Lecture Notes in Computer Science 11943

More information about this series at http://www.springer.com/series/7409

Giuseppe Nicosia · Panos Pardalos ·
Renato Umeton · Giovanni Giuffrida ·
Vincenzo Sciacca (Eds.)

Machine Learning, Optimization, and Data Science

5th International Conference, LOD 2019
Siena, Italy, September 10–13, 2019
Proceedings

Springer

Editors
Giuseppe Nicosia
University of Cambridge
Cambridge, UK

Renato Umeton
Harvard University
Cambridge, MA, USA

Vincenzo Sciacca
Almawave
Rome, Roma, Italy

Panos Pardalos
University of Florida
Gainesville, FL, USA

Giovanni Giuffrida
Università di Catania
Catania, Catania, Italy

ISSN 0302-9743 ISSN 1611-3349 (electronic)
Lecture Notes in Computer Science
ISBN 978-3-030-37598-0 ISBN 978-3-030-37599-7 (eBook)
https://doi.org/10.1007/978-3-030-37599-7

LNCS Sublibrary: SL3 – Information Systems and Applications, incl. Internet/Web, and HCI

This Springer imprint is published by the registered company Springer Nature Switzerland AG
The registered company address is: Gewerbestrasse 11, 6330 Cham, Switzerland

Preface

LOD is the international conference embracing the fields of machine learning, optimization, and data science. The fifth edition, LOD 2019, was organized during September 10–13, 2019, in Certosa di Pontignano (Siena) Italy, a stunning medieval town dominating the picturesque countryside of Tuscany.

The International Conference on Machine Learning, Optimization, and Data Science (LOD) has established itself as a premier interdisciplinary conference in machine learning, computational optimization, and big data. It provides an international forum for the presentation of original multidisciplinary research results, as well as the exchange and dissemination of innovative and practical development experiences.

The LOD Conference Manifesto is the following:

The problem of understanding intelligence is said to be the greatest problem in science today and "the" problem for this century – as deciphering the genetic code was for the second half of the last one. Arguably, the problem of learning represents a gateway to understanding intelligence in brains and machines, to discovering how the human brain works, and to making intelligent machines that learn from experience and improve their competences as children do. In engineering, learning techniques would make it possible to develop software that can be quickly customized to deal with the increasing amount of information and the flood of data around us.
The Mathematics of Learning: Dealing with Data
Tomaso Poggio and Steve Smale

LOD 2019 attracted leading experts from industry and the academic world with the aim of strengthening the connection between these institutions. The 2019 edition of LOD represented a great opportunity for professors, scientists, industry experts, and post-graduate students to learn about recent developments in their own research areas and to learn about research in contiguous research areas, with the aim of creating an environment to share ideas and trigger new collaborations.

As chairs, it was an honor to organize a premiere conference in these areas and to have received a large variety of innovative and original scientific contributions.

During LOD 2019, nine plenary talks were presented:

"Deep Learning on Graphs and Manifolds: Going Beyond Euclidean Data"
Michael Bronstein, Imperial College London, UK

"Backpropagation and Lagrangian Multipliers - New Frontiers of Learning"
Marco Gori, University of Siena, Italy

"A Kernel Critic for Generative Adversarial Networks"
Arthur Gretton, UCL, UK

"Rethinking Planning in Reinforcement Learning"
Arthur Guez, Google DeepMind, UK

"Interactive Multiobjective Optimization in Decision Analytics with a Case Study"
Kaisa Miettinen, University of Jyväskylä, Finland

"Sustainable Interdependent Networks"
Panos Pardalos, Center for Applied Optimization, University of Florida, USA

"Biased Random-Key Genetic Algorithms - Learning Intelligent Solutions from Random Building Blocks"
Mauricio G. C. Resende, Amazon.com Research and University of Washington Seattle, USA

"Building Iride: How to Mix Deep Learning and Ontologies Techniques to Understand Language"
Raniero Romagnoli and Vincenzo Sciacca, Almawave, Italy

"Extending the Frontiers of Deep Learning Using Probabilistic Modelling"
Richard E. Turner, University of Cambridge, UK

LOD 2019 received 158 submissions from 54 countries in 5 continents, and each manuscript was independently reviewed by a committee formed by at least 5 members. These proceedings contain 64 research articles written by leading scientists in the fields of machine learning, artificial intelligence, reinforcement learning, computational optimization, and data science presenting a substantial array of ideas, technologies, algorithms, methods, and applications.

At LOD 2019, Springer LNCS generously sponsored the LOD Best Paper Award. This year, the paper by Sean Tao titled "Deep Neural Network Ensembles" received the LOD 2019 Best Paper Award.

This conference could not have been organized without the contributions of exceptional researchers and visionary industry experts, so we thank them all for participating. A sincere thank you goes also to the Program Committee, formed by more than 450 scientists from academia and industry, for their valuable and essential work of selecting the scientific contributions.

Finally, we would like to express our appreciation to the keynote speakers who accepted our invitation, and to all the authors who submitted their research papers to LOD 2019.

September 2019

Giuseppe Nicosia
Panos Pardalos
Giovanni Giuffrida
Renato Umeton
Vincenzo Sciacca

Organization

General Chairs

Vincenzo Sciacca	Almawave, Italy
Renato Umeton	Dana-Farber Cancer Institute and MIT, USA

Conference and Technical Program Committee Co-chairs

Giovanni Giuffrida	Neodata Group and University of Catania, Italy
Panos Pardalos	University of Florida, USA

Special Sessions Chairs

Giorgio Jansen	University of Cambridge, UK
Salvatore Danilo Riccio	Queen Mary University of London, UK

Tutorial Chair

Vincenzo Sciacca	Almawave, Italy

Publicity Chair

Stefano Mauceri	NCRA, University College Dublin, Ireland

Industrial Session Chairs

Giovanni Giuffrida	Neodata Group and University of Catania, Italy
Vincenzo Sciacca	Almawave, Italy

Organizing Committee

Alberto Castellini	University of Verona, Italy
Piero Conca	Fujitsu, Ireland
Jole Costanza	Italian Institute of Technology, Italy
Giuditta Franco	University of Verona, Italy
Marco Gori	University of Siena, Italy
Giorgio Jansen	University of Cambridge, UK
Kaisa Miettinen	University of Jyväskylä, Finland
Giuseppe Narzisi	New York University Tandon School of Engineering and New York Genome Center, USA
Salvatore Danilo Riccio	Queen Mary University of London, UK

Steering Committee

Giuseppe Nicosia	University of Cambridge, UK
Panos Pardalos	University of Florida, USA

Technical Program Committee

Jason Adair	University of Stirling, UK
Agostinho Agra	Universidade de Aveiro, Portugal
Kerem Akartunali	University of Strathclyde, UK
Richard Allmendinger	The University of Manchester, UK
Paula Amaral	University Nova de Lisboa, Portugal
Aris Anagnostopoulos	Università di Roma La Sapienza, Italy
Davide Anguita	University of Genova, Italy
Alejandro Arbelaez	Cork Institute of Technology, Ireland
Danilo Ardagna	Politecnico di Milano, Italy
Roberto Aringhieri	University of Turin, Italy
Takaya Arita	Nagoya University, Japan
Jason Atkin	University of Nottingham, UK
Martin Atzmueller	Tilburg University, The Netherlands
Chloe-Agathe Azencott	Institut Curie Research Centre, France
Kamyar Azizzadenesheli	University of California at Irvine, USA
Ozalp Babaoglu	Università di Bologna, Italy
Jaume Bacardit	Newcastle University, UK
James Bailey	University of Melbourne, Australia
Marco Baioletti	Università degli Studi di Perugia, Italy
Elena Baralis	Politecnico di Torino, Italy
Xabier E. Barandiaran	University of the Basque Country, Spain
Cristobal Barba-Gonzalez	University of Malaga, Spain
Helio J. C. Barbosa	Laboratório Nacional de Computação Científica, Brazil
Anasse Bari	New York University, USA
Thomas Bartz-Beielstein	IDEA, TH Köln, Germany
Mikhail Batsyn	Higher School of Economics, Russia
Lucia Beccai	Istituto Italiano di Tecnologia, Italy
Aurélien Bellet	Inria Lille, France
Gerardo Beni	University of California at Riverside, USA
Katie Bentley	Harvard Medical School, USA
Peter Bentley	University College London, UK
Heder Bernardino	Universidade Federal de Juiz de Fora, Brazil
Daniel Berrar	Tokyo Institute of Technology, Japan
Adam Berry	CSIRO, Australia
Luc Berthouze	University of Sussex, UK
Martin Berzins	SCI Institute, University of Utah, USA
Manuel Alejandro Betancourt Odio	Universidad Pontificia Comillas, Spain
Rajdeep Bhowmik	Cisco Systems, Inc., USA

Mauro Birattari	IRIDIA, Université Libre de Bruxelles, Belgium
Arnim Bleier	GESIS – Leibniz-Institute for the Social Sciences, Germany
Konstantinos Blekas	University of Ioannina, Greece
Leonidas Bleris	University of Texas at Dallas, USA
Christian Blum	Spanish National Research Council, Spain
Martin Boldt	Blekinge Institute of Technology, Sweden
Flavia Bonomo	Universidad de Buenos Aires, Argentina
Gianluca Bontempi	Université Libre de Bruxelles, Belgium
Ilaria Bordino	UniCredit R&D, Italy
Anton Borg	Blekinge Institute of Technology, Sweden
Paul Bourgine	École Polytechnique Paris, France
Anthony Brabazon	University College Dublin, Ireland
Paulo Branco	Instituto Superior Técnico, Portugal
Juergen Branke	University of Warwick, UK
Alexander Brownlee	University of Stirling, UK
Marcos Bueno	Radboud University, The Netherlands
Larry Bull	University of the West of England, UK
Tadeusz Burczynski	Polish Academy of Sciences, Poland
Robert Busa-Fekete	Yahoo! Research, USA
Adam A. Butchy	University of Pittsburgh, USA
Sergiy I. Butenko	Texas A&M University, USA
Luca Cagliero	Politecnico di Torino, Italy
Stefano Cagnoni	University of Parma, Italy
Yizhi Cai	The University of Edinburgh, UK
Guido Caldarelli	IMT Lucca, Italy
Alexandre Campo	Université Libre de Bruxelles, Belgium
Angelo Cangelosi	University of Plymouth, UK
Mustafa Canim	IBM Thomas J. Watson Research Center, USA
Salvador Eugenio Caoili	University of the Philippines Manila, Philippines
Timoteo Carletti	University of Namur, Belgium
Jonathan Carlson	Microsoft Research, USA
Celso Carneiro Ribeiro	Universidade Federal Fluminense, Brazil
Alexandra M. Carvalho	Universidade de Lisboa, Portugal
Alberto Castellini	University of Verona, Italy
Michelangelo Ceci	University of Bari, Italy
Adelaide Cerveira	INESC-TEC and Universidade de Trás-os-Montes e Alto Douro, Portugal
Uday Chakraborty	University of Missouri at St. Louis, USA
Lijun Chang	The University of Sydney, Australia
Xu Chang	The University of Sydney, Australia
W. Art Chaovalitwongse	University of Washington, USA
Nitesh Chawla	University of Notre Dame, USA
Antonio Chella	Università di Palermo, Italy
Rachid Chelouah	EISTI, Université Paris-Seine, France
Haifeng Chen	NEC Labs, USA

Keke Chen	Wright State University, USA
Steven Chen	University of Pennsylvania, USA
Ying-Ping Chen	National Chiao Tung University, Taiwan
Gregory Chirikjian	Johns Hopkins University, USA
Silvia Chiusano	Politecnico di Torino, Italy
Miroslav Chlebik	University of Sussex, UK
Sung-Bae Cho	Yonsei University, South Korea
Stephane Chretien	National Physical Laboratory, UK
Anders Lyhne Christensen	University of Southern Denmark, Denmark
Stéphan Clémençon	Télécom ParisTech, France
Philippe Codognet	Sorbonne University, France
Carlos Coello Coello	Cinvestav-IPN, Mexico
George Coghill	University of Aberdeen, UK
Sergio Consoli	Philips Research, The Netherlands
David Cornforth	Newcastle University, UK
Luís Correia	University of Lisbon, Portugal
Chiara Damiani	University of Milan-Bicocca, Italy
Thomas Dandekar	University of Würzburg, Germany
Ivan Luciano Danesi	Unicredit Bank, Italy
Christian Darabos	Dartmouth College, USA
Patrick De Causmaecker	KU Leuven, Belgium
Kalyanmoy Deb	Michigan State University, USA
Nicoletta Del Buono	University of Bari, Italy
Jordi Delgado	Universitat Politècnica de Catalunya, Spain
Mauro Dell'Amico	Università degli Studi di Modena e Reggio Emilia, Italy
Ralf Der	MPG, Germany
Clarisse Dhaenens	University of Lille, France
Barbara Di Camillo	University of Padova, Italy
Gianni Di Caro	IDSIA, Switzerland
Luigi Di Caro	University of Torino, Italy
Luca Di Gaspero	University of Udine, Italy
Tom Diethe	Amazon Research Cambridge, UK
Matteo Diez	CNR-INM, National Research Council-Institute of Marine Engineering, Italy
Stephan Doerfel	Micromata GmbH, Germany
Karl Doerner	University of Vienna, Austria
Rafal Drezewski	AGH University of Science and Technology, Poland
Devdatt Dubhashi	Chalmers University, Sweden
Juan J. Durillo	Leibniz Supercomputing Centre, Germany
Omer Dushek	University of Oxford, UK
Nelson F. F. Ebecken	University of Rio de Janeiro, Brazil
Marc Ebner	Ernst Moritz Arndt Universität Greifswald, Germany
Tome Eftimov	Stanford University, USA
Pascale Ehrenfreund	The George Washington University, USA
Gusz Eiben	VU Amsterdam, The Netherlands
Aniko Ekart	Aston University, UK

Talbi El-Ghazali	University of Lille, France
Michael Elberfeld	RWTH Aachen University, Germany
Michael T. M. Emmerich	Leiden University, The Netherlands
Andries Engelbrecht	Stellenbosch University, South Africa
Anton Eremeev	Sobolev Institute of Mathematics, Russia
Harold Fellermann	Newcastle University, UK
Chrisantha Fernando	Queen Mary University, UK
Cèsar Ferri	Universidad Politécnica de Valencia, Spain
Paola Festa	University of Napoli Federico II, Italy
José Rui Figueira	Instituto Superior Técnico, Portugal
Grazziela Figueredo	University of Nottingham, UK
Alessandro Filisetti	Explora Biotech Srl, Italy
Steffen Finck	FH Vorarlberg University of Applied Sciences, Austria
Christoph Flamm	University of Vienna, Austria
Salvador A. Flores	Center for Mathematical Modelling, Chile
Enrico Formenti	Nice Sophia Antipolis University, France
Giuditta Franco	University of Verona, Italy
Piero Fraternali	Politecnico di Milano, Italy
Valerio Freschi	University of Urbino, Italy
Enrique Frias Martinez	Telefonica Research, Spain
Walter Frisch	University of Vienna, Austria
Rudolf M. Füchslin	Zurich University of Applied Sciences, Switzerland
Marcus Gallagher	The University of Queensland, Australia
Claudio Gallicchio	University of Pisa, Italy
Patrick Gallinari	LIP6, University of Paris 6, France
Luca Gambardella	IDSIA, Switzerland
Jean-Gabriel Ganascia	LIP6, Pierre and Marie Curie University, France
Xavier Gandibleux	Université de Nantes, France
Alfredo García Hernández-Díaz	Pablo de Olvide University, Spain
José Manuel García Nieto	University of Málaga, Spain
Paolo Garza	Politecnico di Torino, Italy
Romaric Gaudel	ENSAI, France
Nicholas Geard	The University of Melbourne, Australia
Martin Josef Geiger	Helmut-Schmidt-Universität, Germany
Michel Gendreau	Polytechnique Montréal, Canada
Philip Gerlee	Chalmers University, Sweden
Mario Giacobini	University of Torino, Italy
Onofrio Gigliotta	University of Naples Federico II, Italy
David Ginsbourger	Idiap Research Institute and University of Bern, Switzerland
Giovanni Giuffrida	University of Catania, Italy
Aris Gkoulalas-Divanis	IBM Watson Health, USA
Giorgio Stefano Gnecco	IMT School for Advanced Studies, Lucca, Italy
Christian Gogu	Université Toulouse III, France
Faustino Gomez	IDSIA, Switzerland

Teresa Gonçalves	University of Évora, Portugal
Michael Granitzer	University of Passau, Germany
Alex Graudenzi	University of Milan-Bicocca, Italy
Julie Greensmith	University of Nottingham, UK
Roderich Gross	University of Sheffield, UK
Mario Guarracino	ICAR-CNR, Italy
Francesco Gullo	UniCredit R&D, Italy
Vijay K. Gurbani	Illinois Institute of Technology and Vail Systems, Inc., USA
Steven Gustafson	Maana, Inc., USA
Jin-Kao Hao	University of Angers, France
Simon Harding	Machine Intelligence Ltd., Canada
William Hart	Sandia National Laboratories, USA
Richard Hartl	University of Vienna, Austria
Inman Harvey	University of Sussex, UK
Mohammad Hasan	Indiana University-Purdue University Indianapolis, USA
Geir Hasle	SINTEF Digital, Norway
Verena Heidrich-Meisner	Kiel University, Germany
Eligius M. T. Hendrix	Universidad de Málaga, Spain
Carlos Henggeler Antunes	University of Coimbra, Portugal
Francisco Herrera	University of Granada, Spain
J. Michael Herrmann	The University of Edinburgh, UK
Arjen Hommersom	Radboud University, The Netherlands
Vasant Honavar	Pennsylvania State University, USA
Hongxuan Huang	Tsinghua University, China
Fabrice Huet	University of Nice Sophia Antipolis, France
Hiroyuki Iizuka	Hokkaido University, Japan
Takashi Ikegami	University of Tokyo, Japan
Hisao Ishibuchi	Osaka Prefecture University, Japan
Peter Jacko	Lancaster University Management School, UK
Christian Jacob	University of Calgary, Canada
Hasan Jamil	University of Idaho, USA
Yaochu Jin	University of Surrey, UK
Colin Johnson	University of Kent, UK
Gareth Jones	Dublin City University, Ireland
Laetitia Jourdan	University of Lille, CNRS, France
Narendra Jussien	Ecole des Mines de Nantes, LINA, France
Janusz Kacprzyk	Polish Academy of Sciences, Poland
Theodore Kalamboukis	Athens University of Economics and Business, Greece
Valeriy Kalyagin	National Reaserch University Higher School of Economics, Russia
George Kampis	Eotvos University, Hungary
Jaap Kamps	University of Amsterdam, The Netherlands
Dervis Karaboga	Erciyes University, Turkey
George Karakostas	McMaster University, Canada

Istvan Karsai	ETSU, USA
Zekarias T. Kefato	University of Trento, Italy
Jozef Kelemen	Silesian University, Czech Republic
Graham Kendall	Nottingham University, UK
Navneet Kesher	Facebook, USA
Didier Keymeulen	NASA - Jet Propulsion Laboratory, USA
Michael Khachay	Ural Federal University Ekaterinburg, Russia
Arbaaz Khan	University of Pennsylvania, USA
Daeeun Kim	Yonsei University, South Korea
Lefteris Kirousis	National and Kapodistrian University of Athens, Greece
Zeynep Kiziltan	University of Bologna, Italy
Elena Kochkina	University of Warwick, UK
Min Kong	Hefei University of Technology, China
Erhun Kundakcioglu	Ozyegin University, Turkey
Jacek Kustra	Philips, The Netherlands
C. K. Kwong	The Hong Kong Polytechnic University, Hong Kong, China
Renaud Lambiotte	University of Namur, Belgium
Doron Lancet	Weizmann Institute of Science, Israel
Dario Landa-Silva	University of Nottingham, UK
Pier Luca Lanzi	Politecnico di Milano, Italy
Alessandro Lazaric	Facebook Artificial Intelligence Research (FAIR), France
Sanja Lazarova-Molnar	University of Southern Denmark, Denmark
Doheon Lee	KAIST, South Korea
Eva K. Lee	Georgia Tech, USA
Jay Lee	Center for Intelligent Maintenance Systems - UC, USA
Tom Lenaerts	Universite Libre de Bruxelles, Belgium
Rafael Leon	Universidad Politécnica de Madrid, Spain
Carson Leung	University of Manitoba, Canada
Kang Li	Google, USA
Lei Li	Florida International University, USA
Shuai Li	Cambridge University, UK
Xiaodong Li	RMIT University, Australia
Lukas Lichtensteiger	Zurich University of Applied Sciences, Switzerland
Joseph Lizier	The University of Sydney, Australia
Giosue' Lo Bosco	Universita' di Palermo, Italy
Daniel Lobo	University of Maryland Baltimore County, USA
Fernando Lobo	University of Algarve, Portugal
Daniele Loiacono	Politecnico di Milano, Italy
Jose A. Lozano	University of the Basque Country, Spain
Paul Lu	University of Alberta, Canada
Angelo Lucia	University of Rhode Island, USA
Gabriel Luque	University of Málaga, Spain
Dario Maggiorini	University of Milano, Italy

Gilvan Maia	Universidade Federal do Ceará, Brazil
Donato Malerba	University of Bari, Italy
Anthony Man-Cho So	The Chinese University of Hong Kong, Hong Kong, China
Jacek Mandziuk	Warsaw University of Technology, Poland
Vittorio Maniezzo	University of Bologna, Italy
Marco Maratea	University of Genova, Italy
Elena Marchiori	Radboud University, The Netherlands
Tiziana Margaria	University of Limerick and Lero, Ireland
Magdalene Marinaki	Technical University of Crete, Greece
Yannis Marinakis	Technical University of Crete, Greece
Omer Markovitch	University of Groningen, The Netherlands
Carlos Martin-Vide	Rovira i Virgili University, Spain
Dominique Martinez	Loria, France
Aldo Marzullo	University of Calabria, Italy
Joana Matos Dias	Universidade de Coimbra, Portugal
Matteo Matteucci	Politecnico di Milano, Italy
Stefano Mauceri	University College Dublin, Ireland
Giancarlo Mauri	University of Milano-Bicocca, Italy
Antonio Mauttone	Universidad de la República, Uruguay
Mirjana Mazuran	Politecnico di Milano, Italy
James McDermott	University College Dublin, Ireland
Suzanne McIntosh	NYU Courant, NYU Center for Data Science, USA
Peter Mcowan	Queen Mary University, UK
Gabor Melli	Sony Interactive Entertainment Inc., USA
Jose Fernando Mendes	University of Aveiro, Portugal
Lu Meng	University of Buffalo, USA
Rakesh R. Menon	University of Massachusetts Amherst, USA
David Merodio-Codinachs	ESA, France
Silja Meyer-Nieberg	Universität der Bundeswehr München, Germany
Martin Middendorf	University of Leipzig, Germany
Taneli Mielikäinen	Yahoo!, USA
Orazio Miglino	University of Naples Federico II, Italy
Julian Miller	University of York, UK
Marco Mirolli	ISTC-CNR, Italy
Natasa Miskov-Zivanov	University of Pittsburgh, USA
Carmen Molina-Paris	University of Leeds, UK
Sara Montagna	Università di Bologna, Italy
Marco Montes de Oca	Clypd, Inc., USA
Monica Mordonini	University of Parma, Italy
Mohamed Nadif	Paris Descartes University, France
Hidemoto Nakada	National Institute of Advanced Industrial Science and Technology, Japan
Mirco Nanni	CNR-ISTI, Italy
Sriraam Natarajan	Indiana University, USA
Chrystopher L. Nehaniv	University of Hertfordshire, UK

Michael Newell	Athens Consulting, LLC, USA
Binh P. Nguyen	Victoria University of Wellington, New Zealand
Giuseppe Nicosia	University of Catania, Italy
Sotiris Nikoletseas	University of Patras and CTI, Greece
Xia Ning	IUPUI, USA
Jonas Nordhaug Myhre	The Arctic University of Tromsø, Norway
Wieslaw Nowak	Nicolaus Copernicus University, Poland
Eirini Ntoutsi	Leibniz University Hanover and L3S Research Center, Germany
Michal Or-Guil	Humboldt University Berlin, Germany
Gloria Ortega López	University of Málaga, Spain
Mathias Pacher	Goethe Universität Frankfurt am Main, Germany
Ping-Feng Pai	National Chi Nan University, Taiwan
Wei Pang	University of Aberdeen, UK
George Papastefanatos	IMIS and Athena RC, Greece
Luís Paquete	University of Coimbra, Portugal
Panos Pardalos	Center for Applied Optimization, University of Florida, USA
Rohit Parimi	Bloomberg LP, USA
Andrew J. Parkes	Nottingham University, UK
Konstantinos Parsopoulos	University of Ioannina, Greece
Andrea Patanè	University of Oxford, UK
Joshua Payne	University of Zurich, Switzerland
Jun Pei	Hefei University of Technology, China
Nikos Pelekis	University of Piraeus, Greece
David A. Pelta	Universidad de Granada, Spain
Dimitri Perrin	Queensland University of Technology, Australia
Milena Petkovic	Zuse Institute Berlin, Germany
Koumoutsakos Petros	ETH, Switzerland
Juan Peypouquet	Universidad Técnica Federico Santa María, Chile
Andrew Philippides	University of Sussex, UK
Stefan Pickl	Universität der Bundeswehr München, Germany
Fabio Pinelli	Vodafone, Italy
Joao Pinto	Technical University of Lisbon, Portugal
Vincenzo Piuri	University of Milano, Italy
Alessio Plebe	University of Messina, Italy
Nikolaos Ploskas	University of Western Macedonia, Greece
Valentina Poggioni	University of Perugia, Italy
George Potamias	Institute of Computer Science, Greece
Philippe Preux	Inria, France
Mikhail Prokopenko	The University of Sydney, Australia
Paolo Provero	University of Torino, Italy
Buyue Qian	IBM T. J. Watson, USA
Chao Qian	University of Science and Technology of China, China
Tomasz Radzik	King's College London, UK
Günther Raidl	TU Wien, Austria

Helena Ramalhinho Dias Lourenco	University Pompeu Fabra, Spain
Palaniappan Ramaswamy	University of Kent, UK
Jan Ramon	Inria, France
Vitorino Ramos	Technical University of Lisbon, Portugal
Shoba Ranganathan	Macquarie University, Australia
Zbigniew Ras	University of North Carolina, USA
Jan Rauch	University of Economics, Czech Republic
Cristina Requejo	Universidade de Aveiro, Portugal
Paul Reverdy	University of Arizona, USA
John Rieffel	Union College, USA
Francesco Rinaldi	University of Padova, Italy
Laura Anna Ripamonti	Università degli Studi di Milano, Italy
Humberto Rocha	University of Coimbra, Portugal
Eduardo Rodriguez-Tello	Cinvestav Tamaulipas, Mexico
Andrea Roli	Università di Bologna, Italy
Vittorio Romano	University of Catania, Italy
Pablo Romero	Universidad de la República, Uruguay
Andre Rosendo	University of Cambridge, UK
Samuel Rota Bulò	Mapillary Research, Austria
Arnab Roy	Fujitsu Laboratories of America, USA
Alessandro Rozza	Parthenope University of Naples, Italy
Kepa Ruiz-Mirazo	University of the Basque Country, Spain
Florin Rusu	University of California Merced, USA
Jakub Rydzewski	Nicolaus Copernicus University, Poland
Nick Sahinidis	Carnegie Mellon University, USA
Lorenza Saitta	University of Piemonte Orientale, Italy
Andrea Santoro	Queen Mary University London, UK
Francisco C. Santos	INESC-ID and Instituto Superior Técnico, Lisboa Portugal
Claudio Sartori	University of Bologna, Italy
Fréderic Saubion	Université d'Angers, France
Robert Schaefer	AGH University of Science and Technology, Poland
Andrea Schaerf	University of Udine, Italy
Christoph Schommer	University of Luxemburg, Luxemburg
Oliver Schuetze	Cinvestav-IPN, Mexico
Luís Seabra Lopes	Universidade of Aveiro, Portugal
Natalia Selini Hadjidimitriou	University of Modena and Reggio Emilia, Italy
Alexander Senov	Saint Petersburg State University, Russia
Andrea Serani	CNR-INM, Italy
Roberto Serra	University of Modena and Reggio Emilia, Italy
Marc Sevaux	Université de Bretagne-Sud, France
Nasrullah Sheikh	University of Trento, Italy
Leonid Sheremetov	Mexican Petroleum Institute, Mexico
Ruey-Lin Sheu	National Cheng Kung University, Taiwan

Hsu-Shih Shih	Tamkang University, Taiwan
Kilho Shin	Gakushuin University, Japan
Patrick Siarry	Université Paris-Est Créteil, France
Sergei Sidorov	Saratov State University, Russia
Alkis Simitsis	HP Labs, USA
Alina Sirbu	University of Pisa, Italy
Konstantina Skouri	University of Ioannina, Greece
Johannes Söllner	Sodatana e.U., Austria
Ichoua Soumia	Embry-Riddle Aeronautical University, USA
Giandomenico Spezzano	CNR-ICAR, Italy
Antoine Spicher	LACL, Université Paris-Est Créteil, France
Claudio Stamile	Université de Lyon 1, France
Pasquale Stano	University of Salento, Italy
Thomas Stibor	GSI Helmholtz Centre for Heavy Ion Research, Germany
Catalin Stoean	University of Craiova, Romania
Johan Suykens	KU Leuven, Belgium
Reiji Suzuki	Nagoya University, Japan
Domenico Talia	University of Calabria, Italy
Kay Chen Tan	National University of Singapore, Singapore
Letizia Tanca	Politecnico di Milano, Italy
Charles Taylor	UCLA, USA
Maguelonne Teisseire	Irstea. UMR-TETIS, France
Fabien Teytaud	Université Littoral Côte d'Opale, France
Tzouramanis Theodoros	University of the Aegean, Greece
Jon Timmis	University of York, UK
Gianna Toffolo	University of Padova, UK
Michele Tomaiuolo	University of Parma, Italy
Joo Chuan Tong	Institute of High Performance Computing, Singapore
Jamal Toutouh	Massachusetts Institute of Technology, USA
Jaden Travnik	University of Alberta, Canada
Nickolay Trendafilov	Open University, UK
Sophia Tsoka	King's College London, UK
Shigeyoshi Tsutsui	Hannan University, Japan
Ali Emre Turgut	IRIDIA-ULB, France
Karl Tuyls	The University of Liverpool, UK
Gregor Ulm	Fraunhofer-Chalmers Research Centre for Industrial Mathematics, Sweden
Jon Umerez	University of the Basque Country, Spain
Renato Umeton	Dana-Farber Cancer Institute, USA
Ashish Umre	University of Sussex, UK
Olgierd Unold	Politechnika Wroclawska, Poland
Rishabh Upadhyay	Innopolis University, Russia
Giorgio Valentini	Universita' degli Studi di Milano, Italy
Sergi Valverde	University Pompeu Fabra, Spain
Werner Van Geit	EPFL, Switzerland

Pascal Van Hentenryck	University of Michigan, USA
Ana Lucia Varbanescu	University of Amsterdam, The Netherlands
Carlos Varela	Rensselaer Polytechnic Institute, USA
Iraklis Varlamis	Harokopio University of Athens, Greece
Eleni Vasilaki	University of Sheffield, UK
Richard Vaughan	Simon Fraser University, Canada
Kalyan Veeramachaneni	MIT, USA
Vassilios Verykios	Hellenic Open University, Greece
Herna L. Viktor	University of Ottawa, Canada
Mario Villalobos-Arias	Univesidad de Costa Rica, Costa Rica
Marco Villani	University of Modena and Reggio Emilia, Italy
Susana Vinga	INESC-ID and IST-UL, Portugal
Mirko Viroli	Università di Bologna, Italy
Katya Vladislavleva	Evolved Analytics LLC, Belbium
Stefan Voss	University of Hamburg, Germany
Dean Vucinic	Vesalius College, Vrije Universiteit Brussel, Belgium
Markus Wagner	The University of Adelaide, Australia
Toby Walsh	UNSW Sydney, Australia
Lipo Wang	Nanyang Technological University, Singapore
Liqiang Wang	University of Central Florida, USA
Longshaokan Wang	North Carolina State University, USA
Rainer Wansch	Fraunhofer IIS, Germany
Syed Waziruddin	Kansas State University, USA
Janet Wiles	The University of Queensland, Australia
Man Leung Wong	Lingnan University, Hong Kong, China
Andrew Wuensche	University of Sussex, UK
Petros Xanthopoulos	University of Central Florida, USA
Ning Xiong	Mälardalen University, Sweden
Xin Xu	George Washington University, USA
Gur Yaari	Yale University, USA
Larry Yaeger	Indiana University, USA
Shengxiang Yang	De Montfort University, USA
Xin-She Yang	Middlesex University London, UK
Sule Yildirim-Yayilgan	Norwegian University of Science and Technology, Norway
Shiu Yin Yuen	City University of Hong Kong, Hong Kong, China
Zelda Zabinsky	University of Washington, USA
Ras Zbyszek	University of North Carolina, USA
Hector Zenil	University of Oxford, UK
Guang Lan Zhang	Boston University, USA
Qingfu Zhang	City University of Hong Kong, Hong Kong, China
Rui Zhang	IBM Research, USA
Zhi-Hua Zhou	Nanjing University, China
Tom Ziemke	Linköping University, Sweden
Antanas Zilinskas	Vilnius University, Lithuania

Best Paper Awards

LOD 2019 Best Paper Award
"Deep Neural Network Ensembles"
Sean Tao
Carnegie Mellon University, USA

Springer sponsored the LOD 2019 Best Paper Award with a cash prize of EUR 1,000.

LOD 2018 Best Paper Award
"Calibrating the Classifier: Siamese Neural Network Architecture for End-to-End Arousal Recognition from ECG"
Andrea Patané* and Marta Kwiatkowska*
*University of Oxford, UK

Springer sponsored the LOD 2018 Best Paper Award with a cash prize of EUR 1,000.

MOD 2017 Best Paper Award
"Recipes for Translating Big Data Machine Reading to Executable Cellular Signaling Models"
Khaled Sayed*, Cheryl Telmer**, Adam Butchy*, and Natasa Miskov-Zivanov*
*University of Pittsburgh, USA
**Carnegie Mellon University, USA

Springer sponsored the MOD 2017 Best Paper Award with a cash prize of EUR 1,000.

MOD 2016 Best Paper Award
"Machine Learning: Multi-site Evidence-based Best Practice Discovery"
Eva Lee, Yuanbo Wang, and Matthew Hagen
Eva K. Lee, Professor Director, Center for Operations Research in Medicine and HealthCare H. Milton Stewart School of Industrial and Systems Engineering, Georgia Institute of Technology, USA

MOD 2015 Best Paper Award
"Learning with discrete least squares on multivariate polynomial spaces using evaluations at random or low-discrepancy point sets"
Giovanni Migliorati
Ecole Polytechnique Federale de Lausanne – EPFL, Switzerland

Contents

Deep Neural Network Ensembles

Sean Tao(✉)

Carnegie Mellon University, Pittsburgh, PA 15213, USA
shtao@alumni.cmu.edu

Abstract. Current deep neural networks suffer from two problems; first, they are hard to interpret, and second, they suffer from overfitting. There have been many attempts to define interpretability in neural networks, but they typically lack causality or generality. A myriad of regularization techniques have been developed to prevent overfitting, and this has driven deep learning to become the hot topic it is today; however, while most regularization techniques are justified empirically and even intuitively, there is not much underlying theory. This paper argues that to extract the features used in neural networks to make decisions, it's important to look at the paths between clusters existing in the hidden spaces of neural networks. These features are of particular interest because they reflect the true decision-making process of the neural network. This analysis is then furthered to present an ensemble algorithm for arbitrary neural networks which has guarantees for test accuracy. Finally, a discussion detailing the aforementioned guarantees is introduced and the implications to neural networks, including an intuitive explanation for all current regularization methods, are presented. The ensemble algorithm has generated state-of-the-art results for Wide-ResNets on CIFAR-10 (top 5 for all models) and has improved test accuracy for all models it has been applied to.

Keywords: Deep learning · Interpretability · Ensemble

1 Introduction

Consider a simple feed forward neural network. Define a hidden space corresponding to a hidden layer of a neural network as the space containing the outputs of the hidden nodes at that hidden layer for some input. All hidden spaces are composed of perceptrons with respect to the previous layer, each of which has a hyperplane decision boundary. Points in the previous space are mapped to a constant function of their distance to this plane, and depending on the activation function, compressed or stretched. This stretching and compressing naturally leads to clustering in hidden spaces. The process is repeated for each perceptron comprising the hidden space, where adding a perceptron adds a dimension to the hidden space by projecting the points into a new dimension depending on their distances from the hyperplane of the new perceptron and the activation function. Define a feature to be a measurable characteristic which

G. Nicosia et al. (Eds.): LOD 2019, LNCS 11943, pp. 1–12, 2019.
https://doi.org/10.1007/978-3-030-37599-7_1

a neural network uses to make its classification decision. Unfortunately, these clusters are not features in and of themselves but rather mixtures of features. To extract individual features, then, cluster paths should be examined, where a path is defined per individual input point as the sequence of clusters in the neural network the point belongs to, starting from cluster it belongs to in the input space, then the clusters it belongs to in each hidden space, and finally the cluster it belongs to in the output space.

Intuitively, paths represent features because each path defines a region of the input space which will eventually end up at the same cluster in the output space via the same logic pathway used by the neural network. Thus, it immediately follows that a point on a path must be classified similarly as other points on that path, assuming all points in the output space are classified by cluster. This is another way of stating that points on the same path are indistinguishable from each other to the neural network, and since points in different paths were differentiated in some layer, paths separate out features of neural networks. This process is formalized in the algorithm described below. However, besides merely finding the features in a neural network, the paths serve an even more important purpose–they separate the input into regions of confidence with respect to the output classification. In particular, paths represent specific features, some of which were found because they are truly useful and others due to overfitting.

The process of determining "good" and "bad" data points, formally defined below, attempts to separate data into these two categories. Informally, "good" data come from paths that contain many points that are classified correctly, since these are likely not due to random chance and thus are real features. An ensemble algorithm can then be created to combine different models, where models only vote on their "good" data points. This is equivalent to querying neural networks for points where they are confident in their predictions. Points where the model is unsure can then be classified by other models.

2 Related Work

Much effort has been put into interpreting deep neural networks in the past [12]. A few meta studies summarize the effort of the community as a whole rather well. There are three general types of methods for deep neural networks; namely, by discovering "ways to reduce the complexity of all these operations," "understand[ing] the role and structure of the data flowing through these bottlenecks," and "creat[ing] networks that are designed to be easier to explain" [4]. The difference between the algorithm presented here and the aforementioned attempts at interpretability is that this derives its logic entirely from the network–in particular, paths, by definition, represent the features used in classification.

In addition, clustering has been applied to the hidden spaces of neural networks in prior work, also in the context of interpretability [11]. However, cluster path analysis in deep neural networks could not be found.

Finally, ensembling algorithms of deep neural networks have been attempted before. For instance, algorithms already presented in different contexts have been applied to neural networks [6], such as the Super Learner [9] and AdaBoost [13]. Nevertheless, "the behavior of ensemble learning with deep networks is still not well studied and understood" [6].

3 Algorithm

3.1 Training

Here, the algorithm to train the ensemble is presented; this also contains the process to generate paths. As clarification, the distinction in this paper between training data and a training set is that the training data is split into a training set and a validation set. First, partition the training data into training/validation sets. Repeat this such that there is minimal overlap between points in different training sets. Then, for each of the partitions, conduct the following steps.

Train a neural network on the training set, selecting the best model via validation accuracy. Add this neural network to the set of models named model 1s ("original models"). Determine the optimal number of means in each layer via the "elbow" technique [18] by plotting inertia against the number of means with k-means [16]. Run k-means again with the optimal number of means on the input layer, each of the hidden layers, and output layer for the training set. For each point in the training and validation sets, determine the path of clusters through the neural network. Find optimal values of the following three parameters via grid search, filter out validation points which do not meet the criteria, and calculate the validation accuracy on the remaining points. The three parameters are: maximum distance to a cluster center, minimum number of data points in a split, and minimum accuracy in a split. Define a split to be a partial path of length 2–in other words, how one cluster was split into different clusters in the subsequent layer. Thus, for each point in the validation set, if it is too far from its cluster in any layer, if it is in a split which has too few points, or if it is in a split with too low validation accuracy, filter out the point, and calculate the validation accuracy on the remaining points. Ideally, the parameters which represent the tightest restrictions that filter out the fewest points and achieve the desired validation accuracy should be chosen. This process identifies paths which represent features not generated by overfitting and contain points of high validation, and thus test, accuracy. The idea is to then train other models to focus specifically on the "bad" test points. Using the same parameters found above, separate the training set into "good" and "bad" training points, where "good" data points are points which satisfy the parameters found above. Train another neural network, with the same partition of train and validation sets. However, repeat each "bad" training point in the training set, such that each "bad" point appears twice. Add this network to the set of models named model 2s ("bad 1 models"). Repeat the analysis steps on this new neural network.

3.2 Testing

To run this ensemble on the test set, for each test data point, conduct the following steps. Determine whether or not its a "good" data point in each of the

Algorithm 1. Training the Ensemble

```
Partition the training data into training/validation sets
for each partition do
    Train a neural network
    Add this network to the set of model 1s ("original models")
    Run k-means on all layers
    for each point in training and validation sets do
        Determine the path of clusters
    end for
    Find parameters for filtering clusters
    Separate the training set into "good" and "bad" points
    Train another neural network with oversampling on the "bad" points
    Add this network to the set of model 2s ("bad 1 models")
    Repeat the analysis steps for the new neural network
end for
```

models in the set of model 1s by using the parameters for the respective model. If its a "good" data point in at least half of the models in the set of model 1s, and any of these models agree on a label, return that label. Call these test points "original good" test points. Otherwise, determine whether or not its a "good" test point in each of the models in the set of model 2s. If its a "good" test point in at least half of the models in the set of model 2s, and if any of these models agree on a label, return that label. Call these test points "bad 1" test points. Otherwise, add the output vectors of each of the models in the set of model 2s, and return the label corresponding to the largest value in the resulting sum vector. Call these data points "bad 2" data points.

Algorithm 2. Testing the Ensemble

```
for each test data point do
    if "good" in set of model 1's and agree on label then
        Classify via voting of "good" model 1's ("good" test point)
        continue
    end if
    if "good" in set of model 2's and agree on label then
        Classify via voting of "good" model 2's ("bad 1" test point)
        continue
    end if
    Classify via majority voting of all model 2's ("bad 2" test point)
end for
```

3.3 Larger Models

While this algorithm works to increase test accuracy in smaller, feed forward neural networks, it must be modified to so that it is effective for larger, state-of-the-art models. This is necessary for a few reasons. First, the architecture may not be able to be represented as a simple directed acyclic graph, and thus hidden spaces may not be defined. Second, even if some modern architectures avoid these problems, they often contain such high numbers of hidden nodes that clustering is not meaningful due to the curse of dimensionality. Fortunately, the concept of "good" and "bad" points seem to transcend model architectures and training processes. That is, models all seem to classify "good" points with much higher accuracy than "bad" points, regardless of their architecture, methods of regularization applied, or other training techniques. In particular, larger models seem to be able to accurately classify points which smaller models can classify correctly when trained on any subset of the training data. Then it follows that only "bad 2" test data are inaccurate for larger models. Thus, for larger models, to create an ensemble, conduct the following steps.

First, train one neural network normally ("original model"). This will be used to classify "good" and "bad 1" test data. Then, train another neural network ("bad model"), oversampling all training points which were classified as "bad" in the majority of the set of model 1's, as defined in the training section. Finally, use this model to classify all "bad 2" test data.

Algorithm 3. Training and Testing on Larger Models

Train larger model normally ("original model")
Train larger model with oversampled "bad" training points as found in smaller models ("bad model")
for each test data point **do**
    if "good" or "bad 1" test point **then**
        Classify using "original model"
        continue
    end if
    if "bad 2" test point **then**
        Classify using "bad model"
    end if
end for

4 Results

The path analysis portion of the algorithm was run on a feed forward neural network with Dropout [15] before each of 4 hidden layers and the input layer, all of which utilized sigmoid activations, trained with the Adam optimizer [7] on the MNIST digits dataset [10]. Features were generated from all "good" splits in two manners: first, by averaging all the training points in that split; and second, by finding via backpropagation the input which best generates the cluster center at that layer while the hidden layers' weights remain fixed. No additional regularization is used, demonstrating an improvement over existing techniques, since typically this would just generate noise [12]. The former technique was used on the splits from the first to second hidden layer, since the previous splits had too many images to display, and subsequent layers only had 10 clusters, one per label. Due to lack of resources, cluster split means could not be found, but those could potentially generate clearer features. The latter technique was used on splits from the input to the first hidden layer. For both methods, visual inspection confirms that the features generated separate different digits and, when they exist, different manners of writing each digit. The features are presented below (Figs. 1 and 2).

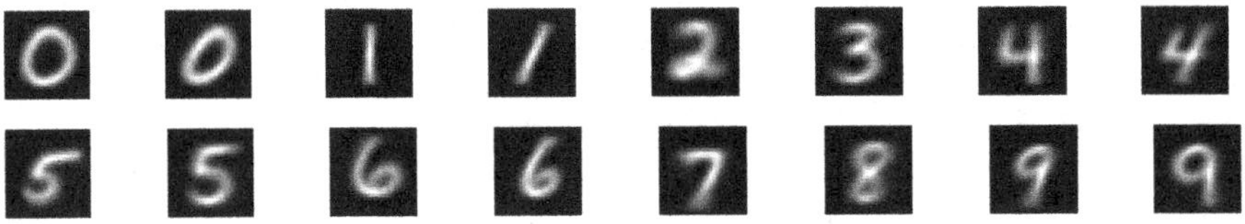

Fig. 1. All "good" features from the first to the second hidden layer via input averaging

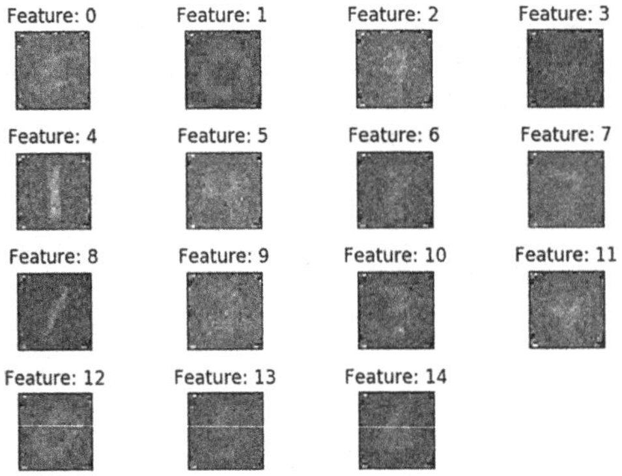

Fig. 2. All "good" features from the input to the first hidden layer via backpropagation

The entire training algorithm was run on CIFAR-10 [8] with an ensemble where the set of model 1's consisted of neural networks created by transfer learning of InceptionResNet [17] trained on ImageNet [2] and the set of model 2's consisted of VGG [14] models specific to CIFAR-10 [3]. The highest test accuracy of any individual of these neural networks when trained on CIFAR-10 is 93.56% [3]. With five validation sets ("Block Partitions"), where each validation set was created by partitioning the training data into contiguous blocks of size 10000, the ensemble achieves 94.63% test accuracy. With a different partition ("Stride 1 Partitions"), where the five validation sets were created by equivalence classes of the training data enumerated and taken modulo 5, the ensemble achieves 94.26% test accuracy. When these two are combined and 10 validation sets are used, the accuracy increases to 94.77%, a 1.21% increase over any the test accuracy of individual network trained on the entire training data [3]. The specific breakdown over the different sets of points are presented below. The difference between the test accuracy of "original good" points, which were "good" on the set of model 1's, and "bad 1" points, which were "good" on the set of model 2's, and "bad 2" points is apparent.

Small model test accuracy breakdown

Ensemble (total num models)	"Original good"	"Bad 1"	"Bad 2"	Overall
Block partitions (10)	98.97%	99.37%	79.65%	94.63%
Stride 1 partitions (10)	99.11%	99.33%	76.71%	94.26%
All partitions (20)	99.06%	99.29%	77.80%	94.77%

Small model total number of points by type breakdown

Ensemble (total num models)	"Original good"	"Bad 1"	"Bad 2"	Overall
Block partitions (10)	5815	1900	2285	10000
Stride 1 partitions (10)	5255	2555	2190	10000
All partitions (20)	5853	2106	2041	10000

Finally, the large model algorithm (Algorithm 3) was applied a Wide-Res Net (WRN) [22] trained with a data augmentation technique, AutoAugment [1]. This ensemble achieved 97.51% test accuracy, 0.18% better than the results for WRN's reported in the AutoAugment paper as of 9 October 2018 and a top 5 result of all papers on CIFAR-10. ShakeDrop [21] with AutoAugment achieved state-of-the-art results on CIFAR-10 when it was published, but the large model algorithm could not be applied to this due to resource limitations.

Large model ensemble test accuracy breakdown

Ensemble (total num models)	"Original good"	"Bad 1"	"Bad 2"	Overall
WRN (2)	99.62%	99.57%	89.31%	97.51%

Large model ensemble total number of points by type breakdown

Ensemble (total num models)	"Original good"	"Bad 1"	"Bad 2"	Overall
WRN (2)	5853	2106	2041	10000

5 Analysis

5.1 "Good" Paths

Bounds on Test Error. Assume that the training data and the test data are independent and identically distributed, and the training data is sufficiently large. For neural networks, there exist potentially disjoint subspaces of the input space which correspond to "good" points in a path–namely, if a point were to fall into a particular subspace, then the point will be "good" in that is does not get filtered, and it will follow the corresponding path. Consider k neural networks each trained on disjoint and insignificant fractions f of the training data, and assume that each discovers some similar subspace. Due to the assumption that the training data is large, each of the k neural networks is effectively trained on some subset of the data sampled from the distribution of all (train and test) data. Moreover, because the samples are disjoint, there is no dependence introduced by overlapping data.

Now, consider the intersection of all k previously discussed subspaces, one found per network; call this s. Assume n points from the respective validation set of each network reside in s, and the maximum classification error of the respective n points for each of the k networks is ϵ'. Define p to be the probability that a training set of size f will produce a neural network that will classify all points in this subspace from the training and test points with error at most ϵ'; in other words, the probability that a neural network will discover s. Noting that the maximum variance for a binary variable is $\frac{1}{4}$ and applying the central limit theorem, a confidence interval can be created to estimate p, since a sample mean is approximately normally distributed with variance at most $\frac{1}{4\sqrt{k}}$, and there exists a single sample mean with value 1. Define ϵ to be the average true error for all points in this subspace of any model trained on the fraction f of the training data. Due to the properties of the normal distribution, the true value of p is almost certainly within 6 standard deviations of this, or $1 - \frac{6}{4\sqrt{k}}$. Then with high confidence, and by applying similar logic as above, the probability ϵ' is approximately normally distributed with mean ϵ and variance at most $\frac{1}{4\sqrt{n}}$ is $(1 - \frac{6}{4\sqrt{k}})$, which converges to 1 as k grows large. More concisely, for any confidence level, by using a sufficiently large ensemble, ϵ' is normally distributed with mean ϵ and variance at most $\frac{1}{4\sqrt{n}}$. This is important because now a confidence interval can be created for ϵ, the average actual error of all points, train and test, in this subspace.

Theorem 1. *Define ϵ to be the average true error for all points in some subspace of any model trained on the fraction f of the training data. Consider k neural networks each trained on disjoint and insignificant fractions f of the training data, and assume that each discovers some similar subspace. Assume n points from the respective validation set of each network reside in s, and the maximum classification error of the respective n points for each of the k networks is ϵ'.*

Then with high confidence, the probability ϵ' is approximately normally distributed with mean ϵ and variance at most $\frac{1}{4\sqrt{n}}$ is $(1 - \frac{6}{4\sqrt{k}})$, which converges to 1 as k grows large.

Implications. The only previous work comparable to this requires either a finite hypothesis space or restrictions on the capabilities of the model–specifically the Vapnik-Chervonenkis (VC) dimension–and this is inapplicable to modern neural networks [20]. This lies in between–it provides bounds on test error for extremely complicated models, but only for certain data points. The intuitive explanation behind the previous section is thus: a neural network trained on training data should perform better than random on unseen test data is because the test data is expected to be drawn from the same distribution as the training data. Any given pattern, which is what a subspace represents, may not be applicable in the test set, but a pattern found in multiple disjoint subsets is most likely to be real, and the probability can be mathematically described via the central limit theorem. When applied on an ensemble of models trained with heavily overlapping training sets, the above analysis does not hold, and the ensemble does not perform too much better than the best model. On the other hand, the above algorithm provides a framework from which useful features can be extracted, and this allows larger models to differentiate between true patterns and overfitting.

The analysis above also serves as a justification of most regularization methods as well as an explanation for Occam's Razor in machine learning. "Simple" models are not preferred because they have a higher prior probability than a complicated one–there is no justification for this. "Simple" models are preferred to complicated ones because, in general, they tend to discover regions which are larger and not particularly convoluted and therefore more easily discovered by other models. This, in turn, implies a better bound on the test error. To see why regularization works, it's important to understand what regularization is doing. Regularization effectively forces models to find more similar solutions; for instance, by placing restrictions on weights via the ridge [5] and the lasso [19]. Thus, models, even when trained on different training sets or with different initializations, are more likely to find similar patterns, and this in turn implies that the patterns which are found are more likely to generalize.

Unfortunately, using regularization means imposing a bias on the model, as per the bias-variance trade-off. These trade-offs are often only backed by intuition and better test error but not justified mathematically. On the other hand, consider an algorithm which follows the general idea as presented above: it trains an ensemble of models; analyzes cluster paths; defines a set of "good" points; and handles "bad" points in some manner, whether by oversampling and training another model or by using some different method. Patterns which were found in smaller models are likely to be discovered by larger models, especially if they are true patterns and are not symptoms of overfitting. These patterns are identified in the larger models, so another model can be trained to focus on the truly bad (the "bad 2") points. This is beneficial for three reasons.

First, this ensemble is effectively a form of regularization on the larger models, since it forces the real patterns to be kept, and this increases test accuracy. There is no added bias since no new assumptions were made. Second, this method can be used iteratively, creating a process to continuously increase test accuracy. Third, this provides a framework for how models can collaborate with each other–they should yield when they are unsure and speak up when they are relatively certain. Combined with feature extraction techniques as outlined above, this could allow for an entirely new field of machine learning: continuous learning with neural networks.

5.2 Bounds on Validation Error of Ensembles

Consider an ensemble consisting of k neural networks, of which all incorrectly classify at most v of the "good" validation points. Then the number of incorrectly classified validation points when applied only to points such that at least a fraction f_1 of the models deem it "good" and of those, at least a fraction f_2 agree, is at most $\frac{v}{f_1 * f_2}$. This establishes a lower bound on the validation accuracy of the ensemble for "good" points. This is necessary because ensembling effectively trades a decrease in validation accuracy for expanding the number of "good" data points.

6 Discussion

6.1 Oversampling

The idea behind the model 2s is this: the model performs significantly better on "good" data points than "bad" data points because "good" data points represent features that the neural network is confident were not created as a result of overfitting. It is natural to continue by creating a new model to focus on the "bad" data points to improve test accuracy. Even if they were classified correctly in the training set, they may still have comparatively high loss. Thus, for any model used to classify "bad" data points, the idea is to increase the weight the model puts on these points in the training set to reduce this loss. Oversampling achieves this effect. On another note, it may be possible perform this process recursively; that is, to continue process for the models in the set of model 2s and create a set of so-called model 3's.

6.2 Partitions

For the training algorithm, the idea of partitioning the training data into a training and validation sets in multiple ways is crucial. This is because the multiple partitions are what create the ability to differentiate between real features and overfitting. Larger models seem to be able to emulate the effects of ensembles of smaller models trained on different portions of the training data. Intuitively, this makes sense–larger models are effectively just smaller models combined together, since neural networks are an iterative model, by design. This is further supported by the fact that for state-of-the-art models, "bad 1" test points are classified with similar accuracy as "original good" test points. Indeed, the critical step occurs in identifying "bad 2" data points, which are outliers in every model, and dealing with these specifically. While the algorithm presented oversamples the corresponding training data, this is not necessarily the optimal approach, and a better method of creating ensembles of models to handle these points should be considered in the future. Finally, it should be noted that the partitions should overlap as little as possible with each other, thus decreasing the probability that the same overfit features would be found by two different models.

6.3 Justification for Parameters Filtering Clusters

There exists an intuitive explanation behind the three parameters which were chosen when determining which data points are "good." The distance to a cluster center is considered because all data must belong to a cluster, and therefore clusters contain data which do not really belong to any single cluster. The idea is to filter out these outliers. The number of points in a split is considered because if this is too small, its impossible to reason about the accuracy of points which follow that path. Intuitively, this is because points in small splits are most likely outliers which do not truly belong, and while they may be classified correctly in the training set, this is most likely due to overfitting. Finally, the accuracy of a split is taken into consideration because if the accuracy of the split is low in the training or the validation set, theres no reason to believe it will be better in the test set.

7 Conclusion

In conclusion, this paper presents an algorithm to analyze features of neural networks and differentiate between useful features and those found due to overfitting. This process is then applied to larger models, generating state-of-the-art results for Wide-ResNets on CIFAR-10. Lastly, an analysis concerning bounds for the test accuracy of these ensembles is detailed, and a theorem bounding the test accuracy of neural networks is presented. Finally, the implications of this theorem, including an intuitive understanding of regularization, are presented.

Acknowledgments. I like to call this project the Miracle Project because it's a true miracle that this project happened...there are too many people to thank. First, thanks to my mom, dad, and brother. Second, thanks to Professors Barnabs Pczos and Majd Sakr. Third, thanks to Theo Yannekis, Michael Tai, Max Mirho, Josh Li, Zach Pan, Sheel Kundu, Shan Wang, Luis Ren Estrella, Eric Mi, Arthur Micha, Rich Zhu, Carter Shaojie Zhang, Eric Hong, Kevin Dougherty, Catherine Zhang, Marisa DelSignore, Elaine Xu, David Skrovanek, Anrey Peng, Bobby Upton, Angelia Wang, and Frank Li. You guys are the real heroes of this project. And, lastly, thanks to you, my dear reader.

References

1. Cubuk, E.D., Zoph, B., Mané, D., Vasudevan, V., Le, Q.V.: Autoaugment: learning augmentation policies from data. CoRR abs/1805.09501 (2018)
2. Deng, J., Dong, W., Socher, R., Li, L.J., Li, K., Fei-Fei, L.: ImageNet: a large-scale hierarchical image database. In: CVPR 2009 (2009)
3. Geifman, Y.: cifar-vgg (2018). https://github.com/geifmany/cifar-vgg
4. Gilpin, L.H., Bau, D., Yuan, B.Z., Bajwa, A., Specter, M., Kagal, L.: Explaining explanations: an overview of interpretability of machine learning. In: 2018 IEEE 5th International Conference on Data Science and Advanced Analytics (DSAA), pp. 80–89 (2018)
5. Hoerl, A.E., Kennard, R.W.: Ridge regression: biased estimation for nonorthogonal problems. Technometrics **42**, 80–86 (2000)
6. Ju, C., Bibaut, A., van der Laan, M.J.: The relative performance of ensemble methods with deep convolutional neural networks for image classification. CoRR abs/1704.01664 (2017)
7. Kingma, D.P., Ba, J.: Adam: a method for stochastic optimization. CoRR abs/1412.6980 (2015)
8. Krizhevsky, A.: Learning multiple layers of features from tiny images (2009)
9. van der Laan, M.J., Polley, E.C., Hubbard, A.E.: Super learner. Stat. Appl. Genet. Mol. Biol. **6**, Article 25 (2007)
10. LeCun, Y., Cortes, C.: MNIST handwritten digit database (2010). http://yann.lecun.com/exdb/mnist/
11. Liu, X., Wang, X., Matwin, S.: Interpretable deep convolutional neural networks via meta-learning. In: 2018 International Joint Conference on Neural Networks (IJCNN), pp. 1–9 (2018)
12. Olah, C., Mordvintsev, A., Schubert, L.: Feature visualization. Distill (2017). https://doi.org/10.23915/distill.00007. https://distill.pub/2017/feature-visualization
13. Schwenk, H., Bengio, Y.: Boosting neural networks. Neural Comput. **12**, 1869–1887 (2000)
14. Simonyan, K., Zisserman, A.: Very deep convolutional networks for large-scale image recognition. CoRR abs/1409.1556 (2015)
15. Srivastava, N., Hinton, G.E., Krizhevsky, A., Sutskever, I., Salakhutdinov, R.R.: Dropout: a simple way to prevent neural networks from overfitting. J. Mach. Learn. Res. **15**, 1929–1958 (2014)
16. Steinhaus, H.: Sur la division des corps matériels en parties. Bull. Acad. Pol. Sci. Cl. III **4**, 801–804 (1957)
17. Szegedy, C., Ioffe, S., Vanhoucke, V.: Inception-v4, inception-resnet and the impact of residual connections on learning. In: AAAI (2016)
18. Thorndike, R.L.: Who belongs in the family? Psychometrika **18**(4), 267–276 (1953). https://doi.org/10.1007/BF02289263
19. Tibshirani, R.: Regression shrinkage and selection via the lasso (1994)
20. Vapnik, V.N.: The nature of statistical learning theory. In: Statistics for Engineering and Information Science (2000)
21. Yamada, Y., Iwamura, M., Akiba, T., Kise, K.: Shakedrop regularization for deep residual learning (2018)
22. Zagoruyko, S., Komodakis, N.: Wide residual networks. CoRR abs/1605.07146 (2016)

Driver Distraction Detection Using Deep Neural Network

Shokoufeh Monjezi Kouchak(✉) and Ashraf Gaffar

Arizona State University, Tempe, AZ, USA
{smonjezi,agaffar}@asu.edu

Abstract. Driver distraction, drunk driving and speed are three main causes of fatal car crashes. Interacting with intricated infotainment system of modern cars increases the driver's cognitive load notably and consequently, it increases the chance of car accident. Analyzing driver behavior using machine learning methods is one of the suggested solutions to detect driver distraction and cognitive load. A variety of machine learning methods and data types have been used to detect driver status or observe the environment to detect dangerous situations. In many applications with a huge dataset, deep learning methods outperform other machine learning approaches since they can extract very intricated patterns from enormous datasets and learn them accurately. We conducted an experiment, using a car simulator, in eight contexts of driving including four distracted and four non-distracted contexts. We used a deep neural network to detect the context of driving using driving data which have collected by the simulator automatically. We analyzed the effect of the depth and width of the network on the train and test accuracy and found the most optimum depth and width of the network. We can detect driver status with 93% accuracy.

Keywords: Deep learning · Driver distraction · Driver safety · Neural network

1 Introduction

Driver distraction is defined as engaging in a secondary task that diverts the driver attention from the primary task of driving [1]. Driver distraction caused 10% of fatal car crashes and 15% of injury crashes in 2015. Based on NHTSA data, which were collected from 50 states, 37,461 people were killed in car crashes on U.S. roads in 2016, an increase of 5.6 percent from the previous year [2]. The fatal crash rate for teenagers is three times more than adults [3]. Driver distraction caused 58% of teenage drivers' car crashes [4]. The results of the most comprehensive research ever conducted using car accidents videos of teenager drivers show that distracted driving is likely much more serious problem than previously known, according to the AAA Foundation for Traffic Safety [4].

There are four types of driver distraction including visual, manual, cognitive and audio distraction. Distractive tasks such as texting, talking on the phone and eating can cause one or multiple types of distraction. Cognitive distraction is the hardest type of distraction to detect and eliminate since it is not visible [5]. Advanced Driver Assistant System, autonomous vehicles and evaluating the driver's behavior are three

G. Nicosia et al. (Eds.): LOD 2019, LNCS 11943, pp. 13–23, 2019.
https://doi.org/10.1007/978-3-030-37599-7_2

predominant solutions to enhance driver's safety and reduce the car crashes that are caused by distracted drivers.

The majority of car accidents happen due to human errors. ADAS goal is helping drivers in the driving process to diminish the possible driving error and improve the driver's safety. Some of them such as adaptive cruise control, adaptive light control and automatic parking automate repetitive or difficult tasks for drivers. Consequently, they make the task of driving easier [6]. On the other hand, the goal of other ADAS systems such as emergency brake, blind spot detection and collision avoidance system is to enhance car safety for everyone on the road by detecting predefined potentially dangerous situations and taking the control of the car for a short time [7]. Although using ADAS reduces driving fatality, it is not the ultimate solution for driver distraction problem since they work in some specific predefined situations and they cannot detect and handle all potentially dangerous situations.

An autonomous vehicle can observe its environment using a variety of sensors, such as camera, Radar and LIDAR and uses the collected data to move from the start point to its destination with minimum or no human input. Although an autonomous car controls the car for the majority of the driving time, the driver still needs to take over the control of the car in emergency situations, so with the current level of automation, autonomous vehicles are not the ultimate solution for driver distraction. Besides, there are many challenges in this area including winning public trust, hardware reliability and service certification, streamlined government regulations. They might solve the car safety challenge in the future when there are only autonomous vehicles on the road and the necessary infrastructures for them will be available in all cities [8–10].

Driver status is an important factor in driver distraction detection. Driver behavior detection using driver status data, driving data, psychological data or combination of them is another proposed solution for estimating driver distraction or the level of the driver's cognitive load. These data can be collected from car data buses or external sensors such as cameras and eye trackers. Deep learning and other machine learning methods use these data to learn driving patterns and driver status [11, 12]. Car infotainment system with car-friendly interface design is another solution that can diminish driver distraction notably [13–15].

In this paper, we used both traditional neural network and deep neural network to detect the driver distraction using driving data that was collected during our experiment. In the experiment, we defined four contexts of driving and two scenarios including distracted driving and non-distracted driving for each context. Then we used a multilayer neural network and a simple neural network to classify collected data in eight groups including four distracted and four non-distracted ones. So, the model detects if the driver is distracted or not and what is the context of driving. In Sect. 2, we compare the performance of the deep neural network (multilayer neural network) and the shallow neural network with one hidden layer. Section 3 is the experiment that we conducted to collect driving data vectors. Section 4 is the deep neural network results and Sect. 5 is the simple neural network results. Section 6 is the conclusion.

2 Deep Learning vs Neural Network

Deep learning is a branch of machine learning that has great flexibility to learn intricated patterns in real-world data by nesting hierarchy of concepts each of which is related to simpler concepts. In deep learning more abstract representation computes in terms of less abstract ones. Deep neural networks learn categories incrementally using their hidden layers and they are powered by an enormous amount of data. Using deep networks decreases the need to domain expertise and dedicated feature selection since the model learns complex patterns in an incremental approach by itself [16–18]. Theoretically, a simple neural network with one hidden layer can estimate almost any function. But in practice shallow neural networks can be inefficient for learning intricated patterns, nonlinear correlations and large datasets [19, 20]. Deep learning and traditional neural networks have been used in many driving safety-related applications, following are some examples.

Li et al. [21] discuss a lane detection method using deep neural networks. Two deep neural networks were used in the model. A multitask deep convolutional network has been used to detect the target considering the region of interest. A recurrent neural network, which is adapted to be used as structured visual detection, is another deep network that's used in this model. The first network models lane structures, and the recurrent neural network detects the lane boundaries without any prior knowledge. Both proposed networks outperform conventional detectors in practical traffic scene. Images of the driving environment were used as inputs of these two deep networks, so it can be used in cars that equipped with cameras such as autonomous cars. This model completely relies on the collected data by external devices [21].

Koesdwiady et al. [22] propose a new three-parts framework for driver distraction detection. A dataset of images with a variety of driver's ethnicity, gender and camera position was used in this framework. The first part is a convolutional deep neural network which is used for high abstracted feature extraction. The second part is a max-pooling layer that decreases the dimension of features, and the last part includes six fully connected layers and a SoftMax layer. The best test accuracy was 90% and the average accuracy was 80% per class. This system uses a multi-class classification process to map the input observation to a driver state. This approach needs some cameras to collect driving data, but our approach detects driver state, distracted and not distracted, only using Can-bus data [22].

Monjezi Kouchak and Gaffar [23] discusses a new non-intrusive model that can predict the driver behavior in distracted driving scenarios using only driving data with more than 90% accuracy. They used a neural network with one hidden layer and a small dataset with 525 driving data vectors. The results show shallow neural networks can extract and learn the patterns of driving in this small driving related dataset with more than 92% accuracy.

In this paper we use both multilayer neural network and simple neural network to classify the driving data in four distracted and four non-distracted driving status and analyze the effect of the depth and width of the network on the training process and training accuracy. In this model, we only used driving data that were collected from car CAN-bus to detect drivers' status.

Based on our collected data and our observations during the experiment and our previous experiment [23], driver distraction doesn't have same effect on the driver's performance in different contexts of driving. Driving in adverse conditions such as fogy weather needs more cognitive attention, so handling a secondary task in these conditions can resulted in more severe driver distraction compared to ideal driving context. As a result, we decided to consider several contexts of driving including day (ideal), Fog (adverse), Night (adverse) and Fog & night (double adverse) in our experiment instead of only one driving context that has been used in other similar works. The output of our classifier is the context of driving and the driver distraction status which gives use more information about the severity level of distraction. For example, in our experiment on average 50.73% of drivers made more errors while driving in adverse and double adverse conditions compared to the ideal mode, so if we detect distraction in fog context, it shows more distraction compared to distraction in day context.

3 Experiment

3.1 Volunteers and Equipment

We conducted an Experiment in the ASU HCaI lab using a Drive Safety Research simulator DS-600 that is fully integrated, high performance, high fidelity driving simulation system which includes multi-channel audio/visual system, a minimum 180° wraparound display, full-width automobile cab (Ford Focus) including windshield, driver and passenger seats, center console, dash and instrumentation as well as real-time vehicle motion simulation. We used an android application, which displays the user interface and it was hosted on the Android v4.4.2 based Samsung Galaxy Tab4 8.0 which was connected to the Hyper Drive simulator.

We invited 97 students including undergrad and grad students of Arizona State University in the range of 20 to 35 years old with at least two years of driving experience to participate in our experiment. Each student took an initial training on the simulator, then they were asked to take an approximately 40-minute simulated drive in eight driving scenarios including four distracted and four non-distracted ones. We defined four driving contexts including Day, Night, Fog and Fog & Night and for each driving context, the drivers tried both distracted and non-distracted driving. We used minimalist design in our interface [13] and in this interface design each application can be accessed with 2,3 or 4 steps of navigation from the main screen. In this experiment, distracting task is defined as reaching a specific application on the interface. We defined three types of distracting tasks including two, three and four steps tasks. As soon as the driver starts each distracted mode's trip, we asked him to do some tasks continuously considering this fact that driving is the primary task and the secondary task has less priority. In non-distracted scenario, we asked the driver to completely focus on driving.

These eight distracted and non-distracted scenarios were used to generate a large number of complex drive situations that will be recorded and analyzed using our deep neural network and traditional neural network. The same road which is an urban road with high traffic has been used in all scenarios. Besides we added some unpredictable events to the road such as some pedestrians to make our experiment more realistic. Figure 1(a) shows the designed road and Fig. 1(b) shows the volunteer's view of the road in night mode from the simulator's driver seat.

Fig. 1. (a) Designed road (b) Simulator night mode

3.2 Data

The simulator collected 14,914,947 data vectors during 776 trips that have been done in this experiment. The sampling rate of the simulator set on the 60 samples per second that is the maximum possible sampling rate of the simulator. First, we preprocessed the dataset and removed vectors with zero accelerating. After that, compressed the dataset and replace every 20 vectors with the average of them since we set the maximum sampling rate and using all collected data was computationally expensive besides it doesn't provide much more information compared to a compressed dataset. The final dataset has 621,088 data vectors. We had same number of driving trips for each of these eight driving scenarios, so almost $1/8^{th}$ of the collected data belongs to each scenario.

Collected data vectors have 53 features, so we used a paired t-test to select most significant ones for our classifier. Based on the t-test results, ten driving features were chosen including Velocity, Lane position, Steering wheel, Speed, Accelerating, Brake, Long Accelerating, Lateral accelerating, Headway Time and Headway distance. Besides, we normalized these features in order to reduce the chance of overfitting while training. The target dataset shows the driving scenario which includes four distracted and four non-distracted scenarios. We used numbers 1 to 8 to represent these scenarios. These input dataset and targets were used in both Deep Neural Network and traditional Neural Network and we analyze the effect of depth and length of the network on the training speed and the network's accuracy.

4 Deep Neural Network Results

We designed a multilayer neural network using TensorFlow DNNClassifier and we used 10 selected driving features as the input of this network. The output is eight driving scenarios of our experiment that each of them shows both the driving context and the driver distraction status. We selected Adam optimizer as the network optimization function since it is recommended as an appropriate optimizer for deep learning and it had better performance compared to other optimizers that we tried. After that we tried a variety of learning rates from 0.1 to 0.0001. Considering the speed and accuracy of training, learning rate 0.0001 was the best choice for our network. We chose 80% of data for training and 20% for testing the model.

The number of hidden neurons is an important factor in the accuracy of deep networks, so we tried large range of hidden neurons from 50 to 5,000. Between 500 to 1,000 neurons was the best range for our application. Figure 2 shows the loss of 10 hidden layers network with 500 neurons in each hidden layer and Fig. 3 shows the loss of the same network with 1,000 hidden neurons.

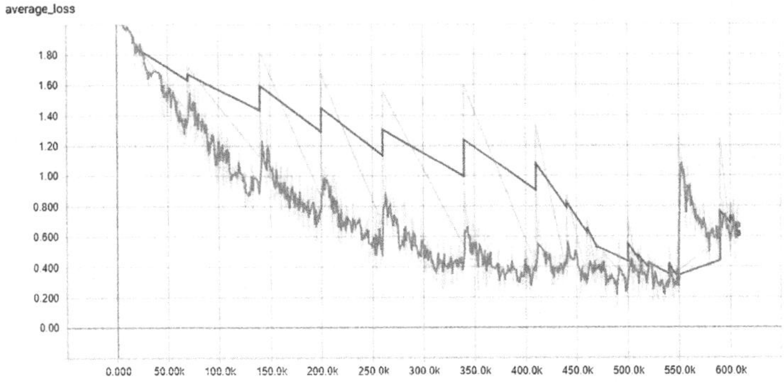

Fig. 2. The loss of 10-hidden layers network with 500 hidden neurons in each layer

The network with 1,000 hidden neurons reached 88.21% train and 86.63% test accuracy after 400,000 epochs. On the other hand, the network with 500 hidden neurons reached 80.78% train and 78.13% test accuracy after 600,000 epochs. In sum, the training process with less hidden neurons was significantly slower and less accurate. We tried more than 1,000 hidden neurons but the accuracy in both test and train was 2% less than the network with 1,000 neurons, so increasing the number of hidden neurons more than 1,000 didn't help us to enhance the network accuracy. In sum, we chose 1,000 as the number of hidden neurons in each hidden layer.

We tried a variety of number of hidden layers from 5 to 15. Table 1 shows some of the achieved results. The range of accuracy for the networks with 7 to 11 hidden layers was between 85% to 87% and the gap between the train and the test accuracy was between 2% to 5%. The network with 13 hidden layers outperforms all shallower networks. Figure 4 shows the accuracy of this network. In the first 250,000 epochs, the accuracy increased fast and between 250,000 to around 450,000 epochs the speed of

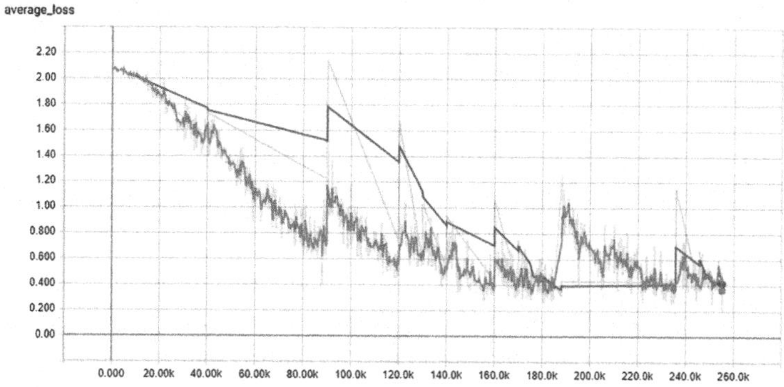

Fig. 3. The loss of 10-hidden layers network with 1000 hidden neurons in each layer

improving accuracy decreased significantly. In sum, the whole process is smooth and there isn't any sudden change and abnormally in the curve. We tried to enhance the network's performance by adding more than 13 hidden layers but both train and test accuracy decreased around 2% compared to 13 hidden layers network when we added one more hidden layer. In sum, the most appropriate network for our application is a network with 13 hidden layers, 1,000 neurons in each hidden layer and Adam optimizer.

Table 1. Results of multilayer neural network

Number of hidden layers	Number of neurons	Training accuracy	Test accuracy
7	1000	87.52	82.42
9	1000	88.08	86.48
10	1000	88.21	86.63
11	1000	85.34	83.58
13	1000	94.06	93.1
14	1000	81.79	79.83
10	500	80.78	78.13
10	2000	86.01	84.98

5 Neural Network Results

We built a neural network with one hidden layer having 1,000 hidden neurons. A variety of learning rates, from 0.1 to 0.0001, have been tried and learning rate 0.001 shows the best performance. After that, we tried different number of hidden neurons from 100 to 10,000. In order to improve the accuracy of the model, we increased the number of epochs and if after a large number of training epochs, we didn't see any improvement, we increased the number of hidden neurons. Table 2 shows a summary of achieved results.

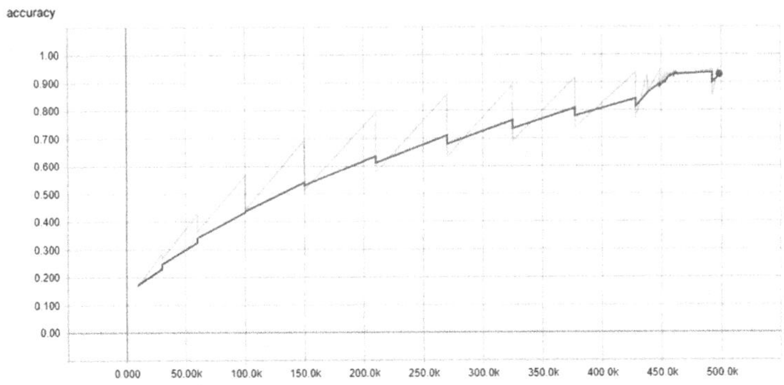

Fig. 4. The 13-hidden layers network's accuracy

Table 2. The summary of the simple neural network's results with the whole dataset

Hidden neurons	Number of epochs	Training accuracy	Testing accuracy
100	500,000	19.71	19
1000	100,000	21.42	20.73
2000	500,000	24.59	23.59
5000	500,000	25.18	23.83
5000	1,000,000	26.28	24.25
10000	1,000,000	26.48	24
10000	10,000,000	29.62	26.88

The network that was trained with 100 neurons didn't show much improvement after 500,000 epochs, so we increased the neurons 10 times and still it didn't help. We kept adding to the number of neurons and epochs. Finally increasing the number of hidden neurons 100 times and the number of epochs 20 times only improved the accuracy of the model 10%, so we decided to use smaller dataset since the network couldn't learn the dataset's patterns efficiently.

We made a dataset with 8,000 data vectors using 1,000 samples from each driving scenario, so the dataset is balanced and much smaller than the first one. We trained the same neural network for 2,500 epochs and with a variety of learning rates. The best results achieved with 0.01 as the learning rate. After that, we started from 100 hidden neurons and increased it when the network overfitted. Table 3 shows some of achieved results.

As we increased the number of hidden neurons, the model overfitted faster. Finally, the network with 500 hidden neurons after 10,000 epochs reached the best results which is 85.81% training accuracy and 75.99% test accuracy. Increasing the number of hidden neurons even eight times more than this didn't enhance the model's performance. The best accuracy of one hidden layer network with a small part of our dataset is not as good as the worst result of our deep neural network with the whole dataset.

When we used the whole dataset, increasing the number of hidden neurons enhanced the speed of training process, but increasing the width of the network is not

Table 3. The summary of the simple neural network's results with small dataset

Hidden neurons	Number of epochs	Training accuracy	Testing accuracy
100	70,000	72.7	63.81
500	15,000	88.03	75
1000	15,000	88.37	74.18
2000	15,000	87.18	75.56
4000	10,000	85.22	73.37

enough to detect and learn intricated patterns of our dataset. Even 10 times wider network, compared to our deep neural network, couldn't reach acceptable accuracy after a long training process. The combination of width and depth of network can help to extract complex and non-linear patterns from the dataset. When we used only 1.2% of our dataset, although the performance enhanced significantly compared to the network that trained with whole dataset the final accuracy was much weaker than the deep neural network. Besides, the network overfitted after few thousand epochs, so we couldn't continue training to improve the network's accuracy.

The novelty of our classifier is that we consider multiple contexts of driving and this model not only can distinguish the normal behavior and distracted driving using driving data, but also can detect the driving context that the input data sample were collected in it. The size of the dataset and the complexity of driving patterns of this dataset are two factors that show a simple neural network is not suitable choice for our model since it cannot extract complex driving patterns from this large dataset. The deep neural network could extract more details about driving patterns in each layer and classify the input data vectors based on the context of driving and distraction status to eight classes accurately.

6 Conclusion

Driver distraction is one of the leading causes of car fatal accidents. Many solutions have been suggested to diminish driver distraction and increase driver safety. Using machine learning and deep learning methods to detect driver behavior is one of these solutions. We conducted an experiment in eight different driving scenarios and the same road to produce an artificial intelligence friendly dataset. After that, we used this dataset to train a traditional neural network and a multilayer neural network. We used these two networks to classify driving data to eight groups, and each group reveals two important information about the driver status including driver distraction status and the driving context. Based on our observation and collected data in this experiment, driving context has effect on the severity of driver distraction. It means, doing a distractive task in adverse conditions resulted in higher level of driver distraction and has more negative effect on the driver performance. As a result, we believe that detecting the context of driving and driver distraction status together provides more valuable information about the cognitive load and distraction level of the driver compared to methods that only detect if the driver is distracted or not.

We compared the results of these two networks to analyze the effect of depth and width of the neural network on the accuracy of our classifier. The results show that shallow neural networks unable to classify the whole dataset and even when we chose around 1% of the dataset, the best accuracy of the shallow neural network was weaker than the weakest performance of our deep neural network. When we have intricated and non-linear patterns, using deep learning is a more suitable choice although the training process is slower and needs more computation power compared to other machine learning methods. We reached 93% test accuracy using deep neural network but the best accuracy of the shallow network with the whole dataset was only 26%.

References

1. Gaffar, A., Monjezi Kouchak, S.: Quantitative driving safety assessment using interaction design benchmarking. In: IEEE Advanced and Trusted Computing (ATC 2017), San Francisco Bay Area, USA, 4–8 August 2017 (2017)
2. NHTSA, USDOT Releases 2016 Fatal Traffic Crash Data. https://www.nhtsa.gov/press-releases/usdot-releases-2016-fatal-traffic-crash-data. Accessed 3 Mar 2018
3. IIHS HDLI: Insurance Institute for Highway Safety Highway Lost Data Institute, Teenagers Driving carries extra risk for them. https://www.iihs.org/iihs/topics/t/teenagers/fatalityfacts/teenagers. Accessed 10 Feb
4. AAA, Distraction and Teen Crashes: Even Worse than We Thought. https://newsroom.aaa.com/2015/03/distraction-teen-crashes-even-worse-thought/. Accessed 5 Jan
5. Gaffar, A., Monjezi Kouchak, S.: Using artificial intelligence to automatically customize modern car infotainment systems. In: Proceedings on the International Conference on Artificial Intelligence (ICAI), pp. 151–156 (2016)
6. Geronimo, D., Lopez, A., Sappa, D., Graf, T.: Survey of pedestrian detection for advanced driver assistance systems. IEEE Trans. Pattern Anal. Mach. Intell. **32**(7), 1239–1258 (2010)
7. Paul, A., Chauhan, R., Srivastava, R., Baruah, M.: Advanced driver assistance systems, SAE Technical Paper 2016-28-0223 (2016). https://doi.org/10.4271/2016-28-0223
8. Monjezi Kouchak, S., Gaffar, A.: Determinism in future cars: why autonomous trucks are easier to design. In: IEEE Advanced and Trusted Computing (ATC 2017), San Francisco Bay Area, USA, 4–8 August 2017 (2017). https://doi.org/10.1109/uic-atc.2017.8397598
9. Maurer, M., Gerdes, J.C., Lenz, B., Winner, H. (eds.): Autonomous Driving: Technical, Legal and Social Aspects. Springer, Heidelberg (2016). https://doi.org/10.1007/978-3-662-48847-8. ISBN 978-3662488454
10. Gaffar, A., Monjezi Kouchak, S.: Undesign: future consideration on end-of-life of driver cars. In: IEEE Advanced and Trusted Computing (ATC 2017), San Francisco Bay Area, USA, 4–8 August 2017 (2017)
11. Lee, J.: Dynamics of driver distraction: the process of engaging and disengaging. Association for Advancement of Automotive Medicine. PMC4001670, pp. 24–32 (2014)
12. Fuller, R.: Towards a general theory of driver behavior. Accid. Anal. Prev. **37**(3), 461–472 (2005)
13. Gaffar, A., Monjezi Kouchak, S.: Minimalist design: an optimized solution for intelligent interactive infotainment systems. In: IEEE IntelliSys, the International Conference on Intelligent Systems and Artificial Intelligence, London, 7th–8th September 2017 (2017)
14. Cellario, M.: Human-centered intelligent vehicles: toward multimodal interface integration. IEEE Intell. Syst. **16**(4), 78–81 (2001)

15. Gaffar, A., Monjezi Kouchak, S.: Using simplified grammar for voice commands to decrease driver distraction. In: The 14th International Conference on Embedded System, pp. 23–28 (2016)
16. Adam, G., Josh, P.: Deep Learning. O'Reilly (2017)
17. Jürgen, S.: Deep learning in neural networks: an overview. Neural Netw. **61**, 85–118 (2015)
18. Graves, A.: Supervised Sequence Labelling with Recurrent Neural Networks. Studies in Computational Intelligence. Springer, Heidelberg (2012). https://doi.org/10.1007/978-3-642-24797-2. ISBN 978-3642247965
19. Stuart, J., Peter, N.: Artificial Intelligence a Modern Approach. Prentice Hall, Upper Saddle River (2010)
20. Bishop, C.: Pattern Recognition and Machine Learning, 1st edn. Springer, New York (2006). ISBN 0-387-31073-8
21. Li, J., Mei, X., Prokhorov, D., Tao, D.: Deep neural network for structural prediction and lane detection in traffic Scene. IEEE Trans. Neural Netw. Learn. Syst. **28**, 14 (2017)
22. Koesdwiady, A., Bedawi, S.M., Ou, C., Karray, F.: End-to-end deep learning for driver distraction recognition. In: Karray, F., Campilho, A., Cheriet, F. (eds.) ICIAR 2017. LNCS, vol. 10317, pp. 11–18. Springer, Cham (2017). https://doi.org/10.1007/978-3-319-59876-5_2
23. Monjezi Kouchak, S., Gaffar, A.: Non-intrusive distraction pattern detection using behavior triangulation method. In: International Conference on Computational Science and Computational Intelligence (CSCI), USA, 14th–16th December 2017 (2017). https://doi.org/10.1109/csci.2017.140

Deep Learning Algorithms for Complex Pattern Recognition in Ultrasonic Sensors Arrays

Vittorio Mazzia[1,2,3(✉)], Angelo Tartaglia[1,2], Marcello Chiaberge[1,2], and Dario Gandini[1,2]

[1] Department of Electronic and Telecommunications Engineering (DET), Politecnico di Torino, Turin, Italy
vittorio.mazzia@polito.it

[2] PIC4SeR - Politecnico Interdepartmental Centre for Service Robotic, Turin, Italy

[3] SmartData@PoliTo - Big Data and Data Science Laboratory, Turin, Italy

Abstract. Nowadays, applications of ultrasonic proximity sensors are limited to a post-processing of the acquired signals with a pipeline of filters and threshold comparators. This article proposes two different and novel processing methodologies, based on machine learning algorithms, that outperform classical approaches. Indeed, noisy signals and presence of thin or soundproofing objects are likely sources of false positive detections that can make traditional approaches useless and unreliable. In order to take advantage of correlations among the data, multiple parallel signals, coming from a cluster of ultrasonic sensors, have been exploited, producing a number of different features that allowed to achieve more accurate and precise predictions for object detection. Firstly, model-based learning as well as instance-based learning systems have been investigated for an independent time correlation analysis of the different signals. Particular attention has been given to the training and testing of the deep fully connected network that showed, since the beginning, more promising results. In the second part, a recurrent neural network, based on long short term memory cells, has been devised. As a result of its intrinsic nature, time correlations between successive samples are not more overlooked, further improving the overall prediction capability of the system. Finally, cutting edge training methodologies and strategies to find the different hyperparameters have been adopted in order to obtain the best results and performance from the available data.

Keywords: Deep learning · Ultrasound sensors · Industrial security

1 Introduction

The usage of ultrasonic sensors is widely diffused in many and different fields. In [1], this type of sensors are used to measure the distance of an object. It is a very

G. Nicosia et al. (Eds.): LOD 2019, LNCS 11943, pp. 24–35, 2019.
https://doi.org/10.1007/978-3-030-37599-7_3

old technology [2], but still new patents [3] and papers [4] are continuously disclosed in a wide range of fields: from distance security measurement, automotive field to robotics applications. Other newest applications use ultrasonic sensors for image reconstruction of organs like liver [4] or heart [5]. These last techniques are also associated with machine and deep learning analysis for post-processing of the captured images. Other machine learning related works with ultrasonic sensors can be found in [6–8].

This paper proposes an alternative methodology, based on machine learning algorithm, to manage and extract results from an ensemble of sensors in order to detect the presence of possible obstacles. Indeed, instead of using a single transducer, correlations between different signals open the possibility to detect targets that are not recognized by a classical methodology. In fact, it is possible to demonstrate that a redundancy of sensors makes the entire system more robust and in addition it opens the possibility to drastically reduce the number of false positive detections. So, based on this new type of data and based on modern technique of machine learning, this paper shows that is possible to design a more accurate and precise system than ones offered by tradition approaches.

2 Case Study

Industrial security is continuously evolving enclosing techniques and methods that take advantage of latest researches. Often classical processing of data is not sufficient to reach the desired level of security for workers and this paper deals with this scenario. The proposed research can be a valuable support for classical methodologies, making a more strong and robust overall system. Possible applications of the enhanced system include obstacle avoidance for automatic guided vehicles (AGV), stacker cranes and open computer numerical control (CNC) machines. The devised test-bed is a replica of the worst condition found in a

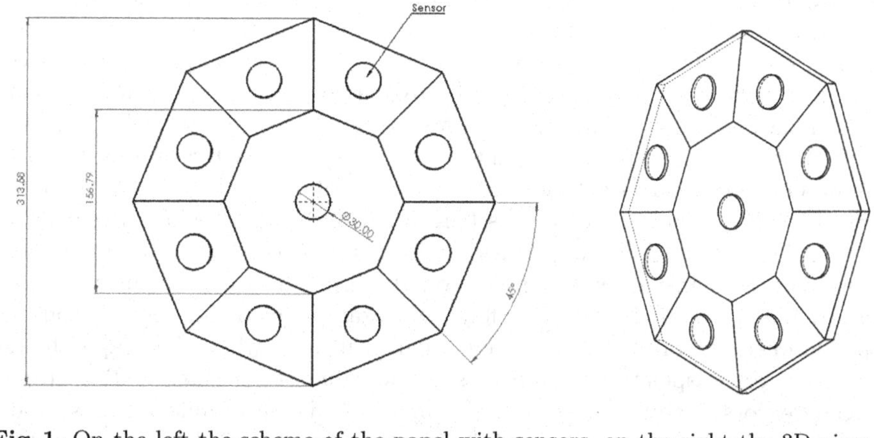

Fig. 1. On the left the scheme of the panel with sensors, on the right the 3D view of overall system.

hypothetical real industrial case and it was used to generate the data set and later for the tests of the final models.

2.1 Dataset Acquisition Bench

The system that has been considered for data acquisition is composed of an array of nine coaxial ultrasonic sensors equally spaced on an octagonal board (Fig. 1). The test bench has been used to detect the presence of a thin rod posed in front of the cluster. The target has been coated with soundproofing material in order to test the system in critical condition reducing the amount of reflected waves. The target has been placed in different positions, orientations and distances trying to emulate the real conditions and to produce a sufficiently large and comprehensive data set for the training of the different models. Figure 1 shows the sensor cluster and the overall system setting. Lateral borders are slightly tilted in order to further open the field of view of the array of sensors. The described testing area has been used for the generation of a very large data set composed by the multiple signals acquired from the nine different sensors. Signals have been recorded as raw data, only discarding the first noisy part of the acquisitions. Indeed, having used coaxial type of sensors (emitter and receiver are on the same axis), the excitation phase produces, for a short period, saturated signals on all transducers that can be easily omitted. Moreover, it is pretty straightforward to automatize this simple procedure removing always the first part of the incoming signals.

3 Dataset Visualization

Firstly, the recorded raw data have been analyzed and visualized with different solutions. Using, principal component analysis (PCA), it has been possible to plot on a tridimensional graph the different acquisition points. In Fig. 2, blue points are samples characterized by the presence of the target and orange points when the object is not present in the field of view. Reflections made by the thin rod are so subtle that they can be easily lost in the background noise. However, even if data points are not well separated, it is interesting to notice how the two classes clustered in specific regions of the space making it clear how correlations between different sensors are a key aspect to exploit for the detection of the target. As already introduced, the main motivation for the use of an ensemble of sensors is that it opens the possibility to exploit correlations between the acquired parallel signals (Fig. 3). A first high-level analysis can be done considering the following considerations: if the correlation between two sensors reaches high values, the points on the graph will be roughly disposed on a straight line and there can be found an example of this in the correlation graph between sensors 9 and 2 (green square); instead when the graph shows a cloud-pattern disposition it indicates a low value of correlation. The graph between sensors 1 and 3 (red square) can be used as example for this case. From those correlations the network is capable to understand information not explicitly given by the sensor itself.

Fig. 2. Projection of the nine signals using PCA. The data is normalized by shifting each point by the centroid and making it unit norm. (Color figure online)

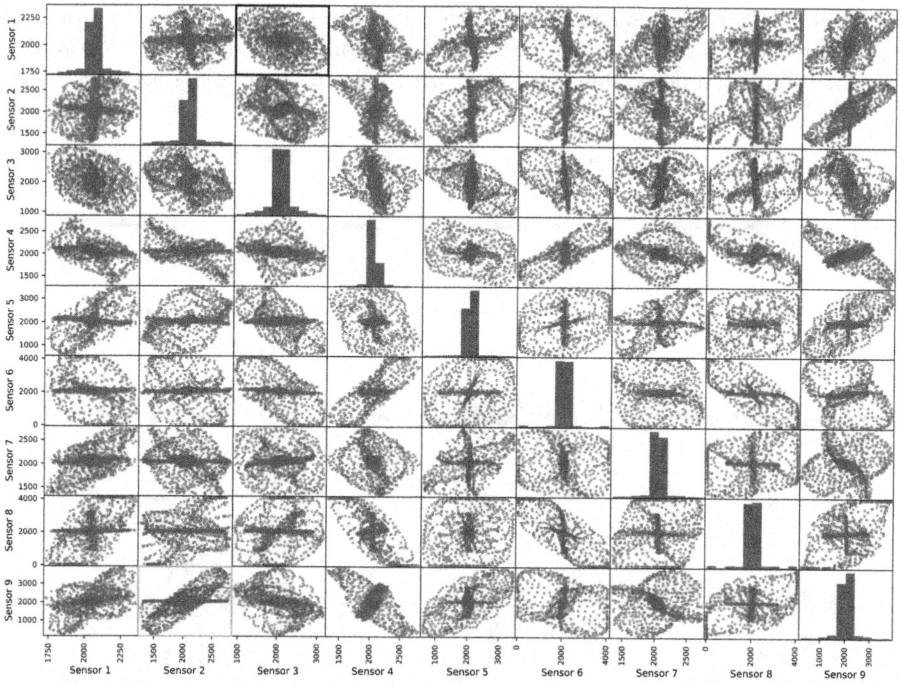

Fig. 3. Correlation matrix of the signals coming from the nine sensors. (Color figure online)

4 Preliminary Analysis

First of all, the acquired raw dataset has been analyzed with a simple classical methodology that can be easily improved with more elaborated processing [6]. However, already in this preliminary part, it has been taken into account the advantage of working with a redundancy of sensors. Indeed, it is obvious from the following results that correlations between the sensors can largely improve the detection accuracy. The nine signals produced by one of the acquisitions are represented in Fig. 4. The detection is easily recognizable with the second sensor response, but it is lost in the background noise in almost all remaining sensors. The envelope curve of each acquisition is highlighted in orange on each sensor signal. Moreover, it is clear how the acquired waves are very dissimilar due to the different location of the sensors on the board.

Firstly, it is possible to perform a simple detection independently on each sensor, simply by comparing the resulting envelops with a manually tuned threshold T_a. However, due to the critical setting of the designed test bench, it is possible to demonstrate that with this simple strategy each sensor produces a huge number of false positive detections.

On the other hand, it is possible to implement a more elaborated processing, exploiting the concurrent signals coming from the nine sensors. As it is represented in Fig. 5, it is pretty straightforward to construct a cumulative function which takes into account how many sensors are above the imposed threshold T_a. At this point, a second threshold T_b has to be set, in order to determine how many sensors should concur to confirm a detection. On Fig. 6, with two different colors are presented

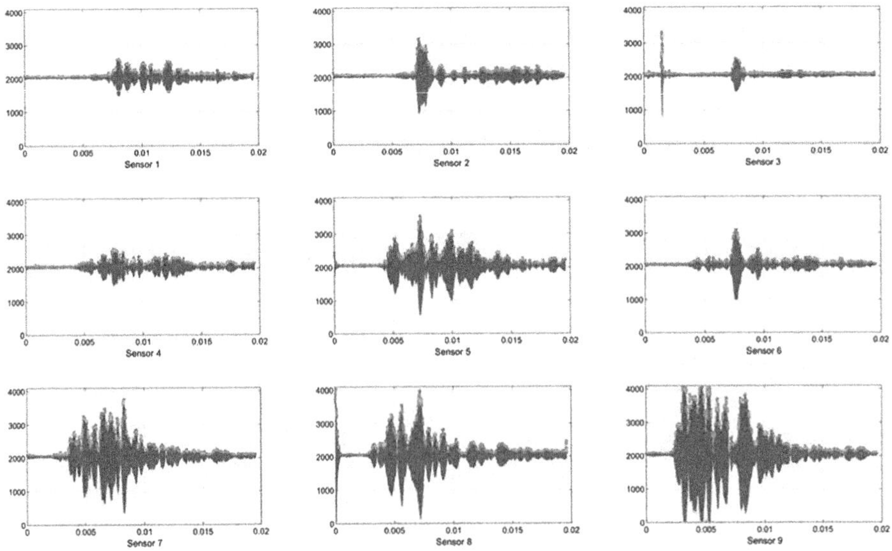

Fig. 4. Waveforms of the nine sensors of a sample acquisition. Envelops are highlighted in orange on each sensors graph. (Color figure online)

two possible outcomes with different values of T_b. In conclusion, it is clear from these preliminary analysis how thresholds T_a and T_b can greatly affect the overall result of the system, generating false positive alarm or miss detections. So, difficulty in selecting optimized thresholds, the manual interpretation of a considerable amount of acquisitions and boost in performance exploiting correlations between the different sensors, are the driving motivations for the presented innovative methodologies.

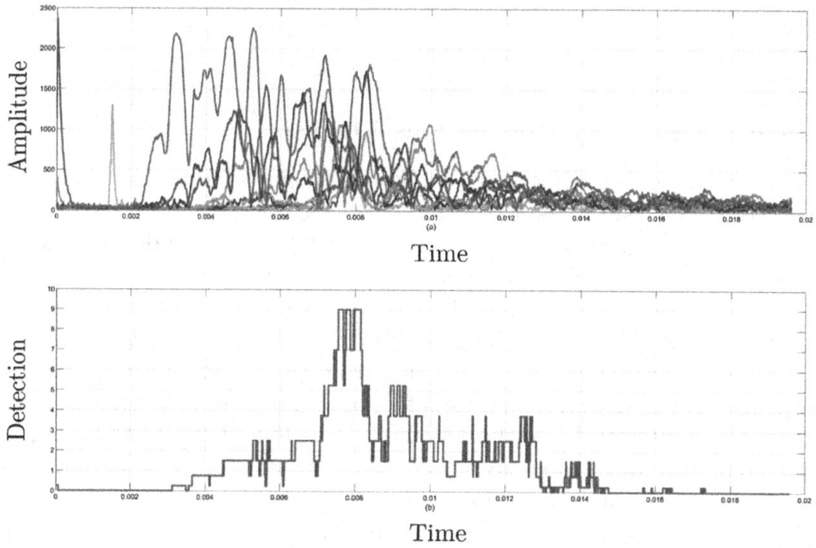

Fig. 5. Above, envelope curves of the acquired signals. Below, cumulative function of the sum of concurrent detections.

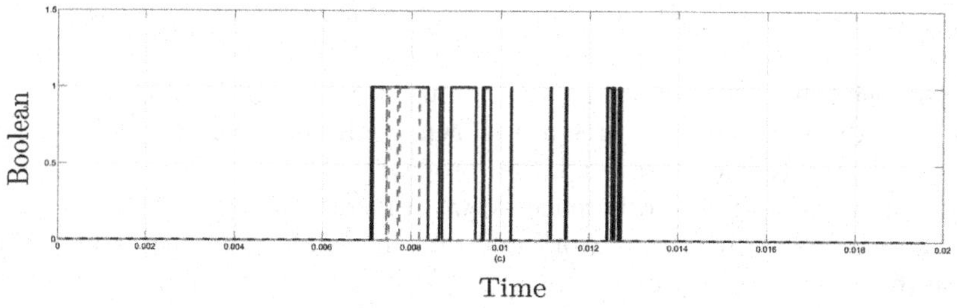

Fig. 6. The solid blue line: detections with concordance of 3 sensors and orange dashed line: concordance of 6. (Color figure online)

5 Proposed Methodologies

The analysis and processing of the ultrasonic waves has been carried out using two distinct approaches. Initially, only samples value range are took into account, neglecting temporal correlations between subsequent instances. However, the contemporary use of multiple sensors and the derived correlations between different signals, opens the possibility to build and train robust models, able to achieve optimal results. Secondly, a much more sophisticated model has been designed to exploit jointly the time information of the different wave forms and their amplitude values. In the next section both methodology are compared and analyzed in view of the results achieved.

5.1 Independent Time Correlations Analysis

Model-based learning as well as instance-based learning systems have been investigated, showing excellent results on both types of binary classifiers. The preprocessing phase is a standard procedure. Firstly, the acquired dataset has been processed with a simple pipeline, removing non representative samples (initial time steps with saturated signals due to the emission of ultrasound waves) and applying standardization to perform feature scaling. Secondly, the resulting training data, with nine attributes, coming from the designed array of sensors, have been manually annotated and used to train and evaluate the different models. In order to better evaluate the different cases, K-fold cross-validation, with k equal to ten, and randomized search have been used in order to evaluate some initial models by varying some hyperparameters. For the sake of completeness, for the first algorithm, K-nearest neighbors (K-NN) [10], 6 is the number of neighbors used with Euclidean as distance reference metric. For the Random Forest [11], 50 is the number of estimators with 3 as minimum samples split (minimum number of samples required to split an internal node) and maximum depth of 15. The selected function to measure the quality of a split was 'entropy' for the information gain. Finally, for the kernel SVM [12], it has been used a radial basis function kernel 'rbf' with 1.5 as value for the penalty parameter. After this first preliminary phase, the analysis has been focused only on the most promising hyperparamenters, fine tuning it with a grid search strategy. In Table 1 the different architectures are shown, with their corresponding results generated by the K-fold cross-validation process [13].

It is possible to see the difference between results achieved by the feed forward fully connected neural network and all other supervised algorithms. That is made possible by the presence of multiple hidden layers, capable of mining deep correlations between the different samples and represent knowledge in a hierarchical structure [14]. After the grid search analysis, the selected configuration accounts five hidden layers with 32, 32, 32, 52 and 62 rectified linear units (ReLU) respectively, for a total of 7,497 trainable parameters. Moreover, a batch of 128 instances has been used with AMSGrad [15] optimizer, an adaptive learning rate method which modifies the basic ADAM optimization algorithm. The overall algorithm update rule, without the debiasing step, is shown in Eq. 1.

The first two equations are the *exponential decay* of the gradient and *gradient squared*, respectively. Instead, with the third equation, keeping an higher v_t term results in a much smaller learning rate, η, fixing the exponential moving average and preventing to converge to a sub-optimal point of the cost function.

$$\begin{aligned} m_t &= \beta_1 m_{t-1} + (1-\beta_1) g_t \\ v_t &= \beta_2 v_{t-1} + (1-\beta_2) g_t^2 \\ \hat{v_t} &= max(\hat{v_{t-1}}, v_t) \\ \theta_{t+1} &= \theta_t - \frac{\eta}{\sqrt{\hat{v_t}} + \varepsilon} m_t \end{aligned} \tag{1}$$

Finally, only Dropout regularization technique has been applied at each layer with a drop rate of 30%. It is also worth pointing out that, due to the online learning nature of neural networks, once a system is trained on the original dataset, it can be further improved with new mini-batches coming from the actual application. That opens the possibility to devise a dynamic setup able to adjust and evolve to the specific application and to new environmental changes.

5.2 Time Correlations Analysis

In this second part, time correlations, between subsequent data samples, are no more overlooked. On the contrary, that additional information is exploited to achieve an even better model, more prone to generalize over new sample data points. That is achieved using a deep learning model based on long short term memory (LSTM) layers that increasingly extracts longer range temporal patterns. Moreover, in order to avoid random interactions between different time-steps and channels (LSTM hidden units), a many-to-many architecture [16] has been adopted. For this purpose, a TimeDistributedDense layer is applied which executes a Dense function across every output over time, using the same set of weights, preserving the multidimensional nature of the processed data. Figure 7 presents a simplified overview of the devised model. The adoption of this strategy, opens the possibility to exploit the temporal correlations between the different samples without leaving out the interactions among the nine signals.

Dataset Pre-processing. For this second case study in the pre-processing pipeline few more steps were needed. Indeed, the devised recurrent neural network (RNN), requires data points to be collected in slice of time series. So, before feeding the model, the samples of each sensor have been collected in chunks of data, forming a three-dimensional array, used for the training of the model. In order to maximize the number of possible new instances of the training dataset the following rule has been adopted:

$$X_{n,:,:} = x^{(i-t):(i)} = (x^{(i-t)}, ..., x^{(i)}) \tag{2}$$

where t is the chosen time-step range, $\underline{X} \in \mathbb{R}^{i \times j \times k}$ is the 3rd-order dataset output tensor and $x^{(i)}$ are the feature vectors of the old dataset which has

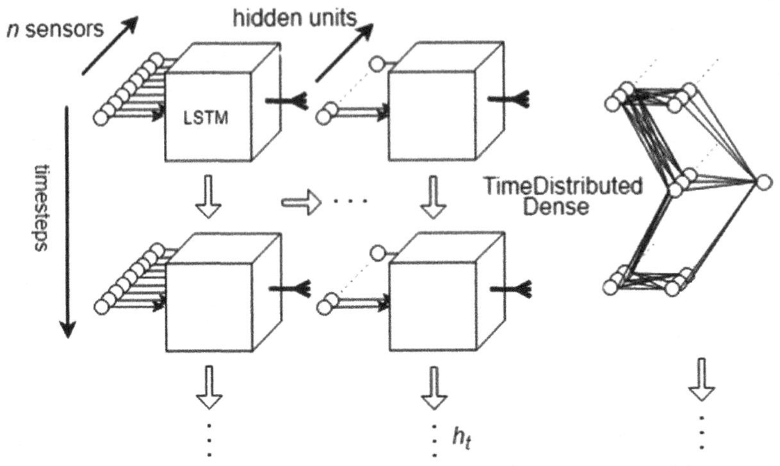

Fig. 7. Overview of the adopted RNN architecture based on LSTM units. X^i samples feed a sequence to sequence model based on LSTM units that gradually extract higher level temporal representations. After that, a TimeDistributedDense layer is applied which executes a Dense function across every output over time using the same set of weights. Finally, a fully connected layer is applied with a sigmoid end point unit to perform the predictions.

$(x^{(i)})^T$ as rows. Finally, for the ground truth array, the following approach has been adopted:

$$Y^{(i)} = H(\sum_{n=i-t}^{i} y^{(n)} - t/l) \quad (3)$$

where $H(x)$ is the Heaviside step function, $Y \in \mathbb{R}^i$ is the new label array, $y^{(i)}$ is the old one and t the time-step value. The proposed approach introduced a new hyperparameters, l, which is required to be set and can affect the reactivity of the system. Figure 8 presents the architecture of the generated data structure. The left cube of data represents the different instances used as input for the RNN. At the centre, there are the feature extracted by the signals of the sensors and finally, on the right, all individual sample data points, generated by the nine sensors in t time steps. After a randomized search, l has been set to 2 and the time-step parameter to 20 that is, feeding the RNN with twenty subsequent acquisitions with each instance. Finally, non-representative samples were removed and standard deviation applied.

Training and Results. The final designed model presents three LSTM layers with 32, 15 and 32 hidden units, respectively. Then, a TimeDistributedDense layer with 52 ReLu neurons and finally, a single sigmoid unit to perform the predictions. In this second case, particular attention has been given to regularization. As before, Dropout has been used at each layer, but in order to prevent over-fitting also weight decay [17] has been exploited. Again, this is a simple fix

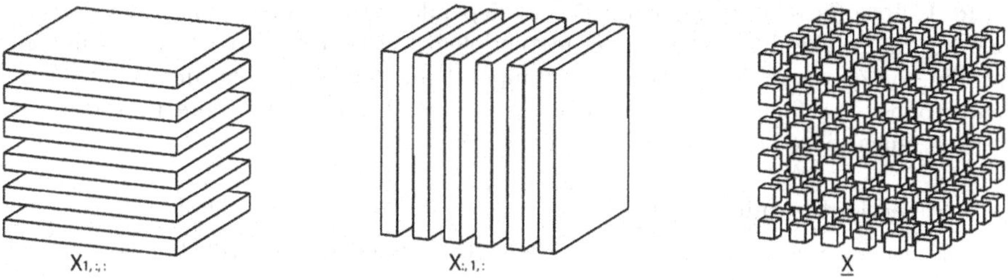

Fig. 8. Overview of the LSTM dataset structure.

to Eq. 1 updating rule, but that has shown, in our case study, for better results than L2 regularization:

$$\theta_{t+1} = \theta_t - \frac{\eta}{\sqrt{\hat{v}_t + \varepsilon}} m_t - \eta w_t \theta_t \tag{4}$$

The new term $\eta w_t \theta_t$ added in Eq. 4 subtracts a little portion of the weights at each step, hence the name decay. On the other hand, L2 regularization is a classic method to reduce over-fitting, and consists in adding to the loss function the sum of the squares of all the weights of the model, multiplied by a given hyper-parameter. So, the two strategies are only the same thing for vanilla stochastic gradient descent (SGD), but as soon as we add momentum, or use a more sophisticated optimizer like Adam or AMSGrad, L2 regularization and weight decay become different giving completely results. After, finding the maximum value of learning rate to start with [18], training the RNN with cosine annealing strategy and evaluating the architecture with K-fold cross-validation, the final model has been shown to achieve over 97% (Table 1) of F_1-score, setting a huge gap between all other tested systems based on a independent time correlations analysis. All training have been carried out on a GPU 2080 Ti with 11 GB of memory exploiting the parallel computing platform CUDA version 10 created by Nvidia. The time spent for the training of the independent time correlation networks was negligible. Instead, the RNN training, due to the complexity of the networks, has required approximately 3 h.

5.3 Future Works

The proposed algorithms could be easily applied in real time applications that would not require the presence of hardware with high computational capabilities. A future development of the proposed methodologies could be an embedded deployment of the overall system, taking advantage of hardware accelerators for the time correlation solution. Indeed, a simple solution could include the usage of hardware like a simple micro-processor for the acquisition of raw data coming from the sensors array, their analysis through the chosen model and generation of warning signals in presence of detected objects. This configuration could solve huge security problems, with a low cost solution, combining two important requirements for present companies.

Table 1. Results of the independent and dependent time correlation analysis kept separate due to the different training methodologies. It is clear, from the last model, how the contemporary exploitation of temporal and instantaneous interactions can greatly boost the final scores. Moreover, it is clear from the F1-score that the last model is much more balanced in performance than others.

Supervised learning alg.	Accuracy	Precision	Recall	F1 score
K-nearest neighbors (K-NN)	0.9280	0.9979	0.8516	0.9190
Random forest	0.9229	0.9798	0.8004	0.8811
Kernel SVM (rbf)	0.9042	0.9703	0.8004	0.8772
Feed forward NN	0.9576	0.9796	0.9311	0.9547
Recurrent neural network (RNN)	0.9615	0.9677	0.9782	0.9729

6 Conclusions

Two novel methodologies have been presented as valuable alternatives to traditional approaches for processing information of ultrasonic proximity sensors. Classical methodologies, in critical situations, struggle to find a correct level of threshold and filtering resulting in miss detections and high false positive rate. Several binary classifiers have been trained in a supervised matter and tested on real case applications where classical techniques struggles to detect the presence of an object. The contemporary usage of multiple sensors, in a cluster disposition, revealed clear correlations between the different signals. These recurrent and marked patterns enables the possibility to build more robust and reliable models able to build their knowledge on these information. Finally, results have shown a significant boost in accuracy exploiting not only the instantaneous correlation of the different signals but also the temporal pattern of the acquired sample. Indeed, the simultaneous observation of multiple samples and their temporal correlation can deeply mine into data, producing more elaborated and articulated predictions.

Acknowledgements. This work has been developed with the contribution of the Politecnico di Torino Interdepartmental Centre for Service Robotics PIC4SeR (https://pic4ser.polito.it) and SmartData@Polito (https://smartdata.polito.it).

References

1. Shrivastava, A.K., Verma, A., Singh, S.P.: Distance measurement of an object or obstacleby ultrasound sensors using P89C51RD2. Int. J. Comput. Theory Eng. **2**(1), 1793–8201 (2010)
2. Houghton, R., DeLuca, F.: Ultrasonicensordevice. Patentinspiration (1964)
3. Li, S.-H.: Ultrasound sensor for distance measurement. Google Patents (2002)
4. Wu, K., Chen, X., Ding, M.: Deep learning based classification of focal liver lesions with contrast-enhanced ultrasound. Optik-Int. J. Light Electron Opt. **125**(15), 4057–4063 (2014)

5. Carneiro, G., Nascimento, J.C., Freitas, A.: The segmentation of the left ventricle of the heart from ultrasound data using deep learning architectures and derivative-based search methods. IEEE Trans. Image Process. **21**(3), 968–982 (2012)
6. Farias, G., et al.: A neural network approach for building an obstacle detection model by fusion of proximity sensors data. Sensors **18**(3), 683 (2018)
7. De Simone, M., Rivera, Z., Guida, D.: Obstacle avoidance system for unmanned ground vehicles by using ultrasonic sensors. Machines **6**(2), 18 (2018)
8. Lee, D., Kim, S., Tak, S., Yeo, H.: Real-time feed-forward neural network-based forward collision warning system under cloud communication environment. IEEE (2018)
9. Parrilla, M., Anaya, J.J., Fritsch, C.: Digital signal processing techniques for high accuracy ultrasonic range measurements. IEEE Trans. Instrum. Meas. **40**(4), 759–763 (1991)
10. Cover, T., Hart, P.: Nearest neighbor pattern classification. IEEE Trans. Inf. Theory **13**(1), 21–27 (1967)
11. Breiman, L.: Random forests. Mach. Learn. **45**(1), 5–32 (2001)
12. Cortes, C., Vapnik, V.: Support-vector networks. Mach. Learn. **20**(3), 273–297 (1995)
13. Tharwat, A.: Classification assessment methods. Appl. Comput. Inf. (2018)
14. LeCun, Y., Bengio, Y., Hinton, G.: Deep learning. Nature **521**(7553), 436–444 (2015)
15. Reddi, S.J., Kale, S., Kumar, S.: On the convergence of adam and beyond (2018)
16. Lipton, Z.C., Berkowitz, J., Elkan, C.: A critical review of recurrent neural networks for sequence learning, arXiv preprint arXiv:1506.00019 (2015)
17. Loshchilov, I., Hutter, F.: Fixing weight decay regularization in adam, arXiv preprint arXiv:1711.05101 (2017)
18. Smith, L.N.: Cyclical learning rates for training neural networks. In: 2017 IEEE Winter Conference on Applications of Computer Vision (WACV) (2017)

An Information Analysis Approach into Feature Understanding of Convolutional Deep Neural Networks

Zahra Sadeghi(✉)

Universitat Autonoma de Barcelona, Barcelona, Spain
zahsade@gmail.com

Abstract. This paper is centered on feature analysis and visualization of pre-trained deep neural networks based on responses of neurons to input images. In order to address this problem, first the information content of learned encodings of neurons is investigated based on the calculation of the salient activation map of each neuron. The salient activation map is considered to be the activation map that has the highest aggregative value over all its cells. Second, neurons are visualized based on their maximum activation response at each location. The results put forward the uncertainty reduction over the stage of deeper layers as well as a decrement pattern in Variation of Information.

Keywords: Information analysis · Deep network understanding · Deep network visualization · Uncertainty and dependency

1 Introduction

The development of Deep Neural Networks (DNNs) has changed the trend in machine learning methods for feature extraction and object recognition problems. The successful structure and mechanism of these networks has encouraged many researchers to propose various DNN architectures [4,5,11]. Despite the advancement and ubiquitous application of DNNs, still the questions about interpretability of the building blocks of these networks are not fully addressed. Dealing with these questions can facilitate research and development of deep networks. In addition, gaining understanding about the underlying mechanism of individual units of deep networks can lead to designing better bio-inspired methods. A very important aspect of DNNs lies within their hierarchical structure. DNNs encompass many connected hidden layers. Each hidden layer itself is composed of a number of hidden units or neurons which act as feature detectors. It has shown that neurons functionality evolves from edge detectors in low hidden layers to shape detectors in the high hidden layers [17]. In addition, each layer achieves a new representation which can be useful for classification of input data at different levels of abstraction [1,12]. Analyzing the activities of neurons within each layer can provide a better understanding about the pattern

G. Nicosia et al. (Eds.): LOD 2019, LNCS 11943, pp. 36–44, 2019.
https://doi.org/10.1007/978-3-030-37599-7_4

of change across layers. Deep network understanding has been a topic of interest to many researchers over the past few years and authors have tackled this problem from different perspectives. One approach is concerned with dissecting pre-trained convolutional networks. To this end, researchers have examined neurons' weights [16] or their responses either based on their reaction to images [3] or by analyzing their activation maps [6]. Activation maximization is an optimization method that has been widely deployed by researchers in order to create images that highly activate neurons [2,9,14] or cause high classification scores [13,18]. Another approach focuses on studying the latent representation of each neuron by leveraging a reverse mechanism. Deconvolution and inversion methods are among the most successful and well-known methods which attempt to reconstruct a representative image for each layer in a backward manner [7,15]. See [10] for review and discussion about feature visualization. However, instead of the growing body of research on DNN visualization (including the interpretation and explanation problems [8]), information analysis about the course of development of internal properties of neurons at each layer and statistical characteristic of learned content of neurons has not received much attention. The contribution of this work is twofold. First, the information content of learned activation maps is analyzed. Second, each neuron is visualized by considering the regions of images with maximum activation responses.

2 Method

2.1 Feature Analysis

The purpose of this section is to investigate the learned pattern of neurons in a pre-trained convolutional deep neural network. To this end, I focus on activation layers of Keras VGG16[1] model which is pre-trained on ILSVRC2012. This network has 13 convolutional layers. In order to make a comparison between neurons, a salient activation map for each neuron at each convolutional layer is created. The salient activation map is considered as the activation map that has the highest summation value over all of its cells and here is called maxsum vector. In this paper, it is assumed that this vector holds the most important information because it incorporates the highest activation map based on the aggregative response of a neuron at all locations. In order to find this vector, all images in the validation set of ImageNet are taken into consideration and the corresponding activation maps from all the units of VGG16 model are computed. This vector is used as the basis of comparison between different neurons and hence, the Mutual Information and Variation of Information between all pairs of neurons at each convolutional layer by utilizing their maxsum vectors are calculated. Mutual information (MI) of two variables reflects the amount of reduction in uncertainty. It measures the amount of shared information content between two random variables. In other words, it quantifies how much information the presence of one variable can provide to decrease the amount of randomness about

[1] I also tried VGG19 model and obtained similar results.

the presence of the other one, or how much knowing X gives information about knowing Y. If MI (X, Y) is zero it means that X and Y are fully independent. Higher values of MI indicate less uncertainty and more dependency between the two variables and perhaps higher degrees of redundancy. On the other hand, Variation of Information (VI) is a metric for evaluating the amount of shared information distance between two variables. This value demonstrates the amount of uncertainty and independence between two variables. Both of the MI and VI notions are linked to entropy of random variables and are quantified in Eqs. 1, 2 and 3 and are illustrated in Fig. 1. Entropy is obtained by measuring the normalized histogram with n (20 in my simulations) bins which is denoted by p in Eq. 4. Using MI and VI notions, the dependency between neurons is assessed.

$$H(X,Y) = H(X) + H(Y) - MI(X,Y) \tag{1}$$

$$MI(X,Y) = H(X) + H(Y) - H(X,Y) \tag{2}$$

$$VI(X,Y) = H(X) + H(Y) - 2MI(X,Y) \tag{3}$$

$$H(X) = -\sum_{i=1}^{n} p_i \log_2 p_i \tag{4}$$

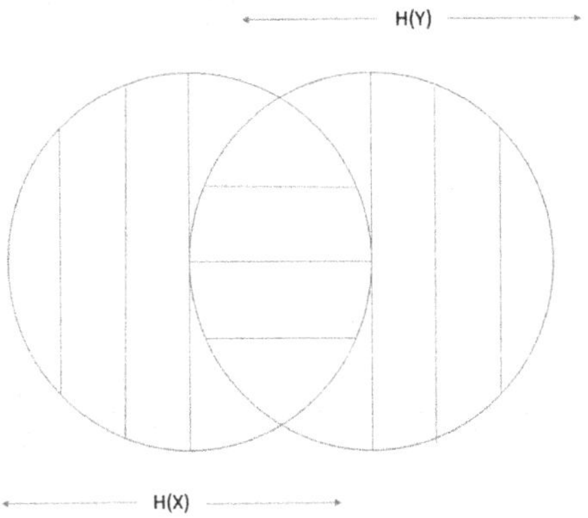

Fig. 1. Mutual Information is represented with parallel horizontal lines and Variation of Information is specified with parallel vertical lines. Entropy of a variable X is denoted by H(X).

2.2 Salient Information Visualization

In this section, I propose a method for visualizing hidden units of convolutional deep neural networks. In previous feature visualization approaches, authors used 9 top images/patches with strongest activation values for each neuron [14,15]. Note that each value in activation map corresponds to the result of a convolution operator that is applied to a patch of input image. Here, I visualize the salient activation maps by means of patches that cause maximum activation in each location for each neuron. This method is based on concatenation of the patches of images that achieve the maximum response (i.e. activation value) at each location. More formally, for each neuron in each layer and for each location in each activation map, first the maximum activation value from all the images of ImageNet validation set is obtained. Then the corresponding patch of the image that caused highest activation value is extracted. Finally, for each neuron, all the top patches are collected and are located at their corresponding position according to their activation map. In this way, a super image as a visualized identity map for each neuron can be acquired.

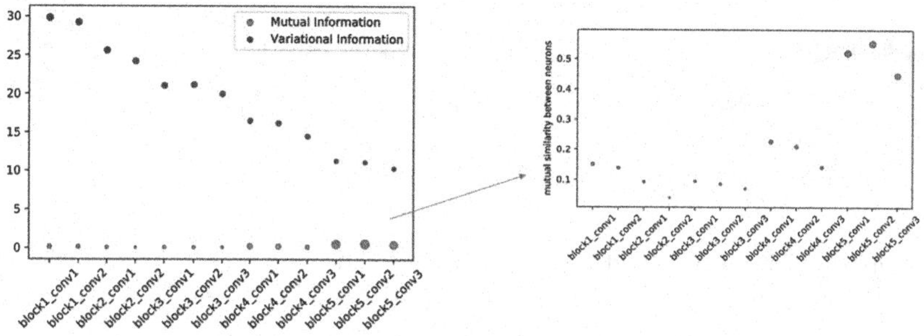

Fig. 2. Mean MI and mean VI for all neurons in each layer.

3 Results and Discussion

In previous section, first the Mutual Information and Variation of Information are calculated for all neurons at all convolutional layers. Mutual information can be interpreted as the similarity of encoding. High Mutual Information indicates strong dependency between two variables, whereas low Mutual Information is a sign of independent variables. On the other hand, Variation of Information incorporates the distance between information content of variables and is served as a dissimilarity measurement. As described before, MI and VI are calculated based on the salient activation map between all neurons. For the sake of interpretation, the mean value of MI/VI between all neuron pairs in each layer is then evaluated and shown in Fig. 2. In order to make values more noticeable and to compensate the scale differences, each dot is shown by a circle with an area that is expanded 100 times greater for MI and 10 times greater for VI. The results show that the mean VI decreases as we go towards the deeper layers, whereas the mean MI indicates an opposite behavior. In addition, I plot the distribution of mean MI/VI between one neuron and all other neurons in each layer in Fig. 3. Each dot corresponds to the mean MI/VI of one neuron in one layer which is specified by one color. Sequential convolutional layers before pooling operators are considered as a block. It is remarkable that there is an overlap in values of mean MI/VI of neurons within the same blocks which posits the similarity in information content of neurons which are subject to consecutive convolutional filters. Moreover, an increasing pattern of MI values and a decreasing trend in VI

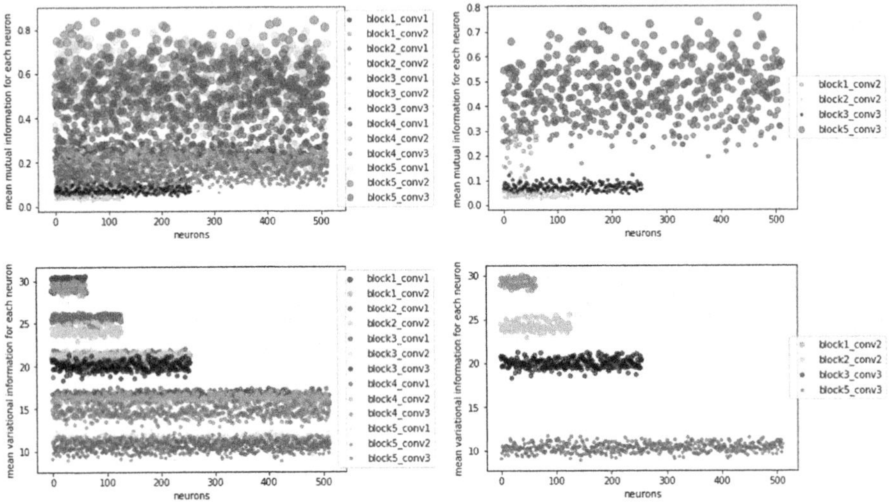

Fig. 3. Distribution of MI (first row) and VI (second row) values for all neurons at each block (first column) and last block of each layer (second column).

values from lower layers towards deeper layers is observable. This can indicate the cooperative behavior and dependency between convolutional filters of higher layers. It also implies an incremental pattern in redundancy of information and a reduction in uncertainty of information in deeper layers. In order to investigate similarities between the functionality of DNN models and human brain in terms of information properties between artificial neurons and the neurons in visual cortex, it is worthwhile that the statistical properties of biological neuron activities along visual pathways be analyzed by using computational neuroscientific approaches. Moreover, a visualization method is proposed in this paper. Figures 4 and 5 exhibit examples of visualization for neurons in each convolutional layer. The results show that the diversity of patterns corresponding to neurons in lower layers is much more compared to the deeper layers. Moreover, in the higher layers, the neighboring patches represent similar textures that belong to similar classes. This result represents the spatial information connectivity in activation maps and indicates a topological ordering in features captured by convolutional and pooling operators. More importantly, visualization maps of neurons suggest that patches from similar classes are captured by different neurons. More specifically, it is observable that the patches that appear in the visualization map of different neurons are not unique and hence it suggests that there is an overlap in information that is carried by neurons in deeper layer. This is in agreement the previous finding in this paper that Mutual Information of the deeper layers is higher (and accordingly Variation of Information is lower) compared to the shallower layers. This method provides a feature visualization at all spatial locations of activation maps which provides a more detailed representation about the patches of input images that activate each neuron. The results of this research

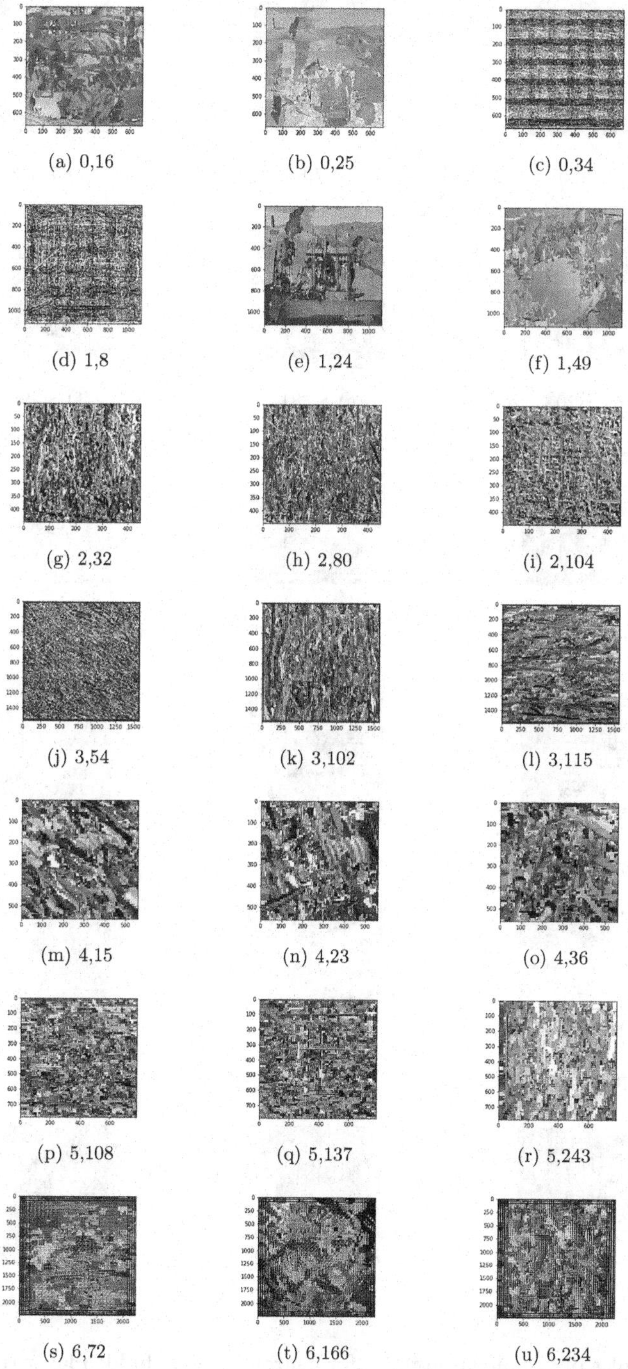

(a) 0,16 (b) 0,25 (c) 0,34

(d) 1,8 (e) 1,24 (f) 1,49

(g) 2,32 (h) 2,80 (i) 2,104

(j) 3,54 (k) 3,102 (l) 3,115

(m) 4,15 (n) 4,23 (o) 4,36

(p) 5,108 (q) 5,137 (r) 5,243

(s) 6,72 (t) 6,166 (u) 6,234

Fig. 4. Examples of visualization maps for neurons in each hidden layer (part 1).

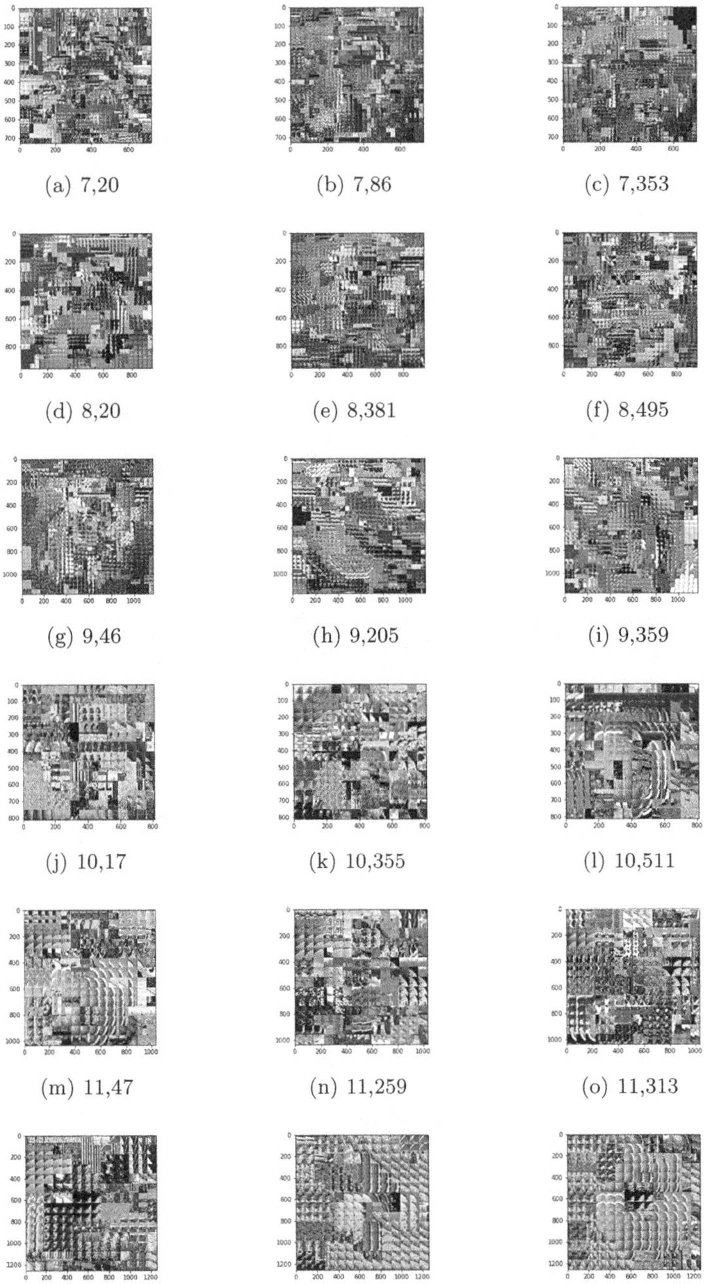

(a) 7,20 (b) 7,86 (c) 7,353

(d) 8,20 (e) 8,381 (f) 8,495

(g) 9,46 (h) 9,205 (i) 9,359

(j) 10,17 (k) 10,355 (l) 10,511

(m) 11,47 (n) 11,259 (o) 11,313

(p) 12,35 (q) 12,174 (r) 12,270

Fig. 5. Examples of visualization maps for neurons in each hidden layer (part 2). From left to right, the first and the second number below each sub-figure indicate the layer number and neuron number correspondingly.

can be useful for uncovering the black box nature of deep neural networks and can be helpful in improving internal representation and building blocks of deep learning models.

Acknowledgment. This research has received funding from the European Union's Horizon 2020 research and innovation program under the Marie Skłodowska-Curie grant agreement No. 665919.

References

1. Bilal, A., Jourabloo, A., Ye, M., Liu, X., Ren, L.: Do convolutional neural networks learn class hierarchy? IEEE Trans. Visual Comput. Graph. **24**(1), 152 (2018)
2. Erhan, D., Bengio, Y., Courville, A., Vincent, P.: Visualizing higher-layer features of a deep network. Univ. Montreal **1341**(3), 1 (2009)
3. Girshick, R., Donahue, J., Darrell, T., Malik, J.: Region-based convolutional networks for accurate object detection and segmentation. IEEE Trans. Pattern Anal. Mach. Intell. **38**(1), 142–158 (2016)
4. He, K., Zhang, X., Ren, S., Sun, J.: Deep residual learning for image recognition. In: Proceedings of the IEEE Conference on Computer Vision and Pattern Recognition, pp. 770–778 (2016)
5. Huang, G., Liu, Z., Van Der Maaten, L., Weinberger, K.Q.: Densely connected convolutional networks. In: Proceedings of the IEEE Conference on Computer Vision and Pattern Recognition, pp. 4700–4708 (2017)
6. Long, J.L., Zhang, N., Darrell, T.: Do convnets learn correspondence? In: Advances in Neural Information Processing Systems, pp. 1601–1609 (2014)
7. Mahendran, A., Vedaldi, A.: Understanding deep image representations by inverting them. In: Proceedings of the IEEE Conference on Computer Vision and Pattern Recognition, pp. 5188–5196 (2015)
8. Montavon, G., Samek, W., Müller, K.R.: Methods for interpreting and understanding deep neural networks. Digit. Signal Proc. **73**, 1–15 (2018)
9. Nguyen, A., Dosovitskiy, A., Yosinski, J., Brox, T., Clune, J.: Synthesizing the preferred inputs for neurons in neural networks via deep generator networks. In: Advances in Neural Information Processing Systems, pp. 3387–3395 (2016)
10. Nguyen, A., Yosinski, J., Clune, J.: Understanding neural networks via feature visualization: a survey. arXiv preprint arXiv:1904.08939 (2019)
11. Sabour, S., Frosst, N., Hinton, G.E.: Dynamic routing between capsules. In: Advances in Neural Information Processing Systems, pp. 3856–3866 (2017)
12. Sadeghi, Z.: Deep learning and developmental learning: emergence of fine-to-coarse conceptual categories at layers of deep belief network. Perception **45**(9), 1036–1045 (2016)
13. Simonyan, K., Vedaldi, A., Zisserman, A.: Deep inside convolutional networks: visualising image classification models and saliency maps. arXiv preprint arXiv:1312.6034 (2013)
14. Yosinski, J., Clune, J., Nguyen, A., Fuchs, T., Lipson, H.: Understanding neural networks through deep visualization. arXiv preprint arXiv:1506.06579 (2015)
15. Zeiler, M.D., Fergus, R.: Visualizing and understanding convolutional networks. In: Fleet, D., Pajdla, T., Schiele, B., Tuytelaars, T. (eds.) ECCV 2014. LNCS, vol. 8689, pp. 818–833. Springer, Cham (2014). https://doi.org/10.1007/978-3-319-10590-1_53

16. Zeiler, M.D., Krishnan, D., Taylor, G.W., Fergus, R.: Deconvolutional networks (2010)
17. Zhou, B., Khosla, A., Lapedriza, A., Oliva, A., Torralba, A.: Object detectors emerge in deep scene CNNs. arXiv preprint arXiv:1412.6856 (2014)
18. Zintgraf, L.M., Cohen, T.S., Welling, M.: A new method to visualize deep neural networks. arXiv preprint arXiv:1603.02518 (2016)

Stochastic Weight Matrix-Based Regularization Methods for Deep Neural Networks

Patrik Reizinger(✉) and Bálint Gyires-Tóth

Budapest University of Technology and Economics, Budapest, Hungary
rpatrik96@sch.bme.hu, toth.b@tmit.bme.hu

Abstract. The aim of this paper is to introduce two widely applicable regularization methods based on the direct modification of weight matrices. The first method, Weight Reinitialization, utilizes a simplified Bayesian assumption with partially resetting a sparse subset of the parameters. The second one, Weight Shuffling, introduces an entropy- and weight distribution-invariant non-white noise to the parameters. The latter can also be interpreted as an ensemble approach. The proposed methods are evaluated on benchmark datasets, such as MNIST, CIFAR-10 or the JSB Chorales database, and also on time series modeling tasks. We report gains both regarding performance and entropy of the analyzed networks. We also made our code available as a GitHub repository (https://github.com/rpatrik96/lod-wmm-2019).

Keywords: Deep learning · Generalization · Regularization · Weight matrix

1 Introduction

The importance of mitigating overfitting, i.e., the reduction of the generalization ability of a deep neural network, has gained more importance with the broadening of the spectrum of using deep learning for solving real-world problems. This is mainly due to the fact that artificial intelligence has become a tool widely used in the academia and industry, and in several cases, no feasible way exists to collect enough or perfectly representative data. The existing pool of methods is rather

The research presented in this paper has been supported by the European Union, co-financed by the European Social Fund (EFOP-3.6.2-16-2017-00013, Thematic Fundamental Research Collaborations Grounding Innovation in Informatics and Infocommunications), by the BME-Artificial Intelligence FIKP grant of Ministry of Human Resources (BME FIKP-MI/SC), by Doctoral Research Scholarship of Ministry of Human Resources (ÚNKP-19-4-BME-189) in the scope of New National Excellence Program, by János Bolyai Research Scholarship of the Hungarian Academy of Sciences. We gratefully acknowledge the support of NVIDIA Corporation with the donation of the Titan Xp GPU used for this research. Patrik Reizinger expresses his gratitude for the financial support of the Nokia Bell Labs Hungary.

G. Nicosia et al. (Eds.): LOD 2019, LNCS 11943, pp. 45–57, 2019.
https://doi.org/10.1007/978-3-030-37599-7_5

numerous, nevertheless, new approaches can bring additional advantages. Before introducing our proposal, it is worth briefly summarizing the aspects of the most popular regularization methods. Traditional constraints on the parameters (like L1 and L2 norms) add a prior to the weight matrix distribution, so the method takes its effect not directly on the weights, as the methods described in Sect. 3. Novel results [11] propose the usage of smaller batch sizes; more robust training and improved generalization performance is reported - while the variance of the minibatch, in general, tends to be higher. Dropout [16] and its generalization, DropConnect [17], are approaching the effect of ensemble networks by partially blocking different parts of the error propagation in the computational graph throughout the training. Batch Normalization [7] affects the gradients, by taking a normalization step improvements are reported both regarding accuracy and rate of convergence.

Another approach of regularization, stemming from the scarceness of available data, is the augmentation of the dataset. For this purpose, data transformations (e.g. random cropping, rotations for images) are applied or noise is added, thus providing more samples with slight differences. However, in the case of working with, e.g. time series or natural language, besides noise, it is rather difficult to find the equivalent transformations which are extensively used in the case of computer vision for data augmentation. For time series classification, there exist methods of similar nature, such as Window Warping [5], which scales a randomly selected slice of a signal. A promising result regarding noise as means of regularization for reinforcement learning is given in [14], the use of additive noise in the parameter space contributed to significant improvements in several tasks. For recurrent networks, another type of regularization method has turned out to be successful: Zoneout [9] can be considered as a variant of Dropout for recurrent neural networks such as LSTM (Long Short-Term Memory, [6]). The main difference compared to Dropout is that Zoneout maintains at least the identity - i.e., instead of dropping specific cells, the activation values from the last step are used in the graph -, thus preserving information flow in the recurrent connections.

This paper is organized as follows: Sect. 2 briefly reviews the motivation for the proposed methods, Sect. 3 includes the detailed overview of the proposed algorithms, the experimental setup is described in Sect. 4, while the results are introduced in Sect. 5 and discussed in Sect. 6.

2 Motivation

Considering the most widely-used approaches and the possibilities, we decided to investigate more thoroughly the direct effects of regularization applied to weight matrices. We refer to all of the tunable parameters of a layer (excluding hyper-parameters) as weight or weight matrix in this paper. While doing so, we also considered that in contrast to the real-world or artificial data, parameters do not possess such a high degree of internal error-resilience. I.e., an improperly designed method can lead to significant degradation of the training, even to complete failure. As weights capture information about the underlying stochastic process in a compact way, this is something which should be considered. Thus, such schemes are needed which do not cause significant parameter degradation - if degradation occurs, it should be corrected before triggering early stopping. This means, the intervention should be local and the change should be small.

Given these considerations and constraints, our goal is to design robust, generally applicable regularization methods based on the direct modification of the weight matrices.

3 Proposed Methods

In this section, we describe both proposed methods, collectively named WMM (Weight Matrix Modification). Each of them modifies weights explicitly, but the two approaches are different. WMM-WR (Weight Reinitialization) can be considered as a way of expressing distrust of backpropagation by discarding some of its parameter updates, while WMM-WS (Weight Shuffling) is a local approach, which constrains its intervention based on an assumption regarding the autocorrelation function of a specific weight matrix.

In this section W denotes a weight matrix, p a probability value, $\mathcal{L}$ the set of affected layers and c a constant (called *coverage*). Thus, p determines the probability of carrying out WMM-WR or WMM-WS and c specifies the size of the window where the method takes effect. Weight matrices are considered two dimensional as it is the general case (for 2D CNNs - Convolutional Neural Networks -, we consider the 2D filters separately – not the filter bank as a whole). Nevertheless, it is done without loss of generalization, hence this property is not exploited in the proposed methods. Please note that both methods are concerning the weight matrices only and not the bias vectors, because our evaluations showed that including the bias cannot bring further advantages.

3.1 Weight Reinitialization (WMM-WR)

As mentioned in Sect. 2, the initial weight distribution can be crucial for good results. Like the weight initialization scheme of [4], or the common choice of

$$W \sim U\left[-\frac{1}{\sqrt{n_i}}, \frac{1}{\sqrt{n_i}}\right], \tag{1}$$

where n_i is the number of columns of the weight matrix W, and U denotes a uniform distribution. From an information theoretical point of view, using uniformly distributed weights corresponds to the principle of maximal entropy, thus a high degree of uncertainty. However, the cited paper from Glorot and Bengio [4], which can improve the convergence with normalizing the gradients, also uses a uniform parameter distribution.

These considerations provide us means for proposing a regularization method with a Bayesian approach. It is common in estimation theory to combine different probability distributions, often weighted with the inverse covariance matrix. This approach can also be used for weight matrices with a slight modification, which is needed to constrain the amount of change to maintain stability during training. For this, we apply a sparse mask. Since we only modify a small amount of the weights, it seemed not feasible to compute the statistics of the whole layer. Different sparse masks are generated in each step, which is also not possible for all combinations without significant speed degradation. Thus, the algorithm either preserves the actual weights or draws a sample from the initial uniform distribution.

Algorithm 1 describes the steps of WMM-WR. It is carried out based on a probabilistic scheme, which is controlled by the p parameter. This approach reduces the probability to introduce a high level of degradation into the network, on the other hand, it also saves computational power, because the method will not be applied in every step. First, a mask M is generated by sampling a uniform distribution U. Then a window position is selected randomly (its size is controlled by the c parameter) – which means that all items outside the window are cleared in M. After that the same p is used as a threshold for ensuring sparsity, thus clearing the majority of the elements within the window. W^* denotes the temporary matrix used for storing the samples from the distribution U_w, the distribution used for initializing the weights at the start of the training. Based on the mask values (which remained within the window after thresholding with p), the elements of the original weight matrix W_i get reinitialized. Generating both M and U_w is linear in the number of the weights, as in the case of memory cost too.

In the case of overfitting, activations of a layer are distributed rather unevenly. As activation functions are in general injective, the distribution of the weights is uneven, too. By reinitializing a random subset of weights, if weights from the peaks of the weight distribution are chosen, overfitting can be reduced with doing a modification step towards maximal entropy, thus smoothing the histogram. On the other hand, reinitializing values near the mean would result in small changes only. If the mean of the weight matrix is not the same as the expected value of the trained weight matrix, then reinitializing a subset of weights expresses our distrust and is opting for maximal entropy, thus for uniform distribution. Independent of the value of the mean of the whole layer, the reinitialized subset itself has an expected value of zero (Eq. 1), thus the method will not result in a significant change, meaning reduced sensitivity. An interesting side-effect of WMM-WR is the fact that it reduces the Kullback-Leibler divergence between the actual distribution of the weights and the initial one. The reason for this is that the same distribution is used for WMM-WR as for initializing the layers. The method increases the entropy (at least momentarily, as a uniform distribution is used), which effect will be investigated in Sect. 5.

3.2 Weight Shuffling (WMM-WS)

As already mentioned in Sect. 1, noise has a wide range of applicability from the dataset to the weights. Instead of using Gaussian white noise, we consider

Algorithm 1. Algorithm for WMM-WR

Input: $\mathcal{L}, p, c$
  for all $W_i \in \mathcal{L}$ **do**
    if $p > U[0;1)$ **then**
      $M \leftarrow U[0;1)$
      $M \leftarrow window(M, c)$
      $M \leftarrow threshold(M, p)$
      $W^* \leftarrow U_w$
      $W_i[M_{jk} == 1] = W^*$
    end if
  end for

using non-white, correlated noise directly for a neighborhood of weights. We refer to a neighborhood of weights as a rectangular window (determined by the parameter c) with dimensions comparably smaller than the size of the weight matrix itself. Because the weights are intended to represent the characteristics of the underlying stochastic process, they should be correlated with each other (i.e. white noise itself is not appropriate for capturing the underlying process). Consequently, by selecting a continuous subset of the weights, the correlation should still hold, hence the process to be modeled stays the same. Shuffling them within the bounds of a local window can be interpreted as using correlated noise for that given subset. Interestingly, this transformation is invariant regarding both the weight distribution and the entropy of the matrix – the latter of which will be investigated in Sect. 5.

A more expressive way is to think about WMM-WS as a type of a 'parameter ensemble', i.e., weights in a neighborhood are forced to fit well a range of input locations. As the ensemble approach, or the approximations of it - such as Dropout -, have been proven to be successful in practice due to the utilization (or simulation) of an averaging step, an improved degree of robustness is expected from WMM-WS as well.

The steps of WMM-WS are described in Algorithm 2. As in the case of Algorithm 1, we have a probabilistic first step which stochastically decides whether WMM-WS will be carried out or not (using parameter p). The mask M is sampled from a Bernoulli-distribution $\mathcal{B}$ directly (as we need a binary mask), thus the density of the mask is independent of the parameter p. For the selection of the local window, c is used as for WMM-WR. Having generated the mask, the weights indexed by M are shuffled for each weight matrix of the layer subset $\mathcal{L}$ (filters of convolutional and gates of recurrent layers are handled independently from each other). The complexity of the algorithm is determined with the shuffle operator, which scales linearly in the number of the weights while selecting the local window has a constant computational cost.

Algorithm 2. Algorithm for WMM-WS

Input: $\mathcal{L}, p, c$
  for all $W_i \in \mathcal{L}$ **do**
    if $p > U[0;1)$ **then**
      $M \leftarrow \mathcal{B}[0;1]$
      $M \leftarrow window(M, c)$
      $shuffle(W_i[M])$
    end if
  end for

4 Experimental Setup

Due to the fact that WMM intervenes into the training loop on the lower levels in the software stack, we have searched for a framework which provides such a low-level interface. PyTorch [13] was chosen to be used for the implementation because it makes possible to modify fine details of the training process. We utilized Ignite [1], a high-level library for PyTorch, which was built by the community to accelerate the implementation of deep learning models by providing a collection of frequently used routines.

The framework of the proposed methods is schematically depicted in Fig. 1: Ignite provides the wrapper which merges the PyTorch model with WMM utilizing its callback interface. Due to the fact that while designing WMM neither architecture- nor task-specific assumptions were made, our approach provides great flexibility, which will be elaborated shortly. For hyper-optimization, a Bayesian approach (tree of Parzen estimators) was used, based on a method shown in [2].

Regarding the parameters of WMM, the following intervals were chosen. For p and c (coverage) a log-uniform distribution was selected to avoid a big value for both parameters at the same time, the intervals were $[0.05; 0.4]$ and $[0.03; 0.35]$, respectively. Each layer was affected by WMM-WS or WMM-WR with an equal probability.

To prove the general applicability of the proposed methods, we set up a testing environment with different model architectures and datasets. Tests were carried out with fully connected, convolutional and LSTM (Long Short-Term Memory, [6]) layers. In case of convolutional layers, regardless of the filterbank representation of the kernels, WMM operations were, of course, restricted to operate only within one filter. For LSTM the effect of WMM on different gates was investigated, the same holds for GRU (Gated Recurrent Unit, [3]) as well.

For evaluation, several datasets were investigated. The MNIST [10] dataset and its sequential (row-by-row) variant were split up into 55,000-5,000-10,000 samples (28×28 pixels, 10 classes, for each handwritten digit) for training, validation, and test purposes. The networks consisted of 2 convolutional and 2 fully connected or 2 LSTM and 1 fully connected layers for the traditional and sequential approaches, respectively. For CIFAR-10 [8] the same splitting policy was used (the samples are $32 \times 32 \times 3$, 10 classes), in that case, the

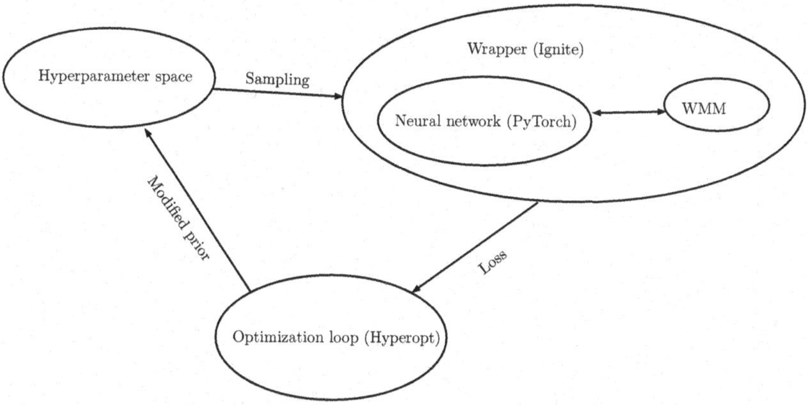

Fig. 1. Structure of the framework for WMM

network consisted of 2 convolutional (with a max pooling between them) and 3 fully connected layers. The JSB Chorales [15] task was organized into 229-76-77 chorales (with a window length of 25, 88 classes) for training, validation, and test purposed. Although being a multilabel problem a one-layer LSTM was sufficient. Besides classification, regression problems were also investigated, for that purpose an artificially generated (synthetic) dataset was used, using also a 55,000-5,000-10,000 data split (window size 50). The synthetic time series were fed into the neural network both with and without colored noise, and consisted of the linear combination of sinusoidal, exponential and logarithmic functions. Our intent with this choice - which is somewhat similar to that of the Phased LSTM [12] paper - was, to begin with simpler datasets and proceed to more complex ones. For both variants of the regression task, a network of 2 LSTM and 1 fully connected layers was used.

5 Results and Discussion

During the analysis of the results, several factors were considered. The three most important factors were performance (accuracy/MSE (Mean Squared Error)), entropy and robustness.

Our analysis intended to identify the tasks where the proposed approach turns out to be advantageous. Nonetheless, we could even make useful conclusions with datasets such as MNIST, where state-of-the-art results are near perfect, using it as a testbed for algorithm stability. For each dataset and each of the proposed methods, we have conducted more than 2000 experiments. The aim of which was to evaluate the behavior of both methods empirically, mainly for describing parameter sensitivity (see Fig. 5 and its discussion below).

To clearly identify the effects of WMM and to prevent overfitting originating from the simple nature of some datasets, the networks used for evaluation were

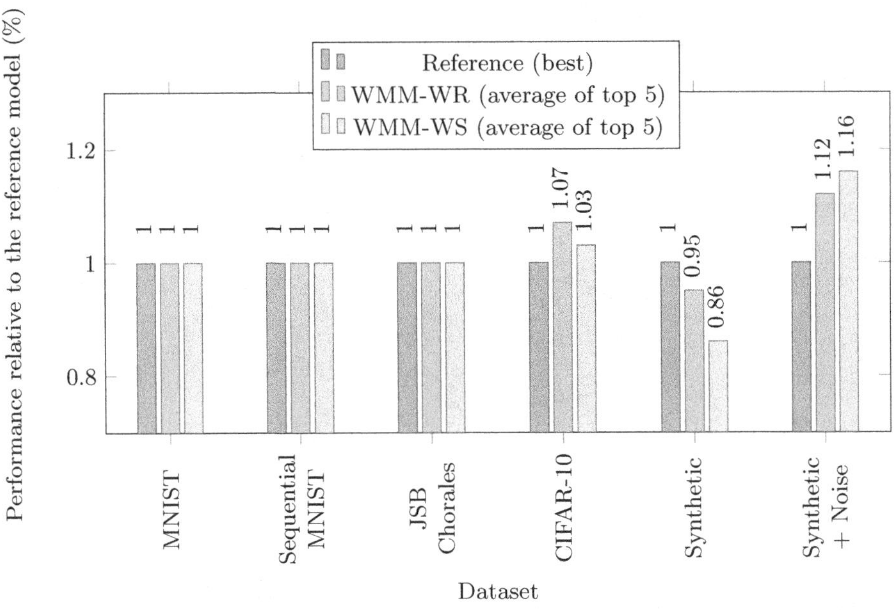

Fig. 2. The effect of WMM on performance metrics (higher values are better)

not very deep (maximum depth was 5 layers). During the evaluation, a reference model (for each task, respectively) was constructed from the best result using L2-regularization, Batch Normalization [7] or Dropout [16], which were chosen as they are currently among the most successful and most widely-used regularization techniques.

The figures included in this section are comparing different aspects of the networks. We compared the average of the top 5 best results utilizing WMM to the best results of other regularization techniques. The effect of the proposed methods regarding performance metrics is depicted in Fig. 2 (higher is better), we converted accuracy and MSE to be able to display them on the same scale (it displays relative change).

Regarding performance, the change is not significant for MNIST (traditional and sequential approach) and JSB Chorales (below 0.1%). In the case of the synthetic dataset, performance suffers even from degradation. On the other hand, for CIFAR-10 WMM-WR brought a performance boost of 7%, while WMM-WS one of 3%. The biggest performance increase (over 10%) was experienced in the case of the synthetic sequential dataset with colored noise.

For the tasks utilizing recurrent networks (sequential MNIST, JSB Chorales, synthetic data) the best results were achieved by different choices of the LSTM gates. In case of WMM-WR the modification of the weights of the forget gate was proven to be the best for the sequential MNIST task, while for the other three problems the cell gate was the best choice. For WMM-WS the output gate

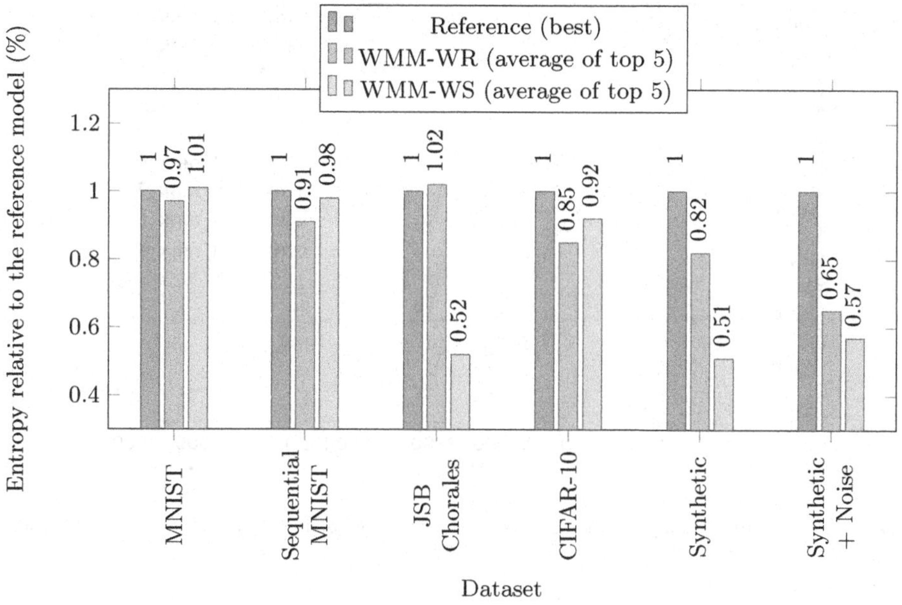

Fig. 3. The effect of WMM onto the entropy of neural networks (lower values are better)

was preferred (synthetic + JSB Chorales) – the sequential MNIST task resulted in modifying the weights of the last, fully connected layer.

Partially due to the fact that the proposed methods have an effect onto the entropy of the networks – WMM-WR is expected to increase (as the distribution used for reinitialization is uniform, thus has maximal entropy), while WMM-WS to retain it – so we decided to compare also entropy values for the different tasks. These results are shown in Fig. 3 (lower is better), where we also compare to the respective reference models (which are the best using other regularization methods).

For MNIST the change is not too significant, but in general, WMM was able to compress the networks by 2–9%, except WMM-WS for the not sequential task. For CIFAR-10, the compression is more significant, both methods rank in the same range.

In the case of the only multilabel dataset, JSB Chorales, the results were rather astonishing: while WMM-WR has not brought a significant change, WMM-WS compressed the network by almost a factor of two. This superiority of WMM-WS compared to WMM-WR regarding data compression is also characteristic of both variants of the synthetic dataset. Nevertheless, in the latter case WMM-WR also reduced entropy for the synthetic data, but there is a difference of 27% between the noiseless and noisy cases.

Generally speaking, WMM-WS with its entropy-invariant approach was able to reduce entropy more than WMM-WR. Despite using a probability

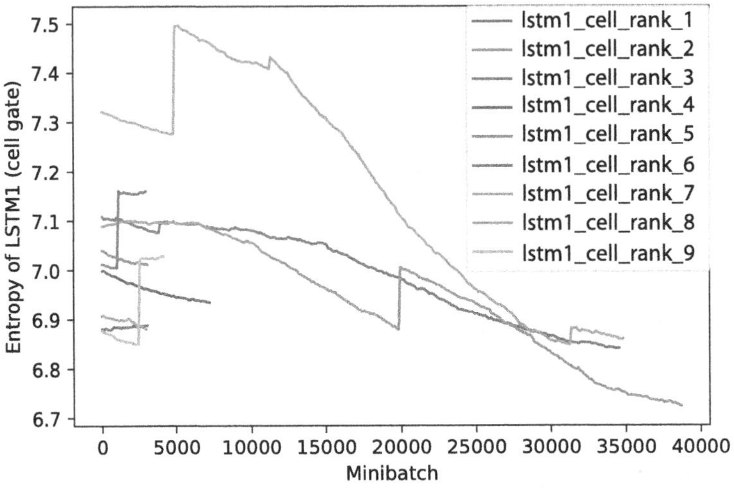

Fig. 4. Temporal entropy change during WMM-WR (JSB Chorales, lower values are better)

distribution according to the principle of maximal entropy, WMM-WR is also able to compress the networks. The analysis of the temporal change of entropy in the case of WMM-WR is shown in Fig. 4 using the best results, where the legend shows the $\mathcal{L}$ parameters of Algorithm 1, also including the gates of the LSTM layer and the ranking according to MSE (the smaller the number, the lower the MSE). Although the entropy momentarily increases due to WMM-WR, in the end, it can help to compress the network, as shown in Fig. 4.

The robustness of the proposed methods is also a crucial factor. Figure 5 shows the distribution of MSE w.r.t $p * c$ (c is the coverage parameter) during Bayesian hyper-optimization. The $\mathcal{L}$ parameter is color-coded in the figures, *none* denotes models without WMM (i.e. the reference). In the case of WMM-WR (Fig. 5a), the plot has a horizontally elongated form, which means smaller sensitivity to parameter change (the less elongation in the y-direction, the better). Although using WMM-WR imposes a higher variance due to its stochastic nature (the scatter plot displays the results of the whole hyper-optimization loop, containing also the not that good results at the beginning), it is clear to see that better results can be achieved with it. A rather unexpected observation was, that in the case of WMM-WS (Fig. 5b) the bounding ellipse is rotated with 90° compared to Fig. 5a, thus showing higher stochastic sensitivity - nevertheless, the Bayesian optimization loop resulted in several good parameter sets.

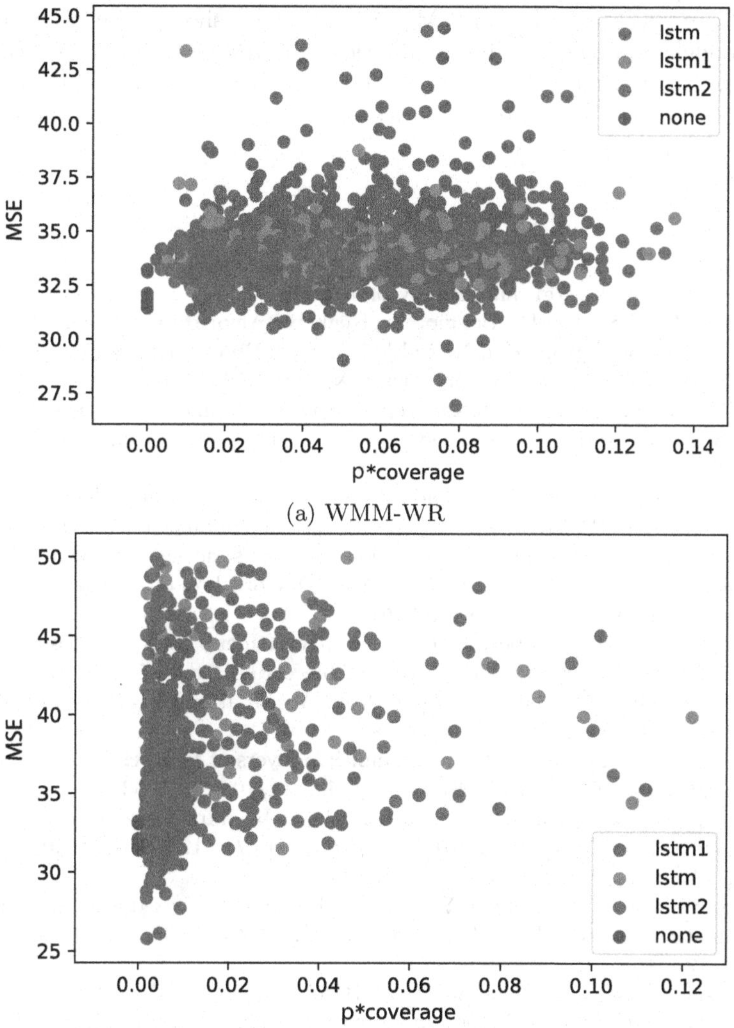

(a) WMM-WR

(b) WMM-WS

Fig. 5. Illustration of the robustness examination for the synthetic dataset (with noise)

6 Conclusions

In this paper two, weight matrix-based regularization methods were proposed and investigated for deep neural network applications. The main advantages of both WMM-WR and WMM-WS are their architecture- and task-independence. Our evaluations resulted in the following conclusions: both methods are capable of reducing overfitting, first of all in the case of noisy data, but also for CIFAR-10 better results were reported for the networks used throughout testing. On the other hand, WMM-WR and WMM-WS also have shown their potential concerning model compression. The capabilities of WMM regarding noise reduction will be investigated more thoroughly in the future, we also intend to introduce an adaptive way (i.e. based on the statistics of the weight matrix) for selecting the by WMM affected areas.

References

1. Ignite (2019). https://github.com/pytorch/ignite
2. Bergstra, J., Bardenet, R., Bengio, Y., Kégl, B.: Algorithms for hyper-parameter optimization. In: Advances in Neural Information Processing Systems (NIPS), pp. 2546–2554 (2011). https://doi.org/2012arXiv1206.2944S
3. Cho, K., et al.: Learning Phrase Representations Using RNN Encoder-Decoder for Statistical Machine Translation. arXiv (2014). https://doi.org/10.1074/jbc.M608066200
4. Glorot, X., Bengio, Y.: Understanding the difficulty of training deep feedforward neural networks. PMLR **9**, 249–256 (2010). https://doi.org/10.1.1.207.2059
5. Guennec, A.L., et al.: Data augmentation for time series classification using convolutional neural networks. In: ECML/PKDD Workshop on Advanced Analytics and Learning on Temporal Data (2016)
6. Hochreiter, S., Schmidhuber, J.: Long short-term memory. Neural Comput. **9**(8), 1735–1780 (1997). https://doi.org/10.1162/neco.1997.9.8.1735
7. Ioffe, S., Szegedy, C.: Batch Normalization: Accelerating Deep Network Training by Reducing Internal Covariate Shift. arXiv (2015)
8. Krizhevsky, A., Hinton, G.: Learning multiple layers of features from tiny images. Technical report, Citeseer (2009). https://doi.org/10.1.1.222.9220
9. Krueger, D., et al.: Zoneout: Regularizing RNNs by Randomly Preserving Hidden Activations. arXiv, pp. 1–11 (2016). https://doi.org/10.1227/01.NEU.0000210260.55124.A4. http://arxiv.org/abs/1606.01305
10. LeCun, Y., Bottou, L., Bengio, Y., Haffner, P.: Gradient-based learning applied to document recognition. Proc. IEEE **86**(11), 2278–2323 (1998). https://doi.org/10.1109/5.726791
11. Masters, D., Luschi, C.: Revisiting Small Batch Training for Deep Neural Networks. arXiv, pp. 1–18 (2018). https://doi.org/10.1016/j.biortech.2007.04.007. http://arxiv.org/abs/1804.07612
12. Neil, D., Pfeiffer, M., Liu, S.C.: Phased LSTM: accelerating recurrent network training for long or event-based sequences. In: Lee, D.D., Sugiyama, M., Luxburg, U.V., Guyon, I., Garnett, R. (eds.) Advances in Neural Information Processing Systems 29, pp. 3882–3890. Curran Associates, Inc. (2016)
13. Paszke, A., et al.: Automatic differentiation in PyTorch. In: Advances in Neural Information Processing Systems 30 (NIPS), pp. 1–4 (2017)
14. Plappert, M., et al.: Parameter Space Noise for Exploration. arXiv, pp. 1–18 (2017). http://arxiv.org/abs/1706.01905
15. Raś, Z., Wieczorkowska, A.: Advances in Music Information Retrieval. Studies in Computational Intelligence, vol. 274. Springer, Heidelberg (2010). https://doi.org/10.1007/978-3-642-11674-2

16. Srivastava, N., Hinton, G., Krizhevsky, A., Sutskever, I., Salakhutdinov, R.: Dropout: a simple way to prevent neural networks from overfitting. J. Mach. Learn. Res. **15**, 1929–1958 (2014). https://doi.org/10.1214/12-AOS1000
17. Wan, L., Zeiler, M., Zhang, S., LeCun, Y., Fergus, R.: Regularization of neural networks using DropConnect. In: ICML, pp. 109–111 (2013). https://doi.org/10.1109/TPAMI.2017.2703082

Quantitative and Ontology-Based Comparison of Explanations for Image Classification

Valentina Ghidini[1(✉)], Alan Perotti[1], and Rossano Schifanella[1,2]

[1] ISI Foundation, Turin, Italy
valentina.ghidini95@gmail.com
[2] University of Turin, Turin, Italy

Abstract. Deep Learning models have recently achieved incredible performances in the Computer Vision field and are being deployed in an ever-growing range of real-life scenarios. Since they do not intrinsically provide insights of their inner decision processes, the field of eXplainable Artificial Intelligence emerged. Different XAI techniques have already been proposed, but the existing literature lacks methods to quantitatively compare different explanations, and in particular the semantic component is systematically overlooked. In this paper we introduce quantitative and ontology-based techniques and metrics in order to enrich and compare different explanations and XAI algorithms.

Keywords: Explainable artificial intelligence · Neural networks · Deep learning · Computer vision

1 Introduction

In the past few years, Artificial Intelligence (AI) has been a subject of intense media hype - coming up in countless articles, often outside of technology-minded publications. This renewed interest is rooted in the new paradigm of Deep Learning (DL), and specifically in the ground-breaking results that convolutional neural networks unlocked in vision-based real-world tasks, spanning from enabling self-driving car technology [14] to achieving super-human accuracy in image-based medical diagnosis [6].

Alas, there is a clear trade-off between accuracy and interpretability, and DL models fall on the far left side of the spectrum: especially for computer vision (CV) tasks, the best performing models are so-called black boxes: they do not provide a human-understandable representation of their encoded knowledge.

Given the pervasive nature of the recent advancements in AI, and the ever-growing application in real-world domains, the debate around ethical issues in

A. Perotti—Acknowledges support from Intesa Sanpaolo Innovation Center. The funder had no role in study design, data collection and analysis, decision to publish, or preparation of the manuscript.

G. Nicosia et al. (Eds.): LOD 2019, LNCS 11943, pp. 58–70, 2019.
https://doi.org/10.1007/978-3-030-37599-7_6

AI technologies and algorithms is more lively than ever: scientists and engineers should be able to ensure that intelligent systems are actually capable of acting in the interest of people's well-being [2].

However, there is no consensus on how to explain a black-box classifier, and not even on what an explanation is in the first place [9]; consequently, there are no quantitative approaches for comparing different eXplainable Artificial Intelligence (XAI) approaches for CV tasks.

Furthermore, image classification is typically framed as a one-vs-all task, whereas we humans rely on structured symbolic knowledge: e.g., given the picture of a Siamese cat, we know that labelling it as another cat breed is not as wrong as labelling it as a vehicle for example. We argue that this factor has to be taken into account when inspecting the classification behaviour of a black-box model.

In this paper, we propose a tool set for quantitatively comparing explanations from black-box image classifiers, factoring in numeric values as well as semantic features of the image-label-explanation tuple.

This paper is structured as follows: in Sect. 2 we provide an overview of Neural Networks and DL models for CV tasks; while in Sect. 3 we describe the black-box problems, what could constitute an explanation, and what are the state-of-the-art XAI algorithms. In Sect. 4 we introduce heatmap-based metrics and provide a visual and quantitative comparison of XAI algorithms. In Sect. 5 we link explanations to an ontology and we introduce a semantics-based metric for explanations. We discuss critical steps in Sect. 6 and we conclude outlining directions for future work in Sect. 7.

2 Deep Learning for Computer Vision

Neural networks are densely connected sets of computational units called artificial neurons [10]. In recent years, more complex models (globally referred to as DL [7]) emerged and started obtaining important results; besides algorithmic advancements, key enabling factors for the rise of DL were the explosive growth and availability of data and the remarkable advancement in hardware technologies [3]. More importantly, Deep Learning became the de facto standard approach for several Computer Vision tasks, such as image classification.

The benchmark for image classification is the ILSVRC challenge, based on the ImageNet dataset (millions of images for one thousand of labels) [4]. The groundbreaking model for Computer Vision was AlexNet, a deep convolutional NN that won the ILSVRC competition in 2013. Increasingly complex models were introduced in later years (most notably the Inception architecture, which we use in this paper) but all deep neural networks for computer vision share the same generic structure with interleaved convolutional and pooling layers followed by fully connected ones.

Since in this paper we focus on explaining already-trained classifiers, we will exploit a pre-trained InceptionV3 [19] model, available within the Keras[1] and Pytorch[2] libraries.

3 Towards eXplainable Artificial Intelligence

3.1 The Black Box Problem

One of the several advantages of deep NNs is their ability to automate the feature extraction process within a completely data-driven framework: for instance, in order to build an image classifier able to distinguish between a stop signal and a tree, it is not necessary to give a formal (machine-runnable) definition of *tree* - it is sufficient to provide a large number of labelled example images. The DL model, during training, builds its own representation of the entities and performs its own feature engineering. The downside of this approach is that the detected features are sub-symbolic (numerical), numerous, and without any attached semantics. It is therefore totally possible to observe the input and the output data of a black box model, and consequently to evaluate its performance, without having any understanding about its internal operations.

The difficulty of inspecting the internal state of a DL model, and therefore understanding *why* it produced a given output, is commonly referred to as the Black Box Problem. This problem becomes crucial when such models are deployed in real-world sensitive scenarios, ranging from default risk prediction to medical diagnosis: there are several reasons why an unexplained black box model can be troublesome.

From a legal viewpoint, AI systems are regulated by law with the General Data Protection Regulation (GDPR) [8] - which includes many regulations regarding algorithmic decision-making. For instance, GDPR states that the decisions *which produces legal effects concerning him or her or of similar importance shall not be based on the data revealing* sensitive information (for example about ethnic origins, political opinions, sexual orientation). Clearly this is impossible to guarantee without opening the black box. Moreover, the GDPR states that the controller [8] must ensure the right for individuals to obtain further information about the decision of any automated system, which is precisely the goal of XAI.

Second, an unexplained DL model might be right for the wrong reasons (e.g. might have picked up a watermark in the images of one class, or recognize an object thanks to the recurrent background in the training set) within the initially provided data and fail spectacularly when presented with new batches of data from different sources (e.g. medical data acquired with a different commercial device). In one almost anecdotal experiment [15], a NN was trained to classify pictures of huskies versus wolves - the resulting classifier was accurate, but XAI techniques showed that the NN was focusing on the snow in the wolf images,

[1] github.com/keras-team/keras.
[2] pytorch.org.

since all the photos of wolves had snow in them, but the husky photos did not. Furthermore, explanations would help gain the trust of domain experts regarding the adoption of new decision support systems, such as medical staff using intelligent ultrasound machines.

3.2 Explanations

To explain (or to interpret) means to provide some meaning in understandable terms. This definition also holds in XAI, where interpreting a model means giving an explanation to the decisions of a certain model, that has to be at the same time an accurate proxy of its decision making process (a property called *fidelity*) and understandable to humans.

One of the most important distinctions among explainability methods is their *local* or *global* interpretability [9]. A global explanation allows to describe the general decision process implemented by the model; on the other hand, local explanations lead to the comprehension of a specific decision on a single data point. Another important feature of explanation methods is how they relate to the model they are trying to explain. In particular, one can have a *model agnostic* explanator, which is the outcome of an algorithm not tied to the architectural details of the specific model being explained. Conversely, *model aware* explanation techniques rely on inspecting inner characteristics of the black box model of interest, such as gradients in a NN: clearly, these approaches are less general as they can be applied only on specific classifiers.

Similarly, an explanation technique can be *data agnostic*, i.e. it can explain any kind of input data (images, texts or tabular), or *data aware*. In CV (and therefore this paper) the input data are always images, and typically explanations are heatmaps, highlighting the most important regions of the data instance for the prediction. This allows the user to visually understand which pixels correlate with the predicted label, and decide whether the DL model focused on a reasonable region of the input image.

3.3 XAI Algorithms

In this paper we compare six state-of-the-art XAI algorithms for image classification: they all provide local explanations as heatmaps.

- The first algorithm is LIME [15], which is a data and model agnostic explanation method. Its application to images involves their initial partition in superpixels, then each one is silenced to test its importance for the output, using a local linear model.
- The idea of RISE [13] is similar, since it still perturbs the input by means of random binary masks (without the need of the segmentation into superpixels) to study the impact on the output. Note that RISE can be applied only to images, but it is still model agnostic.
- Then some saliency masks have been built exploiting the CNN's properties: this is the case of Vanilla Gradient [5], which consists in looking for existing input patterns of bounded norm maximizing the activation of a chosen layer.

- Another model aware technique is Guided Backpropagation [18], which uses a deconvolutional neural network to invert the propagation from the output to the input layer in order to reconstruct numerically the image which maximize the activation of interest.
- Grad-CAM [17] uses the gradients flowing into the final convolutional layer to produce a coarse localization map highlighting the important regions in the input image for a target class of interest.
- Finally, Layerwise Relevant Propagation (LRP) [1] allows to find for each pixel a *relevance score* using a local distribution rule.

4 Heatmap-Based Comparison

4.1 Visual Comparison

As introduced in the previous section, the typical explanation for black box classification algorithms is a heatmap overlayed on the input image, so that each pixel's hotness represents its relevance in the classification - according to the XAI algorithm of interest. As black box classifier we used an Inception-V3 [19] pretrained with ImageNet and obtained the explanation heatmaps for the six XAI algorithms introduced in Sect. 3.

For a first qualitative comparison, we consider one example image (label: *ladybug* and show the explanation heatmaps (for each XAI algorithm) relative to the two top scoring classes (the correct label *ladybug* and a reasonable second *leaf beetle*) and the two classes with lowest predicted probability (*trolley* and *crab*).

All heatmaps are displayed in Fig. 1: the red color is associated to the highest values (the most important regions in the image), the blue color for the less important areas. The saliency masks are normalized for the sake of visualization (red for the 99° percentile, primary blue for the 1st percentile of the values of the mask). Two maps which are visually equivalent may not be the same numerically.

However, the explanations for the correct label (ladybug) are much more concentrated on the correct item, while considering the heatmaps created for wrong classes one can see that they focus on background and surroundings of the item. Moreover, there is a clear difference between the explanation methods: Vanilla Gradient and LRP give a quite sparse heatmap, which is concentrated on the ladybug (for all the labels considered), but also expanding in the surroundings. Guided Backpropagation saliency map is slightly more concentrated than the previous ones but still without any tangible visual difference between the explanations for different labels. LIME and Grad-CAM differentiate in a sharp way the explanations for different classes: the correct label is explained quite precisely (in LIME the most important superpixel is exactly overlayed to the ladybug, while GradCAM is a little less precise). The explanation for RISE is really sparse, and it does not seem to differentiate visually between the different classes, meaning that the important regions are always the same ones, no matter the label to be explained.

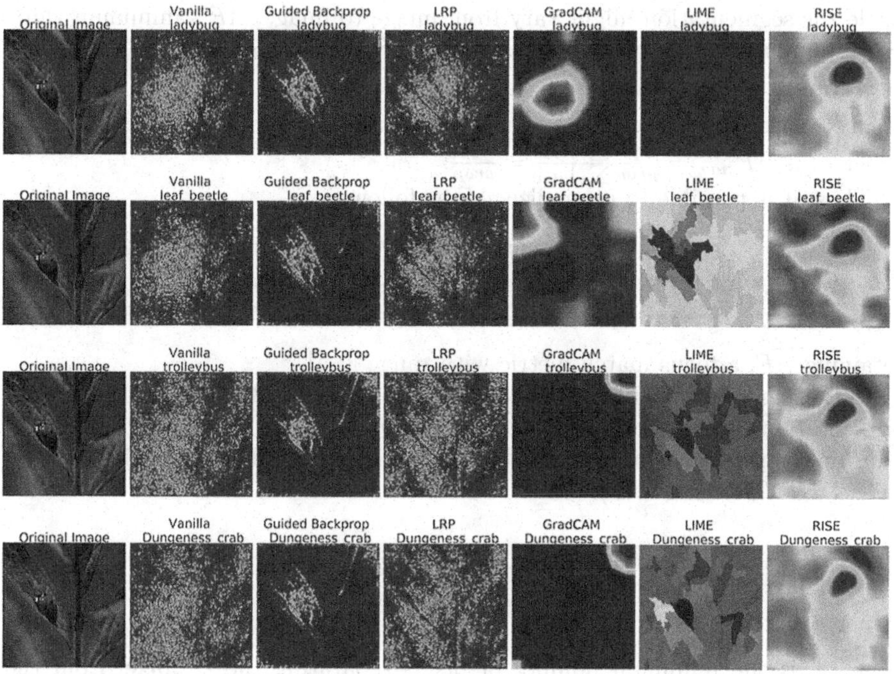

Fig. 1. Visual comparison of explanation heatmaps for different XAI algorithms (Color figure online)

4.2 Metrics

The analysis of a heatmap is twofold. First, one can evaluate how coherent the heatmap is w.r.t. the location of the actual content within an image - that is, whether the explanation focuses onto the same region where a human would. This kind of approach allows to spot *right-for-the-wrong-reasons* scenarios, such as the aforementioned wolf-snow case [15]. Second, one can measure how well the explanation was able to highlight the pixels that contributed most to the correct classification - regardless of their position in the image and coherence with human focus. The first pair of metrics we introduce, *AF* (*Area Focus*) and *BF* (*Border Focus*), measure the match between the hot region in an explanation heatmap and an object segmentation provided as ground truth.

The first (absolute) metric is defined as

$$AF = \frac{1}{prop} \frac{\sum_i m_i^+ - \sum_i m_i^-}{\sum_i |m_i|}$$

where m_i are all pixels in the heatmap, and m_i^+ and m_i^- denote respectively the pixels inside and outside the segmentation contour. $\frac{1}{prop}$ is a normalization factor (with $prop$ being the ratio between the number of pixels in the segmentation area and the total pixel count) introduced to balance the fact that the area inside the segmentation might vary from image to image. AF's minimum value (corresponding to the worst case scenario, where the explanation is actually the reversed segmentation) is equal to $AF_{min} = -\frac{1}{prop}\frac{\sum_i |m_i|}{\sum_i |m_i|} = -\frac{1}{prop}$ and its maximum (in the best case scenario, where segmentation and explanation coincide) is $AF_{max} = \frac{1}{prop}\frac{\sum_i |m_i|}{\sum_i |m_i|} = \frac{1}{prop}$.

Thus, it is possible to normalize AF as follows:

$$\tilde{AF} = \frac{AF - AF_{min}}{AF_{max} - AF_{min}} = \frac{AF + \frac{1}{prop}}{\frac{2}{prop}} = \frac{1}{2}\left(\frac{\sum_i m_i^+ - \sum_i m_i^-}{\sum_i |m_i|} + 1\right)$$

obtaining $\tilde{AF}$, a focus match metric with range $[0, 1]$.

Besides measuring whether the explanation focuses on the inside region of the segmentation, another option is to analyze how well the explanation is able to focus on the general outline of the segmentation. We formalize this with another metric, BF (*Border Focus*):

$$BF = \frac{1}{prop}\frac{\sum_i \frac{m_i^+}{1+d_i} - \sum_i \frac{m_i^-}{1+d_i}}{\sum_i \frac{|m_i|}{d_i+1}}$$

where d_i is the minimum number of pixels separating the ith pixel from the border of the segmentation: this allows to weight more the pixel near the border of the object, such that the explanations highlighting the contour of the image are considered better than the ones farther away from it. BF can be normalized using $BF_{min} = -\frac{1}{prop}\frac{\sum_i \frac{|m_i|}{1+d_i}}{\sum_i \frac{|m_i|}{1+d_i}} = -\frac{1}{prop}$ and $BF_{max} = \frac{1}{prop}\frac{\sum_i \frac{|m_i|}{1+d_i}}{\sum_i \frac{|m_i|}{1+d_i}} = \frac{1}{prop}$, so that $\tilde{BF} = \frac{BF+1}{2}$ is independent of the size of the segmentation area as well.

On the one hand, we argue that it is important to check whether the explanation of the model focuses on the image regions that a human would deem relevant to the classification task. On the other hand, this approach requires the dataset to provide ground truth segmentations defining the exact outline of the objects in picture. Typically this is not an available information for image classification benchmarks - for instance, ImageNet (the world standard dataset for computer vision) does not provide any precise segmentation for its instances.

In order to evaluate the correlation between hotness in the explanation and actual contribution to the correct classification, we include two more metrics [13], namely *insertion* and *deletion*.

The *deletion metric* progressively removes pixels from the image (according to the ranking provided by the explanation heatmap) and measures the decrease in the prediction probability of correct label. For a good explanation, one should observe a sharp drop and thus a low area under the probability curve (AUC) obtained plotting the proportion of deleted pixels versus the probability of belonging to the class of interest for the data instance.

The *insertion metric*, on the other hand, takes a complementary approach, measuring the increase in the probability as more and more pixels are introduced to an empty background according the ranking provided by the heatmap scores, with higher AUC indicative of a better explanation.

These metrics belong to the interval $[0, 1]$, do not require a ground truth segmentation, but need to query the black box classifier several times and to access the classification score.

4.3 Quantitative Comparison

In Fig. 2 we consider again the ladybug example image and compute the four described metrics ($\tilde{AF}$, $\tilde{BF}$, insertion, deletion) for the four labels (top two, bottom two) for the heatmaps provided by six XAI algorithms (Vanilla, Guided Backprop, LRP, GradCAM, LIME, RISE).

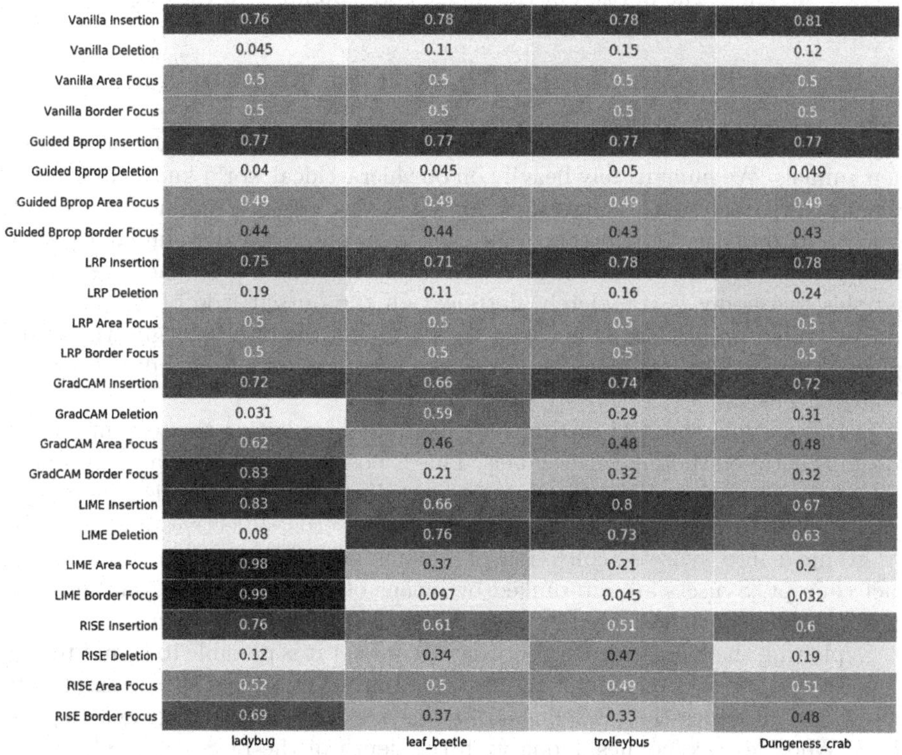

Fig. 2. Assessment of explainability methods

An important observation rising from the discussed metrics is that there is a clear discrepancy between the regions of the image perceived as important from the human eye and the parts which actually give a contribution to the outcome of the model: in particular, according to the deletion and the insertion metrics (which can be considered a proxy of the capability of the explanations to capture important regions purely from the model's perspective), the best performing techniques are model aware (in particular LRP, Guided Backpropagation and Vanilla Gradient).

On the other hand, considering the two metrics involving the segmentation as measures of the accordance between human and model perception of the data instance, all the explanations (except for LIME and GradCAM) seem mediocre, since they are not able to precisely isolate the main object in the image.

Hence, it is possible to conclude that LIME and GradCAM return an explanation which is much closer to the human perception of the input, while they fail to spot important pixels purely for the outcome as well as the model aware techniques, which are better at explaining the system's interpretation of the image.

This observation somehow agrees with the visual heatmaps in Fig. 1, which precisely highlight the portion of the image corresponding to the ladybug.

5 Linking Explanations with Structured Knowledge

Items in real life belong to taxonomies - e.g. Siamese cats are cats, then mammals, then animals. We humans rely heavily on our hierarchical world knowledge when learning and reasoning. Images used for CV tasks visually retain this kind of structured similarity: for instance, all cats species are visually similar to each other, and so on. However, image classification is framed as a one-vs-all task, and this taxonomy is completely flattened on the output side of the learning process, where each class is encoded as a one-hot vector. We argue that this approach is not ideal for labelling items when pairwise ontological distances are heterogeneous.

In this section, the XAI methods are compared exploiting the semantic relationships between the ImageNet labels. This is possible since all ImageNet labels are nodes (called *synsets*) in the WordNet ontology [12]. WordNet is a large lexical database of English language, where nouns, verbs, adjectives and adverbs are grouped into synsets (collection of synonyms), each one expressing a distinct concept. Synsets are interlinked by means of various kinds of relations (in particular, we have considered the *is-a* relationship), creating a network.

Exploiting the hierarchical structure of WordNet it is possible to compute the semantic distance between different classes in ImageNet, using the *wup* similarity [20], which measures the relatedness of two synsets by considering their depths in the WordNet taxonomies, along with the depth of the LCS (Least Common Subsumer).

The setup of this experiment is the following: given one image and an explanation, we progressively mask pixels according to the ranking provided by the explanation heatmap (with the *deletion* approach) and use the black box to classify the masked images. We then compute the wup similarity between the correct label and the one that was obtained by feeding the partially masked image to the black box classifier. We therefore obtain, for each image-explanation pair, a trajectory that described how the predicted label semantically drifts away from the correct one while the image is progressively masked. Examples of such trajectories are visualized in Fig. 3: on the X-axis is the semantic distance from the correct label, and on the Y-axis the percentage of masked pixels: note that the area under this trajectory is in the range [0, 1].

A good explanatory heatmap should highlight first the most relevant pixels for the correct classification; therefore masking pixels in this order should cause a sudden drop in the semantic similarity between the new predicted label and the original one. Consequently, this should yield a relatively horizontal trajectory and consequently a small resulting Area Under Curve (AUC).

Conversely, a bad explanation would highlight non-relevant pixels - producing a trajectory with more vertical segments and a bigger AUC. Thanks to the wup distance we are able to discriminate errors with different semantic distance from the correct label: *tabby cat* - *Siamese cat* from *tabby cat* - *bus*.

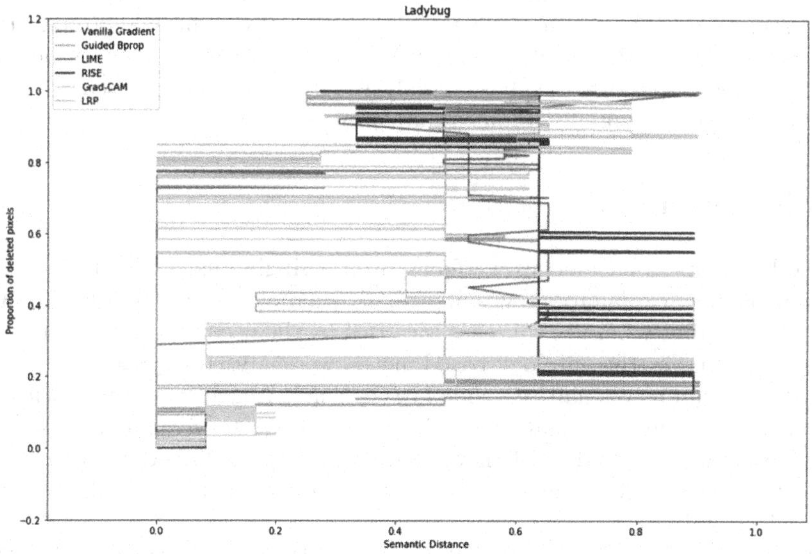

Fig. 3. Example of semantic trajectories, computed for the same image using different XAI techniques

For this experiment we considered all images in the ImageNet label *tabby cat*, linked to the homonym synset in WordNet (∼1000 images). For each image and XAI algorithm we produced an explanation heatmap. We then proceeded to compute each trajectory and determine its AUC. For each XAI method we then aggregated all AUCs (for all the images in the considered synset) and computed their distribution of the AUCs: the best explanation methods should have a mean close to 0, equivalent to a high and positive skewness (since the data range is [0, 1]). The results in Table 1 confirm what has been described in the previous sections: the model-aware methods are better at capturing important pixels for the classification of the model than model agnostic techniques, with the mean of the AUC curves' distribution closer to 0. LRP is performing particularly well, followed by Vanilla Gradient.

Table 1. Skewness of the distribution of AUCs for each XAI technique

XAI method	Vanilla gradient	GradCAM	LIME	Guided backprop	RISE	LRP
Skewness	0.909	0.872	0.590	0.887	0.882	1.032

An important observation is that each trajectory depends on the size of the object represented in the image, but for this analysis we aggregate the trajectories for a whole ImageNet-WordNet synset, thus comparing distributions of a thousand trajectories, over the same images, for each XAI algorithm.

We remark that this experiment does not involve any additional ground truth (such as the segmentation), as we rely on WordNet's structure. We are therefore able to connect a performance-based analysis with a semantics-based approach using benchmark data and without a human in the loop.

6 Discussion

For the heatmap-based metrics we focused on a single image because our goal was to show how explanations visually differ when changing XAI algorithm or label - the scalability limit in this case is caused by the need for ground truth segmentation data. For the ontology-based metric we analyzed a synset as a whole; this approach can be virtually extended to the whole ImageNet dataset, with the sole concern of computational cost - since every trajectory requires classifying multiple partially masked versions of the same image. All XAI algorithms, as well as the pretrained DL model and all data, are publicly available following the provided links.

There is a number of other quantitative comparisons (that we omitted for the sake of brevity) that we performed, such as analyzing how explanations change for different labels, or measuring how noisy or contiguous the hot regions in the heatmaps are. We argue that these analyses are paramount in order to be able to compare explanations heatmaps and XAI algorithms.

The experiments described in the previous sections show a clear discrepancy between XAI techniques: in particular, it is possible to conclude that model aware algorithms are better at discerning important regions for the model, which may not coincide with the human perception of the input. On the contrary, model agnostic methods return explanations which are closer to human common sense, but the corresponding heatmaps highlight regions that correlate poorly with the classification performance. Therefore we argue that these two families of XAI algorithms should be adopted in different settings, according to the priority assigned to fidelity versus human interpretability.

7 Future Work

In the future, we will apply the metrics and comparison techniques defined in the previous sections to new emerging XAI algorithms.

More generally, we aim at further investigate in the direction of semantic explanation, driven by the intuition that human-understandable explanations have to be articulated and therefore require to link the black box outputs with some form of structured symbolic knowledge; in particular, we will keep exploiting the connection between ImageNet and WordNet. For example, for each label-synset (e.g. *tabby cat*) one can find all sibling nodes in WordNet that correspond to ImageNet labels (*tiger cat, Siamese cat*, and so on); setting the corresponding multi-hot output vector in a DL model and applying existing XAI algorithms would allow to obtain a generalized explanation of the *cat* superclass, an ancestor in the WordNet hierarchy but not an existing (and therefore explainable) class for DL models trained on ImageNet. With the same logic counterfactual explanations can be obtained, e.g. providing an explanation of *why an image was classified as cat but specifically as a tabby cat and not a Siamese one.* With the same goal, also *part-of* links can be navigated in order to further enrich the otherwise explanations provided by XAI algorithms.

References

1. Binder, A., Bach, S., Montavon, G., et al.: Layer-wise relevance propagation for deep neural network architecture. In: ICISA 2016 (2016)
2. Chakraborty, S., Tomsett, R., Raghavendra, R., et al.: Interpretability of deep learning models: a survey of results (2017). https://doi.org/10.1109/UIC-ATC.2017.8397411
3. Chollet, F.: Deep Learning with Python. Manning Publishers & Co (2018)
4. Deng, J., Dong, W., Socher, R., et al.: ImageNet: a large-scale hierarchical image database (2009). https://doi.org/10.1109/CVPR.2009.5206848
5. Erhan, D., Bengio, Y., Courville, A., Vincent, P.: Visualizing Higher-Layer Features of a Deep Network (2009). arXiv:1903.02313
6. Esteva, A., Robicquet, A., Ramsundar, B., et al.: A guide to deep learning in healthcare (2019). https://doi.org/10.1038/s41591-018-0316-z
7. Goodfellow, I., Bengio, Y., Courville, A.: Deep Learning. O'Reilly Media (2018)
8. Goodman, B., Flaxman, S.: European Union regulations on algorithmic decision-making and a "right to explanation" (2017). https://doi.org/10.1609/aimag.v38i3.2741
9. Guidotti, R., Monreale, A., Ruggieri, S., et al.: A Survey of Methods for Explaining Black Box Models (2018). arXiv:1802.01933v3
10. Haykin, S.: Neural Networks and Learning Machines. Pearson Prentice Hall (2009). ISBN: 978-0-13-147139-9
11. Li, H., Cai, J., Nguyen, T., Zheng, J.: A benchmark for semantic image segmentation. In: ICME (2013)
12. Miller, G.A.: WordNet: a lexical database for English. Commun. ACM **38**(11), 39–41 (1995)
13. Petsiuk, V., Das, A., Saenko, K.: RISE: Randomized Input Sampling for Explanation of Black-Box Models (2018). arXiv:1806.07421v3

14. Rao, Q., Frtunikj, J.: Deep Learning for Self-driving Cars: Chances and Challenges (2018). https://doi.org/10.1145/3194085.3194087
15. Ribeiro, M.T., Singh, S., Guestrin, C.: Why Should I Trust You? Explaining the Predictions of Any Classifier (2016). arXiv:1602.04938v3
16. Russell, S., Norvig, P.: Artificial Intelligence: A Modern Approach. Prentice Hall Press (1994). ISBN 0136042597 9780136042594
17. Selvaraju, R.R., Cogswell, M., Das, A., Vedantam, R.: Grad-CAM: Visual Explanations from Deep Networks via Gradient-based Localization (2017). arXiv:1610.02391v3
18. Springenberg, J.T., Dosovitskiy, A., Brox, T., Riedmiller, M.: Striving for Simplicity: The All Convolutional Net (2015). arXiv:1412.6806v3
19. Szegedy, C., Vanhoucke, V., Ioffe, S., Shlens, J.: Rethinking the Inception Architecture for Computer Vision (2015). arXiv:1512.00567v3
20. Wu, Z., Palmer, M.: Verbs Semantics and Lexical Selection (2004). https://doi.org/10.3115/981732.981751

About Generative Aspects of Variational Autoencoders

Andrea Asperti(✉)

Department of Informatics: Science and Engineering (DISI),
University of Bologna, Bologna, Italy
andrea.asperti@unibo.it

Abstract. An essential prerequisite for random generation of good quality samples in Variational Autoencoders (VAE) is that the distribution of variables in the latent space has a known distribution, typically a normal distribution $N(0,1)$. This should be induced by a regularization term in the loss function, minimizing *for each data* X, the Kullback-Leibler distance between the posterior inference distribution of latent variables $Q(z|X)$ and $N(0,1)$. In this article, we investigate the marginal inference distribution $Q(z)$ as a Gaussian Mixture Model, proving, under a few reasonable assumptions, that although the first and second moment of $Q(z)$ might indeed be coherent with those of a normal distribution, there is no reason to believe the same for other moments; in particular, its Kurtosis is likely to be different from 3. The actual distribution of $Q(z)$ is possibly very far from a Normal, raising doubts on the effectiveness of generative sampling according to the vanilla VAE framework.

1 Introduction

A large amount of the recent research in deep learning has been driven by the challenging task of building scalable generative models able to describe rich data distributions comprising images, audio and videos. Generative Adversarial Networks (GAN) [7] and Variational Autoencoders (VAE) [11,13] are the two most prominent classes of generative models which have been investigated, each one giving rise to a long series of variants, theoretical in vestigations, and practical applications.

An interesting aspect of Variational Autoencoders is that they couple the generative model (the decoder) with an inference model (the encoder) synthesizing the latent representation of input data: from this point of view, VAEs provide a major tool for the investigation of unsupervised representation learning.

To fix notation, let $P(X|z)$ be the generative model distribution of input X given the latent representation z and $Q(z|X)$ be the encoder distribution, computing the probability of an internal representation z given X.

The relation between $P(X|z)$ and $Q(z|X)$ is given by the following equation (see e.g. [5]), where KL is the Kullback-Leibler divergence:

$$\begin{aligned} log(P(X)) - KL(Q(z|X)||P(z|X)) = \\ \mathbb{E}_{z\sim Q(z|X)}\, log(P(X|z)) - KL(Q(z|X)||P(z)) \end{aligned} \tag{1}$$

G. Nicosia et al. (Eds.): LOD 2019, LNCS 11943, pp. 71–82, 2019.
https://doi.org/10.1007/978-3-030-37599-7_7

Since the Kullback-Leibler divergence is always positive, the term on the right is a lower bound to the loglikelihood $log(P(X))$, known as Evidence Lower Bound (ELBO). Moreover, assuming the encoder $Q(z|X)$ provides a good approximation of the actual model distribution $P(z|X)$, the Kullback-Leibler divergence $KL(Q(z|X)||P(z|X))$ should be small, and the ELBO gives a good approximation of the loglikelihood for X.

The prior $P(z)$ must be a known distribution, we can easily sample from: traditionally, it is assumed to be a spherical Normal distribution $N(1,0)$. The whole point of the VAE is to *induce* the generator to produce a marginal encoding distribution[1] $Q(z) = \mathbb{E}_X Q(z|X)$ close to the prior $P(z)$. Averaging the Kullback-Leibler regularizer $KL(Q(z|X)||P(z))$ on all input data, and expanding the Kullback-Leibler divergence in terms of entropy, we get:

$$\begin{aligned} &\mathbb{E}_X KL(Q(z|X)||P(z)) \\ &= -\mathbb{E}_X \mathcal{H}(Q(z|X)) + \mathbb{E}_X \mathcal{H}(Q(z|X), P(z)) \\ &= -\mathbb{E}_X \mathcal{H}(Q(z|X)) + \mathbb{E}_X \mathbb{E}_{z\sim Q(z|X)} logP(z) \\ &= -\mathbb{E}_X \mathcal{H}(Q(z|X)) + \mathbb{E}_{z\sim Q(z)} logP(z) \\ &= \underbrace{-\mathbb{E}_X \mathcal{H}(Q(z|X))}_{\substack{\text{Avg. Entropy} \\ \text{of } Q(z|X)}} + \underbrace{\mathcal{H}(Q(z), P(z))}_{\substack{\text{Cross-entropy} \\ \text{of } Q(X) \text{ vs } P(z)}} \end{aligned} \tag{2}$$

The cross-entropy between two distributions is minimal when they coincide, so we are pushing $Q(z)$ towards $P(z)$. At the same time, we try to augment the entropy of each $Q(z|X)$; under the usual assumption that $Q(z|X)$ is Gaussian, this amounts to enlarge the variance, further improving the coverage of the latent space, essential for generative sampling (at the cost of more overlapping, and hence more confusion between the encoding of different datapoints).

From the previous equation, we also understand in which sense the Kullback-Leibler term can be understood as a regularizer term: a simple way to drastically reduce the KL distance between an arbitrary distribution $Q(z)$ and a normal distribution is to shift and rescale the former in such a way as to bring its mean to 0 and its variance to 1. This affine deformation of the latent space is easily performed by a neural network and, especially, it has no impact at all on the reconstruction error, since the scaling operation can be immediately compensated by acting on the inputs in the next layer of the network.

The obvious question is if the result of the Kullback-Leibler regularization can be more effective than just pushing the two first moments of the distribution of $Q(z)$ towards those of a normal distribution (also in view of the first term in the r.h.s. of Eq. 2, whose effect is difficult to assess).

To this aim, in this article, we shall try to explicitly investigate the marginal encoding distribution $Q(z) = \mathbb{E}_X Q(z|X)$ as a Gaussian Mixture Model (GMM), with one Gaussian for each data point. This is per se an interesting and informative exploration, providing a more operational and explicit grasp of the actual behaviour of Variational Autoencoders. The main conclusion is that, although

[1] Called by some authors *aggregate posterior distribution* [12].

the two first moments of the GMM may indeed become equal to 0 and 1 by effect of the KL-regularization, there is no major reason to believe that the others moments will agree with those of a Normal distribution. In particular, their Excess Kurtosis are likely to be different.

The structure of the article is the following. In Sect. 2, we address VAEs from the perspective of a Gaussian Mixture Model, and recall the closed form of the KL-divergence traditionally used in implementations (cfr. Eq. 3); we also give, at the end of this Section, a short discussion of the variable collapse phenomenon. In Sect. 3, we derive an interesting law for Variational Auotencoders: for each latent variable, the sum between its variance and the average variance of the Gaussian distributions $Q(z|X)$ should be equal to 1; we also provide experimental evidence for this law in Sect. 3.1. Section 4 is devoted to the moments of the marginal encoding distribution $Q(z)$ (regarded as a GMM): using the law of Sect. 3, we prove that the two first moments should agree with those of a Normal distribution, and argue that there is no reason to believe the same for other moments. A few concluding remarks are given in Sect. 5.

2 A Gaussian Mixture Model

In this section we give a more operational and friendly introduction to autoencoders, stressing the Gaussian assumption about the encoder distribution $Q(z|X)$ and the resulting Gaussian Mixture Model (GMM) obtained by marginalization on all data points.

A traditional autoencoder is not suitable for generative purposes. We can obviously pass any combination of latent values as input to the decoder, but since we have no knowledge about their distribution in the latent space, it is extremely unlikely that the decoder will produce a sensible result. What we would like to obtain is to force the latent space to have a known distribution like, say, a normal spherical distribution. Note that this is sensibly different from simply renormalizing the latent space to have zero mean and unitarian variance (that could be trivially achieved by techniques similar to batch-normalization [10]): we really want to have a uniform coverage of the latent space, without holes or low-density regions, that is an essential prerequisite for random generative sampling (Fig. 1).

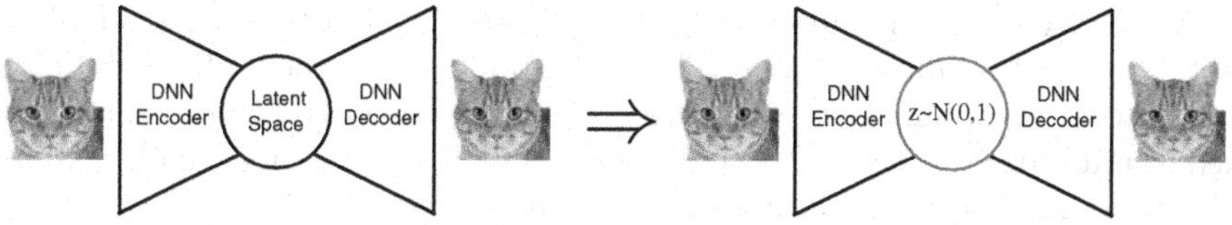

Fig. 1. A traditional autoencoder is not suitable for generative purposes: we need to know the distribution of the latent space, that moreover must be simple enough to allow easy sampling, e.g. a normal distribution.

The route followed by Variational Autoencoders to achieve this objective is a bit contrived.

Let us consider a data X, and its latent encoding z_X. The first observation is that, presumably, all values around X in a suitable neighborhood of it are likely to encode essentially the same information, hence producing a reconstruction close the original. This suggests that we can consider the encoder distribution $Q(z|X)$ as some Gaussian distribution centered in $\mu(X) = z_X$, and with some variance $\sigma^2(X)$. Each point X will have its own, distinct Gaussian distribution $Q(z|X) = N(\mu(X), \sigma^2(X))$ (see Fig. 2).

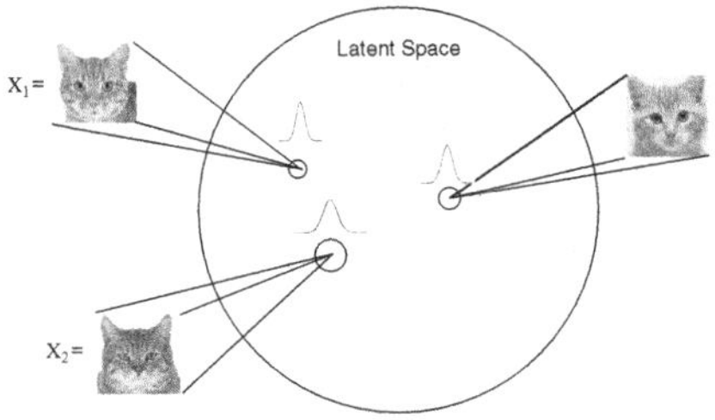

Fig. 2. For each data point X the posterior distribution $Q(z|X)$ is a different Gaussian distribution $N(\mu(X), \sigma^2(X))$.

A traditional autoencoder would just try to compute the expected value $\mu(X)$ by means of a neural network. In the case of VAE, we also try to estimate the variance $\sigma^2(X)$ or, as it is more usual but essentially equivalent, its logarithm. Normally, the two networks share most of the initial layers. The way we estimate $\sigma^2(X)$ has an interest in its own: instead of directly passing $\mu(X)$ to the decoder we sample around $\mu(x)$ using the current estimation of $\sigma(X)$ and pass the resulting point to the decoder. The reconstruction error is used to correctly tune $\sigma^2(X)$ along with $\mu(X)$. Since sampling cannot be directly taken into account during backpropagation, we sample z according to a normal $N(0,1)$ distribution and use $\mu(X)$ and $\sigma(X)$ as rescaling parameters, resulting in the point $z' = z * \sigma(X) + \mu(X)$. In this way, sampling is outside the backpropagation flow for $\sigma^2(X)$ and $\mu(X)$, that can be jointly learned with traditional training: this is the so called reparametrization trick [11,13].

If not contrasted in any way, $\sigma^2(X)$ would naturally tend to go to zero: $\mu(X)$ is the expected encoding, and there is no reason the network should be induced to prefer a different value. This is the point where the Kullaback-Leibler regularization component steps in: we add to the reconstruction error a term aimed to minimize the KL-distance $KL(Q(z|X)||P(z)$ between each $Q(z|X)$ and a *fixed prior* $P(z)$, typically a normal distribution (see Fig. 3).

Since the term $KL(Q(z|X)||P(z)$ is the KL-divergence between two Gaussian distributions $N(\mu(X), \sigma^2(X))$ and $N(1,0)$ it can be computed in closed form:

$$\begin{aligned} &KL(N(\mu(X), \sigma(X)), N(0,1)) = \\ &\quad \tfrac{1}{2}(\mu(X)^2 + \sigma^2(X) - log(\sigma^2(X)) - 1) \end{aligned} \tag{3}$$

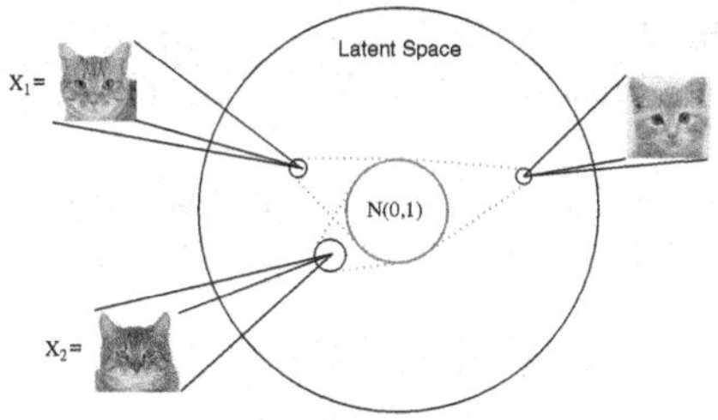

Fig. 3. For each data point X the posterior distribution $Q(z|X)$ is a different Gaussian distribution $N(\mu(X), \sigma^2(X)$.

The quadratic penalty over $\mu(X)$ has a clear interpretation: it will have the final effect to center the marginal distribution $Q(z)$ around the origin bringing its first moment to 0. On the other side, the terms acting on the variance are much less evident and almost counter-intuitive.

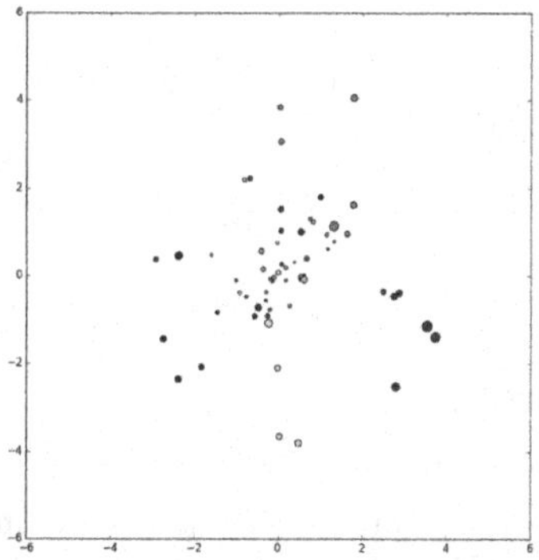

Fig. 4. Position and variance of 60 MNIST digits after 10 epochs of training in a bidimensional latent space. The dimension of the point is proportional to the variance. Observe the Gaussian-like distribution of all points (imagine it with 60000 points instead of 60) and also the really small values of the variance

The actual variance $\sigma^2_\theta(X)$ of the encoding distribution $Q(z|X)$ is typically very small (see Fig. 4), reflecting the fact that only a small region of the latent space around $\mu(X)$ can produce a reconstruction close to X. Pushing $\sigma^2(X)$ towards 1 will cause a large overlaps between the Gaussians relative to different points, augmenting the confusion of the latent representation[2].

A first answer to the previous question is given by Eq. 2: when averaging on all input data, the expected KL-term $\mathbb{E}_X\, KL(Q(z|X)||P(z))$ is just equal

[2] As suggested by β-VAE [3,9], this can also result in learning more disentangled and general features, shared by a large quantity of data.

to the cross entropy $\mathcal{H}(Q(z), P(z))$ between $Q(z)$ and $P(z) = N(0,1)$ (minimal when the two distributions coincide), plus an additional term equal to $-\mathbb{E}_X \mathcal{H}(Q(z|X))$.

In the next Sections we shall do a similar investigation, explicitly addressing the marginal encoding distribution $Q(z) = \mathbb{E}_X Q(z|X)$ as a Gaussian Mixture Model.

Before entering in this topic, let us make a final observation regarding the problem of the collapse of variables in latent spaces of high-dimensions, typical of Variational Autoencoders. The point is that the Kullback-Leibler regularization may, in some pathological but not infrequent cases, prevail over the reconstruction gain provided by the a specific latent variable. In this case, we shall end up in a configuration of the kind described in Fig. 5. Inactive latent variables are characterized by:

- a variance close to 0: the mean value $\mu(X)$ is almost always 0;
- for any X, a computed variance $\sigma^2(X)$ close to 1.

The variance computed by the network is essentially meaningless in this case, since the variable is neglected by the generator.

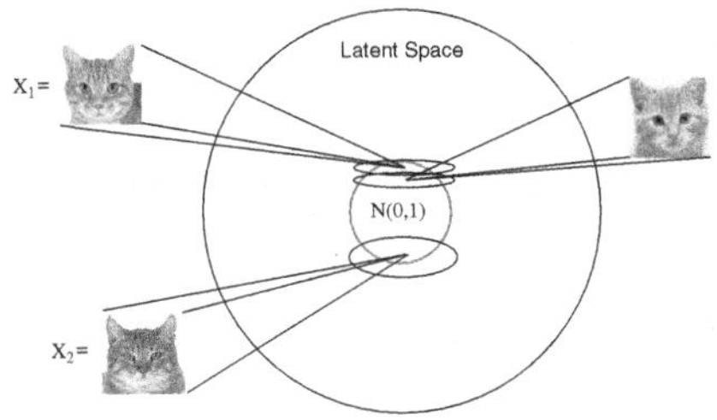

Fig. 5. Effect of the KL-regularization on a weak feature, in this case, x: the variable collapse. For any data X its encoding is 0; the variance of the variable is close to 0, while the computed (but wrong) variance $\sigma^2(X)$ is 1 for each X. This has no effect on the reconstruction, since in any case the variable is ignored by the decoder.

This is sometimes considered as a problem of VAE (overpruning, see e.g. [2,14]), since it causes an under-use of the model capacity. However, the phenomenon only happens when the information associated with the latent variable is really low: in such a situation, it could be wise to renounce to exploit the variable, especially for generative purposes. In other words, the collapse of latent variables can be interpreted as a sparsity phenomenon [1], with all the beneficial effects typically attributed to this property. The self-regularization of the model capacity is not an issue, but a remarkable and astonishing property of VAEs.

3 The Variance Law

Trying to compute relevant statistics for the posterior distribution $Q(z|X)$ of latent variables without some kind of regularization constraint does not make much sense. As a matter of fact, given a network with mean $\mu_z(X)$ and variance $\sigma_z^2(X)$ we can easily build another one having precisely the same behaviour by scaling mean and standard deviation by some constant γ (for all data, uniformly), and then downscaling the generated samples in the next layer of the network: an operation that is easily performed by a neural network.

Let's see how the KL-divergence helps to choose a solution. In the following, we suppose to work on a specific latent variable z.

The expression we must consider is the average, over all data points X of the closed form for the KL-divergence given in Eq. 3, that is:

$$\frac{1}{N}\sum_X \frac{1}{2}(\mu(X)^2 + \sigma^2(X) - log(\sigma^2(X)) - 1) \tag{4}$$

where we used the notation $\widehat{f(X)}$ to abbreviate the average $\frac{1}{N}\sum_X f(X)$ of $f(X)$ on all data X. As we already observed, for the network it is easy to keep a fixed ratio $\rho^2 = \frac{\widehat{\sigma^2(X)}}{\widehat{\mu^2(X)}}$. For similar reasons, we may also assume that latent variable to be centered around zero so that $\widehat{\mu^2(X)}$ is just the (global) variance σ^2 of the latent variable.

Pushing ρ^2 in the previous equation, we get

$$\frac{1}{2}(\widehat{\sigma^2(X)}\frac{1+\rho^2}{\rho^2} - \widehat{log(\sigma^2(X))} - 1)$$

Now we perform a somewhat rough approximation. The average of the logarithms $\widehat{log(\sigma^2(X))}$ is the logarithm of the geometric mean of the variances. If we replace the geometric mean with an arithmetic mean, we get the following much simpler expression:

$$\frac{1}{2}(\widehat{\sigma^2(X)}\frac{1+\rho^2}{\rho^2} - log(\widehat{\sigma^2(X)}) - 1) \tag{5}$$

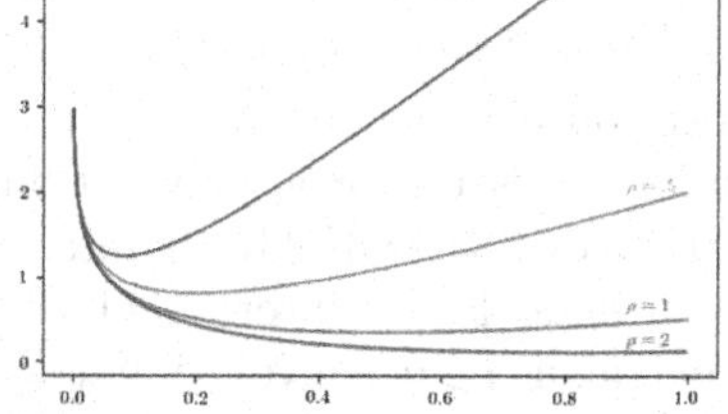

Fig. 6. KL-divergence for different values of ρ: observe the strong minimum for small values of ρ.

In Fig. 6 we plot the previous function in terms of the mean of the variances, for a few given values of ρ. It is easy to derive that we have a minimum for the previous equation at

$$\widehat{\sigma^2(X)} = \frac{\rho^2}{1+\rho^2} \tag{6}$$

that is close to 0 when ρ is small, and close to 1 when ρ is high.

Substituting the definition of $\rho^2(X)$ in Eq. 6, we expect to reach a minimum when

$$\widehat{\sigma^2(X)} = \frac{\widehat{\sigma^2(X)}}{\widehat{\mu^2(X)}} \frac{\widehat{\mu^2(X)}}{\widehat{\mu^2(X)} + \widehat{\sigma^2(X)}}$$

that, by trivial computations, implies the following simple stationary condition:

$$\widehat{\sigma^2(X)} + \widehat{\mu^2(X)} = 1 \tag{7}$$

or simply,

$$\boxed{\widehat{\sigma^2(X)} + \sigma^2 = 1} \tag{8}$$

where we replaced $\widehat{\mu^2(X)}$ with the variance σ^2 of the latent variable in view of the consideration above.

3.1 Experimental Validation

The law in Eq. 8 can be experimentally verified. Actually, we observed it before addressing a theoretical investigation, and it was one of the motivations for this investigation.

As a first example, in Table 1 we provide relevant statistics for a input-256-32-24-16 dense VAE trained over a couple of simple datasets: the first one is MNSIT, and the second one is the *tiles* problem described in [1], consisting in generating small white tiles on a black 28 × 28 background. Observe, by the way, the collapse phenomenon: all variables highlighted in red have collapsed.

For the data in Table 1 the average *sum* of σ^2 and $\widehat{\sigma^2(X)}$ (the two cells in each row) is 0.997 with a variance of 0.00048! Let us remark that condition 8 holds both for active and inactive variables, and not just for the cases when values are close to 0 or 1; for instance, observe the case of variable 12 in the right Table, which has a global variance around 0.797 and a mean local variance $\widehat{\sigma^2(X)}$ around 0.238, almost complementary to 1.

In Table 2 we give similar statistics for a convolutional VAE applied to the same datasets, comprising moreover the CIFAR-10 dataset. The network encoder is composed by five convolutional layers intermixed with batch normalization layers; the decoder is symmetric, using deconvolutional layers (see [1] for a precise definition).

In this case, the average *sum* of σ^2 and $\widehat{\sigma^2(X)}$ is 1.027 with a variance of 0.01045.

Table 1. Latent Space Statistics for a dense VAE relative to MNIST (left) and to the problem of Generating Tiles (right)

no.	variance	mean($\sigma_\theta^2(X)$)
0	8.847272e-05	0.9802212
1	0.00011756	0.9955146
2	6.665453e-05	0.9851733
3	0.97417927	0.0087413
4	0.99131817	0.0061861
5	1.0012343	0.0101425
6	0.94563377	0.0571693
7	0.00015841	0.9820533
8	0.94694275	0.0332076
9	0.00014789	0.9850558
10	1.0040375	0.0181513
11	0.98543876	0.0239957
12	0.000107441	0.9829797
13	4.5068125e-05	0.9989834
14	0.00010853	0.9604088
15	0.9886378	0.0444058

no.	variance	mean($\sigma_\theta^2(X)$)
0	0.89744579	0.08080574
1	5.1123271e-05	0.98192715
2	0.00013507	0.99159979
3	0.99963027	0.016493475
4	6.6830005e-05	1.01567184
5	0.96189236	0.053041528
6	1.01692736	0.012168014
7	0.99424797	0.037749815
8	0.00011436	0.96450048
9	3.2284329e-05	0.97153729
10	7.3369141e-05	1.01612401
11	0.91368156	0.086443416
12	0.79746723175	0.23826576
13	7.9485260e-05	0.9702732
14	0.92481815	0.089715622
15	4.3311214e-05	0.95554572

Table 2. Latent Space Statistics for a convolutional VAE relative to MNIST (left), CIFAR-10 (center) and to the problem of Generating Tiles (right)

no.	variance	mean($\sigma_\theta^2(X)$)
0	3.9236e-05	0.9866523
1	0.9204337	0.0192987
2	0.9943176	0.0073644
3	0.9857711	0.014468
4	1.0453497	0.0048676
5	1.0301798	0.0064604
6	3.0508e-05	0.9371982
7	1.0328044	0.0040635
8	0.9190037	0.0338749
9	0.9930814	0.0121515
10	0.9599534	0.0189501
11	1.3106997	0.0022037
12	1.1483558	0.0025166
13	1.8960e-05	0.9543213
14	1.0768713	0.0077547
15	1.2181355	0.0017144

no.	variance	mean($\sigma_\theta^2(X)$)
0	0.9301896	0.0126611
1	1.0369887	0.0015773
2	2.8245e-05	1.0028544
3	5.4145e-05	0.9690067
4	0.9657640	0.0211818
5	0.9713142	0.0165921
6	0.9378762	0.0275781
7	0.9885434	0.0043563
8	0.9885434	0.0178736
9	0.9782546	0.9782546
10	0.9519605	0.0125327
11	0.9759746	0.0132204
12	0.9555376	0.0157400
13	0.9555232	0.0309539
14	1.0171765	0.0048841
15	0.9341189	0.0387961

no.	variance	mean($\sigma_\theta^2(X)$)
0	0.8974457	0.0808057
1	5.1123e-05	0.9819271
2	0.0001350	0.9915997
3	0.9996302	0.0164934
4	6.6830e-05	1.0156718
5	0.9618923	0.0530415
6	1.0169273	0.0121680
7	0.9942479	0.0377498
8	0.0001143	0.9645004
9	3.22843e-05	0.9715372
10	7.33691e-05	1.0161240
11	0.9136815	0.0864434
12	0.7974672	0.2382657
13	7.94852e-05	0.9702732
14	0.9248181	0.0897156
15	4.33112e-05	0.9555457

4 Moments of the GMM Distribution

Now we compute moments for $Q(z)$ considering it as a Gaussian Mixture Model. The computation is quite standard, and we mostly do it for the sake of completeness.

Our model is a mixture of Gaussian distributions $f(X) = N(\mu(X), \sigma^2(X))$, one for each data X. We suppose all data to be significant samples of the model distribution, hence having the same individual probability. Using our averaging notation, the mixture is just

$$f = \widehat{f(X)}$$

and for any moment k,

$$\mu^{(k)} = \mathbb{E}_f[x^k] = \widehat{\mathbb{E}_{f(X)}[x^k]} = \widehat{\mu^{(k)}(X)} \tag{9}$$

that is to say that each moment of the mixture is just the average of the corresponding moments of its components.

The variance of $f = Q(z)$ is

$$Var(f) = \mu^{(2)} - (\mu^{(1)})^2 = \widehat{\mu^{(2)}(X)} - (\mu^{(1)})^2$$

Under the assumption that $\mu^{(1)} = 0$ (thanks to simple neural network translation operations),

$$Var(f) = \mu^{(2)} = \widehat{\mu^{(2)}(X)} = \widehat{\sigma^2(X)} + \widehat{(\mu^{(1)}(X))^2} = 1$$

due to variance law of Eq. 7.

So, the marginal distribution $Q(z)$ will have indeed a unitarian variance, similarly to the normal prior.

Unfortunately, we know very little about the other moments. Let us consider for instance the Kurtosis, that is usually interpreted as a measure of the "tailedness" of the probability distribution. The standard measure of kurtosis, due to Pearson, is a scaled version of the fourth central moment of the dataset: $k = \frac{\mu_c^{(4)}}{\sigma^4}$.

For our Gaussian Mixture Model, supposing it has zero mean and unitarian variance, the Kurtosis coincides with its fourth moment, that is just the mean of the (non-central) fourth moments of its components, in view of Eq. 9.

For a Gaussian distribution, its non-central fourth moment is

$$mu^{(4)} = \mu^4 + 6\mu^2\sigma^2 + 3\sigma^2$$

Even taking into account the constraint of Eq. 8, the sum over all data of these moments can be arbitrary, and very unlikely to be 3 (the Kurtosis of the Normal). The resulting MGG model could be both platykurtic or leptokurtic. Even worse, while the actual tailedness of the distribution might not be so relevant for many practical purposes, the marginal distribution $Q(z)$ can easily be multimodal, with low density regions where we would not expect them, essentially compromising the possibility of good generative sampling.

5 Conclusions

In this article, we addressed the marginal posterior distribution $Q(z) = \mathbb{E}_X\, Q(z|X)$ as a Gaussian Mixture Model (GMM), with a distinct Gaussian for each data point X, with the aim to investigate the resulting distribution.

As an effect of the Kullback-Leibler regularization component we would expect $Q(z)$ to get close to the prior model distribution $P(z)$, that is traditionally assumed to be a spherical Normal distribution $N(0, 1)$. Considering the moments of the GMM, this implies that for each latent variable z the sum between its variance and the average over all data X of the variance of the constituent Gaussian distributions $Q(z|X)$ should be 1 (variance law). We argue that for a neural network it is relatively easy to ensure this condition by a simple affine transformation of the latent space, and show how the variance law can be derived from the closed form of the KL-divergence. The variance law is a simple sufficient condition ensuring the KL-divergence is properly working, and should be systematically checked when implementing VAEs.

Nevertheless, other moments of $Q(z)$ might still be different form those of a Normal distribution. So, in spite of the Kullback-Leibler regularization, $Q(z)$ could actually sensibly depart from the expected prior $P(z)$, posing serious problems for generative sampling.

This seems to suggest that we need to use more sophisticated architectures, organized with some kind of layered, possibly iterative structure. Examples of this kind are Deep Recurrent Attentive Writers (DRAW) [8] Generative Query Networks [6], or the recent two-stage approach described in [4]. In the latter case, a second VAE is trained to learn an accurate approximation of $Q(z)$; we generate samples of $Q(z)$ starting from a normal distribution, and then fed them to the actual generator of data points. In this way, the dissimilarity between $P(z)$ and $Q(z)$ is not an issue, since we can sample from the latter using the second-stage VAE. Similarly to a vanilla VAE, the approach does not require additional hyperparameters or sensitive tuning and seems to produce high-quality samples, competitive with state-of-the-art GAN models, both in terms of FID score and visual quality.

References

1. Asperti, A.: Sparsity in variational autoencoders. In: Proceedings of the First International Conference on Advances in Signal Processing and Artificial Intelligence, ASPAI 2015, Barcelona, Spain, 20–22 March 2019 (2019)
2. Burda, Y., Grosse, R.B., Salakhutdinov, R.: Importance weighted autoencoders. CoRR, abs/1509.00519 (2015)
3. Christopher, P., et al.: Understanding disentangling in β-VAE (2018)
4. Dai, B., Wipf, D.P.: Diagnosing and enhancing VAE models. In: Seventh International Conference on Learning Representations (ICLR 2019), New Orleans, 6–9 May 2019 (2019)
5. Doersch, C.: Tutorial on variational autoencoders. CoRR, abs/1606.05908 (2016)
6. Eslami, S.M.A., et al.: Neural scene representation and rendering. Science **360**(6394), 1204–1210 (2018)

7. Goodfellow, I.J., et al.: Generative Adversarial Networks. arXiv e-prints, June 2014
8. Gregor, K., Danihelka, I., Graves, A., Rezende, D.J., Wierstra, D.: DRAW: a recurrent neural network for image generation. In: Proceedings of the 32nd International Conference on Machine Learning, ICML 2015, 6–11 July 2015, Lille, France. JMLR Workshop and Conference Proceedings, vol. 37, pp. 1462–1471. JMLR.org (2015)
9. Higgins, I., et al.: β-VAE: learning basic visual concepts with a constrained variational framework (2017)
10. Ioffe, S., Szegedy, C.: Batch normalization: accelerating deep network training by reducing internal covariate shift. CoRR, abs/1502.03167 (2015)
11. Kingma, D.P., Welling, M.: Auto-encoding variational bayes. CoRR, abs/1312.6114 (2013)
12. Makhzani, A., Shlens, J., Jaitly, N., Goodfellow, I.J.: Adversarial autoencoders. CoRR, abs/1511.05644 (2015)
13. Rezende, D.J., Mohamed, S., Wierstra, D.: Stochastic backpropagation and approximate inference in deep generative models. In: Proceedings of the 31st International Conference on Machine Learning, ICML 2014, Beijing, China, 21–26 June 2014. JMLR Workshop and Conference Proceedings, vol. 32, pp. 1278–1286. JMLR.org (2014)
14. Yeung, S., Kannan, A., Dauphin, Y., Fei-Fei, L.: Tackling over-pruning in variational autoencoders. CoRR, abs/1706.03643 (2017)

Adapted Random Survival Forest for Histograms to Analyze NOx Sensor Failure in Heavy Trucks

Ram B. Gurung(✉)

Department of Computer and Systems Sciences,
Stockholm University, Stockholm, Sweden
gurung@dsv.su.se

Abstract. In heavy duty trucks operation, important components need to be examined regularly so that any unexpected breakdowns can be prevented. Data-driven failure prediction models can be built using operational data from a large fleet of trucks. Machine learning methods such as Random Survival Forest (RSF) can be used to generate a survival model that can predict the survival probabilities of a particular component over time. Operational data from the trucks usually have many feature variables represented as histograms. Although bins of a histogram can be considered as an independent numeric variable, dependencies among the bins might exist that could be useful and neglected when bins are treated individually. Therefore, in this article, we propose extension to the standard RSF algorithm that can handle histogram variables and use it to train survival models for a NOx sensor. The trained model is compared in terms of overall error rate with the standard RSF model where bins of a histogram are treated individually as numeric features. The experiment results shows that the adapted approach outperforms the standard approach and the feature variables considered important are ranked.

Keywords: Histogram survival forest · Histogram features · NOx sensor failure

1 Introduction

In heavy duty trucks operation, ensuring availability of a truck is very important. Unexpected failures often result in unnecessary delays and huge expenses. Therefore, a precise warning of impending failure can ensure significant amount of cost savings and customer satisfaction [12]. It is therefore useful to know the current health status of various components in a truck, this information can also be used for scheduling maintenance for individual trucks [1]. Building a physical model (e.g. [3]) to monitor gradual degradation of component health is time consuming, requires domain experts and hence very expensive. As a cost effective alternative, data-driven approaches (e.g. [4,14]) are gaining attention. Operational profiles of a large fleet of trucks can be used to study various failure

G. Nicosia et al. (Eds.): LOD 2019, LNCS 11943, pp. 83–94, 2019.
https://doi.org/10.1007/978-3-030-37599-7_8

patterns and to make useful failure predictions, for example in [12]. This paper focuses on one specific component i.e. NOx sensor, but the overall process is rather generic and can be easily replicated for any other component. The NOx sensor was selected because it is one of the most important components in order for a truck to operate properly as it measures the concentration of NOx content in exhaust gases which are an atmospheric pollutant and therefore enforced by law to keep the concentration below certain level. NOx sensors are usually positioned in the tailpipe, exposed to a very high exhaust temperatures resulting in frequent failures. Therefore, the NOx sensor was selected for this study.

In our previous study [6], we studied the failure of NOx sensor as a classification problem, with a fixed prediction horizon of 3 months into the future. So the basic assumption while training the model was such that, if the sensor survives beyond the three months margin, it is treated as healthy whereas if it fails before that margin, it is treated as failure or near to fail. This assumption was useful as long as the model can accurately predict whether the sensor will survive until next scheduled workshop visit (before three months into the future). From a usability point of view, if the model predicts the sensor to be in danger of breaking down before next scheduled visit, required actions can be taken when the truck is already in the workshop. However, it would be desirable to have a model that does not limit itself to a fixed prediction horizon (e.g. 3 months into future) but rather model the continuous degradation over time. Furthermore, a new model needs to be trained for different prediction horizon. A model that regress over failure times of all trucks would have got the job done. But, not all trucks considered in our study have experienced failure, information about their survival (or failure) is known only until some point. Usually, trucks with no failures constitutes large fraction of all considered trucks, so, they cannot be ignored. Such instances (trucks in our case) can be considered as right-censored instances. Information from censored observations is important and therefore need to be fully taken into account during analysis. Therefore, NOx sensor failure is analyzed using survival analysis methods that can handle such censored data. Survival analysis studies the relationship between feature variables and the expected duration of time until when an event of interest occurs. Such analysis would enable us to learn and predict the survival probabilities of the NOx sensor of a given truck over time.

The Cox regression model [2] and its extensions are popular methods for conducting survival analysis. The popularity of the method can be attributed to its convenience to use. They however impose constrains on the underlying distributions and a Cox regression model does not reduce to ordinary linear regression in absence of censoring [8]. Therefore, other approaches such as tree-based methods like random survival forests [10] are popular. Random forest algorithms for survival analysis are preferred as they can handle high-dimensional feature space and also can easily deal with high-level interactions among features. Further details on the rationale for using tree-based methods for survival analysis can be read in [15]. For our analysis of NOx sensor failure, we choose to use and adapt a popular implementation of random survival forest [10] by Ishwaran et al.

The current implementation of random survival forest [10] can train from data that has simple numeric and/or categorical features. However, training data available in various domains can have features that are rather complex such as histograms. In heavy duty trucks, the operational profile of a truck has many features represented as histograms. For example in [14], authors have used random survival forest for heavy duty truck's battery lifetime prediction and the data they used had many feature variables as histograms. However, they simply considered each bin of a histogram variable as an independent numeric feature while training the model. But we would like to consider histogram variables as a whole while training the model so as to preserve the meta-information and inherent dependencies among histogram bins while training the model. In order to do so, the standard random survival forest algorithm needs to be adapted. Machine learning algorithms that can train on histogram data have not been studied to a major extent. But there exists few examples such as in [5–7,11,13]. Therefore, we intend to extend the current implementation of random survival forest to train on data that has feature variables represented as histogram(s). We use the implementation for random survival forest by Ishwaran et al. [10] as it adapts the standard random forest to survival response. Techniques to handle histogram features in a random forest is similar to one in our earlier study on histogram based random forest for classification [7]. The adapted random survival forest algorithm is then used to learn and predict survival curves for NOx sensor failure. The paper contributes by presenting methods to handle histogram features in a data to learn better survival models to predict imminent failures.

In the next section, a brief description of the random survival forest algorithm [10] and the random forest algorithm for histogram data in classification setting [7] are presented. In Sect. 3, our proposed changes is presented. In Sect. 4, the experimental setup for analysis of NOx sensor failure: data preparation, training model and results are presented. Finally, in Sect. 5, conclusion and direction for future research are pointed out.

2 Background

2.1 Random Survival Forest

The random survival forest is an ensemble of many base survival trees. The algorithm differs from its traditional classification and regression kind in terms of node splitting procedure. Log-rank test is used to evaluate the measure of between-node heterogeneity while splitting the node at each tree. Each tree is grown from a bootstrap samples and terminal nodes are ensured to have at least one event of interest (which is component breakdown in our case). A small subset of candidate feature variables are randomly selected for evaluating a node split. The node is split using the feature variable that maximizes survival differences across child nodes. Eventually, each terminal node in the tree becomes homogeneous with instances that have similar survival pattern. A cumulative hazard function is then determined for the instances at each terminal node [9].

Let τ be a set of terminal nodes in a tree. Let $(T_{1,h}, \delta_{1,h}), ..., (T_{n(h),h}, \delta_{n(h),h})$ be the survival times and censoring indicator (right censored) for instances in a terminal node $h \in \tau$. For instances i, $\delta_{i,h} = 0$ or $\delta_{i,h} = 1$ respectively indicates censored or event occurred at $T_{i,h}$. Let, $t_{1,h} < t_{2,h} < ... < t_{N(h),h}$ be the $N(h)$ distinct event times (failure times). At time $t_{l,h}$, if $d_{l,h}$ and $Y_{l,h}$ be the number of failed instances and instances at risk of failure respectively, the cumulative hazard function (CHF) using Nelson-Aalen estimator for h is

$$\hat{H}_h(t) = \sum_{t_{l,h} \leq t} \frac{d_{l,h}}{Y_{l,h}} \tag{1}$$

For an instance i with feature variables x_i that ends up in terminal node h, CHF estimate is

$$H(t|x_i) = \hat{H}_h(t) \tag{2}$$

An ensemble cumulative hazard estimate $H_e(t|x_i)$ is calculated by combining information from all trees. This hazard estimate can then be converted into survival function as

$$e^{-H_e(t|x_i)} \tag{3}$$

2.2 Random Forest Classifier for Histogram Data

Histogram data has one or more feature variables that are histograms. The standard random forest algorithm when trained on such data would consider each bin individually as a separate numeric variable while evaluating a split of a node in a base tree. However, we believe that, since the essence of the feature lies in whole histogram, the algorithm should consider all the bins (or multiple bins) simultaneously at each node split.

As described in [7], when a histogram variable is selected for evaluation of a node split in a base tree, groups of subset of bins of the histogram are generated first. The number of bins to be bundled in a subset is a model hyper-parameter. For simplicity, a sliding window method is used to bundle consecutive bins as shown in Fig. 1. The size of a sliding window (number of bins) can be tuned. Once a possible subsets of bins are generated, a certain number (usually squared root of total number of possible subsets, subject to yet another possible hyper-parameter value) of subsets are randomly selected. The randomly selected subsets of bins are then evaluated for splitting the node.

The purpose of using multiple bins simultaneously for a node split is to make sure that the split utilizes relevant information from all the bins. As explained in [7], given a subset of bins to be evaluated for a node split, a principal component analysis (PCA) is applied on the bins. Each principal component is a linear combinations of the original bins and therefore expected to contain information about the (linear) dependencies among such bins. The algorithm then performs evaluation of splitting the node on these subsequent principal components which are treated as a standard numeric feature variables as shown in Fig. 2.

3 Methods

The objective of this paper is to perform survival analysis of NOx sensor in heavy duty trucks by using adapted version of standard random survival forest for histogram variables. Although the current implementation of the RSF algorithm can be used to train on data with histogram features by treating each bin as an individual numeric feature, we adapt the algorithm to make it utilize histogram variables as a whole. We have investigated two approaches, first one uses multiple bins of a chosen histogram to split tree nodes and the second approach uses weighted probability of selecting bin in a chosen histogram to evaluate a node split. Both approaches are explained in detail below.

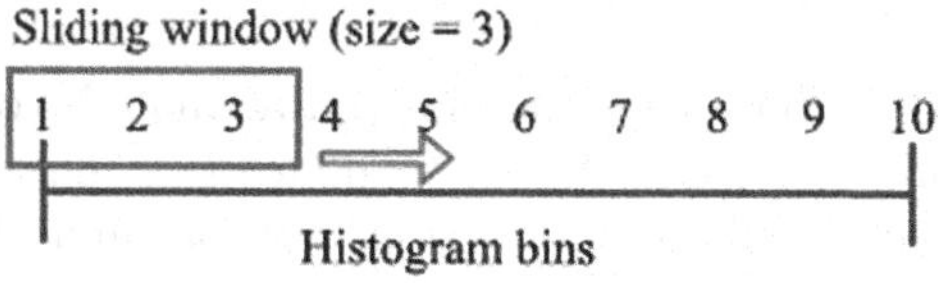

Fig. 1. Sliding window method to group bins

3.1 Using Multiple Bins Together for Evaluating Node Split

This approach derives from [7] where information from multiple bins of histogram is used to split a tree node. A high level description of adapted random survival forest algorithm is as follows:

1. Draw B bootstrap samples from the original training data.
2. Grow a histogram tree from each bootstrap sample. At each node of the tree randomly select a subset of p feature variables for splitting.
   (a) If a feature variable is numeric or categorical, find the cutoff value or best category that maximizes survival difference across child nodes.
   (b) If the feature variable is a histogram, randomly select $\sqrt{n}$ of the n possible subsets of bins from the histogram using a sliding window. For each selected subset of bins, apply PCA and evaluate each principal components for the best cutoff value that maximizes survival difference between child nodes. A bin subset that gives the best split is then selected for the histogram variable that is being considered.
3. Grow the survival tree to a full size until each terminal node has d unique deaths.
4. Compute the ensemble cumulative hazard estimate by combining information from all trees.
5. For a given test example, the cumulative hazard is estimated by dropping it down the root node of each tree. For a histogram variable, PCA rotation coefficients are used to transform the bin set of the test example and the cutoff value of the best principal component to route the test example to correct child node.

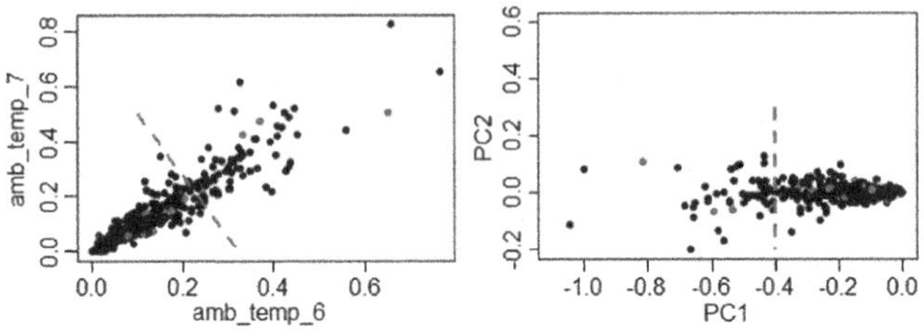

Fig. 2. Principal component analysis on group of bins

3.2 Weighted Probability of Getting Selected for Split Evaluation Assigned to a Bin

This is a simple modification of how the bins (individually) would be selected for evaluating a node split. If we were to use standard random forest approach, all the bins of all histograms would be simply considered as numeric feature variables and the algorithm would randomly pick small subset of all variables (numeric and categorical) to evaluate a node split. Instead, in this approach, selection of features to be evaluated for node split is done in two steps. In the first step, all numeric, categorical and histogram variables are considered for random selection and have equal probability of being selected. If a histogram variable is selected, in the next step, a small subset of bins of this histogram are again randomly selected. In the second step, instead of assigning equal probability for each bins to be selected, we assign a weighted probability. We assume failure as a case of anomaly such that breakdown occurs when a truck's operation deviates from a regular norm. An operation at extreme conditions are usually recorded along either ends of an operational histogram feature. Therefore, bins towards either ends of a histogram could have more relevant information regarding breakdowns. So, such bins should be more frequently evaluated. Histogram bin with a smaller average value (not frequent) will get higher probability of getting selected for split evaluation. However, bins with average zero value will not be selected at all whereas maximum probability is clipped at 0.5 to prevent heavy selection bias. For example, as shown in Fig. 3, for the Ambient Temperature histogram feature variable, bins around third and ninth have higher probabilities of getting selected based on the average value they contain in a given node.

4 Empirical Evaluation

The modifications in the algorithm described in Sect. 2 was implemented in the R programming language. The survival analysis model for the NOx sensor was built using operational data from large fleet of heavy duty trucks built by one of the truck manufacturing company in Sweden.

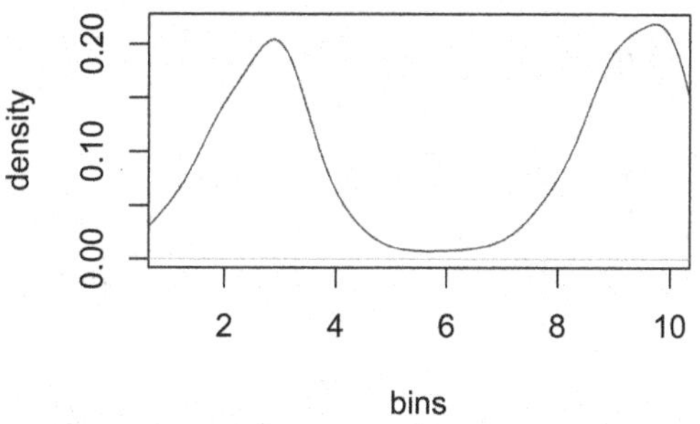

Fig. 3. Example showing which bins of Ambient Temperature histogram has more chances of getting selected for evaluating node split given that the feature Ambient Temperature gets randomly selected

4.1 Data Preparation

The data set has information about heavy duty trucks and their operations. Various sensors in the truck record data which are related to truck's operation and are stored on-board the truck. These data are extracted when the truck visits workshop. In order to limit our analysis to a homogeneous fleet, only trucks built for long haulage operation are considered. For our experiment, we only considered a failure of one specific component in a truck, i.e. NOx sensor. This is an important and sensitive component in the truck and it tends to fail rather frequently. Each data extract from a truck contains information corresponding to various feature variables associated to operational behavior of the trucks such as ambient temperature, fuel temperature, fuel consumption etc. Most of the operational variables are represented as histograms. For example, the ambient temperature variable can be a histogram with certain predefined number of bins that carries information (frequency count) about various ranges of ambient temperatures that the truck had operated under. Some trucks could have multiple failures of the same component over its lifetime, in such cases, we take into account only the first failure for the sake of simplicity. For our experiment, 10,000 trucks are considered with only 1000 of them having recorded NOx sensor failure. This experimental dataset considers nine histogram feature variables for the operational profile and six variables corresponding to the technical specification of the trucks from among many possible features. These features were selected after consulting with domain experts in the company.

4.2 Experiment Setup and Results

The survival model from the experimental data was generated using both the standard random survival forest approach and the new proposed algorithms with suggested changes for histogram features. The comparison of the performance for all approaches was done in terms of error rate. Experiments were conducted in the R programming environment. 5-fold cross validation was performed and

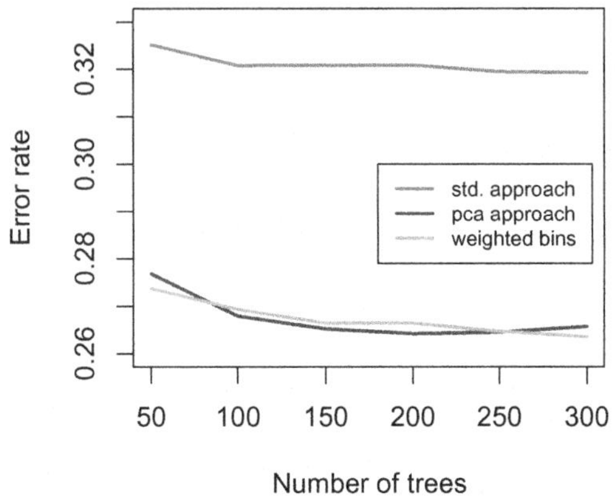

Fig. 4. Error rate vs Number of trees

error rates were averaged over all five folds. 5-fold cross validation was performed instead of 10-folds as the data set is already skewed heavily and the training of the model is also time consuming. The number of trees in both random forest models was set to 300. As shown in Fig. 4, performance gets stable well before using all 300 trees therefore more trees were not considered. Each tree in a model was grown to make sure that a leaf node has at least one failure (death) examples (trucks). Limit on the number of examples in a node to become a leaf node is set to be seven after which splitting stops. For the histogram approach that implements PCA using sliding window (as shown in Fig. 1 to group bins together before evaluating node split, size of the window was set to a modest value of 3. For a 2 dimensional histogram, sliding window of a block size of 2×2 was set. Different window sizes could have been examined but for this experiment only a modest setup was deployed, leaving further exploration for later. The results are shown in the Table 1. The columns of the table are approaches used to train the model, average cross-validation error rate, model training time relative to the standard approach and average number of leaf nodes obtained using different approaches respectively.

As shown in the result Table 1, the histogram approach using PCA and the weighted bin selection approach of training random survival forest outperforms the standard approach as the average error rate are quite lower for both approaches.

The model predicted survival curves for a randomly selected 100 new test examples are shown in the Fig. 5. Two out of these 100 trucks had experienced NOx failure. For the purpose of clarity, only 100 examples were plotted or else the survival curves would not be clearly visible. As shown in the figure, the red curves belong to these two faulty trucks. Faulty trucks are expected to have steeper curve along the time axis indicating their higher risk of failure. Such trend is evident in case of survival models trained using PCA and weighted binned approaches. It is evident from all three models that the survival curves are steeper

Table 1. Results for NOx sensor survival models

Survival forest models	Error rates	Training time (relative)	Avg. leaf nodes
Standard approach	0.321	1	288.5
PCA approach	0.265	1.72	276.4
Weighted bin prob	0.263	1.14	271.7

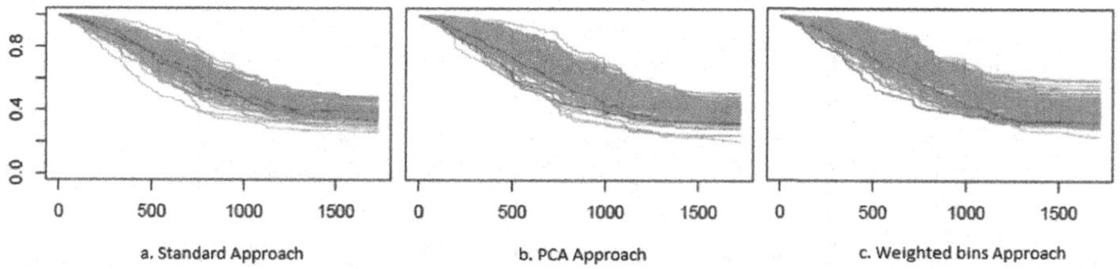

Fig. 5. Model predicted survival curves for 100 randomly selected test examples, two of which had NOx Sensor failure at some point in time.

in the beginning and then eventually gets stable after a while. The variables that were considered important by the trained models are shown in the Fig. 6. Standard approach on the left shows only top 20 important features. Importance score is obtained as a decrease in average accuracy when particular variable is permuted. *Age* seems to be the most significant variable. It is a number of days since a truck is delivered until the day when the information from the truck was extracted. *ambient.temp* records the temperature of the surroundings where the truck operated. So, the temperature of the region where the trucks operates more often might have influence on it's NOx Sensor failure. *production.type* also seem to be an important feature which carries information about batch and location where the truck was manufactured. In the the standard approach list on the left of the figure, the number after a variable name represents the bin of that histogram.

An intuition of variable's importance is also obtained from where in the base tree it is used to split a tree node. If it is used early on, towards the root node, it is considered important. Based on this, minimal depth (distance from root node to a node where it is used for the first time) of each variable is computed across all trees in the forest and are averaged. The minimal depth rank for all feature variables of the trained models are shown the Fig. 7. *torque.map* and *fuel.consumption* are the two largest histogram variables in our dataset. The PCA approach uses multiple bins simultaneously to make splitting decision, because of large histogram size, they have more options to choose from and hence the better chance of finding a good split. Therefore, they might have appeared on top of the list. The problem with weighted bin probability approach however is that, if the counts (value) in the bins are almost evenly distributed, only very few bins at extreme ends are more likely to be selected for evaluation and hence a small chance of finding better split. Perhaps, this approach is not very useful for such large histograms.

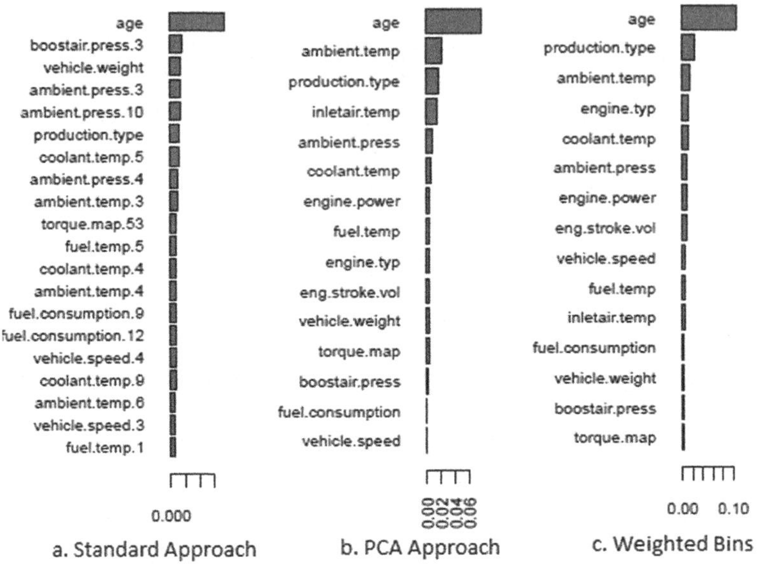

Fig. 6. Variable importance rank

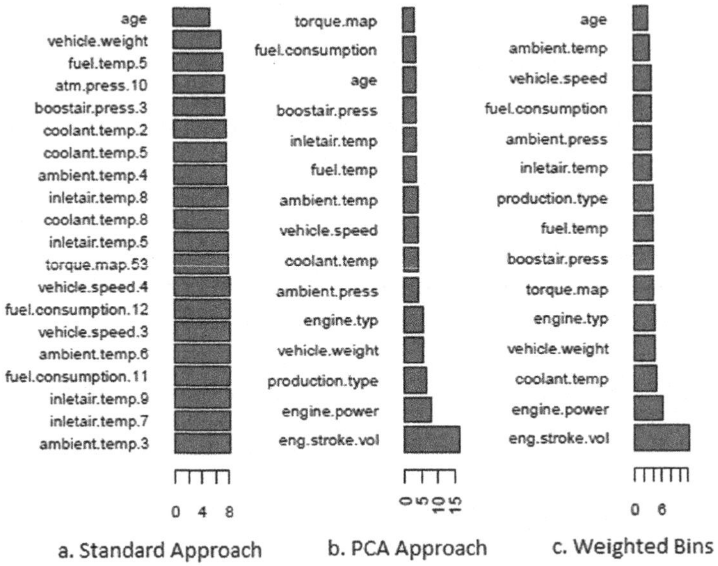

Fig. 7. Minimal depth rank

5 Conclusion and Future Work

In this paper, NOx sensor failure in heavy duty trucks was analyzed. The random survival forest (RSF) algorithm was used to train on operational data from a large fleet of trucks. The operational data had many features represented as histograms. Since the standard random survival forest is not built to handle histogram features, an approach to adapt and train it better on data with histogram feature variables were discussed. Two approaches were examined, first by using multiple bins to make a node split decision and second by assigning more weights to bins at extreme ends to ensure their frequent selection for evaluating a split. The intuition behind first approach is to use information of a local region (multiple neighboring bins simultaneously for split evaluation) in histogram to get better splits while the second approach consider failures (component breakdowns) as an abnormal case and expects to find relevant information about such breakdown in bins that capture extreme operating conditions (which usually are along the extreme ends of histogram where operational feature variables are represented as histograms). The adapted algorithms were used to train a survival model and to predict survival curves for NOx sensor failure in heavy duty trucks. According to the results obtained from the experimental evaluation, the proposed approach outperformed the standard approach in terms of error rate. For NOx sensor failure analysis in the considered experimental data, age of a truck seems to be important. Also, the information about ambient temperature of regions where truck was operated seems to carry useful information related to the NOx sensor failure.

In the current implementation of the histogram based approach, missing values in the features are simply ignored, hence better ways of handling missing values in histogram variables can be explored in future. More elaborate comparison with other standard methods in terms of performance metrics such as brier score and mortality can be performed in future.

Acknowledgment. This work has been funded by the Vinnova program for Strategic Vehicle Research and Innovation (FFI) Transport Efficiency.

References

1. Biteus, J., Lindgren, T.: Planning flexible maintenance for heavy trucks using machine learning models, constraint programming, and route optimization. SAE Int. J. Mater. Manf. **10**, 306–315 (2017)
2. Cox, D.R.: Regression models and life-tables. J. Roy. Stat. Soc.: Ser. B (Methodol.) **34**(2), 187–220 (1972)
3. Daigle, M.J., Goebel, K.: A model-based prognostics approach applied to pneumatic valves. Int. J. Progn. Health Manag. **2**, 84 (2011)
4. Frisk, E., Krysander, M., Larsson, E.: Data-driven lead-acid battery prognostics using random survival forests. In: Annual Conference of the Prognostics and Health Management Society 2014, pp. 92–101 (2014)
5. Gurung, R., Lindgren, T., Bostrom, H.: Learning decision trees from histogram data using multiple subsets of bins. In: Proceedings of the 29th International Florida Artificial Intelligence Research Society Conference (FLAIRS), pp. 430–435 (2016)
6. Gurung, R., Lindgren, T., Boström, H.: Predicting NOx sensor failure in heavy duty trucks using histogram-based random forests. Int. J. Progn. Health Manag. **8**(008) (2017)

7. Gurung, R., Lindgren, T., Boström, H.: Learning random forest from histogram data using split specific axis rotation. Int. J. Mach. Learn. Comput. **8**(1), 74–79 (2018)
8. Hothorn, T., Bühlmann, P., Dudoit, S., Molinaro, A., Van Der Laan, M.J.: Survival ensembles. Biostatistics **7**(3), 355–373 (2006)
9. Ishwaran, H., Kogalur, U.: Random survival forests for R. Rnews **7**(2), 25–31 (2007)
10. Ishwaran, H., Kogalur, U., Blackstone, E., Lauer, M.: Random survival forests. Ann. Appl. Statist. **2**(3), 841–860 (2008)
11. Le-Rademacher, J., Billard, L.: Principal component analysis for histogram-valued data. Adv. Data Anal. Classif. **11**(2), 327–351 (2017)
12. Prytz, R., Nowaczyk, S., Rögnvaldsson, T., Byttner, S.: Predicting the need for vehicle compressor repairs using maintenance records and logged vehicle data. Eng. Appl. Artif. Intell. **41**, 139–150 (2015)
13. Rivoli, L., Verde, R., Irpino, A.: The median of a set of histogram data. In: Alleva, G., Giommi, A. (eds.) Topics in Theoretical and Applied Statistics. STAS, pp. 37–48. Springer, Cham (2016). https://doi.org/10.1007/978-3-319-27274-0_4
14. Voronov, S., Frisk, E., Krysander, M.: Data-driven battery lifetime prediction and confidence estimation for heavy-duty trucks. IEEE Trans. Reliab. **67**(2), 623–639 (2018)
15. Zhou, Y., McArdle, J.J.: Rationale and applications of survival tree and survival ensemble methods. Psychometrika **80**, 811–833 (2015)

Incoherent Submatrix Selection via Approximate Independence Sets in Scalar Product Graphs

Stéphane Chrétien[1,2,3](✉) and Zhen Wai Olivier Ho[2]

[1] National Physical Laboratory, Teddington TW11 0LW, UK
stephane.chretien@npl.co.uk
[2] Laboratoire de Mathématiques de Besançon, 16 Route de Gray, 25030 Besançon, France
[3] FEMTO-ST, 15B avenue des Montboucons, 25030 Besançon, France
https://sites.google.com/view/stephanechretien/home

Abstract. This paper addresses the problem of extracting the largest possible number of columns from a given matrix $X \in \mathbb{R}^{n\times p}$ in such a way that the resulting submatrix has an coherence smaller than a given threshold η. This problem can clearly be expressed as the one of finding a maximum cardinality stable set in the graph whose adjacency matrix is obtained by taking the componentwise absolute value of X^tX and setting entries less than η to 0 and the other entries to 1. We propose a spectral-type relaxation which boils down to optimising a quadratic function on a sphere. We prove a theoretical approximation bound for the solution of the resulting relaxed problem.

1 Introduction

The goal of the present paper is to address the problem of incoherent submatrix extraction. Recall that the coherence of a matrix X with ℓ_2-normalised columns is the maximum absolute value of the scalar product of any two different columns of X. It is usually denoted by $\mu(X)$. Controlling the coherence of a matrix is of paramount importance in statistics [10,13,14,20], signal processing, compressed sensing and image processing [2,4,7,11,12,15,17,19], etc. Incoherence is associated with interesting questions in combinatorics [18]. Efficient recovery of incoherent matrices, an important problem in dictionnary learning, was addressed by the computer science community in [3]. Incoherence is a key assumption behind the current approaches of sparse estimation based on convex optimisation [8].

In many real life problems from statistics, and signal and image processing, the incoherence assumption breaks down [1,2,9], etc. Several approaches helping to get around this problem have been proposed in the literature but, to the best of our knowledge, the problem of extracting a sufficiently incoherent submatrix from a given matrix has not yet been studied in the literature. On the other hand, incoherent submatrix extraction is an important problem and a computationally efficient method for it will definitely allow to select sufficiently different features

G. Nicosia et al. (Eds.): LOD 2019, LNCS 11943, pp. 95–105, 2019.
https://doi.org/10.1007/978-3-030-37599-7_9

from data and make high dimensional sparse representation of the data possible in difficult settings where naive use of ℓ_1 penalised estimation was previously doomed to slow learning rates [5,6].

In the present paper, we propose a new approach to address the incoherent submatrix selection problem seen as a weighted version of the independent set problem in a graph. More precisely, we propose a new spectral-type estimator for the column subset selection problem and study its performance by bounding its scalar product with the indicator vector of the best column selection, in the case it is unique.

The plan of the paper is as follows. In Sect. 2, we reformulate the problem as an independent set computation problem. Section 3 presents a new spectral estimator of the weighted stable set and provides a theoretical error bound.

2 Incoherent Submatrix Extraction as an Approximate Independent Set Computation

The problem of extracting the largest submatrix with coherence less than a given threshold η, by appropriate column selection, can be expressed as an instance of the maximum stable set in graph theory. In order to achieve this, we associate with our matrix X a graph $\mathcal{G} = (V, E)$ as follows:

- $V = \{1, \ldots, p\}$
- E is defined by

$$(j, j') \in E \text{ if and only if } |\langle X_j, X_{j'} \rangle| > \eta. \tag{1}$$

Then, clearly, finding the largest stable set in this graph will immediately provide a submatrix with coherence less than η.

The main difficulty with the independent set approach is that it belongs to the class of NP-hard problems.

3 Relaxing on the Sphere: A New Extraction Approach

In this section, we present a new spectral-type estimator for the independent set.

3.1 The Spectral Estimator

We start from the following relaxed problem

$$x^* \leftarrow \max_{x \in \{0,1\}^p} \quad e^t x \quad \text{s.t.} \quad x^t M x \leq r. \tag{2}$$

The approach of the previous section was addressing the special case corresponding to the value $\lambda = 0$ and M chosen as e.g. the adjacency matrix of $\mathcal{G}$. As in the previous section, we can reformulate this problem using a binary ±1 variable z as follows:

$$\max_{z\in\{-1,1\}^p} \quad e^t\frac{z+e}{2} \quad \text{s.t.} \quad \frac{1}{4}\,(z+e)^t M(z+e) \leq r. \tag{3}$$

The spectral estimator is the vector obtained as the solution of the following problem.

$$\max_{\|z\|_2^2=p} \quad e^t\frac{z+e}{2} \quad \text{s.t.} \quad \text{s.t.} \quad \frac{1}{4}\,(z+e)^t M(z+e) \leq r. \tag{4}$$

It is equivalent to the penalised version

$$\hat{z} \leftarrow \min_{\|z\|_2^2=p} \quad -e^t\frac{z+e}{2} + \lambda\,\left(\frac{1}{4}\,(z+e)^t M(z+e) - r\right) \tag{5}$$

for some Lagrange multiplier $\lambda > 0$[1]. Set

$$\hat{x} = \frac{\hat{z}+e}{2}. \tag{6}$$

3.2 Theoretical Guarantees

In this section, we provide our main theoretical result concerning the performance of our approach.

Theorem 1. *Let ρ^* denote the indicator vector of the maximal independent set defined by*

$$\rho^* \leftarrow \max_{\rho\in\{0,1\}^p} \quad e^t\rho \quad s.t. \quad \rho^t M\rho = 0.$$

Let $\delta > 0$ be such that $M_\delta = M + \delta I$ is positive definite. Let λ_1 be the smallest eigenvalue of M_δ and $\phi_1,\ldots,\phi_p$ be the pairwise orthogonal, unit-norm eigenvectors of M_δ. Set

$$q_1 = \frac{1}{\sqrt{p}} M_\delta e \quad and \quad q_2 = -\frac{1}{\sqrt{p}}\Big(\frac{(1+\delta)}{\lambda}\; e - M_\delta e\Big)$$

and set

$$\gamma_{k,i} = \phi_i^t q_k$$

for $k = 1, 2$ and $i = 1,\ldots,p$. Then,

$$\|x^* - \rho^*\|_\infty \leq \sqrt{p}\left(\frac{(1+\delta)}{\lambda(\lambda_1-\mu_2)} + \frac{\|\gamma_1\|_2\;\nu^*}{(\lambda_1-\mu_1)(\lambda_1-\mu_2)}\right)^2,$$

[1] Here, positivity is trivial.

with ν^* *given by*

$$\nu^* = (\lambda_p - \mu_1)\phi\left(p\ \frac{\gamma_{1,\max}^2}{\gamma_{1,\min}^2}\ \frac{\frac{(1+\delta)^2}{\lambda^2} + \frac{2}{\sqrt{p}}\frac{1+\delta}{\lambda}e^t M_\delta e}{2\ \|\gamma_2\|_2^2}\right)$$

where ϕ *denotes the inverse function of* $x \mapsto x/(1+x)^3$.

Proof. The proof will consist of three steps. The first step re-expresses the problem as the one of minimising the distance to the oracle plus a linear penalisation term. The second step identifies the oracle as the solution to a perturbed problem. The third step then uses a perturbation result proved in the Appendix.

First Step. Let us now note that since $\|z\|_2^2 = p$, we may incorporate any multiple of the term $z^t z - p$ into the objective function and obtain

$$\min_{\|z\|_2^2=p} \quad -e^t\frac{z+e}{2} + \lambda\left(\frac{1}{4}\ (z+e)^t M(z+e) - r\right) + \frac{1}{4}\ \delta(z^t z - p),$$

without changing the solution. Using the fact that

$$\begin{aligned} z^t z - p &= (z+e-e)^t(z+e-e) - p \\ &= (z+e)^t(z+e) - 2e^t(z+e) \end{aligned}$$

we obtain the equivalent problem

$$\min_{\|z\|_2^2=p} \quad -(1+\delta)\ e^t\frac{z+e}{2} + \frac{\lambda}{4}\ (z+e)^t(M+\delta I)(z+e). \tag{7}$$

Set $M_\delta = M + \delta$. We can now expand the term

$$\begin{aligned} \frac{1}{4}\ (z+e)^t M_\delta(z+e) &= \left(\frac{z+e}{2} - \rho^* + \rho^*\right)^t M_\delta \left(\frac{z+e}{2} - \rho^* + \rho^*\right) \\ &= \left(\frac{z+e}{2} - \rho^*\right)^t M_\delta \left(\frac{z+e}{2} - \rho^*\right) \\ &\quad + 2\left(\frac{z+e}{2}\right)^t M_\delta \rho^*, \end{aligned}$$

where we used the fact that ${\rho^*}^t M\rho^* = 0$ in the last equality. Therefore, (7) is equivalent to

$$\min_{\|z\|_2^2=p} \quad \left(2M_\delta\rho^* - \frac{(1+\delta)}{\lambda}\ e\right)^t\left(\frac{z+e}{2}\right) + \left(\frac{z+e}{2} - \rho^*\right)^t M_\delta\left(\frac{z+e}{2} - \rho^*\right). \tag{8}$$

Second Step. When $1/\lambda = 0$, the solution to (9) is readily seen to be equal to the oracle ρ^*.

Third Step. We will now use a perturbation result proved in Lemma 3. For this purpose, we first make a change of variable in order to transform the problem into an optimisation problem on the unit sphere. Let $\tilde{z} = 1/\sqrt{p}z$. Then problem (8) is equivalent to

$$\min_{\|\tilde{z}\|_2^2=1} \left(2M_\delta \rho^* - \frac{(1+\delta)}{\lambda} e\right)^t \left(\frac{\sqrt{p}\tilde{z}+e}{2}\right) + \left(\frac{\sqrt{p}\tilde{z}+e}{2} - \rho^*\right)^t M_\delta \left(\frac{\sqrt{p}\tilde{z}+e}{2} - \rho^*\right).$$

This is equivalent to solving the problem

$$\min_{\|\tilde{z}\|_2=1} \quad \frac{1}{2}\tilde{z}^t Q \tilde{z} - q^t \tilde{z}. \tag{9}$$

with

$$Q = M_\delta \quad \text{and} \quad q = -\frac{1}{\sqrt{p}}\Big(\frac{(1+\delta)}{\lambda} e - M_\delta e\Big).$$

Let $\lambda_1 \leq \ldots \leq \lambda_p$ be the eigenvalues of Q and $\phi_1, \ldots, \phi_p$ be associated pairwise orthogonal, unit-norm eigenvectors. Set

$$q_1 = \frac{1}{\sqrt{p}} M_\delta e \quad \text{and} \quad q_2 = -\frac{1}{\sqrt{p}}\Big(\frac{(1+\delta)}{\lambda} e - M_\delta e\Big)$$

and set

$$\gamma_{k,i} = \phi_i^t q_k$$

for $k = 1, 2$ and $i = 1, \ldots, p$. Thus, we have

$$\|\gamma_1 - \gamma_2\|_2 = \frac{(1+\delta)}{\lambda}.$$

and

$$\|\gamma_1^2 - \gamma_2^2\|_1 \leq \frac{(1+\delta)^2}{\lambda^2} + \frac{2}{\sqrt{p}} \frac{1+\delta}{\lambda} e^t M_\delta e.$$

Combining these bounds with Lemma 3, we obtain the announced result.

4 Conclusion and Future Works

In the present paper, we proposed an alternative approach to the problem of column selection, viewed as quadratic binary maximisation problem. We studied the approximation error of the solution to the easier spherically constrained quadratic problem obtained by relaxing the binary constraints.

Future work will consist in further exploring the quality of the bound obtained in Theorem 1. In particular, we will try to clarify for which types of graphs the error bound can be small, i.e. $|\lambda_1 - \mu_1|$ is bounded from below by an appropriate function of p. Our next plans will also address practical assessment of the efficiency of the method.

A Minimizing Quadratic Functionals on the Sphere

A.1 A Semi-explicit Solution

The following result can be found in [16].

Lemma 1. *For $Q \in \mathbb{S}_p$ and $q \in \mathbb{R}^p$, consider the following quadratic programming problem over the sphere:*

$$\min_{\|x\|_2=1} \quad \frac{1}{2}x^t Q x - q^t x. \tag{10}$$

Let $\lambda_1 \leq \ldots \leq \lambda_p$ be the eigenvalues of Q and $\phi_1,\ldots,\phi_p$ be associated pairwise orthogonal, unit-norm eigenvectors. Let $\gamma_{k,i} = q^t\phi_i$, $i = 1,\ldots,p$. Let $\mathcal{E}_1 = \{i \text{ s.t. } \lambda_i = \lambda_1\}$ and $\mathcal{E}_+ = \{i \text{ s.t. } \lambda_i > \lambda_1\}$. Then, x^* is a solution if and only if

$$x^* = \sum_{i=1}^{p} c_i^* \phi_i$$

and

1. degenerate case: *If $\gamma_i = 0$ for all $i \in \mathcal{E}_1$ and*

$$\sum_{i \in \mathcal{E}_+} \frac{\gamma_i^2}{(\lambda_i - \lambda_1)^2} \leq 1.$$

   then $c_i^* = \gamma_i/(\lambda_i - \lambda_1)$, $i \in \mathcal{E}_1$ and c_i^*, $i \in \mathcal{E}_1$ are arbitrary under the constraint that $\sum_{i \in \mathcal{E}_1} c_i^{*^2} = 1 - \sum_{i \in \mathcal{E}_+} c_i^{*^2}$.
2. nondegenerate case: *If not in the degenerate case, $c_i^* = \gamma_i/(\lambda_i - \mu)$, $i = 1,\ldots,n$ for $\mu > -\lambda_1$ which is a solution of*

$$\sum_{i=1,\ldots,n} \frac{\gamma_i^2}{(\lambda_i - \mu)^2} = 1. \tag{11}$$

Moreover, we have the following useful result.

Corollary 1. *If Q is positive definite, and $\sum_{i=1,\ldots,p} \gamma_i^2/\lambda_i^2 < 1$, then $0 < \mu < \lambda_1$.*

Proof. This follows immediately from the intermediate value theorem.

A.2 Bounds on μ

From (11), we can get the following easy bounds on μ.

Lemma 2. *Let* $\gamma_{\min} = \min_{i=1}^{p} \gamma_i$ *and* $\gamma_{\max} = \max_{i=1}^{p} \gamma_i$. *Then, we have*

$$p\gamma_{\max}^2 \geq \max_{i=1}^{p} \{(\lambda_i - \mu)^2\} \geq p\gamma_{\min}^2. \tag{12}$$

and

$$\gamma_{\min}^2 \leq \min_{i=1}^{p} \{(\lambda_i - \mu)^2\} \leq \|\gamma\|_2^2 \tag{13}$$

Proof. The proof is divided into three parts, corresponding to each (double) inequality.

Proof of (12): We have

$$\begin{aligned} \max_{i=1}^{p} \frac{\gamma_{\max}^2}{(\lambda_i - \mu)^2} &\geq \max_{i=1}^{p} \frac{\gamma_i^2}{(\lambda_i - \mu)^2} \\ &\geq \frac{1}{p} \sum_{i=1}^{p} \frac{\gamma_i^2}{(\lambda_i - \mu)^2} \\ &= \frac{1}{p}. \end{aligned}$$

This immediately gives $p\gamma_{\max} \geq \max_{i=1}^{p} \{(\lambda_i - \mu)^2\}$. On the one hand, we have

$$1 = p \sum_{i=1,\ldots,p} \frac{\gamma_i^2}{(\lambda_i - \mu)^2} \geq \frac{p\gamma_{\min}^2}{\max_{i=1}^{p}\{(\lambda_i - \mu)^2\}}.$$

Therefore, we get $\max_{i=1}^{p}\{(\lambda_i - \mu)^2\} \geq p\, \gamma_{\min}^2$. On the other hand, we have

Proof of (13):

$$\frac{\gamma_i^2}{(\lambda_i - \mu)^2} \leq 1$$

which gives

$$(\lambda_i - \mu)^2 \geq \gamma_i^2$$

for $i = 1, \ldots, p$. Thus, the lower bound follows. For the other bound, since

$$\sum_{i=1}^{p} \frac{\gamma_i^2}{(\lambda_i - \mu)^2} = 1, \tag{14}$$

we get

$$1 \leq \sum_{i=1}^{p} \frac{\gamma_i^2}{(\lambda_i - \mu)^2} \leq \frac{\|\gamma\|_2^2}{\min_{i=1}^{p} (\lambda_i - \mu)^2}$$

and the proof in completed.

A.3 ℓ_∞ Perturbation of the Linear Term

We now consider the problem of controlling the solution under perturbation of q.

Lemma 3. *Consider the two quadratic programming problems over the sphere:*

$$\min_{\|x\|_2=1} \quad \frac{1}{2}x^tQx - q_k^t x, \tag{15}$$

for $k = 1, 2$. Assume that the solution to (15) is non-degenerate in both cases $k = 1, 2$ and let x_1^* and x_2^* be the corresponding solutions. Assume further that $\sum_{i=1,\dots,n} \gamma_{k,i}^2/\lambda_i^2 < 1$, $k = 1, 2$. Let ϕ denote the inverse function of $x \mapsto x/(1+x)^3$. Then, we have

$$\|x_1^* - x_2^*\|_\infty \leq \sqrt{p}\left(\frac{\|\gamma_1 - \gamma_2\|_2}{(\lambda_1 - \mu_2)} + \frac{\|\gamma_1\|_2 \ \nu^*}{(\lambda_1 - \mu_1)(\lambda_1 - \mu_2)}\right)^2,$$

with r^* given by

$$\nu^* = (\lambda_p - \mu_1)\phi\left(p \ \frac{\gamma_{1,\max}^2}{\gamma_{1,\min}^2} \frac{\|\gamma_1^2 - \gamma_2^2\|_1}{2 \ \|\gamma_2\|_2^2}\right)$$

Proof. Let Φ denote the matrix whose columns are the eigenvectors of A. More precisely, $\lambda_1 \leq \cdots \leq \lambda_p$ and let ϕ_i be an eigenvector associated with λ_i, $i = 1, \dots, p$. Let $\gamma_i = q^t\phi_i$, $i = 1, \dots, p$. Let c_1^* (resp. c_2^*) be the vector of coefficients of x_1^* (resp. x_2^*) in the eigenbasis of A. For each $k = 1, 2$, there exists a real μ_k such that

$$c_{k,i}^* = \frac{\gamma_{k,i}}{(\lambda_i - \mu_k)},$$

$i = 1, \dots, p$ for $\mu_k > -\lambda_1$ which is a solution of

$$\sum_{i=1}^p \frac{\gamma_{k,i}^2}{(\lambda_i - \mu)^2} = 1.$$

Now, apply Neuberger's Theorem 2 to obtain an estimation of $|\mu_1 - \mu_2|$ as a function of γ_1 and γ_2. For this purpose, set

$$F(\mu) = \sum_{i=1}^p \frac{\gamma_{2,i}^2}{(\lambda_i - \mu)^2} - 1, \ i.e. \quad F'(\mu) = 2\sum_{i=1}^p \frac{\gamma_{2,i}^2}{(\lambda_i - \mu)^3}.$$

Now, we need to find the smallest value of ν such that, for all $\mu \in B(\mu_1, \nu)$, we need to find a number $h \in \bar{B}(0, \nu)$ such that

$$h = F'(\mu)^{-1} \ F(\mu_1)$$

We therefore have that

$$h = \frac{\sum_{i=1}^{p} \frac{\gamma_{2,i}^2}{(\lambda_i-\mu_1)^2} - 1}{2\sum_{i=1}^{p} \frac{\gamma_{2,i}^2}{(\lambda_i-\mu)^3}} = \frac{\sum_{i=1}^{p} \frac{\gamma_{1,i}^2}{(\lambda_i-\mu_1)^2} - 1 + \sum_{i=1}^{p} \frac{\gamma_{2,i}^2-\gamma_{1,i}^2}{(\lambda_i-\mu_1)^2}}{2\sum_{i=1}^{p} \frac{\gamma_{2,i}^2}{(\lambda_i-\mu)^3}}$$

and since

$$\sum_{i=1}^{p} \frac{\gamma_{1,i}^2}{(\lambda_i - \mu_1)^2} = 1,$$

we have

$$h \leq \frac{(\min_{i=1}^{p} \{(\lambda_i - \mu_1)^2\})^{-1} \, \|\gamma_1^2 - \gamma_2^2\|_1}{2\, \|\gamma_2\|_2^2 \, (\max\{(\lambda_i - \mu)^3\})^{-1}}$$

where $\cdot^2$ is to be understood componentwise. Moreover, since $\sum_{i=1,\ldots,p} \gamma_{k,i}^2/\lambda_i^2 < 1$, $k = 1, 2$,

$$\max\{(\lambda_i - \mu)^3\} = (\lambda_p - \mu_1 + r)^3 \text{ and } \min_{i=1}^{p}\{(\lambda_i - \mu_1)^2\} = (\lambda_1 - \mu_1)^2.$$

Thus, for $\nu > 0$ such that

$$\nu \geq \frac{\|\gamma_1^2 - \gamma_2^2\|_1 \, (\lambda_p - \mu_1 + \nu)^3}{2\, \|\gamma_2\|_2^2 \, (\lambda_1 - \mu_1)^2},$$

we get from Theorem 2 that there exists a solution to the equation $F(u) = 0$ inside the ball $\bar{B}(\mu_1, \nu)$. Make the change of variable

$$\nu = \alpha(\lambda_p - \mu_1)$$

and obtain that we need to find $\alpha \in (0, 1)$ such that

$$\frac{\alpha}{(1+\alpha)^3} \geq \frac{\|\gamma_1^2 - \gamma_2^2\|_1 \, (\lambda_n - \mu_1)^2}{2\, \|\gamma_2\|_2^2 \, (\lambda_1 - \mu_1)^2}.$$

Lemma 2 now gives

$$\frac{(\lambda_n - \mu_1)^2}{(\lambda_1 - \mu_1)^2} \leq p \, \frac{\gamma_{1,\max}^2}{\gamma_{1,\min}^2}$$

from which we get that the value ν^* of ν given by

$$\nu^* = (\lambda_p - \mu_1)\phi\left(p \, \frac{\gamma_{1,\max}^3}{\gamma_{1,\min}^2} \frac{\|\gamma_1^2 - \gamma_2^2\|_1}{2\, \|\gamma_2\|_2^2}\right)$$

is admissible, for $\|\gamma_1^2 - \gamma_2^2\|_1$ such that the term involving ϕ is less than one.

$$\frac{\gamma_{1,i}}{(\lambda_i - \mu_1)} - \frac{\gamma_{2,i}}{(\lambda_i - \mu_2)} = \frac{\gamma_{1,i}(\lambda_i - \mu_1 + \mu_1 - \mu_2) - \gamma_{2,i}(\lambda_i - \mu_1)}{(\lambda_i - \mu_1)(\lambda_i - \mu_2)}$$
$$= \frac{(\gamma_{1,i} - \gamma_{2,i})}{\lambda_i - \mu_2} + \frac{\gamma_{1,i}(\mu_1 - \mu_2)}{(\lambda_i - \mu_1)(\lambda_i - \mu_2)}.$$

Therefore,

$$\|c_1^* - c_2^*\|_2^2 \leq \left(\frac{\|\gamma_1 - \gamma_2\|_2}{(\lambda_1 - \mu_2)} + \frac{\|\gamma_1\|_2 \; |\mu_1 - \mu_2|}{(\lambda_1 - \mu_1)(\lambda_1 - \mu_2)} \right)^2 .$$

Finally, using that $|\mu_1 - \mu_2| \leq \nu^*$, we get

$$\|c_1^* - c_2^*\|_2 \leq \left(\frac{\|\gamma_1 - \gamma_2\|_2}{(\lambda_1 - \mu_2)} + \frac{\|\gamma_1\|_2 \; \nu^*}{(\lambda_1 - \mu_1)(\lambda_1 - \mu_2)} \right)^2 ,$$

which gives

$$\|x_1^* - x_2^*\|_\infty \leq \sqrt{p} \left(\frac{\|\gamma_1 - \gamma_2\|_2}{(\lambda_1 - \mu_2)} + \frac{\|\gamma_1\|_2 \; \nu^*}{(\lambda_1 - \mu_1)(\lambda_1 - \mu_2)} \right)^2 ,$$

as announced.

A.4 Neuberger's Theorem

In this subsection, we recall Neuberger's theorem.

Theorem 2. *Suppose that $r > 0$, that $x \in R^p$, and that F is a continuous function from $\bar{B}(x, r)$ to R^m with the property that for each y in $B(x, r)$, there is an h in $\bar{B}(0, r)$ such that*

$$\lim_{t \to 0+} \frac{(F(y + th) - F(y))}{t} = -F(x). \tag{16}$$

Then, there exists u in $\bar{B}(x, r)$ such that $F(u) = 0$.

References

1. Adcock, B., Hansen, A.C., Poon, C., Roman, B.: Breaking the coherence barrier: a new theory for compressed sensing. Forum Math. Sigma **5**, 84 (2017)
2. Adcock, B., Hansen, A.C., Poon, C., Roman, B., et al.: Breaking the coherence barrier: asymptotic incoherence and asymptotic sparsity in compressed sensing. Preprint (2013)
3. Arora, S., Ge, R., Moitra, A.:. New algorithms for learning incoherent and overcomplete dictionaries. In: Conference on Learning Theory, pp. 779–806 (2014)
4. Baraniuk, R.G.: Compressive sensing [lecture notes]. IEEE Signal Process. Mag. **24**(4), 118–121 (2007)
5. Bellec, P.C.: Localized Gaussian width of m-convex hulls with applications to lasso and convex aggregation. arXiv preprint arXiv:1705.10696 (2017)
6. Bühlmann, P., Van De Geer, S.: Statistics for High-dimensional Data: Methods, Theory and Applications. Springer, Heidelberg (2011). https://doi.org/10.1007/978-3-642-20192-9
7. Candes, E., Romberg, J.: Sparsity and incoherence in compressive sampling. Inverse Prob. **23**(3), 969 (2007)

8. Candès, E.J.: Mathematics of sparsity (and a few other things). In: Proceedings of the International Congress of Mathematicians, Seoul, South Korea, vol. 123. Citeseer (2014)
9. Candes, E.J., Eldar, Y.C., Needell, D., Randall, P.: Compressed sensing with coherent and redundant dictionaries. Appl. Comput. Harmonic Anal. **31**(1), 59–73 (2011)
10. Candès, E.J., Plan, Y.: Near-ideal model selection by l1 minimization. Ann. Stat. **37**(5A), 2145–2177 (2009)
11. Candès, E.J., Wakin, M.B.: An introduction to compressive sampling. IEEE Signal Process. Mag. **25**(2), 21–30 (2008)
12. Cevher, V., Boufounos, P., Baraniuk, R.G., Gilbert, A.C., Strauss, M.J.: Near-optimal Bayesian localization via incoherence and sparsity. In: International Conference on Information Processing in Sensor Networks, IPSN 2009, pp. 205–216. IEEE (2009)
13. Chrétien, S., Darses, S.: Invertibility of random submatrices via tail-decoupling and a matrix Chernoff inequality. Stat. Probab. Lett. **82**(7), 1479–1487 (2012)
14. Chrétien, S., Darses, S.: Sparse recovery with unknown variance: a Lasso-type approach. IEEE Trans. Inf. Theory **60**(7), 3970–3988 (2014)
15. Foucart, S., Rauhut, H.: A Mathematical Introduction to Compressive Sensing, vol. 1. Birkhäuser, Basel (2013)
16. Hager, W.W.: Minimizing a quadratic over a sphere. SIAM J. Optim. **12**(1), 188–208 (2001)
17. Mallat, S.: A Wavelet Tour of Signal Processing: The Sparse Way. Academic Press, Cambridge (2008)
18. Nelson, J.L., Temlyakov, V.N.: On the size of incoherent systems. J. Approximation Theory **163**(9), 1238–1245 (2011)
19. Romberg, J.: Imaging via compressive sampling. IEEE Signal Process. Mag. **25**(2), 14–20 (2008)
20. Van De Geer, S.A., Bühlmann, P., et al.: On the conditions used to prove oracle results for the Lasso. Electron. J. Stat. **3**, 1360–1392 (2009)

LIA: A Label-Independent Algorithm for Feature Selection for Supervised Learning

Gail Gilboa-Freedman(✉), Alon Patelsky, and Tal Sheldon

The Interdisciplinary Center Herzliya, Herzliya, Israel
gail.gilboa@idc.ac.il

Abstract. The current study considers an unconventional framework of unsupervised feature selection for supervised learning. We provide a new unsupervised algorithm which we call *LIA*, for Label-Independent Algorithm, which combines information theory and network science techniques. In addition, we present the results of an empirical study comparing LIA with a standard supervised algorithm (MRMR). The two algorithms are similar in that both minimize redundancy, but MRMR uses the labels of the instances in the input data set to maximize the relevance of the selected features. This is an advantage when such labels are available, but a shortcoming when they are not. We used cross-validation to evaluate the effectiveness of selected features for generating different well-known classifiers for a variety of publicly available data sets. The results show that the performance of classifiers using the features selected by our proposed label-independent algorithm is very close to or in some cases better than their performance when using the features selected by a common label-dependent algorithm (MRMR). Thus, LIA has potential to be useful for a wide variety of applications where dimension reduction is needed and the instances in the data set are unlabeled.

Keywords: Machine learning · Feature selection · Symmetric uncertainty · Supervised learning · Classifiers · Algorithm · Community detection · Louvain algorithm · Performance evaluation

1 Introduction

Feature selection is the task of identifying the most informative features in any body of data. Traditionally, feature selection algorithms for supervised learning are designed to work with known classification problems, selecting the subset of features that promises to provide the most accurate classification for a given problem. Yet rapid advances in data storage and sharing technologies now enable the collection and warehousing of vast quantities of high-dimensional data, often before it is known precisely how those data will be used. Very often, feature selection must take place already during the pre-processing phase [7] to enable effective ETL (Extract, Transform, and Load) processes. However, the data at

G. Nicosia et al. (Eds.): LOD 2019, LNCS 11943, pp. 106–117, 2019.
https://doi.org/10.1007/978-3-030-37599-7_10

this phase are often unlabeled, and therefore supervised feature selection algorithms cannot be applied. The motivation for the present research derives from this challenge.

Our novel approach to feature selection assumes that the selection is unsupervised, but the application is in supervised learning. This framework is in accordance with many real-life situations. For instance, patient data collected by hospitals (physical measurements, current diagnosis, clinical history, etc.) are routinely supplied to researchers for various applications, including generating predictive models aimed at (for example) optimizing medical procedures or personalizing treatments. Maintaining such databases may be costly, time-consuming, and reliant on compliance by medical staff. Given the inefficiencies of storing redundant data, and of re-selecting information for each potential application, hospitals would like to minimize the number of features in the data they store while keeping it informative. We hypothesize that in many such cases, including our medical data example, analysis of internal relations between predictive features (e.g., patients' "blood-pressure" and "weight" in our example) may suffice for selecting a feature subset that would be efficient when integrated into a variety of classifiers. This raises the interesting and seldom-studied question of whether unsupervised methods for feature selection can achieve performance comparable to supervised feature selection methods.

It is natural to assume that an algorithm which specifies a different feature subset for each classification problem would perform better than an algorithm which takes a unsupervised approach and selects the same feature subset for all problems. The present paper explores the extent of this advantage. We suggest an algorithm that analyzes the mutual relations between predictive features and exploits these relations to identify redundancies in the data set. We call our algorithm *LIA*, for Label-Independent Algorithm. LIA has three steps: partition, initialization, and greedy selection. In the first step, partition, the algorithm represents the data set as a large network, where each node or vertex stands for a feature. Pairs of nodes are connected by edges with weights that represent the mutual dependency between the corresponding features. Then, LIA detects communities in this feature network, using the familiar Louvain's algorithm [3]. In the second step, initialization, LIA initializes the set of selected features with one feature from each community. In the third step, greedy selection, LIA adds new features in each iteration so as to minimize the mutual information between new features and the features selected thus far.

LIA is an unsupervised method and therefore is used prior to knowing which classifier uses the selected features. The effect of this scenario is double-sided. On the one hand, it improves many aspects of data maintenance, and avoids the need for selecting a different data set for each classification purpose. On the other hand, it reduces users' ability to identify the subset of features that is most relevant to any specific classification problem. Which effect is more dominant? The answer, of course, depends on the costs of each (e.g., the cost of storing the data vs. the cost of false classification). However, to begin with, it is interesting to examine the influence of using an unlabeled data set regardless of these costs.

For this purpose, we conducted an empirical study, comparing LIA with the standard label-dependent feature selection algorithm, MRMR.

Our main finding is that LIA not only reduces the costs of handling high-dimensional data (our primary motivation for presenting this algorithm), but remains competitive in terms of the accuracy of classifiers that use its output features. This result is worth noting, because label-dependent algorithms are based on knowledge about the relevance of different features to different classes, and therefore should be expected to perform significantly better than LIA. The results point to the huge potential of using LIA, or other unsupervised algorithms, in cases where it is costly to maintain large data sets that are not yet specified to a certain classification problem.

2 Literature Review

The feature selection problem has attracted substantial attention in the computer science literature (see [6] and [21] for reviews). The problem is often formalized as an initial step in building machine-learning models, in both supervised and unsupervised learning (see reviews in [39] and [15], respectively). Solutions proposed for feature selection have proven effective for many real-world applications [18] where stored data is potentially highly redundant, for example in the realms of network intrusion detection [37], face recognition [43] and gene expression [9].

Feature selection algorithms for supervised learning aim to choose the subset of features that is optimal for building an effective classifier. By classifiers, we refer to classification models such as those widely used in machine learning [19], whose purpose is to predict a class value from particular predictive features. Thus, theses algorithms aim to minimize redundancy while maximizing the relevance of the selected features [18,22], all in a time-efficient manner. These algorithms usually fall into one of four categories: filter, wrapper, embedded, and hybrid (see [41] for a review). The proposed LIA falls into the filter category as it does not employ a classifier [5].

Notably, our proposed algorithm differs from "class-independent feature selection" algorithms, so named because they select the same feature subsets for different classes. While they select the same subset for each label-value, these algorithms are not label-independent like ours. For example, ReliefF [33] selects features in order to separate instances from different classes.

The proposed LIA algorithm follows the information-based approach prevalent in the feature selection literature. Many real-world data sets include pairwise dependencies between features [13], and it is common to quantify these mutual dependencies using information measures (see [38] for a review). Prominent feature selection algorithms that employ the mutual information measure include MRMR [27], TMI-FX [26], and MIFS-ND [16]. LIA uses *symmetric uncertainty* (see [36]), which is a normalization of the mutual information measure to the *entropy* (see [32] for definition) of the features. We selected this measure because it is well-established in the feature selection literature and has proven to be effective [13,36].

LIA employs network analysis and represents the features in the input data set as nodes in a network. A similar approach has been shown to be effective for feature selection in previous work [8,36]. However, in contrast with previous applications of clustering analysis, LIA quantify feature–feature relations in a manner that is independent of their relevance to any given class. LIA partitions this network, using Louvain's algorithm [3], which is based on modularity optimization. The maximization of modularity is an NP-complete problem [4], but Louvain's algorithm achieves good partitions reasonably quickly by taking a heuristic greedy approach. Some eminent examples of other community detection algorithms include Newman-Girvan [14], InfoMap [34], stochastic block models [1], and label propagation algorithms [23]. The curious reader is referred to the comparative analyses of community detection algorithms in [25] and [20].

LIA follows a greedy approach that is similar to the greedy approach used in other feature selection algorithms like MRMR [27] or MIFS-ND [16], but is also different for having a criterion that uses only feature–feature computations.

Finally, the concept of unsupervised feature selection has been widely addressed in the literature, but mostly in the context of unsupervised learning. In this context, features are selected during the pre-processing phase as part of the construction of *clustering models*, which are widely used in machine learning (see [12] and [17] for reviews). In the bulk of the literature on unsupervised feature selection, it is common to evaluate algorithms by comparing their performance when employed in different unsupervised models. Moreover, unsupervised algorithms often consider these performances as part of their process (for examples see [11]). It is not a common approach to compare an unsupervised algorithm with a supervised one [42]. The current study takes this uncommon approach.

3 Formalizing the Problem

In many real-world situations there are advantages to reducing the dimensions of collected data at an early stage, prior to labeling the data, even to the point of compromising on the accuracy of the classifiers that might use these data in the future (see the hospital example described in the introduction). Motivated by these situations, we formulate the problem as follows:

The Unsupervised Feature Selection Problem for Supervised Learning: Given a data set D of dimension n with a feature set $F = \{f_1, f_2, \ldots, f_n\}$ and a fixed integer K, what is the optimal subset of features $S = \{s_1, s_2, \ldots, s_K\}$ that are informative but non-redundant?

4 The Label-Independent Algorithm (LIA)

The proposed LIA uses a combination of information theory and network science techniques. Inspired by the principle of identifying features that demonstrate minimal redundancy, LIA computes the feature–feature *symmetric uncertainty* measure to quantify feature–feature similarity. The symmetric uncertainty of two random variables measures the degree to which knowing the value of either random variable reduces uncertainty regarding the value of the other [29]. It is derived by normalizing the *information gain* to the *entropies* of the random variables (see [32] for definitions of these measures). Normalization induces values in the range [0,1] and assures symmetry. A value of zero means that the two variables are independent. Thus, higher (vs. lower) values mean that knowing the value of either variable is more (vs. less) useful for predicting the value of the other. LIA has three steps, described next.

4.1 Step 1: Network-Based Partition

Given a data set D of dimension n with a feature set $F = \{f_1, f_2, ..., f_n\}$, we represent it by a network (complete graph) $G = (V, E)$, in which the vertices (nodes) $V = \{f_1, f_2, ..., f_n\}$ represent the features and the edges are $E = \{(f_i, f_j)|i, j \in [1, n], i \neq j\}$. In an adjacency matrix A, all the edges are weighted to reflect how informative its vertices (i.e., pairs of features) are in relation to one another, such that $A_{i,j}$ is the symmetric uncertainty between the ith and j^{th} feature. For formal definition of the symmetric uncertainty of two features X, Y see [36]. Then, we partition the network G into m parts or communities $P = \{P_1, P_2, ..., P_m\}$, applying the Louvain algorithm [3]. Each community includes a feature subset such that features within any given community are more informative toward one another than toward features in any other community.

4.2 Step 2: Initialization

In the second step, we use the feature–feature symmetric uncertainty values and the partition P (all computed in step 1) to select a subset of K features. We first calculate the *internal average symmetric uncertainty (IASU)* for each feature f_i. For each feature, we compute this as the average of the symmetric uncertainty values between that feature and the other features in its community. We denote the set of selected features by S, and initialize it by inserting the feature from each community with the highest $IASU$ value.

4.3 Step 3: Greedy Selection

In the third step, for each non-selected feature, we calculate the *average symmetric uncertainty (ASU)* as the average of the symmetric uncertainty values between that feature and the features in S. We then add $K - 1$ features to S, one by one, in a greedy manner. In each iteration, we add the feature with the minimal ASU to S, and update the ASU values of all non-selected features accordingly. The process ends when S includes K features.

4.4 Time Complexity Analysis

The time complexity of LIA depends on the number of features in the input data set n. The most time-consuming stage is the first step of the algorithm. This

Algorithm 1. LIA
Input: Data set D with dimension n and features $F = \{f_1, f_2, \ldots, f_n\}$, integer K
Output: Set of K selected features S

```
// Step 1: Network-Based Partition
Initialize G=NULL
for {f_i, f_j} in F do
    Add f_i and/or f_j to the set of the vertices V(G)
    Add (f_i, f_j) to the set of the edges E(G)
    A_{i,j} = SU(f_i, f_j) as the weight of the edge
end for
P = Louvain (G)
// Step 2: Initialization
for P_i in P do
    for f_j in P_i do
        IASU_j = Average of SU(f_j, f_l) for f_l ∈ P_i
    end for
    S = ∪s_i, where s_i = feature with highest IASU in P_i
end for
// Step 3: Greedy selection
if |S| < K then
    F' = F\S
    for f_j in F' do
        ASU_j = Average of SU(f_j, f_l) for f_l ∈ S
    end for
    for k = |S| + 1 to K do
        s_k = feature with lowest ASU
        Transform s_k from F' to S
        if k < K then
            Update ASU
        end if
    end for
end if
return S
```

includes the initial $O(n^2)$ computations of all the pair-wise symmetric uncertainty values; the construction of a feature network with time complexity $O(n^2)$ (as it is linear with the number of elements in the network); and Louvain's community detection, which is also $O(n^2)$ as it is linear with the number of network edges [40]. In the second step, LIA performs a greedy process with $O(n)$ computations in each iteration (for updating the average symmetric uncertainty of each non-selected feature with the features selected thus far). The number of iterations is lower than K, which is lower than n. In total the time complexity of the greedy process is $O(Kn)$, which is bounded by $O(n^2)$. The overall time complexity of LIA is $O(n^2)$.

5 Symmetric Uncertainty

Our algorithm uses the symmetric uncertainty measure. Formally the symmetric uncertainty of two features X, Y is computed as $SU(X,Y) = \frac{2 \times IG(X|Y)}{H(X)+H(Y)}$, where $IG(X|Y)$ is the information gain, or additional information about Y provided by X and is computed by $IG(X|Y) = H(X) - H(X|Y)$; $H(X)$ is the entropy of X, which quantifies its uncertainty, computed by $H(X) = -\sum_{x \in X} p(x) log_2 p(x)$; and $H(X|Y)$ is the *conditional entropy*, which quantifies the remaining uncertainty of a random variable X given that the value of Y is known, computed by $H(X|Y) = -\sum_{y \in Y} p(y) \sum_{x \in X} p(x|y) log_2$. For detailed description see [32,36].

6 Empirical Study

We evaluate the performance of LIA by comparing its results to those achieved using MRMR [27], a standard label-dependent feature selection algorithm that is similar to LIA in being both information-based and greedy.

Simulations were carried out on a workstation with a 2.3 GHz Intel Xeon E5 2686 v4 processor (Ubuntu 18.04.1 bionic). The algorithm was implemented using Python 3.6.6 software, with packages from the data analysis library *Panda* [31], with the GitHub information theory toolkit [10], and with the network analysis packages *Networkx* [24] and *Community* [30]. The implementation of MRMR was taken from a publicly available resource (see [28]). We use four common classifiers, namely the decision tree, K-nearest neighbor (KNN), random forest, and support vector machine (SVM), all implemented in scikit-learn Python packages (see [35]). We use 10-fold cross-validation and compare the accuracy and F-score measures for all classifiers and for data sets with categorical features taken from the UCI Machine Learning Repository [2]. These vary in size (number of instances), dimension (number of features), and application domain (see Table 1).

Table 1. Data sets examined.

Data Set	Domain	Instances	Features
WINE	INTRUSION	178	13
CONGRESS VOTING	LIFE SCIENCE	435	16
LYMPH	SOCIAL SCIENCE	148	18
MUSHROOM	LIFE SCIENCE	8124	21
SPECT	LIFE SCIENCE	267	22
PHISHING WEBSITES	SECURITY	11055	30
KR-VS-KP	GAMES	3916	36
CONNECT-4	GAMES	67557	42
COIL2000	SOCIAL SCIENCE	9821	85
LUNG	BIOLOGY	73	325
COLON	BIOLOGY	62	2000

7 Results

We present the results of our empirical study comparing LIA with MRMR for a variety of classification problems and classifiers. The study's main result is that in most cases LIA's accuracy is similar to that of MRMR. This is a non-intuitive result, considering that LIA ignores the labels in the input data set, while MRMR strongly relies on analyzing the relationships between the features and labels.

7.1 Performance Evaluation for Different Iterations of LIA

To verify that LIA is competitive with a standard label-dependent algorithm for a variety of data sets and classifiers, we evaluated the efficiency of LIA over the course of its iterations for each of the data sets in our study. By way of example, the graphs in Figs. 1 and 2 illustrate the results for the SPECT data set.

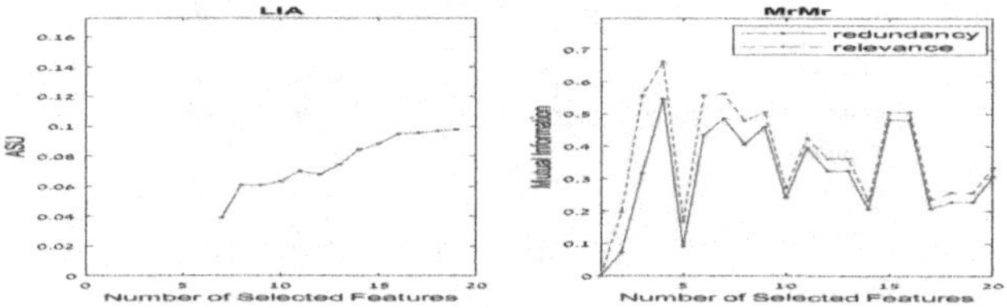

Fig. 1. Criteria of the selected features (SPECT data set). The left-hand graph shows the average symmetric uncertainty (*ASU*) over the iterations of LIA for the SPECT data set. The right-hand graph shows the redundancy and relevance criteria of the selected features over the iterations of MRMR. It is worth reiterating here that LIA is label-independent and therefore does not take relevance into account.

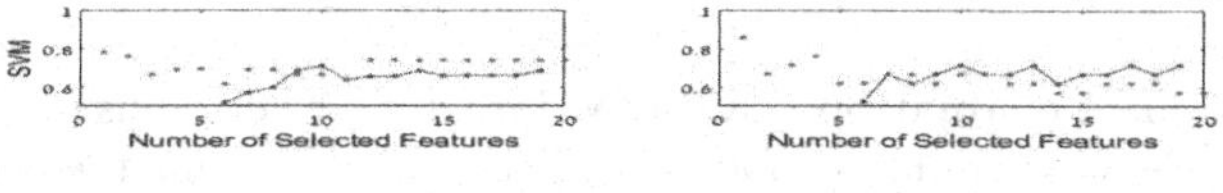

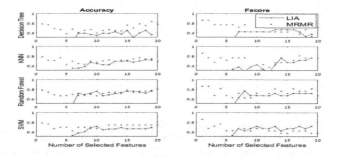

Fig. 2. Performance evaluation (SPECT data set). The graphs present the accuracy (left) and F-scores (right) of four classifiers which contain the set of features selected by LIA (solid line) vs. MRMR (dashed line) over the algorithms' iterations.

7.2 Performance Evaluation for the First Iteration of LIA

The table in Fig. 3 shows the performance measures (accuracy and F-score) under a predefined stopping rule such that LIA selects one feature from each community (without the following greedy part of the algorithm, and without having K as an input). Performance measures are shown for LIA in comparison with MRMR when it selects the same number of features, and when all features are selected. As can be seen in the table, for most of the data sets in our study, LIA's performance is similar to that of MRMR even when a stopping condition is predefined. Figure 4 emphasizes that the efficiency of LIA improves when the third step is executed.

		Performance Measures			
Data Set	Communities	Decision Tree	KNN	Random Forest	SVM
WINE	4	0.81(0.88) 0.86(0.70) 0.92(0.92)	0.64(0.56) 0.65(0.65) 0.67(0.66)	0.88(0.83) 0.89(0.71) 0.98(0.98)	0.68(0.61) 0.46(0.38) 0.44(0.44)
CONGRESS VOTING	5	0.94(0.94) 0.95(0.94) 0.94(0.93)	0.94(0.91) 0.96(0.93) 0.91(0.91)	0.95(0.94) 0.95(0.94) 0.98(0.98)	0.96(0.96) 0.96(0.96) 0.94(0.94)
LYMPH	5	0.77(0.62) 0.77(0.74) 0.77(0.87)	0.78(0.60) 0.79(0.72) 0.76(0.79)	0.81(0.66) 0.85(0.76) 0.85(0.85)	0.77(0.66) 0.79(0.74) 0.77(0.83)
MUSHROOM	6	0.99(0.99) 0.99(0.99) 0.99(0.99)	0.99(0.99) 0.99(0.99) 0.99(0.99)	0.99(0.99) 0.99(0.99) 0.99(0.99)	0.99(0.99) 0.99(0.99) 0.99(0.99)
SPECT	5	0.79(0.77) 0.80(0.74) 0.77(0.70)	0.80(0.78) 0.79(0.74) 0.79(0.77)	0.80(0.80) 0.81(0.76) 0.79(0.76)	0.78(0.79) 0.78(0.79) 0.89(0.79)
PHISHING WEBSITES	5	0.88(0.88) 0.61(0.61) 0.95(0.94)	0.88(0.87) 0.49(0.59) 0.93(0.91)	0.88(0.88) 0.61(0.61) 0.96(0.95)	0.88(0.88) 0.61(0.60) 0.94(0.93)
KR-VS-KP	7	0.69(0.69) 0.93(0.93) 0.99(0.99)	0.65(0.67) 0.93(0.93) 0.99(0.99)	0.69(0.69) 0.95(0.94) 0.99(0.99)	0.69(0.69) 0.94(0.94) 0.99(0.99)
CONNECT4	8	0.65(0.64) 0.67(0.66) 0.69(0.69)	0.57(0.49) 0.59(0.61) 0.71(0.68)	0.65(0.64) 0.67(0.66) 0.78(0.76)	0.65(0.65) 0.66(0.66) 0.67(0.67)
COIL2000	10	0.93(0.92) 0.93(0.92) 0.89(0.88)	0.93(0.92) 0.93(0.93) 0.93(0.92)	0.98(0.98) 0.93(0.93) 0.92(0.92)	0.93(0.98) 0.93(0.93) 0.93(0.93)
LUNG	9	0.52(0.52) 0.54(0.28) 0.59(0.52)	0.56(0.44) 0.60(0.36) 0.69(0.44)	0.60(0.44) 0.58(0.44) 0.65(0.56)	0.58(0.48) 0.50(0.48) 0.54(0.48)
COLON	6	0.39(0.43) 0.69(0.71) 0.75(0.85)	0.55(0.52) 0.68(0.76) 0.66(0.71)	0.50(0.52) 0.72(0.71) 0.77(0.8)	0.52(0.52) 0.72(0.62) 0.63(0.61)

Fig. 3. Performance of LIA when it stops after its second step. Each cell in the table represents a specific combination of data set and classifier. Efficiency using features selected by LIA (first line in each cell) is compared with efficiency using the same number of features selected by MRMR (second line), and with efficiency using the complete set of all features (third line) as a benchmark. Efficiency is measured by accuracy (and Fscore).

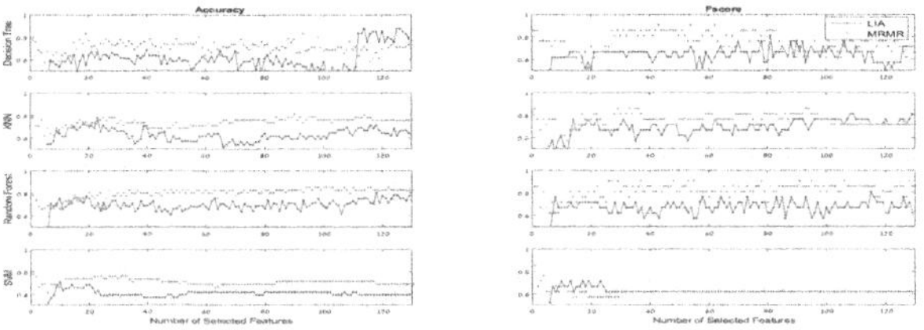

Fig. 4. Performance evaluation (COLON data set). The graphs present the accuracy (left) and F-scores (right) of four classifiers which contain the set of features selected by LIA (solid line) vs. MRMR (dashed line) over the algorithms' iterations.

8 Conclusions and Future Work

The current study sheds light on situations where the available data is still unlabeled, meaning that the benefits of dimension reduction for data maintenance are achievable only with a feature selection algorithm that is unsupervised. The performance evaluation of our proposed algorithm shows that our unsupervised approach is attractive in today's Big Data era, when data collectors must sift through massive quantities of data features and decide which to keep and share – often before the (possibly many) applications of the data are known.

As the main result of this study, we show that our label-independent algorithm (LIA) is competitive with a standard label-dependent algorithm (MRMR) in terms of the accuracy of classifiers that use their outputs. This is despite the fact that LIA discounts the relevance of the features chosen to the (eventual) classification problem. As a second result, we show that the first iteration of LIA is also competitive with MRMR. This result leads to the practical recommendation that for some data sets, LIA may be used for dimension reduction of non-labeled data to a size that does not need to be chosen by the data collector.

We recognize the importance of time efficiency in feature selection processes, and we are aware of the advantage of algorithms that do not need to perform the time-consuming analysis of the mutual relations between pairs of features required by LIA. However, in the wider context of Big Data, where dimension reduction is impossible with label-dependent algorithms, LIA induces efficiency.

We also recognize that the unsupervised approach taken in this article will not be appropriate for all settings. Indeed, in some cases the performance of LIA may be poor in comparison with label-dependent algorithms, either demonstrating low accuracy or requiring more features to reach an optimal value. However, based on the variety of data sets, properties, and classifiers chosen for our empirical study, we believe that any such poor performance will likely arise from specific characteristics of the input data set that do not exist in the examples chosen for this study. More precisely, poor performance of LIA may be expected in cases where the input data sets are characterized by two properties: first, where the features are relatively independent of one another, such that there is no clear underlying pattern of relations between them; and second, where the variance of mutual relations is low. Together, these properties generate ambiguity in patterns of feature community. When selecting features in such cases, redundancy can be a less dominant factor than relevance (for predicting the value of the label).

In another study, we are also applying the techniques developed in the current paper to the feature selection problem for **un**supervised learning. Specifically, we apply the information-based partition of the features to the problem of feature selection for clustering models.

References

1. Abbe, E., Sandon, C.: Community detection in general stochastic block models: fundamental limits and efficient recovery algorithms. In: Proceedings of the IEEE 56th Annual Symposium on Foundations of Computer Science, pp. 670–688 (2015)
2. Bache, K., Lichman, M.: UCI machine learning repository (2013). http://archive.ics.uci.edu/ml
3. Blondel, V.D., Guillaume, J.L., Lambiotte, R., Lefebvre, E.: Fast unfolding of communities in large networks. J. Stat. Mech.: Theory Exp. **10**, P10008 (2008)
4. Brandes, U., et al.: Maximizing modularity is hard (2006)
5. Cadenas, J., Garrido, M., Martínez, R.: Feature subset selection filter-wrapper based on low quality data. Expert Syst. Appl. **40**(16), 6241–6252 (2013)
6. Chandrashekar, G., Sahin, F.: A survey on feature selection methods. Comput. Electr. Eng. **40**(1), 16–28 (2014)
7. Chen, M., Mao, S., Liu, Y.: Big data: a survey. Mobile Netw. Appl. **19**(2), 171–209 (2014)
8. Dhillon, I.S., Mallela, S., Kumar, R.: A divisive information theoretic feature clustering algorithm for text classification. J. Mach. Learn. Res. **3**, 1265–1287 (2003)
9. Ding, C., Peng, H.: Minimum redundancy feature selection from microarray gene expression data. J. Bioinform. Comput. Biol. **3**(2), 185–205 (2005)
10. Dit: dit: Python package for information theory. https://github.com/dit/dit
11. Dy, J.G., Brodley, C.E.: Feature selection for unsupervised learning. J. Mach. Learn. Res. **4**, 845–889 (2004)
12. Everitt, B.S., Landau, S., Leese, M.: Cluster Analysis. Wiley, New York (2001). isbn 9780340761199
13. Gaidhani, J.P., Natikar, S.B.: Implementation on feature subset selection using symmetric uncertainty measure. Int. J. Sci. Eng. Res. **5**(7), 1396–1399 (2014)
14. Girvan, M., Newman, M.E.: Community structure in social and biological networks. Proc. Nat. Acad. Sci. **99**(12), 7821–7826 (2002)
15. Guyon, I., Gunn, S., Ben-Hur, A., Dror, G.: Result analysis of the NIPS 2003 feature selection challenge. In: Advances in Neural Information Processing Systems (2005)
16. Hoque, N., Bhattacharyya, D.K., Kalita, J.K.: MIFS-ND: a mutual information-based feature selection method. Expert Syst. Appl. **41**(14), 6371–6385 (2014)
17. Jain, A., Dubes, R.: Algorithms for Clustering Data. Prentice-Hall, Englewood Cliffs (1988). ISBN 978-0130222787
18. Jovic, A., Brkić, K., Bogunović, N.: A review of feature selection methods with applications. In: Proceedings of the 38th International Convention on Information and Communication Technology, Electronics and Microelectronics (MIPRO), pp. 1200–1205 (2015)
19. Kotsiantis, S.B., Zaharakis, I., Pintelas, P.: Supervised machine learning: a review of classification techniques. In: Proceedings of the 2007 Conference on Emerging Artificial Intelligence Applications in Computer Engineering, pp. 3–24. IOS Press, Amsterdam (2007)
20. Lancichinetti, A., Fortunato, S.: Community detection algorithms: a comparative analysis. Phys. Rev. E **80**(5), 056117 (2009)
21. Liu, H., Motoda, H.E.: Computational Methods of Feature Selection. CRC Press, Boca Raton (2007)
22. Liu, H., Yu, L.: Toward integrating feature selection algorithms for classification and clustering. IEEE Trans. Knowl. Data Eng. **17**(4), 491–502 (2005)

23. Liu, X., Murata, T.: Advanced modularity-specialized label propagation algorithm for detecting communities in networks. Phys. A **389**(7), 1493–1500 (2010)
24. NetworkX: Software for complex networks. https://networkx.github.io/
25. Orman, G.K., Labatut, V., Cherifi, H.: Qualitative comparison of community detection algorithms. In: Cherifi, H., Zain, J.M., El-Qawasmeh, E. (eds.) DICTAP 2011. CCIS, vol. 167, pp. 265–279. Springer, Heidelberg (2011). https://doi.org/10.1007/978-3-642-22027-2_23
26. Oveisi, F., Oveisi, S., Efranian, A., Patras, I.: Tree-structured feature extraction using mutual information. IEEE Trans. Neural Netw. Learn. Syst. **23**, 127–137 (2012)
27. Peng, H., Long, F., Ding, C.: Feature selection based on mutual information criteria of max-dependency, max-relevance, and min-redundancy. IEEE Trans. Pattern Anal. Mach. Intell. **27**(8), 1226–1238 (2005)
28. Peng, H.: MRMR implementation. http://home.penglab.com/proj/mRMR/
29. Press, W.H., Flannery, B.P., Teukolsky, S.A., Vetterling, W.T.: Numerical Recipes in C. Cambridge University Press, Cambridge (1988)
30. Python: Python louvain package. https://pypi.org/project/python-louvain/
31. Python: Python pandas package. https://pypi.org/project/pandas/
32. Renyi, A.: On measures of entropy and information. In: Proceedings of the Fourth Berkeley Symposium on Mathematical Statistics and Probability, Volume 1: Contributions to the Theory of Statistics, pp. 547–561. University of California Press, Berkeley, CA (1961). https://projecteuclid.org/euclid.bsmsp/1200512181
33. Robnik-Šikonja, M., Kononenko, I.: Theoretical and empirical analysis of RELIEFF and RRELIEFF. Mach. Learn. **53**(1–2), 23–69 (2003)
34. Rosvall, M.: Maps of information flow reveal community structure in complex networks. Proc. Nat. Acad. Sci. **105**(4), 1118–1123 (2007)
35. Scikit-Learn: Machine learning in python. https://scikit-learn.org/stable/
36. Song, Q., Ni, J., Wang, G.: A fast clustering-based feature subset selection algorithm for high-dimensional data. IEEE Trans. Knowl. Data Eng. **25**(1), 1–14 (2013)
37. Stein, G., Chen, B., Wu, A.S., Hua, K.A.: Decision tree classifier for network intrusion detection with GA-based feature selection. In: Proceedings of 43rd Annual Southeast Regional Conference (ACM-SE 43), vol. 2, pp. 136–141. ACM (2005)
38. Steuer, R., Kurths, J., Daub, C.O., Weise, J., Selbig, J.: The mutual information: detecting and evaluating dependencies between variables. Bioinformatics **18**(2), 231–240 (2002)
39. Tang, J., Alelyani, S., Liu, H.: Feature selection for classification: a review. Data Classif.: Algorithms Appl. (2014)
40. Traag, V.: Faster unfolding of communities: speeding up the Louvain algorithm. Phys. Rev. E **92**(3), 032801 (2015)
41. Wang, L., Zhou, N., Chu, F.: A wrapper approach to selection of class-dependent features. IEEE Trans. Neural Netw. **19**(7), 1267–1278 (2008)
42. Wei, X., Xie, S., Cao, B., Yu, P.S.: Rethinking unsupervised feature selection: from pseudo labels to pseudo must-links. In: Ceci, M., Hollmén, J., Todorovski, L., Vens, C., Džeroski, S. (eds.) ECML PKDD 2017. LNCS (LNAI), vol. 10534, pp. 272–287. Springer, Cham (2017). https://doi.org/10.1007/978-3-319-71249-9_17
43. Yang, A.Y., Wright, J., Ma, Y., Sastry, S.S.: Feature selection in face recognition: a sparse representation perspective. Technical report, EECS Department, University of California, Berkeley, Technical Report No. UCB/EECS-2007-99 (2007)

Relationship Estimation Metrics for Binary SoC Data

Dave McEwan(✉) and Jose Nunez-Yanez

University of Bristol, Bristol, UK
{dave.mcewan,eejlny}@bristol.ac.uk

Abstract. System-on-Chip (SoC) designs are used in every aspect of computing and their optimization is a difficult but essential task in today's competitive market. Data taken from SoCs to achieve this is often characterised by very long concurrent bit vectors which have unknown relationships to each other. This paper explains and empirically compares the accuracy of several methods used to detect the existence of these relationships in a wide range of systems. A probabilistic model is used to construct and test a large number of SoC-like systems with known relationships which are compared with the estimated relationships to give accuracy scores. The metrics $\dot{\text{Cov}}$ and $\dot{\text{Dep}}$ based on covariance and independence are demonstrated to be the most useful, whereas metrics based on the Hamming distance and geometric approaches are shown to be less useful for detecting the presence of relationships between SoC data.

Keywords: Binary time series · Bit vector · Correlation · Similarity · System-on-chip

1 Introduction

SoC designs include the processors and associated peripheral blocks of silicon chip based computers and are an intrinsic piece of modern computing, owing their often complex design to lifetimes of work by hundreds of hardware and software engineers. The SoC in a RaspberryPi [11] for example includes 4 ARM processors, memory caches, graphics processors, timers, and all of the associated interconnect components. Measuring, analysing, and understanding the behavior of these systems is important for the optimization of cost, size, power usage, performance, and resiliance to faults.

Sampling the voltage levels of many individual wires is typically infeasible due to bandwidth and storage constraints so sparser event based measurements are often used instead; E.g. Observations like "`cache_miss` @ 123 ns". This gives rise

This project is supported by the Engineering and Physical Sciences Research Council (EP/I028153/ and EP/L016656/1); the University of Bristol and UltraSoC Technologies Ltd.

G. Nicosia et al. (Eds.): LOD 2019, LNCS 11943, pp. 118–129, 2019.
https://doi.org/10.1007/978-3-030-37599-7_11

to datasets of very long concurrent streams of binary occurrence/non-occurrence data so an understanding of how these event measurements are related is key to the design optimization process. It is therefore desirable to have an effective estimate of the connectedness between bit vectors to indicate the existence of pairwise relationships. Given that a SoC may perform many different tasks the relationships may change over time which means that a windowed or, more generally, a weighted approach is required. Relationships between bit vectors are modelled as boolean functions composed of negation (NOT), conjunction (AND), inclusive disjunction (OR), and exclusive disjunction (XOR) operations since this fits well with natural language and has previously been successfully applied to many different system types [5]; E.g. Relationships of a form like "`flush` occurs when `filled` AND `read_access` occur together".

This paper provides the following novel contributions:

- A probabilistic model for SoC data which allows a large amount of representative data to be generated and compared on demand.
- An empirical study on the accuracy of several weighted correlation and similarity metrics in the use of relationship estimation.

A collection of previous work is reviewed, and the metrics are formally defined with the reasoning behind them. Next, assumptions about the construction of SoC relationships are explained and the design of the experiment is described along with the method of comparison. Finally results are presented as a series of Probability Density Function (PDF) plots and discussed in terms of their application.

2 Previous Work

An examination of currently available hardware and low-level software profiling methods is given by Lagraa [9] which covers well known techniques such as using counters to generate statistics about both hardware and software events – effectively a low cost data compression. Lagraa's thesis is based on profiling SoCs created specifically on Xilinx MPSoC devices, which although powerful, ensures it may not be applied to data from non-Field Programmable Gate Array (FPGA) sources such as designs already manufactured in silicon which is often the end goal of SoC design. Lo et al. [10] described a system for describing behavior with a series of statements using a search space exploration process based on boolean set theory. While this work has a similar goal of finding temporal dependencies it is acknowledged that the mining method does not perform adequately for the very long traces often found in real-world SoC data. Ivanovic et al. [7] review time series analysis models and methods where characteristic features of economic time series are described such as drawn from noisy sources, high auto-dependence and inter-dependence, high correlation, and non-stationarity. SoC data is expected to have these same features, together with full binarization and much greater length. The expected utility approach to learning probabilistic models by Friedman and Sandow [3] minimises the Kullbach-Leibler distance between observed data and a model, attempting to fit that data using an iterative method. As noted in Friston et al. [4] fully learning all parameters of a Bayesian network through empirical observations is an intractable analytic problem which simpler non-iterative measures can only roughly approximate. The approach of modelling relationships as boolean functions has been used for measuring complexity and pattern detection in a variety of fields including complex biological systems from the scale of proteins to groups of animals [20].

'Correlation' is a vague term which has several possible interpretations [18] including treating data as high dimensional vectors, sets, and population samples. A wide survey of binary similarity and distance measures by Choi et al. [1] tabulates 76 methods from various fields and classify them as either distance, non-correlation, or correlation based. A similarity measure is one where a higher result is produced for more similar data, whereas a distance measure will give a higher results for data which are further apart, i.e, less similar. The distinction between correlation and similarity can be shown with an example: If it is noticed over a large number of parties that the pattern of attendance between Alice and Bob is similar then it may be inferred that there is some kind of relationship connecting them. In this case the attendance patterns of Alice and Bob are both similar and correlated. However, if Bob is secretly also seeing Eve it would be noticed that Bob only attends parties if either Alice or Eve attend, but not both at the same time. In this case Bob's pattern of attendance may not be similar to that of either Alice or Eve, but will be correlated with both. It can therefore be seen that correlation is a more powerful approach for detecting relationships, although typically involves more calculation.

In a SoC design the functionallity is split into a number of discrete logical blocks such as a timer or an ARM processor which communicate via one or more buses. The configuration of many of these blocks and buses is often specified with a non-trivial set of parameters which affects the size, performance, and cost of the final design. The system components are usually a mixture of hardware and software which should all be working in harmony to achieve the designer's goal and the designer will usually have in mind how this harmony should look. For example the designer will have a rule that they would like to confirm "software should use the cache efficiently" which will be done by analysing the interaction of events such as `cache_miss` and `enter_someFunction`. By recording events and measuring detecting inter-event relationships the system designer can decide if the set of design parameters should be kept or changed [14], thus aiding the SoC design optimization process.

3 Metrics

A measured stream of events is written as f_i where i is an identifier for one particular event source such as `cache_miss`. Where $f_i(t) = 1$ indicates event i was observed at time t, and $f_i(t) = 0$ indicating i was not observed at time t. A windowing or weighting function w is used to create a weighted average of each measurement to give an expectation of an event occurrence.

$$\mathbb{E}[f_i] = \frac{1}{\sum_t w(t)} \sum_t w(t) * f_i(t) \quad \in [0,1] \tag{1}$$

Bayes theorem may be rearranged to find the conditional expectation.

$$\Pr(X|Y) = \frac{\Pr(Y|X)\Pr(X)}{\Pr(Y)} = \frac{\Pr(Y \cap X)}{\Pr(Y)}, \quad \text{if } \Pr(Y) \neq 0 \tag{2}$$

$$\mathbb{E}[f_x|f_y] := \begin{cases} \text{NaN} & : \mathbb{E}[f_y] = 0 \\ \dfrac{\mathbb{E}[f_x \odot f_y]}{\mathbb{E}[f_y]} & : \text{otherwise} \end{cases} \tag{3}$$

It is not sufficient to look only at conditional expectation to determine if X and Y are related. For example, the result $\Pr(X|Y) = 0.9$ may arise from X's relationship with Y, but may equally arise from the case $\Pr(X) = 0.9$.

A naïve approach might be to estimate how similar a pair of bit vectors are by counting the number of matching bits. The expectation that a pair of corresponding bits are equal is the Hamming Similarity [6], as shown in Eq. (4). Where X and Y are typical sets [12] this is equivalent to $|\mathbb{E}[X] - \mathbb{E}[Y]|$. The absolute difference $|X - Y|$ may also be performed on binary data using a bitwise XOR operation.

$$\dot{\text{H}}\text{am}(f_x, f_y) := 1 - \mathbb{E}[|f_x - f_y|] \tag{4}$$

The dot in the notation is used to show that this measure is similar to, but not necessarily equivalent to the standard definition. Modifications to the standard definitions may include disallowing NaN, restricting or expanding the range to $[0,1]$, or reflecting the result. For example, reflecting the result of $\mathbb{E}[|f_x - f_y|]$ in the definition of $\dot{\text{H}}\text{am}$ a metric is given where 0 indicates fully different and 1 indicates exactly the same.

A similar approach is to treat a pair of bit vectors as a pair of sets. The Jaccard index first described for comparing the distribution of alpine flora [8], and later refined for use in general sets is defined as the ratio of size the intersection to the size of the union. Tanimoto's reformulation [19] of the Jaccard index shown in Eq. (5) was given for measuring the similarity of binary sets.

$$J(X,Y) = \frac{|X \cap Y|}{|X \cup Y|} = \frac{|X \cap Y|}{|X| + |Y| - |X \cap Y|}, \quad |X \cup Y| \neq \varnothing \tag{5}$$

$$\dot{\text{T}}\text{mt}(f_x, f_y) := \frac{\mathbb{E}[f_x \odot f_y]}{\mathbb{E}[f_x] + \mathbb{E}[f_y] - \mathbb{E}[f_x \odot f_y]} \tag{6}$$

Treating measurements as points in bounded high dimensional space allows the Euclidean distance to be calculated, then reflected and normalized to $[0,1]$ to show closeness rather than distance. This approach is common for problems where the alignment of physical objects is to be determined such as facial detection and gene sequencing [2].

$$\dot{\text{Cls}}(f_x, f_y) := 1 - \sqrt{\mathbb{E}\left[|f_x - f_y|^2\right]} \tag{7}$$

It can be seen that this formulation is similar to using the Hamming distance, albeit growing quadratically rather than linearly as the number of identical bits increases. Another geometric approach is to treat a pair of measurements as bounded high dimensional vectors and calculate the angle between them using the cosine similarity as is often used in natural language processing [15] and data mining [17].

$$\text{CosineSimilarity}_{X,Y} = \frac{X \cdot Y}{|X|\,|Y|},\ X, Y \neq 0 \quad \in [-1, 1] \tag{8}$$

$$\dot{\text{Cos}}(f_x, f_y) := \frac{\mathbb{E}[f_x \odot f_y]}{\sqrt{\mathbb{E}[f_x^2]}\sqrt{\mathbb{E}[f_y^2]}} \quad \in [0, 1] \tag{9}$$

The strict interval of the measured bit vectors $f_x, f_y \in [0, 1]$ mean that $\dot{\text{Cos}}$ is always positive.

The above metrics attempt to uncover relationships by finding pairs of bit vectors which are similar to each other. These may be useful for simple relationships of forms similar to "X leads to Y" but may not be useful for finding relationships which incorporate multiple measurements via a function of boolean operations such as "A AND B XOR C leads to Y". Treating measurement data as samples from a population invites the use of covariance or the Pearson correlation coefficient as a distance metric. The covariance, as shown in Eq. (10), between two bounded-value populations is also bounded, as shown in Eq. (11). This allows the $\dot{\text{Cov}}$ metric to be defined, again setting negative correlations to 0. For binary measurements with equal weights $\dot{\text{Cov}}$ can be shown to be equivalent to the Pearson correlation coefficient.

$$\text{cov}(X, Y) = \mathbb{E}[(X - \mathbb{E}[X])\,(Y - \mathbb{E}[Y])] = \mathbb{E}[XY] - \mathbb{E}[X]\mathbb{E}[Y] \tag{10}$$

$$X, Y \in [0, 1] \implies \frac{-1}{4} \leqslant \text{cov}(X, Y) \leqslant \frac{1}{4} \tag{11}$$

$$\dot{\text{Cov}}(f_x, f_y) := 4\left|\mathbb{E}[f_x \odot f_y] - \mathbb{E}[f_x]\mathbb{E}[f_y]\right| \quad \in [0, 1] \tag{12}$$

Using this definition it can be seen that if two random variables are independent then $\dot{\text{Cov}}(X, Y) = 0$, however the reverse is not true in general as the covariance of two dependent random variables may be 0. The definition of independence in Eq. (13) may be used to define a metric of dependence.

$$X \perp\!\!\!\perp Y \iff \Pr(X) = \Pr(X|Y) \tag{13}$$

$$\dot{\text{Dep}}(f_x, f_y) := \left|\frac{\mathbb{E}[f_x|f_y] - \mathbb{E}[f_x]}{\mathbb{E}[f_x|f_y]}\right|, \quad \text{if } \mathbb{E}[f_x] \leqslant \mathbb{E}[f_x|f_y] \tag{14}$$

Normalizing the difference in expectation $\mathbb{E}[f_x|f_y] - \mathbb{E}[f_x]$ to the range $[0, 1]$ allows this to be rearranged showing that $\dot{\text{Dep}}(X, Y)$ is an undirected similarity, i.e. the order of X and Y is unimportant.

$$\dot{\text{Dep}}(f_x, f_y) = \frac{\mathbb{E}[f_x|f_y] - \mathbb{E}[f_x]}{\mathbb{E}[f_x|f_y]} = 1 - \frac{\mathbb{E}[f_x]\mathbb{E}[f_y]}{\mathbb{E}[f_x \odot f_y]} = \dot{\text{Dep}}(f_y, f_x) \tag{15}$$

The metrics defined above $\dot{\text{Ham}}$, $\dot{\text{Tmt}}$, $\dot{\text{Cls}}$, $\dot{\text{Cos}}$, $\dot{\text{Cov}}$, and $\dot{\text{Dep}}$ all share the same codomain $[0, 1]$ where 1 means the strongest relationship. In order to compare these correlation metrics an experiment has been devised to quantify their effectiveness, as described in Sect. 4.

4 Experimental Procedure

This experiment constructs a large number of SoC-like systems according to a probabilistic structure and records event-like data from them. The topology of each system is fixed which means relationships between bit vectors in each system are known in advance of applying any estimation metric. The metrics above are then applied to the recorded data and compared to the known relationships which allows the effectiveness of each metric to be demonstrated empirically.

The maximum number of measurement nodes $2n_{\text{maxm}}$ is set to 100 to keep the size of systems within reasonable limits. Each system is composed of a number of measurement nodes $e_{i \in [1,m]}$ such that $m = m_{\text{src}} + m_{\text{dst}}$ of either type 'src' or 'dst' arranged in a bipartite graph as shown in Fig. 1. In each system the numbers of measurement nodes are chosen at random $m_{\text{src}}, m_{\text{dst}} \sim \text{U}(1, n_{\text{maxm}})$. Src nodes are binary random variables with a fixed densitity $\sim \text{Arcsin}(0, 1)$ where the approximately equal number of high and low density bit vectors represents equal importance of detecting relationships and anti-relationships. The value of each dst node is formed by combining a number of edges $\sim \text{Lognormal}(0, 1)$ from src nodes. There are five types of systems which relate to the method by which src nodes are combined to produce the value at a dst node. One fifth of systems use only AND operations ($\wedge$) to combine connections to each dst node, another fifth uses only OR ($\vee$), and another fifth uses only XOR ($\oplus$). The fourth type of system uniformly chooses one of the $\wedge$, $\vee$, $\oplus$ methods to give a mix of homogeneous functions for each dst node. The fifth type gets the values of each dst node by applying chains of operations $\sim \text{U}(\{\wedge, \vee, \oplus\})$ combine connections, implemented as Left Hand Associative (LHA). By keeping different connection strategies separate it is easier to see how the metrics compare for different types of relationships.

The known relationships were used to construct an adjacency matrix where $K_{ij} = 1$ indicates that node i is connected to node j, with 0 otherwise. The diagonal is not used as these tautological relationships will provide a perfect score with every metric without providing any new information about the metric's accuracy or effectiveness. Each metric is applied to every pair of nodes to construct an estimated adjacency matrix E. Each element E_{ij} is compared with

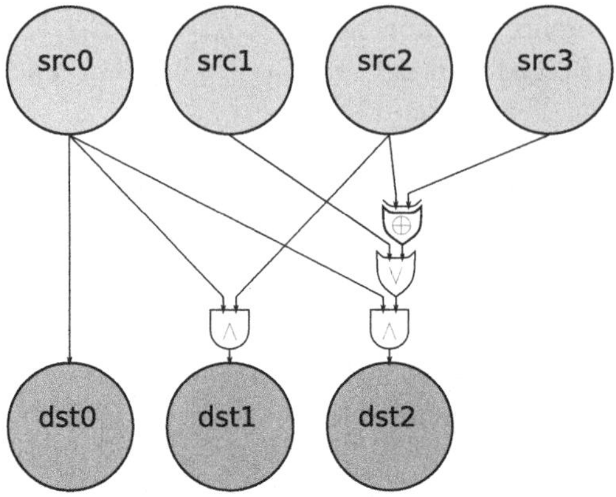

Fig. 1. Example system with src and dst nodes connected via binary operations.

K_{ij} to give an amount of True-Positive (TP) and False-Negative (FN) where $K_{ij} = 1$ or an amount of True-Negative (TN) and False-Positive (FP) where $K_{ij} = 0$. For example if a connection is known to exist ($K_{ij} = 1$) and the metric calculated a value of 0.7 then the True-Positive and False-Negative values would be 0.7 and 0.3 respectively, with both True-Negative and False-Positive equal to 0. Alternatively if a connection is know to not exist ($K_{ij} = 0$) then True-Negative and False-Positive would be 0.3 and 0.7, with True-Positive and False-Negative equal to 0. These are used to construct the confusion matrix and subsequently give scores for the True Positive Rate (Sensitivity) (TPR), True Negative Rate (Specificity) (TNR), Positive Predictive Value (Precision) (PPV), Negative Predictive Value (NPV), Accuracy (ACC), Balanced Accuracy (BACC), Book-Maker's Informedness (BMI), and Matthews Correlation Coefficient (MCC).

$$\mathrm{TP} = \sum_{i,j} \min(K_{ij}, E_{ij}) \qquad \mathrm{FN} = \sum_{i,j} \min(K_{ij}, 1 - E_{ij})$$

$$\mathrm{FP} = \sum_{i,j} \min(1 - K_{ij}, E_{ij}) \qquad \mathrm{TN} = \sum_{i,j} \min(1 - K_{ij}, 1 - E_{ij})$$

$$\mathrm{TPR} = \frac{\mathrm{TP}}{\mathrm{TP} + \mathrm{FN}} \qquad \mathrm{TNR} = \frac{\mathrm{TN}}{\mathrm{TN} + \mathrm{FP}}$$

$$\mathrm{PPV} = \frac{\mathrm{TP}}{\mathrm{TP} + \mathrm{FP}} \qquad \mathrm{NPV} = \frac{\mathrm{TN}}{\mathrm{TN} + \mathrm{FN}}$$

$$\mathrm{ACC} = \frac{\mathrm{TP} + \mathrm{TN}}{\mathrm{TP} + \mathrm{FN} + \mathrm{TN} + \mathrm{FP}} \qquad \mathrm{BACC} = \frac{\mathrm{TPR} + \mathrm{TNR}}{2}$$

$$\mathrm{BMI} = \mathrm{TPR} + \mathrm{TNR} - 1$$

$$\mathrm{MCC} = \frac{\mathrm{TP} \times \mathrm{TN} - \mathrm{TP} \times \mathrm{TN}}{\sqrt{(\mathrm{TP} + \mathrm{FP})(\mathrm{TP} + \mathrm{FN})(\mathrm{TN} + \mathrm{FP})(\mathrm{TN} + \mathrm{FN})}}$$

To create the dataset 1000 systems were generated, with 10000 samples of each node taken from each system. This procedure was repeated for each metric for each system and the PDF of each metric's accuracy is plotted using Kernel Density Estimation (KDE) to see an overview of how well each performs over a large number of different systems.

5 Results and Discussion

The metrics defined in Sect. 3 function as binary classifiers therefore it is reasonable to compare their effectiveness using some of the statistics common for binary classifiers noted above. The TPR measures the proportion of connections which are correctly estimated and the TNR similarly measures the proportion of non-connections correctly estimated. The PPV and NPV measures the proportion of estimates which are correctly estimated to equal the known connections and non-connections. ACC measures the likelihood of an estimation matching a known connection or non-connection. For imbalanced data sets ACC is not necessarily a good way of scoring the performance of these metrics as it may give an overly optimistic score. Normalizing TP and TN by the numbers of samples gives the Balanced Accuracy [16] which may provide a better score for large systems where the adjacency matrices are sparse. Matthews Correlation Coefficient finds the covariance between the known and estimated adjacency matrices which may also be interpreted as a useful score of metric performance. Youden's J statistic, also known as Book-Maker's Informedness similarly attempts to capture the performance of a binary classifier by combining the sensitivity and specifitiy to give the probability of an informed decision.

Each statistic was calculated for each metric for each system. Given the large number of systems of various types, PDFs of these statistics are shown in Fig. 2 where more weight on the right hand side towards 1.0 indicates a better metric.

Figure 2a shows that $\dot{\mathrm{C}}\mathrm{ov}$ and $\dot{\mathrm{D}}\mathrm{ep}$ correctly identify the existence of around 25% of existing connections and other metrics identify many more connections. However, Fig. 2b shows that $\dot{\mathrm{C}}\mathrm{ov}$ and $\dot{\mathrm{D}}\mathrm{ep}$ are much more likely to correctly identify non-connections than other metrics, especially $\dot{\mathrm{H}}\mathrm{am}$ and $\dot{\mathrm{C}}\mathrm{ov}$.

For a metric to be considered useful for detecting connections the expected value of both PPV and NPV must be greater than 0, and ACC must be greater than 0.5. It can be seen in Fig. 2d that all metrics score highly for estimating negatives; I.e. when a connection does not exist they give a result close to 0. On its own this does not carry much meaning as a constant 0 will always give a correct answer. Similarly, a constant 1 will give a correct answer for positive links so the plots in the middle and right columns must be considered together with the overall accuracy to judge the usefulness of a metric.

Given that ACC is potentially misleading for imbalanced data sets such as this one it is essential to check against BACC. $\dot{\mathrm{H}}\mathrm{am}$ usually has ACC of close to 0.5 which alone indicates than it is close to useless for detecting connections in binary SoC data. The wider peaks of $\dot{\mathrm{C}}\mathrm{os}$ and $\dot{\mathrm{T}}\mathrm{mt}$ in both ACC and BACC indicate that these metrics are much more variable in their performance than

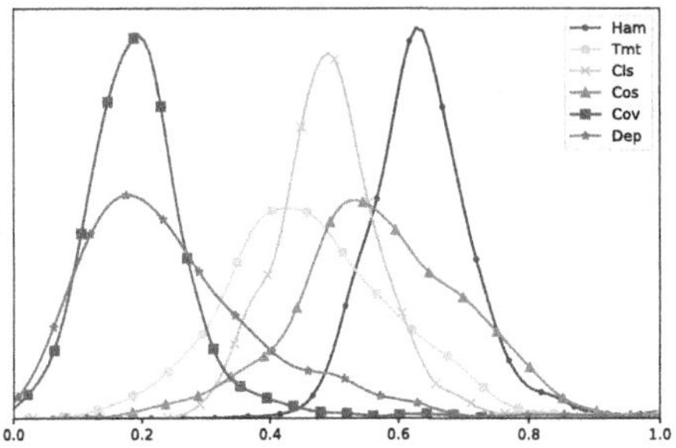

(a) True Positive Rate (Sensitivity).

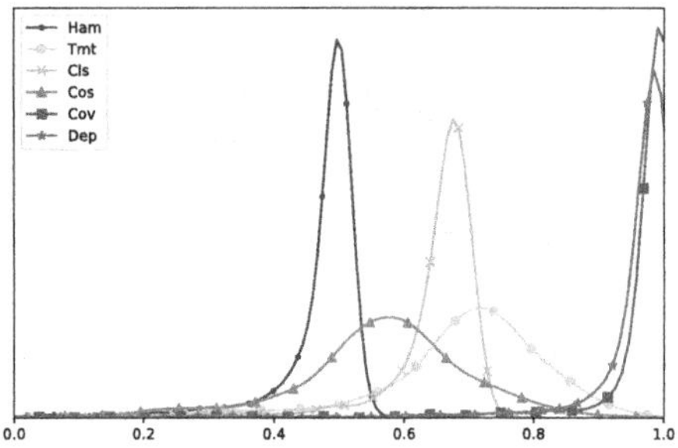

(b) True Negative Rate (Specificity).

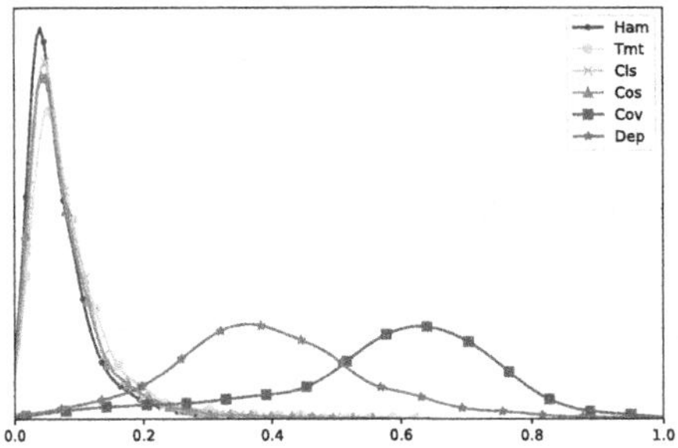

(c) Positive Predictive Value (Precision).

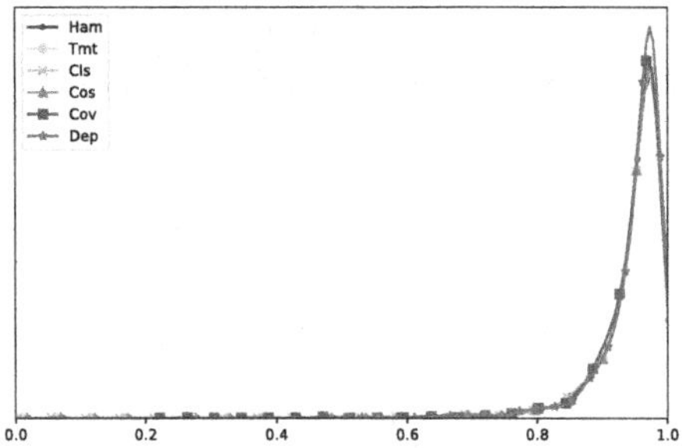

(d) Negative Predictive Value.

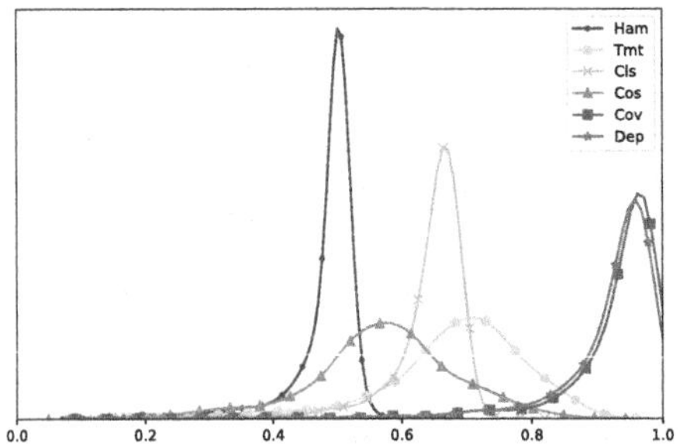

(e) Accuracy.

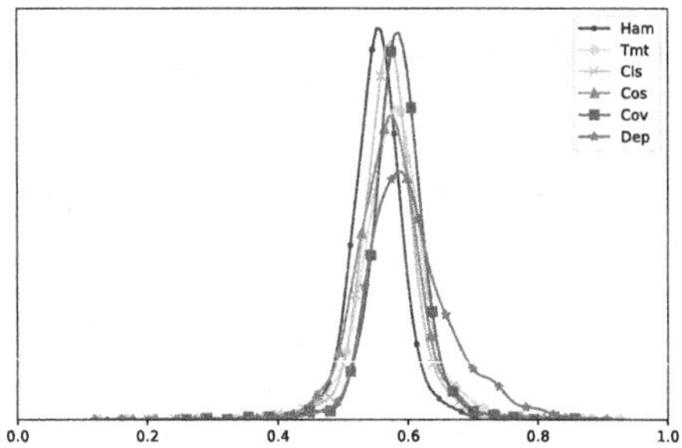

(f) Balanced Accuracy.

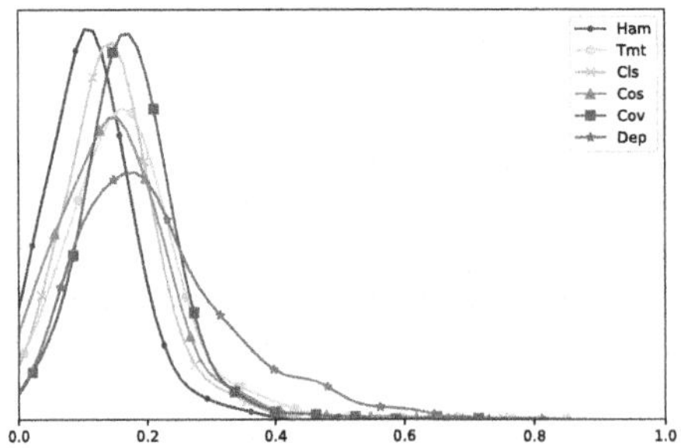

(g) Book-Maker's Informedness.

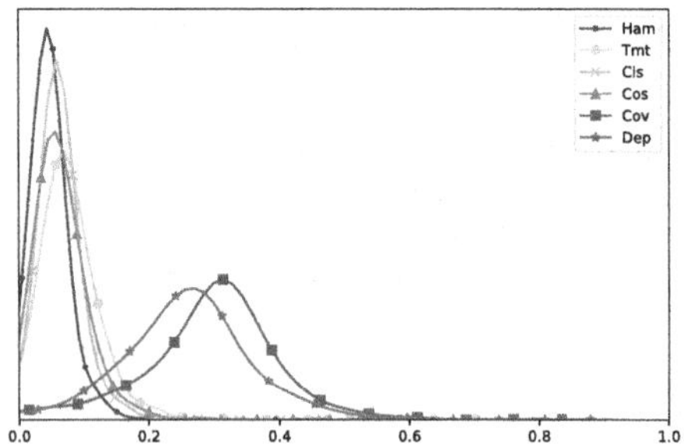

(h) Matthews Correlation Coefficient.

Fig. 2. KDE plots of score PDFs averaged across all system types. More weight on the right hand side is always better.

the likes of $\dot{\text{Cls}}$, $\dot{\text{Cov}}$, and $\dot{\text{Dep}}$. In this pair of plots where $\dot{\text{Cov}}$ and $\dot{\text{Dep}}$ both have much more weight towards the right hand side this indicates that these metrics are more likely to give a good estimate of connectedness.

Finally, using Fig. 2h and g as checks it can be see again that $\dot{\text{Cov}}$ and $\dot{\text{Dep}}$ outperform the other metrics. MCC actually has an interval of $[-1, 1]$ though the negative side is not plotted here, and given that all metrics have weight on the positive side this shows that all of the defined metrics contain at least some information on the connectedness.

The overall results indicate that $\dot{\text{Ham}}$, $\dot{\text{Tmt}}$, $\dot{\text{Cos}}$ and $\dot{\text{Cls}}$ are close to useless for detecting connections in datasets resembling the SoC data model described above.

A characteristic feature employed by both $\dot{\text{Tmt}}$ and $\dot{\text{Cos}}$ is the intersection $f_x \odot f_y$, whereas $\dot{\text{Ham}}$ and $\dot{\text{Cls}}$ employ an absolute difference $|f_x - f_y|$. The best performing metrics $\dot{\text{Cov}}$ and $\dot{\text{Dep}}$ have consistently higher accuracy scores and employ both the intersection, and the product of expectations $\mathbb{E}[f_x]\mathbb{E}[f_y]$.

The simplicity of these metrics allows hints about the system function to be found quickly in an automated manner, albeit without further information about the formulation or complexity of the relationships. Any information which can be extracted from a dataset about the workings of its system may be used to ease the work of a SoC designer. For example, putting the results into a suitable visualization provides an easy to consume presentation of how related a set of measurements are during a given time window. This allows the SoC designer to make a more educated choice about the set of design parameters in order to provide a more optimal design for their chosen market.

6 Conclusion

The formulation and rationale behind six methods of measuring similarity or correlation to estimate relationships between weighted bit vectors has been given. The given formulations may also be applied more generally to bounded data in the range $[0, 1]$, though this is not explored in this paper and may be the subject of future work. Other directions of future work include testing and comparing additional metrics or designing specialized metrics for.

It has been shown that using methods which are common in other fields such as the Hamming distance, Tanimoto distance, Euclidean distance, or Cosine similarity are not well suited to low-cost relationship detection when the relationships are potentially complex. This result highlights a potential pitfall of not considering the system construction for data scientists working with related binary data streams.

The metrics $\dot{\text{Cov}}$ and $\dot{\text{Dep}}$ are shown to consistently estimate the existance of relationships in SoC-like data with higher accuracy than the other metrics. This result gives confidence that detection systems may employ these approaches in order to make meaningful gains in the process of optimizing SoC behavior. By using more accurate metrics unknown relationships may be uncovered giving

SoC designer the information they need to optimize their designs and sharpen their competitive edge.

The Python code used to perform the experiments is available online [13].

References

1. Choi, S.S., Cha, S.H., Tappert, C.C.: A survey of binary similarity and distance measures. Syst. Cybern. Inform. **8**(1), 43–48 (2010)
2. Frey, B.J., Dueck, D.: Clustering by passing messages between data points. Science **315**, 972–976 (1950)
3. Friedman, C., Sandow, S.: Learning probabilistic models: an expected utility maximization approach. J. Mach. Learn. Res. **4**, 257–291 (2003)
4. Friston, K., Parr, T., Zeidman, P.: Bayesian model reduction. The Welcome Centre for Human Neuroimaging, Institute of Neurology, London (2018). https://arxiv.org/pdf/1805.07092.pdf
5. Gheradi, M., Rotondo, P.: Measuring logic complexity can guide pattern discovery in empirical systems. Complexity **21**(S2), 397–408 (2018)
6. Hamming, R.W.: Error detecting and error correcting codes. Bell Syst. Tech. J. **29**(2), 147–160 (1950)
7. Ivanovic, M., Kurbalija, V.: Time series analysis and possible applications. In: 39th International Convention on Information and Communication Technology, Electronics and Microelectronics, pp. 473–479 (2016). https://doi.org/10.1109/MIPRO.2016.7522190
8. Jaccard, P.: The distribution of the flora in the alpine zone. New Phytol. **11**(2), 37–50 (1919). http://nph.onlinelibrary.wiley.com/doi/abs/10.1111/j.1469-8137.1912.tb05611.x
9. Lagraa, S.: New MP-SoC profiling tools based on data mining techniques. Ph.D. thesis, L'Université de Grenoble (2014). https://tel.archives-ouvertes.fr/tel-01548913
10. Lo, D., Khoo, S.C., Liu, C.: Mining past-time temporal rules from execution traces. In: ACM Workshop On Dynamic Analysis, pp. 50–56, July 2008
11. Loo, G.V.: BCM2836 ARM Quad-A7 (2014). https://www.raspberrypi.org/documentation/hardware/raspberrypi/bcm2836/QA7_rev3.4.pdf
12. MacKay, D.J.C.: Information Theory, Inference and Learning Algorithms. Cambridge University Press, Cambridge (2002)
13. McEwan, D.: Relest: relationship estimation (2019). https://github.com/DaveMcEwan/dmppl/blob/master/dmppl/experiments/relest.py
14. McEwan, D., Hlond, M., Nunez-Yanez, J.: Visualizations for understanding SoC behaviour. In: 2019 15th Conference on Ph.D. Research in Microelectronics and Electronics (PRIME), July 2019. https://doi.org/10.1109/PRIME.2019.8787837. https://arxiv.org/abs/1905.06386
15. Mikolov, T., Chen, K., Corrado, G., Dean, J.: Efficient estimation of word representations in vector space. arXiv e-prints arXiv:1301.3781, January 2013
16. Mower, J.P.: PREP-Mt: predictive RNA editor for plant mitochondrial genes. BMC Bioinform. **6**(1), 96 (2005)
17. Rahutomo, F., Kitasuka, T., Aritsugi, M.: Semantic cosine similarity. In: Proceedings of the 7th ICAST 2012, Seoul, October 2012
18. Rodgers, J.L., Nicewander, W.A.: Thirteen ways to look at the correlation coefficient. Am. Stat. **42**, 59–66 (1988)

19. Rogers, D.J., Tanimoto, T.T.: A computer program for classifying plants. Science **132**(3434), 1115–1118 (1960). http://science.sciencemag.org/content/132/3434/1115/tab-pdf
20. Tkacik, G., Bialek, W.: Information processing in living systems. Complexity **21**(S2), 397–408 (2018)

Network Alignment Using Graphlet Signature and High Order Proximity

Aljohara Almulhim(✉), Vachik S. Dave, and Mohammad Al Hasan

Department of Computer and Information Sciences,
Indiana University Purdue University, Indianapolis, IN 46202, USA
aalmulhi@iu.edu, {vsdave,alhasan}@iupui.edu

Abstract. Network alignment problem arises in graph-based problem formulation of many computer science and biological problems. The alignment task is to identify the best one-to-one matching between vertices for a pair of networks by considering the local topology or vertex attributes or both. Existing algorithms for network alignment uses a diverse set of methodologies, such as, Eigen-decomposition of a similarity matrix, solving a quadratic assignment problem through subgradiant optimization, or heuristic-based iterative greedy matching. However, these methods are either too slow, or they have poor matching performance. Some existing methods also require extensive external node attributes as prior information for the purpose of node matching. In this paper, we develop a novel topology-based network alignment approach which we call *GraphletAlign*. The proposed method uses graphlet signature as node attributes and then uses a bi-partite matching algorithm for obtaining an initial alignment, which is later refined by considering higher-order matching. Our results on large real-life networks show the superiority of *GraphletAlign* over the existing methods; specifically, *GraphletAlign*'s accuracy improvement is up to 20%–72% compared to existing network alignment methods over six large real-life networks.

Keywords: Network alignment · Higher order proximity

1 Introduction

Network is the natural representation of many real-world data, examples include social interaction among people (email networks, social networks, collaboration networks), molecular-level interaction among biological entities, such as proteins, or genes (phylogenetic networks, protein-interaction network), and infrastructure-related information for transmitting data, power, or people (road networks, power-grid networks, computer networks). Increasingly, many such data are being stored as networks and higher level data mining and analysis algorithms are being built on these networks jointly. However, as network data grows in number, it has become more likely that a collection of node entities are residing in many networks with possibly a different node identifier in each

G. Nicosia et al. (Eds.): LOD 2019, LNCS 11943, pp. 130–142, 2019.
https://doi.org/10.1007/978-3-030-37599-7_12

network; for instance, a person may be in both Facebook, and LinkedIn network with a different Id; a protein may be in various interaction networks of multiple species but its name differ in different species; or a router node may be part of multiple network traffic graphs with different identifiers that were generated randomly for preserving anonymity. This problem is specifically ubiquitous when we have multiple snapshots of the same network for different time, or for different external conditions. As a consequence of this problem, performing knowledge discovery over multiple data sources jointly is becoming increasingly difficult, as one needs to match (align) the node entities in different graphs before building the design matrix of any learning algorithm. This task of aligning vertices of multiple networks is commonly known as network alignment, which is the focus of this work.

Given a pair of networks, a network alignment method maps the vertices of a source network to the vertices of a destination network with the assumption that the number of nodes in the destination network is equal or larger. On some occasions, the number of nodes are identical in both networks, then the mapping is mutual or one-to-one. In terms of matching criteria, in some application scenarios of network alignment, such as, biology [15], ontology matching [11], and database schema matching [9], prior information regarding the vertices is available which acts as main features for matching the vertices. On the contrary, in some of the application domains, like, computer vision [2,13], and social networks [10] no prior information regarding the vertices is available so the matching objective is to align vertices that are very similar in their local topology. A combination of node attributes and local topology based matching are also considered in some tasks. Among the different variants of vertex matching criteria, the topology based matching is the most general and difficult, which we investigate in our paper.

For topology-based network alignment, the objective is to map the vertex-set of a source network to the vertex-set of a destination network, such that local topology (adjacency and non-adjacency) of the aligned vertices are maximally matched. Note that this task is different from the well-known subgraph isomorphism problem (a known NP-Complete problem), where the objective is to exactly match the topology of all vertices of the source graph with the same of a subgraph of the destination graph. When the source and destination graphs are identical except that the vertex identifiers of the graphs are from a permutation group, the graph alignment problem then becomes a graph isomorphism problem whose complexity status is yet unknown [3]. Nevertheless, in real-life graph alignment task, the source and destination graphs are very similar, but not identical, which makes it a maximum matching optimization problem.

In existing literature researchers designed diverse techniques for solving network alignment [4,14,17]. These works can be categorized into two major groups: the works in the first group [7,14] build node level features and then match the nodes in the two networks through a greedy criterion. The works in the second group [1,6,17] cast the network alignment task as a global optimization problem which is solved by optimization methodologies. The second group, more often, assumes that the vertices and/or edges are labeled and the matching task is required with respect to those labels. In a very recent work [4], representation learning framework has also been used for solving graph alignment task, where the node level features are learned automatically, then greedy matching criterion is used for solving the alignment task. Our experiments with the existing methodologies reveal that the performance of the existing methods critically depends on the node-level features. Unless the node-level features are very expressive, the alignment task fails to obtain a matching which is substantially accurate.

In this work, we propose *GraphletAlign*, a topology-based network alignment method, which uses local graphlet frequency as node-level feature and obtains an initial alignment through efficient bi-partite matching. The matching is later refined by considering higher-order matching. Experiments on six large real-life networks show that the accuracy of *GraphletAlign* is up to 20%–72% better compared with the best performance of the existing network alignment methods over six large real-life networks. Besides, *GraphletAlign*'s runtime performance is also favourable over the competing algorithms.

2 Related Works

The network alignment problem has been studied by different communities, including database, bioinformatics, computer vision, and social sciences. One of the earliest solution for this problem is IsoRank proposed by Singh et al. [14]. Staring from an initial similarity matrix (user provided), IsoRank obtains a higher-level similarity matrix via iterative matrix-vector multiplication like PageRank. Then the nodes are matched by obtaining matched pair greedily based on the higher-level similarity value. GRAAL (GRAph ALigner) [7] is probably the most well-known among the existing network alignment algorithms. It uses graphlet degree based features for each nodes and then uses a greedy node matching for solving graph alignment.

In recent years, mathematical optimization methods are being used for solving network alignment. Klau [6] formulated this as an integer linear programming task and solved it by using Lagrangian relaxation algorithm. Bayati et al. [1] proposed NetAlign, which casts the network alignment problem as a quadratic optimization task. To speed up the computation, the authors of NetAlign have also proposed several approximation approaches. Recently, Zhang and Tong [17] proposed FINAL algorithm, which also uses mathematical optimization framework for aligning attributed networks. REGAL [4] is one of the recent works which solves the network alignment task by finding a latent representation of nodes and then matching the nodes through the similarity of nodes in the latent space.

3 Problem Definition

In this paper, scalar values are denoted by lowercase letters (e.g., n). Vectors are represented by boldface lowercase letters (e.g., $\mathbf{x}$). Bold uppercase letters (e.g.,

$\mathbf{X}$) denote matrices, the ith row of a matrix $\mathbf{X}$ is denoted as $\mathbf{x}_i$ and j^{th} element of the vector $\mathbf{x}_i$ is represented as x_i^j. Calligraphic uppercase letter (e.g., $\mathcal{X}$) is used to denote a set and $|\mathcal{X}|$ is used to denote the cardinality of the set $\mathcal{X}$.

An undirected network is represented as $G_i = (\mathcal{V}_i, \mathcal{E}_i)$, where $\mathcal{V}_i$ is the set of vertices and $\mathcal{E}_i$ is the set of edges. For any vertex $u \in \mathcal{V}_i$, all (direct) neighbors of vertex u are represented as $\Gamma_u^{\langle 1 \rangle}$. The higher order (multi-hop) neighbors of u are defined as $\Gamma_u^{\langle i \rangle}$, which consists of all nodes that are reachable from u by a walk less than or equal to i.

Given two undirected networks G_1 and G_2, our objective is to find a unique mapping function $\phi : \mathcal{V}_1 \rightarrow \mathcal{V}_2$ such that this mapping maximizes the local topological alignment of networks G_1 and G_2.

4 *GraphletAlign* Based Network Alignment

In this section, we discuss the proposed framework *GraphletAlign*, which uses graphlet signature vector and order function ψ to efficiently solve the graph alignment problem. Graphlet first calculates a base cost matrix $\mathbf{D}^{\langle 1 \rangle}$ from graphlet signature vector and then the *GraphletAlign* finds a base mapping $\phi^{\langle 1 \rangle}$. Next, we define an order function ψ which takes in a lower order cost matrix $\mathbf{D}^{\langle 1 \rangle}$ and corresponding mapping $\phi^{\langle 1 \rangle}$ to generate higher order cost matrix $\mathbf{D}^{\langle 2 \rangle}$ and higher order mapping $\phi^{\langle 2 \rangle}$. This order function can be used recursively to obtain higher order cost matrices and higher order mappings.

4.1 Obtaining Base Mapping

The primary objective of the network alignment problem is to obtain pair-wise alignment among vertices so that the structural differences between two aligned vertices is minimized. For that, *GraphletAlign* first quantifies the structural features of a vertex in an undirected network. There are many traditional network features to represent a vertex, such as a degree, node centrality, clustering coefficient etc. However, these features are not very informative and inadequate for complex task of network alignment. On the other hand, graphlet frequency is proved to be effective network feature for network classification [12]. Hence, *GraphletAlign* uses graphlet frequency vector (GFV) as a node feature to solve the network alignment problem.

Graphlet Signature

We use local graphlet frequencies of a vertex which gives structural information from the local neighborhood of a node. Since we are interested in local neighborhood information, we limit the graphlet size upto 4. Also, we discriminate the graphlets based on the node orbits to obtain node centric features i.e. same structural graphlets with the target node in different orbits are considered as different graphlets. For example, in Fig. 1(a), graphlets $g6$ and $g7$ are structurally same (star topology) but the target node u is in different orbit i.e. the target node u is a tail node in graphlet $g6$, while u is an internal node in graphlet $g7$.

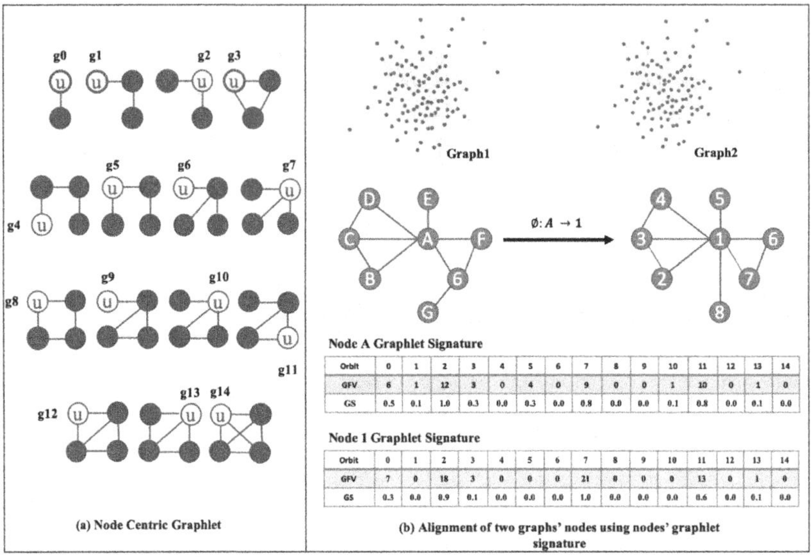

Fig. 1. Example of using node centric graphlets size 4 to align nodes in two graphs

After orbit based separation, we have total 15 local graphlets as shown in Fig. 1. For each node of networks G_1 and G_2, we calculated the number of occurrence of each of these graphlets with respect to the target node using Orca method [5]. After calculating graphlet frequency vector (GFV) for each node, we normalize the GFV such that the values in the vector are in the range $[0, 1]$, this normalized vector is called graphlet signature (GS) of the node. For any node $u \in \mathcal{V}_1$ corresponding graphlet signature (GS) is represented as $\mathbf{f}_u \in \mathbb{R}^{15}$.

Base Cost Matrix

These graphlet signatures are representation of vertices, so the cost between pair of vertices can be calculated as a distance in the vector space. We build a cost matrix $\mathbf{D}^{\langle 1 \rangle} \in \mathbb{R}^{|\mathcal{V}_1| \times |\mathcal{V}_2|}$, where rows of the matrix represent vertices of network G_1 and columns are vertices of network G_2. We calculate values of $\mathbf{D}^{\langle 1 \rangle}$ matrix using Euclidean distance; for example, any two vertices $i \in \mathcal{V}_1$ and $j \in \mathcal{V}_2$ corresponding matrix element $d^{\langle 1 \rangle}{}_i{}^j = \|\mathbf{f}_i - \mathbf{f}_j\|_2$.

Notice that, to find each value of the cost matrix we need to calculate all-node pair distance between two networks. However, for large networks, calculating this full distance matrix can be inefficient. To solve this problem, we calculate distance vector of each node using only a subset of the other network; i.e. given a vertex we limit our search space to only top-p similar (lowest distance) vertices from the other network. This can be done efficiently using k-d tree data structure. For small $p \ll |\mathcal{V}_2|$, this approach reduces the distance calculation cost significantly.

Base Mapping
For isomorphic networks G_1 and G_2, the ideal network alignment is to find the permutation mapping function $\phi : u \rightarrow v$, $u \in \mathcal{V}_1$ and $v \in \mathcal{V}_2$, such that ϕ converts G_1 to G_2. However, in practice no two networks are isomorphic and hence the objective of the network alignment is to reduce the mapping cost. Hence, our objective function is defined as

$$\Theta(G_1, G_2, \mathbf{D}) = \min_{\phi} \sum_{u \in \mathcal{V}_1, v \in \mathcal{V}_2} d_u^v \times \phi(u, v) \tag{1}$$

Where, the mapping function $\phi(u, v)$ returns 1 if node u of G_1 is mapped to node v of G_2, 0 otherwise; $\mathbf{D}$ is input cost matrix such that d_u^v is the cost (distance) between u and v by using graphlet feature.

To find the optimum base mapping $\phi^{\langle 1 \rangle}$, we use base cost matrix $\mathbf{D}^{\langle 1 \rangle}$ as an input. We solve the objective $\Theta(G_1, G_2, \mathbf{D})$ by using the popular Hungarian assignment method [8].

4.2 Higher Order Calculation

It has been shown [16] that higher order (multi-hop) information is useful to obtain a better proximity function between nodes. We use this fact to obtain a refined similarity function from a base alignment. The key observation is if two nodes (say, u, and v) are aligned correctly their corresponding neighbors need to be aligned correctly. So, we can detect a wrong mapping by introducing a cost reflecting the aggregated distance between the neighbor-sets of u and v. This cost refinement entails a higher order proximity function between a pair of nodes, which enhances the performance of network alignment.

To infuse this higher order information in the proposed *GraphletAlign*, we define an order function $\psi : \left(\mathbf{D}^{\langle \mathbf{t} \rangle}, \phi^{\langle t \rangle}\right) \rightarrow \mathbf{D}^{\langle \mathbf{t+1} \rangle}$, which takes a cost matrix $(\mathbf{D}^{\langle \mathbf{t} \rangle})$ and an existing mapping $(\phi^{\langle t \rangle})$ as inputs and provides a new $(t+1)^{th}$ order cost matrix $(\mathbf{D}^{\langle \mathbf{t+1} \rangle})$ as an output. ψ uses direct and indirect neighbors of a vertex along with input distance matrix to calculate new distance between a pair of vertices. The $(t+1)^{th}$ order distance between vertex $i \in \mathcal{V}_1$ and $j \in \mathcal{V}_2$ is a sum of their t^{th} order distance and t^{th} order neighboring distance. Mathematically,

$$d^{\langle t+1 \rangle}{}_i^j = d^{\langle t \rangle}{}_i^j + n^{\langle t \rangle}{}_i^j \tag{2}$$

where, $n^{\langle t \rangle}{}_i^j$ is the tth order neighboring distance, which is calculated from t^{th} order neighbors of vertices i and j. To calculate neighboring distance between $i \in \mathcal{V}_1$ and $j \in \mathcal{V}_2$, we use the input mapping $\phi^{\langle t \rangle}$. For $i \in \mathcal{V}_1$'s neighbor set, $\Gamma_i^{\langle t \rangle}$, $\phi^{\langle t \rangle}$ provides the following mapped neighbor set of $j \in \mathcal{V}_2$:

$$\phi^{\langle t \rangle}_{\Gamma_i^{\langle t \rangle}} = \bigcup_{k \in \Gamma_i^{\langle t \rangle}} \phi^{\langle t \rangle}(k)$$

Given these two sets, a simple approach to calculate neighboring distance is to count the number of mis-matches in the neighbors. Hence, if there are more

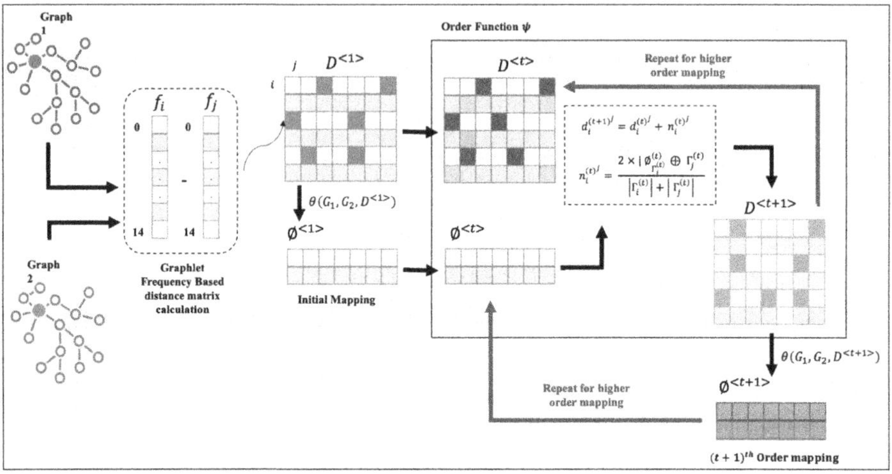

Fig. 2. Graphlet frequency and higher order distance based proposed framework.

mis-matched neighbors the neighboring distance is high otherwise it is low. We use normalized cardinality of exclusive disjoint set of $\phi^{\langle t \rangle}_{\Gamma_i^{\langle t \rangle}}$ and $\Gamma_j^{\langle t \rangle}$, i.e.

$$n^{\langle t \rangle}{}_i^j = \frac{2 * |\phi^{\langle t \rangle}_{\Gamma_i^{\langle t \rangle}} \oplus \Gamma_j^{\langle t \rangle}|}{|\Gamma_i^{\langle t \rangle}| + |\Gamma_j^{\langle t \rangle}|}$$

Here, we have 2 in the numerator because for each mis-match we have a pair of vertices which are wrongly mapped.

Lastly, after calculating $(t+1)^{th}$ order distance matrix $\mathbf{D}^{\langle \mathbf{t+1} \rangle}$, we minimize the objective in Eq. 1 with the new cost matrix, i.e. $\Theta(G_1, G_2, \mathbf{D}^{\langle \mathbf{t+1} \rangle})$, to generate a new mapping $\phi^{\langle t+1 \rangle}$.

4.3 *GraphletAlign* Framework

We illustrate our proposed framework *GraphletAlign* in the Fig. 2, which has two parts. The first is the base model that calculated base cost matrix $\mathbf{D}^{\langle 1 \rangle}$ from graphlet signatures ($\mathbf{f}_i$, $\mathbf{f}_j$) and then find the lowest cost mapping function $\phi^{\langle 1 \rangle}$ optimizing Eq. 1. The second part is used to compute higher order cost matrix $\mathbf{D}^{\langle \mathbf{t+1} \rangle}$ from lower order matrix $\mathbf{D}^{\langle \mathbf{t} \rangle}$ using neighboring information and pass it to the order function to get a higher order mapping $\phi^{\langle t+1 \rangle}$. As shown in the Fig. 2, we can repeatedly use the order function to get higher order mapping.

5 Experiments and Results

We perform several experiments to validate the performance of *GraphletAlign*in noisy graph setting using real-world datasets. We also compare our method with various baseline methods in term of accuracy and running time.

Table 1. Statistics of six real-world datasets

Dataset	# Nodes	# Edges	Description
Moreno	1870	2277	Protein-Protein Interaction
Hamsterster	1858	12,534	Social Network
Power-Grid	4941	6594	Infrastructure Network
AS-Com	6474	13,895	Communication Network
PGP	10,680	24,316	Online Contact Interaction
Astro-Ph	18,771	198,050	Co-authorship Network

Datasets. We use six real-world datasets in our experiment. All datasets are collected from KONECT (The Koblenz Network Collection). In Table 1, we present basic statistics and short description of each network. They are varied in size and dense as well in domain and application. More details of each dataset is given below: *Moreno Protein-Protein (Moreno)* is a protein-protein network preserves only the high degree protein in yeast in which nodes refer to protein and links refer to their interactions. *Hamsterster* is a social network depicts friendships between users of hamsterster.com website. *Power-Grid* is an infrastructure network where nodes refer to either a generator, a transformator or a substation and links refer to power supply lines. *Autonomous Systems (AS-Com)* is a network of autonomous systems with autonomous systems as nodes and their communication as links. *Pretty Good Privacy (PGP)* is an online contact network that captures the interaction between users nodes. *Astro-Ph* is the authors of scientific papers collaboration network which has authors as nodes and collaborations as links.

Experimental Setup. For each real network G_1, a new network G_2 and the ground-truth alignment table are generated by randomly shuffling the edges and nodes of G_1. Then, we remove edges with a probability prb: 0.01, 0.03, and 0.05 to add a structure noise preserving the graph connectivity. Moreover, we calculate a prior alignment information using nodes degree similarity which is required as an input for all baseline methods except ours and REGAL. We extract the graphlets signature for nodes up to size four using Orca tool [5]. The value of the parameter p which is required in constructing the kd-tree is set up to $0.25 \times |\mathcal{V}_2|$.

Baseline Methods. We compare against the following five baseline methods:

1. **IsoRank** [14] computes nodes' similarity score by solving the alignment as eigenvelaue problem with relaxed constraints.
2. **Klau's** [6] is an iterative process to solve the task as quadratic assignment problem using a combination of lagrangian relaxation constraints and subgradient optimization.
3. **GRAAL** [7] incorporates the node graphlet signature in computing the cost of node similarity matrix to match nodes across graphs.

Table 2. Quantitative results of alignment accuracy between our proposed method and other baselines for the graph alignment task using graphlet count on different noisy networks

Moreno						
Noise	IsoRank	Klau's	GRAAL	FINAL	REGAL	OurMethod
1%	0.4088	**0.6945**	0.5093	0.4347	0.5760	$0.6914_{\pm 0.0027}$
3%	0.4922	**0.6940**	0.4613	0.4152	0.5748	$0.6896_{\pm 0.0048}$
5%	0.3711	**0.6895**	0.5055	0.2994	0.5678	$0.6894_{\pm 0.0047}$
Hamsters						
Noise	IsoRank	Klau's	GRAAL	FINAL	REGAL	OurMethod
1%	0.2517	0.7315	0.5763	0.2425	0.7945	$\mathbf{0.8166}_{\pm 0.0024}$
3%	0.2821	0.7429	0.5826	0.3365	0.8045	$\mathbf{0.8190}_{\pm 0.0006}$
5%	0.3290	0.7413	0.5736	0.2174	0.7833	$\mathbf{0.8141}_{\pm 0.0029}$
Power-Grid						
Noise	IsoRank	Klau's	GRAAL	FINAL	REGAL	OurMethod
1%	0.4252	0.7475	0.6098	0.6830	0.4602	$\mathbf{0.7921}_{\pm 0.0040}$
3%	0.2970	0.7496	0.6147	0.5187	0.4566	$\mathbf{0.7896}_{\pm 0.0034}$
5%	0.6554	0.7583	0.6140	0.7209	0.4623	$\mathbf{0.7933}_{\pm 0.0015}$
AS-Com						
Noise	IsoRank	Klau's	GRAAL	FINAL	REGAL	OurMethod
1%	0.3851	0.5347	0.4035	0.3645	0.5110	$\mathbf{0.5374}_{\pm 0.0010}$
3%	0.2052	0.5139	0.4405	0.1782	0.5051	$\mathbf{0.5361}_{\pm 0.0014}$
5%	0.3407	0.5344	0.4227	0.2873	0.5095	$\mathbf{0.5370}_{\pm 0.0005}$
PGP						
Noise	IsoRank	Klau's	GRAAL	FINAL	REGAL	OurMethod
1%	0.5834	0.6070	–	0.6795	0.5867	$\mathbf{0.7272}_{\pm 0.0005}$
3%	0.4500	0.6506	–	0.5609	0.5849	$\mathbf{0.7260}_{\pm 0.0016}$
5%	0.4671	0.7186	–	0.6210	0.5869	$\mathbf{0.7270}_{\pm 0.0004}$
Astro-Ph						
Noise	IsoRank	Klau's	GRAAL	FINAL	REGAL	OurMethod
1%	0.2160	0.8009	–	0.4333	0.7997	$\mathbf{0.8219}_{\pm 0.0013}$
3%	0.1425	0.7967	–	0.3742	0.8029	$\mathbf{0.8224}_{\pm 0.0009}$
5%	0.6191	0.8092	–	0.3961	0.8111	$\mathbf{0.8230}_{\pm 0.0010}$

4. **FINAL** [17] solves the alignment method as optimization method by leveraging the attributes information into the topology structure.
5. **REGAL** [4] is proposed as node representation learning-based graph alignment framework to match nodes in two graphs.

Metric. We measure the performance of our method using network alignment accuracy. Network alignment accuracy is calculated as the number of correct alignments over the total number of possible alignments. We consider a hard assignment to calculate the alignment accuracy in which each node in the first network is assigned to one and only one corresponding node in the second network. Both the mean accuracy score and the standard deviation are reported for five independent trails for each noisy network. Another criteria we consider for our method evaluation is the average runtime. Code and implementation details can be found on https://github.com/aalmulhi/GraphletAlign.

5.1 Results and Discussion

The primary results of comparison experiments are shown in Table 2, where each column represents different competing methods and rows are grouped into seven parts for each real-world dataset. Each group has alignment accuracy values for 1%, 3% and 5% noise in corresponding cells.

From Table 2, we can see that the proposed *GraphletAlign* outperforms all five competing methods in all datasets, except for Moreno dataset Klau's method performs very similar i.e. it performs $\leq 0.6\%$ better than *GraphletAlign*. Among competing methods IsoRank performs really poor, while state of the art Klau's method and REGAL performs better than other three methods. However, proposed *GraphletAlign* consistently performs better than both Klau's method and REGAL. For example, our proposed *GraphletAlign* improves accuracy of Kalu's method up to 20.00%, while *GraphletAlign* improves accuracy of REGAL up to 72.00%. We believe, the main reason for superior performance of the proposed *GraphletAlign* is its ability to exploit higher order proximity information.

Furthermore, FINAL shows bad performance due to its sensitivity toward noise because the attribute choice of node's degree. GRAAL algorithm integrates the node degree and the graphlet signature in computing their cost matrix. Thus, noise affects the performance badly. We have not reported the GRAAL results for PGP and Astro-Ph datasets because they take more than three days for running therefore we kill the processes.

In addition to the accuracy, we report the average run-time in Table 3. REGAL is the fastest method among competitors since it is an embedding-based method. However, our method is better than REGAL in accuracy. The second best is *GraphletAlign* and IsoRank. IsoRank is better than us in three datasets: Hamesters, As-Com, and Astro-Ph since their approach has fast convergence rate. Despite that, their accuracy results are poor. GRAAL is the worst among competing methods in term of runtime. For example, GRAAL is 1000 times slower than *GraphletAlign* on As-Com dataset.

5.2 Importance of the Order Cell

We conducted another important experiment to verify our assumption regardig the importance of the higher order proximity information. In this experiment, we show two accuracy results with and without considering the order cell proximity. The results of this experiment are depicted in the Table 4, where rows represent different datasets and three group of columns are for different noise level of each dataset. There are two columns in each column group, in which the first column (Acc.1) is showing assignment accuracy of the *GraphletAlign* without higher

Table 3. Average of Running Time in seconds on graphs with different noise settings

Moreno						
Noise	IsoRank	Klau's	GRAAL	FINAL	REGAL	OurMethod
1%	10.27	39.22	116.32	68.35	**3.65**	**9.28**
3%	14.17	38.51	118.97	55.93	**3.75**	**9.27**
5%	13.57	40.62	117.20	58.57	**3.79**	**9.37**
Hamsters						
Noise	IsoRank	Klau's	GRAAL	FINAL	REGAL	OurMethod
1%	**12.15**	39.22	1324.17	152.11	**7.52**	31.29
3%	**15.20**	38.51	1338.59	138.82	**7.42**	31.35
5%	**13.24**	40.62	1351.07	136.80	**7.43**	31.32
Power-Grid						
Noise	IsoRank	Klau's	GRAAL	FINAL	REGAL	OurMethod
1%	177.10	1019.41	2031.27	1101.34	**14.34**	**111.74**
3%	124.03	1032.08	2159.75	1160.94	**14.82**	**110.38**
5%	112.56	1006.60	2003.62	1096.79	**14.34**	**108.88**
AS-Com						
Noise	IsoRank	Klau's	GRAAL	FINAL	REGAL	OurMethod
1%	**141.60**	1839.78	234272.57	1992.69	**36.01**	224.02
3%	**150.50**	2448.72	203669.85	2130.37	**34.59**	235.37
5%	**154.80**	1956.47	199289.25	1893.58	**35.68**	233.63
PGP						
Noise	IsoRank	Klau's	GRAAL	FINAL	REGAL	OurMethod
1%	2002.68	1935.50	–	5889.18	**39.03**	**609.93**
3%	1203.40	1692.83	–	5625.91	**39.16**	**607.59**
5%	1201.66	1694.65	–	5469.67	**38.27**	**615.17**
Astro-Ph						
Noise	IsoRank	Klau's	GRAAL	FINAL	REGAL	OurMethod
1%	**3319.16**	4045.31	–	29231.52	**126.49**	4139.57
3%	**3176.66**	4277.25	–	30145.99	**137.09**	4135.04
5%	**3445.26**	4089.80	–	29158.77	**162.89**	4041.43

order information (order cell), while values in the second column (Acc.2) represents accuracy of the proposed *GraphletAlign* (with order cell). From the table, we can observe that higher order proximity improves the assignment accuracy for four datasets with different noise level. For instance, in Power-Grid dataset the higher order proximity information improves the assignment accuracy up to 38.41%.

Table 4. Assignment accuracy on datasets before and after considering the high order approximation.

	Noise 1%		Noise 3%		Noise 5%	
	Acc1.	Acc2.	Acc1.	Acc2.	Acc1.	Acc2.
Moreno	0.6187	**0.6914**	0.6173	**0.6896**	0.6180	**0.6894**
Hamsters	0.8143	**0.8166**	0.8126	**0.8190**	0.8140	**0.8141**
Power-Grid	0.5723	**0.7921**	0.5703	**0.7896**	0.5734	**0.7933**
AS-Com	0.5331	**0.5374**	0.5312	**0.5361**	0.5329	**0.5370**
PGP	0.6404	**0.7272**	0.6360	**0.7260**	0.6402	**0.7270**
Astro-Ph	0.8099	**0.8219**	0.8189	**0.8224**	0.8190	**0.8230**

6 Conclusion

We show that *GraphletAlign* use the graphlet signature as a novel-feature for network alignment problem proves that the structure of the network alone with no other support node/edge features gives a high accuracy with the second order information. The assumption of relying on the network topology for network matching task is valid and helps especially in the case of networks with no attributes assign to them.

References

1. Bayati, M., Gleich, D.F., Saberi, A., Wang, Y.: Message-passing algorithms for sparse network alignment. Tran. Knowl. Discov. Data **7**(1), 1–31 (2013)
2. Conte, D., Foggia, P., Sansone, C., Vento, M.: Thirty years of graph matching in pattern recognition. Int. J. Pattern Recognit Artif Intell. **18**(03), 265–298 (2004)
3. Garey, M.R., Johnson, D.S.: Computers and Intractability: A Guide to the Theory of NP-Completeness. W. H. Freeman & Co., New York (1979)
4. Heimann, M., Shen, H., Safavi, T., Koutra, D.: Regal: representation learning-based graph alignment. In: ACM CIKM, pp. 117–126 (2018)
5. Hočevar, T., Demšar, J.: A combinatorial approach to graphlet counting. Bioinformatics **30**(4), 559–565 (2014)
6. Klau, G.W.: A new graph-based method for pairwise global network alignment. BMC Bioinform. **10**(1), S59 (2009)
7. Kuchaiev, O., Milenković, T., Memišević, V., Hayes, W., Pržulj, N.: Topological network alignment uncovers biological function and phylogeny. J. R. Soc. Interface **7**(50), 1341–1354 (2010)
8. Kuhn, H.W.: The hungarian method for the assignment problem. Naval Res. Logistics Q. **2**(1–2), 83–97 (1955)
9. Lacoste-Julien, S., Taskar, B., Klein, D., Jordan, M.I.: Word alignment via quadratic assignment. In: Conference of the North American Chapter of the Association of Computational Linguistics, pp. 112–119 (2006)
10. Liu, S., Wang, S., Zhu, F., Zhang, J., Krishnan, R.: Hydra: large-scale social identity linkage via heterogeneous behavior modeling. In: ACM SIGMOD International Conference on Management of Data, pp. 51–62 (2014)

11. Melnik, S., Garcia-Molina, H., Rahm, E.: Similarity flooding: a versatile graph matching algorithm and its application to schema matching. In: Proceedings 18th International Conference on Data Engineering, pp. 117–128. IEEE (2002)
12. Rahman, M., Bhuiyan, M.A., Hasan, M.A.: Graft: an efficient graphlet counting method for large graph analysis. IEEE TKDE **26**(10), 2466–2478 (2014)
13. Schellewald, C., Schnörr, C.: Probabilistic subgraph matching based on convex relaxation. In: Rangarajan, A., Vemuri, B., Yuille, A.L. (eds.) EMMCVPR 2005. LNCS, vol. 3757, pp. 171–186. Springer, Heidelberg (2005). https://doi.org/10.1007/11585978_12
14. Singh, R., Xu, J., Berger, B.: Pairwise global alignment of protein interaction networks by matching neighborhood topology. In: Speed, T., Huang, H. (eds.) RECOMB 2007. LNCS, vol. 4453, pp. 16–31. Springer, Heidelberg (2007). https://doi.org/10.1007/978-3-540-71681-5_2
15. Srinivasan, B.S., Novak, A.F., Flannick, J.A., Batzoglou, S., McAdams, H.H.: Integrated protein interaction networks for 11 microbes. In: Apostolico, A., Guerra, C., Istrail, S., Pevzner, P.A., Waterman, M. (eds.) RECOMB 2006. LNCS, vol. 3909, pp. 1–14. Springer, Heidelberg (2006). https://doi.org/10.1007/11732990_1
16. Tang, J., Qu, M., Wang, M., Zhang, M., Yan, J., Mei, Q.: Line: large-scale information network embedding. In: WWW 2015, pp. 1067–1077 (2015)
17. Zhang, S., Tong, H.: Final: fast attributed network alignment. In: ACM SIGKDD, pp. 1345–1354 (2016)

Effect of Market Spread Over Reinforcement Learning Based Market Maker

Abbas Haider[1](✉), Hui Wang[1], Bryan Scotney[2], and Glenn Hawe[1]

[1] Ulster University, Shore Road, Newtownabbey, UK
{haider-a, h.wang, gi.hawe}@ulster.ac.uk
[2] Ulster University, Cromore Road, Londonderry, UK
bw.scotney@ulster.ac.uk

Abstract. Market Making (also known as liquidity providing service) is a well-known trading problem studied in multiple disciplines including Finance, Economics and Artificial Intelligence. This paper examines the impact of Market Spread over the market maker's (or liquidity provider's) convergence ability through testing the hypothesis that "Knowledge of market spread while learning leads to faster convergence to an optimal and less volatile market making policy". Reinforcement Learning was used to mimic the behaviour of a liquidity provider with Limit Order Book using historical Trade and Quote data of five equities, as the trading environment. An empirical study of results obtained from experiments (comparing our reward function with benchmark) shows significant improvement in the magnitude of returns obtained by a market maker with knowledge of market spread compared to a market maker without such knowledge, which proves our stated hypothesis.

Keywords: Market making · Market spread · Reinforcement learning · Reward function

1 Introduction

Market Spread can be defined as the mean difference between the prices quoted by buyers and sellers for an asset traded in one or more financial markets (Eq. 1). In other words, it is the quantitative difference between the amount paid while buying an asset (stocks, bonds etc.) and the amount received while selling. Limit Order Book (LOB), that has two sides (buy and sell), was used to imitate the actual financial markets where traders (buyers and sellers) quote their limit orders. LOB contained price and size of limit orders at multiple levels on both sides. Orders are arranged in increasing sequence on buy side and in decreasing sequence on sell side. Order which lies at the top of ask book is known as best-ask whereas the topmost bid book order is known as best-bid, and the difference between best-ask and best-bid prices is called as bid-ask spread. Level 5 LOB contains 5 ask orders on the ask side and 5 bid orders on the bid side and averaging the prices of these asks and bids together gives $Mean_{sellprice}$ on the ask side or ask book and $Mean_{buyprice}$ on the bid side or bid book. This work examines the influence of market spread on market making, a fundamental trading problem, using the reinforcement learning (RL) paradigm by modifying the existing benchmark reward function.

G. Nicosia et al. (Eds.): LOD 2019, LNCS 11943, pp. 143–153, 2019.
https://doi.org/10.1007/978-3-030-37599-7_13

$$Market_{Spread} = Mean_{sellprice} - Mean_{buyprice} \tag{1}$$

Our hypothesis here is "Knowledge of market spread while learning leads to faster convergence to an optimal and less volatile market making policy" which will be tested by developing a market maker learning agent. An agent interacts with an LOB environment consisting of multiple technical indicators *(such as inventory position, ask level, bid level, book imbalance, market sentiment etc.)* and a reward function that decides the behaviour to be learned. This paper compares two different versions of the reward function to show how our version leads to the faster learning of RL agent in the light of market spread with less volatility against that of a benchmark approach. Like the benchmark method our approach also prevents speculative trading but also makes the agent perform well in less fluid environment (markets having higher value of market spread) unlikely to the benchmark method and successfully learns the behaviour of liquidity provision to illiquid markets as well. Section 2 describes the types of financial markets that currently exist including the one used in this work namely continuous double action market along with market making and related work, Sect. 3 focuses on problem modelling in a RL context and Sect. 4 presents comparative results and discussion to prove the stated hypothesis. Finally, Sect. 5 summarizes our conclusions.

2 Background and Related Work

2.1 Financial Markets

There are two distinct types of financial markets which exist worldwide, and some operate as a combination of known types. **Dealer** markets are trading venues where investors or traders cannot trade directly with each other but only through the dealers who are responsible for placing buy and sell orders. Another one is **Auction** markets where investors can trade directly and can see all the quotes placed by other traders without the intervention of dealers in **pure auction markets** and with dealers' presence in **double auction markets**. In this work, the continuous double auction market was used which allows traders or investors to place their orders in either direction along with the presence of market makers to provide the liquidity to less liquid or completely illiquid markets. In particular, New York Stock Exchange (NYSE), Chicago Board Options Exchange (CBOE) and European Options Exchange (EOE) operate as continuous double auction markets and can be considered as an example of hybrid markets—a mixture of two described above. A LOB, simulation was used to a mimic real market environment, and contained limit orders from investors and market maker to make the learning agent capable of trading in competing financial markets. [1] states that there is a high inter-market competition of quote placement which decreases market liquidity. Market spread has an inverse relationship with market liquidity therefore as market spread increases liquidity decreases which results in increased trading costs. Moreover, liquidity variation influences traders through trade execution cost (price paid by a trader when his order gets processed) and desired stock position. This concretizes the fact that market spread plays a vital role in profits of interested traders hence market makers.

2.2 Market Making

Market making (or liquidity providing) is a facilitation service provided by liquidity providers to ensure smooth flow of order arrivals and executions in financial markets. In other words, market making helps to keep the market sufficiently liquid to preserve the trading interest of its participants. In compensation they charge their quoted spread (difference between their ask and bid quotes) as profit. Within the finance literature [2] studied inventory effect by deriving optimal asset price which they called "True Price" which was later used in empirical study of AMEX by [3]. Then, [4] approached this problem as optimal bid and ask quotes placement in LOB using probabilistic knowledge where order arrivals and executions follow their model. [5] studied the impact of market making on price formation and modeled exploration vs exploitation trade off as price vs profit. However, this work is concentrated on price prediction and stability rather than on enhancing quality of market measured by market spread. Recently, [6] argued that high frequency market makers provide more liquidity as volatility increases with the increment of risk of arbitrageurs picking their quotes. Within the AI domain, [7] used RL to develop the first practical model of market making. They investigated the influence of uninformed market participants on the nature of placing quotes by market makers and argued that RL policies converge successfully along with maintaining balance between profit and spread. Next, [8] used online learning approach to design a market maker and empirically assess it presuming market has enough liquidity. Now recently, [9] developed a market maker agent that studied the problem on realistic grounds. We use this work as a benchmark to study the impact of market spread and compare their reward function with the one proposed here proving the hypothesis stated in Sect. 1.

3 Problem Modelling

[10] states that RL is a computational method of learning through trial and error with the goal of maximizing reward received while an agent interacts with complex and uncertain environment. In other words, it is a type of machine learning method, which aims to mimic the behavior of a real-world entity (e.g. humans or animals) through a direct interaction with a surrounding environment which generates a feedback signal that serves as the backbone of the agent's learning.

3.1 Environment

Our RL based market maker agent requires a realistic surrounding to interact with and to receives feedback signal from to learn intended behavior of providing liquidity to illiquid markets. [11] states that LOBs are being used, in more than half of financial markets around the world, as a means of executing buy and sell orders. They also argued that LOB simulation provides inner view of actual market functioning. Therefore, we used LOB simulation as a trading environment for our learning agent, It contains a bid and an ask book which further contain buy and sell limit orders respectively. Limit orders have been arranged in ascending and descending order, in ask and bid books respectively, based on limit price and timestamp. Environment states are only partially observable as the agent doesn't have any information about other traders hence it follows partially observable markov decision process with the environment being model free i.e. no knowledge of transition probabilities (*or environment dynamics*) between states. In this problem setting, a LOB snapshot at time t serves as a state which changes at time $t + \delta$ and so on might lead to infinite number of environment states. Hence, traditional learning algorithms such as maintaining a state-action mapping in a table won't work therefore function approximation technique (*such as TileCodings*) can solve the problem of state dimensionality reduction.

3.1.1 States

States can be defined as the combination of multiple technical indicators known as state variables. State variables include inventory position, ask level, bid level, book imbalance, strength volatility index, and market sentiment as mentioned in Sect. 1. Each variable gets calculated using a defined statistical formula e.g. book imbalance uses formula denoted by Eq. 2. In this way, the state vector in Eq. 3 represents state's value which are then mapped to feature vectors denoting actions. As mentioned in Sect. 3.1, states are partially observable because only 7 state variables (containing real values) are used to represent the entire state. State vector has infinite possible combinations of values of state variables hence the problem has a continuous state space.

$$imb = total_{ask_{volume}} - total_{bid_{volume}} \tag{2}$$

$$State = \sum\nolimits_{i=1}^{n} \lambda_i . Var_i \tag{3}$$

As discussed in Sect. 1, a LOB simulation serves as simulated financial market and provides crucial information to a decision maker (agent) in terms of technical and market indicators as state variables denoted by *Var*. Equation 3 denotes the value of state in terms of weighted financial indicators (λ_i are the weights) and this gives flexibility of preferring some indicators over others depending on the market context. This way of representation makes environment states multi-dimensional therefore making overall information more expressive for an agent to learn a policy.

3.2 Learning

Optimal policy denoted by π^* can be identified by determining optimal state-value function (Eq. 4) or optimal action-value function (Eq. 5). State value function *V(s)* quantifies the total future expected return from *s* till the terminal state is encountered in a finite MDP. Similarly, the Action-Value function *Q (s, a)* provides the estimation of total future discounted reward when an action *a* is taken from *s*. π^* denotes the optimal state-action mapping ensuring that agent prefers the best action in every state of the environment.

$$\pi^* = argmax(V^*(s)) \tag{4}$$

Table 1. Action space

Count	Action
1	Quote (1, 1)
2	Quote (2, 2)
3	Quote (3, 3)
4	Quote (1, 2)
5	Quote (0, 1)
6	Quote (1, 0)

$$\pi^* = argmax(Q^*(s, a)) \tag{5}$$

As mentioned in Sect. 3.1.1 this RL problem has continuous state-space, but discrete action-space is being taken to maintain the simplicity of the solution. The agent is supposed to choose actions from a predefined action set (see Table 1) and every action has an associated feature vector [9]. Each state requires the calculation of a best action using epsilon-greedy policy (see [10] for more detail) and agent places bid and ask quotes in lob for execution.

$$\begin{aligned} \psi_{update} &= \alpha\left(r + \gamma Q\left(s', a'\right) - Q(s, a)\right) \\ Q(s, a) &= Q(s, a) + \psi_{update} \end{aligned} \tag{6}$$

A discrete set of actions has been quantified (Eq. 6) Ψ_{update} serves as one step Temporal Difference (TD) or TD error to be minimized, to determine the values of selected features representing every action in every state with α being the learning rate of the algorithm and γ is the discount factor. An action-value function *Q (s, a)* used a function approximation method to learn features of discrete actions as the Q table approach was not feasible due to infinite amount of environment states. Tile Coding function approximation method suitable for multi-dimensional environment state problem was used to quantize states and map them to the set of features representing actions as in [9].

3.3 Reward Function

Reward Function is an indispensable component of RL framework that generates feedback from environment against each action taken. The goal is to maximize the total future reward obtained after each interaction of an agent with its environment. An action that produces positive reward value is considered as a good action in a certain state whereas negative reward informs the agent to never pick that action in that certain state. A scalar value generated by reward function is being used by Eq. 6 to obtain temporal difference update denoted by Ψ_{update} then Feature vector of each action gets amended by $\Psi_{update.}$ Designing a reward function is not an easy task to accomplish and may cause fatal behavior learned by the agent. Especially in this problem setting an agent ought to place quotes and expect to earn profits hence the reward function should contain relevant information pertinent to financial markets. We may conclude that if we provide more and more information then it will lead to higher profits, but it may cause agent to lose money. Providing more and more information in reward function does not always improve performance and it will be proved by graphical results in later sections.

$$R = \phi_{profit/loss} - \lambda * max((Inv * TP), 0) \tag{7}$$

Reward function developed by [9] (Eq. 7) contains two main components which are obvious for this problem domain i.e. profit/loss obtained by agent and inventory generated in each trade. These components are left unchanged as they are necessary to get the desired behavior i.e. gain profit. However, this behavior can be reinforced more to let the agent make more profitable trades by providing more relevant market detail to the reward function. It will be shown that adding any other non-relevant information can cause performance degradation but only adding market spread leads to improvement. This paper focuses on incorporating market spread to test our hypothesis "Knowledge of market spread while learning leads to faster convergence to an optimal and less volatile market making policy", where m_{sp} stands for *market spread.*

$$R = \phi_{profit/loss} - \lambda * \left[max((Inv * TP), 0) - m_{sp}\right] \tag{8}$$

Our stated hypothesis can be proved by establishing contradictory hypothesis i.e. "Knowledge of market spread, and market volatility together leads to loss incurring market making policy". Contradictory hypothesis is described by Eq. 9 where *vol* represents market volatility. Results obtained using Eq. 9 as a reward function confirms that market spread increases profit levels whereas adding other information which is relevant to the problem but not suitable to be included in reward function leads to performance degradation.

$$R = \phi_{profit/loss} - \lambda * \left[max((Inv * TP), 0) - m_{sp} - vol\right] \tag{9}$$

Rationale behind using m_{sp} term in reward function ensures that the agent must learn to select an optimal action to accomplish the goal of providing liquidity to less or completely illiquid markets. Reward value generated using market spread *(m_{sp})* component of reward function adds more importance to those actions which transform less liquid or completely illiquid market into an interesting trading environment. Reward function denoted by Eq. 8 idealizes the true and optimal behavior of providing liquidity to every financial market along with the capability of refraining itself from learning speculative trading nature. Table 2 shows the pseudocode of the actual program used for conducting experiments and collecting results.

Table 2. Pseudocode

Step	Description
1	Load data from file
2	Read a row
3	Populate Limit Order Book
4	Calculate values of all state variables
5	Initialize feature vectors of all actions
6	Calculate values of each action using *epsilon-greedy* policy
7	Execute action using trades data
8	Collect reward value generated by reward function using Eqs. 7, 8 and 9.
9	Collect profit/loss generated
10	Update features of all actions using Eq 6.
11	Repeat steps 2–10 until *episode_count* < *total_episodes.*

4 Experimental Results

4.1 Training Data

Chicago Board of Options Exchange (CBOE) US equities data service was used to obtain real time bid/ask quotes of depth five and trades data. This data service provides reference quotes and trades prices through four different exchanges namely BZX, BYX, EDGX, EDGA. Dataset collected belong to BZX equities such as Vodafone (VOD), GlaxoSmithKline (GSK), American Airlines (AAL), Nvidia Inc. (NVDA) and Apple Inc. (APPL) and was used to simulate LOB environment where market maker places bid and ask orders, up to depth level five, to boost market liquidity. Date Range was from 12-Jan-2017 to 12-Mar-2017 with approximately 38 million milliseconds historical quotes and trades records per equity. Quotes data included top five bids (buys) and asks (sells) per record whereas trades data provided price and volume of shares per trade. Each equity had separate excel file for its historical records of quotes placed and trades executed. As quoted limit orders get executed against market orders profit/loss received, market spread (difference between best bid and best ask) and inventory position all together form the reward signal value for the agent from learning environment.

4.2 Results and Discussion

This section deals with comparative study of results obtained during training phase of RL based market maker on five different BZX equities datasets as mentioned in Sect. 4.1 with the settings summarized in Table 3. RL algorithm *Sarsa* was used to estimate action-value function denoted by $Q\ (s,\ a)$ for all states and actions using temporal difference method. *Sarsa* algorithm guarantees convergences to an optimal action-value function and policy if all pairs $(s,\ a)$ are visited infinite times and when it converges in the limit of greedy policy [10]. Hence, epsilon greedy was used to select actions that gives maximum total discounted return with probability *1-epsilon* and choose action randomly with probability *epsilon/number of actions* in consensus with exploitation vs exploration tradeoff. The agent was trained on five equities dataset which are VOD, GSK, AAL, NVDA, APPL for 2000 episodes (or iterations) to show consistency among results obtained using Eq. 7 against the consistent results received using Eq. 8 whereas backtesting was done on sliced datasets (80% training, 20% testing) for 20 episodes just to evaluate the learned policy. Discount factor decides the nature of decision maker (learner) whether it be short sighted or long sighted, hence a non-myopic market maker was used in this implementation with its value closer to the highest value.

Table 3. Configuration settings

Configuration	Value
RL algorithm	Sarsa
Policy used	Epsilon greedy
Number of training and backtest episodes	2000 and 20
Discount Factor	0.9

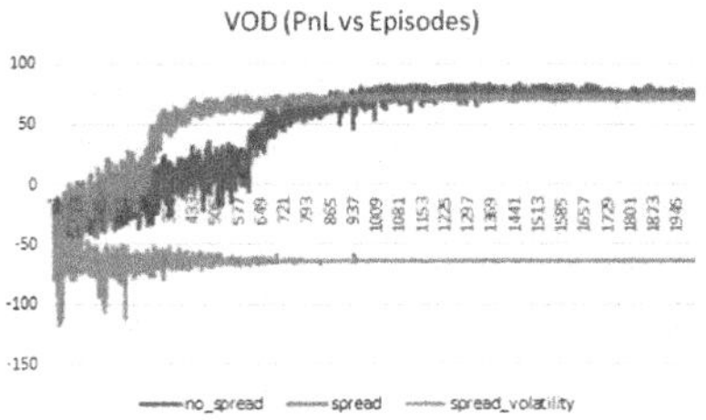

Fig. 1. Vodafone training curve (Color figure online)

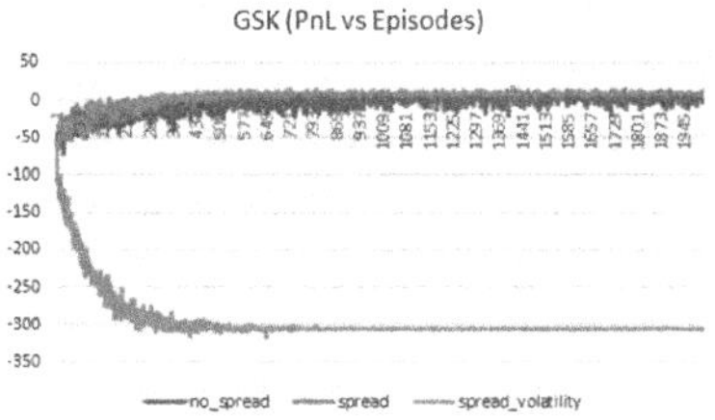

Fig. 2a. GlaxoSmithKline training curve (Color figure online)

Figure 1, compares the performance of a RL agent using benchmark (blue), our reward function (orange) and contradictory hypothesis reward function (gray). It is evident from the curves that spread based reward function leads to the faster convergence of decision maker towards an optimal policy, moreover smooth- ness and sleekness of orange curve depicts less volatility and more stability in learning.

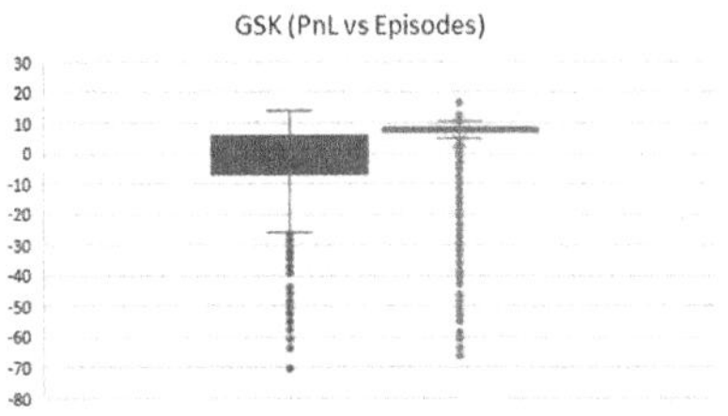

Fig. 2b. GlaxoSmithKline box-whisker plot

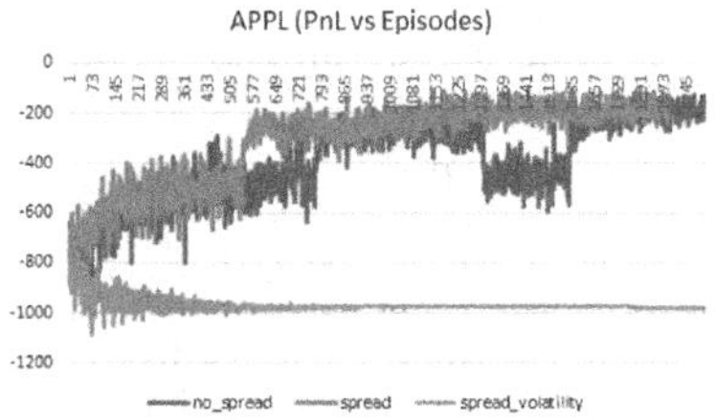

Fig. 3. Apple training curve

However, returns obtained using Eq. 9 (gray curve) are not comparable with either benchmark or with our reward function (Eq. 8) proving contradictory hypothesis.

Similarly, in Fig. 2 orange curve shows similar traits depicted in Fig. 1. The orange curve lies over blue curve therefore slope value during range of episode 217–361 is higher confirming faster convergence and higher sleekness observed in orange curve. Moreover, it reaches higher return value than blue curve for GSK dataset as shown in Fig. 2a hence supporting our hypothesis. Gray curve supports contrary hypothesis as well strengthening our stated hypothesis. Apple dataset has shown consistency with Vodafone and GlaxoSmithKline, orange curve converges faster than blue during episode range 577–649. Moreover, blue curve degrades during 1297–1585 range as compared to orange counterpart. However, again gray curve's performance is consistently poor supporting contrary hypothesis which emboldens our main hypothesis.

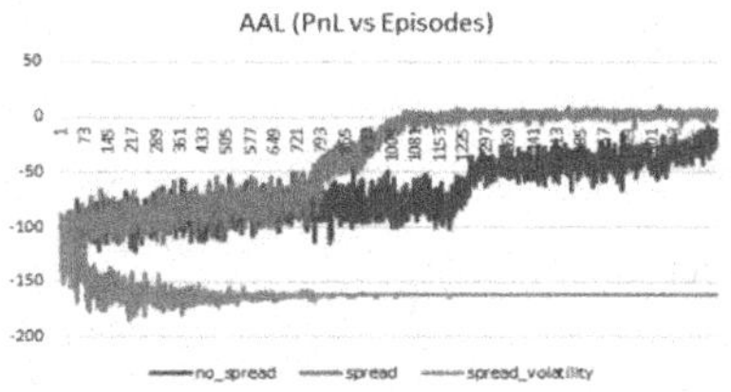

Fig. 4. American Airlines training curve (Color figure online)

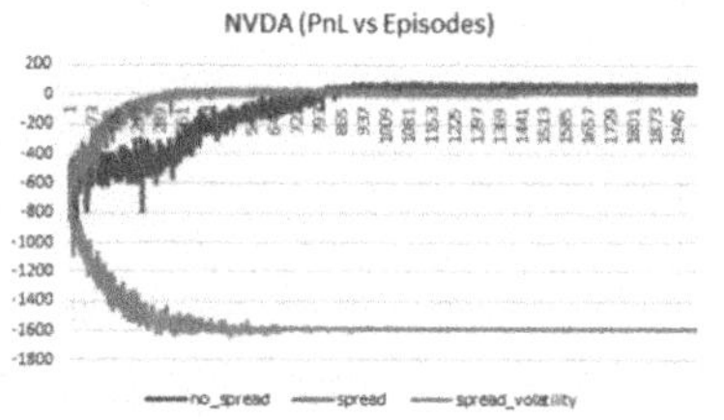

Fig. 5. Nvidia training curve (Color figure online)

In Fig. 4, American Airlines dataset symbolizes consistently improved returns depicted by orange curve as compared to blue curve. Higher slope value of orange curve confirms faster convergence and visible blank region between blue and orange graphs during episode range 793–2000 proves our hypothesis again. Whereas, gray curve's performance evidently proves contradiction in consensus with Figs. 1, 2a and 3. Nvidia dataset see Fig. 5 supports our main hypothetical statement in a crystal-clear fashion. However, maximum return value for orange curve is slightly lower than blue counterpart but a visible blank region between them during episode range 73–577 proves it. Moreover, contrary hypothesis proved by consistent poor performance of gray curve finally confirms the validity of target hypothesis. Training graphs shown above

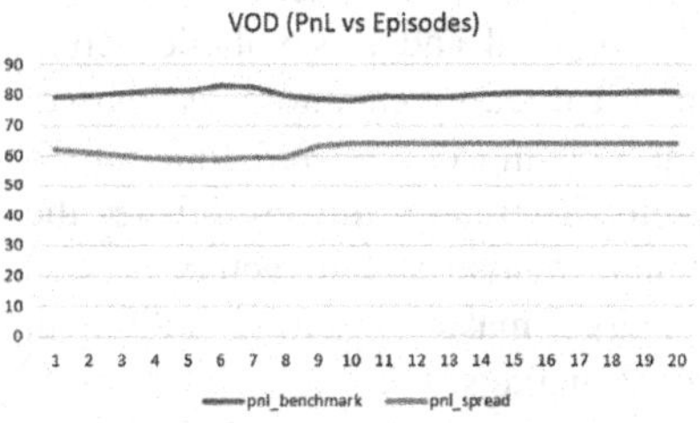

Fig. 6. Vodafone Test curve

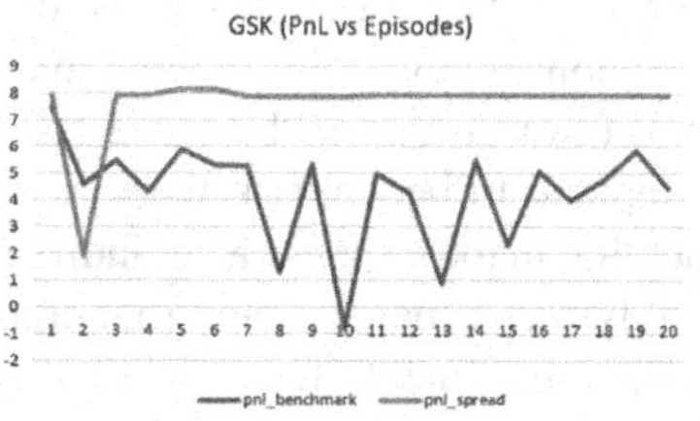

Fig. 7. GlaxoSmithKline Test curve

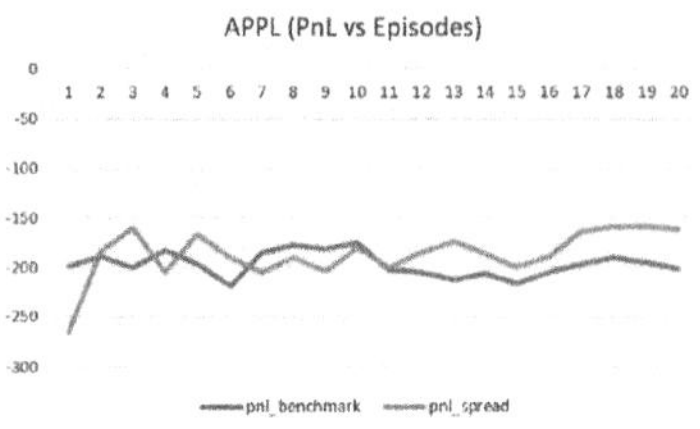

Fig. 8. Apple Test curve

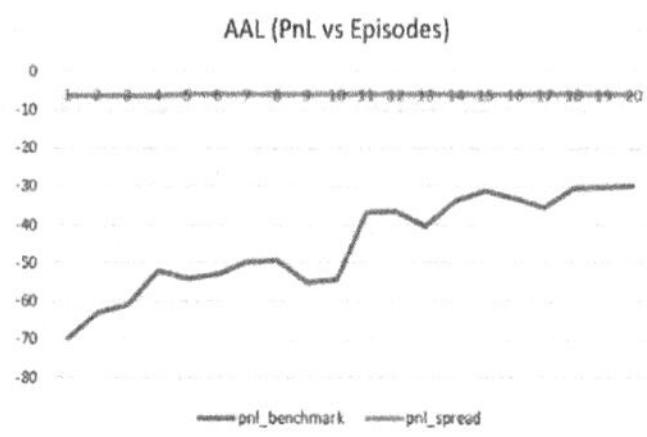

Fig. 9. American Airlines Test curve

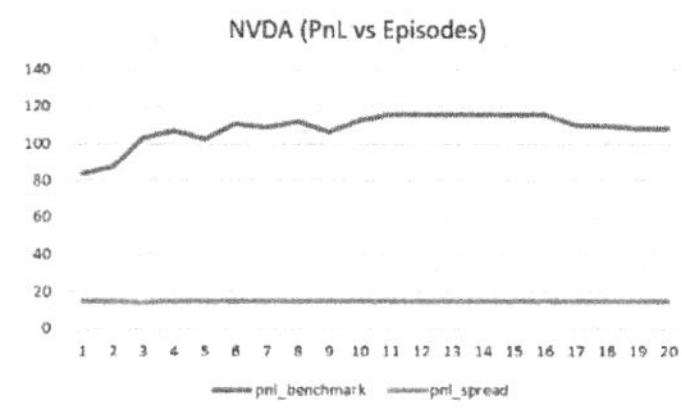

Fig. 10. Nvidia Test curve

completely support our main hypothesis and contradictory hypothesis also justifies that only market spread component leads to higher return values whereas including one more component "market volatility" caused huge loss.

Consistency in return values of our reward function (orange curve) as compared to benchmark (blue curve), across all datasets except apple where volatility level is almost similar among both curves, supports the main hypothesis (Figs. 6, 7, 8, 9 and 10). Moreover, higher returns in 3 out 5 datasets also showed that market making policy learned using Eq. 8 is more profitable and less volatile as stated.

5 Conclusion

Market spread has a profound impact over market maker's action policy while executing its intended goal of providing immediacy especially to illiquid markets and therefore prevents losing investor's interest in the equities listed there. The results obtained have proven our stated hypothesis which states that "Knowledge of market spread to the market maker helps in identifying an optimal and less volatile action selection strategy". Moreover, an action taking policy learned by an automated market maker was found to be more stable and consistent in using our reward function as compared to benchmark work as shown in test results. Therefore, this establishes the fact that market spread is an indispensable component of market making activity and is highly recommended being incorporated by every market maker system or individual traders responsible for providing liquidity to financial markets.

References

1. Upson, J., Van Ness, R.A.: Multiple markets, algorithmic trading, and market liquidity. J. Financ. Markets **32**, 49–68 (2017)
2. Ho, T., Stoll, H.: Optimal dealer pricing under transactions and return uncertainty. J. Financ. Econ. **9**, 47–73 (1981)
3. Ho, T., Macris, R.: Dealer bid-ask quotes and transaction prices: an empirical study of some AMEX options. J. Finance **39**(1), 23–45 (1984)
4. Avellaneda, M., Stoikov, S.: High-frequency trading in a limit order book. Quant. Financ. **8**(3), 217–224 (2008)
5. Das, S.: The effects of market-making on price dynamics. In: International Foundation for Autonomous Agents and Multiagent Systems, Richland, vol. 2, pp. 887–894 (2008)
6. Ait-Sahalia, Y., Sağlam, M.: High Frequency Market Making: Optimal Quoting (2017). https://ssrn.com/abstract=2331613 or http://dx.doi.org/10.2139/ssrn.2331613
7. Chan, N.T., Shelton, C.: An Electronic Market-Maker (2001). http://hdl.handle.net/1721.1/7220
8. Abernethy, J., Kale, S.: Adaptive market making via online learning. In: Neural Information Processing Systems (NIPS), 5–10 December 2013. Curran Associates Inc., Lake Tahoe, Nevada, USA, vol. 2, pp 2058–2066 (2013)
9. Spooner, T., Fearnley, J., Savani, R., Koukorinis, A.: Market making via reinforcement learning. In: International Conference on Autonomous Agents and Multiagent Systems (AAMAS), 10–15 July 2018, Stockholm, Sweden, SIGAI, IFAAMAS, International Foundation for Autonomous Agents and Multiagent Systems, Richland, vol. 17, pp. 434–442 (2018)
10. Sutton, R.S., Barto, A.G.: Reinforcement Learning: An Introduction, 2nd edn. MIT Press, Cambridge (2018)
11. Gould, M., Porter, M., Williams, S., McDonald, M., Fenn, D., Howison, S.: Limit order books. Quant. Financ. **13**, 1709–1742 (2013)

A Beam Search for the Longest Common Subsequence Problem Guided by a Novel Approximate Expected Length Calculation

Marko Djukanovic[1(✉)], Günther R. Raidl[1], and Christian Blum[2]

[1] Institute of Logic and Computation, TU Wien, Vienna, Austria
{djukanovic,raidl}@ac.tuwien.ac.at
[2] Artificial Intelligence Research Institute (IIIA-CSIC), Campus UAB, Bellaterra, Spain
christian.blum@iiia.csic.es

Abstract. The longest common subsequence problem (LCS) aims at finding a longest string that appears as subsequence in each of a given set of input strings. This is a well known $\mathcal{NP}$-hard problem which has been tackled by many heuristic approaches. Among them, the best performing ones are based on beam search (BS) but differ significantly in various aspects. In this paper we compare the existing BS-based approaches by using a common BS framework making the differences more explicit. Furthermore, we derive a novel heuristic function to guide BS, which approximates the expected length of an LCS of random strings. In a rigorous experimental evaluation we compare all BS-based methods from the literature and investigate the impact of our new heuristic guidance. Results show in particular that our novel heuristic guidance leads frequently to significantly better solutions. New best solutions are obtained for a wide range of the existing benchmark instances.

Keywords: String problems · Expected value · Beam search

1 Introduction

We define a *string* s as a finite sequence of $|s|$ characters from a finite alphabet Σ. Strings are widely used for representing sequence information. Words, and even whole texts, are naturally stored by means of strings. In the field of bioinformatics, DNA and protein sequences, for example, play particularly important roles. A frequently occurring necessity is to detect similarities between several strings in order to derive relationships and possibly predict diverse aspects of some strings. A *sequence* of a string s is any sequence obtained by removing an arbitrary number of characters from s. A natural and common way to compare two or more strings is to find *common subsequences*. More specifically, given a set of m input strings $S = \{s_1, \ldots, s_m\}$, the *longest common subsequence* (LCS)

G. Nicosia et al. (Eds.): LOD 2019, LNCS 11943, pp. 154–167, 2019.
https://doi.org/10.1007/978-3-030-37599-7_14

problem [16] aims at finding a subsequence of maximal length which is common for all the strings in S. Apart from applications in computational biology [13], this problem appears, for example, also in data compression [1,19], text editing [14], and the production of circuits in field programmable gate arrays [6].

The LCS problem is $\mathcal{NP}$-hard for an arbitrary number (m) of input strings [16]. However, for fixed m polynomial algorithms based on dynamic programming (DP) are known [10]. Standard dynamic programming approaches run in $O(n^m)$ time, where n is the length of the longest input string. This means that these exact methods become quickly impractical when m grows and n is not small. In practice, heuristics optimization techniques are therefore frequently used. Concerning simple approximate methods, the expansion algorithm [5] and the Best-Next heuristic [9,11] are well known construction heuristics.

A break-through in terms of both, computation time and solution quality was achieved by the *Beam Search* (BS) of Blum et al. [3]. This algorithm is an incomplete tree search which relies on a solution construction mechanism based on the Best-Next heuristic and exploits bounding information using a simple upper bound function to prune non promising solutions. The algorithm was able to outperform all existing algorithms at the time of its presentation. Later, Wang et al. [21] proposed a fast A*-based heuristic utilizing a new DP-based upper bound function. Mousavi and Tabataba [17] proposed a variant of the BS which uses a probability-based heuristic and a different pruning mechanism. Moreover, Tabataba et al. [20] suggested a hyper-heuristic approach which makes use of two different heuristics and applies a beam search with low beam width first to make the decision about which heuristic to use in a successive BS with higher beam width. This approach is currently state-of-the-art for the LCS problem.

Recently, a *chemical reaction optimization* [12] was also proposed for the LCS and the authors claimed to achieve new best results for some of the benchmark instances. We gave our best to re-implement their approach but were not successful due to many ambiguities and mistakes found within the algorithm's description and open questions that could not be resolved from the paper. The authors of the paper also were not able to provide us the original implementation of their approach or enough clarification. Therefore, we exclude this algorithm from further consideration in our experimental comparison in this article.

In conclusion, the currently best performing heuristic approaches to solve also large LCS problem instances are thus based on BS guided by different heuristics and incorporate different pruning mechanisms to omit nodes that likely lead to weaker suboptimal solutions. More detailed conclusions are, however, difficult as the experimental studies in the literature are limited and partly questionable: On the one hand, in [3,20] mistakes in earlier works have been reported, whose impact on the final solution quality are not known. On the other hand, the algorithms have partly been tested on different LCS instance sets and/or the methods were implemented in different programming languages and the experiments were performed on different machines.

In our work we re-implemented all the mentioned BS-based methods from the literature with their specific heuristics and pruning mechanisms within a common BS framework in order to rigorously compare the methods on the same sets of existing benchmark instances. Furthermore, we derive a novel heuristic function for guiding BS that computes an approximate expected length of an LCS. This function is derived from our previous work done for the *palindromic* version of the LCS problem [7]). The experimental evaluation shows that this heuristic is for most of the benchmark sets a significantly better guidance than the other so far used heuristic functions, and we obtain new best known results on several occasions. An exception is only one benchmark set were the input strings are artificially created in a strongly correlated way.

The rest of the paper is organized as follows. Section 2 describes the state graph to be searched for solving the LCS problem. In Sect. 3 we present our united BS framework which covers all the considered BS-based algorithms from the literature. We further propose our novel heuristic function for approximating the expected length for the LCS here. Section 4 reports the main results of our rigorous experimental comparison. We finally conclude with Sect. 5 where also some directions for promising future work are sketched.

2 State Graph

We start this section by introducing some common notation. Let n, as already stated, be the maximum length of the strings in S. The j-th letter of a string s is denoted by $s[j]$, $j = 1, \ldots, |s|$, and let $s_1 \cdot s_2$ denote the concatenation obtained by appending string s_2 to string s_1. By $s[j, j']$, $j \le j'$, we denote the continuous subsequence of s starting at position j and ending at position j'; if $j > j'$, $s[j, j']$ is the empty string ε. Finally, let $|s|_a$ be the number of occurrences of letter $a \in \Sigma$ in string s. Henceforth, a string s is called a (valid) *partial solution* concerning input strings $S = \{s_1, \ldots, s_m\}$, if s is a common subsequence of the strings in S.

Let $p^{\mathrm{L}} \in \mathbb{N}^m$ be an integer valued vector such that $1 \le p^{\mathrm{L}}_i \le |s_i|$, for $i = 1, \ldots, m$. Given such vector p^{L}, the set $S[p^{\mathrm{L}}] := \{s_i[p^{\mathrm{L}}_i, |s_i|] \mid i = 1, \ldots, m\}$ consists of each original string's continuous subsequence from the position given in p^{L} up to the end. We call vector p^{L} *left position vector*, and it represents the LCS subproblem on the strings $S[p^{\mathrm{L}}]$. Note that the original problem can be denoted by $S[(1, \ldots, 1)]$. We are, however, not interested in all possible subproblems but just those which are *induced* by (meaningful) partial solutions. A partial solution s induces the subproblem $S[p^{\mathrm{L}}]$ for which

- $s_i[1, p^{\mathrm{L}}_i - 1]$ is the minimal string among strings $s_i[1, x], x = 1, \ldots, p^{\mathrm{L}} - 1$ such that it contains s as a subsequence, for all $i = 1, \ldots, m$.

For example, if $S = \{\texttt{abcbacb}, \texttt{accbbaa}\}$, then for the partial solution $s = \texttt{acb}$ the induced subproblem is represented by $p^{\mathrm{L}} = (5, 5)$.

The state graph of the LCS problem is a directed acyclic graph $G = (V, A)$, where a node $v \in V$ is represented by the respective left position vector $p^{\mathrm{L},v}$ and by the length l_v of the partial solution which induces the corresponding subproblem $S[p^{\mathrm{L},v}]$, i.e. $v = (p^{\mathrm{L},v}, l_v)$. An arc $a = (v_1, v_2) \in A$ with a label $\ell(a) \in \Sigma$ exists if l_{v_2} is by one larger than l_{v_1} and if the partial solution obtained by appending letter $\ell(a)$ to the partial solution of v_1 induces the subproblem $S[p^{\mathrm{L},v_2}]$. The root node r of the state graph G corresponds to the original problem, which can be said to be induced by the empty partial solution ε, i.e., $r = ((1,\ldots,1),0)$. In order to derive the successor nodes of a node $v \in V$ in G, we first determine all the letters by which partial solutions inducing $S[p^{\mathrm{L},v}]$ can be feasibly extended, i.e., all letters $a \in \Sigma$ appearing at least once in each string in $S[p^{\mathrm{L},v}]$. Let Σ_v denote this set of feasible extension letters w.r.t. node v. For each letter $a \in \Sigma_v$, the position of the first occurrence of a in $s_i[p_i^{\mathrm{L},v}, |s_i|]$ is denoted by $p_{i,a}^{\mathrm{L},v}$, $i = 1,\ldots,m$. We can in general reduce Σ_v by disregarding *dominated* letters. We say letter $a \in \Sigma_v$ dominates letter $b \in \Sigma_v$ iff $p_{i,a}^{\mathrm{L},v} \leq p_{i,b}^{\mathrm{L},v}$ for all $i = 1,\ldots,m$. Dominated letters can safely be ignored since they always lead to suboptimal solutions as they "skip" some other letter that can be used before. Let Σ_v^{nd} be the obtained set of feasible and non-dominated letters for node v. For each letter $a \in \Sigma_v^{\mathrm{nd}}$ we derive a successor node $v' = (p^{\mathrm{L},v'}, l_v + 1)$, where $p_i^{\mathrm{L},v'} = p_{i,a}^{\mathrm{L},v'} + 1$, $i = 1,\ldots,m$. Each node that receives no successor node, i.e., where $\Sigma_v^{\mathrm{nd}} = \emptyset$, is called a *complete* node. Now, note that any path from the root node r to any node in V represents the feasible partial solution given by the sequence of labels of the traversed arcs. Any path from r to a complete node represents a common subsequence of S that cannot be further extended, and an LCS is therefore given by a longest path from r to any complete node.

3 Beam Search Framework

In the literature for the LCS problem, the so far leading algorithms to approach larger instances heuristically are all based on Beam Search (BS). This essentially is an incomplete tree search which expands nodes in a breadth-first search manner. A collection of nodes, called the *beam*, is maintained. Initially, the beam contains just the root node r. In each major iteration, BS expands all nodes of the beam in order to obtain the respective successor nodes at the next level. From those the $\beta > 0$ most promising nodes are selected to become the beam for the next iteration, where β is a strategy parameter called beam width. This expansion and selection steps are repeated level by level until the beam becomes empty. We will consider several ways to determine the most promising nodes to be kept at each step of BS in Sect. 3.1. The BS returns the partial solution of a complete node with the largest l_v value discovered during the search. The main difference among BS approaches from the literature are the heuristic functions used to evaluate LCS nodes for the selection of the beam and pruning mechanisms to recognize and discard dominated nodes. A general BS framework for the LCS is shown in Algorithm 1.

ExtendAndEvaluate(B, h) derives and collects the successor nodes of all $v \in B$ and evaluates them by heuristic h, generating the set of extension nodes V_{ext} ordered according to non-increasing h-values. Prune($V_{\mathrm{ext}}, ub_{\mathrm{prune}}$) optionally removes any dominated node v for which $l_v + ub_{\mathrm{prune}}(v) \leq |s_{\mathrm{lcs}}|$, where $ub_{\mathrm{prune}}(\cdot)$ is an upper bound function for the number of letters that may possibly still be appended, or in other words an upper bound for the LCS of the

Algorithm 1. BS framework for the LCS problem

```
Input: an instance (S, Σ), heuristic function h to evaluate nodes; upper bound
   function ub_prune to prune nodes; parameter k_best to filter nodes (non-dominance
   relation check); β: beam size (and others depending on the specific algorithm)
Output: a feasible LCS solution
B ← {r}
s_lcs ← ε
while B ≠ ∅ do
    V_ext ← ExtendAndEvaluate(B, h)
    update s_lcs if a complete node v with a new largest l_v value reached
    V_ext ← Prune(V_ext, ub_prune)  // optional
    V_ext ← Filter(V_ext, k_best)  // optional
    B ← Reduce(V_ext, β)
end while
return s_lcs
```

corresponding remaining subproblem, and $|s_{\mathrm{lcs}}|$ is the length of the so far best solution. Filter$(V_{\mathrm{ext}}, k_{\mathrm{best}})$ is another optional step. It removes nodes corresponding to dominated letters as defined in Sect. 2, but in a possibly restricted way controlled by parameter k_{best} in order to limit the spent computing effort. More concretely, the dominance relationship is checked for each node $v \in V_{\mathrm{ext}}$ against the k_{best} most promising nodes from V_{ext}. Last but not least, Reduce$(V_{\mathrm{ext}}, \beta)$ returns the new beam consisting of the β best ranked nodes in V_{ext}.

3.1 Functions for Evaluating Nodes

In the literature, several different functions are used for evaluating and pruning nodes, i.e., for h and ub_{prune}. In the following we summarize them.

Fraser [9] used as upper bound on the number of letters that might be further added to a partial solution leading to a node v—or in other words the length of an LCS of the induced remaining subproblem $S[p^{\mathrm{L},v}]$–by $\mathrm{UB}_{\min}(v) = \mathrm{UB}_{\min}(S[p^{\mathrm{L}}]) = \min_{i=1,\ldots,m}(|s_i| - p_i^{\mathrm{L},v} + 1)$.

Blum et al. [3] suggested the upper bound $\mathrm{UB}_1(v) = \mathrm{UB}_1(S[p^{\mathrm{L},v}]) = \sum_{a \in \Sigma} c_a$, with $c_a = \min_{i=1,\ldots,m} |s_i[p_i^{\mathrm{L},v}, |s_i|]|_a$, which dominates $\mathrm{UB}_{\min}$. While UB_1 is efficiently calculated using smart preprocessing in $O(m \cdot |\Sigma|)$, it is still a rather weak bound. The same authors proposed the following ranking function Rank(v) to use for heuristic function h. When expanding a node v, all the successors v' of v are ranked either by $\mathrm{UB}_{\min}(v')$ or by $g(v, v') = \left(\sum_{i=1}^{m} \frac{p_i^{\mathrm{L},v'} - p_i^{\mathrm{L},v} - 1}{|s_i| - p_i^{\mathrm{L},v}}\right)^{-1}$. If v' has the largest $\mathrm{UB}_{\min}(v')$ (or $g(v, v')$) value among all the successors of v, it receives rank 1, the successor with the second largest value among the successors receives rank 2, etc. The overall value Rank(v) is obtained by summarizing all the ranks along the path from the root node to the node corresponding to the partial solution. Finally, the nodes in V_{ext} are sorted according to non-increasing values Rank(v) (i.e., smaller values preferable here).

Wang et al. [21] proposed a DP-based upper bound using the LCS for two input strings, given by

$$\mathrm{UB}_2(v) = \mathrm{UB}_2(S[p^{\mathrm{L},v}]) = \min_{i=1,\ldots,m-1} |\mathrm{LCS}(s_i[p_i^{\mathrm{L},v}, |s_i|], s_{i+1}[p_{i+1}^{\mathrm{L},v}, |s_{i+1}|])|.$$

Even if not so obvious, UB_2 can be calculated efficiently in time $O(m)$ by creating appropriate data structures in preprocessing. By combining the two upper bounds we obtain the so far tightest known bound $\mathrm{UB}(v) = \min(\mathrm{UB}_1(v), \mathrm{UB}_2(v))$ that can still efficiently be calculated in $O(m \cdot |\Sigma|)$ time. This bound will serve in Prune() of our BS framework, since it can filter more non-promising nodes than when UB_1 or UB_2 are just individually applied.

Mousavi and Tabataba [17,20] proposed two heuristic guidances. The first estimation is derived by assuming that all input strings are uniformly at random generated and that they are mutually independent. The authors derived a recursion which determines the probability $\mathcal{P}(p,q)$ that a uniform random string of length p is a subsequence of a string of length q. These probabilities can be calculated during preprocessing and are stored in a matrix. For some fixed k, using the assumption that the input strings are independent, each node is evaluated by $\mathrm{H}(v) = \mathrm{H}(S[p^{\mathrm{L},v}]) = \prod_{i=1}^{m} \mathcal{P}(k, |s_i| - p_i^{\mathrm{L},v} + 1)$. This corresponds to the probability that a partial solution represented by v can be extended by k letters. The value of k is heuristically chosen as $k := \max\left(1, \left\lfloor \frac{1}{|\Sigma|} \cdot \min_{v \in V_{\mathrm{ext}}, i=1,\ldots,m}(|s_i| - p_i^{\mathrm{L},v} + 1)\right\rfloor\right)$. The second heuristic estimation, the so called *power* heuristic, is proposed as follows:

$$\mathrm{Pow}(v) = \mathrm{Pow}(S[p^{\mathrm{L},v}]) = \left(\prod_{i=1}^{m}(|s_i| - p_i^{\mathrm{L},v} + 1)\right)^q \cdot \mathrm{UB}_{\min}(v),\ q \in [0,1).$$

It can be seen as a generalized form of $\mathrm{UB}_{\min}$. The authors argue to use smaller values for q in case of larger m and specifically set $q = a \times \exp(-b \cdot m) + c$, where $a, b, c \geq 0$ are then instance-independent parameters of the algorithm.

3.2 A Heuristic Estimation of the Expected Length of an LCS

By making use of the matrix of probabilities $\mathcal{P}(p,q)$ from [17] for a uniform random string of length p to be a subsequence of a random string of length q and some basic laws from probability theory, we derive an approximation for the expected length of a LCS of $S[p^{\mathrm{L},v}]$ as follows. Let Y be the random variable which corresponds to the length of an LCS for a set S of uniformly at random generated strings. This value cannot be larger than the length of the shortest string in S, denoted by $l_{\max} = \min_{i=1,\ldots,m}(|s_i| - p_i^{\mathrm{L},v} + 1)$. Let us enumerate all sequences of length k over alphabet Σ; trivially, there are $|\Sigma|^k$ such sequences. We now make the simplifying assumption that for any sequence of length k over Σ, the event that the sequence is a common subsequence of all strings in S is independent of the corresponding events for other sequences. Let $Y_k \in \{0,1\}$, $k = 0, \ldots, l_{\max}$, be a binary random variable which denotes the event that S has a common subsequence of length at least k. If S has a common subsequence of length $k+1$, this implies that it has a common subsequence of length k as well. Consequently, we get that $\Pr[Y=k] = \mathbb{E}[Y_k] - \mathbb{E}[Y_{k+1}]$ and

$$\mathbb{E}[Y] = \sum_{k=1}^{l_{\max}} k \cdot \Pr[Y=k] = \sum_{k=1}^{l_{\max}} k \cdot (\mathbb{E}[Y_k] - \mathbb{E}[Y_{k+1}]) = \sum_{k=1}^{l_{\max}} \mathbb{E}[Y_k]. \quad (1)$$

The probability that the input strings of S have no common subsequence of length k is equal to $1 - \mathbb{E}[Y_k]$. Due to our assumption of independence, this probability is equal to $(1 - \prod_{i=1}^{m} \mathcal{P}(k, |s_i|))^{|\Sigma|^k}$. Finally, for any node $v \in V$, we obtain the approximate expected length

$$\mathrm{EX}(v) = \mathrm{EX}(S[p^{\mathrm{L},v}]) = \sum_{k=1}^{l_{\max}} \left(1 - \left(1 - \prod_{i=1}^{m} \mathcal{P}(k, |s_i| - p_i^{\mathrm{L},v} + 1)\right)^{|\Sigma|^k}\right). \quad (2)$$

Since a direct numerical calculation would yield too large intermediate values for a common floating point arithmetic when k is large, we use the decomposition

$$A^{|\Sigma|^k} = \left(\underbrace{\left(\cdots (A)^{|\Sigma|^p} \cdots\right)^{|\Sigma|^p}}_{\lfloor k/p \rfloor}\right)^{|\Sigma|^{(k \bmod p)}}$$

for any expression A with $p = 20$. Moreover, the calculation of (2) can in practice be efficiently done in $O(m \log(n))$ time by exploiting the fact that the terms under the sum present a decreasing sequence of values within $[0, 1]$ and many values are very close to zero or one: We apply a divide-and-conquer principle for detecting the values $\left(1 - \prod_{i=1}^{m} \mathcal{P}(k, |s_i| - p_i^{\mathrm{L},v} + 1)\right)^{|\Sigma|^k} \in (\epsilon, 1-\epsilon)$ with the threshold $\epsilon = 10^{-6}$. Furthermore, note that if the product which appears in (2) is close to zero, this might cause further numerical issues. These are resolved by replacing the term $\left(1 - \prod_{i=1}^{m} \mathcal{P}(k, |s_i| - p_i^{\mathrm{L},v} + 1)\right)^{|\Sigma|^k}$ with an approximation derived from the Taylor expansion of $(1-x)^\alpha$; details are described in [7].

3.3 Expressing Existing Approaches in Terms of Our Framework

All BS-related approaches from the literature (see Sect. 1) can be defined as follows within our framework from Algorithm 1.

BS by Blum et al. [3]. The heuristic function h is set to $h = \mathrm{Rank}^{-1}$. Function $\mathrm{Prune}(V_{\mathrm{ext}}, ub_{\mathrm{prune}})$ uses $ub_{\mathrm{prune}} = \mathrm{UB}_1$. Moreover, all nodes that are not among the $\mu \cdot \beta$ most promising nodes from V_{ext} are pruned. Hereby, $\mu \geq 1$ is an algorithm-specific parameter. Finally, with a setting of $k_{\mathrm{best}} \geq \beta \cdot |V_{\mathrm{ext}}|$ in function $\mathrm{Filter}(V_{\mathrm{ext}}, k_{\mathrm{best}})$, the original algorithm is obtained. Instead of testing this original algorithm, we study here an improved version that uses $ub_{\mathrm{prune}} = \mathrm{UB}$. Moreover, during tuning (see below) we will consider also values for k_{best} such that $k_{\mathrm{best}} < \beta \cdot |V_{\mathrm{ext}}|$. The resulting method is henceforth denoted by BS-BLUM.

Heuristic by Wang [21]. $h = \text{UB}_2$ is used as heuristic function. Moreover, a priority queue to store extensions is used instead of the standard vector structure used in the implementation of other algorithms. Function $\text{Prune}(V_{\text{ext}}, ub_{\text{prune}})$ removes all those nodes from V_{ext} whose h-values deviate more than $W \geq 0$ units from the priority value of the most promising node of V_{ext}. Filtering is not used. Instead of $h = \text{UB}_2$ (as in the original algorithm) we use here $h = \text{UB}$, which significantly improves the algorithm henceforth denoted by Bs-Wang.

BS Approaches by Mousavi and Tabataba [17,20]. The first approach, denoted by Bs-H, uses $h = \text{H}$, whereas in the second one, denoted by Bs-Pow, $h = \text{Pow}$ is used. No pruning is done. Finally, a restricted filtering ($k_{\text{best}} > 0$) is applied.

The Hyper-heuristic Approach by Mousavi and Tabataba [20]. This approach, henceforth labeled Hh, combines heuristic functions H and Pow as follows. First, Bs-H and Bs-Pow are both executed using a low beam width $\beta_h > 0$. Based on the outcome of these two executions, either Bs-H or Bs-Pow will be selected as the final method executed with a beam width $\beta \gg \beta_h$. The result is the best solution found in both phases.

4 Experimental Evaluation

The presented BS framework was implemented in C++ and all experiments were performed in single-threaded mode on an Intel Xeon E5-2640 with 2.40 GHz and 16 GB of memory. In addition to the five approaches from the literature as detailed in Sect. 3.3, we test our own approach, labeled Bs-Ex, which uses $h = \text{EX}$, no pruning, and a restricted filtering.

The related literature offers six different benchmark sets for the LCS problem. The `ES` benchmark, introduced by Easton and Singireddy [8], consists of 600 instances of different sizes in terms of the number and the length of the input strings, and in terms of the alphabet size. A second benchmark consists of three groups of 20 instances each: `Random`, `Rat` and `Virus`. It was introduced by Shyu and Tsai [18] for testing their ant colony optimization algorithm. Hereby, `Rat` and `Virus` consist of sequences from rat and virus genomes. The `BB` benchmark of 80 instances was generated by Blum and Blesa [2] in a way such that a large similarity between the input strings exists. Finally, the `BL` instance set [4] consists of 450 problem instances that were generated uniformly at random.

The final solution quality produced by any of the considered BS methods is largely determined by the beam size parameter β. We decided to test all algorithms with a setting aiming for a low computation time ($\beta = 50$) and with a second setting aiming for a high solution quality ($\beta = 600$). The first setting is henceforth called the *low-time* setting, and the second one the *high-quality* setting. Note that when using the same value of β, the considered algorithms expand a comparable number of nodes. The remaining parameters of the algorithms are tuned by *irace* [15] for the *high-quality* setting. Separate tuning runs with a budget of 5000 algorithm applications are performed for benchmark instances in which the input strings have a random character (ES, Random, Rat, Virus, BL),[1] and for the structured instances from set BB. 30 training instances are used for the first tuning run and 20 for the second one.

The outcome reported by *irace* for random instances is as follows. Bs-Blum makes use of function $g(.,.)$ within $h = \text{Rank}^{-1}$. Moreover, $\mu = 4.0$ and $k_{\text{best}} = 5$ are used. Bs-Wang uses $W = 10$. For Bs-H we obtain $k_{\text{best}} = 50$, for Bs-Pow we get $k_{\text{best}} = 100$, $a = 1.677$, $b = 0.054$, and $c = 0.074$. Finally, Hh uses $\beta_h = 50$, and for Bs-Ex we get $k_{\text{best}} = 100$.

For the structured instances from set BB *irace* reports the following. Bs-Blum makes use of $\text{UB}_{\min}$ within $h = \text{Rank}^{-1}$. Moreover, it uses $\mu = 4.0$ and $k_{\text{best}} = 1000$. Bs-Wang uses $W = 10$, Bs-H needs $k_{\text{best}} = 100$, and Bs-Pow requires $k_{\text{best}} = 100$, $a = 1.823$, $b = 0.112$, and $c = 0.014$. Finally, Hh uses $\beta_h = 50$ and for Bs-Ex $k_{\text{best}} = 100$. At this point we want to emphasize that we made sure that the five re-implemented competitor algorithms obtain equivalent (and often even better) results on all benchmark sets than those reported in the original papers.

We now proceed to study the numerical results presented in Tables 1, 2, 3, 4 and 5. In each table, the first three columns describe the respective instances in terms of the alphabet size ($|\Sigma|$), the number of input strings (n), and the maximum string length (m). Columns 4–8 report the results obtained with the *low-time* setting, while columns 9–13 report on the results of the *high-quality* setting. The first three columns of both blocks provide the results of the best performing algorithm among the five competitors from the literature. Listed are for each instance (or instance group) the (average) solution length, the respective (average) computation time, and the algorithm that achieved this result. The last two columns of both blocks present the (average) solution length and the (average) computation time of our new Bs-Ex. The overall best result of each comparison is indicated in bold font, and an asterisk indicates that this result is better than the so-far best known one from the literature. These results allow to make the following observations.

- Concerning the *low-time* setting of the algorithms, the approaches from the literature compare as follows. Bs-H and Bs-Pow seem to outperform the other approaches in the context of benchmarks Rat and Virus, with Bs-Blum and Bs-Wang gaining some terrain when moving towards the alphabet size of $|\Sigma| = 20$. Furthermore, Hh and—to some extent—Bs-H dominate the remaining approaches in the context of benchmarks ES and BL. Concerning the structured instances from set BB the picture is not so clear. Here, the oldest BS approach (Bs-Blum) is able to win over the other approaches in three out of seven cases.

[1] Note that even instance sets Rat and Virus contain sequences that are close to random strings.

Table 1. Results on benchmark set `Rat`.

			low-time, literature			*low-time*, Bs-Ex		*high-quality*, literature			*high-quality*, Bs-Ex	
$\|\Sigma\|$	n	m	$\|\bar{s}_{\text{best}}\|$	$\bar{t}_{\text{best}}$	Algo.	$\|\bar{s}_{\text{best}}\|$	$\bar{t}[s]$	$\|\bar{s}_{\text{best}}\|$	$\bar{t}_{\text{best}}$	Algo.	$\|\bar{s}_{\text{best}}\|$	$\bar{t}[s]$
4	600	10	**201**	0.09	Bs-Pow	198	0.22	204	1.18	Bs-Pow	***205**	3.09
4	600	15	**182**	0.10	Bs-Pow	**182**	0.18	184	0.62	Bs-H	***185**	2.65
4	600	20	**169**	0.05	Bs-Pow	168	0.15	170	0.94	Bs-Pow	***172**	2.25
4	600	25	166	0.12	Bs-Pow	**167**	0.18	168	1.01	Bs-Pow	***170**	2.71
4	600	40	**151**	0.04	Bs-H	146	0.15	150	1.02	Bs-Pow	**152**	1.81
4	600	60	149	0.10	Bs-Pow	**150**	0.17	151	1.16	Bs-Pow	***152**	2.27
4	600	80	**137**	0.05	Bs-H	**137**	0.17	139	0.67	Bs-H	***142**	2.47
4	600	100	**133**	0.07	Bs-Pow	131	0.14	135	0.47	Bs-H	***137**	2.50
4	600	150	125	0.06	Bs-H	**127**	0.13	126	0.91	Bs-Pow	***129**	1.97
4	600	200	**121**	0.09	Bs-Pow	**121**	0.17	***123**	0.70	Bs-Pow	***123**	2.65
20	600	10	**70**	0.09	Bs-H	**70**	0.37	***71**	1.86	Bs-H	***71**	3.44
20	600	15	61	0.15	Bs-Pow	**62**	0.28	62	1.40	Bs-H	***63**	2.55
20	600	20	**53**	0.12	Bs-Pow	**53**	0.20	**54**	1.15	Bs-H	**54**	2.45
20	600	25	**50**	0.22	Bs-Wang	**50**	0.21	51	1.09	Bs-H	***52**	2.94
20	600	40	**48**	0.09	Bs-H	47	0.19	**49**	1.15	Bs-Blum	**49**	2.97
20	600	60	**46**	0.09	Bs-H	**46**	0.20	**47**	1.61	Bs-Pow	46	2.42
20	600	80	**43**	0.18	Bs-Blum	41	0.21	***44**	1.14	Bs-H	43	2.64
20	600	100	**38**	0.11	Bs-Pow	**38**	0.23	39	0.96	Bs-H	***40**	2.54
20	600	150	**36**	0.32	Bs-Blum	**36**	0.14	**37**	5.11	Bs-Wang	**37**	2.03
20	600	200	**34**	0.10	Bs-Pow	**34**	0.18	**34**	2.62	Bs-Blum	**34**	2.74

- The results obtained by Bs-Ex with the *low-time* setting are comparable to the best results obtained by the methods from the literature. More specifically, Bs-Ex produces comparable results for `Virus` and `Rat` and is able to outperform the other approaches in the context of `ES` and `BL`. As could be expected, for the `BB` instance set, in which the input strings have a strong relation to each other, the EX guiding function cannot successfully guide the search. This is because EX assumes the input strings to be random strings, that is, to be independent of each other.
- Concerning the results obtained with the *high-quality* setting, the comparison of the algorithms from the literature can be summarized as follows. For all benchmark sets (apart from `BB`) the best performance is shown by Bs-H and/or Bs-Pow. For benchmark set `BB` the picture is, again, not so clear, with Bs-Blum gaining some terrain.
- Bs-Ex with the *high-quality* setting outperforms the other approaches from the literature on all benchmark sets except for `BB`, the latter again due to the strong correlation of the strings. In fact, in 48 out of 67 cases (concerning benchmarks `Rat`, `Virus`, `ES` and `BL`) Bs-Ex is able to obtain a new best-known result. Moreover, in most of the remaining cases, the obtained result is equal to the so-far best known one.
- Concerning the run times of the approaches, the calculation of EX is done in $O(m \log n)$ time and, therefore, is a bit more expensive when compared to the simpler UB, H, or Pow. However, this is not a significant issue since almost all runs completed within rather short times of usually a few seconds up to less than two minutes.

– We performed Wilcoxon signed-rank tests with an error level of 5% to check the significance of differences in the results of the approaches. These indicate that the solutions of the *high-quality* Bs-Ex are in the expected case indeed significantly better than those obtained from the *high-quality* state-of-the-art approaches from the literature for all except for the `Virus` benchmark, where no conclusion can be drawn, and the `BB` benchmark, where the Bs-Ex results are significantly worse due to the strong relationship among the sequences.

Table 2. Results on benchmark set `Virus`.

			low-time, literature			*low-time*, Bs-Ex		*high-quality*, literature			*high-quality*, Bs-Ex	
$\lvert\Sigma\rvert$	n	m	$\lvert\overline{s}_{\text{best}}\rvert$	$\overline{t}_{\text{best}}$	Algo.	$\lvert\overline{s}_{\text{best}}\rvert$	$\overline{t}[s]$	$\lvert\overline{s}_{\text{best}}\rvert$	$\overline{t}_{\text{best}}$	Algo.	$\lvert\overline{s}_{\text{best}}\rvert$	$\overline{t}[s]$
4	600	10	**225**	0.04	Bs-H	223	0.21	226	0.68	Bs-H	***227**	2.88
4	600	15	200	0.04	Bs-H	**201**	0.23	204	0.71	Bs-H	***205**	2.24
4	600	20	186	0.05	Bs-H	**188**	0.18	190	0.69	Bs-H	***192**	2.69
4	600	25	**191**	0.06	Bs-H	**191**	0.20	***194**	0.68	Bs-H	***194**	2.20
4	600	40	165	0.04	Bs-H	**167**	0.17	***170**	1.21	Bs-Pow	***170**	2.24
4	600	60	**163**	0.04	Bs-H	162	0.27	***166**	0.69	Bs-H	***166**	2.38
4	600	80	157	0.04	Bs-H	**158**	0.19	159	0.72	Bs-H	***163**	2.70
4	600	100	153	0.07	Bs-H	**156**	0.19	**158**	0.90	Bs-H	**158**	2.31
4	600	150	**154**	0.06	Bs-H	**154**	0.22	**156**	0.66	Bs-H	**156**	2.37
4	600	200	**153**	0.09	Bs-H	152	0.39	***155**	1.22	Bs-H	154	2.63
20	600	10	**75**	0.15	Bs-Pow	74	0.28	***77**	2.38	Bs-Pow	76	2.86
20	600	15	**63**	0.16	Bs-Pow	**63**	0.24	***64**	1.57	Bs-H	***64**	2.91
20	600	20	**59**	0.13	Bs-H	**59**	0.29	**60**	1.58	Bs-H	**60**	2.68
20	600	25	**55**	0.11	Bs-Pow	54	0.20	**55**	1.10	Bs-H	**55**	2.65
20	600	40	**49**	0.08	Bs-H	**49**	0.20	***50**	0.85	Bs-H	***50**	2.85
20	600	60	**47**	0.16	Bs-Pow	46	0.19	47	1.43	Bs-Blum	***48**	3.34
20	600	80	44	0.18	Bs-Blum	**46**	0.30	**46**	1.39	Bs-H	**46**	2.60
20	600	100	**44**	0.14	Bs-H	**44**	0.27	44	2.04	Bs-Blum	***45**	2.33
20	600	150	**45**	0.11	Bs-H	**45**	0.24	**45**	2.94	Bs-Blum	**45**	2.75
20	600	200	**43**	0.17	Bs-H	**43**	0.28	**44**	1.69	Bs-H	43	3.17

Table 3. Results on benchmark set `ES` (averaged over 50 instances per row).

			low-time, literature			*low-time*, Bs-Ex		*high-quality*, literature			*high-quality*, Bs-Ex	
$\lvert\Sigma\rvert$	n	m	$\lvert\overline{s}_{\text{best}}\rvert$	$\overline{t}_{\text{best}}$	Algo.	$\lvert\overline{s}_{\text{best}}\rvert$	$\overline{t}[s]$	$\lvert\overline{s}_{\text{best}}\rvert$	$\overline{t}_{\text{best}}$	Algo.	$\lvert\overline{s}_{\text{best}}\rvert$	$\overline{t}[s]$
2	1000	10	608.52	0.31	Hh	**609.80**	0.39	614.2	1.42	Bs-Pow	***615.06**	4.43
2	1000	50	533.16	0.33	Hh	**535.02**	0.42	536.46	1.05	Bs-H	***538.24**	4.43
2	1000	100	515.94	0.11	Bs-H	**517.38**	0.46	518.56	1.33	Bs-H	***519.84**	4.82
10	1000	10	199.10	0.53	Hh	**199.38**	0.47	202.72	2.52	Bs-Pow	***203.10**	5.64
10	1000	50	133.86	0.46	Hh	**134.74**	0.35	135.52	2.12	Bs-Pow	***136.32**	3.94
10	1000	100	121.28	0.50	Hh	**122.10**	0.40	122.40	1.50	Bs-H	***123.32**	4.32
25	2500	10	**230.28**	2.33	Hh	223.00	1.57	***235.22**	10.45	Bs-Pow	231.12	19.10
25	2500	50	136.6	1.69	Hh	**137.90**	1.24	138.56	7.23	Bs-Pow	***139.50**	14.51
25	2500	100	120.3	1.74	Hh	**121.74**	1.32	121.62	7.29	Bs-Pow	***122.88**	15.97
100	5000	10	**141.86**	16.12	Hh	139.82	6.98	***144.90**	75.88	Bs-Pow	144.18	91.87
100	5000	50	70.28	9.16	Hh	**71.08**	4.79	71.32	39.11	Bs-Pow	***71.94**	53.54
100	5000	100	59.2	8.71	Hh	**60.04**	4.75	60.06	36.03	Bs-Pow	***60.66**	53.67

Table 4. Results on benchmark BL (averaged over 10 instances per row, $|\Sigma| = 4$).

			low-time, literature			*low-time*, Bs-Ex		*high-quality*, literature			*high-quality*, Bs-Ex	
$\|\Sigma\|$	n	m	$\|\overline{s}_{\text{best}}\|$	$\overline{t}_{\text{best}}$	Algo.	$\|\overline{s}_{\text{best}}\|$	$\overline{t}[s]$	$\|\overline{s}_{\text{best}}\|$	$\overline{t}_{\text{best}}$	Algo.	$\|\overline{s}_{\text{best}}\|$	$\overline{t}[s]$
4	100	10	**34.0**	0.01	Bs-Pow	**34.0**	0.02	***34.1**	0.14	Bs-Pow	***34.1**	0.39
4	100	50	23.7	0.03	Hh	**23.8**	0.02	***24.2**	0.08	Bs-H	***24.2**	0.30
4	100	100	**21.5**	0.03	Hh	**21.5**	0.02	***22.0**	0.23	Bs-Wang	***22.0**	0.27
4	100	150	**20.2**	0.03	Hh	**20.2**	0.02	***20.5**	0.12	Bs-Pow	***20.5**	0.31
4	100	200	**19.8**	0.01	Bs-H	19.5	0.02	***19.9**	0.14	Bs-H	***19.9**	0.31
4	500	10	**182.0**	0.24	Hh	181.2	0.25	***184.1**	1.03	Bs-Pow	184.0	2.41
4	500	50	138.6	0.21	Hh	**139.1**	0.19	140.1	0.90	Bs-Pow	***141.0**	2.13
4	500	100	129.2	0.06	Bs-H	**129.7**	0.18	130.2	1.01	Bs-Pow	***130.8**	2.10
4	500	150	124.7	0.07	Bs-H	**125.5**	0.19	125.9	0.79	Bs-H	***126.4**	2.38
4	500	200	122.6	0.07	Bs-H	**123.0**	0.22	123.2	0.83	Bs-H	***123.7**	2.61
4	1000	10	368.3	0.35	Hh	**368.5**	0.42	373.2	1.80	Bs-Pow	***374.6**	5.22
4	1000	50	284.2	0.36	Hh	**286.2**	0.35	287.0	1.69	Bs-Pow	***288.6**	4.43
4	1000	100	267.5	0.11	Bs-H	**268.8**	0.41	269.5	1.36	Bs-H	***270.6**	4.56
4	1000	150	259.5	0.14	Bs-H	**261.2**	0.47	261.5	1.38	Bs-H	***262.8**	5.30
4	1000	200	254.9	0.17	Bs-H	**256.0**	0.52	256.5	1.81	Bs-H	***257.6**	6.31

Table 5. Results on benchmark set BB (averaged over 10 instances per row).

			low-time, literature			*low-time*, Bs-Ex		*high-quality*, literature			*high-quality*, Bs-Ex	
$\|\Sigma\|$	n	m	$\|\overline{s}_{\text{best}}\|$	$\overline{t}_{\text{best}}$	Algo.	$\|\overline{s}_{\text{best}}\|$	$\overline{t}[s]$	$\|\overline{s}_{\text{best}}\|$	$\overline{t}_{\text{best}}$	Algo.	$\|\overline{s}_{\text{best}}\|$	$\overline{t}[s]$
2	1000	10	**662.9**	0.33	Hh	635.1	0.44	***676.5**	1.16	Bs-H	673.5	5.49
2	1000	100	**551.0**	0.54	Hh	525.1	0.50	***560.7**	2.10	Bs-Pow	536.6	6.05
4	1000	10	**537.8**	0.43	Hh	453.0	0.48	***545.4**	1.73	Bs-H	545.2	6.24
4	1000	100	**371.2**	0.24	Bs-Pow	318.6	0.53	***388.8**	2.86	Bs-Pow	329.5	5.85
8	1000	10	**462.6**	0.27	Bs-Blum	338.8	0.53	***462.7**	7.93	Bs-Blum	***462.7**	7.90
8	1000	100	**260.9**	0.87	Bs-Blum	198.0	0.67	***272.1**	18.43	Bs-Blum	210.6	8.00
24	1000	10	**385.6**	0.67	Bs-Blum	**385.6**	1.04	385.6	13.14	Bs-Blum	**385.6**	16.24
24	1000	100	**147.0**	0.66	Bs-Pow	95.8	0.98	***149.5**	8.01	Bs-Pow	113.3	12.45

Overall, the numerical results clearly show that EX is a better guidance for BS than the heuristics and upper bounds used in former work, as long as (near-) independence among the input strings is given. Finally, note that the result for Random and for the instances with $|\Sigma| > 4$ of set BL are not provided due to page limitations. Full results can be found at https://www.ac.tuwien.ac.at/files/resources/instances/LCS/LCS-report.zip.

5 Conclusions and Future Work

This paper presents a beam search (BS) framework for the longest common subsequence (LCS) problem that covers all BS-related approaches that were proposed so-far in the literature. These approaches are currently state-of-the-art for the LCS problem. A second contribution consists of a new heuristic function for BS that is based on approximating the expected length of an LCS, assuming that the input strings are randomly created and independent of each other. Our experimental evaluation showed that our new heuristic guides the search in a much better way than the heuristics and upper bounds from the literature. In particular, we were able to produce new best-known results in 48 out of 67 cases.

On the other side, the experimental evaluation has shown that the new guiding function does not work so well in the context of instances in which the input strings are strongly correlated. This, however, comes with no surprise. In future work we therefore aim at combining our new heuristic with the ones available in the literature in order to exploit the advantages of each of them. For example, it could be considered to evaluate the extensions of the nodes in the current beam by applying several heuristics and keep those that are ranked promising by any of the heuristics. Alternatively, algorithm selection techniques might be applied in order to chose the most promising heuristic based on features of the problem instance. Moreover, note that the general search framework proposed in this paper also naturally extends towards exact search algorithms for the LCS, such as an A^* algorithm, for example. Solving small instances to proven optimality and returning proven optimality gaps for larger instances is also our future goal.

Acknowledgments. We gratefully acknowledge the financial support of this project by the Doctoral Program "Vienna Graduate School on Computational Optimization" funded by the Austrian Science Foundation (FWF) under contract no. W1260-N35. Moreover, Christian Blum acknowledges the support of LOGISTAR, a proyect from the European Union's Horizon 2020 research and innovation programme under grant agreement No. 769142.

References

1. Beal, R., Afrin, T., Farheen, A., Adjeroh, D.: A new algorithm for "the LCS problem" with application in compressing genome resequencing data. BMC Genom. **17**(4), 544 (2016)
2. Blum, C., Blesa, M.J.: Probabilistic beam search for the longest common subsequence problem. In: Stützle, T., Birattari, M., H. Hoos, H. (eds.) SLS 2007. LNCS, vol. 4638, pp. 150–161. Springer, Heidelberg (2007). https://doi.org/10.1007/978-3-540-74446-7_11
3. Blum, C., Blesa, M.J., López-Ibáñez, M.: Beam search for the longest common subsequence problem. Comput. Oper. Res. **36**(12), 3178–3186 (2009)
4. Blum, C., Festa, P.: Longest common subsequence problems. In: Metaheuristics for String Problems in Bioinformatics, chapter 3, pp. 45–60. Wiley (2016)
5. Bonizzoni, P., Della Vedova, G., Mauri, G.: Experimenting an approximation algorithm for the LCS. Discrete Appl. Math. **110**(1), 13–24 (2001)
6. Brisk, P., Kaplan, A., Sarrafzadeh, M.: Area-efficient instruction set synthesis for reconfigurable system-on-chip design. In: Proceedings of the 41st Design Automation Conference, pp. 395–400. IEEE press (2004)
7. Djukanovic, M., Raidl, G., Blum, C.: Anytime algorithms for the longest common palindromic subsequence problem. Technical Report AC-TR-18-012, TU Wien, Vienna, Austria (2018)
8. Easton, T., Singireddy, A.: A large neighborhood search heuristic for the longest common subsequence problem. J. Heuristics **14**(3), 271–283 (2008)
9. Fraser, C.B.: Subsequences and Supersequences of Strings. Ph.D. thesis, University of Glasgow, Glasgow, UK (1995)

10. Gusfield, D.: Algorithms on Strings, Trees, and Sequences. Computer Science and Computational Biology. Cambridge University Press, Cambridge (1997)
11. Huang, K., Yang, C., Tseng, K.: Fast algorithms for finding the common subsequences of multiple sequences. In: Proceedings of the IEEE International Computer Symposium, pp. 1006–1011. IEEE press (2004)
12. Islam, M.R., Saifullah, C.M.K., Asha, Z.T., Ahamed, R.: Chemical reaction optimization for solving longest common subsequence problem for multiple string. Soft Comput. **23**(14), 5485–5509 (2018). In press
13. Jiang, T., Lin, G., Ma, B., Zhang, K.: A general edit distance between RNA structures. J. Comput. Biol. **9**(2), 371–388 (2002)
14. Kruskal, J.B.: An overview of sequence comparison: time warps, string edits, and macromolecules. SIAM Rev. **25**(2), 201–237 (1983)
15. López-Ibáñez, M., Dubois-Lacoste, J., Pérez Cáceres, L., Stützle, T., Birattari, M.: The irace package: iterated racing for automatic algorithm configuration. Oper. Res. Perspect. **3**, 43–58 (2016)
16. Maier, D.: The complexity of some problems on subsequences and supersequences. J. ACM **25**(2), 322–336 (1978)
17. Mousavi, S.R., Tabataba, F.: An improved algorithm for the longest common subsequence problem. Comput. Oper. Res. **39**(3), 512–520 (2012)
18. Shyu, S.J., Tsai, C.-Y.: Finding the longest common subsequence for multiple biological sequences by ant colony optimization. Comput. Oper. Res. **36**(1), 73–91 (2009)
19. Storer, J.: Data Compression: Methods and Theory. Computer Science Press, Rockville (1988)
20. Tabataba, F.S., Mousavi, S.R.: A hyper-heuristic for the longest common subsequence problem. Comput. Biol. Chem. **36**, 42–54 (2012)
21. Wang, Q., Korkin, D., Shang, Y.: A fast multiple longest common subsequence (MLCS) algorithm. IEEE Trans. Knowl. Data Eng. **23**(3), 321–334 (2011)

An Adaptive Parameter Free Particle Swarm Optimization Algorithm for the Permutation Flowshop Scheduling Problem

Yannis Marinakis(✉) and Magdalene Marinaki

School of Production Engineering and Management,
Technical University of Crete, Chania, Greece
marinakis@ergasya.tuc.gr, magda@dssl.tuc.gr

Abstract. The finding of suitable values for all parameters of a Particle Swarm Optimization (PSO) algorithm is a crucial issue in the design of the algorithm. A trial and error procedure is the most common way to find the parameters but, also, a number of different procedures have been applied in the past. In this paper, an adaptive strategy is used where random values are assigned in the initialization of the algorithm and, then, during the iterations the parameters are optimized together and simultaneously with the optimization of the objective function of the problem. This approach is used for the solution of the Permutation Flowshop Scheduling Problem. The algorithm is tested in 120 benchmark instances and is compared with a number of algorithms from the literature.

Keywords: Particle swarm optimization · Variable neighborhood search · Path relinking · Permutation flowshop scheduling problem

1 Introduction

Particle Swarm Optimization (PSO) is a population-based swarm intelligence algorithm that was originally proposed by Kennedy and Eberhart [7]. The finding of the optimal set of parameters is not a new idea. In the literature and for the various variants of the PSO algorithm, authors have proposed different ways for calculating the main parameters of the algorithms. The most common way is to use the parameters that most researchers have used in the literature. Another way is to test for some instances a number of different sets of parameters, find the best values of the parameters for these instances and, then, use these values for the rest of the instances. Nowadays, a number of algorithms have been proposed for automated tunning of the parameters inside the procedure. For analytical presentations of these algorithms please see [12].

In this paper, a new adaptive parameter free strategy denoted as **Hybrid Adaptive Particle Swarm Optimization (HAPSO)** algorithm is presented.

G. Nicosia et al. (Eds.): LOD 2019, LNCS 11943, pp. 168–179, 2019.
https://doi.org/10.1007/978-3-030-37599-7_15

All parameters of the Particle Swarm Optimization algorithm (acceleration coefficients, iterations, local search iterations, upper and lower bounds of the velocities and of the positions and number of particles) are optimized during the procedure and, thus, the algorithm works independently and without any interference from the user. All parameters are randomly initialized and, afterwards, during the iterations the parameters are adapted using three different conditions, the first is used for all parameters except the number of particles, the second is used for the increase of the number of particles and the third is used for the decrease of the number of particles. We applied the idea using as main algorithm the classic Constriction Particle Swarm Optimization algorithm [2].

The algorithm is tested for the solution of the Permutation Flowshop Scheduling Problem (PFSP). PFSP is an interesting NP-hard problem for which there is a large number of benchmark instances (120 instances) in the literature that could be used for testing the algorithm. There is, also, a large number of publications in the literature for solving this problem with heuristic, metaheuristic and evolutionary algorithms and, thus, it was easy to compare our results with the results of other algorithms from the literature (please see [1,8,10,11,13,17]).

A brief description of the PFSP is the following: in a flowshop scheduling problem [6], there is a set of n jobs, tasks or items to be processed in a set of m machines or processors in the same order while at any time each job can be processed on at most one machine and each machine can process at most one job. It should be noted that once a job is processed on a machine, it cannot be terminated before completion [10,11]. Also, all jobs are independent and are available for processing at time 0, the set-up times of the jobs on machines are negligible and, therefore, can be ignored and the machines are continuously available. The objective is to find a sequence for the processing of the jobs in the machines so that a given criterion is optimized. In the literature, the most common criterion is the minimization of the makespan (C_{max}), i.e the minimization of the maximum completion time [13].

The rest of the paper is organized as follows: In the next section the proposed algorithm, the Hybrid Adaptive Particle Swarm Optimization (HAPSO), is presented and analyzed in detail. Computational results are presented and analyzed in the third section while in the last section conclusions and future research are given.

2 Hybrid Adaptive Particle Swarm Optimization

Initially, a set of particles is created randomly where each particle corresponds to a possible solution. Each particle, denoted i, has a position (x_{ij} where j is the dimension of the problem) in the space of solutions and moves with a given velocity (v_{ij}). Each particle is represented via the permutation π of jobs. For example, if we have a particle (solution) with ten jobs, a possible permutation representation is the following:

$$2\ 3\ 8\ 5\ 4\ 10\ 9\ 6\ 1\ 7$$

As the calculation of the velocity of each particle is performed by Eq. (1) (see below), the above mentioned representation should be transformed appropriately. We transform each element of the solution into a floating point in the interval (0,1], calculate the velocities and the positions of all particles and, then, convert back the particles' positions into the integer domain using relative position indexing [9]. Its performance is evaluated on the predefined fitness function (makespan). The velocity and position equations are updated as follows (constriction PSO) [2]:

$$v_{ij}(t+1) = \chi(v_{ij}(t) + c_1 rand_1(pbest_{ij} - x_{ij}(t)) + c_2 rand_2(gbest_j - x_{ij}(t))) \quad (1)$$

$$x_{ij}(t+1) = x_{ij}(t) + v_{ij}(t+1) \quad (2)$$

where

$$\chi = \frac{2}{|2 - c - \sqrt{c^2 - 4c}|} \text{ and } c = c_1 + c_2, c > 4 \quad (3)$$

t is the iterations' counter, $pbest_{ij}$ is the personal best position of each particle and $gbest_j$ is the global best solution of the whole swarm, c_1 and c_2 are the acceleration coefficients, $rand_1$ and $rand_2$ are two random variables in the interval (0, 1), and c is equal to the sum of the c_1 and c_2 as it is mentioned in [2]. A modified version of a local search strategy based on the Variable Neighborhood Search (VNS) algorithm [4] is applied [11] in each particle in the swarm in order to improve the solutions produced from the PSO algorithm. Finally, a modified version of a Path Relinking strategy [3] with starting solution the best particle and target solution one of the other particles of the swarm is applied [11]. In each iteration of the algorithm, the optimal solution of the whole swarm and the optimal solution of each particle are kept, as they are both needed for the calculation of the $gbest_j$ the first one and the $pbest_{ij}$ the second one, respectively.

The most important and novel part of the algorithm is the optimization of the parameters during the iterations of the algorithm. Initially, random values of the parameters are given. The parameters that should be optimized are the number of particles, the number of iterations, the number of local search iterations, the coefficients c_1 and c_2, the upper and lower bounds of the positions and of the velocities. As the algorithm should have a stopping criterion, the only parameter, that was selected by us, is the number of consecutive iterations with no improvement in the results of the best solution (this value is selected to be equal to 20).

After the initialization of the particles, the fitness function of each particle in addition to the best particle are calculated. The initial velocities are set equal to zero. It is, also, calculated the average values in the fitness function of all particles and the average values of the best solutions of all particles. In the first iteration, these values are the same for all particles. The initial random values of all parameters are the best values so far, and, the algorithm proceeds as a classic constriction PSO. The best values for each parameter are updated when the average values of the best solutions of all particles in a specific iteration are less than the best average values of the best solutions of all particles. Three different conditions are controlling the parameters during the iterations of the algorithm. In the first one, if for a consecutive number of iterations the best solution has not been improved, the values of the number of local search iterations, the c_1 and c_2, the upper and lower bounds of the positions and of the velocities are updated as follows: $c_1 = c_{1opt} + \frac{c_1 - c_{1opt}}{c_{1opt}}$, $c_2 = c_{2opt} + \frac{c_2 - c_{2opt}}{c_{2opt}}$, $u_{positions} = UB + \frac{u_{positions} - UB}{UB}$, $l_{positions} = -u_{positions}$, $u_{velocities} = V + \frac{l_{positions} - V}{V}$, $l_{velocities} = -u_{velocities}$, $Local\ Search\ iter = LS + \frac{Local\ Search\ iter - LS}{LS}$, where c_{1opt}, c_{2opt}, UB, V and LS are the optimum values for the c_1, c_2, upper bounds in positions, upper bounds in velocities and local search iterations, respectively.

In the second condition, the increase of the number of particles is performed. If for a consecutive number of iterations, the best solution and the average best solution of all particles are not improving, then, a number of new particles are created. All new particles are initialized from random values and they use as initial values of the parameters the current values, $Particles = NP + \frac{Particles - NP}{NP}$, where NP is the number of particles in which the best so far solution has been produced.

Finally, in the third condition, the decrease of the number of particles is performed. If for a consecutive number of iterations the best solution has not been improved and the best value of a particle is more than 5% of the best value of the swarm, then, this particle is deleted from the swarm. The number of consecutive iterations was set equal to $abs(Initial\ Number\ of\ Particles - Particles)$ if the number of particles has been changed at least one time during the iterations, otherwise the value was set equal to $Initial\ Number\ of\ Particles - \frac{Max\ iterations}{abs(Max\ iterations - Local\ Search\ iter)}$. When the algorithm converges, except of the best solution, the best parameters have, also, been calculated.

3 Results and Discussion

The algorithm (implemented in Fortran 90) was tested on the 120 benchmark instances of Taillard [14]. In these instances, there are different sets having 20, 50, 100, 200 and 500 jobs and 5, 10 or 20 machines. There are 12 sets with 10 problems inside every size set (for example 20×5 means 20 jobs and 5 machines). The efficiency of the HAPSO algorithm is measured by the quality of the produced solutions. The quality is given in terms of the relative deviation from the best known solution, that is $\omega = \frac{(c_{HAPSO} - c_{BKS})}{c_{BKS}}\%$, where c_{HAPSO} denotes the cost of the solution found by HAPSO and c_{BKS} is the cost of the best known solution. To test the performance of the proposed algorithm we applied HAPSO (and the other algorithms used in the comparisons) 10 times to each test instance.

In Table 1 for each of the 12 sets we have averaged the quality of the 10 corresponding instances that belong to each one of the 12 sets. The Table is divided in three parts. In the first part, the results concerning the quality of the solutions are presented. For the first two parts of the Table, the results are referring to the best run out of the ten runs for each set. More precisely, in the first row (denoted by $HAPSO$) the average best solution in each of the 12 sets is presented. In the second row (AP), the average solution of each particle in each of the 12 sets is presented while in the third row (AO), the average best solution of each particle in each of the 12 sets is presented. In rows four to six the respective qualities of the three previously mentioned values are presented (Q is the quality for the average best solution, QP is the quality for the average solution of each particle and QA is the quality for the average best solution of each particle). In row seven, the number of optimum values based on BKS from the literature is presented while in rows eight and nine the lower (LVI) and upper (UVI) values of the qualities of the algorithm in each instance for each set are presented. In the second part the results concerning the optimization of the parameters of the PSO algorithm are presented (Parameters optimization). In the first row of the second part the average iteration number ($Iter$) where the algorithm converged for each set is presented while in the second row the average optimized number of particles (NP) needed for each set is presented. In the third and fourth rows the optimum values of c_1 and c_2 (c_{1opt} and c_{2opt}) are presented, respectively, while in fifth and sixth rows the optimum values for the upper bound (UB) for the positions and the velocities (V) are presented, respectively. The lower bounds for the positions and for the velocities are the negative values of the upper bounds. In the last row of the second part, the optimum number of local search (LS) iterations is presented. Finally, in the third part of the Table, the average values of the ten runs are presented (Ten Runs). In the first two rows, the average values ($A10$) and the average qualities ($AQ10$) of the ten runs in each set are presented while in the third and fourth rows the standard deviation ($stdev$) and the variance (var) of the ten runs are presented, respectively. Finally, in the last row the average median (M) values are presented. In Figs. 1 and 2, the optimization of the parameters in two different instances, the one is the first instance in the fist set of Taillard and it has 20 jobs and 5 machines (Fig. 1) and the other is the eight instance in the fourth set of Taillard and it has 50 jobs and 5 machines (Fig. 2), are presented.

As it can be seen the algorithm finds the best known solutions for thirty three instances, i.e. in 27.5% of all cases, for twenty six instances (in 21.66% of all cases) the quality of the solutions is between 0 and 1, for sixteen instances (in 12.8% of all cases) the quality of the solutions is between 1 and 2.5, for thirty two instances (in 25.6% of all cases) the quality of the solutions is between 2.5 and 4 and, finally, for thirteen instances (in 10.8% of all cases) the quality of the solutions is between 4 and 5.22. The instances can be divided in 5 different categories based on the number of jobs. The first category contains the instances in which the number of jobs is equal to 20 and the number of machines varies between 5 to 20. In these instances, the quality of the solutions is between 0 and 0.40 with average quality equal to 0.05. In the second category, the number of jobs is equal to 50 and the number of machines varies between 5 to 20. In these instances, the quality of the solutions is between 0 and 3.78 with average quality equal to 1.75. In the third category, the number of jobs is equal to 100 and the number of machines varies between 5 to 20. In these instances, the quality of the solutions is between 0 and 4.78 with average quality equal to 1.76. In the fourth category, the number of jobs is equal to 200 and the number of machines varies between 10 to 20. In these instances, the quality of the solutions is between 0.90 and 5.22 with average quality equal to 2.67. And, finally, in the last category, the number of jobs is equal to 500 and the number of machines is equal to 20. In these instances, the quality of the solutions is between 2.63 and 4.24 with average quality equal to 3.43.

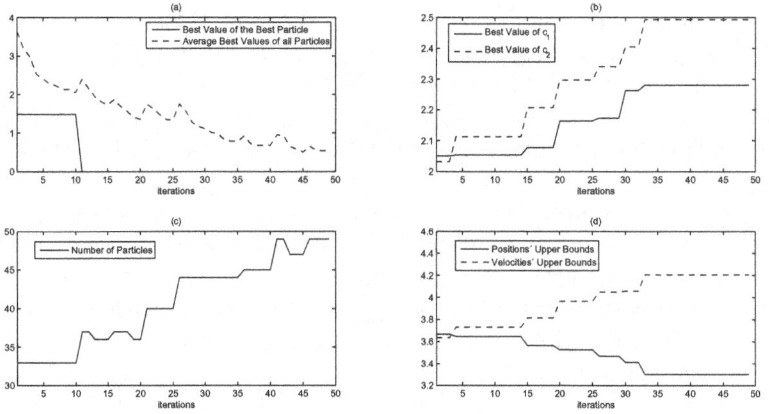

Fig. 1. Optimization of the parameters for the first instance of Taillard

The instances can, also, be divided in categories based on the number of machines. Thus, in the first category, there are the instances with number of machines equal to 5 and the number of jobs varies between 20 to 100. In these instances, the quality of the solutions is between 0.00 and 0.29 with average quality equal to 0.06. In the second category, there are the instances with number of machines equal to 10 and the number of jobs varies between 20 to 200. In these instances, the quality of the solutions is between 0.00 and 2.79 with average quality equal to 1.09. Finally, in the last category, the number of machines is equal to 20 and the number of jobs varies between 20 to 500. In these instances, the quality of the solutions is between 0.00 and 5.22 with average quality equal to 2.99. In the third part of the Table, the average values of the ten runs are presented. The values of the average qualities, the standard deviation and the variance prove that the algorithm is stable and finds in most runs very good and efficient solutions. The most important values of Table 1 are the values of the second part where the optimized parameters for each of the main parameters of the algorithm are presented. As it can be seen, the convergence of the algorithm is succeeded using between 72.6 iterations in the set 500×20 to 183.7 iterations in the set 50×20. The optimum number of particles varies between 72.6 to 150.9. The c_{1opt} and the c_{2opt} vary between 2.52 to 2.80 and 2.57 to 2.83, respectively. The upper bounds of the positions and the velocities vary between 2.48 to 3.26 and 4.75 to 5.70, respectively. Finally, the values of the local search iterations vary between 94.2 to 149.7.

In Table 2, except of the results of the proposed algorithm, the results of other five algorithms based on Particle Swarm Optimization are presented. The first algorithm, the *HybPSO* [10], is a classic PSO algorithm that uses a global neighborhood topology and a constriction factor in the velocity equation. The other four algorithms are variants of the classic Particle Swarm Optimization algorithm with different velocity equations presented and analyzed in [11]. The *Random Global Particle Swarm Optimization (RGPSO)* uses a global neighborhood topology and a random selection of one out of five different velocities' equations for each particle. The *Local Neighborhood Particle Swarm Optimization (LNPSO)*, *Random Local Neighborhood Particle Swarm Optimization (RLNPSO)* and *Particle Swarm Optimization with Expanding Neighborhood Topology (PSOENT)*, use local neighborhood topology with a constant neighborhood in LNPSO, a random neighborhood in RLNPSO and an expanding neighborhood in PSOENT. The reason that the proposed algorithm is compared analytically with these algorithms is that they use, also, exactly the same metaheuristic algorithms with the same way as in the proposed algorithm but without the optimization of the parameters presented in this paper and, thus, a comparison with them is essential. For analytical description of the algorithms and of how their parameters were selected please see [10,11].

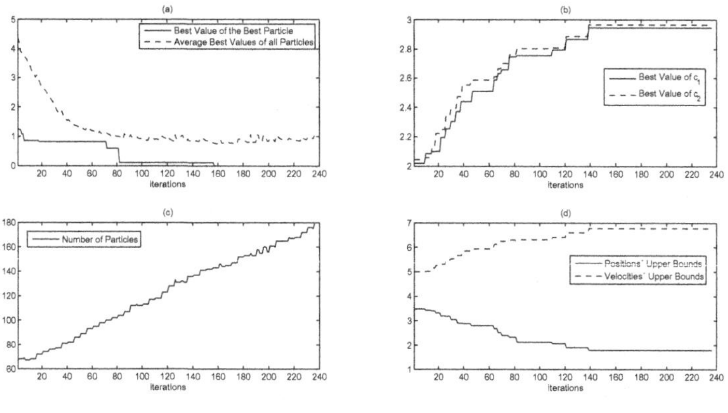

Fig. 2. Optimization of the parameters for the thirty eight instance of Taillard

In Table 2, the average quality (Q) and the average CPU times (T) in seconds of the proposed algorithm and of the other five algorithms are presented. In the first set, all the algorithms found the optimum in all instances. For the other sets, the proposed algorithm gave better results in 6 out of 11 sets as it is presented with bold letters in Table 2. The PSOENT found the best results in 6 out of 11 sets but in one set (200×20) the average results are equal to the ones found by the proposed algorithm. The most important comparison is the comparison with the HybPSO algorithm as both algorithms use the same local search algorithms, the same velocities equations (constriction factor) and the

Table 1. Analytical presentation of the results of the proposed algorithm for each set in Taillard benchmark instances for the PFSP

	20 × 5	20 × 10	20 × 20	50 × 5	50 × 10	50 × 20	100 × 5	100 × 10	100 × 20	200 × 10	200 × 20	500 × 20
Solutions and qualities optimization												
$HAPSO$	1222.4	1514.9	2236.5	2737.7	3043.3	3828.5	5251.8	5693.7	6548.4	10783.2	11737.1	27266.7
AP	1294.37	1682.38	2390.50	2927.44	3321.96	4232.97	5384.20	6159.66	7251.93	11355.29	12263.18	27752.20
AO	1232.78	1534.35	2266.48	2754.42	3096.55	3905.47	5275.99	5758.18	6650.33	10883.23	11893.24	27519.58
Q	0.00	0.09	0.07	0.05	2.01	3.20	0.14	1.17	3.97	1.06	4.27	3.43
QP	6.00	11.05	6.98	6.96	11.38	14.09	2.67	9.54	15.20	6.41	8.95	5.27
QA	0.86	1.36	1.41	0.66	3.80	5.27	0.60	2.32	5.60	2.00	5.66	4.39
$BKSN$	10	7	6	7	0	0	3	0	0	0	0	0
LVI	0.00	0.00	0.00	0.00	0.55	2.37	0.00	0.26	2.92	0.46	3.55	2.99
UVI	0.00	0.40	0.26	0.22	2.79	3.78	0.29	2.02	4.78	1.50	5.22	4.24
Parameters optimization												
$Iter$	102.5	160.3	150.1	154.3	179.3	183.7	136.2	150.7	157	131.7	111.4	72.6
NP	72.6	89.9	110.8	142.5	109.9	98.3	148.1	140.5	118.1	150.9	118.8	112.8
c_{1opt}	2.52	2.68	2.76	2.80	2.74	2.77	2.63	2.80	2.76	2.69	2.66	2.36
c_{2opt}	2.57	2.71	2.73	2.75	2.74	2.80	2.68	2.74	2.83	2.75	2.66	2.39
UB	3.26	2.75	2.80	2.56	2.61	2.59	2.64	2.68	2.48	2.58	2.66	3.20
V	4.93	5.34	5.43	5.42	5.54	5.58	5.32	5.36	5.40	5.34	5.70	4.75
LS	112.3	142.8	140.9	144.5	147.8	146.4	141.3	148.7	149.7	147.1	137.9	94.2
Ten runs												
$A10$	1222.43	1516.01	2238.05	2739.6	3045.42	3830.48	5254.03	5696.05	6551.7	10786.46	11740.58	27270.42
$AQ10$	0.00	0.16	0.14	0.12	2.08	3.25	0.18	1.21	4.03	1.09	4.30	3.44
$stdev$	0.07	1.11	1.39	1.54	1.50	1.49	1.66	1.77	2.38	2.44	2.81	3.04
var	0.05	1.33	1.98	2.40	2.29	2.24	2.82	3.18	5.80	5.96	7.97	9.52
M	1222.4	1515.75	2237.85	2739.8	3045.45	3830.55	5254.05	5696.15	6551.75	10786.4	11740.45	27270.3

Table 2. Comparisons of the results (average qualities) of HAPSO with the other five algorithms in Taillard benchmark instances for the PFSP

	HybPSO		RGPSO		LNPSO		RLNPSO		PSOENT		HAPSO	
	Q	T	Q	T	Q	T	Q	T	Q	T	Q	T
20 × 5	**0.00**	4.15	**0.00**	4.08	**0.00**	4.05	**0.00**	4.02	**0.00**	3.45	**0.00**	*1.58*
20 × 10	0.15	15.15	0.15	18.12	0.08	16.35	0.11	15.5	**0.07**	15.25	0.09	*11.21*
20 × 20	0.31	24.85	0.08	24.5	0.09	24.65	0.11	25.15	0.08	24.52	**0.07**	*18.35*
50 × 5	0.20	8.25	0.07	7.45	0.05	7.52	0.06	7.39	**0.02**	6.15	0.05	*5.24*
50 × 10	2.20	23.45	2.34	23.54	2.05	24.18	2.29	26.21	2.11	23.55	**2.01**	*20.15*
50 × 20	3.81	45.28	3.59	44.15	3.50	46.5	3.51	48.5	3.83	44.25	**3.20**	*39.22*
100 × 5	0.19	22.15	0.13	22.58	0.11	23.15	0.11	24.18	**0.09**	22.85	0.14	*18.12*
100 × 10	1.33	65.25	1.21	64.35	1.23	62.28	1.29	61.28	1.26	60.35	**1.17**	*51.11*
100 × 20	4.36	125.35	4.37	135.48	4.55	140.24	4.21	139.28	4.37	131.15	**4.13**	*118.21*
200 × 10	1.37	122.85	1.12	125.48	1.15	126.35	1.20	125.44	**1.02**	124.18	1.06	*111.21*
200 × 20	4.62	258.46	4.37	255.24	4.45	259.35	4.66	261.28	**4.27**	255.42	**4.27**	*242.34*
500 × 20	3.21	410.18	2.95	408.25	2.74	415.35	2.97	412.24	**2.73**	409.54	3.43	*380.31*
Average	1.81	93.78	1.70	94.43	1.67	95.83	1.71	95.87	1.65	93.38	**1.63**	*84.75*

same global neighborhood topology. The proposed algorithm gave better results in 10 out of 12 sets, in one they both gave equal results and in 1 the HybPSO gave better results. The improvement in the quality of solutions is between 0.05 to 0.61 in the specific 10 sets that the proposed algorithm gave better results.

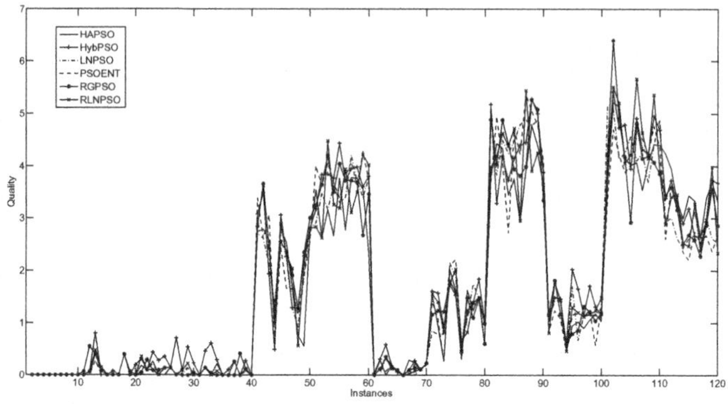

Fig. 3. Quality of solutions of the 6 algorithms for the 120 benchmark instances

The fact that the proposed algorithm performs better than the HybPSO algorithm lead us to the conclusion that the optimization of the parameters inside the algorithm will improve the results of a constriction PSO algorithm. When the proposed algorithm is compared with the RGPSO, the proposed algorithm gave better results in nine sets and worst results in two sets, while when it is compared with LNPSO, the proposed algorithm gave better results in seven sets, worst results in three sets and in one set the two algorithms gave the same average solution. Finally, when the proposed algorithm is compared with RLNPSO, the proposed algorithm gave better results in nine sets and worst results in two sets. Another interesting comparison is with the PSOENT algorithm, which is one of the most efficient algorithms for the solution of the PFSP. In this case, the proposed algorithm gave better results in five sets, worst results in five sets and in one set both algorithms gave equal average solutions. This final comparison proves that the proposed algorithm can compete a very efficient algorithm with a very strong neighborhood and in some instances to improve the results of this algorithm. Thus, the main goal of the paper which was to show if the optimization of the parameters inside the PSO algorithm can produce a very efficient algorithm was achieved. In this table, also, the average computational time for each set are presented. In the algorithms presented in the comparisons [10,11], a maximum computational time had been selected. In this algorithm the maximum computational time parameter is not used at all. However, as it is presented in the Table in every set the algorithm converged faster than every other algorithm used in the comparisons. In Fig. 3, the quality for all algorithms in the 120 instances is presented. The fluctuations in the results are due to the number of jobs, but, mainly to the number of machines. When the number of machines increases, the quality of the solutions deteriorates. In Table 3, a comparison with other algorithms from the literature is performed. There is a number of heuristic and metaheuristic algorithms that have been applied for finding of the makespan in a PFSP. Table 3 presents the average quality of the solutions of the proposed

Table 3. Comparisons of the results of HAPSO with other algorithms from the literature

Problems	HAPSO	PSOENT	NEHT	SGA	MGGA	ACGA	SGGA
20×5	0.00	0.00	3.35	1.02	0.81	1.08	1.1
20×10	0.09	0.07	5.02	1.73	1.4	1.62	1.9
20×20	0.07	0.08	3.73	1.48	1.06	1.34	1.6
50×5	0.05	0.02	0.84	0.61	0.44	0.57	0.52
50×10	2.01	2.11	5.12	2.81	2.56	2.79	2.74
50×20	3.20	3.83	6.26	3.98	3.82	3.75	3.94
100×5	0.14	0.09	0.46	0.47	0.41	0.44	0.38
100×10	1.17	1.26	2.13	1.67	1.5	1.71	1.6
100×20	4.13	4.37	5.23	3.8	3.15	3.47	3.51
200×10	1.06	1.02	1.43	0.94	0.92	0.94	0.8
200×20	4.27	4.27	4.41	2.73	3.95	2.61	2.32
500×20	3.43	2.73	2.24	-	-	-	-
Average	1.63	1.65	3.35	1.93	1.82	1.84	1.85
Problems	DDE	CPSO	GMA	PSO2	GA-VNS	ACS	PSO1
20×5	0.46	1.05	1.14	1.25	0	1.19	1.75
20×10	0.93	2.42	2.3	2.17	0	1.7	3.25
20×20	0.79	1.99	2.01	2.09	0	1.6	2.82
50×5	0.17	0.9	0.47	0.47	0	0.43	1.14
50×10	2.26	4.85	3.21	3.6	0.77	0.89	5.29
50×20	3.11	6.4	4.97	4.84	0.96	2.71	7.21
100×5	0.08	0.74	0.42	0.35	0	0.22	0.63
100×10	0.94	2.94	1.96	1.78	0.08	1.22	3.27
100×20	3.24	7.11	4.68	5.13	1.31	2.22	8.25
200×10	0.55	2.17	1.1	-	0.11	0.64	2.47
200×20	2.61	6.89	3.61	-	1.17	1.3	8.05
500×20	-	-	-	-	0.63	1.68	-
Average	1.37	3.40	2.35	2.40	0.4	1.28	4.01

algorithm (HAPSO) and the average quality of other 12 algorithms from the literature. The first one is the most important heuristic algorithm, the NEHT which is the classic NEH algorithm together with the improvement that was presented by Taillard [14]. A Simple Genetic Algorithm (SGA) [1], a Mining Genetic Algorithm [1], an Artificial Chromosome with Genetic Algorithms [1], a Self Guided Genetic Algorithm (SGGA) [1], four versions of Particle Swarm Optimization algorithm, (PSO1) [15], (CPSO) [5], (PSOENT) [11] and (PSO2) [8], a Discrete Differential Evolution (DDE) algorithm [13], a hybridization of Genetic Algorithm with Variable Neighborhood Search (GA-VNS) [17] and an

Ant Colony Optimization algorithm (ACS) [16] are given. The proposed algorithm finds the best known solution in the first set in all instances. This happens only in two more algorithm, in the GA-VNS and in PSOENT. Both algorithms have as local search phase a Variable Neighborhood Search (VNS) algorithm which lead us to the conclusion that the combination of a population based algorithm (like PSO in the proposed algorithm and in PSOENT and a genetic algorithm in GA-VNS) with a very strong local search technique like VNS increases both the exploration and exploitation abilities of the algorithm. The proposed algorithm performs better in 10 out of the 13 other algorithms used for the comparisons. The most important comparison is the one with the other four Particle Swarm Optimization algorithms. The results of the proposed algorithm in every set and the average results of all sets are better from the three implementations of PSO. The proposed algorithm performs equally well with the ACS and the PSOENT and performs slightly inferior from the GA-VNS.

4 Conclusions

In this paper, a Hybrid Adaptive Particle Swarm Optimization (HAPSO) algorithm was presented for the solution of the Permutation Flowshop Scheduling Problem. The algorithm was tested on 120 benchmark instances and was compared with other Particle Swarm Optimization algorithms and other algorithms from the literature. In this algorithm, all the parameters are optimized during the iterations together and simultaneously with the optimization of the objective function of the problem. The proposed algorithm was proved very efficient as it gave better results and faster convergence compared to other algorithms which use a set of constant parameters for all instances and in all iterations. Future research will be focused on the solution of other problems using the HAPSO and on the application of this technique in other evolutionary algorithms in order to see if this technique can be proved efficient not only for a PSO variant but, also, for any other evolutionary optimization algorithm.

References

1. Chen, S.-H., Chang, P.-C., Cheng, T.C.E., Zhang, Q.: A self-guided genetic algorithm for permutation flowshop scheduling problems. Comput. Oper. Res. **39**, 1450–1457 (2012)
2. Clerc, M., Kennedy, J.: The particle swarm: explosion, stability and convergence in a multi-dimensional complex space. IEEE Trans. Evol. Comput. **6**, 58–73 (2002)
3. Glover, F., Laguna, M., Marti, R.: Scatter search and path relinking: advances and applications. In: Glover, F., Kochenberger, G.A. (eds.) Handbook of Metaheuristics, pp. 1–36. Kluwer Academic Publishers, Boston (2003)
4. Hansen, P., Mladenovic, N.: Variable neighborhood search: principles and applications. Eur. J. Oper. Res. **130**, 449–467 (2001)
5. Jarboui, B., Ibrahim, S., Siarry, P., Rebai, A.: A combinatorial particle swarm optimisation for solving permutation flow shop problems. Comput. Ind. Eng. **54**, 526–538 (2008)

6. Johnson, S.: Optimal two-and-three stage production schedules with setup times included. Naval Res. Logist. Q. **1**, 61–68 (1954)
7. Kennedy, J., Eberhart, R.: Particle swarm optimization. In: Proceedings of 1995 IEEE International Conference on Neural Networks, vol. 4, pp. 1942–1948 (1995)
8. Liao, C.-J., Tseng, C.-T., Luarn, P.: A discrete version of particle swarm optimization for flowshop scheduling problems. Comput. Oper. Res. **34**, 3099–3111 (2007)
9. Lichtblau, D.: Discrete optimization using Mathematica. In: Callaos, N., Ebisuzaki, T., Starr, B., Abe, J.M., Lichtblau, D. (eds.) World Multi-conference on Systemics, Cybernetics and Informatics (SCI 2002). International Institute of Informatics and Systemics, vol. 16, pp. 169–174 (2002)
10. Marinakis, Y., Marinaki, M.: A hybrid particle swarm optimization algorithm for the permutation flowshop scheduling problem. In: Migdalas, A., et al. (eds.) Optimization Theory, Decision Making, and Operational Research Applications. Springer Proceedings in Mathematics and Statistics, vol. 31, pp. 91–101 (2013)
11. Marinakis, Y., Marinaki, M.: Particle swarm optimization with expanding neighborhood topology for the permutation flowshop scheduling problem. Soft. Comput. **17**(7), 1159–1173 (2013)
12. Marinakis, Y., Marinaki, M., Migdalas, A.: A multi-adaptive particle swarm optimization for the vehicle routing problem with time windows. Inf. Sci. **481**, 311–329 (2019)
13. Pan, Q.-K., Tasgetiren, M.F., Liang, Y.-C.: A discrete differential evolution algorithm for the permutation flowshop scheduling problem. Comput. Ind. Eng. **55**, 795–816 (2008)
14. Taillard, E.: Benchmarks for basic scheduling problems. Eur. J. Oper. Res. **64**, 278–285 (1993)
15. Tasgetiren, M., Liang, Y., Sevkli, M., Gencyilmaz, G.: A particle swarm optimization algorithm for makespan and total flow time minimization in the permutation flowshop sequencing problem. Eur. J. Oper. Res. **177**, 1930–1947 (2007)
16. Ying, K.C., Liao, C.J.: An ant colony system for permutation flow-shop sequencing. Comput. Oper. Res. **31**, 791–801 (2004)
17. Zobolas, G.I., Tarantilis, C.D., Ioannou, G.: Minimizing makespan in permutation flow shop scheduling problems using a hybrid metaheuristic algorithm. Comput. Oper. Res. **36**, 1249–1267 (2009)

The Measure of Regular Relations Recognition Applied to the Supervised Classification Task

Yuriy Mikheev(✉)

Saint Petersburg State University, Saint Petersburg, Russia
yuri.mikheev@gmail.com

Abstract. The probability measure of regularities recognition in an information stream is introduced in the paper. The measure allows for the creation of machine-learning models without a supervisor. The experiment described in the paper empirically proves that the measure allows the recognition of regularities and helps to find out regular relations between the values of variables.

The machine learning model finds out regular relations in data set and by the model allow reconstructing unknown values of the classification variable. The classification algorithm on the basis of the probability measure of regularities recognition is described in the paper. The measure connection with entropy is demonstrated and mutual information is used to optimise the algorithm's performance. The accuracy of the algorithm matches the accuracy of well-known supervised machine learning algorithms and also exceeds them.

Keywords: Cognition · Machine-learning · Classification · Supervised classification algorithm · Prediction · Regularity · ZClassifier · Association rules analysis

1 Human and Machine Cognition

Before computers and computer science were developed, the research cognition activity could be conducted only by studying human beings. From the middle of the 20th-century, scientists had been able to make computer models of cognition and then test the models [1].

Now, because of the computers power growth and growth in the availability of datasets, it becomes easier to make and test various cognition activity models. There are many machine-learning algorithms that are widely used today [1–4, 7, 8, 14, 15, 18, 19].

Cognition is an essential factor that influences survival. Prediction is a result of the cognition activity, and prediction accuracy is the measure of the quality of cognition. So, if the model cognition activity makes a good prediction, it could be considered as realistic and functionally relevant to the original cognition subject.

In the paper, cognition is considered as a process of regularities recognition in the input information stream. The cognition system finds the regularities, memorises it and then uses it for prediction.

G. Nicosia et al. (Eds.): LOD 2019, LNCS 11943, pp. 180–191, 2019.
https://doi.org/10.1007/978-3-030-37599-7_16

Entropy as a measure of certainty is widely used in supervised machine learning [10, 15]. Here we use entropy for unsupervised machine learning to select the regularities needed for prediction.

2 Probability Measure of Regular Relations Recognition

Let us use the following operational definition - regularity is an often-observed repeated relation between symbols in an information stream. A symbol in the information stream is the simplest item of the information stream, for example, the attribute value in a dataset, a letter in a text or user action in website monitoring data.

Relations are considered in the same way as in the set theory, for example, relations of order, similarity, identity, etc. In this paper, only the similarity relation is used, but the same approach could be easily applied to any relations. The considered measure could be applied to any relation or set of relations between symbols in the information stream.

There are simple regularities between two symbols and more complex among three and more symbols. We should consider that regularity exists if the probability of observing the relation between symbols is not random. We title the measure of deference from random probability as "The measure of regularities recognition".

Symbols in the information stream have the empirical probability of appearing p_s, where $s \in S$ is a subset of all the possible input symbols or the input dictionary. $s_z \in S$ is the subset of the symbols included in regularity. p_{s_z} is defined as the probability of the regularity between symbols in s_z.

The measure of regularities recognition is:

$$\frac{p_{s_z}}{p'_{s_z}}$$

And in the logarithm form:

$$Z = \log\left(\frac{p_{s_z}}{p'_{s_z}}\right)$$

Where $p_{s_z} = \frac{f(s_z)}{m}$, $f(s_z)$ is a frequency of the relation appearance, m is the number of observations in a data stream, p'_{s_z} is the probability of appearance $s_z \in S$ as if they are independent. From the theorem of probabilities, multiplication follows that the probability of the independent events is the multiplication of each of the probabilities:

$$p'_{s_z} = \prod_{s \in s_z} p_s$$

Then:

$$Z = \log\left(\frac{p_{s_z}}{\prod_{s \in s_z} p_s}\right) \tag{1}$$

The measure shows how big the difference between the regular and the random appearance of the symbols s_z is. The function (1) grows faster when $N = |s_z|$ and slower in case of p_{s_z}.

The measure is closely related to function lift in the association rule algorithms. In association rule works the lift function is used as one of apriori criteria for selecting rules [12–14], we are using it as measure sufficiency of information for prediction (Fig. 1).

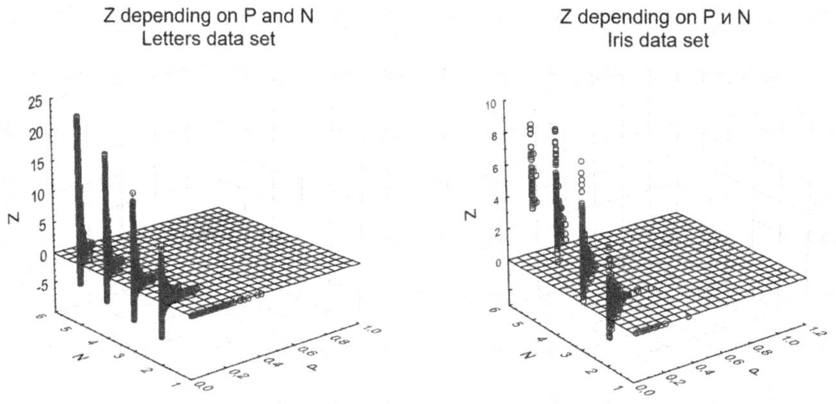

Fig. 1. Z depending on p_{s_z} and $N = |s_z|$

Expression (1) in the form of conditional probabilities:

$$\begin{aligned} Z &= \log\left(\frac{p_{s_z}}{\prod_{s \in s_z} p_s}\right) = \log\left(\frac{p(s_1|C_{-1} = \{s_2, s_3, \ldots, s_n\}) * p(C_{-1})}{\prod_{s \in s_z} p_s}\right) \\ &= \log\left(\frac{p(s_1|C_{-1}) * p(s_2|C_{-2}) * \ldots * p(s_{n-1}|s_n) * p(s_n)}{\prod_{s \in s_z} p_s}\right) \\ &= \log\left(\frac{p(s_1|C_{-1}) * p(s_2|C_{-2}) * \ldots * p(s_{n-1}|s_n)}{\prod_{s \in s_z \cap \overline{s_n}} p_s}\right) \end{aligned} \tag{1.1}$$

Where $C_{-1} = \{s_2, s_3, \ldots, s_n\}$ is a subset s_z excluding s_1, that is $s \in s_z \cap \overline{s_1}$. $C_{-2} = \{s_3, s_4, \ldots, s_n\}$ is a subset s_z excluding s_2 that is $s \in s_z \cap \overline{s_1} \cap \overline{s_2}$, and so on.

So, Z depends on the conditional probabilities $s \in s_z$. Therefore, Z could help select a set of symbols that have the strongest influence on each other. (1.1) is a very important property for the optimisation of the algorithms and selection of the most important symbols for predicting.

If Z < 0 then:

$$\frac{p_{s_z}}{\prod_{S_z} p_s} < 1$$

Regularity is not found because the probability regularity appearance is less than the random appearance of the symbols in relation s_z.

If Z $\geq$ 0 then:

$$\frac{p_{s_z}}{\prod_{S_z} p_s} \geq 1 \tag{2}$$

Regularity is recognised in the input data stream.

Thus:

$Z \in (-\infty, 0)$ – Random relation of symbols
$Z \in [0, \infty)$ – Regular relation s_z

2.1 Related Works

Association rules mining is mostly related to the paper [22]. Many works were published in the field of association rules analysis for the classification problem, for example [9, 11, 18, 21, 22]. The major problem in those works was reducing the number of rules in the model and selecting the most useful rules for the classification rules selection. We dealt with a similar problem in this paper. Unlike the mentioned works, entropy and the amount of mutual information for rule selection are used in the paper.

In association rules, approaches unusual for the probabilistic theory terms are widely used. The most popular of them are support (X), which is the same as the probability of X in the dataset, and confidence (X => Y), which is a conditional probability. In the paper, the classical probability theory terms are used in order to link with the information theory and informational entropy.

The entropy approach in the form of information gain is widely used in decision tree algorithms and regression [5, 8, 10, 15]. All decision tree algorithms are supervised machine learning kind of algorithms. The way of using the entropy approach for unsupervised learning is presented here. In the paper, a large amount of mutual information is used for calculating values of the classification variable.

2.2 Classification Task

In a classification task, we have the dataset D containing m records with observations of variables $x \in X$.

Let us be sure that the mathematical expectation of Z of $c \in D$ in all the observed combinations of the values of the variables in the dataset has full mutual information of the variables in the dataset.

Full mutual information between two systems [6, 17]:

$$I_{X \leftrightarrow Y} = E\left[\log \frac{P(X, Y)}{P(X)P(Y)}\right]$$

In this case, the systems are subsets of the variables in the dataset. If we put the variables from X in this expression we get:

$$I_{X \leftrightarrow X} = E\left[\log \frac{P(x_1, x_2, \ldots, x_n)}{\prod_{x_i \in X} P(x_i)}\right] = E[Z] \tag{3}$$

From the attributes of the full mutual information [12, 20] we get:

$$I_{X \leftrightarrow X} = E[Z] = \sum_{x \in X} H(x) - H(C),$$

Where C is a set of relations on the set of values of variables $x \in X$, $H(x) = \sum_{x_i \in X} p(x_i) \log p(x_i)$ is the entropy of the variable x, $H(C)$ is the entropy of relations values of x.

Let's express it by conditional entropy. We know:

$$H(X, Y) = H(X) + H(Y|X)$$

Thus:

$$H(C) = \sum_{x \in X} H(x) + \sum_{x \in X} \sum_{r \in X} H(x|r),$$

$$I_{X \leftrightarrow X} = E[Z] = \sum_{x \in X} H(x) - H(C) = \sum_{x \in X} H(x) - \sum_{x \in X} H(x) - \sum_{x \in X} \sum_{r \in X} H(x|r)$$

$$E[Z] = -\sum_{x \in X} \sum_{r \in X} H(x|r) = I_{X \leftrightarrow X} \tag{3.1}$$

So, if we calculate E[Z] of the variables or any set of symbols included in C, we'll have an amount of information concentrated in the set of the variables or set of symbols. The amount of mutual information shows how strong the interdependent values of the variables are and to what degree could they be used for reinstating one of them, for example, classification variable.

The following hypothesis should be tested:

1. Expression (1) could be used for recovering relations between the values of the variables in datasets and reinstating the unknown values investigated variable.
2. Using (2), we can create algorithms that recognise and use regularities for prediction.

If we succeed in it, we empirically prove the truth of (1) and usefulness of (2).

Let's be sure of it and make the algorithm on the basis of (1) and (2). We title the algorithm "ZClassifier".

3 Experimental Testing of the Measure of Regularities Recognition

In order to test the measure of regularities recognition, we'll solve the problem of supervised classification for published datasets. If ZClassifier does accurate classifications, then we will empirically prove the correctness of the measure of regularities recognition.

Using (1) we can calculate how strong the interconnected values of variables in the input dataset are.

Let data in the dataset be structured as a record $r_k \in D$. Every record is a tuple that contains values of the variables x_i, $i \in 1, \ldots, n$, n is the number of variables in the dataset.

Let S be the set of values of the variables $x_i \in X$, s_{x_i} is the set of values of the variables x_i. S is a dictionary of input dataset and $s^l_{x_i} \in S$ is a symbol from the dictionary. Thus, each record $r_k \in D$ is a tuple of s_{x_i} and every record in the input dataset contains a set of relations between symbols of the records c_k, which is the powerset $\left(\left\{s^l_{x_i} | s^l_{x_i} \in r_k\right\}\right)$. $\forall c_k \in C$, C is the set of all relations acquired in the dataset.

For each combination, we can calculate the probability $p(c_k)$ and expression (1) becomes:

$$Z(c_k) = \log\left(\frac{p(c_k)}{\prod_{s_{x_i}} p_{s^l_{x_i}}}\right) \tag{4}$$

$p_{s^l_{x_i}}$ is a probability appearance of $s^l_{x_i}$ in the dataset.

Z (4) shows how strong are the variables values' interdependencies.

The direct algorithm of the making model is:
Input: Training dataset

1. for each record from the training dataset do
2. for each relation c in the record
3. increment appearance frequency $f(c_k)$
4. Save c_k and $f(c_k)$ to the model C

The back index algorithm of the making model is:
Input: Training dataset D, I_t is the mutual information threshold

1. create back index of the training dataset
2. for all values s_x of variables $x \in D$ save s_x in list TC
3. for all relations c of the values of variables $x, y \in D$ calculate mutual information $I_c = p_c Z(c)$ and save c in list TC
4. for all $c_i, c_j \in TC$ do
5. Make $c_n = c_i \cup c_j$ and calculate $I_{c_n} = p_{c_n} Z(c_n)$ by back index
6. If ($I_{c_n} > I_t$) save c_n in TC and save c_n to the model C
7. repeat till $TC \neq \emptyset$

The classification variable recovering algorithm is:
Input: Testing Dataset, $s_{\acute{x}}$ values (symbols) of $\acute{x}$ is investigated variable, C model containing all found relations

1. for each record r from the test dataset
2. Select all relations among $s^l{}_{x_i} \in r$ and save it in C_0
3. for all $c_k \in C_0$ and each $s^l{}_{\acute{x}} \in S$ create relation $c_k \cup s^l{}_{\acute{x}}$, and calculate $Z(c_k \cup s^l{}_{\acute{x}})$,
4. if $Z(c_k \cup s^l{}_{\acute{x}}) > 0$ then save $c_k \cup s^l{}_{\acute{x}}$ in $C_{\acute{x}}$
5. $\acute{x} = argmax(Pf(s^l{}_{\acute{x}}))$, where $Pf(s_{\acute{x}})$ is one of the following functions:
   a. $Pf(s^l{}_{\acute{x}}) = \max_{c_k \in C_0} (Z(c_k \cup s^l{}_{\acute{x}})^a \, p'(c_k \cup s^l{}_{\acute{x}})^b)$
   b. $Pf(s^l{}_{\acute{x}}) = \sum_{c_k \in C_0} (Z(c_k \cup s^l{}_{\acute{x}})^a \, p'(c_k \cup s^l{}_{\acute{x}})^b)$
   c. $Pf(s^l{}_{\acute{x}}) = \frac{1}{|C_0|} \sum_{c_k \in C_0} (Z(c_k \cup s^l{}_{\acute{x}})^a \, p'(c_k \cup s^l{}_{\acute{x}})^b)$,

a and b are experimentally fit constants, $p' = \frac{p(c_k \cup s^l{}_{\acute{x}})}{p(c_k \in C_0)}$ is the conditional probability of $c_k \cup s^l{}_{\acute{x}}$, if c_k was appeared

The algorithm helps to find out the most probable $s^l_{\acute{x}}$ by the max (a) or sum (b) or mean (c) of Z function for relations containing $s^l_{\acute{x}}$ where $s^l_{\acute{x}}$ is the value of classification variable, or in other words, the number of class that best matched the record in the dataset.

We are choosing a prediction function *Pf* by experiments in order to maximise the prediction accuracy of $\acute{x}$.

If we deal with numeric values of x_i (not categorical values), we should define intervals in order to decrease the amount of $s_{x_i}^l$. In that case, accuracy is dependent on the interval borders.

3.1 Direct Classifier Algorithm Characteristics and Optimisation

3.1.1 Working Time

The making model algorithm is sensitive to the number of variables in the dataset and the true complexity of the data. Algorithm complexity is O $(2^n * m)$, where n is the number of variables and m is the number of records. Time and memory usage exponentially grow with the growth in the number of variables.

Restricting the maximum amount of symbols in relations sufficiently decreases the working time. In most cases, 5–7 symbols max in relation are enough for excellent accuracy. Complexity in those cases is O $\left(C_i^7 * m\right)$, which is significantly less than O $(2^n * m)$ where $C_n^k = \frac{n!}{(n-k)!k!}$ – is Binomial coefficient.

Another way to optimise the time performance is by reducing the number of variables in the dataset. In practice, to make a prediction, we only need variables that correlate with the classification variable. We can estimate how the strong values of variables influence each other by using (1.1). The variable value influence is the input for the amount of mutual information of relations containing the variable value.

Variable value influence estimation algorithm:
Input: Model C is the set of all relations acquired in the dataset
1. for all $c \in C$
2. for all s from c
3. $influence(s) = p_c Z(c) - p_{c \cap \bar{s}} Z(c \cap \bar{s})$

The influence shows how important the variable value is for prediction and helps to decide if it is worth using the variable in the model or not. Variables could be selected by influence.

3.1.2 Model Size

The result of the training stage is a huge amount of relations. To reduce the model size, we can use (3.1) and select only the informative relations at the pruning stage.

Pruning algorithm:
Input: Model C is the set of all relations acquired in the dataset
$minInformation$ – minimal information of c
1. for all $c \in C$ (C is the set of all relations acquired in the dataset)
2. if $p_c Z(c) < minInformation$ then remove c from C

Pruning allows a reduction in the model size by several times without any losses in accuracy (Fig. 2).

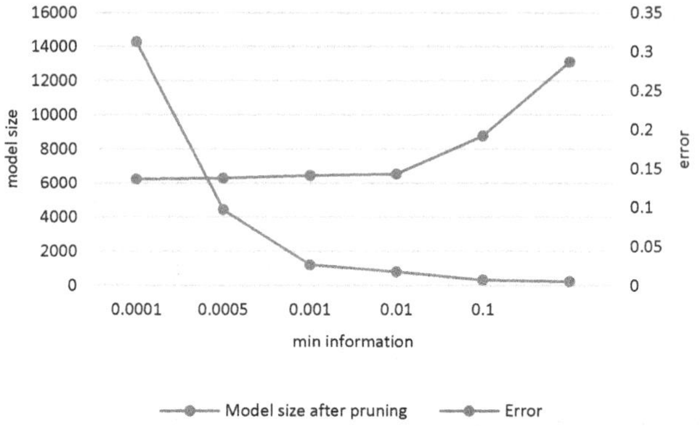

Fig. 2. Reducing model size and error growth after pruning (Adult dataset)

3.2 Back Index ZClassifier Algorithm Complexity

Formally, the algorithm complexity is O $(2^n * m * l^2)$, where n is the number of variables, m is the number of records and l is the number of relations in the sorted list. That is, the direct algorithm will have lower complexity than the back index algorithm.

In practice, the back index ZClassifier algorithm works faster with datasets that have a larger amount of variables because the back index algorithm deals with "good" relations having a high amount of mutual information. A model improbable is needed to have 2^n relations for accurate prediction.

For example, the Home Credit Default Risk competition dataset from Kaggle has 123 variables in the main data table. Direct ZClassifier can create models that contain relations with a max length of 3 symbols. Back index ZClassifier creates a model without restrictions on the relation length for less than 30 min. The size comparison of the algorithms models is provided in Table 2 and the accuracy is provided in Table 3.

4 The Results of the Experiments

ZClassifier was applied to several UCI datasets. Accuracy was estimated by kFold with the k most appropriate to the dataset. Datasets are described in Table 1.

The meanings of k in kFold were chosen in order to provide enough size of training data. So, for the small datasets Zoo and Iris, k was equal to 2 and we performed the experiment 5 times.

Compare the ZClassifier results to other supervised classification algorithms [10] (Table 2):

From the results, we can conclude that classification on the basis (1) has a comparable or better accuracy [13].

Gradient boosting [17] shows better results but the boosting process takes a lot of time for iterations. ZClassifier helps in finding out a good decision by one iteration and could also be boosted in a similar way as other classification algorithms. This task is planned to be solved in the future.

Table 1. UCI dataset in the experiment

Dataset	Variables	Direct ZClassifier		Back index ZClassifier		Records	kFold's k
		Used variables	Working time (sec)	Used variables	Working time (sec)		
Mushrooms	17	17	28,9	17	8,6	8 125	10
Adult	14	4	3,4	4	2,5	48 842	10
Census	40	4	17,2	4	9,8	299 291	10
Letters	17	15	2 920	15	20,8	20 000	10
Zoo	17	17	122	17	2,5	160	5 times k = 2
Iris	4	4	1,5	4	1,5	149	5 times k = 2
Segments	19	13	60,03	13	9,3	2310	10

Table 2. Direct and Back Index algorithms complexity comparison

Dataset	Direct ZClassifier		Back index ZClassifier	
	Relations in model	Relations after pruning	Relations in model	Relations after pruning
Mushrooms	53 168	5 108	3 737	2 634
Adult	12 920	1 622	3 024	2 120
Census	49 199	5 804	19 692	5 507
Letters	51 304 560	46 890 748	34 355	22 823
Zoo	1 040 992	–	1 060	–
Iris	608	608	679	–
Segments	6 379 889	–	11 132	–

Table 3. The comparison of the resulting accuracy

Dataset	NB-Tree	Naïve Bayes	C4.5	Gradient boosting	CMAR [21]	Direct ZClassifier	Back index ZClassifier
Mushrooms	100	95,52	100	100		100 ± 0	99,17 ± 0,56
Adult	85,9	83,88	85,54	87,53		86,18 ± 0,91	86,1 ± 0,75
Census		76,8	95,2	95,4		94,87 ± 0,16	94,82 ± 0,2
Letter	85,66	64,07	88,03	93,4		91,42 ± 0,40	32,52 ± 1,35
Zoo	93,07	94,97	92,61		97,1	95,12 ± 0,48	95,0 ± 3,0
Iris	95,4	95,53	94,73	95.83	94	92,21 ± 0,75	92,8 ± 2,84
Segments	95,34	80,17	96,79	99,9		98,11 ± 0,24	79,81 ± 2,16

5 Conclusions

The measure of regularities recognition is introduced in the paper. ZClassifier is an algorithm created in order to demonstrate that the measure of regularities recognition is working and that the measure shows good results in the classification task.

The hypotheses that the expression (1) could be used for recovering relations between values of the variables in datasets and reinstating unknown values investigated variable and that (2), we can create algorithms that recognise and use regularities for prediction are proven.

The experiment shows that condition (2) decreases the algorithm work time. Also pruning, based on the most informative relations selection, allows a sufficient reduction in the model size.

The measure of regularities recognition could be applied for regularities recognition in data streams having various structures of data. For example, the author created the algorithm for recognition regularities in English and Russian texts. The algorithm can distinguish words and regular phrases from the texts.

An example of the entropy approach to unsupervised machine learning was demonstrated in the paper. The measure of regularities recognition could be used in other algorithms. In the future, on the basis of the measure of regularities recognition, we are going to create unsupervised machine-learning algorithms for various types of data.

References

1. Ben-Hur, A., et al.: Support vector clustering. J. Mach. Learn. Res. **2**, 125–137 (2001)
2. Bertsekas, D.P., Bertsekas, D.P.: Dynamic Programming and Optimal Control: Approximate Dynamic Programming, vol. II. Athena Scientific (2012)
3. Hawkins, J., Blakeslee, S.: On Intelligence. Times Books (2004)
4. Brown, J.D.: Principal components analysis and exploratory factor analysis – definitions, differences and choices. Shiken: JALT Test. Eval. SIG Newsl. **13**, 26–30 (2009)
5. Cai, X., Sowmya, A.: Level learning set: a novel classifier based on active contour models. In: Kok, J.N., Koronacki, J., Mantaras, R.L., Matwin, S., Mladenič, D., Skowron, A. (eds.) ECML 2007. LNCS (LNAI), vol. 4701, pp. 79–90. Springer, Heidelberg (2007). https://doi.org/10.1007/978-3-540-74958-5_11
6. Ventcel, E.S.: Teoria veroytnostey. Вентцель, Е. С. (1999). Теория вероятностей: учебник для вузов. - Moscow (1999)
7. Estivill-Castro, V.: Why so many clustering algorithms – a position paper. ACM SIGKDD Explor. Newsl. **4**(1), 65–75 (2002)
8. Freedman, D.A.: Statistical Models: Theory and Practice. Cambridge University Press, Cambridge (2005)
9. Hahsler, M., Hornik, K., Reutterer, T.: Implications of probabilistic data modeling for rule mining. Research Report Series, Department of Statistics and Mathematics, 14 (2005)
10. Jiang, L., Li, C.: Scaling up the accuracy of decision-tree: a Naive-Bayes combination. J. Comput. **6**(7), 1325–1331 (2011)
11. Kliegr, T.: Quantitative CBA: Small and Comprehensible Association Rule Classification Models. School of Electronic Engineering and Computer Science, Queen Mary University of London, United Kingdom, and Faculty of Informatics and Statistics, VSE, Czec (2017)
12. MacKay, D.J.C.: Information Theory, Inference, and Learning Algorithms, 1st edn, p. 34. Cambridge University Press, Cambridge (2003)
13. Caruana, R., Niculescu-Mizil, A.: An Empirical Comparison of Supervised Learning Algorithms. Department of Computer Science, Cornell University, Ithaca, NY 14853 USA (2006)
14. Rish, I.: An Empirical Study of the Naïve Bayes Classifier. In: IJCAI Work Empirical Methods Artificial Intelligence, vol. 3 (2001)

15. Rokach, L., Maimon, O.: Data Mining with Decision Trees. Theory and Applications. Ben-Gurion University of the Negev, Tel-Aviv University, Israel (2007)
16. Shannon, C., Weaver, W.: The mathematical. Theory communication. Bell Syst. Tech. J. **27**, 379–423, 623–656 (1948)
17. Sigrist, F.: Gradient and Newton. Boosting for Classification. Lucerne University of Applied Sciences and Arts (2019)
18. Srikant, R., Agrawal, R.: Fast algorithms for mining association rules. In: 20th International Conference on Very Large Data Bases, Santiago, Chile, pp. 487–499 (1994)
19. Russell, S.J., Norvig, P.: Artificial Intelligence: A Modern Approach. Prentice Hall (2009)
20. Vapnik, V.: The Nature of Statistical Learning Theory. Springer, New York (1999)
21. Li, W., Han, J., Pei, J.: CMAR: accurate and efficient classification based on multiple class-association rules. School of Computing Science, Simon Fraser Universit. IEEE International Conference on Data Mining (2001)
22. Yin, X., Han, J.: CPAR: Classification Based on Predictive Association Rules. University of Illinois at Urbana-Champaign (2003)

Simple and Accurate Classification Method Based on Class Association Rules Performs Well on Well-Known Datasets

Jamolbek Mattiev[1] and Branko Kavšek[1,2](✉)

[1] University of Primorska, Glagoljaška 8, 6000 Koper, Slovenia
jamolbek.mattiev@famnit.upr.si, branko.kavsek@upr.si
[2] Jožef Stefan Institute, Jamove cesta 39, 1000 Ljubljana, Slovenia

Abstract. Existing classification rule learning algorithms use mainly greedy heuristic search to find regularities in datasets for classification. In recent years, extensive research on association rule mining was performed in the machine learning community on learning rules by using exhaustive search. The main objective is to find all rules in data that satisfy the user-specified minimum support and minimum confidence constraints. Although the whole set of rules may not be used directly for accurate classification, effective and efficient classifiers have been built using these, so called, classification association rules.

In this paper, we compare "classical" classification rule learning algorithms that use greedy heuristic search to produce the final classifier with a class association rule learner that uses constrained exhaustive search to find classification rules on "well known" datasets. We propose a simple method to extract class association rules by simple pruning to form an accurate classifier. This is a preliminary study that aims to show that an adequate choice of the "right" class association rules by considering the dependent (class) attribute distribution of values can produce a compact, understandable and relatively accurate classifier. We have performed experiments on 12 datasets from UCI Machine Learning Database Repository and compared the results with well-known rule-based and tree-based classification algorithms. Experimental results show that our method was consistent and comparative with other well-known classification algorithms. Although not achieving the best results in terms of classification accuracy, our method is relatively simple and produces compact and understandable classifiers by exhaustively searching the entire example space.

Keywords: Attribute · Frequent itemset · Minimum support · Minimum confidence · Class association rules (CAR) · Associative classification

1 Introduction

Huge amounts of data are being generated and stored every day in corporate computer database systems. Mining association rules from transactional data is becoming a popular and important knowledge discovery technique [14]. For example, association rules (ARs) of retail data can provide valuable information on customer buying behavior. The number of rules discovered in a "real-life" dataset can easily exceed 10,000.

G. Nicosia et al. (Eds.): LOD 2019, LNCS 11943, pp. 192–204, 2019.
https://doi.org/10.1007/978-3-030-37599-7_17

To manage this knowledge, rules have to be pruned and grouped, so that only a reasonable number of rules have to be inspected and analyzed.

Classification rule mining aims at discovering a small set of rules in the database that forms an accurate classifier. Association rule mining finds all the rules existing in the database that satisfy some minimum support and minimum confidence constraints.

The classification problem has been extensively studied by the research community. Various types of classification approaches have been proposed (e.g., KNN [19], Bayesian classifiers [3], decision trees [17], neural networks [20], and associative classifiers [15, 16]). In recent years, associative classification has been investigated widely. It integrates association rule mining algorithm and classification. Associative classification induces a set of classification association rules from the training dataset which satisfies certain user-specified frequency and confidence. Then it selects a small set of high quality association classification rules and uses this rule set for classification. The experimental results from [7, 15] indicate that associative classification achieves higher accuracy than some traditional classification approaches such as decision trees. In comparison with some traditional rule-based classification approaches, associative classification has two characteristics: (1) it generates a large number of association classification rules and (2) the measure support and confidence are used for evaluating the significance of classification association rules. However, associative classification has some weaknesses. First, it often generates a very large number of classification association rules in association rule mining, especially when the training dataset is large and dense. It takes great efforts to select a set of high quality classification rules among them. Second, the accuracy of associative classification depends on the setting of the minimum support and the minimum confidence. Although associative classification has some drawbacks, it achieves higher accuracy than rule and tree based classification algorithms on most of real life datasets.

In this work, we propose a classification method that selects a reasonable number of rules for classification. We have performed experiments on 12 datasets from UCI Machine Learning Database Repository [10] and compare the experimental result with 8 well-known classification algorithms (Naïve Bayes [3], Decision Table [13], OneR [12], Ripper [9], PART [11], C4.5 [17], Prism [6], Random Forest [5]).

The rest of the paper is organized as follows. Section 2 highlights the related work to our research work. Preliminary concepts are provided in Sects. 3. Section 4 gives the details of problem statement. Generating the class association rules are given in Sect. 5. In Sect. 6, our proposed method is described. Section 7 provides with details (about datasets, results) of Experimental Evaluation. Conclusions and future plans are given in Sect. 8. Acknowledgement and References close the paper.

2 Related Work

Our proposed method is partially similar to CPAR [18] method but the only difference is that CPAR adopts a greedy algorithm to generate association rules directly from training dataset, instead of generating a large number of candidate rules as in associative classification methods, while our proposed method generates strong association rules from frequent itemsets.

CMAR [15] associative classification method also uses multiple association rules for classification. This method extends FP-growth (an efficient frequent pattern mining method) to mine large datasets. The difference between our method and CMAR is that CMAR stores and retrieves mined association rules by CR-tree structure and prunes rules based on *confidence*, *correlation* and *database coverage*. In our method rules are pruned based on minimum *support* and *confidence* constraints and our method uses "Apriori" algorithm to generate frequent itemsets, instead of FP-tree.

Our proposed method uses strong class association rules for classification and is different from existing classification rule based or tree based methods, such as C4.5, PART, CN2 [8], Prism and Ripper. All of these methods use heuristic search to learn the regularities in dataset to build a classifier, while our method uses exhaustive search to learn the rules and aims to find all rules in the dataset. Some classification method's rule selection technique in classifier building is related to the traditional *covering method*, which is used in [6, 9]. The covering method works as follows: A rule set is built in a greedy way, one rule at a time. After a rule is found, all examples that are covered by the rule are removed. This process is repeated until there are no training examples left. The key difference between our proposed method and the covering methods is in the rules that they use. Our method learns all rules from the whole training dataset by using exhaustive search, while each rule in the covering methods is learnt using a heuristic search from the remaining dataset after all examples covered by previous rules are removed.

In data mining [4], association rule miner is used to generate high confident rules (greater than 90%). Association rule miner is applied to find rules that can describe individual classes in [2]. In both works, researchers did not intend to build a classifier.

3 Preliminary Concepts

An association rule has two parts, an antecedent (if) and a consequent (then). An antecedent is an item found in the data. A consequent is an item that is found in combination with the antecedent. Association rules are created by analyzing data for frequent if/then patterns and using the criteria *support* and *confidence* to identify the most important relationships. *Support* is an indication of how frequently the items appear in the dataset. *Confidence* indicates the number of times the if/then statements have been found to be true.

Let D be a dataset with n attributes $\{A_1, A_2, \ldots, A_n\}$ that are classified into M known classes and $|D|$ objects. Let $Y = \{y_1, y_2 \ldots . y_m\}$ be a list of class labels. A specific value of an attribute A_i and class Y is denoted by lower-case letters a_{in} and y_j respectively.

Definition 1. An itemset is a set of some pairs of attributes and a specific value, denoted $\{(A_{i1}, a_{i1}), (A_{i2}, a_{i2}), \ldots, (A_{in}, a_{in})\}$.

Definition 2. A class association rule R has the form $\{(A_{i1}, a_{i1}), .., (n, a_{in})\} \rightarrow y_j$ where $\{(A_{i1}, a_{i1}), .., (A_{in}, a_{in})\}$ is an itemset and $y_j \in Y$ is a class label.

Definition 3. The actual occurrence *ActOcc(R)* of a rule *R* in *D* is the number of records of *D* that match *R*'s antecedent (left-hand side).

Definition 4. The support of rule *R*, denoted by *Supp(R)*, is the number of records of *D* that match *R*'s antecedent and are labeled with *R*'s class.

Definition 5. The confidence of rule R, denoted by *Conf(R)*, is defined as follows:

$$Conf(R) = Supp(R)/ActOcc(R). \quad (1)$$

4 Problem Statement

Our proposed research assumes that the dataset is a normal relational table, which consists of *N* cases described by *L* distinct attributes. These *N* cases have been classified into *M* known classes. An attribute can be a categorical (or discrete) or a continuous (or numeric) attribute. In this work, we treat all the attributes uniformly. For a categorical attribute, all the possible values are mapped to a set of consecutive positive integers. For a continuous attribute, its value range is discretized into intervals (bins), and the intervals are also mapped to consecutive positive integers. With these mappings, we can treat a dataset as a set of (attribute, integer-value) pairs and a class label. We call each (attribute, integer-value) pair an item. Discretization of continuous attributes will not be discussed in this paper as there are many existing algorithms in the machine learning literature that can be used. Let *D* be the dataset and $I = \{x_1, x_2, \ldots, x_n\}$ be a set of items in *D*. A set $X \subseteq I$ is called an itemsets and $Y = \{y_1, y_2, \ldots, y_m\}$ be the set of class labels. We say that a data case $d \in D$ contains $X \subseteq I$, a subset of items, if $X \subseteq d$. A class association rule is an implication of the form $X \rightarrow y_j$, where $X \subseteq I$, and $y_j \in Y$. A rule $X \rightarrow y_j$ holds in *D* with confidence *conf* if *conf* % of cases in *D* that contain *X* are labeled with class y_j. The rule $X \rightarrow y_j$ has support *supp* in *D* if *supp*% of the cases in *D* contain *X* and are labeled with class y_j.

Our objectives are (1) to generate the complete set of CARs that satisfy the user-specified minimum support and minimum confidence constraints, and (2) to extract the reasonable number of strong CARs by grouping to build a simple and accurate classifier.

5 Generating the Complete Set of CARs

Association rule generation is usually split up into two separate steps:

1. First, minimum support is applied to find all frequent itemsets in a database.
2. Second, these frequent itemsets and the minimum confidence constraint are used to form rules.

5.1 Finding the Frequent Itemsets Using Candidate Generation

While the second step is straight forward, the first step needs more attention. Apriori is a seminal algorithm described in [1] for mining frequent itemsets for association rules. The name of the algorithm is based on the fact that the algorithm uses *prior* knowledge of frequent itemset properties, as we shall see following. Apriori employs an iterative approach known as a *level-wise* search, where k-itemsets are used to explore $(k + 1)$-itemsets. First, let L be a set of frequent itemsets, the set of frequent 1-itemsets is found by scanning the database to accumulate the count for each item and collecting those items that satisfy minimum support. The resulting set is denoted L_1. Next, L_1 is used to generate L_2, the set of frequent 2-itemsets, which is used to generate L_3, and so on, until no more frequent k-itemsets can be found. The finding of each L_k requires one full scan of the database.

Finding all frequent itemsets in a database is difficult since it involves searching all possible itemsets (item combinations). The set of possible itemsets is the power set over L and has size $2^n - 1$ (excluding the empty set which is not a valid itemset). Although the size of the powerset grows exponentially in the number of items n in L, efficient search is possible using the downward-closure property of support (also called anti-monotonicity) which guarantees that for a frequent itemset, all its subsets are also frequent and thus for an infrequent itemset, all its supersets must also be infrequent.

Algorithm 1: Apriori algorithm for finding the frequent itemsets

Input: *D*, a database of transactions, *min_sup*, the minimum support count threshold
Output: *L*, frequent itemsets in *D*

```
1:   L_1 = find_frequent_1-itemsets(D);
2:   for (k= 2; L_{k-1} !=∅; k++) do begin
3:       C_k = apriori_gen( L_{k-1} );
4:       for each transaction t ∈ D do begin // scan D for counts
5:           C_t =subset( C_k ,t); // get the subset of t that are candidates
6:           for each candidate c ∈ C_t do begin
7:               c.count++; // Increment the count of all candidates in C_k that
8:               are contained in t
9:           end
10:      L_k ={ c ∈ C | c.count ≥ min_sup };
11:      end
12:  end
13:  return L=∪_k L_k ;
```

Line 1 represents that Apriori finds the frequent 1-itemsets, L_1. In line 2–12, L_{k-1} is used to generate candidate C_k in order to find L_k for $k \geq 2$. The apriori_gen procedure generates the candidates and then uses to eliminate those having a subset that is not frequent (line 3). Once all of the candidates generated, the database is scanned (line 4). For each transaction, a subset function is used to find all subsets of the transaction that are candidates (line 5), and the count for each of those candidates satisfying minimum support (line 6–10) form the set of frequent itemsets, L (line 13).

5.2 Generating the Class Association Rules from Frequent Itemsets

Once the frequent itemsets from transactions in a database have been found, it is straightforward to generate strong class association rules from them (where strong CARs satisfy both minimum support and minimum confidence). This can be done using following equation for confidence:

$$confidence(A \rightarrow B) = \frac{support\_count(A \cup B)}{support\_count(A)}. \quad (2)$$

The Eq. (2) is expressed in terms of itemsets support count, where A is premises (itemsets that is left-hand side of the rule), B is consequence(class label that is right-hand side of the rule), *support_count*($A \cup B$) is the number of transactions containing the itemsets $A \cup B$, and *support_count*(A) is the number of transactions containing the itemsets A. Based on this equation, CARs can be generated as follows:

- For each frequent itemsets L and class label C, generate all nonempty subsets of L.
- For every nonempty subset S of L, output the rule "$S \rightarrow C$" if $\frac{support\_count(L)}{support\_count(S)} \geq$ min_*conf*, where min_*conf* is the minimum confidence threshold.

6 Definition of Our Proposed Method

Once the class associations rules are generated in 5.2 based on minimum support and minimum confidence thresholds, then, we will form a simple and accurate classifier by extracting the reasonable number of strong CARs. Our proposed method is outlined in Algorithm 2. First line finds all frequent itemsets in the dataset, strong class association rules that are satisfied the minimum support and confidence constrains are generated from frequent itemsets in line 2. In line 3, CARs are sorted in *confidence* and *support* descending order as follow:

Given two rules R_1 and R_2, R_1 is said having higher rank than R_2,, denoted as $R_1 > R_2$,

- If and only if, $conf(R_1) > conf(R_2)$; or
- If $conf(R_1) = conf(R_2)$ but, $supp(R_1) > supp(R_2)$: or
- If $conf(R_1) = conf(R_2)$ and $supp(R_1) = supp(R_2), R_1$ has fewer attribute values in its left-hand side than R_2 does;

Line 4 illustrates the grouping of class association rules by their class labels (for example, if the class has three values, then, rules are clustered into three groups). From line 5–8, we extract the reasonable number of rules per class that are equal to *numrules* to form a simple and accurate classifier. These set of rules become our final classifier. In line 9–12, classification is performed by extracted CARs in line 5–8, if the rule can classify the example correctly, then, we increase the corresponding class count by one and store the it. By line 13–16, If none of the rules can classify the example correctly, then, algorithm returns the majority class value in training dataset. Otherwise, it returns the majority class value of correctly classified rules.

Algorithm 2: (Simple and accurate classification algorithm)

Input: a set of CARs with their *support* and *confidence* constraints
Output: a subset of rules for classification

```
1:    F= frequent_itemsets(D);
2:    R= genCARs(F);
3:    R= sort(R, minconf, minsup);
4:    G=Group(R);
5:    for (k=1; k≤ numClass; k++) do begin
6:        X= extract(class[k], numrules);
7:        Y= Y.add(X);
8:    end
9:      for each rule y ∈ Y do begin
10:         if y classify new_example then
11:             class_count[y.class]++;
12:     end
13:   if max(class_count)==0 then
14:         predicted_class=majority_class(D);
15:   else predicted_class= index_of_max(class_count);
16:   return predicted_class
```

Example 2. Let us assume that we have following class association rules (satisfied the user specified minimum support and confidence thresholds) with support and confidence which were generated in previous section:

CARs	Support	Confidence
{al = 1, a3 = 2, a4 = 1, a5 = 3} → {1}	4	72%
{al = 1, a2 = 5, a5 = 2} → {3}	7	100%
{al = 1, a3 = 5, a4 = 4} → {3	9	95%
{a2 = 3, a3 = 4, a4 = 4, a5 = 4} → {1}	3	100%
{al = 2, a2 = 3, a3 = 1} → {2}	11	85%
{a3 = 1, a4 = 1, a5 = 2} → {1}	8	90%
{al = 1, a2 = 5, a3 = 4, a5 = 2} → {2}	11	85%
{al = 1, a2 = 2, a3 = 1, a4 = 3, a5 = 1} → {3}	4	64%
{a2 = 2, a5 = 5} → {2}	6	69%

In the first step, we obtain the following result by sorting the CARs in *confidence* and *support* descending order:

CARs	Support	Confidence
{al = 1, a2 = 5, a5 = 2} → {3}	7	100%
{a2 = 3, a3 = 4, a4 = 4, a5 = 4} → {1}	3	100%
{al = 1, a3 = 5, a4 = 4} → {3	9	95%
{a3 = 1, a4 = 1, a5 = 2} → {1}	8	90%
{al = 2, a2 = 3, a3 = 1} → {2}	11	85%
{al = 1, a2 = 5, a3 = 4, a5 = 2} → {2}	11	85%
{al = 1, a3 = 2, a4 = 1, a5 = 3} → {1}	4	72%
{a2 = 2, a5 = 5} → {2}	6	69%
{al = 1, a2 = 2, a3 = 1, a4 = 3, a5 = 1} → {3}	4	64%

In the next step, CARs are distributed into groups that are equal to number of classes. We can achieve this result by sorting the CARs on class label.

CARs	Support	Confidence
{a2 = 3, a3 = 4, a4 = 4, a5 = 4} → {1}	3	100%
{a3 = 1, a4 = 1, a5 = 2} → {1}	8	90%
{al = 1, a3 = 2, a4 = 1, a5 = 3} → {1}	4	72%
{al = 2, a2 = 3, a3 = 1} → {2}	11	85%
{al = 1, a2 = 5, a3 = 4, a5 = 2} → {2}	11	85%
{a2 = 2, a5 = 5} → {2}	6	69%
{al = 1, a2 = 5, a5 = 2} → {3}	7	100%
{al = 1, a3 = 5, a4 = 4} → {3	9	95%
{al = 1, a2 = 2, a3 = 1, a4 = 3, a5 = 1} → {3}	4	64%

Last but not least, we extracted the highly qualitative rules (already sorted by support and confidence) that are equal to intended user specified *numrules* parameter for our classifier, let us apply *numrules* is equal to 2:

CARs	Support	Confidence
1. {a2 = 3, a3 = 4, a4 = 4, a5 = 4} → {1}	3	100%
2. {a3 = 1, a4 = 1, a5 = 2} → {1}	8	90%
3. {al = 2, a2 = 3, a3 = 1} → {2}	11	85%
4. {al = 1, a2 = 5, a3 = 4, a5 = 2} → {2}	11	85%
5. {al = 1, a2 = 5, a5 = 2} → {3}	7	100%
6. {al = 1, a3 = 5, a4 = 4} → {3}	9	95%

Now, we classify the following example;

$$\{A1 = 1, a2 = 5, a3 = 5, a4 = 4, a5 = 2\} \rightarrow ?$$

So, this example is classified by 4, 5, 6 rules. Class label of the rules that classified the example correctly are 2, 3, 3. Our classifier predicts that the class of new example is 3 (because third class value has max count).

7 Experimental Evaluation

To evaluate the accuracy of our method (SA), we have performed two experiments on 12 datasets from UCI Machine Learning Database Repository and compared the results with 8 classification algorithms namely, Naive Bayes (NB), Decision Tree (DT), C4.5, PART (PT), 1R, Prism (PR), Random Forest (RF) and JRip (JR). The Weka software is used to explore all 8 classification algorithms.

***Experiment 1*:** In the first experiment we set up the same default parameters for all datasets and all classification algorithms. For all classification algorithms and our method, we used percentage split (75% is for learning, 25% is for testing, preserved order of split) method to split the dataset into learning and test sets. All classification algorithms parameters are default values. For our method, we set up the following default values for the first experiment: *minsup = 0.5%, minconf = 50%, num-rules = 10*. The results of first experiment are shown in Table 1.

Table 1. The comparison between our method and some classification algorithms

Dataset	# attr	# Cls	# recs	Accuracy (%)								
				NB	DT	C4.5	PT	1R	JR	PR	RF	SA
Breast cancer	10	2	286	78	79	69	68	69	78	62	71	81
Balance	5	3	625	89	67	64	76	57	69	59	80	76
Car.Evn	7	4	1728	84	59	66	77	59	62	59	64	59
Vote	17	2	435	91	96	95	94	98	97	95	99	95
Tic-Tac	10	2	958	73	77	86	93	74	99	95	96	86
Contact lenses	5	3	24	83	83	83	83	66	83	83	83	83
Nursery	9	5	12960	92	94	96	99	71	96	96	98	80
Mushroom	23	2	8124	94	100	100	100	98	100	100	100	92
Hayes	6	3	160	81	54	75	78	54	84	72	84	72
Lymp	19	4	148	89	64	72	81	72	70	72	75	72
Monks	7	2	554	73	100	100	100	73	100	99	100	90
Spect.H	23	2	267	80	93	93	89	93	93	80	97	89
Average accuracy (%)				83.9	80.5	83.3	86.5	73.7	85.9	81	87.3	81.3

Table 1 shows that in the first experiment, our proposed method achieved better average accuracy than Decision Table, OneR and Prism algorithms with 81.3%, while Random Forest method gained the highest average accuracy.

Experiment 2: All classification algorithms parameters are also set up as default values in the second experiment and it is used percentage split (66% is for learning, 34% is for testing, preserved order of split) method to split the dataset into learning and test sets, it is also used the randomize function on some datasets to apply percentage split method with preserved order function. We test both decision tree methods and rule based method. We report the accuracy of both methods.

For our method, we set up different default parameters for different datasets to obtain intended number of rules per class and to increase the accuracy. We also take into consideration the class distribution to keep balance when we are extracting the rules, for example, if the class has three values that distributed 60%, 20%, 20% respectively, then, our classifier should have 20, 5, 5 rules approximately for each class and we apply *numrules* higher to achieve intended number of rules for each class. *#attr* (number of attributes), *#cls* (number of class values), #recs (number of records) are same as Table 1. Results are shown in Table 2.

Table 2. The results of second experiment (with different default parameters)

Data set	Min Sup	Min Conf	#Rules per Class	Accuracy (%)								
				NB	DT	C4.5	PT	1R	JR	PR	RF	SA
Breast cancer	1%	75%	10	77	74	75	71	71	70	60	69	80
Balance	0.3%	60%	15	90	65	65	76	57	71	60	76	74
Car.Evn	0.5%	60%	72	72	62	65	72	59	61	61	64	63
Vote	1%	80%	15	92	94	95	97	96	96	91	97	94
Tic-Tac	1%	90%	30	70	73	79	95	73	98	92	94	92
Contact lenses	1%	80%	6	75	75	75	75	63	75	75	75	75
Nursery	2%	60%	55	89	93	95	98	71	93	96	98	89
Mushroom	25%	90%	100	94	100	100	100	98	100	100	100	94
Hayes	0.5%	50%	20	86	57	71	64	51	91	73	82	71
Lymp	5%	80%	18	81	76	76	74	72	74	68	74	74
Monks	1%	70%	30	74	100	92	100	74	100	94	98	92
Spect.H	3%	60%	25	81	92	92	89	92	92	70	95	88
Average accuracy (%)				81.8	80.1	81.7	84.3	73.1	85.1	78.3	85.2	82.2

As can be seen from the results, our proposed method outperforms Naive Bayes, OneR and Prism on accuracy, more specifically, it wins more than 50% of datasets among these three algorithms and our method has similar result with Decision Table method, it wins half of the datasets on accuracy. Even though our method gained lower average accuracy than Naive Bayes and C4.5 algorithms (81.3, 83.9 and 83.3 respectively) in the first experiment, it got higher average accuracy than these algorithms in the second experiment (82.2, 81.8 and 81.7 respectively). Although our algorithm loses around 70% of the datasets to C4.5, PART, JRip and Random Forest algorithms, it achieves the best accuracy on "Breast Cancer" dataset in both experiments. We could obtain slightly higher accuracy in the second experiment comparing to first one with 82.16 and 81.25 respectively, because, we modified the default parameters and *minsup, minconf* constraints.

We tested our method with different *confidence* and *numrules* thresholds on all datasets and here, it is reported the experiments on "Nursery" dataset to show that how we analyzed and selected the best *numrules* and *confidence* thresholds which achieves the highest accuracy. In some datasets, higher confidence and higher number of rules achieve higher confidence (most probably in larger datasets). In some datasets, lower confidence achieves higher accuracy depending on the class distribution. Results are shown in Fig. 1.

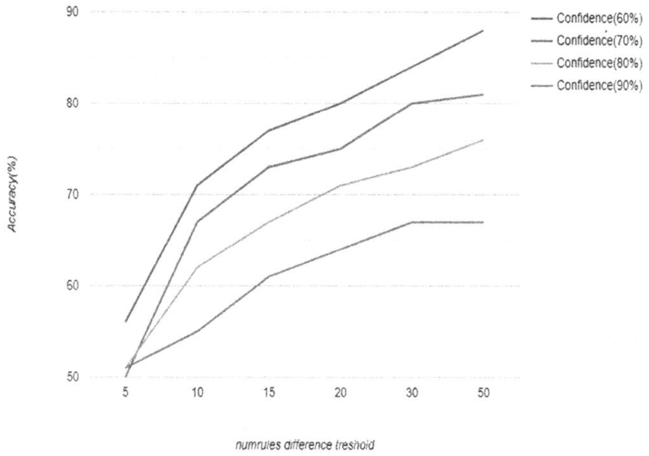

Fig. 1. The effect of *confidence* and *numrules* thresholds to the accuracy on "Nursery" dataset

Figure 1 illustrates that the peaks of accuracy are achieved at *numrules* >= 50 of all curves. The best accuracy (89%) in "Nursery" dataset is gained when *numrules* is 55 and *confidence* is 60%, that is, there is optimal settings for both thresholds.

8 Conclusion and Future Work

Accuracy and efficiency are crucial factors in classification tasks in data mining. Associative classification gets higher accuracy than some traditional rule-based classification approaches in some cases. However, it generates a large number of association classification rules. Therefore, the efficiency of associative classification is not high when the minimum support is set to be low and the training dataset is large.

Our experiments on 12 databases from the UCI machine learning repository show that our method is consistent, accurate, and comparative with other well-known classification methods. It achieved the fourth highest average accuracy (82.2%) among all classification methods on the selected UCI datasets.

Since, this is just a preliminary research and a step to our final goal of producing a more comprehensive and interactively explorable classifier using class association rules and clustering, 10-fold cross-validation was not used at this stage – it would probably not give much different results on the selected datasets. Significance testing was also omitted in this preliminary stage. All those are already being implemented for our "final" algorithm. Our plans for future work are thus to include statistical testing of results, using measures such as, ROC, precision, recall, and measuring running times as well as test the updated approach on larger "real-life" datasets.

Acknowledgement. The authors gratefully acknowledge the European Commission for funding the InnoRenew CoE project (Grant Agreement #739574) under the Horizon2020 Widespread-Teaming program and and the Republic of Slovenia (Investment funding of the Republic of Slovenia and the European Union of the European Regional Development Fund).

References

1. Agrawal, R., Srikant, R.: Fast algorithms for mining association rules. In: VLDB 1994 Proceedings of the 20th International Conference on Very Large Data Bases, Chile, pp. 487–499 (1994)
2. Ali, K., Manganaris, S., Srikant, R.: Partial classification using association rules. In: Proceedings of KDD-1997, U.S.A., pp. 115–118 (1997)
3. Baralis, E., Cagliero, L., Garza, P.: A novel pattern-based Bayesian classifier. IEEE Trans. Knowl. Data Eng. **25**(12), 2780–2795 (2013)
4. Bayardo, R.J.: Brute-force mining of high-confidence classification rules. In: Proceedings of the Third International Conference on Knowledge Discovery and Data Mining, U.S.A., pp. 123–126 (1997)
5. Breiman, L.: Random Forests. Mach. Learn. **45**(1), 5–32 (2001)
6. Cendrowska, J.: PRISM: an algorithm for inducing modular rules. Int. J. Man-Mach. Stud. **27**(4), 349–370 (1987)
7. Chen, G., Liu, H., Yu, L., Wei, Q., Zhang, X.: A new approach to classification based on association rule mining. Decis. Support Syst. **42**(2), 674–689 (2006)
8. Clark, P., Niblett, T.: The CN2 induction algorithm. Mach. Learn. **3**(4), 261–283 (1989)
9. Cohen, W.W.: Fast effective rule induction. In: ICML 1995 Proceedings of the Twelfth International Conference on Machine Learning, Tahoe City, California, pp. 115–123 (1995)
10. Dua, D., Graff, C.: UCI Machine Learning Repository. University of California, Irvine (2019)
11. Frank, E., Witten, I.: Generating accurate rule sets without global optimization. In: Fifteenth International Conference on Machine Learning, USA, pp. 144–151 (1998)
12. Holte, R.: Very simple classification rules perform well on most commonly used datasets. Mach. Learn. **11**(1), 63–91 (1993)
13. Kohavi, R.: The power of decision tables. In: 8th European Conference on Machine Learning, Heraclion, Crete, Greece, pp. 174–189 (1995)

14. Lent, B., Swami, A., Widom, J.: Clustering association rules. In: ICDE 1997 Proceedings of the Thirteenth International Conference on Data Engineering, England, pp. 220–231 (1997)
15. Li, W., Han, J., Pei, J.: CMAR: accurate and efficient classification based on multiple class-association rules. In: Proceedings of the 1st IEEE International Conference on Data Mining (ICDM 2001), San Jose, California, USA, pp. 369–376 (2001)
16. Liu, B., Hsu, W., Ma, Y.: Integrating classification and association rule mining. In: Proceedings of the 4th International Conference on Knowledge Discovery and Data Mining (KDD 1998), New York, USA, pp. 80–86 (1998)
17. Quinlan, J.: C4.5: Programs for machine learning. Mach. Learn. **16**(3), 235–240 (1993)
18. Xiaoxin, Y., Jiawei, H. CPAR: classification based on predictive association rules. In: Proceedings of the SIAM International Conference on Data Mining, San Francisco, U.S.A., pp. 331–335 (2003)
19. Zhang, M., Zhou Z.: A k-nearest neighbor based algorithm for multi-label classification. In: Proceedings of the 1st IEEE International Conference on Granular Computing (GrC 2005), Beijing, China, vol. 2, pp. 718–721 (2005)
20. Zhou, Z., Liu, X.: Training cost-sensitive neural networks with methods addressing the class imbalance problem. IEEE Trans. Knowl. Data Eng. **18**(1), 63–77 (2006)

Analyses of Multi-collection Corpora via Compound Topic Modeling

Clint P. George[1,3], Wei Xia[2], and George Michailidis[1,2](✉)

[1] Informatics Institute, University of Florida, Gainesville, USA
gmichail@ufl.edu
[2] Department of Statistics, University of Florida, Gainesville, USA
[3] School of Mathematics and Computer Science,
Indian Institute of Technology Goa, Veling, India

Abstract. Popular probabilistic topic models have typically centered on one single text collection, which is deficient for comparative text analyses. We consider a setting where we have partitionable corpora. Each subcollection shares a single set of topics, but there exists relative variation in topic proportions among collections. We propose the compound latent Dirichlet allocation (cLDA) model that encourages generalizability, depends less on user-input parameters, and includes any prior knowledge corpus organization structure. For parameter estimation, we study Markov chain Monte Carlo (MCMC) and variational inference approaches extensively and suggest an efficient MCMC method. We evaluate cLDA using both synthetic and real-world corpora and cLDA shows superior performance over the state-of-the-art models.

Keywords: Statistical inference · Unsupervised learning · Text analysis

1 Introduction

Newspapers, magazines, scientific journals, and social media messages being composed in daily living produce routinely an enormous volume of text data. The corresponding content comes from diverse backgrounds and represent distinct themes or ideas; modeling and analyzing such heterogeneity in large-scale is crucial in any text mining frameworks. Typically, text mining aims to extract relevant and interesting information from the text by the process of structuring the written text (e.g. via semantic parsing, stemming, lemmatization), inferring hidden patterns within the structured data, and finally, deciphering the results. To address these tasks in an unsupervised manner, numerous statistical methods such as TF-IDF [20], latent semantic indexing [5, LSI], and probabilistic topic models, e.g. probabilistic LSI [12, pLSI], latent Dirichlet allocation [4, LDA], have emerged in the literature.

Supported by grants NIH #7 R21 GM101719-03 to G. Michailidis and Institute of Education Sciences #R305160004, and the University of Florida Informatics Institute.

G. Nicosia et al. (Eds.): LOD 2019, LNCS 11943, pp. 205–218, 2019.
https://doi.org/10.1007/978-3-030-37599-7_18

Topic models such as LDA are developed to capture the underlying semantic structure of a corpus (i.e. a document collection) based on co-occurrences of words, assuming documents as bags of words—i.e. the order of words in a document discarded. LDA (Fig. 2) assumes that (a) a topic is a latent (i.e., unobserved) distribution on the corpus vocabulary, (b) each document in the corpus is described by a latent mixture of topics, and (c) each observed word in a document, there is a latent variable representing a topic from which the word is drawn. This model is suitable for documents coming from a single collection. This assumption is insufficient for comparative analyses of text, especially, for partition-able text corpora, as we describe next.

Suppose, we deal with corpora of (i) articles accepted in various workshops or consecutive proceedings of a conference or (ii) blogs/forums from people different countries. It may be of interest to explore research topics across multiple/consecutive proceedings or workshops of a conference or cultural differences in blogs and forums from different countries or languages [23,27]. Moreover, category labels, workshop names, article timestamps, or geotags can provide significant prior knowledge regarding the unique structure of these corpora. We aim to include these prior structures and characteristics of the corpus into a probabilistic model adding less computational burden, for expert analyses of the corpus, collection, and document-level characteristics.

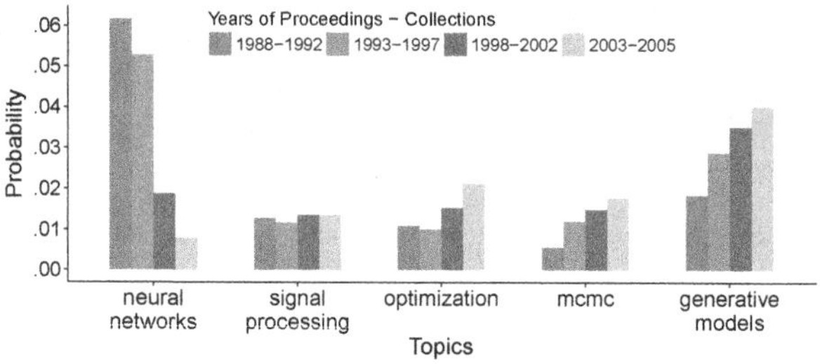

Fig. 1. Estimated topic proportions of five topics for the four time spans of the NIPS proceedings.

For example, Fig. 1 gives the results of an experiment on a real document corpus employing the proposed model in this paper. The corpus consists of articles accepted in the proceedings of the NIPS conference for the years from 1988 to 2005. We wish to analyze topics that evolve over this timespan. We thus partitioned the corpus into four collections based on time (details in Sect. 4). The plot shows estimates of topic proportions for all four collections. Evidently, some topics got increased attention (e.g. Markov chain Monte Carlo (MCMC), generative models) and some topics got decreased popularity (e.g. neural networks) over the years of the conference. For some topics, the popularity is relatively constant (e.g. signal processing) from the beginning of NIPS. The algorithm used is unsupervised, and only takes documents, words, and their collection labels as input.

Assuming a flat structure to all documents in a corpus, LDA or its variants is not well suited to the multi-collection corpora setting: (1) Ignoring organization of collections may produce topics that describe only some, not all of the collections. (2) No direct way exists to define topics that describe the shared information across collections and a topic that describes the information specific to a particular collection. As a solution, we can crudely consider each collection as a separate corpus and fit an LDA model for each corpus. However, we must consider (i) the nontrivial alignment of topics in each model, for any useful comparison of topics among collections (and, topics inferred from individual corpus partitions and the whole corpus can themselves be different) and (ii) information loss due to modeling collections separately, especially for small datasets (Sect. 4).

We introduce the compound latent Dirichlet allocation (cLDA, Sect. 2) model that incorporates any prior knowledge on the organization of documents in a corpus into a unified modeling framework. cLDA assumes a shared, collection-level, latent mixture of topics for documents in each collection. This collection-level mixture is used as the base measure to derive the (latent) topic mixture for each document. All collection-level and document-level variables share a common set of topics at the corpus level, which enables us to perform exciting inferences, as shown in Fig. 1. cLDA can thus aid visual thematic analyses of large sets of documents, and include corpus, collection, and document specific views.

The parameters of interest in cLDA are hidden and exact posterior inference is intractable in cLDA. Non-conjugate relationships in the model further make approximate inference complex. Popular approximate posterior inference methods in topic models are MCMC and variational methods. MCMC methods enable us to sample hidden variables from the intractable posterior with convergence guarantees. We consider two MCMC methods for cLDA: (a) one uses the traditional auxiliary variable updates within Gibbs sampling, and (b) the other uses Langevin dynamics within Gibbs sampling, a method that received recent attention. Our experimental evidence suggests that the former method gives superior performance with only a little computational overhead compared to the collapsed Gibbs sampling algorithm for LDA [10, CGS] (details in Sects. 3.2, 4, [7]), and is the main focus here. Variational methods are often used in topic modeling as they give fast implementations by construction. Although they converge rather quickly, their solutions are suboptimal compared to the results of the other two MCMC schemes (Sect. 3.2 and [7]).

Our contributions are three-fold: (i) we propose a probabilistic model (i.e. cLDA) that can capture the topic structure of a corpus including organization hierarchy of documents (Sect. 2), (ii) we study efficient methods for posterior inference in cLDA (Sect. 3), and (iii) we perform an empirical study of the real-world applicability of the cLDA model—for example, (a) analyzing topics that evolve over time, (b) analyzing patterns of topics on customer reviews, and (c) summarizing topic structure of a corpus—via three real world corpora (Sect. 4). Moreover, the inference about collection-level topic mixtures are of interest to the general issue of posterior sampling on the probability simplex in statistics.

2 A Compound Hierarchical Model

There is a vocabulary $\mathcal{V}$ of V terms in the corpus; in general, $\mathcal{V}$ is considered as the union of all the word tokens in all the documents of the corpus, after removing stop-words and normalizing tokens (e.g. stemming). The number of topics K is assumed to be known. (Selection in practice, Sect. 4.) By definition, a topic is a distribution over $\mathcal{V}$, i.e., a point in the V-1 dimensional simplex $\mathbb{S}_V$. We will form a $K \times V$ matrix $\boldsymbol{\beta}$, whose k^{th} row is the kth topic (how $\boldsymbol{\beta}$ is formed will be described shortly). Thus, the rows of $\boldsymbol{\beta}$ are vectors $\boldsymbol{\beta}_1, \ldots, \boldsymbol{\beta}_K$, all lying in $\mathbb{S}_V$. There are J collections in the corpus and for $j = 1, 2, \ldots, J$, collection j has D_j documents. For $d = 1, \ldots, D_j$, document d in collection j (i.e. document jd) has n_{jd} words, $w_{jd1}, \ldots, w_{jdn_{jd}}$. Each word is represented by the index or id of the corresponding term from the vocabulary. We represent document jd by the vector $\boldsymbol{w}_{jd} = (w_{jd1}, \ldots, w_{jdn_{jd}})$, collection j by the concatenated vector $\boldsymbol{w}_j = (\boldsymbol{w}_{j1}, \ldots, \boldsymbol{w}_{jD_j})$, and the corpus by the concatenated vector $\boldsymbol{w} = (\boldsymbol{w}_1, \ldots, \boldsymbol{w}_J)$.

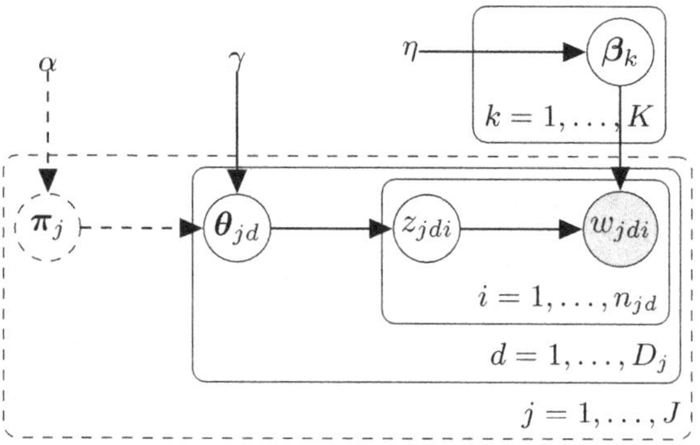

Fig. 2. Graphical model of the LDA model (the inner structure) and cLDA model (the outer dashed-structure is the extension to LDA): Nodes denote random variables, shaded nodes denote observed variables, edges denote conditional dependencies, and plates denote replicated processes.

We use $\text{Dir}_L(a\omega_1, \ldots, a\omega_L)$ to denote the finite-dimensional Dirichlet distribution on the (L-1)-dimensional simplex. This has two parameters, a scale (concentration) parameter a and a base measure $(\omega_1, \ldots, \omega_L)$ on the (L-1)-dimensional simplex. Thus $\text{Dir}_L(a, \ldots, a)$ denotes an L-dimensional Dirichlet distribution with the constant base measure $(1, \ldots, 1)$. $\text{Mult}_L(b_1, \ldots, b_L)$ to represents the multinomial distribution with number of trials equal to 1 and probability vector $(b_1, \ldots, b_L)$. Let $h = (\eta, \alpha, \gamma) \in (0, \infty)^3$ be the hyperparameters in the model. We formally define the cLDA model as (see Fig. 2).

$$\boldsymbol{\beta}_k \overset{\text{iid}}{\sim} \text{Dir}_V(\eta, \ldots, \eta), \text{ for topic } k = 1, \ldots, K \tag{1}$$

$$\boldsymbol{\pi}_j \overset{\text{iid}}{\sim} \text{Dir}_K(\alpha, \ldots, \alpha), \text{ for collection } j = 1, \ldots, J \tag{2}$$

$$\boldsymbol{\theta}_{jd} \overset{\text{iid}}{\sim} \text{Dir}_K(\gamma\pi_{j1}, \ldots, \gamma\pi_{jK}), \text{ for document } jd \tag{3}$$

$$z_{jdi} \overset{\text{iid}}{\sim} \text{Mult}_K(\boldsymbol{\theta}_{jd}), \text{ for each word } w_{jdi} \tag{4}$$

$$w_{jdi} \overset{\text{ind}}{\sim} \text{Mult}_V(\boldsymbol{\beta}_{z_{jdi}}) \tag{5}$$

The distribution of $z_{jd1}, \ldots, z_{jdn_{jd}}$ will depend on the document-level variable $\boldsymbol{\theta}_{jd}$ that represents a distribution on the topics for document jd. A single $\boldsymbol{\theta}_{jd}$ for document jd encourages different words in document jd to share the same document level characteristics. The distribution of $\boldsymbol{\theta}_{j1}, \ldots, \boldsymbol{\theta}_{jD_j}$ will depend on the collection-level variable $\boldsymbol{\pi}_j$ which indicates a distribution on the topics for collection j. A single $\boldsymbol{\pi}_j$ for collection j encourages different documents in collection j to have the same collection level properties. A single $\boldsymbol{\beta}$ is shared among all documents, which encourages documents in various collections in the corpus to share the same set of topics. The standard LDA model [4] is a special case of the proposed cLDA model, where there exist single parameters $\boldsymbol{\pi}_j$ and $\boldsymbol{\beta}$ that fail to capture potential heterogeneity amongst the predefined collections, an objective that cLDA is designed for.

Related Work

Zhai et al. [27] refer to the problem of multi-collection corpora modeling as comparative text mining (CTM), which uses pLSI as a building block. Comparing with CTM model, cLDA employs an efficient and generalizable LDA-based framework that has several advantages over pLSI—for example, LDA incorporates Dirichlet priors for document topic structures in a natural way to deal with newly encountered documents. Furthermore, cLDA combines collection-specific characteristics in a natural probabilistic framework enabling efficient posterior sampling, depending less on user-defined parameters as in CTM.

The cLDA model shares some similarities, but also exhibits differences from: Hierarchical Dirichlet Process [21] and Nested Chinese Restaurant Process [3], which are introduced to learn document and topic hierarchies from the data nonparametrically. In nonparametric models, as we observe more and more data, the data representations grow structurally. Instead of manually specifying the number of topics K, these nonparametric topic models infer K from the data by assuming a Dirichlet process prior for document topic distributions and topics. A major hurdle in these frameworks is that inference can be computationally challenging for large datasets. Our experimental evidence also shows that HDP produces too many fragmented topics, which may lead the practitioner to bear the additional burden of post processing (Sect. 4). Here, similar to LDA, cLDA is a parametric model, i.e., the data representational structure is fixed and does not grow as more data are observed. cLDA assumes that K is fixed and can be inferred from the data directly (e.g. empirical Bayes methods [6]) or by cross-validation. We thus have simple Dirichlet priors in cLDA without adding much burden to the model and inference (Sect. 3).

In light of adding supervision, several modifications to LDA model have been proposed in the literature. Supervised LDA [14, sLDA] is an example; for each document d, sLDA introduces a response variable y_d that is assumed to be generated from document d's empirical topic mixture distribution. In practice, the posterior inference in such a setting can be inefficient due to the high non-linearity of the discrete distribution on the empirical parameters [28]. cLDA, on the other hand, proposes a generative framework incorporating the collection-level characteristics, without much computational burden (Sect. 4). Also, the objectives of these related models are different from the focus of this paper.

We can view the hierarchical model of cLDA as a special instance of [13], which extends LDA to have multiple layers in the hierarchy, for corpora with many categories and subcategories. In cLDA, we are interested in corpora with one layer of top-level categories, which is adequate for real-world corpora such as books with many chapters or conference proceedings spanning over the years. Moreover, in our experience, variational inference (the inference scheme in [13]) for the two-layered hierarchical model is suboptimal compared to MCMC methods for inference presented in this study.

3 Posterior Sampling

The parameters of interest in cLDA: (a) corpus-level topics, (b) collection-level mixture of topics, (c) document-level mixture of topics, and (d) topic indices of words are hidden. We identify the hidden variables given the observed word statistics and document organization hierarchy in the corpus via posterior inference.

Let $\boldsymbol{\theta} = (\boldsymbol{\theta}_{11}, \ldots, \boldsymbol{\theta}_{1D_1}, \ldots, \boldsymbol{\theta}_{J1}, \ldots, \boldsymbol{\theta}_{JD_J})$, $\boldsymbol{z}_{jd} = (z_{jd1}, \ldots, z_{jdn_{jd}})$ for $d = 1, \ldots, D_j$, $j = 1, \ldots, J$, $\boldsymbol{z} = (\boldsymbol{z}_{11}, \ldots, \boldsymbol{z}_{1D_1}, \ldots, \boldsymbol{z}_{J1}, \ldots, \boldsymbol{z}_{JD_J})$, and $\boldsymbol{\pi} = (\boldsymbol{\pi}_1, \ldots, \boldsymbol{\pi}_J)$. We will then use $\boldsymbol{\psi}$ to denote the latent variables $(\boldsymbol{\beta}, \boldsymbol{\pi}, \boldsymbol{\theta}, \boldsymbol{z})$ in the cLDA model. For any given h, (1)–(5) in the hierarchical model induce a prior distribution $p_h(\boldsymbol{\psi})$ on $\boldsymbol{\psi}$. Equation (5) gives the likelihood $\ell_{\boldsymbol{w}}(\boldsymbol{\psi})$. The words $\boldsymbol{w}$ and their document and collection labels are observed. We are interested in $p_{h,\boldsymbol{w}}(\boldsymbol{\psi})$, the posterior distribution of $\boldsymbol{\psi}$ given $\boldsymbol{w}$ corresponding to the prior $p_h(\boldsymbol{\psi})$. Applying Bayes rule and using (1)–(5), we write the posterior distribution $p_{h,\boldsymbol{w}}$ of $\boldsymbol{\psi}$ as [7]

$$
\begin{aligned}
p_{h,\boldsymbol{w}}(\boldsymbol{\psi}) \propto & \left[\prod_{j=1}^{J}\prod_{d=1}^{D_j} \frac{\prod_{k=1}^{K} \theta_{jdk}^{n_{jdk}+\gamma\pi_{jk}-1}}{\prod_{k=1}^{K} \Gamma(\gamma\pi_{jk})}\right] \left[\prod_{j=1}^{J}\prod_{k=1}^{K} \pi_{jk}^{\alpha-1}\right] \\
& \left[\prod_{k=1}^{K}\prod_{v=1}^{V} \beta_{kv}^{\sum_{j=1}^{J}\sum_{d=1}^{D_j} m_{jdkv}+\eta-1}\right]
\end{aligned}
\qquad (6)
$$

where n_{jdk} is the number of words in document d in collection j that are assigned to topic k, and m_{jdkv} is the number of words in document d in collection j for which the latent topic is k and the index of the word in the vocabulary is v. These count statistics depend on both $\boldsymbol{z}$ and $\boldsymbol{w}$. Note that the constants in the Dirichlet normalizing constants are absorbed into the overall constant of proportionality. Unfortunately, the normalizing constant of the posterior $p_{h,w}(\boldsymbol{\psi})$, is the likelihood of the data with all latent variables integrated out, is a non-trivial integral. This makes exact inference difficult in cLDA.

Popular methods for approximate posterior inference in topic models are Markov chain Monte Carlo [10, MCMC] and variational methods [4, VEM]. Although variational methods may give a fast and scalable approximation for the posterior, due to optimizing the proxy lower-bound, it may not produce optimal solutions as the MCMC methods in practice [22]. That is the case in cLDA, as our experimental evidence suggests [7].

3.1 Inference via Markov Chain Monte Carlo Methods

According to the hierarchical model (1)–(5), $\boldsymbol{\theta}_{jd}$'s and $\boldsymbol{\beta}_k$'s are independent, and by inspecting the posterior (6), given $(\boldsymbol{\pi}, \boldsymbol{z})$, we get:

$$\begin{aligned} \boldsymbol{\theta}_{jd} &\sim \text{Dir}_K\Big(n_{jd1} + \gamma\pi_{j1}, \ldots, n_{jdK} + \gamma\pi_{jK}\Big), \\ \boldsymbol{\beta}_k &\sim \text{Dir}_V\Big(m_{k1} + \eta, \ldots, m_{kV} + \eta\Big), \end{aligned} \tag{7}$$

where $m_{kv} = \sum_{j=1}^{J}\sum_{d=1}^{D_j} m_{jdkv}$ and $d = 1, \ldots, D_j, j = 1, \ldots, J, k = 1, \ldots, K$. Note that (7) implicitly dependent on the observed data $\boldsymbol{w}$.

We can integrate out $\boldsymbol{\theta}$'s and $\boldsymbol{\beta}$'s to get the marginal posterior distribution of $(\boldsymbol{\pi}, \boldsymbol{z})$ (up to a normalizing constant) as

$$\begin{aligned} p_{h,w}(\boldsymbol{\pi}, \boldsymbol{z}) \propto & \left[\prod_{j=1}^{J}\prod_{d=1}^{D_j}\prod_{k=1}^{K} \frac{\Gamma(\gamma\pi_{jk} + n_{jdk})}{\prod_{k=1}^{K}\Gamma(\gamma\pi_{jk})}\right] \\ & \left[\prod_{j=1}^{J}\prod_{k=1}^{K}\pi_{jk}^{\alpha-1}\right]\left[\prod_{k=1}^{K}\frac{\prod_{v=1}^{V}\Gamma(m_{..kv} + \eta)}{\Gamma(m_{..k.} + V\eta)}\right] \end{aligned} \tag{8}$$

Let the vector $\boldsymbol{z}^{(-jdi)}$ be the topic assignments of all words in the corpus except for word w_{jdi}. And, we define $n_{jd} := \sum_{k=1}^{K} n_{jdk}$ and $m_k := \sum_{v=1}^{V} m_{kv}$. By inspecting (8), we obtain a closed form expression for the conditional posterior distribution for z_{jdi}, given $\boldsymbol{z}^{(-jdi)}$ and π_{jk}, as

$$p_{h,w}\left(z_{jdik} = 1 \,|\, .\right) \propto \frac{\gamma\pi_{jk} + n_{jdk}^{(-jdi)}}{\gamma + n_{jd}^{(-jdi)}} \frac{\eta + m_{kv}^{(-jdi)}}{V\eta + m_k^{(-jdi)}} \tag{9}$$

where the superscript $(-jdi)$ for the count statistics n_{jdk}, n_{jd}, m_{kv}, and m_k means that we discard the contribution of word w_{jdi} for counting [7]. This enables us to build a Gibbs sampling chain on $\boldsymbol{z}$, by sampling z_{jdi}, given π_{jk} and $\boldsymbol{z}^{(-jdi)}$.

Given $\boldsymbol{z}_j$ and the observed data $\boldsymbol{w}_j$, we have the unnormalized posterior probability density function for $\boldsymbol{\pi}_j$ as

$$\tilde{p}_{\boldsymbol{w}_j}(\boldsymbol{\pi}_j \mid \boldsymbol{z}_j) \propto \prod_{d=1}^{D_j} \prod_{k=1}^{K} \frac{\Gamma(\gamma\pi_{jk} + n_{jdk})}{\Gamma(\gamma\pi_{jk})} \prod_{k=1}^{K} \pi_{jk}^{\alpha-1} \tag{10}$$

Here we use the fact that $\boldsymbol{\pi}_j$'s are independent of $\boldsymbol{\beta}_k$'s. We wish to sample $\boldsymbol{\pi}_j$'s from this distribution, however the normalized density function $p_{\boldsymbol{w}_j}(\boldsymbol{\pi}_j \mid \boldsymbol{z}_j)$ is computationally intractable: the density $p_{\boldsymbol{w}_j}(\boldsymbol{\pi}_j \mid \boldsymbol{z}_j)$ has a non-conjugate relationship between $\boldsymbol{\pi}_j$'s and $\boldsymbol{\theta}_{dj}$'s, which makes its normalizer intractable. We developed two Markov chain Monte Carlo schemes that enable us to sample from this posterior density: one uses the auxiliary variable sampling, and other uses the Metropolis-Hastings algorithm [11,15, MH] with Langevin dynamics [8]. The latter scheme has become popular for sampling on the simplex recently [19]. However, we shall only focus on the first scheme due to its superior performance in our numerical experiments. We denote the Markov chain on $(\boldsymbol{\pi}, \boldsymbol{z})$ induced by the latter scheme by [7, MGS].

Auxiliary Variable Sampling The idea of auxiliary variable sampling is that one can sample from a distribution $f(x)$ for variable x by sampling from some augmented distribution $f(x, s)$ for variable x and auxiliary variable s, such that the marginal distribution of x is $f(x)$ under $f(x, s)$. One can build a Markov chain using this idea in which, auxiliary variable s is introduced temporarily and discarded, only leaving the value of x. Since $f(x)$ is the marginal distribution of x under $f(x, s)$, this update for x will leave $f(x)$ invariant [17]. Suppose $\mathrm{S}(n_{jdk}, s)$ denotes the unsigned Stirling number of the first kind. We can then get the expression for augmented sampling by plugging in the factorial expansion [1]

$$\frac{\Gamma(\gamma\pi_{jk} + n_{jdk})}{\Gamma(\gamma\pi_{jk})} = \sum_{s=0}^{n_{jdk}} \mathrm{S}(n_{jdk}, s)(\gamma\pi_{jk})^{s}, \tag{11}$$

into the marginal posterior density (10) as

$$\tilde{p}_{\boldsymbol{w}_j}(\boldsymbol{\pi}_j \mid .) \propto \prod_{d=1}^{D_j} \prod_{k=1}^{K} \mathrm{S}(n_{jdk}, s_{jdk})(\gamma\pi_{jk})^{s_{jdk}} \prod_{k=1}^{K} \pi_{jk}^{\alpha-1} \tag{12}$$

([18] and [21] used a similar idea in a different hierarchical model.) The expression (12) introduces auxiliary variable s_{jdk}. By inspecting (12), we get a closed form expression for sampling $\boldsymbol{\pi}_j$, for $j = 1, \ldots, J$:

$$\boldsymbol{\pi}_j \sim \mathrm{Dir}_K \left(\sum_{d=1}^{D_j} s_{jd1} + \alpha, \ldots, \sum_{d=1}^{D_j} s_{jdK} + \alpha \right), \tag{13}$$

and we update the auxiliary variable s_{jdk} by the Antoniak sampling scheme [18, Appendix A], i.e., the Chinese restaurant process [2, CRP] with concentration parameter $\gamma\pi_{jk}$ and the number of customers n_{jdk}. The auxiliary variable s_{jdk} is typically updated by drawing n_{jdk} Bernoulli variables as

$$s_{jdk} = \sum_{l=1}^{n_{jdk}} s^{(l)}, \quad s^{(l)} \overset{\text{iid}}{\sim} \text{Bernoulli}\left(\frac{\gamma\pi_{jk}}{\gamma\pi_{jk} + l - 1}\right) \tag{14}$$

This augmented update for collection level parameters will add a low computational overhead to the collapsed Gibbs sampling chain on $\boldsymbol{z}$, and is easy to implement in practice. We denote the Markov chain on $(\boldsymbol{\pi}, \boldsymbol{z})$ based on the auxiliary variable update within Gibbs sampling by [7, AGS].

We can easily augment AGS and MGS chains on $(\boldsymbol{\pi}, \boldsymbol{z})$ to a Markov chain on $\boldsymbol{\psi}$ with invariant distribution $p_{h,w}(\boldsymbol{\psi})$ by using the conditional distribution of $(\boldsymbol{\beta}, \boldsymbol{\theta})$, given by (7). The augmented chain will hold the convergence properties of the chain on $(\boldsymbol{\pi}, \boldsymbol{z})$.

3.2 Empirical Evaluation of Samples $\boldsymbol{\pi}$

We consider a synthetic corpus by simulating (1)–(5) of the cLDA hierarchical model with the number of collections $J = 2$ and the number of topics $K = 3$. We did this solely so that we can visualize the results of the algorithms. We also took the vocabulary size $V = 40$, the number of documents in each collection $D_j = 100$, and the hyperparameters $h_{\text{true}} = (\alpha, \gamma, \eta) = (.1, 1, .25)$. Collection-level Dirichlet sampling via (2) with α produced two topic distributions $\boldsymbol{\pi}_1^{\text{true}} = (.002, \epsilon, .997)$ and $\boldsymbol{\pi}_2^{\text{true}} = (.584, .386, .030)$, where ϵ denotes a small number. We study the ability of algorithms AGS, MGS, and VEM for cLDA to recover parameters $\boldsymbol{\pi}_1^{\text{true}}$ and $\boldsymbol{\pi}_2^{\text{true}}$, by comparing $\boldsymbol{\pi}_j$ samples from the AGS and MGS chains with variational estimates of $\boldsymbol{\pi}_j$ from VEM iterations[1]. We initialized $\boldsymbol{\pi}_1^{(0)} = \boldsymbol{\pi}_2^{(0)} = (.33, .33, .33)$ and use hyperparameters h_{true} for all three algorithms. Using the data $\boldsymbol{w}$, we ran both chains AGS and MGS for 2000 iterations, and algorithm VEM converged after 45 EM iterations.

Table 1 gives the values of $\boldsymbol{\pi}_1$ and $\boldsymbol{\pi}_2$ on the 2-dimensional simplex from the last iteration of all three algorithms. We can see that both AGS and MGS chains were able to recover the values $\boldsymbol{\pi}_1^{\text{true}}$ and $\boldsymbol{\pi}_2^{\text{true}}$ reasonably well, although AGS chain has an edge. To converge to optimal regions $\|\boldsymbol{\pi}_1^{\text{true}} - \boldsymbol{\pi}_1^{(s)}\| \leq .003$ and $\|\boldsymbol{\pi}_2^{\text{true}} - \boldsymbol{\pi}_2^{(s)}\| \leq .07$, the AGS chain took 42 and 23, respectively. ($\|.\|$ denotes L-1 norm on the 2-dimensional simplex.) On the other hand, the MGS chain required 248 and 139 cycles, respectively. Finally, at convergence the VEM algorithm converged to solutions that are 0.08 far from $\boldsymbol{\pi}_1^{\text{true}}$ and 0.18 far from $\boldsymbol{\pi}_2^{\text{true}}$. Here, the reported values are after the necessary alignment of topics with the true set of topics, for all the three methods. Trace plots of values of $\boldsymbol{\pi}$ on the 2-dimensional simplex for various iterations of AGS, MGS, and VEM also show the superior performance of the AGS algorithm [7]. Hence, it is used in the experimental analyses next.

[1] Implementation available at https://github.com/clintpgeorge/clda (R package).

Table 1. Estimated values of π via algorithms AGS, MGS, and VEM.

Method	$\hat{\pi}_1$	$\hat{\pi}_2$	Iterations
AGS	$(\epsilon, .996, .003)$	$(.347, .015, .636)$	2,000
MGS	$(.001, .997, \epsilon)$	$(.379, .005, .615)$	2,000
VEM	$(.057, .935, .006)$	$(.258, .155, .585)$	45

4 Experimental Analyses

We use three document corpora based on: (i) the NIPS 00-18 dataset [9], (ii) customer reviews from Yelp[2], and (iii) the 20Newsgroups dataset[3] in our experiments. We applied standard corpus preprocessing, which involves tokenizing text and discarding standard stop-words, numbers, dates, and infrequent words from the corpus. The NIPS 00-18 dataset consists of papers published in proceedings 0 to 18 (i.e. years from 1988 to 2005) of the Neural Information Processing Systems (NIPS) conference. This corpus has 2,741 articles and 9,156 unique words. A question of interest is to see how various research topics evolve over the years 1988–2005 of NIPS proceedings. Typically, new topics do not emerge in consecutive years. So, we partition the NIPS 00-18 corpus into four collections based on time periods 1988–1992, 1993–1997, 1998–2002, and 2003–2005 in our analyses.

Similarly, each Yelp customer review (i.e. a document) is associated with a customer rating on the scale 1 to 5, where 5 being the best and 1 being the worst. This corpus has 24,310 restaurant reviews, and 9,517 unique words. According to the subject matter of articles, we took a subset of the 20Newsgroups dataset with subject groups computers, recreation, science, and politics with 16 newsgroups (16Newsgroups). 16Newsgroups has 10,764 articles and 9,208 unique words.

Comparisons with Models in the Prior Work: Wallach et al. [25, Sect. 3] proposed *non-uniform* base measures instead of the typical *uniform* base measures used in the Dirichlet priors over document-level topic distributions $\boldsymbol{\theta}_d$'s in LDA. The purpose was to get a better asymmetric Dirichlet prior for LDA, which is generally used for corpora with a flat organization hierarchy for documents. We can consider one such model as a particular case of the cLDA model proposed here: their model is closely related to a cLDA model with a single collection. To illustrate, we perform a comparative study of perplexity scores of cLDA with single and multi-collection in our experiments.

One can consider (conceptually) one-collection cLDA model as a two-level Dirichlet Process (DP) mixture model [21]. We thus compare the results of one-collection cLDA and LDA with the results of two-level DP[4] using the popular Chinese Restaurant Franchise (CRF) sampling scheme [21].

[2] http://uilab.kaist.ac.kr/research/WSDM11/.

[3] http://qwone.com/~jason/20Newsgroups.

[4] Code by Wang and Blei (2010): https://github.com/blei-lab/hdp.

Criteria for Evaluation: We use the perplexity score—a popular scheme to evaluate the performance of topics models—as specified as in [19,26] for evaluation. Exact computation of this score is intractable for LDA and cLDA. We estimate this score via MCMC or variational methods for both cLDA and LDA models [7].

Mimno et al. [16] suggested a couple of generic evaluation scores based on human coherence judgments of estimated topics via topic models such as LDA. *Topic size* is one, which is the number of words assigned to a topic in the corpus. We estimate it by samples from the posterior of the topic latent variable $\boldsymbol{z}$, given observed words $\boldsymbol{w}$ (e.g. via Markov chains: collapsed Gibbs sampling (CGS) for LDA [10], AGS, and CRF). We found that LDA models have uniform topic sizes compared to cLDA models. Another option is the topic *coherence score* [16], which is computed based on the most probable words in an estimated topic for a corpus [7]. The intuition behind this score is that group of words belonging to a topic possibly co-occur with in a document. Our analysis shows that coherence scores of cLDA topics are better than LDA topics except for corpus 16Newsgroups, which we think, is partially due to having relatively easily separable topics compared to the other two complex corpora. Moreover, both AGS (with a reasonable number of collections) and CGS chains have comparable computational cost for all three corpora (details in [7]).

Comparing Performance of cLDA and LDA Models: We first look at the criterion perplexity scores for both LDA and cLDA models with various values of the number of topics keeping each model hyperparameter fixed (and default). We ran the Markov chains AGS and CGS for 1000 iterations. Figure 3 gives (average) per-word perplexities for the held-out (test) words using algorithms CGS and AGS (with single collection, i.e., $J = 1$ and multi-collection, $J > 1$) for corpora NIPS 00-18 and 16Newsgroups. From the plots, we see that cLDA outperforms LDA well in terms of predictive power, except for corpus NIPS 00-18, for which cLDA has a slight edge over LDA. We believe the marginal performance for corpus NIPS 00-18 is partly due to the nature of partitions defined in the corpus; collections in this corpus share many common topics, compared to corpus 16Newsgroups. It suggests that cLDA suits well for corpora with separable collections. Similarly, cLDA models show a small improvement with multi-collection over single collection. And, cLDA surpass LDA even with a single collection. Note that cLDA has a better selection of priors as well as the ability to incorporate corpora document hierarchy into the modeling framework.

Selecting K and Hyperparameters α, γ, η: Figure 3 also gives some insights to select K for each corpus. Perplexities of cLDA models go down quickly with increase in the number of topics initially, and then the rate of decrease go steady after reasonable number of topics (e.g. $K = 90$ for NIPS 00-18). LDA, on the other hand, has a "U" curve with increase in perplexity after $K = 30$ or $K = 40$ for 16Newsgroups (and Yelp [7]). However, for corpus NIPS 00-18, LDA shows a behavior similar to cLDA. We thus pick $K = 90$ for corpus NIPS 00-18 and $K = 30$ for corpus 16Newsgroups (and $K = 40$ for corpus Yelp), in our analyses. The hyperparameters (α, γ, η) in cLDA control the prior and posterior distributions of parameters ψ and should be selected carefully. To estimate (γ, η) in

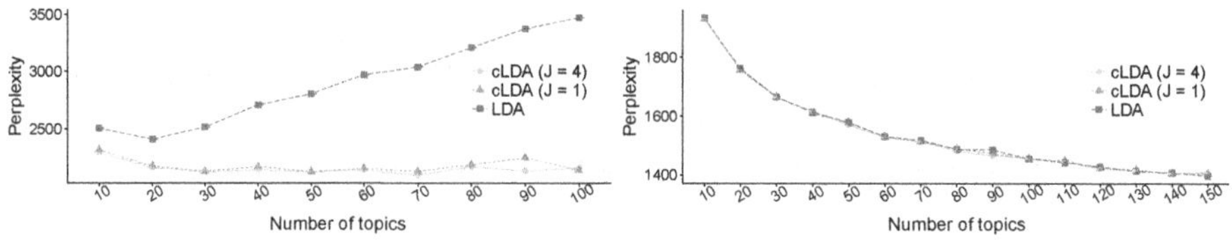

Fig. 3. Perplexities for LDA and cLDA with one-collection ($J = 1$) and multi-collection ($J > 1$) corpus and various values of K, for corpora 16Newsgroups (left) and NIPS 00-18 (right).

cLDA, we use the Gibbs-EM algorithm [7] for all three corpora with the selected K and constant $\alpha = 1$. Estimated $(\hat{\eta}, \hat{\gamma})$'s for cLDA models at convergence for corpora 16Newsgroups, NIPS 00-18, and Yelp are $(.05, 2.07)$, $(.026, 6.72)$, and $(.034, 3.44)$, respectively. Similarly, estimated $(\hat{\eta}, \hat{\alpha})$ for LDA [7,24] for corpora 16Newsgroups, NIPS 00-18, and Yelp are $(.053, .048)$, $(.027, .06)$, and $(.029, .12)$, respectively.

Comparing Performance of cLDA with LDA and HDP: We ran all three chains AGS, CGS, and CRF for 1,000 iterations for corpus NIPS 00-18. We used the last sample from each chain for analysis. We set $K = 90$ for AGS and CGS, but CRF inferred 204 topics from the corpus. To evaluate the quality of topics learned for each method, we applied the `hclust` algorithm on topic distributions. `hclust` computes a topic-to-topic similarity matrix based on the *manhattan* distance, and then builds a hierarchical tree in a bottom-up approach. The hierarchies formed by cLDA and LDA are comparable. Looking closely at topic word distributions shows that cLDA produces better topics compared to LDA, exploiting the asymmetric hierarchical Dirichlet prior on documents. Considering the estimated topic sizes of the three models, HDP found too many redundant (similar) topics [7].

Figure 4 gives boxplot statistics (i.e. median, lower hinge, and upper hinge) of silhouette widths of `hclust` clusters of topics (i.e. $\boldsymbol{\beta}_j$s) and documents (i.e. $\boldsymbol{\theta}_d$s), for different number of clusters (a user-specified parameter in `hclust`). We typically favor methods with high silhouette widths. Overall, cLDA topics outperform HDP topics, and cLDA topics are comparable or better than LDA topics (left). Silhouette widths of clusters (right) based on HDP are relatively constant with different values of the number of clusters. The large set of minute topics ($K = 204$) estimated by HDP may have helped clustering documents. cLDA performs better than LDA and is comparable with HDP or better than HDP with the right number of `hclust` clusters (e.g. from 50 to 150).

One can use trained cLDA models for tasks such as classifying text documents and summarizing document collections and corpora. cLDA also gives excellent options to visualize and browse documents. Our extensive usability study is given in [7] due to space limitations.

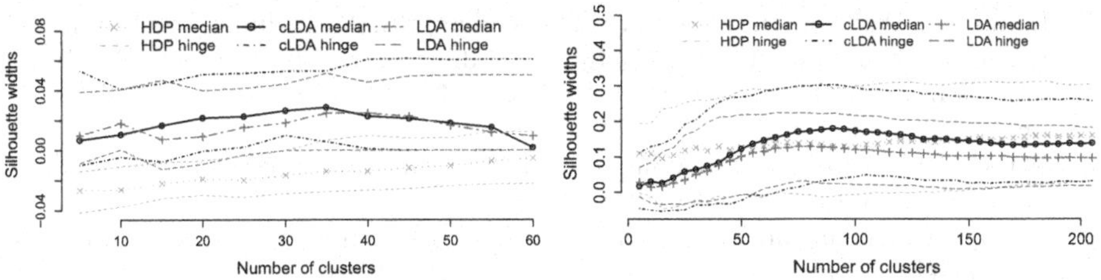

Fig. 4. Boxplot statistics of silhouette widths on topics' (left) and documents' (right) `hclust` clusters with various number of clusters for NIPS 00-18

References

1. Abramowitz, M.: Handbook of Mathematical Functions, With Formulas, Graphs, and Mathematical Tables. Dover Publications, Incorporated (1974)
2. Aldous, D.J.: Exchangeability and related topics. In: Hennequin, P.L. (ed.) École d'Été de Probabilités de Saint-Flour XIII — 1983. LNM, vol. 1117, pp. 1–198. Springer, Heidelberg (1985). https://doi.org/10.1007/BFb0099421
3. Blei, D.M., Griffiths, T.L., Jordan, M.I., Tenenbaum, J.B.: Hierarchical topic models and the nested Chinese restaurant process. In: Advances in Neural Information Processing Systems, p. 2003. MIT Press (2004)
4. Blei, D.M., Ng, A.Y., Jordan, M.I.: Latent Dirichlet allocation. J. Mach. Learn. Res. **3**, 993–1022 (2003)
5. Deerwester, S., Dumais, S.T., Furnas, G.W., Landauer, T.K., Harshman, R.: Indexing by latent semantic analysis. J. Am. Soc. Inf. Sci. **41**(6), 391 (1990)
6. George, C.P.: Latent Dirichlet allocation: hyperparameter selection and applications to electronic discovery. Ph.D. thesis, University of Florida (2015)
7. George, C.P., Xia, W., Michailidis, G.: Analyses of multi-collection corpora via compound topic modeling. Technical report, arXiv (2019)
8. Girolami, M., Calderhead, B.: Riemann Manifold Langevin and Hamiltonian Monte Carlo methods. J. R. Stat. Soc.: Ser. B (Stat. Methodol.) **73**(2), 123–214 (2011)
9. Globerson, A., Chechik, G., Pereira, F., Tishby, N.: Euclidean embedding of co-occurrence data. J. Mach. Learn. Res. **8**, 2265–2295 (2007)
10. Griffiths, T.L., Steyvers, M.: Finding scientific topics. Proc. Natl. Acad. Sci. **101**, 5228–5235 (2004)
11. Hastings, W.K.: Monte Carlo sampling methods using Markov chains and their applications. Biometrika **57**, 97–109 (1970)
12. Hofmann, T.: Probabilistic latent semantic indexing. In: Proceedings of the 22nd Annual International ACM SIGIR Conference on Research and Development in Information Retrieval, SIGIR 1999, pp. 50–57. ACM, New York (1999)
13. Kim, D.k., Voelker, G., Saul, L.K.: A variational approximation for topic modeling of hierarchical corpora. In: Proceedings of the 30th International Conference on Machine Learning (ICML 2013), pp. 55–63 (2013)
14. Mcauliffe, J.D., Blei, D.M.: Supervised topic models. In: Advances in Neural Information Processing Systems, pp. 121–128 (2008)
15. Metropolis, N., Rosenbluth, A.W., Rosenbluth, M.N., Teller, A.H., Teller, E.: Equations of state calculations by fast computing machines. J. Chem. Phys. **21**, 1087–1091 (1953)

16. Mimno, D., Wallach, H.M., Talley, E., Leenders, M., McCallum, A.: Optimizing semantic coherence in topic models. In: Proceedings of the Conference on Empirical Methods in Natural Language Processing, pp. 262–272. Association for Computational Linguistics (2011)
17. Neal, R.M.: Markov chain sampling methods for Dirichlet process mixture models. J. Comput. Graph. Stat. **9**, 249–265 (2000)
18. Newman, D., Asuncion, A.U., Smyth, P., Welling, M.: Distributed algorithms for topic models. J. Mach. Learn. Res. **10**, 1801–1828 (2009)
19. Patterson, S., Teh, Y.W.: Stochastic gradient Riemannian Langevin dynamics on the probability simplex. In: Advances in Nueral Information Processing Systems 26 (Proceedings of NIPS), pp. 1–10 (2013)
20. Salton, G., Wong, A., Yang, C.S.: A vector space model for automatic indexing. Commun. ACM **18**(11), 613–620 (1975)
21. Teh, Y.W., Jordan, M.I., Beal, M.J., Blei, D.M.: Hierarchical Dirichlet processes. J. Am. Stat. Assoc. **101**, 1566–1581 (2006)
22. Teh, Y.W., Newman, D., Welling, M.: A collapsed variational Bayesian inference algorithm for latent Dirichlet allocation. In: Schölkopf, B., Platt, J.C., Hoffman, T. (eds.) Advances in Neural Information Processing Systems, vol. 19, pp. 1353–1360. MIT Press (2007)
23. Vulić, I., De Smet, W., Tang, J., Moens, M.F.: Probabilistic topic modeling in multilingual settings: an overview of its methodology and applications. Inf. Process. Manag. **51**(1), 111–147 (2015)
24. Wallach, H.M.: Topic modelling: beyond bag-of-words. In: Proceedings of the International Conference on Machine Learning, pp. 977–984 (2006)
25. Wallach, H.M., Mimno, D., McCallum, A.: Rethinking LDA: why priors matter. Adv. Neural Inf. Process. Syst. **22**, 1973–1981 (2009)
26. Wallach, H.M., Murray, I., Salakhutdinov, R., Mimno, D.: Evaluation methods for topic models. In: Proceedings of the 26th Annual International Conference on Machine Learning, pp. 1105–1112. ACM (2009)
27. Zhai, C., Velivelli, A., Yu, B.: A cross-collection mixture model for comparative text mining. In: Proceedings of the Tenth ACM SIGKDD International Conference on Knowledge Discovery and Data Mining, KDD 2004, pp. 743–748. ACM, New York (2004)
28. Zhu, J., Xing, E.P.: Discriminative training of mixed membership models (2014)

Text Mining with Constrained Tensor Decomposition

Elaheh Sobhani[1,2(✉)], Pierre Comon[1], Christian Jutten[1], and Massoud Babaie-Zadeh[2]

[1] GIPSA-Lab, University of Grenoble Alpes, CNRS, 38000 Grenoble, France
{elaheh.sobhani,pierre.comon,christian.jutten}@gipsa-lab.grenoble-inp.fr
[2] Sharif University of Technology, Tehran, Iran
mbzadeh@sharif.edu

Abstract. Text mining, as a special case of data mining, refers to the estimation of knowledge or parameters necessary for certain purposes, such as unsupervised clustering by observing various documents. In this context, the topic of a document can be seen as a hidden variable, and words are multi-view variables related to each other by a topic. The main goal in this paper is to estimate the probability of topics, and conditional probability of words given topics. To this end, we use non negative Canonical Polyadic (CP) decomposition of a third order moment tensor of observed words. Our computer simulations show that the proposed algorithm has better performance compared to a previously proposed algorithm, which utilizes the Robust tensor power method after whitening by second order moment. Moreover, as our cost function includes the non negativity constraint on estimated probabilities, we never obtain negative values in our estimated probabilities, whereas it is often the case with the power method combined with deflation. In addition, our algorithm is capable of handling over-complete cases, where the number of hidden variables is larger than that of multi-view variables, contrary to deflation-based techniques. Further, the method proposed therein supports a larger over-completeness compared to modified versions of the tensor power method, which has been customized to handle over-complete case.

Keywords: Data mining · Learning · Latent variable · Multi-view · Non negative · Tensor · Cp decomposition · Eigenvalue

1 Introduction

Modeling. Data mining covers a wide range of methods and tools utilized for discovering knowledge from data [17]. In the context of multi-modal data (*e.g.* text, audio and video for the same event), mining can be considered as the estimation of latent variables that are present in several modalities [6,24].

Supported by Gipsa-Lab.

G. Nicosia et al. (Eds.): LOD 2019, LNCS 11943, pp. 219–231, 2019.
https://doi.org/10.1007/978-3-030-37599-7_19

Multi-view models are useful tools to represent the relationships in this framework [27]. The graphical model we assume is depicted in Fig. 1; a set of multi-view variables, $\{x_1, x_2, \ldots, x_L\}$, and their corresponding latent variable, h, have been represented. This general model does not impose any particular distribution (Gaussian, Dirichlet, etc), and is very common in several tasks of data (e.g. text) mining. Text mining as a special case of data mining is the center of attention of this paper. With this in mind, words in a document are considered as the multi-view variables $\{x_1, x_2, \ldots, x_L\}$ and their topic as the hidden variable h, as in [5].

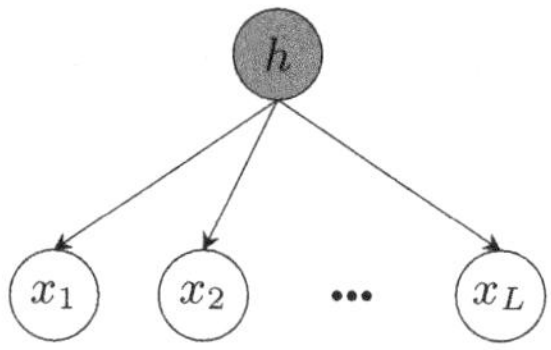

Fig. 1. Multi-view variables and their corresponding hidden variable.

Related Works. Daily increasing amount of text data available online have raised many challenging tasks such as unsupervised learning or clustering, which can be investigated by text mining [17]. Among plenty of algorithms customized for these purposes, we focus on some specific recent methods [2–4], which employ tensor factorization for learning latent variables and of course text mining as a special case. The key is that the latter tensor decomposition is unique, and hence is able to provide an estimation of latent parameters. There exist some other works, such as [9], employing tensors in a similar context. However, those aim at measuring word similarity rather than learning parameters of an underlying graphical model. Therefore, they may resort to the Tucker decomposition, which is not unique, but permits low *multilinear rank* approximation. Unlike [2–4] and the present contribution, which can be applied to other problems than text mining, [9] is hence dedicated to text mining, and does not aim at learning parameters of a model such as that of Fig. 1.

Resorting to moments for estimating parameters has a long history since 1894 till now in various fields such as mixture model [1,5,22], Blind Source Separation [10,11], and machine learning [1–5], just to name a few. For instance, in [5], the graphical model shown in Fig. 1 has been considered, and authors tried to estimate its parameters (probability of h and conditional probability of x_ℓ given h) by the means of diagonalization of two moment matrices.

In the same vein, several recent works have extended the idea of [5] to the tensor framework. To be more precise, [2] is the first contribution in which the Robust tensor power method is used to estimate the above mentioned parameters. One of the limitation of [2] is that it can handle only under-complete cases (*i.e.* when the number of hidden variables is smaller than that of multi-view). In order to overcome this limitation, authors of [3,4] proposed to use other decompositions such as Tucker [12], or enhanced the tensor power method [15,16].

Contribution. In this paper, we use the same formulation as in [2], relating moments and desired parameters. But one important difference lies in assuming a more appropriate cost function incorporating non negative constraints, which we minimize by Alternating-Optimization Alternating Direction Method of Multipliers (AO-ADMM) [18]. This results in better performance than that of [2]. We shall show some figures in Sect. 5.2 demonstrating that our method, unlike the Robust tensor power method [2], does not return negative values for estimated probabilities. Moreover, our method can handle over-complete cases up to higher levels than [3].

Notation. Vectors, matrices, and tensors will be denoted with bold lowercases (*e.g.* $\boldsymbol{a}$), bold uppercases (*e.g.* $\boldsymbol{A}$) and bold calligraphic letters (*e.g.* $\boldsymbol{\mathcal{T}}$), respectively. The contraction symbol[1] $\bullet_k$ indicates a summation on the kth tensor index, and $\otimes$ the tensor (outer) product.

2 Problem Formulation

Although the graphical model in Fig. 1 can be used for a wide range of learning problems [21], we customize the formulation of this section for the particular problem of text mining. Let L be the number of words in a given document, $\boldsymbol{x}_\ell$ the observed words, $\ell \in \mathscr{L} = \{1, 2, \ldots, L\}$, and h a topic encoded into a discrete variable taking K possible integer values, say in $\mathscr{H} = \{1, 2, \ldots, K\}$ with probability $\varphi(k) = Prob(h = k)$. All the words belong to a known encoded dictionary $\Omega = \{\boldsymbol{u}_1, \ldots, \boldsymbol{u}_D\}$ of cardinality D. Put in other words, we can consider a mapping γ (generally not injective) from $\mathscr{L}$ to Ω such that $\boldsymbol{x}_\ell = \boldsymbol{u}_{\gamma(\ell)}$. In the context of text mining, D would be the number of words and K the number of topics. The number of documents, N, is generally larger than D.

Besides the probability of topics, the conditional probability of each word given the topic is also an important parameter in the text mining task. We denote the conditional probability of each word $\boldsymbol{u}_d$ of dictionary Ω, given a particular topic, $h = k$, by $f_k(d) = Prob(\boldsymbol{x} = \boldsymbol{u}_d | h = k)$. Therefore, the joint distribution of $\boldsymbol{X} = [\boldsymbol{x}_1, \ldots, \boldsymbol{x}_L]$ can be written as:

$$p_{\boldsymbol{X}}(\boldsymbol{u}_{\gamma(1)}, \ldots, \boldsymbol{u}_{\gamma(L)}) = \sum_{k=1}^{K} \varphi(k)\, f_k(\gamma(1)) \ldots f_k(\gamma(L)), \qquad (1)$$

which is referred to as the *naive Bayes* model. The main task is to estimate the quantities appearing in the right-hand side of (1) from realizations of $\boldsymbol{X}$. We assume the same assumptions as in [2,5]:

[1] mode-k product between tensors.

- conditional probabilities do not depend on the order of words (exchangablility),
- words are conditionally independent given the topic,
- words have the same conditional distribution given the topic.

In the sequel, we will need the second and third order moments:

$$\boldsymbol{P} \stackrel{\text{def}}{=} \mathbb{E}\{\boldsymbol{x}_p \otimes \boldsymbol{x}_q\} \tag{2}$$

$$\boldsymbol{\mathcal{T}} \stackrel{\text{def}}{=} \mathbb{E}\{\boldsymbol{x}_p \otimes \boldsymbol{x}_q \otimes \boldsymbol{x}_r\}, \tag{3}$$

where $\boldsymbol{P}$ is a $D \times D$ symmetric matrix and $\boldsymbol{\mathcal{T}}$ a $D \times D \times D$ symmetric tensor. These moments do not depend on $\{p, q, r\}$ provided these 3 integers are all different, which ensures the conditional independence assumed in (1). But note that $\{\gamma(p), \gamma(q), \gamma(r)\}$ may not be different because γ is not injective.

As in [2], we encode $\boldsymbol{u}_d$ into the columns of the $D \times D$ identity matrix. Because of this choice made for $\boldsymbol{u}_d$, these moments exhibit the following relations:

$$\boldsymbol{P} = \sum_{k=1}^{K} \varphi_k \, \boldsymbol{a}_k \otimes \boldsymbol{a}_k \tag{4}$$

$$\boldsymbol{\mathcal{T}} = \sum_{k=1}^{K} \varphi_k \, \boldsymbol{a}_k \otimes \boldsymbol{a}_k \otimes \boldsymbol{a}_k, \tag{5}$$

where $\boldsymbol{a}_k$ denotes the kth column of matrix $\boldsymbol{A}$. Note that $\boldsymbol{a}_k$ contains the values of $f_k(d)$ for all d and each k. Rewriting (1) reveals the nice property is that arrays $\boldsymbol{P}$ and $\boldsymbol{\mathcal{T}}$ are actually *joint probabilities* of observations, *i.e.*, $P_{ij} = Prob\{\boldsymbol{x}_p = \boldsymbol{u}_i, \boldsymbol{x}_q = \boldsymbol{u}_j\}$ and $\mathcal{T}_{ijk} = Prob\{\boldsymbol{x}_p = \boldsymbol{u}_i, \boldsymbol{x}_q = \boldsymbol{u}_j, \boldsymbol{x}_r = \boldsymbol{u}_k\}$.

3 The Robust Tensor Power Method

We describe in this section the approach of [2]. We shall compare this method with ours in Subsect. 5.2. In [2], the authors propose to use the two moments defined in (2) and (3) with exactly the same encoding explained in Sect. 2. Therefore, these moments enjoy relations (4) and (5).

Matrix $\boldsymbol{P}$ is theoretically positive semi-definite, since φ_k are positive numbers in (4). Hence in a similar way as had been done for Blind Source Separation [10,11], there exists a "whitening" matrix $\boldsymbol{W}$ such that $\boldsymbol{W}^\mathsf{T}\boldsymbol{P}\boldsymbol{W} = \boldsymbol{I}$, where $\boldsymbol{I}$ is the identity matrix; $\boldsymbol{W}$ can theoretically be easily obtained from the EigenValue Decomposition (EVD) $\boldsymbol{P} = \boldsymbol{U}\boldsymbol{D}\boldsymbol{U}^\mathsf{T}$, $\boldsymbol{U}^\mathsf{T}\boldsymbol{U} = \boldsymbol{I}$, by setting $\boldsymbol{W} = \boldsymbol{U}\boldsymbol{D}^{-1/2}$. Therefore,

$$\sum_{k=1}^{K} \tilde{\boldsymbol{a}}_k \tilde{\boldsymbol{a}}_k^\mathsf{T} = \boldsymbol{I}$$

if $\tilde{\boldsymbol{a}}_k \stackrel{\text{def}}{=} \sqrt{\varphi_k}\, \boldsymbol{W}^\mathsf{T}\boldsymbol{a}_k$. In other words, the matrix $\tilde{\boldsymbol{A}} \in \mathbb{R}^{K \times K}$ containing $\tilde{\boldsymbol{a}}_k$ as its columns is orthogonal, *i.e.* $\tilde{\boldsymbol{A}}\tilde{\boldsymbol{A}}^\mathsf{T} = \boldsymbol{I}$. As proposed in [2], in practice, matrix

---

Algorithm 1. The Robust tensor power method [2]

Input: $\mathcal{T}, \boldsymbol{P}$
Output: $\boldsymbol{A}, \varphi$
1: Whitening: $\boldsymbol{P} = \boldsymbol{U}\boldsymbol{\Sigma}\boldsymbol{V}^{\mathsf{T}}$; $\boldsymbol{W} = \boldsymbol{U}\boldsymbol{\Sigma}^{-1/2}$; $\tilde{\mathcal{T}} \stackrel{\text{def}}{=} \mathcal{T} \bullet_1 \boldsymbol{W} \bullet_2 \boldsymbol{W} \bullet_3 \boldsymbol{W}$
2: **for** $k = 1, \ldots, K$ **do**
3: **for** $m = 1, \ldots, M_1$ **do**
4: 1) draw an initial vector $\boldsymbol{\theta}_k$ of unit 2-norm for $\hat{\boldsymbol{a}}_k$
2) compute M_2 power iteration updates in norm 2, *i.e.*:
$\boldsymbol{t} = \tilde{\mathcal{T}} \bullet_2 \hat{\boldsymbol{a}}_k \bullet_3 \hat{\boldsymbol{a}}_k$; $\hat{\varphi}_k \leftarrow \|\boldsymbol{t}\|_2$; $\hat{\boldsymbol{a}}_k \leftarrow \dfrac{\boldsymbol{t}}{\hat{\varphi}_k}$
5: **end for**
6: Pick the trial (m) having the largest $\hat{\varphi}_k$
7: Refine $\hat{\boldsymbol{a}}_k, \hat{\varphi}_k$ by M_2 extraneous iterations
8: Deflation: $\tilde{\mathcal{T}} \leftarrow \tilde{\mathcal{T}} - \hat{\varphi}\, \hat{\boldsymbol{a}}_k \otimes \hat{\boldsymbol{a}}_k \otimes \hat{\boldsymbol{a}}_k$
9: **end for**
10: De-whitening: $\boldsymbol{B} = (\boldsymbol{W}^T)^{\dagger}$; $\boldsymbol{a}_k = \hat{\varphi}_k \boldsymbol{B} \hat{\boldsymbol{a}}_k$; $\varphi = \dfrac{1}{\hat{\varphi}_k^2}$ for $k = 1, \ldots, K$
11: Back to norm 1: $\alpha = \|\boldsymbol{a}_k\|_1$; $\boldsymbol{a}_k \leftarrow \dfrac{\boldsymbol{a}_k}{\alpha}$; $\varphi_k \leftarrow \dfrac{\varphi_k}{\alpha}$ for $k = 1, \ldots, K$ (our suggestion)

$\boldsymbol{P}$ can be estimated via the mere averaging of samples (7), and consequently, $\boldsymbol{P}$ may have negative eigenvalues because of estimation errors. This issue was not investigated in [2], but they used left singular vectors of $\boldsymbol{P}$ instead of eigenvectors to construct $\boldsymbol{W}$ (cf. Algorithm 1). Next, this whitening matrix is applied to tensor $\mathcal{T}$ to yield:

$$\tilde{\mathcal{T}} \stackrel{\text{def}}{=} \mathcal{T} \underset{1}{\bullet} \boldsymbol{W} \underset{2}{\bullet} \boldsymbol{W} \underset{3}{\bullet} \boldsymbol{W}$$

which have the following element-wise definition [12]:

$$\tilde{\mathcal{T}}(r,t,s) = \sum_{r',t',s'} \mathcal{T}(r',t',s') \boldsymbol{W}(r,r') \boldsymbol{W}(t,t') \boldsymbol{W}(s,s').$$

This new tensor enjoys the relationship

$$\tilde{\mathcal{T}} = \sum_{k=1}^{K} \varphi_k^{-1/2}\, \tilde{\boldsymbol{a}}_k \otimes \tilde{\boldsymbol{a}}_k \otimes \tilde{\boldsymbol{a}}_k.$$

The conclusion is that $\tilde{\mathcal{T}}$ ideally admits an *orthogonal* Canonical Polyadic (CP) decomposition; see *e.g.* [11,12] for an introduction. Many algorithms have been devised for this kind of decomposition, including the pair-sweeping CoM algorithm, or Joint Approximate Diagonalization (JAD) algorithms [10]. But authors in [2] utilized the tensor power iteration [15] to extract the dominant "eigenvector[2]", $\hat{\boldsymbol{a}}$, and dominant "eigenvalue", $\hat{\varphi}$, of $\tilde{\mathcal{T}}$ and then proceeded by deflation, *i.e.* $\tilde{\mathcal{T}} \leftarrow \tilde{\mathcal{T}} - \hat{\varphi}\, \hat{\boldsymbol{a}} \otimes \hat{\boldsymbol{a}} \otimes \hat{\boldsymbol{a}}$, to get the remaining ones.

---

[2] Recall that there exist several definitions of tensor eigenvectors [19,23]. The definition used in [2] – and hence here – is $\mathcal{T} \bullet \boldsymbol{v} \bullet \boldsymbol{v} = \lambda \boldsymbol{v}$, which has the undesirable property that λ depends on the norm of $\boldsymbol{v}$. In fact, if $(\lambda, \boldsymbol{v})$ is an eigenpair, then so is $(\alpha\lambda, \alpha\boldsymbol{v})$ for any nonzero α. This is analyzed in *e.g.* [23].

Unfortunately, except the tensor power iteration [15], there is no pseudo code in [2] describing the entire algorithm. Our description in Algorithm 1 hopefully fills this lack; we used this algorithm in our subsequent computer experiments in Sect. 5.2.

4 Proposed Method

In this section, we detail our contributions, namely a non negative Canonical Polyadic (CP) decomposition executed on tensor (3). Therefore, it should not be considered as a modified version of the Robust Tensor Power Method, since it is a completely different way of computing the unknown parameters. It is clear that $\boldsymbol{\mathcal{T}}$ in (5) corresponds to a CP decomposition structure. As we seek $\boldsymbol{A}$ and $\boldsymbol{\varphi}$, which represent probability values, it would be inevitable to consider non negativity constraint during the decomposition process. However, at least theoretically, we do not need to impose symmetry during the course of the CP decomposition, even if (5) reveals that $\boldsymbol{\mathcal{T}}$ is a symmetric tensor, since it will be expected to be obtained automatically.

Generally, computing a tensor low-rank approximation is ill-posed, except if some constraints, such as *non negativity* or *orthogonality*, are imposed [12,20]. Taking this into account, in order to obtain reliable estimated quantities, $\boldsymbol{A}$ and $\boldsymbol{\varphi}$, we consider the following minimization in $\mathbb{R}_+$:

$$\min_{\boldsymbol{A},\boldsymbol{\varphi}} \|\boldsymbol{\mathcal{T}} - \sum_{k=1}^{K} \varphi(k)\, \boldsymbol{a}_k \otimes \boldsymbol{a}_k \otimes \boldsymbol{a}_k\|_2 \qquad (6)$$
$$\text{s.t.} \quad \boldsymbol{A} \in \mathbb{R}_+^{D\times K},\ \boldsymbol{\varphi} \in \mathbb{R}_+^{K}.$$

Since (6) is a non convex problem, Alternating Optimization (AO) is usually executed for such a problem [13,18,26]. AO is also known as Block Coordinate Descent (BCD) whose convergence has been discussed in [7]. When the cost function is simply least squares with no regularization, AO converted into the well-known method Alternating Least Squares (ALS)[18]. A recent modification named AO-ADMM [18] has been done to improve AO by the means of Alternating Direction Method of Multipliers (ADMM) [8], so that AO can handle constraints as well. Algorithm 2 shows the details of applying AO-ADMM to our problem (6).

It should be noted that in order to ensure the uniqueness of the CP decomposition in Algorithm 2, the rank of $\boldsymbol{\mathcal{T}}$, K, should be less than the *expected rank* R^o, or than R^o_s in the symmetric case [12], with:

$$R^o = \left\lceil \frac{D^3}{3D-2} \right\rceil,\ R^o_s = \left\lceil \frac{(D+1)(D+2)}{6} \right\rceil,$$

Therefore, because the proposed algorithm only uses the third order moment tensor $\boldsymbol{\mathcal{T}}$, it is capable of handling over-complete cases ($K > D$) for estimating probabilities of hidden multi-view variables, up to $K = O(D^2)$.

Algorithm 2. AO-ADMM for Problem (6)

Input: $\mathcal{T}, K, \rho, Q$
Output: $\boldsymbol{A}, \boldsymbol{\varphi}$
for $q = 1 : Q$ (loop on starting points) **do**
  Initialize $\boldsymbol{H}_1, \boldsymbol{H}_2, \boldsymbol{H}_3$
  Initialize $\boldsymbol{U}_1, \boldsymbol{U}_2, \boldsymbol{U}_3$
  repeat
    for $n = 1, 2, 3$ **do**
      $\boldsymbol{T} = \mathcal{T}_{(n)}$ (Unfold mode n)
      $\boldsymbol{Z} = \odot_{j \neq n} \boldsymbol{H}_j$
      Compute the Cholesky factor $\boldsymbol{L}$ of $(\boldsymbol{Z}^{\mathsf{T}}\boldsymbol{Z} + \rho \boldsymbol{I})$
      Update $\boldsymbol{H}_n$ as in Alg.1 of [18]:
      (a) $\boldsymbol{G}_n \leftarrow \boldsymbol{L}^{-\mathsf{T}}\boldsymbol{L}^{-1}[\boldsymbol{Z}^{\mathsf{T}}\boldsymbol{T} + \rho(\boldsymbol{U}_n + \boldsymbol{H}_n)^{\mathsf{T}}]$ # Auxiliary variables
      (b) $\boldsymbol{H}_n \leftarrow \max[\boldsymbol{0}, \boldsymbol{G}_n^{\mathsf{T}} - \boldsymbol{U}_n]$ # Primal variables
      (c) $\boldsymbol{U}_n \leftarrow \boldsymbol{U}_n + \boldsymbol{H}_n - \boldsymbol{G}_n^{\mathsf{T}}$ # Dual variables
    end for
  until some termination criterion
  for $k = 1 : K$ **do** $\boldsymbol{a}_k = H_1(:,k)/\|H_1(:,k)\|_1$; $\varphi_k = \prod_n \|H_n(:,k)\|_1$ **end for**
end for
Pick the best among the Q estimates.

5 Simulations

In this section we describe computer simulations aiming at comparing the proposed method with that of [2] in terms of performance in estimating the probability of topics and the conditional probabilities of words given the topics. In Subsect. 5.1, we explain how one can generate synthetic data according to the graphical model in Fig. 1. The relative error e_{φ} (resp. $e_{\boldsymbol{A}}$) in estimating $\boldsymbol{\varphi}$ (resp. $\boldsymbol{A}$) made by both methods are reported in Subsect. 5.2. If $\hat{\boldsymbol{\varphi}}$ is an estimate of $\boldsymbol{\varphi}$, it is defined as:

$$e_{\varphi} = \frac{\|\boldsymbol{\varphi} - \hat{\boldsymbol{\varphi}}\|_2}{\|\boldsymbol{\varphi}\|_2}$$

5.1 Data Generation

In this subsection we explain how one can generate a set of synthetic encoded words that are related to the same topic (cf. Fig. 1), with these two properties simultaneously:

- conditionally independent given the topic
- having the same conditional distribution

The first step to generate a synthetic data set is to define the arbitrary distributions $\varphi(k)$ and $f_k(d)$ for all $(k, \boldsymbol{u}_d) \in \mathscr{H} \times \Omega$. As mentioned before, the values of $\varphi(k)$ and $f_k(d)$ are stored in a K-dimensional vector $\boldsymbol{\varphi}$ and in a $D \times K$ matrix $\boldsymbol{A}$ respectively.

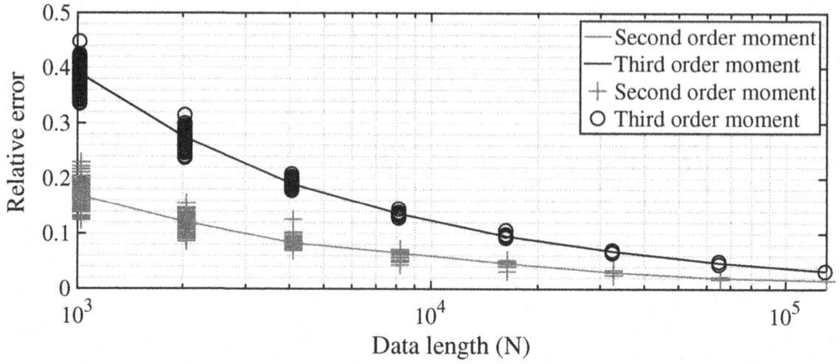

Fig. 2. Consistency of moment estimates. In this figure, the larger the data length, the fewer the number of realizations, to keep the total block length (and hence the computational load) constant.

In order to generate synthetic data according to graphical model in Fig. 1, it is actually useful to use their cumulative distributions, $\Phi(k) = Prob(h \leq k)$ and $F_k(d) = Prob(\gamma \leq d \,|\, h = k)$. $\boldsymbol{\Phi}$ is a $K \times 1$ vector constructed by summing[3] the entries of $\boldsymbol{\varphi}$. Similarly, matrix $\boldsymbol{F}$ is a $D \times K$ matrix obtained by summing the entries of matrix $\boldsymbol{A}$. Our generative algorithm is as follows:

- draw $z \in [0,\, 1]$, and pick $k = \Phi^{-1}(z)$, by selecting the first entry in $\boldsymbol{\Phi}$ that is larger than z.
- for each $\ell \in \{1, 2, \ldots, L\}$,
  - draw $z_\ell \in [0,\, 1]$, and pick $\gamma(\ell) = F_k^{-1}(z_\ell)$, by selecting the first entry in the kth column of $\boldsymbol{F}$ that is larger than z_ℓ.
  - set $\boldsymbol{x}_\ell = \boldsymbol{u}_{\gamma(\ell)}$.

In the second step, we utilize the averages of large number of realizations, say $N = 2^{14}$, to obtain an acceptable approximation of $\boldsymbol{P}$ and $\boldsymbol{\mathcal{T}}$. In each realization, we draw a random encoded topic, k, in the way described above. Then, according to the chosen topic, we draw three random encoded words, $\{\boldsymbol{x}_p = \boldsymbol{u}_{\gamma(p)}, \boldsymbol{x}_q = \boldsymbol{u}_{\gamma(q)}, \boldsymbol{x}_r = \boldsymbol{u}_{\gamma(r)}\}$, such that they satisfy the two properties mentioned above. At the end, by using the following averages, we have an *empirical* approximation of second and third order moments with sample moments below:

$$\boldsymbol{P}_e = \frac{1}{N} \sum_{n=1}^{N} \boldsymbol{u}_{\gamma(p)} \otimes \boldsymbol{u}_{\gamma(q)} \tag{7}$$

$$\boldsymbol{\mathcal{T}}_e = \frac{1}{N} \sum_{n=1}^{N} \boldsymbol{u}_{\gamma(p)} \otimes \boldsymbol{u}_{\gamma(q)} \otimes \boldsymbol{u}_{\gamma(r)} \tag{8}$$

[3] With this goal, the matlab function `cumsum()` can be used.

Moment Consistency. As mentioned in Sect. 2, it can be easily shown that (2) and (3) are respectively equal to (4) and (5). Therefore, the above sample estimates of (2) and (3), *i.e.* (7) and (8), should converge to the true moments $\boldsymbol{P}$ and $\boldsymbol{\mathcal{T}}$ defined in (4) and (5). Hence, it seems essential to check the consistency of generated data in order to make sure that sample moments indeed converge to the true joint probabilities. To this end, we generated N realizations of our synthetic data set, $\boldsymbol{x}_\ell = \boldsymbol{u}_{\gamma(\ell)}$, (in a way explained above) with the following parameter values: $K = 4, D = 6, N = 2^{17}$. We chose a uniform distribution for φ, and K arbitrary distributions (cf. Sect. 5.2) to build a synthetic matrix $\boldsymbol{A}$. Lastly, we calculated sample moments in (7) and (8), and compared them to the true ones (4) and (5). Figure 2 reports the discrepancy between both in terms of Euclidean norm. As can be seen in Fig. 2, the generated data in the way described above are consistent for $N > 10000$.

5.2 Performance

Results as a function of N. As mentioned before, the main goal is the estimation of $\boldsymbol{A}$ and $\boldsymbol{\varphi}$. We report in this section the comparison results obtained with only one dictionary size, $D = 8$, with a fixed number of topics, $K = 4$, and for various data lengths up to $N = 2^{17}$. The power method is run with with $M_1 = 10$ and $M_2 = 5$, and the proposed algorithm with $Q = 100$. Moreover, the results are obtained for one particular realization[4] of matrix $\boldsymbol{A}$ and vector $\boldsymbol{\varphi}$:

$$\boldsymbol{A} = \begin{bmatrix} 0.162 & 0.211 & 0.000 & 0.000 \\ 0.082 & 0.130 & 0.000 & 0.000 \\ 0.022 & 0.235 & 0.000 & 0.403 \\ 0.196 & 0.423 & 0.000 & 0.038 \\ 0.174 & 0.000 & 0.276 & 0.119 \\ 0.104 & 0.000 & 0.133 & 0.439 \\ 0.113 & 0.000 & 0.119 & 0.000 \\ 0.147 & 0.000 & 0.473 & 0.000 \end{bmatrix}, \quad \boldsymbol{\varphi} = \begin{bmatrix} 0.256 \\ 0.163 \\ 0.201 \\ 0.380 \end{bmatrix}.$$

In order to compare the performance of both methods, we run the above simulation for increasing data lengths (note that the larger the data length, the fewer the number of realizations, to keep a computational load constant). To be more precise, by keeping the values of parameters $D = 8, K = 4$, we increase exponentially the value of N from 2^{14} to 2^{17}. For each value of N, we generated several synthetic data sets of size N in the way that has been explained in Subsect. 5.1, and then the methods have been applied to the obtained data to compare relative errors as a function of data length. The results are reported in Fig. 3 and 4; each black circle (resp. red cross) corresponds to the error of power method (resp. proposed method) in estimating $\boldsymbol{\varphi}$ or $\boldsymbol{A}$ for a specific realization of a particular data length, N. To ease comparison, the median (resp. standard deviation) among realizations is also plotted in solid (resp. dotted) line for every data length. As Figs. 3 and 4 show, the error of the proposed method converges

[4] Results on other particular realizations may be found in [25].

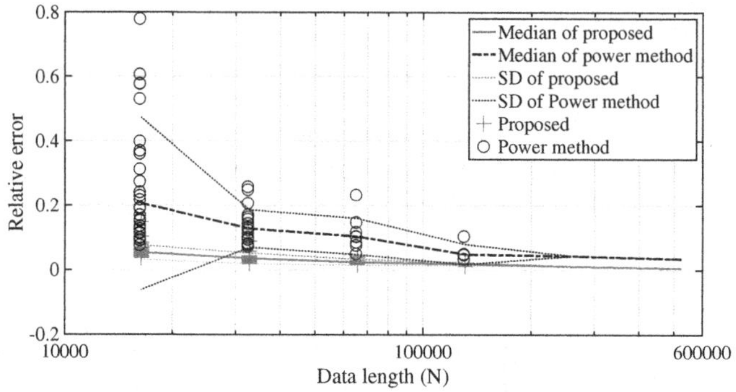

Fig. 3. Performance (relative error) in estimating $\boldsymbol{A}$

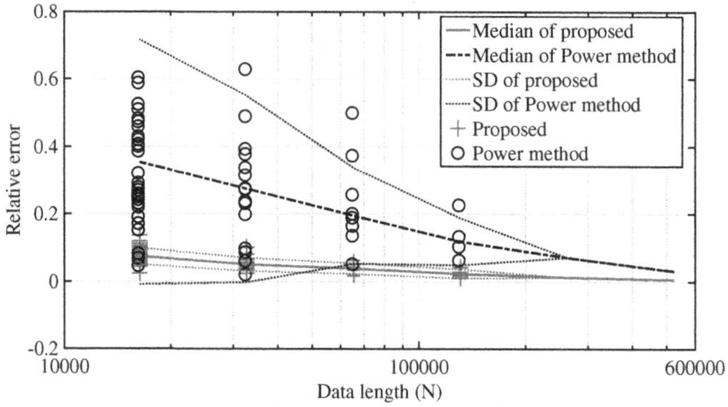

Fig. 4. Performance (relative error) in estimating $\boldsymbol{\varphi}$

faster to zero as the data length increases, and unlike the Robust tensor power method [2], it appears to be much more robust, *i.e.* smaller standard deviation of the relative error.

Results for random $\boldsymbol{A}$ *and* $\boldsymbol{\varphi}$. When running extensive computer simulations, one can assume a uniform distribution for the topics, and K uniform distributions for the conditional probability of words given the topic, which are actually the columns of $\boldsymbol{A}$. Once $\boldsymbol{A}$ and $\boldsymbol{\varphi}$ are obtained, one can proceed exactly the same way as explained in Subsect. 5.1 to generate synthetic datasets and sample moments. For space reasons, we only report in Table 1 the results obtained for a few values of (D, K), averaged over 20 independent trials. These extensive computer experiments confirm the robustness and superiority of the proposed algorithm compared to the power method.

Table 1. Median (left) and standard deviation (right) of error in estimating $\boldsymbol{A}$ and φ, with data length $N = 2^{17} \approx 1.2\,10^6$, and $K = 4$. In each cell, top: proposed method, bottom: power method.

Relative error in estimating $\boldsymbol{A}$										
D	6	7	8	9	10	6	7	8	9	10
Proposed	0.087	0.077	0.070	0.104	0.073	0.100	0.123	0.083	0.104	0.120
Robust power method	0.382	0.178	0.243	0.181	0.118	0.313	0.314	0.386	0.327	0.235
Relative error in estimating φ										
D	6	7	8	9	10	6	7	8	9	10
Proposed	0.145	0.105	0.115	0.096	0.128	0.223	0.118	0.213	0.285	0.315
Robust power method	0.657	0.283	0.310	0.287	0.228	0.545	0.524	0.591	0.407	0.399

6 Conclusion

In this paper, we focus on text mining as a special case of data mining: given the observation of words of several documents, we aim at estimating the probabilities of topics, φ, (as hidden variable cf. Fig. 1) as well as the conditional probabilities of words given topics, $\boldsymbol{A}$. For this purpose, we use (5), which is the same equation as in [2] relating the third order moment, $\mathcal{T}$, of observed words to desired probabilities. However, our cost function is more appropriate since non negativity is imposed. In fact, the Robust tensor power method [2] sometimes delivers negative values for estimated probabilities [25]. This difference appears in our computer results, which show that the performance of the proposed method is better compared to [2]. In addition, our algorithm only requires $\mathcal{T}$ as input, unlike the algorithm in [2], which permits to address the over-complete regime where number of hidden variables is larger than that of multi-view variables.

Further improvements of the power method. As mentioned in Sect. 3, the Robust tensor power method [2], combined with deflation, attempts to estimate eigenvectors and eigenvalues of the whitened tensor. One known problem with deflation, which has been observed even for matrices [14], is that the extracted eigenvectors may loose orthogonality because of rounding errors. The consequence is that the same (dominant) eigenvector may show up several times, especially if the array dimensions become large. This problem can be fixed by a re-orthogonalization of initial vectors with previously found vectors. Suppose that $\boldsymbol{V}(i) \stackrel{\text{def}}{=} [\tilde{\boldsymbol{a}}_1, \ldots \tilde{\boldsymbol{a}}_i]$ is the matrix of the eigenvectors already obtained at iteration i. At the next iteration, instead of choosing simply a random vector $\boldsymbol{\theta}$ as in [2], we propose to pick it in the subspace orthogonal to $\boldsymbol{V}(i)$: $\tilde{\boldsymbol{\theta}} = \boldsymbol{\theta} - \boldsymbol{V}\boldsymbol{V}^{\mathsf{T}}\boldsymbol{\theta}$ is the next initial vector. In even larger dimensions, it may be necessary to also re-orthogonalize the intermediate iterates generated by the tensor power method, because of rounding errors. Another improvement consists in symmetrizing the sample estimates $\mathcal{T}_e$ and $\boldsymbol{P}_e$.

Over-complete regime. We previously mentioned the ability of our algorithm to address the so-called over-complete regimes, where $K > D$, up to $K = O(D^2)$.

In this context, deflation is not possible, and if K eigenvectors are desired, the tensor power method must be run from K initial values, located close enough to the true eigenvectors. But the latter are unknown, which imposes to draw a very large number (growing as a Dth power) of initial vectors. We believe this solution is not workable. The sufficient condition requested in [3] is not only inaccurately defined, but is also very unlikely to be satisfied in practice.

Perspectives. This work must be continued to allow the application of the algorithm we have proposed to real databases. The implementation of certain operations must be revised in large dimensions to allow firstly to save memory space, secondly to limit rounding errors, and thirdly to reduce biases specific to large dimensions.

References

1. Anandkumar, A., Foster, D.P., Hsu, D., Kakade, S.M., Liu, Y.: A spectral algorithm for latent Dirichlet allocation. NIPS **25**, 1–9 (2012)
2. Anandkumar, A., Ge, R., Hsu, D., Kakade, S.M., Telgarsky, M.: Tensor decompositions for learning latent variable models. J. Mach. Learn. Res. **15**, 2773–2832 (2014)
3. Anandkumar, A., Ge, R., Janzamin, M.: Analyzing tensor power method dynamics in overcomplete regime. J. Mach. Learn. Res. **18**, 1–40 (2017)
4. Anandkumar, A., Hsu, D., Janzamin, M., Kakade, S.: When are overcomplete topic models identifiable? Uniqueness of tensor Tucker decompositions with structured sparsity. J. Mach. Learn. Res. **16**, 2643–2694 (2015)
5. Anandkumar, A., Hsu, D., Kakade, S.M.: A method of moments for mixture models and hidden Markov models. In: 25th Conference on Learning Theory, vol. 23, pp. 33.1–33.34 (2012)
6. Atrey, P.K., Hossain, M.A., El Saddik, A., Kankanhalli, M.S.: Multimodal fusion for multimedia analysis: a survey. Multimed. Syst. **16**(6), 345–379 (2010)
7. Beck, A., Tetruashvili, L.: On the convergence of block coordinate descent type methods. SIAM J. Optim. **23**(4), 2037–2060 (2013)
8. Boyd, S., Parikh, N., Chu, E., Peleato, B., Eckstein, J.: Distributed optimization and statistical learning via the alternating direction method of multipliers. Found. Trends Mach. Learn. **3**(1), 1–122 (2011)
9. Chang, K.W., Yih, W.T., Meek, C.: Multi-relational latent semantic analysis. In: Empirical Methods in Natural Language Proceedings, Seattle, pp. 1602–1612, October 2013
10. Comon, P., Jutten, C. (eds.): Handbook of Blind Source Separation. Academic Press, Oxford, Burlington (2010)
11. Comon, P.: Independent component analysis, a new concept? Sig. Proc. **36**(3), 287–314 (1994)
12. Comon, P.: Tensors: a brief introduction. IEEE Sig. Proc. Mag. **31**(3), 44–53 (2014)
13. Comon, P., Luciani, X., De Almeida, A.L.: Tensor decompositions, alternating least squares and other tales. J. Chemom. **23**(7–8), 393–405 (2009)
14. Cullum, J., Willoughby, R.: Lanczos Algorithms, vol. I. Birkhauser, Basel (1985)
15. De Lathauwer, L., Comon, P., et al.: Higher-order power method, application in Independent Component Analysis. In: NOLTA, Las Vegas, pp. 91–96, December 1995

16. De Lathauwer, L., De Moor, B., Vandewalle, J.: On the best rank-1 and rank-(r 1, r 2, ..., rn) approximation of higher-order tensors. SIAM J. Matrix Anal. Appl. **21**(4), 1324–1342 (2000)
17. Han, J., Pei, J., Kamber, M.: Data Mining: Concepts and Techniques. Elsevier, Amsterdam (2011)
18. Huang, K., Sidiropoulos, N., et al.: A flexible and efficient algorithmic framework for constrained matrix and tensor factorization. IEEE Trans. Sig. Proc. **64**(19), 5052–5065 (2016)
19. Lim, L.H.: Singular values and eigenvalues of tensors: a variational approach. In: IEEE SAM Workshop, Puerto Vallarta, Mexico, 13–15 December 2005
20. Lim, L.H., Comon, P.: Nonnegative approximations of nonnegative tensors. J. Chemom. **23**, 432–441 (2009)
21. Murphy, K.P.: Machine Learning. A Probabilistic Perspective. MIT Press, London (2012)
22. Pearson, K.: Contributions to the mathematical theory of evolution. Philos. Trans. R. Soc. Lond. A **185**, 71–110 (1894)
23. Qi, L., Luo, Z.: Tensor Analysis. SIAM, Spectral Theory and Special Tensors (2017)
24. Ramage, D., Manning, C.D., Dumais, S.: Partially labeled topic models for interpretable text mining. In: Proceedings of the 17th ACM SIGKDD International Conference on Knowledge Discovery and Data Mining, pp. 457–465. ACM (2011)
25. Sobhani, E., Comon, P., Babaie-Zadeh, M.: Data mining with tensor decompositions. In: Gretsi. Lille, 26–29 August 2019
26. Sobhani, E., Sadeghi, M., Babaie-Zadeh, M., Jutten, C.: A robust ellipse fitting algorithm based on sparsity of outliers. In: 2017 25th European Signal Processing Conference (EUSIPCO), pp. 1195–1199. IEEE (2017)
27. Whittaker, J.: Graphical Models in Applied Multivariate Statistics. Wiley, Hoboken (2009)

The Induction Problem: A Machine Learning Vindication Argument

Gianluca Bontempi(✉)

Machine Learning Group, Computer Science Department,
Faculty of Sciences, ULB, Université Libre de Bruxelles, Brussels, Belgium
gbonte@ulb.ac.be

Abstract. The problem of induction is a central problem in philosophy of science and concerns whether it is sound or not to extract laws from observational data. Nowadays, this issue is more relevant than ever given the pervasive and growing role of the data discovery process in all sciences. If on one hand induction is routinely employed by automatic machine learning techniques, on the other most of the philosophical work criticises induction as if an alternative could exist. But is there indeed a reliable alternative to induction? Is it possible to discover or predict something in a non inductive manner?

This paper formalises the question on the basis of statistical notions (bias, variance, mean squared error) borrowed from estimation theory and statistical machine learning. The result is a justification of induction as rational behaviour. In a decision-making process a behaviour is rational if it is based on making choices that result in the most optimal level of benefit or utility. If we measure utility in a prediction context in terms of expected accuracy, it follows that induction is the rational way of conduct.

1 Introduction

The process of extraction of scientific laws from observational data has interested the philosophy of science during the last two centuries. Though not definitely settled, the debate is more relevant than ever given the pervasive and growing role of data discovery in all sciences. The entire science, if not the entire human intellectual activity, is becoming data driven. Data science, or the procedure of extracting regularities from data on the basis of inductive machine learning procedures [3], is nowadays a key ingredient of the most successful research and applied technological enterprises.

This may appear paradoxical if we consider that the process of induction, from Hume's [4] work on, has been accused of having no rational foundation. Induction is the inferential process in which one takes the past as grounds for beliefs about the future, or the observed data as grounds for beliefs about the unobserved. In an inductive inference, where the premises are the data (or observations) and the conclusions are referred to as hypothesis (or models), three main properties hold [1]:

G. Nicosia et al. (Eds.): LOD 2019, LNCS 11943, pp. 232–243, 2019.
https://doi.org/10.1007/978-3-030-37599-7_20

1. The conclusions follow *non-monotonically* from the premises. The addition of an extra premise (i.e. more data) might change the conclusion even when the extra premise does not contradict any of the other premises. In other terms $D_1 \subset D_2 \not\Rightarrow (h(D_2) \Rightarrow h(D_1))$ where $h(D)$ is the inductive consequence of D.
2. The truth of the premises is not enough to guarantee the truth of the conclusion as there is no correspondent to the notion of deductive validity.
3. There is an information gain in induction since a hypothesis asserts more than data alone.

In other words an inductive inference is *ampliative*, i.e. it has more content in the conclusion than in the premises, and contrasts with the mathematical and logical reasoning which is deductive and non ampliative. As a consequence, inductive inference is unsafe: no conclusion is a guaranteed truth and so it can dissolve even if no premise is removed.

According to Hume *all reasonings concerning nature are founded on experience, and all reasonings from experience are founded on the supposition that the course of nature will continue uniformly the same* or in other words that the future will be like the past. Any attempt to show, based on experience, that a regularity that has held in the past will or must continue to hold in the future will be circular. It follows that the entire knowledge discovery process from data is relying on shaky foundations. This is well known as the Hume's problem and the philosopher C. D. Broad's defined induction as "the glory of science and the scandal of philosophy". The serious consequences of such result was clear to Hume himself who never discouraged scientists from inductive practices. In fact, in absence of a justification, he provided an explanation which was more psychological than methodological. According to Hume, we, humans, expect the future will be like the past since this is part of our nature: we have inborn inductive habits (or instincts) but we cannot justify them. The *principle of uniformity of nature* is not a priori true, nor it can be proved empirically, and there is no reason beyond induction to justify inductive reasoning. Thus, Hume offers a naturalistic explanation of the psychological mechanism by which empirical predictions are made without any rational justification for this practice.

The discussion about the justification of induction started in the late 19th century. A detailed analysis of the responses to the Hume problem are contained in the Howson book [4]. Let us review some of the most interesting arguments.

First, the Darwinian argument which claims that the inductive habit was inherited as a product of evolutionary pressures. This explanation is a sort of "naturalized epistemology" but can be accused of circularity since we assume inductive science to justify induction in science.

Bertrand Russell suggests instead to accept the insolubility of the problem and proposes to create an "inductive principle" that should be accepted as sound ground for justifying induction.

Another well known argument is "Rule-circularity" which states: "Most of the inductive inferences humans made in the past were reliable. Therefore the majority of inductive inferences are right". Cleve insisted that the use of induction to prove induction is rule circular but not premise circular and, as such, acceptable. Criticisms about this argument, considered to be anyway question-begging, are detailed in [4].

"The No-Miracles" argument states: "If an hypothesis h predicts independently a sufficiently large and precise body of data, it would be a miracle that it would be false then we can reasonably infer the approximate truth of the hypothesis". Though this argument has been often raised in philosophy of science, common wisdom in machine learning may be used to refute it [3]. It is not the degree of fitting of an hypothesis to historical data which makes it true. There is nothing extraordinary (in fact no miracle at all) in predicting a large body of data if the hypothesis is complex enough and built ad hoc. A well know example is an overfitting hypothesis, i.e. a too complex hypothesis which interpolates past data yet without any generalization power. The quality of an hypothesis derives from the quality of the learning procedure used to build it and from the correctness of its implicit assumption, not from the fact of predicting correctly a large (or very large) number of outcomes. This is made explicit in statistical learning by the notions of bias and variance of an estimator [2] which we will use later to establish our argument.

Popper's [9] reaction to Hume's problem was to simply deny induction. According to him, humans or scientists do not make inductions, they make conjectures (or hypothesis) and test them (or their logical consequences obtained by deduction). If the test is successful the conjecture is corroborated but never verified or proven. Confirmation is a myth. No theory or belief about the world can be proven: it can be only submitted to test for falsification and, in the best case, be confirmed by the evidence for the time being. Though the argument of Popper got an immense credit among scientists, it is so strict to close any prospect of automatic induction. If induction does not exist and the hypothesis generation is exclusively a human creative process, any automatic and human independent inductive procedure (like the ones underlying all the successful applications in data science) should be nonsense or at least ineffective. As data scientists who are assisting to an incredible success of inductive practices in any sort of predictive task, we reject the denial of induction and we intend to use arguments from statistical machine learning to justify its use in science.

Let us assume that induction is a goal-directed activity, whose goal is to generalise from observational data. As Howson stresses [4] there are two ways to justify its usage: an internal and an external way. The internal way consists in showing that the structure of the procedures itself inevitably leads to the goal, in other words that the goal (here the hypothesis) necessarily follows from the input (the data). As stressed by Hume and confirmed by well-known results in machine learning (notably the "no free lunch" theorems [14]), induction does not necessary entail truth. There is no gold procedure that given the data returns necessarily the optimal or the best generalization.

The second way is to show that the inductive procedure achieves the goal most of the time, but the success or not of the enterprise depends on factors external to it. There, though there is no necessity in the achievement of the objective, the aim is to show that the inductive procedure is the least incorrect one.

This paper takes this second approach by getting inspiration from the Hans Reichenbach's *vindication* of induction [10], where it is argued that if predictive success is possible, only inductive methods will be able to pursue it. *Consequently we have everything to gain and nothing to lose by employing inductive methods* [11]. In this paper we take a similar pragmatic approach to justify induction as the most rational practice in front of experimental observations. In detail, we extend and formalize the Reichenbach argument by adopting notation and results from the estimation theory in order to assess and measure the quality of an estimator, or more in a general, of any inductive algorithm. Note that in our approach we replace the long-run approach of Reichenbach, which was criticized in terms of limiting values, by a short run or finite sample approach which quantifies the generalization accuracy of any inductive practice using a finite number of observations. What emerges is that, though no guarantee of correct prediction is associated to inductive practices, induction is the most rational choice (in the sense of lowest generalization error) if we have to choose between an inductive or a non inductive approach.

Note that our decision theory arguments differs form the "rule-circularity" argument since we make no assumption that past successes of induction necessarily extrapolate to future ones. Each application domain is different and there is no guarantee that what worked in other contexts (or times) will be useful in our, too. In our argument induction is not perfect, but it is the lesser evil and it has to be preferred whatever is the degree of regularity of the natural phenomenon: in other words if a regularity exists, induction is less error prone than non induction while, in absence of regularity, induction is as weak as non induction.

Another aspect of our approach is that it applies whatever is the adopted inductive procedure (e.g. regression, classification, prediction). This allows us to extend the conventional discourse about induction to other domains than simply induction by enumeration. Finally this paper aims to corroborate the idea that the interaction between machine learning and philosophy of science can play a beneficial role in improving the grasp of the induction process (see [6,12] and other papers in the same special issue).

2 The Machine Learning Argument

We define first what we intend by inductive process and more specifically how we can assess in a quantitative manner its accuracy. In what follows we will represent variability by having recourse to a stochastic notation. We will use the bold notation to denote random variables and the normal font to denote their value. So by $\mathbf{x}$ we denote the random variable while by x we refer to a single realization of the r.v. $\mathbf{x}$. We will have resort to the terminology of the estimation theory [7] where an estimator is defined as any algorithm taking as input a finite dataset D_{tr} made of N_{tr} samples and returning as output an estimation (or prediction) $\hat{\theta}_N$ of an unknown quantity θ. Note that this definition is extremely general since it encompasses many statistical modeling procedures as well as plenty of supervised machine learning and data mining algorithms.

Suppose we are interested in some phenomena characterized by a set of variables for which we have a set of historical records. In particular our interest concerns some properties of these variables, like a parameter of the distribution, correlation or more in general their dependency. We use the target quantity θ to design the object of our estimation. For instance in regression θ could denote the expected value of $\mathbf{y}$ for an observed value x while, in case of binary classification, θ could design the probability that $\mathbf{y} = 1$ for a given x.

Since induction looks for regularity, but there is no logical necessity for regularity in nature, we should assess the benefits of induction (with respect to no induction) by taking into account that either regular or irregular phenomena can occur. We assume that the observed data are samples of a generative model characterized by a parametric probability distribution with parameter $\boldsymbol{\theta}$ where the parameter is random, and no assumption is made about the nature of the distribution. For the sake of simplicity, we will assume that $\boldsymbol{\theta}$ is scalar and with variance $\mathcal{V}$, though the results may be generalized to the multivariate case.

We distinguish between two settings: *regularity* (or uniformity of nature) and *irregularity*. A regular phenomenon is described by a distribution where the parameter θ_{tr} is unknown but constant. An irregular phenomenon is described by a distribution where the parameter is random. In this paper the degree of randomness of $\boldsymbol{\theta}$ is used to denote the degree of irregularity of the phenomenon.

In a regular setting two subsequent observations (or datasets) are i.i.d realizations of the same probability distribution, i.e. characterized by the same parameter θ_{tr}) In an irregular setting two subsequent observations are realizations of two different distributions, for example a training one with parameter θ_{tr} and a test one with parameter θ_{ts}.

Hence after, θ will denote both the parameter of the data distribution and the target of the prediction: for instance θ could be the conditional mean (in a regression task) or the a posteriori probability (in a classification task).

Machine learning decomposes the induction process into three major steps: the collection of a finite dataset D_{tr} made of N samples, the generation of a family $\mathcal{E}$ of estimators (or hypothesis) and the model selection[1]. We call estimator [7] any algorithm[2] taking as input a training set D_{tr} and returning as output an estimation (or prediction) $\hat{\theta}_{tr}$ of the target quantity θ_{tr}.

An estimator is then the main ingredient of any inductive process, since it makes explicit the mapping $\hat{\theta}_{tr} = \hat{\theta}(D_{tr})$ between data and estimation. Since data are variable (or more formally the dataset $\mathbf{D}_{tr}$ is random), the output of the estimation process is the random variable $\hat{\boldsymbol{\theta}}_{tr}$. It follows that we cannot talk about the accuracy of an inductive process without taking into account this variability. This is the reason why, if the goal is to predict θ we cannot assess the quality of an estimator (or more in general of any inductive procedure) in terms of a single prediction but more properly in terms of statistical average of the prediction error.

[1] for simplicity we will not consider here the case of combining estimators.
[2] here we will consider only deterministic algorithms.

The accuracy of $\hat{\boldsymbol{\theta}}_{tr} \in \mathcal{E}$ may be expressed in terms of the Mean Squared Error

$$\text{MSE}(\hat{\boldsymbol{\theta}}_{tr}) = E_{\hat{\boldsymbol{\theta}}_{tr}}[(\theta_{tr} - \hat{\boldsymbol{\theta}}_{tr})^2]$$

which can be notoriously decomposed in the sum

$$(E_{\hat{\boldsymbol{\theta}}_{tr}}[\hat{\boldsymbol{\theta}}_{tr}] - \theta_{tr})^2 + \text{Var}\left[\hat{\boldsymbol{\theta}}_{tr}\right] = B_{tr}^2 + V_{tr}$$

where B_{tr} and V_{tr} denote the bias and the variance, respectively. While the bias is a sort of systematic error, typically due to the family of estimators taken into consideration, the variance measures the sensitivity of the estimator to the training dataset and decreases with the increase of the number N of observations.

What is remarkable here is that those quantities formalize the reliability and the accuracy of any inductive process. On one hand they show that it does not make sense to assume a perfect induction since any induction depends on data and, being data variable, induction is variable too. On the other hand, though perfect induction is illusory, it is possible to have degree of reliability according to the property of the target quantity, of the observed data and the estimator. In what follows we will use these quantities to show that no rational alternative to induction exists if our aim is to perform prediction or more in general extract information from data.

Let

$$\hat{\boldsymbol{\theta}}^* = \arg\min_{\hat{\theta}_{tr} \in \mathcal{E}} \text{MSE}(\hat{\boldsymbol{\theta}}_{tr}) \tag{1}$$

be the estimator in the family $\mathcal{E}$ with the lowest mean-squared-error. The aim of the selection step is to assess the MSE of each estimator in the family $\mathcal{E}$ and return the one with the lowest value, i.e.

$$\hat{\boldsymbol{\theta}}_{tr}^* = \arg\min_{\hat{\theta}_{tr} \in \mathcal{E}} \widehat{\text{MSE}}(\hat{\boldsymbol{\theta}}_{tr}) \tag{2}$$

where $\widehat{\text{MSE}}$ is the estimation of MSE returned by validation procedures (e.g. cross-validation, leave-one-out or bootstrap).

In order to compare inductive with non inductive practice we need to specify what we mean by alternative to induction. In qualitative terms, as illustrated by Salmon [11], *we might make wild guesses, consult a crystal gazer or believe what is found in Chinese fortune cookies.* In our estimation framework, a non inductive practice corresponds to a special kind of estimator which on purpose uses no data (i.e. $N = 0$), i.e. is data independent.

Let us now quantify the expected error of inductive and no-inductive procedures in the two contexts: regularity and no regularity.

Let us first consider the regular setting (denoted by the superscript $^{(r)}$) where the training and test datasets are i.i.d. samples of the $\boldsymbol{\theta} = \theta_{tr}$ distribution. The error of the non inductive process is

$$\text{MSE}_0^{(r)} = E_{\hat{\boldsymbol{\theta}}_0}[(\theta - \hat{\boldsymbol{\theta}}_0)^2] = B_0^2 + V_0$$

where B_0 and V_0 denote the bias and the variance of the non inductive estimator $\hat{\boldsymbol{\theta}}_0$. According to (2) the error of the inductive process is $\text{MSE}_I^{(r)} = \text{MSE}(\hat{\boldsymbol{\theta}}^*_{tr})$.

Now, if we consider that a non inductive procedure is just a specific instance of estimator which is using no data, we can include $\hat{\boldsymbol{\theta}}_0$ in the family $\mathcal{E}$ by default[3]. It follows that the probability that MSE_I is bigger than MSE_0 amounts to the probability of wrong selection of $\hat{\boldsymbol{\theta}}^*$ due to the error in estimating the MSE terms . Since it is well known that this probability can be made arbitrarily small by increasing the number N of samples, we can conclude that the an inductive process cannot be outperformed by a non inductive one in the regular setting.

In the irregular setting the training set and the test set are generated by two different distributions with parameters θ_{tr} and θ_{ts}. In particular we assume that θ_{ts} is a realization of $\boldsymbol{\theta}_{ts}$ whose mean is θ_{tr} and whose variance is $\mathcal{V}$. The Mean Squared Error of the inductive process is now obtained by averaging over all possible training sets and test parameters:

$$\begin{aligned}
\text{MSE}_I^{(irr)} &= E_{\hat{\boldsymbol{\theta}}^*_{tr},\boldsymbol{\theta}_{ts}}[(\boldsymbol{\theta}_{ts} - \hat{\boldsymbol{\theta}}^*_{tr})^2] \\
&= E_{\hat{\boldsymbol{\theta}}^*_{tr},\boldsymbol{\theta}_{ts}}[(\boldsymbol{\theta}_{ts} - \theta_{tr} + \theta_{tr} - \hat{\boldsymbol{\theta}}^*_{tr})^2] \\
&= E_{\boldsymbol{\theta}_{ts}}[(\boldsymbol{\theta}_{ts} - \theta_{tr})^2] - 2E_{\hat{\boldsymbol{\theta}}^*_{tr},\boldsymbol{\theta}_{ts}}[(\boldsymbol{\theta}_{ts} - \theta_{tr})(\hat{\boldsymbol{\theta}}^*_{tr} - \theta_{tr})] + E_{\hat{\boldsymbol{\theta}}^*_{tr}}[(\theta_{tr} - \hat{\boldsymbol{\theta}}^*_{tr})^2] \\
&= \mathcal{V} - 2\Gamma + E_{\hat{\boldsymbol{\theta}}^*_{tr}}[(\boldsymbol{\theta}_{tr} - \hat{\boldsymbol{\theta}}^*_{tr})^2] = \mathcal{V} - 2\Gamma + \text{MSE}_I^{(r)} \qquad (3)
\end{aligned}$$

In the equation above $\mathcal{V}$ quantifies the variability (or irregularity) of the phenomenon and

$$\Gamma = E_{\hat{\boldsymbol{\theta}}^*_{tr},\boldsymbol{\theta}_{ts}}[(\boldsymbol{\theta}_{ts} - \theta_{tr})(\hat{\boldsymbol{\theta}}^*_{tr} - \theta_{tr})] \qquad (4)$$

denotes the covariance between the estimator and $\boldsymbol{\theta}_{ts}$. This term can be different from zero only if the learning procedure incorporates some knowledge (also called inductive bias) about the $\boldsymbol{\theta}_{ts}$ distribution. If no knowledge about $\boldsymbol{\theta}_{ts}$ is available, $\Gamma = 0$.

Note that the Eq. (3) decomposes the testing error in an irregular setting into three terms: a term depending on the variability of phenomenon, a term representing the impact of inductive bias and a term denoting the MSE estimated on the basis of the training set only.

Analogously, for the non inductive case, we have

$$\text{MSE}_0^{(irr)} = \mathcal{V} + \text{MSE}_0^{(r)} \qquad (5)$$

Since model selection ensures $\text{MSE}_I^{(r)} \leq \text{MSE}_0^{(r)}$ for every θ_{tr}, from (3) and (5) we obtain that the error of the inductive process is not larger than the non inductive one. So in the irregular setting too, the inductive approach cannot be outperformed by the non inductive one. Though in the irregular case the resulting error is definitely much larger than in the regular case (notably if

[3] it is indeed a common practice to add random predictors in machine learning pipelines and to use them as null hypothesis against which the generalization power of more complex candidate algorithms is benchmarked.

$\mathcal{V}$ is large), it still holds that an inductive approach is on average more accurate that a non inductive one. This can be explained by the fact that an irregular setting is analogous to an estimation task where a single observation (in this case θ_{tr}, or better its proxy $\hat{\theta}^*_{tr}$) is available about the target distribution (i.e. $\boldsymbol{\theta}$). Though a single sample is not much for returning an accurate estimation, it is nevertheless recommended to use it rather than discard it. A rational agent is therefore encouraged to choose, whatever is its assumption about the reality, an inductive technique to take into account the observed data.

Table 1. MSE for induction and non inductive processes in the case of regular and irregular nature.

MSE	Regular setting $\boldsymbol{\theta} = \theta_{tr}$	Irregular setting $\boldsymbol{\theta}$
Inductive process	$\min_{\mathcal{E}}[B^2_{tr} + V_{tr}]$	$\mathcal{V} + [\min_{\mathcal{E}}[B^2_{tr} + V_{tr}]]$
Non inductive process	$B^2_0 + V_0$	$\mathcal{V} + [B^2_0 + V_0]$

The reasoning is summarized in Table 1 which is on purpose remindful of Table I in [11]. We extended the "nothing to lose" vindicationist argument presented by Reichenbach by using the notion of mean-squared error to quantify the short-run prediction accuracy. Given the impossibility of assured success in obtaining knowledge from data, it is nevertheless possible to show quantitatively that inductive policies are preferable to any competitor. Induction is preferable to soothsaying since it will work if anything will.

2.1 Example

In order to illustrate the result (3) let us consider a simple task where the goal is to learn the expectation of a distribution. Let us suppose that we are in an irregular setting and that $\boldsymbol{\theta}_{ts}$ is distributed according to a Normal distribution with mean θ_{tr} and variance $\mathcal{V} = 1$. Suppose that we can observe a dataset D_N of size $N = 50$ distributed according to a Normal distribution with mean θ_{tr} and variance 1. Let us compare an inductive approach which simply computes a simple average ($\hat{\theta}^*_{tr} = \frac{\sum_{i=1}^N z_i}{N}$) with a number of noninductive strategies $\hat{\theta}_0$ which differ in term of inductive bias (or a priori knowledge), since $\hat{\theta}_0$ is distributed according to a Normal distribution with mean zero and standard deviation $\sigma_0 \in [0.05, 2]$. Figure 1 illustrates $\text{MSE}_I^{(irr)}$ (upper dotted horizontal line) and $\text{MSE}_0^{(irr)}$ for different values of σ_0^2. It appears that for all values of σ_0^2 the inductive approach outperforms (i.e. lower generalization error) the non inductive one. As far as the a priori becomes less informative, the accuracy of the non inductive approach deteriorates due to the increasing of $\text{MSE}_0^{(irr)}$ (continuous increasing line).

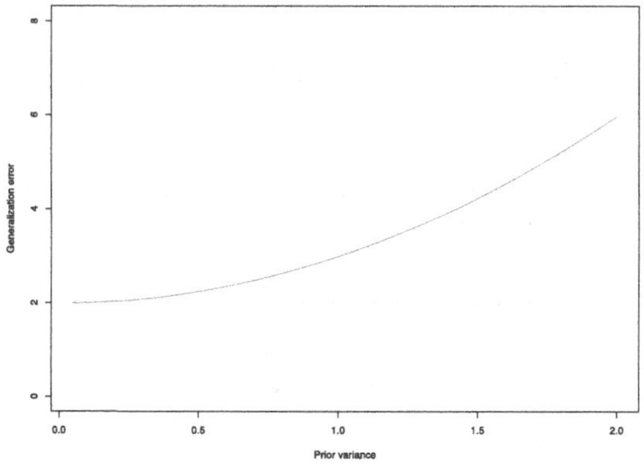

Fig. 1. Generalization error in the irregular setting for inductive and non inductive case: x-axis represents the amount of variance of the noninductive estimator. The horizontal line represents $\mathrm{MSE}_I^{(irr)}$ while the increasing line represent $\mathrm{MSE}_0^{(irr)}$.

3 Discussion

The results of the previous section leave open a set of issues which are worth to be discussed.

- Practical relevance: though the results discussed in the above section appear of no immediate application, there is a very hot domain in machine learning which could take advantage of such reasoning. This is the domain of transfer learning [8] whose aim is to transfer the learning of a *source* model θ_{tr} to a *target* model θ_{ts}, e.g. by using a small amount of target labeled data to leverage the source data. The transfer setting may be addressed in the derivation (3) by making the hypothesis that the distribution of θ_{ts} is no more centered on θ_{tr} but on a value $\mathcal{T} + \theta_{tr}$ where $\mathcal{T}$ denotes the transfer from the source to the target. In this case the result (3) becomes

$$\mathrm{MSE}_I^{(irr)} = \mathcal{V} + \mathcal{T}^2 - 2\mathcal{T}B_{tr} - 2\Gamma + \mathrm{MSE}_I^{(r)}$$

  The difficulty of generalization is made then harder by the presence of additional terms depending on the transfer $\mathcal{T}$. At the same time this setting confirms the superiority of an inductive approach. The availability of a (however small) number of samples about θ_{ts} can be used to estimate the $\mathcal{T}$ term and then reduce the error of a data driven approach with respect to a non inductive practice which would have no manner of accounting for the transfer.
- Bayesianism: the Bayesian formalism has been more and more used in the last decades by philosophers of science to have a formal and quantitative interpretation of the induction process. If one hand, Bayesian reasoning shed light on some famous riddles of induction (see [5]), on the other hand Wolpert [13] showed that conventional Bayesian formalism fails to distinguish target functions from hypothesis functions, and is then incapable of addressing the generalization issue. This limitation is also present in the conventional estimation

formalism which implicitly makes the assumption of a constant invariable target behind the observations (like in the conventional definition of Mean Squared Error). In order to overcome this limitations Wolpert introduces the Extended Bayesian Formalism (EBF). Our work is inspired to Wolpert results and can be seen as a sort of Extended Frequentist Formalism (EFF) aiming to stress the impact on the generalization error of uncoupling the distribution of the estimator and the one of the target.

- The distinction between regular and irregular settings: for the sake of the presentation we put regular and irregular settings in two distinct classes. However as it appears from Table 1, there is a continuum between the two situations. The larger is the randomness of the target (i.e. the larger $\mathcal{V}$) the least accurate is the accuracy of the inductive estimation.
- How to build an estimator: in the previous sections we introduced and discussed the properties of an estimator, intended as a mapping between a dataset and an estimation. An open issue remains however: how to build such mapping? Are there better ways to build this mapping? All depends on the (unknown) relation between the target and the dataset. Typically there are two approaches in statistics: a parametric approach where it is assumed the knowledge of the parametric link between the target and the dataset distribution and a nonparametric approach where no parametric assumption is made. However, it is worth to remark that though nonparametric approaches are distribution free, they are dependent on a set of hyperparameters (e.g. the kernel bandwidth or number of neighbors in nearest neighbour) whose optimal value, unknown a priori, depends on the data distribution. In other words any rule for creating estimators introduce a bias (in parametric or non parametric form) which has a major impact on the final accuracy. This aspect reconciles our vision with the denial of induction made by Popper. It is indeed the case that there is no *tabula rasa* way of making induction and that each induction procedure has its own bias [1]. Choosing a family of estimators is in some sense an arbitrary act of creation which can be loosely justified a priori and that can only be assessed by data. Machine learning techniques however found a way to escape this indeterminacy by implementing an automatic version of the hypotetico-deductionist approach where a number of alternative (nevertheless predefined) families of estimators are generated, then assessed and eventually ranked. This is accomplished by the model selection step whose goal is (i) to assess alternatives on the basis of data driven procedures (e.g. cross-validation or bootstrap), (ii) prune weak hypothesis and (iii) preserve the least erroneous ones.
- Why is inductive machine learning successful? the results of the previous section aim to show that the inductive process is not necessarily correct but necessarily better than non inductive alternatives. Its degree of success depends on the degree of regularity of the nature. Each time we are in front of regular phenomena, or better we formulate prediction problems which have some degree of regularity, machine learning can be effective. At the same time, there are plenty of examples where machine learning and prediction have very low reliability or whose accuracy is slightly better than random guess: this

is the case of long term prediction of chaotic phenomena (e.g. weather forecasting), nonstationary and nonlinear financial/economical problems as well as prediction tasks whose number of features is much larger than the number of samples (notably bioinformatics).

- Undetermination of theory by evidence: the bias/variance formulation of the induction problem supports the empiricist thesis of "undetermination of theory by evidence", i.e. the data do not by themselves choose among competing scientific theories. The bias/variance decomposition shows that it is possible to attain similar prediction accuracy with very different estimators: for instance a low bias large variance estimator could be as accurate as a large bias but low variance one. It is then possible that different machine learning algorithms generate alternative estimators which are compatible with the actual evidence yet for which prediction accuracy provides no empirical ground for the choice. In practice, other (non predictive) criteria can help in disambiguating the situation: think for instance to criteria related to the computational or storage cost as well as criteria related to the interpretability of the model (e.g. decision trees vs. neural networks).
- Practical relevance of those results: though the results discussed in the above section appear of no practical use, there is a very hot domain in machine learning which could take advantage of this reasoning. It is the domain of transfer learning where the issue is indeed to transfer the learning of a source model θ_{tr} to a target model θ_{ts}, for instance thanks to few samples characterizing the target tasks. This aspect could be taken into account in the derivation (3) by making the hypothesis that the distribution of θ_{ts} is not centered on θ_{tr} but on a value $\mathcal{T} + \theta_{tr}$ where $\mathcal{T}$ denotes the transfer from the source to the target. This setting it appears

4 Conclusion

Machine learning aims to extract predictive knowledge from data and as such it is intimately linked to the problem of induction. Its everyday usage in theoretical and applied sciences raises additional pragmatic evidence in favour of induction. But as Hume stressed, past successes of machine learning are no guarantee for the future. Hume's arguments have been for more than 250 years treated as arguments for skepticism about empirical science. However, Hume himself considered that inductive arguments were reasonable. His attitude was that he had not yet found the right justification for induction, not that there was no justification for it.

Reichenbach agreed with Hume that it is impossible to prove that induction will always yield true conclusions (validation). However, a pragmatic attitude is possible by showing that induction is well suited to the achievement of our aim (vindication).

This paper extends the "nothing to lose" vindicationist argument presented by Reichenbach, by using the notion of mean-squared error to quantify the short-run prediction accuracy and by claiming the rationality of induction in a decision making perspective. Given the impossibility of assured success in obtaining knowledge from data, it is nevertheless possible to show quantitatively that inductive policies are preferable to any competitor. Induction is preferable to soothsaying since it will work if anything will.

Humans are confronted from the very first day to an incessant stream of data. They have only two options: make use of them or discard them. Neglecting data in no way can be better than using data. Inductive learning from data has therefore no logical guarantee of success but it is nonetheless the only rational behavior we can afford. Paraphrasing G. Box, induction is wrong but sometimes it is useful.

References

1. Bensusan, H.N.: Automatic bias learning: an inquiry into the inductive basis of induction. Ph.D. thesis, University of Sussex (1999)
2. Geman, S., Bienenstock, E., Doursat, R.: Neural networks and the bias/variance dilemma. Neural Comput. **4**(1), 1–58 (1992)
3. Hastie, T., Tibshirani, R., Friedman, J.: The Elements of Statistical Learning. SSS. Springer, New York (2009). https://doi.org/10.1007/978-0-387-84858-7
4. Howson, C.: Hume's Problem: Induction and the Justification of Belief. Clarendon Press, Oxford (2003)
5. Howson, C., Urbach, P.: Scientific Reasoning: The Bayesian Approach. Open Court, Chicago (2006)
6. Korb, K.B.: Introduction: machine learning as philosophy of science. Mind Mach. **14**(4), 433–440 (2004)
7. Lehmann, E.L., Casella, G.: Theory of Point Estimation. STS. Springer, New York (1998). https://doi.org/10.1007/b98854
8. Pan, S.J., Yang, Q.: A survey on transfer learning. IEEE Trans. Knowl. Data Eng. **22**(10), 1345–1359 (2010)
9. Popper, K.R.: Conjectures and Refutations. Basic Books, New York (1962)
10. Reichenbach, H.: The Theory of Probability. University of California Press (1949)
11. Salmon, W.C.: Hans Reichenbach's vindication of induction. Erkenntnis **27**(35), 99–122 (1991)
12. Williamson, J.: A dynamic interaction between machine learning and the philosophy of science. Minds Mach. **14**(4), 539–549 (2004)
13. Wolpert, D.H.: On the connection between in-sample testing and generalization error. Complex Syst. **6**, 47–94 (1992)
14. Wolpert, D.H.: The lack of a priori distinctions between learning algorithms. Neural Comput. **8**, 1341–1390 (1996)

Geospatial Dimension in Association Rule Mining: The Case Study of the Amazon Charcoal Tree

David N. Prata(✉), Marcelo L. Rocha(✉), Leandro O. Ferreira(✉), and Rogério Nogueira(✉)

Programa de Pós-graduação em Modelagem Computacional de Sistemas, UFT, Palmas, TO, Brazil
{ddnprata,mlisboa,roger}@uft.edu.br, leferfsa@gmail.com

Abstract. Despite the potential of association rules to extract knowledge in datasets, the selection of valuable rules has been the specialists' big challenge because of the great number of rules generated by association rules' algorithms. Unfortunately, for Species Distribution Modeling, researchers and analysts alike has been discourage regarding the efficacy and relative merits to the studies due to the conflicting of results. In this case study, we integrate geoprocessing techniques with association rules to analyze the potential distribution of the charcoal tree species of Amazon, Tachigali Subvelutina. This integration allowed the exploration of a new method to select valuable association rules based not only by frequent item sets, like confidence, support and lift. But also considering the geospatial distribution of the species occurrences. The application of one more dimensional to the measures of the frequent item sets provides a more effective understanding and adequate amount of relevance to represent significant association rules. Further, this new approach can support experts to: identify effective rules, i.e., rules that can be effective used with regression techniques to predict species in geospatial environments; identify variables that play a decisive role to discover the occurrence of species; and evaluate the effectivity of the datasets used for the study to answer the specified scientific questions.

Keywords: Geospatial data mining · Modeling · Statistical methods for data mining

1 Introduction

Forest management plans help to program territorial regulations and to systematize the organization of activities in protected areas. Sustainable forest management requires spatial, biotic, and abiotic data, to be analyzed for government decision-making. The charcoal tree of Amazon, whose scientific name is *Tachigali Subvelutina*, from now on called simply *TS*, is a promising tree species in Brazilian economy to energetic plantation [1]. A geospatial mining analysis for the potential specie distribution of *TS* is

G. Nicosia et al. (Eds.): LOD 2019, LNCS 11943, pp. 244–258, 2019.
https://doi.org/10.1007/978-3-030-37599-7_21

imperative to determine areas for reintroduction, and to discover potential areas of conservation and forest management.

In this study, we model the potential distribution by the use of geographic data, structured in geographic datasets, and manipulated by geographic information systems. For this reason, a large amount of data was preprocessed, explored and analyzed, using geoprocessing techniques and tools. We collected and manipulated geospatial themes of the physical environment. After all, data mining techniques were applied in order to discover interesting rules and patterns.

Despite the range of tools accessible to apply in Species Distribution Modeling, the efficacy regarding the studies is discouraging to researchers and analysts alike due to the conflicting of results [2]. To mitigate this problem, this work close attention to three important concerns to support researches results:

- identify effective rules, i.e., rules that can be effective used with regression techniques to predict species in geospatial environments. Not only in respect to their statistics' correlations and efficiency to frequent item sets, but also because of their role play in the context of the data analysis in question;
- identify variables that play a decisive role to discover the occurrence of species;
- evaluate the effectiveness of the datasets to answer the specified scientific questions.

The purpose of this work was to answer the following scientific question for the case study of the charcoal tree: What are the combinations of abiotic features, which we can most easily find the charcoal tree species for the study area? To present this research study, the work is organized as follows: Sect. 2 summarize special topics of the literature related to the theme. Section 3 presents methods and materials used to collect and to analyze the data. The results are presented in Sect. 4, the discussion is subdivided in aspects of data mining and geoprocessing issues. Finally, Sect. 5 draws our conclusions about this research.

2 Spatial Rule Mining

Association rule is a method used in machine learning for discovering valuable relations in large transaction datasets. The well-known Apriori algorithm [3] had proposed to select an association rule by a minimum support threshold and by a minimum confidence threshold. For prediction, a good measure is the Lift. Lift measures the performance of a model, in this case, the association rule, to predict the population of a classified dataset.

Statistical interpretation models can be used for estimation, description, and prediction. Point process is a model for the spatial distribution of the points, like trees in a forest. Some potentials applications to answer the scientific questions using spatial data mining include co-location pattern discovery, spatial clustering, and spatial hotspot analysis [4].

Spatial dataset usually describes spatial features objects at different locations in the geographic space. Co-location pattern mining discover features located in spatial proximity. Spatial co-location pattern discovery and association rule mining based approaches are used to integrate spatial data transactions over the geographic space and then the Apriori algorithm is used for rules generation [5]. Spatial clustering is the method of grouping spatially point process into clusters. Density-based clustering algorithms are clusters based on the density of data points in a region. Hotspot is a clustered pattern distributed in hotspot regions with high similarity. Differently of spatial clustering, the spatial correlation of attribute values in a hotspot are usually high and drops dramatically at boundaries.

3 Methods and Materials

3.1 Geoprocessing

The study area consists of a stretch of two irregular polygons of Tocantins river basin, located in the north region of Brazil, between latitudes 08°00′00″S and 11°00′0″S, and from longitude 48°54′41.360′W and 47°27′11,395″W, Fig. 1. Its area covers 26,536.5 square kilometers, and has a circumference of about 1,226,084 meters. The entropized area corresponds to 10.045,4 km^2. Tocantins River flows to the mouth of Amazonia River.

The geospatial themes of physical environment of this study include: geology (rocks), hydrogeology, geomorphology (relief), pedology (soil), declivity, erodibility, hypsometry, hydrology, and vegetation. The geological data, geomorphology and pedology, originated from the Settlement Project of the Amazon Protection System Databases, scale 1:250,000, WGS84 reference system. The geospatial vector data of this project were obtained from the original drafts, interpreted in 1: 250,000 scale, from surveys of natural resources conducted by Radambrasil projects. Thereafter, these data were reinterpreted with the support of semi controlled mosaics Radar and Landsat TM-5 images, information of other institutions, and fieldwork.

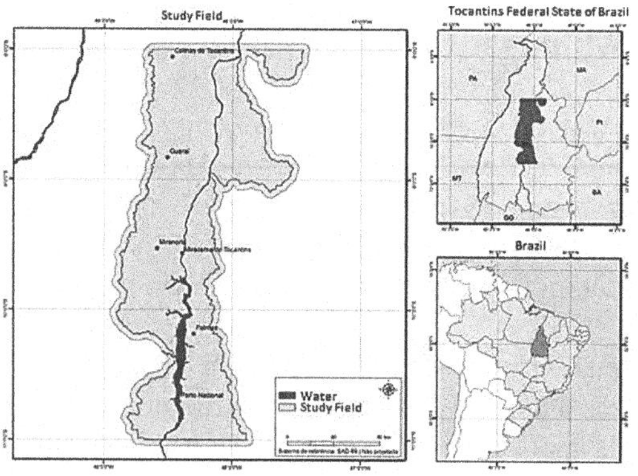

Fig. 1. Study area.

The information plans of Tocantins boundary issues, municipal centers and hydrography, were obtained through the Continuous Digital Cartographic Base of

Tocantins State. These themes were generated in 1:100,000 scale. The hydrogeology theme was obtained through the Geographic Database of Tocantins State - update 2012, 1:1,000,000 scale. To obtain this information, a plan called "State Plan of Water Resources from Tocantins State" was implemented. The Geological Charter of Brazil to the Mineral Resources Research Company and the map of hydrogeological domains and subdomains from Brazil provided the information in 1:2,500,000 scale.

The watershed information plan contains the 30 major river basins of the river systems from Tocantins and Araguaia rivers [6]. This product was created within the agro ecological zoning project with support from topographic charts of IBGE (Brazilian Institute of Geography and Statistics), 1:250,000 scale (generating a digital elevation from level contours), and satellite images from the National Department of Water and Power. Later, we refined this information by the support of Continuous Digital Cartographic Base with CBERS2 and Landsat 5 satellite images, 1:100,000 scale. After all, the limits of the river systems from Araguaia and Tocantins rivers were revised.

The information plans of declivity and potential erodibility of soils were produced in 1:250,000 scale, and obtained through the Digital Atlas of Tocantins - Spatial Data Base. The information plan of declivity was obtained from vector data of topographic charts, 1:250,000 scale (developed by IBGE), semi-controlled mosaics images of radar, scale 1:250,000, of Radambrasil projects. The declivity classes were generated from a digital elevation model, obtained from the level contours, using equidistance of 100 meters. The fitting of lines and contours was possible due to the use of semi-controlled mosaics of radar and altimetry charts. Where relevant, associations of declivity were created from the basic classes, in agreement with the scale of the charters and the equidistance of the level contours. The information plan of potential erodibility soil was obtained through the establishment and digital manipulation of declivity classes and the erosive potential of the soil. The plan considered the interpretation of the following sources:

- Topographic charters of IBGE, 1: 250,000 scale and draft pages of thematic interpretation from soil, geology and geomorphology, 1: 250,000 scale.
- Multispectral images of Landsat TM in bands: 3, 4 and 5; 1: 250,000 scale (1996).
- Semi-controlled radar of mosaics of images, 1: 250,000 scale, by Radambrasil projects.
- Reports of pedology, geomorphology, and geology, by Radam and Radambrasil projects.
- Geoenvironmental Map of the State of Tocantins, 1:1,000,000 scale, produced by the staff of Geosciences Division of the Midwest, department of Geosciences/IBGE in 1995.

The information plan of coverage and land use was produced in 1:100,000 scale, and made available through the Geographic Database of the Dynamic Coverage and Land Use of Tocantins State. The information contained on this basis were obtained by techniques of visual and automated interpretation of Landsat TM-5 satellite images, supported by fieldwork and secondary statistical information. The information plan of hypsometry (altitude) was generated from the polygonising of level contours of Continuous Digital Cartographic Base of Tocantins State.

3.2 Data Mining

The first stage consisted of evaluating data mining objectives [7] in agreement with our data set, which was constructed in order to achieve the purpose of the work. The exploratory research of *TS* species may include an understanding of the relationships between the vegetation and the environmental variables, in order to determine areas for reintroduction, conservation, and forest management. In this case, the data mining of potential geographic distribution of *TS* should identify the relationships between biotic and abiotic factors.

Apriori, an association rule algorithm [8], is capable to find patterns among data properties, by the identification of unknown relationships between them. The association rule has the form A ==> B, where A, called the antecedent, and B, called the consequent, are sets of items or transactions, and the rule can be read as: the attribute A often implies B. To evaluate the generated rules, we used some measures of interest: the measures most used are well established, such as support and confidence in the work of [9]. The authors [10] performed a survey of other metrics and suggested strategies for selecting appropriate measures for certain areas and requirements, as follows:

- Support: The support of a rule is defined as the fraction of the items, which satisfy the set A and B of the rule.
- Confidence: It is a measure of the strength of support to the rules and corresponds to statistical significance. The probability of the rule in finding B so that B also contain A.
- Lift: Used to find dependencies, it indicates how much more common it becomes B when A occurs. It can vary between 0 and ∞.

4 Results and Discussion

(a) **Data Mining Analysis**

Association rules are an exploratory technique to generate rules describing the most relevant patterns found in a selected dataset. These patterns represent the possible combinations of variables occurring with a particular frequency, composed by an antecedent and a consequent. The antecedent contains a subset of attributes and their values, and the consequent has a subset of attributes resulted of the antecedent.

In this work, the purpose for generating association rules is to discover patterns of potential areas where the *TS* specie occurred. Our hypothesis is that the association rules can help us to answer the scientific question cited in this work.

Therefore, the antecedent is composed by abiotic factors of the environment, and the consequent by the biotic factor of *TS* specie. For the data association, we selected out 6 spatial themes of physical geology to be analyzed: geomorphology, erodibility, water body away, soils up to the 2nd level of classification and the vegetation, Table 1.

The generation procedure of association rules gave a total of more than 1,000 association rules, for the *TS* specie as unique consequent, with many different relations extracted from a dataset of 6,803 (six thousand and eighty three hundred) observations. However, we only selected 22 classification rules that best represented the six physical environment themes where *TS* species were found as consequent. The 22 classification rules were selected based on their Confidence index and by the quantity of compound attributes. The selected instances represent a total of 274 individuals of *TS* specie which were occasionally found in the 96 plots scattered along 144,000 m^2, of the collected field. All the plots have the same area 30 m $\times$ 50 m.

To facilitate the analysis of the results, the data is organized by rule groups called: A, B, C, D and E. The first group refers to the rules with only one attribute for the antecedent. The second group has two attributes, and so on, until the sixth group of rules with up to six attributes representing the six physical environment themes. Each rule has different levels of physical themes enrolled, and the interpretation, e.g., rule 508, Table 2.

Table 1. Twenty-two classification rules representing the occurrence of *TS* species (the consequent). 33 of the 96 plots hold *TS* species from the collected field.

RL	Condition (antecedent)	RG	TO	CR	SPTR	Lift
1	Geom_Simb = Dt	A	239	0.05	0.61	1.40
2	Geol_Let = Dp		163	0.05	0.40	1.45
4	CotaFisico = 200		130	0.03	0.57	0.82
6	Erod_Clas = Ligeira		109	0.05	0.29	1.33
8	Dist_Agua = 400		104	0.05	0.25	1.46
10	EB_Nivel_2 = RQo		89	0.05	0.23	1.36
15	Erod_Clas = Muito Forte		87	0.04	0.26	1.18
3	Geol_Let = Dp Geom_Simb = Dt	B	146	0.08	0.25	2.06
7	Geom_Simb = Dt Erod_Clas = Ligeira		107	0.06	0.26	1.49
12	Geom_Simb = Dt EB_Nivel_2 = RQo		89	0.06	0.21	1.52
14	Geol_Let = Dp Dist_Agua = 400		100	0.10	0.14	2.55
19	Geol_Let = Dp Erod_Clas = Muito forte		81	0.12	0.09	3.01
25	Geol_Let = Dp Geom_Simb = Dt Erod_Clas = Muito forte	C	81	0.12	0.09	3.01
29	Geom_Simb = Dt Erod_Clas = Ligeira CotaFisico = 200		81	0.05	0.20	1.46
35	Geom_Simb = Dt CotaFisico = 200 EB_Nivel_2 = RQo		79	0.06	0.17	1.62
41	Geol_Let = Dp Geom_Simb = Dt EB_Nivel_2 = FFc		75	0.08	0.13	2.08
828	Geol_Let = Cpo Geom_Simb = Dt Erod_Clas = Ligeira		36	0.07	0.07	1.83
60	Geol_Let = Dp Geom_Simb = Dt Erod_Clas = Muito forte Dist_Agua = 400	D	73	0.14	0.07	3.70
124	Geol_Let = Dp Geom_Simb = Dt Erod_Clas = Muito forte CotaFisico = 400		67	0.12	0.07	3.10
421	Geol_Let = Dp Geom_Simb = Dt Erod_Clas = Muito forte CotaFisico = 400 Dist_Agua = 400	E	59	0.16	0.05	4.10

(continued)

Table 1. (*continued*)

RL	Condition (antecedent)	RG	TO	CR	SPTR	Lift
508	Geol_Let = Dp Geom_Simb = Dt Erod_Clas = Muito forte CotaFisico = 400 Dist_Agua = 400 EB_Nivel_2 = FFc	F	59	0.16	0.05	4.10
1055	Geol_Let = CPpi Geom_Simb = Dt Erod_Clas = Ligeira CotaFisico = 200 Dist_Agua = 500 EB_Nivel_2 = RQo		34	0.20	0.02	5.08

RL: Number of the rule for the Apriori result; RG: Rules' Group; TO: Total of Occurrences; SPTR: Support of the rule; CR: Confidence of the rule for the Charcoal tree.

The best Support was the rule 1 "Geom_Simb = Dt ==> Acronimo = Scle pani". This type of Geomorphology was associated with 87% of the *TS* specie found in the study area. The rule 2 "Geol_Let = Dp ==> Acronimo = Scle pani" had the third best Support of 0.40, considering 8 degrees of freedom.

Considering 10 degrees of freedom, the rule 8 "Dist_Agua = 400 ==> Acronimo = Sclepani" had Support of 0.25, and the highest Lift of the group, 1.46, with Confidence of 0.05. On the other hand, the quota of altitude between 100 and 200 meters correspond to 47% of the *TS* specie found in the study area, with the worst Lift of the group, 0.82, and the lowest Confidence of 0.03.

As we observed, these rules of A group were significantly mined for the study by the support index. The support of the rules decay according to their groups, from A to F. Conversely, the confidence increases with the rise of the rule groups.

The Lift also behaves in contrast, as the Support declines from simple to complex rules. In general, the Lift increases with the growth of complexity of the rules. The rule goes more Lift's effective as far as the model is going more complexity. All the rules had Lift above 1, promising in some degree (proportionally to the Lift) that those rules are potentially useful to predict the consequent. The only exception was the CotaFisico = 200. The spotlight in this case is group F, with rules 508 and 1055.

Analyzing attributes features for intragroups, rule 12 "Geom_Simb = Dt EB_Nivel_2 = RQo ==> Acronimo = Scle pani", from B group, showed an almost totally dependency between Geomorphology, structural tabular surface composed by dissecting processes, topographic with elevation of 200 meters, and the soil with presence of órticos quartizic neossolos, to the presence of *TS* specie. This type of soil alone, rule 10, had support of 0.23, and in conjunction with this type of Geomorphology (Dt) resulted in about the sameness support of 0.21. These are multicollinearity cases.

This is also the case for the rule 7 "Geom_Simb = Dt Erod_Clas = Ligeira ==> Acronimo = Scle pani" where the Erodibility class "Ligeira" has Support of 0.29 alone, and in conjunction with the Geomorphology (Dt), the Support is about the sameness of 0.26. We can also find a strong dependence among instanced attributes for the rules of groups C, D, E and F, where the same instanced attributes are repeated from groups A and B.

Table 2. Interpretation of rule 508.

Enrolled attributes	Geology, Geomorphology, Pedology (Soil), Erodibility, Hypsometry (altitude), and Hydrology (distance from water bodies)
Rule denotation	Geol_Let = Dp Geom_Simb = Dt Erod_Clas = Muito_forte CotaFisico = 400 Dist_Agua = 400 EB_Nivel_2 = FFc ==> Acronimo = Scle pani
Rule interpretation	If geology is from Pimenteiras formation, and geomorphology has structural tabular surface composed by dissecting processes; and erodibility with shallow and very shallow soils, with presence of rock outcroppings, predominant relief ranging from hilly to steep, and slopes greater than or equal to 45%. Quota of altitude between 300 and 400 meters, with a distance from water bodies from 400 to 500 meters; and the soil composed by concretionary Petric Plinthosols; then there is the presence of the *Tachigali Subvelutina* specie

(b) **Geoprocessing Analysis – Effective Rules**

For the geoprocessing analysis, we create Table 3 depicting the study area for geospatial dimensions, considering the preserved land area. For this study, all the ploted area is preserved land area. However, to calculate the preserved land area, we need to subtract the total land area from the human used area. We also defined 1,500 m^2 as a unit area for this study. This corresponds to the area of one plot.

To calculate the territorial significance of the rules, we formulate the geospatial support, the geospatial confidence, and the geospatial lift, as follows.

- Geospatial Support: GSupport = f(X, Y)/PL. The support of a rule is defined as the fraction of the total preserved land from dataset, which satisfy the set intersection between X and Y of the rule.
- Geospatial Confidence: $GConf(X \rightarrow Y) = Gsupp(X \cup Y)/Gsupp(X)$. It is a measure of the strength of support to the rules and corresponds to statistical significance. The probability of the rule in finding Y for the collected area so that these collected areas also contain X;
- Geospatial Lift: $GLift(X \rightarrow Y) = Gsupp(X \cup Y)/Gsupp(X) \times Gsupp(Y)$. Used to find dependencies, it indicates how much more common becomes Y when X occurs. It can vary between 0 and ∞.

The density was calculated as the fraction of the total quantity of plots where the charcoal tree occurred for the total number of charcoal trees observations that were found in these plots. The α index was calculated as the remaining of the subtraction by one, for the fraction of the total quantity of plots of the rule, which satisfy the total quantity of plots where the charcoal tree occurred. The new approach shows proportional differences between indexes in Tables 1 and 3.

The Support for geospatial dimension still decaying by groups of rules, Pearson r correlation between *sup* and *gsup* is equal to 0,989, Figs. 3 and 4. However, for Figs. 4, 5 and 6, there is no correlation between Confidence and GConfidence. The GConfidence did not presented a pattern for intergroup compared to Confidence (Tables 1 and 3), i.e., Confidence increases as the growth of instanced attributes in intergroups. However, for GConfidence, it does not occur. This fact shows that GConfidence can best classify effective rules independent of their complexity (Fig. 2).

The rules 14 and 828, respectively from groups B and C, have high GConfidence of 0.83. In addition, the 1055 rule has GConfidence of 1 (one). These rules also have high GLifts of 2.42 and 2.90, Figs. 7 and 8. These are the best examples of the potential capability of the GConfidence and GLift to balance the tradeoff of a complexity model in favor of accurately, but also with low bias, i.e., to choose a generalized model as possible to better perform with new data.

Table 3. Twenty-two classification rules representing the occurrence of *TS* species (the consequent) for the 33 plots of the collected field where *TS* species occurred.

RL	Condition (antecedent)	RG	QPO	GS	GC	Den	α	GL
1	Geom_Simb = Dt	A	27	0.56	0.5	8.85	0.5	1.45
2	Geol_Let = Dp		12	0.34	0.36	13.5	0.6	1.05
4	CotaFisico = 200		20	0.57	0.36	6.50	0.6	1.05
6	Erod_Clas = Ligeira		13	0.27	0.5	8.38	0.5	1.45
8	Dist_Agua = 400		6	0.25	0.25	17.3	0.7	0.72
10	EB_Nivel_2 = RQo		8	0.19	0.42	11.1	0.5	1.22
15	Erod_Clas = Muito Forte		5	0.27	0.19	17.4	0.8	0.55
3	Geol_Let = Dp Geom_Simb = Dt	B	9	0.17	0.52	16.2	0.4	1.54
7	Geom_Simb = Dt Erod_Clas = Ligeira		12	0.21	0.57	8.92	0.4	1.66
12	Geom_Simb = Dt EB_Nivel_2 = RQo		8	0.15	0.53	11.1	0.4	1.55
14	Geol_Let = Dp Dist_Agua = 400		5	0.06	0.83	20.0	0.1	2.42
19	Geol_Let = Dp Erod_Clas = Muito forte		4	0.06	0.66	20.2	0.3	1.93
25	Geol_Let = Dp Geom_Simb = Dt Erod_Clas = Muito forte	C	4	0.06	0.66	20.2	0.3	1.93
29	Geom_Simb = Dt Erod_Clas = Ligeira CotaFisico = 200		10	0.17	0.58	8.10	0.4	1.71
35	Geom_Simb = Dt CotaFisico = 200 EB_Nivel_2 = RQo		7	0.13	0.53	11.2	0.4	1.56
41	Geol_Let = Dp Geom_Simb = Dt EB_Nivel_2 = FFc		3	0.09	0.33	25.0	0.6	0.96
828	Geol_Let = Cpo Geom_Simb = Dt Erod_Clas = Ligeira		5	0.06	0.83	7.20	0.1	2.42
60	Geol_Let = Dp Geom_Simb = Dt Erod_Clas = Muito forte Dist_Agua = 400	D	2	0.03	0.66	36.5	0.3	1.93
124	Geol_Let = Dp Geom_Simb = Dt Erod_Clas = Muito forte CotaFisico = 400		3	0.05	0.6	22.3	0.4	1.74
421	Geol_Let = Dp Geom_Simb = Dt Erod_Clas = Muito forte CotaFisico = 400 Dist_Agua = 400	E	1	0.02	0.5	59.0	0.5	1.45
508	Geol_Let = Dp Geom_Simb = Dt Erod_Clas = Muito forte CotaFisico = 400 Dist_Agua = 400 EB_Nivel_2 = FFc	F	1	0.02	0.5	59.0	0.5	1.45
1055	Geol_Let = CPpi Geom_Simb = Dt Erod_Clas = Ligeira CotaFisico = 200 Dist_Agua = 500 EB_Nivel_2 = RQo		1	0.01	1	34.0	0	2.90

The GConfidence of 1.0 (one) with the GLift of 2.90 for the rule 1055 were the best measures of GConfidence and GLift. This fact demonstrates a coherence for these indexes measures compared to the Confidence and Lift indexes of the frequent item set. For this new evaluation, the best significance rule for the geospatial occurrences of the *TS* specie was rule 1055. This rule has low confidence of 0.21. This is because of the intersection of six variables enrolled. In consequence, the preserved land area is too small, 1,500 m^2. This fact gave conditions to the GConfidence be able to be very high, 1.0.

The CORR Procedure

2 Variables:	GSUP SUP

Simple Statistics

Variable	N	Mean	Std Dev	Sum	Minimum	Maximum
GSUP	22	0.17330	0.15930	3.81250	0.01042	0.57292
SUP	22	0.20893	0.15939	4.59635	0.02440	0.61943

Pearson Correlation Coefficients, N = 22
Prob > |r| under H0: Rho=0

	GSUP	SUP
GSUP	1.00000	0.98984 <.0001
SUP	0.98984 <.0001	1.00000

Fig. 2. Pearson correlation between the measures of Support and GSupport.

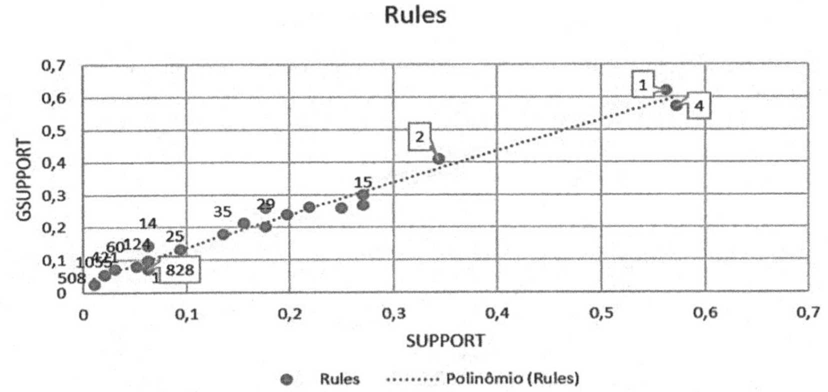

Fig. 3. Polinomial regression for the strong correlation between Support and GSupport.

The CORR Procedure

2 Variables:	GCONFIDENCE CONFIDENCE

Simple Statistics						
Variable	N	Mean	Std Dev	Sum	Minimum	Maximum
GCONFIDENCE	22	0.54325	0.19138	11.95150	0.19231	1.00000
CONFIDENCE	22	0.34638	0.16336	7.62044	0.12409	0.87226

Pearson Correlation Coefficients, N = 22 Prob > |r| under H0: Rho=0		
	GCONFIDENCE	CONFIDENCE
GCONFIDENCE	1.00000	-0.38506 0.0768
CONFIDENCE	-0.38506 0.0768	1.00000

Fig. 4. Pearson correlation for Confidence and GConfidence.

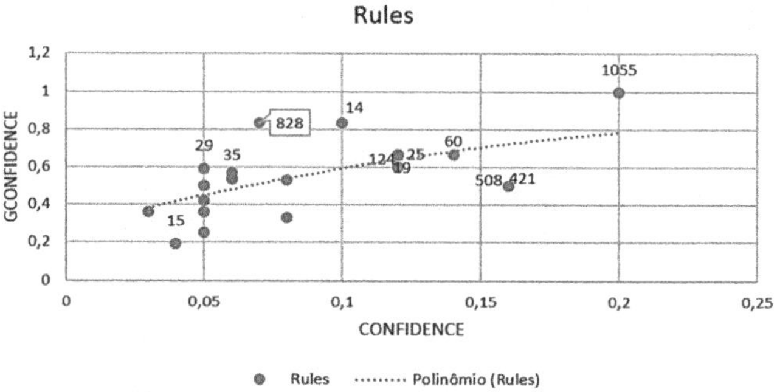

Fig. 5. Polinomial regression for the negative correlation between Confidence and GConfidence.

The CORR Procedure

2 Variables:	LIFT GLIFT

Simple Statistics						
Variable	N	Mean	Std Dev	Sum	Minimum	Maximum
LIFT	22	2.26511	1.15687	49.83243	0.82996	5.08535
GLIFT	22	1.58036	0.55673	34.76801	0.55944	2.90909

Pearson Correlation Coefficients, N = 22 Prob > |r| under H0: Rho=0		
	LIFT	GLIFT
LIFT	1.00000	0.58021 0.0046
GLIFT	0.58021 0.0046	1.00000

Fig. 6. Pearson correlation between the measures of Lift and GLift.

Despite the correlation between Lift and GLift, GLift shows some better models (rules with higher indexes) in groups B and C than in groups D and E. Glift and Lift are correlated with r equals to 0.58, Fig. 7.

The two-dimensional results for Lift and GLift are illustrated in Fig. 8.

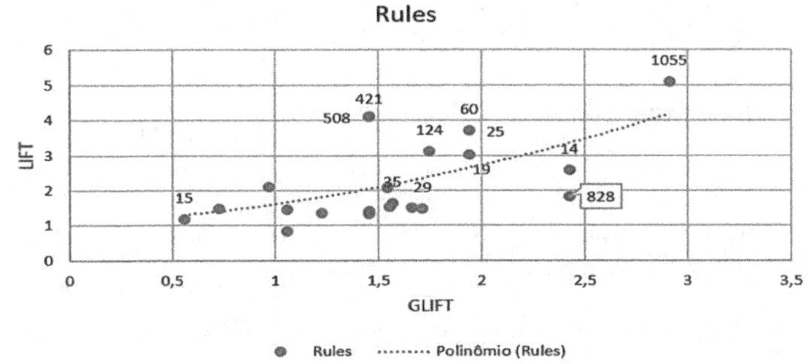

Fig. 7. The two-dimensional classification rules for Lit and Glift.

In Fig. 8, the rules with position at the right are more geospatial representative. In contrast, the rules higher to the left are more representative only for the frequent item set of species. GLift presented three rules under the measure of 1, where Lift presented only one. Some models were not considered effective when considering the geospatial dimension. The rules 14, 19, 25, 60 and 1055, are the best rules selected by GLift and Lift in conjunction. Considering the two dimensions' classification, the best rules to represent areas for reintroduction, and potential areas for conservation and forest management, are rules that belongs to groups B, C, D and F. These rules were picked up from each rule group, without having to follow any group pattern.

Although, we are considering the spatial distribution, the density was not determinant to the best rules. This happened because of the α factor in Table 3. The calculus of the GLift considers GSupport and GConfidence, which intrinsically means α index as fraction of the density. Based on Figs. 9 and 10, we conclude that the data behavior results in this case study for GLift is normally distributed. Nevertheless, the results for Lift did not pass the tests for normality (Fig. 11).

All these results lead us to hypothesize that GLift is more appropriate than Lift to analyze association rules when the database have geospatial dimensions, at least in this case study.

Tests for Normality				
Test	Statistic		p Value	
Shapiro-Wilk	W	0.966405	Pr < W	0.6281
Kolmogorov-Smirnov	D	0.137875	Pr > D	>0.1500
Cramer-von Mises	W-Sq	0.063863	Pr > W-Sq	>0.2500
Anderson-Darling	A-Sq	0.36157	Pr > A-Sq	>0.2500

Fig. 8. GLift test for normality.

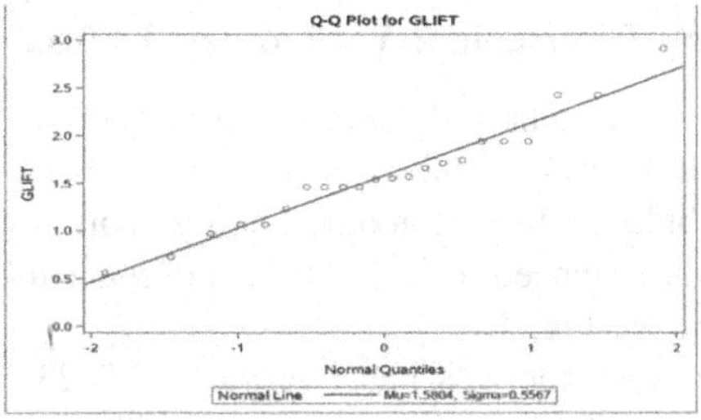

Fig. 9. Q-Q Plot for GLift.

Tests for Normality				
Test	Statistic		p Value	
Shapiro-Wilk	W	0.869657	Pr < W	0.0077
Kolmogorov-Smirnov	D	0.210751	Pr > D	0.0120
Cramer-von Mises	W-Sq	0.219193	Pr > W-Sq	<0.0050
Anderson-Darling	A-Sq	1.212618	Pr > A-Sq	<0.0050

Fig. 10. Lift test for normality.

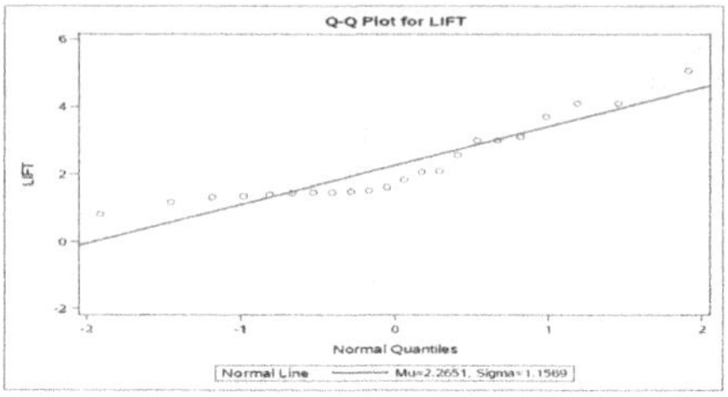

Fig. 11. Q-Q Plot for Lift.

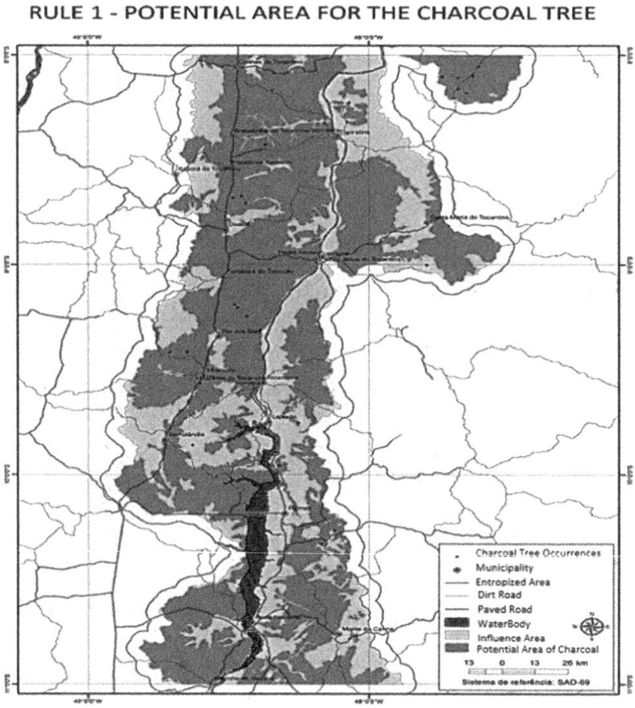

Fig. 12. Potential geographic distribution area for the charcoal tree (Color figure online).

(c) **Determinant Variables and Effective Datasets**

The instanced variable Geom_Simb = Dt is presented in 27 of the 33 plots, in other words, 81% of the occurrences of the charcoal tree were encountered with a structural tabular surface of geomorphology composed by dissecting processes. We can consider this instanced attribute as determinant to find the charcoal tree in the study area (in green), Fig. 12.

The rule 828 had five plots (22, 23, 24, 38, and 46) with the occurrence of the charcoal tree and one plot (25) without the occurrence. At this moment, an issue has been raised: what are the determinant variables to answer why the charcoal tree did not appear in one of the plots?

After analyzing the dataset, we did not find any instanced variables in our dataset that has different values between plot 25 and the others five plots, which are in the same rule.

We can conclude that this dataset does not support the answer for the question about why the charcoal tree did not appear in plot 25.

5 Conclusions

In this work, we developed a geospatial mining analysis for the potential distribution of *TS* specie, with the aim to determine areas for reintroduction and to discover potential areas of conservation and forest management. Specially, we take into consideration the goal for answering the following scientific question: What are the combinations of abiotic features, which we can most easily find in nature the charcoal tree species for the study area? The answer was constructed by the data analysis and it is the instanced attributes of rule 1055, for this case study.

To perform this purpose, we collected and manipulated geospatial themes of physical environment including geology (rocks), hydrogeology, geomorphology (relief), pedology (soil), declivity, erodibility, hypsometry, hydrology, and the vegetation. Applying statistical techniques, we could validate the quality of the data and perform an analysis combining data mining and geoprocessing techniques for the *TS* specie.

The applied geospatial method to select the best classification rules considered three principles. First, the categorization of the classification rules in groups (A, B, C, D, E and F) based on their quantity of compound attributes. It was important to understand the role of each attribute and the composition of the subsets for the rules' analysis. Second, the conception of indexes by the addition of the geospatial dimension. The combination of the spatial dimension with the frequent item set association was valuable to facilitate finding the best significant rules for the geospatial occurrences of the *TS* specie. Third, the comparison of the results of geospatial indexes on this case study showed to be essential to better understand the analysis of geospatial data.

The strategies to enhance the efficacy regarding the studies of Species Distribution Modeling, closed attention to three important issues:

- identify effective rules, i.e., rules that can be effective used with regression techniques to predict species in geospatial environments. Not only in respect to their statistics' correlations and efficiency to frequent item sets, but also because of their role play in the context of the data analysis in question.
- identify variables that played a decisive role to discover the occurrence of species;
- evaluate the effectivity of the datasets used for the study to answer the specified scientific questions.

These goals were reached and described as follows:

- the conception of geospatial measures for support, confidence, and lift.

- the comparison of the created indexes with the indexes of frequent item sets showed more coherence when applied in spatial datasets, at least for this case study. For instance, we conclude that the data behavior results in this case study for GLift was normally distributed, and Lift did not pass the tests for normality.
- we showed that support, confidence, and lift, of frequent item set behave in ascending or descending order relative to complex models. This is an ascertainment to the necessity of these indexes to have a balance between the tradeoff of a complexity model in favor of accurately, but also with low bias, i.e., to choose a generalized model as possible as it can better perform with new data. In this case study, GConfidence and GLift showed their potential capability to deal with this effective model tradeoff.
- the conception of the α index to flag the effective of the rule promotes the analysis of determinant rules for the occurrence or not occurrence of a specie. And also help to analyze the effectivity of the dataset to model the species distribution.

References

1. Prata, D., Sousa, P., Pinheiro, R., Kneip, A.: Geospatial analysis of tree species for ecological economics. Int. Proc. Econ. Dev. Res. **85**, 125–130 (2015)
2. Franklin, J.: Mapping Species Distributions: Spatial Inference and Prediction. Cambridge University Press, Cambridge (2009)
3. Agrawal, R., Imielinski, T., Swami, A.: Database mining: a performance perspective. IEEE Trans. Knowl. Data Eng. **5**, 914–925 (1993). Special issue on Learning and Discovery in Knowledge Based Databases
4. Shekhar, S., Evans, M.R., Kang, J.M., Mohan, P.: Identifying patterns in spatial information: a survey of methods. Data Min. Knowl. Disc. **1**, 193–214 (2011)
5. Koperski, K., Han, J.: Discovery of spatial association rules in geographic information databases. In: Egenhofer, M.J., Herring, J.R. (eds.) SSD 1995. LNCS, vol. 951, pp. 47–66. Springer, Heidelberg (1995). https://doi.org/10.1007/3-540-60159-7_4
6. de Silva, M.A., Trevisan, D.Q., Prata, D.N., Marques, E.E., Lisboa, M., Prata, M.: Exploring an ichthyoplankton database from a freshwater reservoir in legal amazon. In: Motoda, H., Wu, Z., Cao, L., Zaiane, O., Yao, M., Wang, W. (eds.) ADMA 2013. LNCS (LNAI), vol. 8347, pp. 384–395. Springer, Heidelberg (2013). https://doi.org/10.1007/978-3-642-53917-6_34
7. Tetko, I.V., et al.: Benchmarking of linear and nonlinear approaches for quantitative structure − property relationship studies of metal complexation with ionophores. J. Chem. Inf. Modeling **46**, 808–819 (2006)
8. Han, J., Kamber, M., Pei, J.: Data Mining: Concepts and Techniques, 3rd edn. Morgan Kaufmann Publishers Inc., San Francisco (2011)
9. Hall, M.A.: Correlation-based feature selection for machine learning. Ph.D thesis, Waikato University, Hamilton, NZ (1998)
10. Geng, L., Hamilton, H.J.: Interestingness measures for data mining: a survey. ACM Comput. Surv. **38**(3), 9 (2006). Article 9

On Probabilistic k-Richness of the k-Means Algorithms

Robert A. Kłopotek[1(✉)] and Mieczysław A. Kłopotek[2]

[1] Faculty of Mathematics and Natural Sciences. School of Exact Sciences, Cardinal Stefan Wyszyński University in Warsaw, Warsaw, Poland
r.klopotek@uksw.edu.pl

[2] Computer Science Fundamental Research Institute, Polish Academy of Sciences, Warsaw, Poland
mieczyslaw.klopotek@ipipan.waw.pl

Abstract. With Kleinberg's axiomatic system for clustering, a discussion has been initiated, what kind of properties clustering algorithms should have and have not. As Ackerman et al. pointed out, the static properties studied by Kleinberg and other are not appropriate for clustering algorithms with elements of randomness. Therefore they introduced the property of probabilistic k-richness and claimed, without a proof that the versions of k-means both with random initialisation and k-means++ initialization have this property. We prove that k-means++ has the property of probabilistic k-richness, while k-means with random initialisation for well separated clusters does not. To characterize the latter, we introduce the notion of weak probabilistic k-richness and prove it for this algorithm. For completeness, we provide with a constructive proof that the theoretical k-means has the (deterministic) k-richness property.

Keywords: k-means · k-means++ · k-richness · Probabilistic k-richness · Weak probabilistic k-richness

1 Introduction

The outcome of a clustering algorithm is a function of two factors: the data structure and the limitations of the clustering algorithm. For example, algorithms from the k-means family (see [15] for an overview of k-means family) will produce a partition of the data into exactly k clusters, whatever the intrinsic structure of the data is. Are there other limitations? This paper focuses on the richness property, that is how many distinct clusterings can be considered as a possible outcome of the algorithm. We demonstrate that the theoretical k-means has the property of *k-richness*, that is that any partition into k clusters may be obtained. We show that k-means++ is *probabilistically k-rich*, that is by appropriate selection of the data set one can approach the k-richness with any desired probability. We establish that k-means with random initialization

Supported by Polish government fundamental research funds.

G. Nicosia et al. (Eds.): LOD 2019, LNCS 11943, pp. 259–271, 2019.
https://doi.org/10.1007/978-3-030-37599-7_22

does not have the property of probabilistic k-richness, but it has only a weaker property of *weak probabilistic k-richness* which means that the k-richness can be approximated by repeated runs of the algorithm only.

In Sect. 2 we introduce the basic concepts. Section 3 presents positive results for k-richness of three types of k-means algorithms. Section 4 contains the proof of the negative result, that k-means-random is not probabilistically k-rich. Numerical experiments on these issues are reported in Sect. 5. Section 6 relates our results to the literature. Section 7 concludes the paper.

2 Background and Preliminaries

Probably the most cited axiomatic framework for clustering is the Kleinberg's system [12][1]. Kleinberg [12, Section 2] defines clustering function as

Definition 1. *"A* clustering function *is a function f that takes a distance function d on [set] $\mathbf{X}$ [of size $n \geq 2$] and returns a partition Γ of $\mathbf{X}$. The sets in Γ will be called its* clusters*"*

Definition 2. *"With the set $\mathbf{X} = \{1, 2, \ldots, n\}$ [...] we define a distance function to be any function $d : S \times S \to \mathbb{R}$ such that for distinct $i, j \in S$ we have $d(i,j) \geq 0, d(i,j) = 0$ if and only if $i = j$, and $d(i,j) = d(j,i)$."*

Kleinberg [12] introduced an axiomatic system for clustering functions, including the so-called richness axiom/property:

Property 1. *Let Range(f) denote the set of all partitions Γ such that $f(d) = \Gamma$ for some distance function d. If Range(f) is equal to the set of all partitions of $\mathbf{X}$, then f has the* richness *property.*

One of the most popular clustering algorithms, the k-means, does not possess this property, as it splits data in (exactly) k clusters. Nor other k-clustering methods do. Therefore, a modified property was proposed, called k-richness:

Property 2 (Zadeh and Ben-David [17]). *If for any partition Γ of the set $\mathbf{X}$ consisting of exactly k clusters there exists such a distance function d that the clustering function $f(d)$ returns this partition Γ, then f has the k-*richness *property.*

Let us define *the k-means-ideal algorithm* as one that produces a clustering Γ_{opt} attaining the minimum of the cost function $Q(\Gamma)$.

$$Q(\Gamma) = \sum_{i=1}^{m}\sum_{j=1}^{k} u_{ij}\|\mathbf{x}_i - \boldsymbol{\mu}_j\|^2 = \sum_{j=1}^{k}\frac{1}{n_j}\sum_{\mathbf{x}_i,\mathbf{x}_l \in C_j}\|\mathbf{x}_i - \mathbf{x}_l\|^2 \quad (1)$$

where $\mathbf{X}$ is the clustered dataset, Γ is its partition into the predefined number k of clusters, and u_{ij} is an indicator of the membership of data point $\mathbf{x}_i$ in the cluster C_j having the centre at $\boldsymbol{\mu}_j$. As k-means-ideal is NP-hard, the following algorithm is used in practice:

[1] Google Scholar lists about 400 citations.

1. Initialize k cluster centres $\boldsymbol{\mu}_1^{(0)}, \ldots, \boldsymbol{\mu}_k^{(0)}$. Set $t := 0$.
2. Assign each data element $\mathbf{x}_i$ to the cluster $C_j^{(t)}$ identified by the closest $\boldsymbol{\mu}_j^{(t)}$.
3. Update $t := t + 1$. Compute a new $\boldsymbol{\mu}_j^{(t)}$ of each cluster as the gravity centre of the data elements in $C_j^{(t-1)}$.
4. Repeat steps 2 and 3 until reaching a stop criterion (no change of cluster membership, or no sufficient improvement of the objective function, or exceeding some maximum number of iterations, or some other criterion).

If step 1 is performed as random uniform sampling from the set of data points (without replacement), then we will speak about *k-means-random* algorithm. The *k-means++* algorithm is a special case of k-means where the initial guess of cluster centres proceeds as follows. $\boldsymbol{\mu}_1^{(0)}$ is set to be a data point uniformly sampled from $\mathbf{X}$. The subsequent cluster centres are data points picked from $\mathbf{X}$ with probability proportional to the squared distance to the closest cluster centre chosen so far. For details check [4]. Only k-means-ideal is k-rich. k-richness is problematic for randomized algorithms, like the k-means-random or k-means++, as their output is not deterministic. Therefore Ackerman et al. [1, Definition 3 (k-Richness)] introduce the concept of *probabilistic k-richness.*

Property 3. *If for any partition Γ of the set $\mathbf{X}$ into exactly k clusters and every $\epsilon > 0$ there exists such a distance function d that the clustering function $f(d)$ returns this partition Γ with probability exceeding $1 - \epsilon$, then f has the* probabilistic k-richness *property.*

They postulate (omitting the proof) that *probabilistic k-richness* is possessed by k-means-random algorithm (see their Fig. 2). We question this claim - see Theorems 2 and 3. But we show that probabilistic k-richness *is possessed* by k-means++ algorithm and that a weakened version (weak probabilistic k-richness, defined as Property 4) is possessed by k-means-random - see Theorem 1.

Property 4. *A clustering method is said to have* weak probabilistic k-richness *property if there exists a function $pr(k) > 0$ ($k \in \mathbb{N}$) independent of the sample size and distance that for any partition Γ of the set $\mathbf{X}$ consisting of exactly k clusters, then there exists such a distance function d that the clustering function returns this partition Γ with probability exceeding $pr(k)$.*

$pr(k)$ is a minimum probability that the algorithm returns the required partition. It depends only on k and not on the structure of the clustered data set. This property means that we can approximate the k-richness with any probability by re-running the clustering algorithm, with a-priori known number of re-runs.

3 k-Richness Properties of k-Means Variants

In this section we explore the various types of k-richness property for the k-means-ideal, k-means-random and k-means++ algorithms.

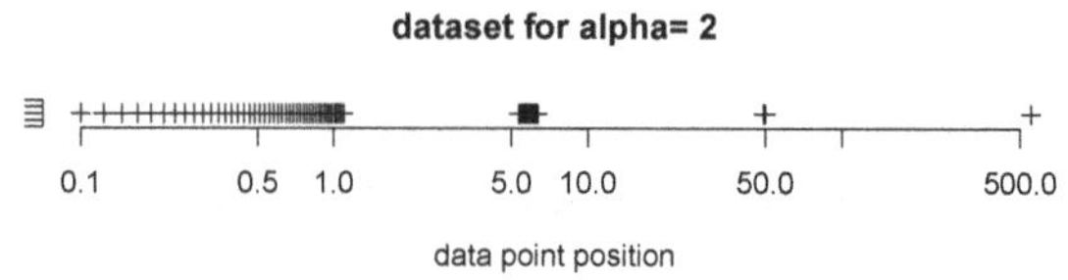

Fig. 1. A one-dimensional sample data set with clusters of cardinalities (from left to right) 45, 40, 35, 30 illustrating Theorem 1. X axis is in log scale.

Theorem 1. *k-means-ideal algorithm is k-rich. k-means-random algorithm is weakly probabilistically k-rich. k-means++ algorithm is probabilistically k-rich.*

Proof. We proceed by constructing a data set for each required partition in one-dimensional space. The construction is valid in higher dimensional space. Let us consider n data points arranged on a straight line and we want to split them into k clusters fitting a concrete partition Γ_0. For this purpose arrange the clusters on the line (left to right) as follows. Elements of each cluster shall be uniformly distributed over a line segment of unit length. Clusters of higher cardinality than a given cluster shall lie to the left of it. The space between the clusters (distance between closest elements of ith and $(i+1)$st cluster) should be set as follows: For $i = 1, \dots, k-1$ and $j < i$ let $dce(j,i)$ denote the distance between the most extreme data points of clusters j and i, $cardc(j,i)$ shall denote the combined cardinality of clusters $j, j+1, \dots, i$. The distance between closest elements of clusters i and $i+1$ shall be then set to

$$dist(i, i+1) = \alpha \cdot dce(1,i) \frac{cardc(1,i) + cardc(i+1,i+1)}{cardc(i+1,i+1)}$$

with $\alpha = 2$ for k-means-ideal and k-means-random. Assume we have a cluster centre μ_{i+1} within the tight interval containing elements of cluster $i+1$ and another one $\mu_{1:i}$ within the tight interval containing clusters $1, \dots, i$. Wherever they will be located, there is no possibility that an element of the cluster $i+1$ is closer to $\mu_{1:i}$ than to μ_{i+1}, and no element of the clusters $1, \dots, i$ is closer to μ_{i+1} than to $\mu_{1:i}$, because the distance between the clusters is chosen to be large enough. What is more, if there is a cluster containing the whole cluster $i+1$ and the whole or parts of the clusters $1, \dots, i$ then the cluster centre will fall closer to cluster $i+1$ than i. An example of cluster spacing and cluster arrangement by cluster size is visible in Fig. 1.

Consider *k-means-ideal.* Let A be the most right cluster of a partition Γ into k clusters different from Γ_0, containing the *gap between clusters i and $i+1$ of* Γ_0. Assume Γ is the optimal partition. Construct a new partition Γ' from Γ: Let us split A into two parts, A_L and A_R to the left and to the right of this *gap between clusters.* If the cluster centre of A was closer to A_L than to A_R, let A_L be a new cluster, and let A_R be attached to the closest right cluster of Γ. Elements of A_L will have smaller sum of squares to cluster centre of A_L than they had to the cluster centre of A. The distance of elements of A_R to cluster centre of A was at least $dce(1,i)$, and now it will be reduced to at most 1. If the cluster centre of A was closer to A_R than to A_L, let A_R be a new cluster, and let A_L be attached to the closest left cluster of Γ. Elements of A_R will have smaller sum of squares to cluster centre of A_R than they had to the cluster centre of A. The distance of elements of A_L to cluster centre of A was at least $dce(1,i)$, and now it will be reduced to at most $dce(1,i)$. In both cases Γ' be more optimal than Γ. This contradiction shows that Γ_0 is a more optimal partition than any other possible partition of the data set. Hence k-means-ideal will choose Γ_0 as the partition. As we can define Γ_0 for any split into k clusters, so we can achieve any partition with this construction.

Consider now *k-means-random*. Assume that each of the clusters of Γ_0 is seeded in the first step. The border between neighbouring clusters, due to the large spaces between clusters of Γ_0, will lie in the gap between the clusters, because it is twice as big as any of these clusters. In the next step of computing new cluster centres, no cluster centre will be moved outside of the area between boundaries of the given cluster. Hence the algorithm will stop as no change of cluster membership occurs.

Now consider a different initial seeding such that there are i seeds in i leftmost clusters of Γ_0 for each $i = 1, \ldots, k$. Consider the case that after the random initialization (or at any later step) we get a partition Γ with cluster centres $\mu_1, \ldots, \mu_k$ such that there is a cluster C_i of Γ_0 that has not been seeded (does not contain a μ_i in its range). Take the most left such cluster. Because of the big gap, no cluster of Γ with centre to the right of C_i would contain any data element from C_i, because the distance from any element of C_i to any element to its left is at most $dce(1,i)$, while to any to the right is at least $2 \cdot dce(1,i)$. So C_i is contained in a cluster B of Γ, not containing any point to the right of C_i.

Let us compute the centre of all the data from clusters C_1 to C_i. Even if all the data from $C_1, \ldots, C_{i-1}$ would be concentrated in the leftmost point of C_1, and all from C_i in the leftmost point of C_i the weight centre would lie at

$$2 \cdot dce(1,i-1)\frac{cardc(1,i-1)+cardc(i,i)}{cardc(i,i)} \cdot \frac{cardc(i,i)}{cardc(1,i-1)+cardc(i,i)}$$
$$= 2 \cdot dce(1,i-1)$$

from the leftmost point of C_1 that is in the gap between the clusters C_{i-1} and C_i at a distance of $dce(1,i-1)$ from the border of C_{i-1}. Therefore a cluster centre update step of k-means-random will locate there or more to the right the cluster centre of the cluster B of Γ. During the update step on class membership, no element to the left of C_i will be included, because elements of $C_1, \ldots, C_{i-1}$ have distances between them of at most $dce(1,i-1)$, and they are at least $dce(1,i-1)$ away from the cluster centre of B. Therefore B will become identical with C_i and in the next step the cluster centre of B will become identical with that of C. The same will happen to all the clusters of Γ_0 that had no seed at a point in time, given that there is a seed to the left of them. And the mentioned initialisation condition (*i seeds in i leftmost clusters*) guarantees this.

However, if the initial seeding does not match the condition *"i seeds in i leftmost clusters"*, we have no guarantee that the iterative steps of k-means-random will manage to move the cluster centres to the clusters of Γ_0.

So let us compute the probability that the initial seeding fulfills the condition *"i seeds in i leftmost clusters"*.

As the clusters are sorted non-increasingly, the probability of hitting the first cluster is at least $\frac{1}{k}$, that of first or second $\frac{2}{k}$, that of first, or second, or,. . . ,or ith is $\frac{i}{k}$. This results in the estimation of appropriate initialisation of $pr(k) = k!/k^k$.

In this way we obtained an estimate of probability of detecting a predefined clustering Γ_0 that is positive, dependent only on k and independent of the sample size. So we can conclude that k-means-random is weakly probabilistically k-rich.

Furthermore, the targeted clustering is the absolute minimum of the k-means-ideal, hence we can run k-means-random multiple time in order to achieve the desired probability of k-richness. E.g. if we need 95% certainty, we need to rerun k-means-random r times with r such that $1 - (1 - k!/k^k)^r \geq 95\%$.

As *k-means++* behaves similarly to k-means-random after initial seeding, the same effect of spreading cluster centres to Γ_0 centres will be reached if the initialisation fulfils the requirement *"i seeds in the i leftmost clusters"*. Let us compute the probability that each cluster gets a seed during the seeding phase. When the first seed is distributed, like in k-means, we have the assurance that an unhit cluster will be hit. The probability, that a cluster is hit during the seeding step after the first one, is proportional to the sum of squared distances of cluster elements to the closest seed assigned earlier. Consider the ith cluster (from the left) that was not hit so far. The closest hit cluster to the left can lie at least a distance

$$\alpha \cdot dce(1, i-1)\frac{cardc(1, i-1) + cardc(i, i)}{cardc(i, i)}$$

and that to the right at

$$\alpha \cdot dce(1, i)\frac{cardc(1, i) + cardc(i+1, i+1)}{cardc(i+1, i+1)}$$

(the first cluster has no left neighbor, and the kth - no right neighbor). The contribution of the ith cluster to the sum of squares estimation for hitting probability in a current state amounts to at least the smaller number of the following two:

$$\text{(i)} \qquad cardc(i, i)\left(\alpha \cdot dce(1, i-1)\frac{cardc(1, i-1) + cardc(i, i)}{cardc(i, i)}\right)^2$$

$$\text{(ii)} \qquad cardc(i, i)\left(\alpha \cdot dce(1, i)\frac{cardc(1, i) + cardc(i+1, i+1)}{cardc(i+1, i+1)}\right)^2$$

The first expression is the smaller one so we will consider it only.

$$cardc(i,i)\left(\alpha \cdot dce(1,i-1)\frac{cardc(1,i-1)+cardc(i,i)}{cardc(i,i)}\right)^2$$
$$= \alpha^2 \cdot dce(1,i-1)^2 \frac{cardc(1,i)^2}{cardc(i,i)}$$

Due to the non-increasing order of cluster sizes, $cardc(1,i) \geq i \cdot cardc(i,i)$, and $cardc(1,i) \geq i\frac{n}{k}$. Therefore

$$\alpha^2 \cdot dce(1,i-1)^2 \frac{cardc(1,i)^2}{cardc(i,i)} \geq \alpha^2 \cdot dce(1,i-1)^2 i \frac{n}{k}$$

Furthermore, $dce(1,1) = 1$, and

$$dce(1,i) = dce(1,i-1) + 1 + \alpha dce(1,i-1)\frac{cardc(1,i-1)+cardc(i,i)}{cardc(i,i)}$$
$$\geq dce(1,i-1) \cdot (1+i\alpha)$$

Hence $dce(1,i) \geq (1+2\alpha)^{i-1}$. Therefore

$$\alpha^2 \cdot dce(1,i-1)^2 i \frac{n}{k} \geq \alpha^2 (1+2\alpha)^{2(i-1)} i \frac{n}{k}$$

After s seeds were distributed the sum of squared distances to the closest seed for hit clusters amounts to at most the combined cardinality of the clusters with seeds times 1 so this does not exceed n. Therefore the probability of hitting an unhit cluster after s seeds were already distributed and hit different clusters is not bigger than

$$\frac{\alpha^2(1+2\alpha)^2\frac{n}{k} + \sum_{i=2}^{k-s} \alpha^2(1+2\alpha)^{2(i-1)} i\frac{n}{k}}{n + \alpha^2(1+2\alpha)^2\frac{n}{k} + \sum_{i=2}^{k-s} \alpha^2(1+2\alpha)^{2(i-1)} i\frac{n}{k}}$$

Hence the probability of hitting all clusters during seeding exceeds:

$$pr(k,\alpha) = \prod_{s=1}^{k-1} \frac{\alpha^2(1+2\alpha)^2 + \sum_{i=2}^{k-s} \alpha^2(1+2\alpha)^{2(i-1)} i}{k + \alpha^2(1+2\alpha)^2 + \sum_{i=2}^{k-s} \alpha^2(1+2\alpha)^{2(i-1)} i}$$

For a fixed α value k-means++ is weakly probabilistically k-rich. In probabilistic sense, as $pr(k,\alpha)$ grows with α and $\lim_{\alpha \to \infty} pr(k,\alpha) = 1$, for a predefined $\epsilon > 0$, one can find α (and hence distances) such that $1 - \epsilon < pr(k,\alpha)$. Therefore k-means++ is also probabilistically k-rich.

4 *k*-Means-Random Not Probabilistically Rich

As already mentioned, we disagree with Ackerman et al. [1] on probabilistic k-richness. In particular, we demonstrate below that given the intrinsic clusters are well separated (with large gaps), it is impossible for k-means-random to attain the probabilistic k-richness.

First we show this for the easier case of data embedded in one-dimensional space (Theorem 2) and thereafter for multidimensional space (Theorem 3). Define the *enclosing radius of a cluster* as the distance from cluster centre to the furthest point of the cluster.

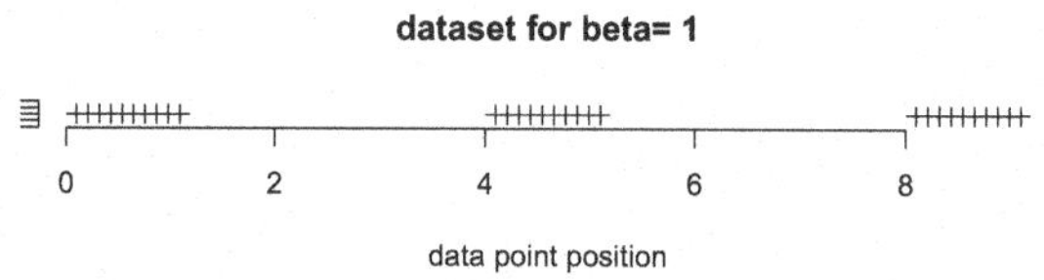

Fig. 2. A one-dimensional sample data set with clusters of cardinalities (from left to right) 10, 10, 10 illustrating Theorem 2 for $\beta = 1, r = 1$. See Table 2 for meaning of β.

Theorem 2. *In one-dimensional space, for $k \geq 3$, when distances between cluster centres exceed 6 times the largest enclosing radius r, k-means-random is not probabilistically k-rich.*

Proof. Let C_1, C_2, C_3 be three intrinsic clusters of identical large cardinality (at least 2) that will be arranged on a line in that order (from the left to the right). Let the distance d_1 between the cluster centre of C_1, C_2 be larger than d_2 between cluster centres C_2, C_3. Let the initial seeding be: one seed $\boldsymbol{\mu}_1^{(0)}$ as the leftmost member of C_1, one seed $\boldsymbol{\mu}_2^{(0)}$ as the rightmost member of C_1, one seed $\boldsymbol{\mu}_3^{(0)}$ as the rightmost member of C_2. Then in the first iterative step of k-means-random the last cluster centre $\boldsymbol{\mu}_3^{(1)}$ will lie in the middle between centres of C_2, C_3, while the two other $\boldsymbol{\mu}_1^{(1)}, \boldsymbol{\mu}_2^{(1)}$ will stay in the cluster C_1. The distance from the first two cluster centres $\boldsymbol{\mu}_1^{(1)}, \boldsymbol{\mu}_2^{(1)}$ to the leftmost point of C_2 will amount to at least $d_1 - 2r$, while the distance to this point from the last centre $\boldsymbol{\mu}_3^{(1)}$ will not exceed $d_2/2 + r$. As $d_1 > 6r$, and $d_1 \geq d_2$, so $d_1 + (d_1 - d_2) > 6r$. Hence, $2d_1 > d_2 + 6r$, $d_1 > d_2/2 + 3r$, $d_1 - 2r > d_2/2 + r$. This means that the last cluster centre $\boldsymbol{\mu}_3^{(t)}$ $(t = 1, 2, \ldots)$ will always keep clusters C_2, C_3, while the first two $\boldsymbol{\mu}_1^{(t)}, \boldsymbol{\mu}_2^{(t)}$ will divide between themselves C_1. And there is no way to change the situation by any manipulation of distances between cluster elements. Though the probability of the mentioned seeding is low, it is finite and cannot be reduced as required by Ackerman's definition of probabilistic k-richness. Hence k-means-random is not probabilistically k-rich for $k = 3$.

In case of more than 3 intrinsic clusters let $C_1, C_2, \ldots, C_k$ be the equal size (at least 2) clusters arranged in this way from left to right such that the smallest distance between leftmost cluster elements is between those of C_i, C_{i+1} , $1 < i < k$, amounting to $d_0 > 5r$. The initial seeding should be as follows: the seeds $j = i+2, \ldots, k$ should be at the rightmost data point of cluster C_j, and the seeds $\boldsymbol{\mu}_j^{(0)}$, $j = 2, \ldots, i-1$ at the rightmost data point of C_{j-1} and the seed $\boldsymbol{\mu}_1^{(0)}$ at the leftmost data point of C_1. The seed $\boldsymbol{\mu}_i^{(0)}$ at the leftmost data point of C_{i-1}. The seed $\boldsymbol{\mu}_{i+1}^{(0)}$ at the leftmost data point of C_{i+1}. No seed occurs at the cluster C_i. Hence, seed $\boldsymbol{\mu}_1^{(0)}$ will get only points from C_1, seeds $\boldsymbol{\mu}_2^{(0)}, \dots, \boldsymbol{\mu}_{i-1}^{(0)}$ will get only points from $C_1, \dots, C_{i-1}$ resp., seeds $\boldsymbol{\mu}_{i+2}^{(0)}, \dots, \boldsymbol{\mu}_k^{(0)}$ from $C_{i+2}, \dots, C_k$ resp. Seed $\boldsymbol{\mu}_{i+1}^{(0)}$ from both C_i, C_{i+1}, $\boldsymbol{\mu}_i^{(0)}$ from C_{i-1}.

In the next step the cluster centres $\boldsymbol{\mu}_j^{(1)} j = i+2, \dots, k$ will relocate to the centres of clusters C_j, the cluster centre $\boldsymbol{\mu}_{i+1}^{(1)}$ will relocate to the midpoint between cluster centres $C_i, C_{i+1}.\boldsymbol{\mu}_i^{(1)}$ will relocate to the centre of C_{i-1}. Centres $\boldsymbol{\mu}_1^{(1)}$ and $\boldsymbol{\mu}_2^{(1)}$ will stay in C_1, and the other cluster centres $\boldsymbol{\mu}_j^{(1)} j = 3, \dots, i-1$ will stay in clusters $C_2, \dots, C_{i-2}$ resp. Now the cluster centre $\boldsymbol{\mu}_{i+1}^{(1)}$ will be at most $d_0/2 + r$ away from any point from $C_i \cup C_{i+1}$. The cluster centre $\boldsymbol{\mu}_i^{(1)}$ will be at least $d_0 - r$ away from point from $C_i \cup C_{i+1}$. The cluster centre $\boldsymbol{\mu}_{i+2}^{(1)}$ will be at least $d_0 - 2r$ away from point from $C_i \cup C_{i+1}$. So neither $\boldsymbol{\mu}_i^{(1)}$ nor $\boldsymbol{\mu}_{i+2}^{(1)}$ will take over any point from $C_{i+1}^{(0)}$. Therefore, there will be no future change of point assignments to cluster centres at any later point in time t.

Consider now the case that the smallest distance between leftmost cluster elements is between those of C_i, C_{i+1} for $i = 1$. In this case the initial seed should be: Seeds $\boldsymbol{\mu}_j^{(0)}$ $j = 1, \dots, k-2$ at rightmost points of C_{j+1}, and seeds $\boldsymbol{\mu}_{k-1}^{(0)}, \boldsymbol{\mu}_k^{(0)}$ at the leftmost and rightmost points of C_k. Seed $\boldsymbol{\mu}_1^{(0)}$ captures then the clusters C_1, C_2, seeds $\boldsymbol{\mu}_j^{(0)}$ $j = 2, \dots, k-2$ capture clusters C_{j+1}, and seeds $\boldsymbol{\mu}_{k-1}^{(0)}, \boldsymbol{\mu}_k^{(0)}$ split the cluster C_k. Again by similar argument after the computation of new cluster centres $\boldsymbol{\mu}_j^{(1)}$ no change of boundaries of detected clusters will occur.

Theorem 3. *For* $k \geq mV_{ball,m,R}/V_{simplex,m,R-4r}$ *where* $V_{ball,m,R} = \frac{\pi^{\frac{m}{2}}}{\Gamma(\frac{m}{2}+1)} R^m$, $V_{simplex,m,R-4r} = \frac{\sqrt{n+1}}{n!\sqrt{2^n}} (R-4r)^m$, *where* m *is the dimension of space,* $S_m(R) = \frac{2\pi^{(m+1)/2}}{\Gamma(\frac{m+1}{2})} R^m$, *when distances between cluster centres exceed 10 times the largest enclosing radius* r, *and* $R = 14r$, *k-means-random is not probabilistically k-rich.*

Proof. Let $C_1, C_2, \dots, C_k$ be sets of equal size (at least 2). Without restricting the generality, let set centres of C_1 and C_2 be the closest ones, at the distance of $d_0 > 10r$. Consider the (finite) set $\mathcal{C}_{12}$ of sets that have set centres closer to C_1 centre than $d_0 + 4r$. Initialize k-means in such a way that:

- there is no seed from the set C_1,
- there is one seed $\boldsymbol{\mu}_j^{(0)}$ from each of the sets $C_j \in \mathcal{C}_{12}$, in each case being the closest point to any point of C_1, and
- there is one seed $\boldsymbol{\mu}_j^{(0)}$ in each set C_j outside of these sets ($C_j \notin \mathcal{C}_{12}$) except for one which is seeded with two seeds, whereby the furthest points from C_1 should be selected; we will show subsequently, that if k is sufficiently large, there exists at least one set C_j outside of $\mathcal{C}_{12}$.

In the first iteration of k-means the clusters that are formed will consist only of points of the set C_j from which each seed stemmed ($C_j \subseteq C_j^{(0)}$ plus some points from the set C_1, because of the large distance between set centres.

Only seeds $\boldsymbol{\mu}_j^{(0)} \in C_j \in \mathcal{C}_{12}$ have a chance to capture elements from C_1, because the distance of C_2 seed $\boldsymbol{\mu}_2^{(0)}$ to any point of the cluster C_1 will not exceed $d_0 + 2r - \epsilon$ (ϵ small positive number, while the distances of seeds from outside of $\mathcal{C}_{12}$ will amount to at least $d_0 + 4r - 2r > d_0 + 2r - \epsilon$. The distances of seeds from outside of $\mathcal{C}_{12}$ to points in sets from $\mathcal{C}_{12}$ will be at least $d_0 - 2r$. But the distances of seeds $\boldsymbol{\mu}_j^{(0)} \in C_j \in \mathcal{C}_{12}$ to elements of their own sets C_j will amount to at most $2r$.

In the second iteration, after the new cluster centres $\boldsymbol{\mu}_j^{(1)}$ are computed, the cluster centres $\boldsymbol{\mu}_j^{(1)}$ for $C_j \notin \mathcal{C}_{12}$ will lie within the area of their respective sets C_j (they will be their set centres). Therefore, again their distances to points in $\mathcal{C}_{12}$ will be at least $d_1 \geq d_0 - 2r$, while to C_1 at least $d_0 + 2r$.

The distances of centres $\boldsymbol{\mu}_j^{(1)}$ for $C_j \in \mathcal{C}_{12}$ of clusters originating from seeds of sets from $\mathcal{C}_{12}$ to their own sets will amount to $d_2 < (d_0 + 4r)/2 + r$. This formula results from the following: The new cluster $C_j^{(0)}$ formed from a seed $\boldsymbol{\mu}_j^{(0)}$ from a set C_j consists of the points in C_j, with the gravity centre at the centre of C_j, and some points from C_1 the gravity centre of which is closer to the centre of C_1 than $d_0 + 4r$. As the sets are of same cardinality, there are more points from C_j than from C_1 so that the new cluster centre $\boldsymbol{\mu}_j^{(1)}$ is closer to the gravity centre of C_j than to C_1. Hence the formula. From elementary geometry we see also, that for a new cluster centre originating in $C_i \in \mathcal{C}_{12}$ to elements of another set $C_j \in \mathcal{C}_{12}$ is not smaller than $\sqrt{0.75}d_0$ so that the sets from $\mathcal{C}_{12}$ are not split in the process between new cluster centres. Also $d_1 - d_2 > d_0 - 2r - (d_0 + 4r)/2 - r = d_0/2 - 5r > 5r - 5r = 0$. Hence, after the iterative step, no cluster centre $\boldsymbol{\mu}_j^{(1)}$ originating from sets C_j outside of $\mathcal{C}_{12}$ will capture points from clusters originating from $\mathcal{C}_{12}$. The distance of cluster centre originating from C_2 to any point of the cluster C_1 will not exceed $d_0 + r$ which is smaller than $d_0 + 2r$. This means stagnation - no chance to get a point from $\mathcal{C}_{12}$ or C_1 by outsiders, no chance of splitting elements of $\mathcal{C}_{12}$, no chance of getting outsider points by clusters originating at $\mathcal{C}_{12}$ so that the set C_1 will never become a cluster. The assumption is that the set of outsiders is non-empty.

As the distance of centres originating at $\mathcal{C}_{12}$ from C_1 is at most $d_0 + 4r$, so they are contained in a ball of radius $d_0 + 4r$ around the centre of C_1. The volume of this ball is V_{ball,m,d_0+4r}. As their distance is at least d_0, the cluster centres originating from $\mathcal{C}_{12}$ may be arranged into simplexes (as corner points of these simplexes) of volume of at least $V_{simplex,m,d_0}$. The quotient $V_{ball,m,d_0+4r}/V_{simplex,m,d_0}$ tells us at most how many simplexes fit the ball. As each simplex has m corner points, so $mV_{ball,m,d_0+4r}/V_{simplex,m,d_0}$ tells us at most how many cluster centres may originate from $\mathcal{C}_{12}$. So if k exceeds this figure, there are clusters not from $\mathcal{C}_{12}$.

5 Experiments

In order to illustrate the difference between the property of probabilistic k-richness and weak probabilistic k-richness, experiments were performed for data sets described in Theorems 1 and 2. Table 1 shows the behaviour of the algorithms k-means-random and k-means++ under the extreme conditions of constructive proof of Theorem 1. Table 2 illustrates this behaviour under more practical settings of the data sets mentioned in the proof of Theorem 2.

Table 1. Number of errors of k-means and k-means++ for the cluster sizes from Fig. 1 and for various values of α on 100,000 repetitions of the algorithms.

Value of α	2	4	6	8	10	12	14	16
Errors of k-means-random	29267	29321	29245	29374	29527	29497	29072	29275
Errors of k-means++	973	312	136	87	69	41	26	23

Table 2. Number of errors committed by k-means and k-means++ algorithms for the cluster sizes from Fig. 2 and for various values of distances $\beta \cdot 6r$ on 100,000 repetitions of the algorithm. rs - number of restarts.

Value of β	$1 \cdot 6r$	$2 \cdot 6r$	$3 \cdot 6r$	$4 \cdot 6r$	$5 \cdot 6r$	$6 \cdot 6r$	$7 \cdot 6r$	$8 \cdot 6r$
Errors of k-means-random	27590	28839	28752	28959	28940	29091	28997	29052
Errors of k-means-random rs = 2	7425	8430	8485	8442	8383	8453	8497	8344
Errors of k-means-random rs = 4	546	731	718	760	720	752	751	760
Errors of k-means++	668	208	102	57	41	26	24	12
Errors of k-means++ rs = 2	4	0	0	0	0	0	0	0

The probabilistic k-richness of k-means++ is visible as increasing distances (via α) decreases probability of not recognizing the intrinsic clustering. k-means-random does not improve its performance with increasing distances between clusters, as it is not probabilistically k-rich. Both algorithms possess the weak probabilistic k-richness as the repetitions decrease the error probability.

6 Discussion

The seminal paper of Kleinberg [12] sparked a continuous interest in exploring axiomatic properties of clustering algorithms, and in particular the richness properties of clustering [6,10,16,17] and related processes, like similarity measurement [11]. It was demonstrated by Ackerman et al. [2], that k-richness is a problematic issue as a useful property of stability of clusters under malicious addition of data points holds only for balanced clusters. A number of such papers directly refer to the Ackerman's et al. publication [1]. Regrettably, no interest is shown in the faithfulness of the results presented there, especially with respect to probabilistic k-richness of k-means algorithm family. Instead, they just mention that clustering is an ill-defined problem [3,14], or that formal methods of clustering algorithm selection require axiomatization [9], or as an alternative method of algorithm characterization [5,7,8,13].

7 Concluding Remarks

In this paper we have demonstrated, that contrary to claims of Ackerman et al. [1], the k-means-random is not probabilistically k-rich. In order to characterize the k-richness properties of the k-means-random algorithm, we introduced the concept of weak probabilistic k-richness and proved that k-means-random is weakly probabilistically k-rich. Additionally, we proved in a constructive manner that k-means++ is probabilistically k-rich and k-means-ideal is k-rich. To our best knowledge, this was not proven constructively so far.

References

1. Ackerman, M., Ben-David, S., Loker, D.: Towards property-based classification of clustering paradigms. In: Advances in Neural Information Processing Systems, vol. 23, pp. 10–18. Curran Associates, Inc. (2010)
2. Ackerman, M., Ben-David, S., Loker, D., Sabato, S.: Clustering oligarchies. In: Proceedings of 16th International Conference on Artificial Intelligence and Statistics, AISTATS 2013, Scottsdale, AZ, USA, 29 April – 1 May 2013, pp. 66–74 (2013)
3. Aluc, G.: Workload matters: A robust approach to physical RDF database design (2015). http://hdl.handle.net/10012/9774
4. Arthur, D., Vassilvitskii, S.: k-means++: the advantages of careful seeding. In: Proceedings of the Eighteenth Annual ACM-SIAM Symposium on Discrete Algorithms, pp. 1027–1035, SODA 2007. SIAM, New Orleans, Louisiana, USA, 7–9 Jan 2007
5. Bouveyron, C., Hammer, B., Villmann, T.: Recent developments in clustering algorithms. In: ESANN (2012)
6. Cohen-Addad, V., Kanade, V., Mallmann-Trenn, F.: Clustering redemption-beyond the impossibility of kleinberg's axioms. In: Advances in Neural Information Processing Systems, vol. 31, pp. 8517–8526. Curran Associates, Inc. (2018)
7. Cornuéjols, A., Wemmert, C., Gançarski, P., Bennani, Y.: Collaborative clustering: why, when, what and how. Inf. Fusion **39**, 81–95 (2018)
8. Hennig, C.: Clustering strategy and method selection. arXiv 1503.02059 (2015)
9. Hennig, C.: What are the true clusters? Pattern Recogn. Lett. **64**(C), 53–62 (2015). https://doi.org/10.1016/j.patrec.2015.04.009
10. Hopcroft, J., Kannan, R.: Computer science theory for the information age (2012), chapter 8.13.2. A Satisfiable Set of Axioms. page 272ff
11. Kang, U., Bilenko, M., Zhou, D., Faloutsos, C.: Axiomatic analysis of co-occurrence similarity functions. Technical report CMU-CS-12-102 (2012)
12. Kleinberg, J.: An impossibility theorem for clustering. Proc. NIPS **2002**, 446–453 (2002). http://books.nips.cc/papers/files/nips15/LT17.pdf
13. van Laarhoven, T., Marchiori, E.: Axioms for graph clustering quality functions. J. Mach. Learn. Res. **15**, 193–215 (2014)

14. Luksza, M.: Cluster statistics and gene expression analysis. Ph.D. thesis, Fachbereich Mathematik und Informatik der Freien Universität Berlin (2011)
15. Wierzchoń, S., Kłopotek, M.: Modern Clustering Algorithms. Studies in Big Data, vol. 34. Springer Verlag (2018)
16. Zadeh, R.B.: Towards a principled theory of clustering. In: Thirteenth International Conference on Artificial Intelligence and Statistics. Proceedings of Machine Learning Research Press (2010)
17. Zadeh, R.B., Ben-David, S.: A uniqueness theorem for clustering. In: Proceedings of the Twenty-Fifth Conference on Uncertainty in Artificial Intelligence, UAI 2009, pp. 639–646. AUAI Press, Arlington, Virginia, USA (2009)

Using Clustering for Supervised Feature Selection to Detect Relevant Features

Christoph Lohrmann(✉) and Pasi Luukka

LUT University, Yliopistonkatu 34, 53850 Lappeenranta, Finland
{christoph.lohrmann,pasi.luukka}@lut.fi

Abstract. In many applications in machine learning, large quantities of features and information are available, but these can be of low quality. A novel filter method for feature selection for classification termed COLD is presented that uses class-wise clustering to reduce the dimensionality of the data. The idea behind this approach is that if a relevant feature would be removed from the set of features, the separation of clusters belonging to different classes will deteriorate. Four artificial examples and two real-world data sets are presented on which COLD is compared with several popular filter methods. For the artificial examples, only COLD is capable to consistently rank the features according to their contribution to the separation of the classes. For the real-world Dermatology and Arrhythmia dataset, COLD demonstrates the ability to remove a large number of features and improve the classification accuracy or, at a minimum, not degrade the performance considerably.

Keywords: Feature selection · Dimensionality reduction · Clustering · Classification · Filter method · Machine learning

1 Introduction

A large number of features, meaning a high dimensionality of a dataset, can lead to severe disadvantages for the analysis of data sets such as computational cost, performance of an algorithm deployed on the data and a lack of generalizability of the results obtained [2,6]. Feature selection is an approach that selects a subset of the existing features in a dataset to reduce the dimension of the data [13]. The objective of feature selection is to discard irrelevant and redundant features [10]. In this paper, we focus on supervised feature selection, which refers to feature selection in the context of a classification task. Since feature selection attempts to select a suitable feature subset, it has to be emphasized that the choice of the subset of the k most relevant features is not necessarily the same as taking the k features that are each by themselves the most relevant ones [5]. This appears obvious when e.g. two features are redundant [5]. However, Elashoff, Elashoff and Goldman [8] show that also in case of independent features, the subset of the k best features is not necessarily the union of the k best single features. This finding is extended by Toussaint [25], who finds that the best subset of

G. Nicosia et al. (Eds.): LOD 2019, LNCS 11943, pp. 272–283, 2019.
https://doi.org/10.1007/978-3-030-37599-7_23

k features does not need to contain the single best feature. As a consequence, a feature selection algorithm should not evaluate features without taking into account the dependence to other features for the selection of the best subset.

2 Objectives

The objective of this research is to introduce a heuristic feature ranking method for supervised feature selection that deploys an intuitive approach via clustering. There has been some research on using clustering in the context of unsupervised [19] and supervised feature selection [4,22,24]. The research on supervised feature selection using clustering evolves around the idea to cluster features and keep for each cluster one (or more) representative feature(s) instead of all features in that cluster [4,22,24].

In contrast to that, the approach suggested in this paper has three distinct differences to these approaches. First, the clustering is not focused on the features but on the observations in the data. Second, the clustering is conducted for each class separtely to identify groups of observations within each class and not within the data as a whole. The focus of the clustering is on classes since the novel algorithm aims to measure the separation among classes. In order to measure how well classes are separated, one or more representative points and information on the covariance can be used. Clustering is deployed for each class, since classes that have more complex structures or groupings can be better represented by multiple representative points (cluster centroids) and the corresponding cluster covariance matrices than by a single representative and covariance matrix. Third, the clusters are not used to discard features that are similar (belonging to the same cluster), (highly) correlated or do not fit any cluster [4,24]. Instead, the approach presented in this paper focuses solely on each feature's ability to contribute to the separation of clusters of different classes in the data. In order to be able to capture each feature's relevance not just on a stand-alone basis (univariate) but also jointly with one or more other features (multivariate), the proposed COLD algorithm compares the cluster separation between the clusters of each class with all features with the cluster separation when a certain feature is removed. In this way, the contribution of a feature to the set of the remaining features can be measured. The following sections will discuss the methods (Sect. 3), the data as well as the training procedure (Sect. 4), the results (Sect. 5) and the conclusion (Sect. 6).

3 Methods

3.1 K-Medoids Clustering and Determining the Number of Clusters

In this paper, we will use a generalization of the k-means algorithm, called the k-medoids clustering [1]. K-means requires the features to be numerical while K-medoids also works with categorical variables [11] and the K-means algorithm is sensitive to the existence of outliers while K-medoids is more robust with respect to outliers and noise present in the data [11,23]. In the scientific literature several methods exist that estimate the optimal number of clusters. In this paper, we will use the Silhouette method due to, in our view, its intuitive idea, and the fact that the optimal number of clusters depends on the partition of the observations and not on the clustering algorithm [21].

3.2 COLD Algorithm

In this paper, we suggest a novel supervised filter method, called COLD, which ranks the features/variables according to their relevance. The name stands for "Clustering One Less Dimension" and is intended as a reference to the way this algorithm functions. The idea is to cluster each class separately to find groupings of observations for each class and then determine how each feature contributes to the separation of the clusters of each class to the clusters of all other classes, where the separation is measured in terms of the Mahalanobis distance. This cluster separation is compared between using the entire feature set and a feature subset that does not contain a certain feature. If clusters of different classes are closer to each other, so less well separated if a certain feature is not included, then this feature is by itself or jointly with other features relevant. The novel algorithm can be illustrated in a five step process:

Step 1: Pre-processing of the Data. The first step in the pre-processing is the scaling of the data into the compact interval $[0, 1]$. Also, those features that are linearly independent of each other remain in the data, whereas all others are considered redundant and are removed. The linear independent features are determined with a rank-revealing QR factorization [3].

Step 2: Clustering and the Choice of the Optimal Number of Clusters for Each Class. To determine the number of clusters for each class, the Silhouette method is used with clustering partitions for each number of clusters k from the minimum value of k specified (e.g. 1) to the maximum value of k (e.g. 10 or based on the sample size). In this paper, the Mahalanobis distance is chosen as distance measure for the k-medoids clustering and for the Silhouette index since it accounts for the covariance of the clusters (groups). To calculate the Mahalanobis distance during the clustering, the covariance matrix of each cluster has to be invertible. However, the clustering algorithm can suggest a cluster in which the covariance matrix is not invertible. If the covariance matrix is rank deficient, it will be regularized via ridge regularization [26] to be invertible by:

$$\Sigma_k = (1 - \alpha)\Sigma_k + \alpha I \tag{1}$$

where I is the identity matrix. The parameter α is controlling how much of the identity matrix I is added to the (rank deficient) original covariance matrix Σ_k to create the regularized version. The outcome of Step 2 is the optimal number of clusters O, and the best clustering result - including cluster memberships, cluster centres and the (regularized) covariance matrices.

Step 3: Calculate the Distance Among Clusters. Let us denote the cluster indices obtained in Step 2 by $o = 1$ to O. We denote with C_o the set of cluster indices that belong to the same class as cluster o and with $\overline{C}_o$ the complement of C_o, so all cluster indices that belong to any other class than the one o belongs to. The same logic is applied to the features. D denotes the number of features in the data set. The index d denotes a feature, whereas $\overline{d}$ denotes the complement containing all features from 1 to D except d itself. The cluster centre of the o-th cluster is denoted by v_o. Then the Mahalanobis distance between two cluster centres v_o and v_m, where $m \in \overline{C}_o$ can be denoted by:

$$S(v_o, v_m) = (v_o - v_m)^T \Sigma_o^{-1}(v_o - v_m) \tag{2}$$

where v_o and v_m are the centres of two clusters that belong to different classes. The Mahalanobis distance between each cluster centre v_o and the cluster centres corresponding to $\overline{C}_o$ is calculated.

Step 4: Calculate the Distance Among Clusters Without One Feature. This step is similar to Step 3 but does not use all features in the calculation. In particular, the Mahalanobis distance between two cluster centres v_o and v_m without the d-th feature can be expressed as follows:

$$S(v_{o,\overline{d}}, v_{m,\overline{d}}) = (v_{o,\overline{d}} - v_{m,\overline{d}})^T \Sigma_{o,\overline{d}}^{-1}(v_{o,\overline{d}} - v_{m,\overline{d}}) \tag{3}$$

where $v_{o,\overline{d}}$ denotes the o-th cluster centre with all the features except for the d-th one. Similar to Step 3, $v_{o,\overline{d}}$ and $v_{m,\overline{d}}$ do not belong to the same class. The covariance matrix $\Sigma_{o,\overline{d}}$ of the o-th cluster without the d-th feature is simply Σ_o without the d-th row and column. This calculation is conducted for each feature between each cluster centre v_o and the cluster centres corresponding to $\overline{C}_o$.

Step 5: Determine the Rank and Score for Each Feature. A ratio of the Mahalanobis distance between the cluster centres v_o and v_m without and with feature d is calculated. This ratio is denoted as:

$$r_{o,m,d} = \frac{S(v_{o,\overline{d}}, v_{m,\overline{d}})}{S(v_o, v_m)} \tag{4}$$

This ratio is $r_{o,m,d} \geq 0$ since the distance between the clusters v_o and v_m measured by Mahalanobis distance cannot take a negative value. The value for $r_{o,m,d}$ will be larger than 1, showing that $S(v_{o,\overline{d}}, v_{m,\overline{d}})$ is larger than $S(v_o, v_m)$, when the distance between clusters increases after the removal of feature d. This implies that the d-th feature is irrelevant or even noise given that it is easier to discriminate between the clusters without it. In contrast to that, $r_{o,m,d}$ is smaller than 1, when $S(v_{o,\overline{d}}, v_{m,\overline{d}})$ is smaller than $S(v_o, v_m)$. Hence, removing the d-th feature decreases the distance between clusters v_o and v_m, making it more difficult to discriminate between them. The values of $r_{o,m,d}$ provide an indication whether the removal of a feature d is supporting the discrimination between two clusters o and m. But, all clusters $\overline{C}_o$ that do not belong to the same class as v_o, have to be taken into account, and they are not equally important. This is premised on the consideration that the change of the distance between cluster v_o to its closest cluster of any other class has likely more impact on the ability to separate classes than any more distant cluster. Hence, the aggregation of $r_{o,m,d}$ over all other clusters $\overline{C}_o$ is weighted according to their distance to v_o.

$$w_{o,m} = \begin{cases} 1 & \text{if } card(\overline{C}_o = 1) \\ 1 - \frac{S(v_o, v_m)}{\sum_{m \in \overline{C}_o} S(v_o, v_m)} & \text{otherwise} \end{cases} \tag{5}$$

The weight defined in this way will be $w_{o,m} \in [0, 1]$. First, the formula considers the case with cardinality $card(\overline{C}_o) = 1$, which simply means that there is only one cluster from another class and the weight of it is set to 1. If there is more than one cluster of any other class, then the weight for the clusters that are closer to v_o are higher. The corresponding weighted average is calculated as:

$$r_{o,d} = \frac{\sum_{m \in \overline{C}_o} r_{o,m,d} * w_{o,m}}{\sum_{m \in \overline{C}_o} w_{o,m}} \tag{6}$$

Since the sum over $w_{o,m}$ is often not 1, the numerator is divided by the sum of the weights. A value of $r_{o,d}$ larger than 1 indicates that the removal of feature d on average separates the cluster o better from clusters of other classes than using all features including d. In contrast to that, a value equal to or smaller than 1 implies that the clusters of other classes are on average as distant or closer to cluster o without feature d. The values of $r_{o,d}$ can simply be averaged by summation and division by O, the number of clusters. To obtain a more intuitive result, the average over $r_{o,d}$ is subtracted from 1, as is stated below:

$$COLD_d = 1 - \frac{\sum_o r_{o,d}}{O} \tag{7}$$

If the $COLD_d$ feature score (being in general ≤ 1) takes a positive value, feature d is relevant for the separation between classes. In contrast to that, if the score is close to zero or negative, this indicates that the feature leads to almost no improvement or on average even an increase in the distances among clusters of different classes, which characterizes a feature that is irrelevant or even noisy. Subsequently, the features can be ranked according to their relevance from the highest feature score to the lowest. In general, at least all features with a COLD score of zero or smaller can be discarded since they deteriorate the separation among the clusters of different classes.

4 Data and Training Procedure

4.1 Artificial Examples

Four artificial classification examples were created with different data structures in order to test the new algorithm and compare it to several popular filter algorithms for feature selection. The first artificial example (Fig. 1) is a classification task with two classes. Each class is characterized by three normal distributed features without any covariance, so the variables are uncorrelated. The first two features resemble a simple numerical XOR problem with two distinct decision regions for each of the classes. The observations of each class's clusters are normally distributed. The third feature overlaps for both classes moderately. The first two features are each on their own overlapping to a large extent, which makes them almost irrelevant by themselves, but they are together able to linearly separate the two classes. The third feature that overlaps only moderately for both classes is by itself the best feature. However, the best feature subset contains only the first two features since they together can linearly separate the two classes without misclassification.

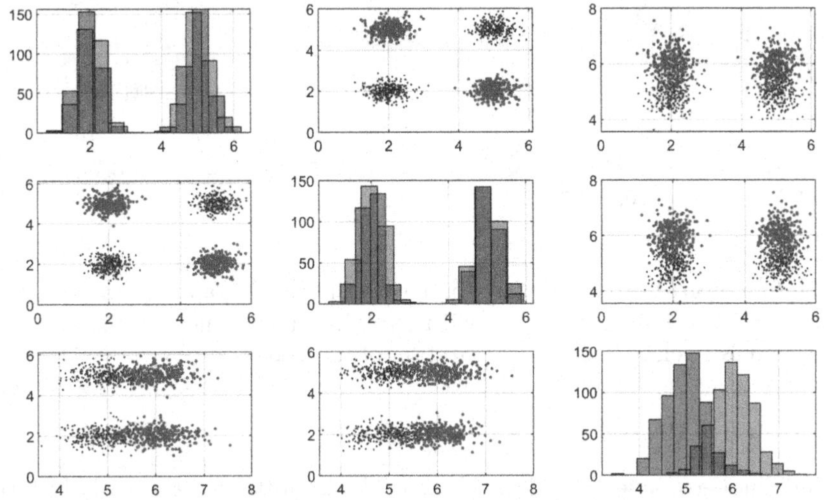

Fig. 1. First artificial example: $\mu_1 = (2, 2, 5)$, $\mu_2 = (5, 5, 5)$, $\mu_3 = (2, 5, 6)$, $\mu_4 = (5, 2, 6)$, $\Sigma = (0.1, 0, 0; 0, 0.1, 0; 0, 0, 0.2)$.

The second artificial example is based on the first example but contains a fourth normal distributed feature with the same mean and variance for both classes, which makes it clearly irrelevant for the separation of the classes.

The third artificial example is another binary class problem with three dimensions. Both classes are drawn from a normal distribution and have exactly the same covariance structure but a different mean. Feature one and two are on their own overlapping to a moderate to large extent, which makes them almost useless by themselves, but they are jointly able to linearly separate the two classes. The third feature overlaps for both classes slightly and is uncorrelated with the other two features. Only considering each dimension on its own suggests that the third feature is the best feature. This example is highlighted in Fig. 2 and is related to the fourth example in [10].

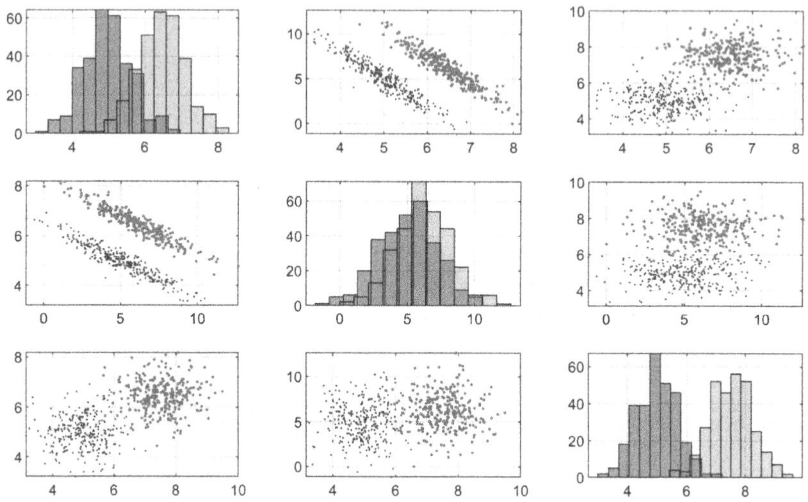

Fig. 2. Second artificial example: $\mu_1 = (5, 5, 5)$, $\mu_2 = (6.5, 6, 7.5)$, $\Sigma = (0.5, -1.5, 0; -1.5, 5, 0; 0, 0, 0.5)$.

The fourth artificial example is based on the third example but contains a fourth normal distributed feature with the same mean and variance for both classes so it is irrelevant for the separation of the classes.

4.2 Training Procedure

For the Silhouette index as well as the k-medoids clustering, the best clustering (in terms of within-sum-of-squares) out of 100 random initializations of the initial cluster centres is selected. In the Silhouette index, the minimum number of clusters is set to 1 and the maximum number of clusters for a class is set to the rounded down quotient of the number of observations in that class divided by 100. This approach is chosen to ensure that there are on average at least 100 observations in each cluster. The parameter α is selected for each cluster separately to be the maximum of a small positive constant 0.000001 and the minimum variance on the diagonal of the covariance matrix that is larger than 0 divided by 2. The idea behind the second part is to keep the ordering of the variances in the covariance matrix the same, but, also to ensure that this value is not too small, but a suitable small positive constant. For the artificial examples, 300 observations for each cluster of a class are generated.

The feature ranking of the COLD algorithm is compared to ReliefF [15] with 10 nearest hits/misses [14] and with 70 nearest hits/misses and a decay factor $s = 20$ [20], the similarity- and entropy-based feature selection approach by Luukka [18], the 'fuzzy similarity and entropy' (FSAE) feature selection [17], the Laplacian score [12], and the Fisher score [7]. All feature selection algorithms are run 500 times to determine how volatile the feature ranking of each method is.

All calculations were implemented with MATLAB^{TM}- software. The code for the Laplacian score was taken from the Matlab toolbox'Feature Selection Library' [9]. The MATLAB^{TM} code for the COLD algorithm can be send via e-mail upon request.

5 Results

Most methods, including the new COLD algorithm, can solve the first artificial example (see Table 1) and only algorithms such as FSAE and the Fisher Score fail to identify that the subset of feature 1 and 2 can be used to better discriminate among classes than any subset of the same size containing feature 3. All feature selection algorithms such as COLD that rank the third feature last, keep or even improve their classification accuracy with the K-nearest-neighbor classifier (k=10), decision tree (Minimum leave size=10) and support vector machines (SVM, with radial basis function) of about 100%. This result represents a highly significant (p-value$<$0.01) outperformance to the 87.9% obtained using the Fisher Score, FS Luukka (2011) or the FSAE, which evaluate each feature only by itself. It is obvious, that when only a single feature is considered and two features have to be discarded, the situation reverses and these three approaches outperform the remaining feature selection algorithms highly significantly. This highlights once more that algorithms such as COLD can find for this example the subset of two features that leads to the best class separation and highest classification accuracy, but also that this subset does not need to contain the single best feature. The set of methods that rank the features as desired remain unaltered after a fourth irrelevant normal distributed feature is added. Moreover, all methods with exception of FS Luukka (2011), FSAE and the Fisher Score correctly identify the fourth feature as the least relevant one.

Table 1. Results first artificial example.

Filter method	Feature 1	Feature 2	Feature 3
Actual structure	Rank 1 & 2		Rank 3
ReliefF (10 h/m)	Rank 1	Rank 2	Rank 3
ReliefF (70 h/m, s = 20)	Rank 1	Rank 2	Rank 3
FS Luukka (2011)	Rank 3	Rank 1	Rank 2
FSAE (l = 1)	Rank 3	Rank 1	Rank 2
COLD	Rank 1	Rank 2	Rank 3
Laplacian score	Rank 2	Rank 1	Rank 3
Fisher score	Rank 3	Rank 1	Rank 2

For the third example, methods that rely on Euclidean distance or simple distance measures such as similarity, that do not account for the cluster covariance, face difficulties. As highlighted in Table 2, only the COLD algorithm and the method of Luukka (2011) classify this example correctly. All other methods, including ReliefF, fail to account for the dependency of the first two features.

Hence, after the first feature removal the mean classification accuracy of COLD remains at the same level or even improves closer to 100% for the three classification algorithms. This is once again a highly significant outperformance (p-value<0.01) of the feature subset containing 2 features suggested by COLD (and the method by Luukka (2011)) compared to the remaining feature selection algorithms. Only the Fisher Score and FSAE have selected the single best feature after two removals, which leads to a performance of about 96%. This is inferior in mean performance to the best subset of two features obtained by COLD, but the best performing subset of a single feature.

Table 2. Results third artificial example.

Filter method	Feature 1	Feature 2	Feature 3
Actual structure	Rank 1 & 2		Rank 3
ReliefF (10 h/m)	Rank 1	Rank 3	Rank 2
ReliefF (70 h/m, s = 20)	Rank 1	Rank 3	Rank 2
FS Luukka (2011)	Rank 2	Rank 1	Rank 3
FSAE (l = 1)	Rank 3	Rank 1	Rank 2
COLD	Rank 1	Rank 2	Rank 3
Laplacian score	Rank 1	Rank 3	Rank 2
Fisher score	Rank 3	Rank 1	Rank 2

For the fourth artificial example again almost all methods leave their ranking unchanged and recognize the fourth feature as the least relevant one. Only FS Luukka (2011), FSAE and the Laplacian Score change their ranking, the first two in an adverse way, and the third now correctly ranks the four features. For each feature selection algorithm for all 500 iterations and each artificial example, the feature ranking obtained is the same and unchanged.

In addition, the feature selection algorithms were tested on two real-world datasets, the "Dermatology Data Set" (358 observations of 34 features) and the "Arrhythmia Data Set" (420 observations of 258 non-constant features) from the UCI Machine Learning Repository [16]. The computational time required to generate the feature ranking for these data sets is presented in Table 3.

Table 3. Computational time of feature selection methods (in seconds).

Data set	ReliefF(10)	ReliefF(70,20)	Luukka	FSAE	COLD	Laplacian	Fisher
Dermatology	0.6	0.7	0.3	0.4	1.6	0.3	0.1
Arrhythmia	6.6	7.3	1.9	2.1	75.4	0.5	0.2

It is apparent, that the COLD algorithm requires more computational time to determine the feature ranking than the other six feature ranking methods. This is the result of deploying clustering, in particular the rather computationally expensive but more noise-resistant k-medoids clustering algorithm, for a multivariate feature ranking that attempts to account for feature interactions.

The results for the mean performance for the three classifiers in this study on each set of features is illustrated in Fig. 3.

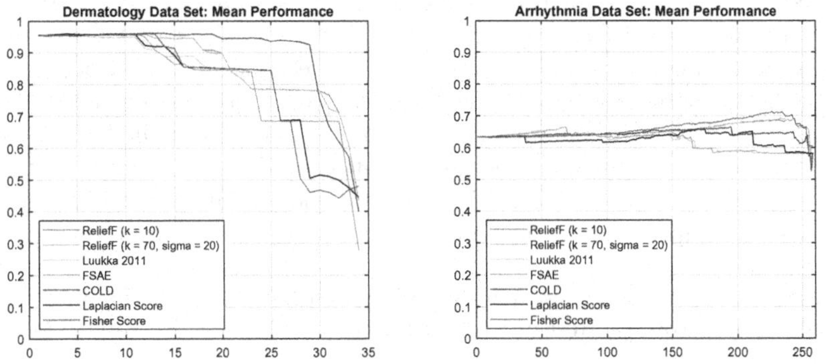

Fig. 3. Mean performance with the KNN, Decision Tree and SVM classifier

The results clearly indicate that for the Dermatology data set, the COLD algorithm determines a feature ranking that outperforms the remaining six feature selection methods between 15 to 29 feature removals with a considerable margin. Until the removal of the 30th out of 34 features, the mean performance of COLD is consistently over 92%. The highest performance with COLD is achieved with the K-nearest-neighbor classifier with 97.9% for the removal of 12 features. The one-sided Welch's test with unequal variance demonstrates that for the K-nearest-neighbor algorithm, the positive difference in the mean accuracy achieved with COLD is highly significant (p-value<0.01) for 7 to 29 removed features compared to all of the other 6 feature selection methods. The same is the case for the decision tree classifier for 17 to 29 discarded features, and with the SVM for 14 to 29 eliminated features. For the Arrhythmia data set, the Fisher Score obtains overall the highest mean accuracies. In general, COLD receives competitive results to the remaining feature selection algorithms for the Arrhythmia data set and can also improve the classification accuracy several percentage points with about 200 features removed and can obtain a similar performance than with all features when discarding more than 250 out of 258 features.

6 Conclusion

In this research, we presented the novel supervised feature selection algorithm COLD. The algorithm is premised on clustering data class-wise and determining clusters within classes and their covariance structure. The algorithm aims to evaluate the contribution of features to the separation among clusters with and without each feature. It only considers linearly independent features, so that redundant features are automatically ranked last. Using multiple artificial examples, it was demonstrated that COLD can correctly rank features according to their joint relevance for the classification and not just by each feature's standalone characteristics. For the artificial examples, only the novel algorithm COLD consistently ranked the features according to their joint relevance. It also always correctly detected the irrelevant feature and ranked it last. On the real-world Dermatology and Arrhythmia data sets, COLD demonstrates that it can in both cases improve the mean classification accuracy for the K-nearest-neighbor, decision tree and SVM classifier and even outperform the remaining six feature ranking methods for the majority of the feature subset sizes for the Dermatology data set. In the future research, the behaviour of the COLD algorithm can be tested for higher dimensional data and additional data structures and improvements of the algorithms complexity can be investigated.

Acknowledgements. This research would like to acknowledge the funding received from the Finnish Strategic Research Council, grant number 313396/MFG40 Manufacturing 4.0.

References

1. Bishop, C.M.: Pattern Recognition and Machine Learning. Springer ScienceBusiness Media, New York (2006)
2. Caruana, R., Freitag, D.: Greedy attribute selection. In: Cohen, W., Hirsh, H. (eds.) Proceedings of the 11th International Conference on Machine Learning (ICML 1994), pp. 28–36. Morgan Kaufmann, New Brunswick (1994)
3. Chan, T.F.: Rank revealing QR factorizations. Linear Algebra Appl. **88–89**, 67–82 (1987)
4. Chormunge, S., Jena, S.: Correlation based feature selection with clustering for highdimensional data. J. Electr. Syst. Inf. Technol. **5**, 542–549 (2018)
5. Cover, T.M.: The best two independent measurements are not the two best. IEEE Trans. Syst. Man Cybern. **4**(1), 116–117 (1974)
6. Dessì, N., Pes, B.: Similarity of feature selection methods. An empirical study across data intensive classification tasks. Expert Syst. Appl. **42**(10), 4632–4642 (2015)
7. Duda, R., Hart, P., Stork, D.: Pattern Classification. John Wiley and Sons, New York (2012)
8. Elashoff, J.E., Elashoff, R.M., Goldman, G.E.: On the choice of variables in classification problems with dichotomous variables. Biometrika **54**(3), 668–670 (1967)
9. Ruffo, G.: Matlab Toolbox: Feature selection library. https://se.mathworks.com/matlabcentral/fileexchange/56937-feature-selection-library. Accessed 1 Dec 2018
10. Guyon, I., Elisseeff, A.: An introduction to variable and feature selection. J. Mach. Learn. Res. **3**, 1157–1182 (2003)
11. Hastie, T., Tibshirani, R., Friedman, J.: Data Mining, Inference, and Prediction. Springer Series in Statistics. Springer, New York (2009). https://doi.org/10.1007/978-0-387-84858-7

12. He X., Cai, D., Niyogi, P.: Laplacian score for feature selection. In: Proceedings of the 18th International Conference on Neural Information Processing Systems (NIPS 2005), pp. 507–514. MIT Press, Cambridge (2005)
13. Kittler, J., Mardia, K.V.: Statistical pattern recognition in image analysis. J. Appl. Stat. **21**(1–2), 61–75 (1994)
14. Kononenko, I.: Estimating attributes: analysis and extensions of RELIEF. In: Bergadano, F., De Raedt, L. (eds.) ECML 1994. LNCS, vol. 784, pp. 171–182. Springer, Heidelberg (1994). https://doi.org/10.1007/3-540-57868-4_57
15. Kononenko, I., Simec, E., Robnik-Sikonja, M.: Overcoming the myopia of inductive learning algorithms with RELIEFF. Appl. Intell. **7**, 39–55 (1997)
16. Lichman, M.: UCI Machine Learning Repository. https://archive.ics.uci.edu/ml/index.php. Accessed 20 June 2019
17. Lohrmann, C., Luukka, P., Jablonska-Sabuka, M., Kauranne, T.: Supervised feature selection with a combination of fuzzy similarity measures and fuzzy entropy measures. Expert Syst. Appl. **110**, 216–236 (2018)
18. Luukka, P.: Feature selection using fuzzy entropy measures with similarity classifier. Expert Syst. Appl. **38**, 4600–4607 (2011)
19. Mitra, P., Murthy, C.A., Pal, S.K.: Unsupervised feature selection using feature similarity. IEEE Trans. Pattern Anal. Mach. Intell. **24**(3), 301–312 (2002)
20. Robnik-Šikonja, M., Kononenko, I.: Theoretical and empirical analysis of ReliefF and RReliefF. Appl. Intell. **53**(1–2), 23–69 (2003)
21. Rousseeuw, P.J.: A graphical aid to the interpretation and validation of cluster analysis. J. Comput. Appl. Math. **20**, 53–65 (1987)
22. Sahu, B., Dehuri, S., Jagadev, A.K.: Feature selection model based on clustering and ranking in pipeline for microarray data. Inf. Med. Unlocked **9**, 107–122 (2017)
23. Sammut, C., Webb, G.I.: Encyclopedia of Machine Learning and Data Mining, 2017th edn. Springer Science+Business Media, New York (2017)
24. Sotoca, J.M., Pla, F.: Supervised feature selection by clustering using conditional mutual information-based distances. Pattern Recogn. **43**, 2068–2081 (2010)
25. Toussaint, G.T.: Note on optimal selection of independent binary-valued features for pattern recognition. IEEE Trans. Inf. Theory **17**(5), 618 (1971)
26. Warton, D.I.: Penalized normal likelihood and ridge regularization of correlation and covariance matrices. J. Am. Stat. Assoc. **103**(481), 340–349 (2008)

A Structural Theorem for Center-Based Clustering in High-Dimensional Euclidean Space

Vladimir Shenmaier(✉)

Sobolev Institute of Mathematics, 4 Koptyug Avenue, 630090 Novosibirsk, Russia
shenmaier@mail.ru

Abstract. We prove that, for any finite set X in Euclidean space of any dimension and for any fixed $\varepsilon > 0$, there exists a polynomial-cardinality set of points in this space which can be constructed in polynomial time and which contains $(1+\varepsilon)$-approximations of all the points of space in terms of the distances to every element of X. The proved statement allows to approximate a lot of clustering problems which can be reduced to finding optimal cluster centers in high-dimensional space. In fact, we construct a polynomial-time approximation-preserving reduction of such problems to their discrete versions, in which the desired centers must be selected from a given finite set.

Keywords: Geometric clustering · Partial clustering · Outliers · Euclidean space · High dimensions · k-Means · k-Median

1 Introduction

We prove a general geometric property of finite sets of points in Euclidean space which may be useful for developing approximation algorithms with performance guarantees for center-based clustering problems.

This property concerns the following concept. We say that a point y in space $\mathbb{R}^d$ is a $(1+\varepsilon)$-approximation of another point x with respect to some finite set $X \subset \mathbb{R}^d$ if the Euclidean distance from y to each element of X is at most $1+\varepsilon$ of that from x. We show that, for any n-element set X in any-dimensional space and any $\varepsilon \in (0, 1]$, there exists an $\mathcal{O}\big((n/\varepsilon)^{\frac{2}{\varepsilon}\log\frac{2}{\varepsilon}}\big)$-element set of points which can be constructed in polynomial time and which contains $(1+\varepsilon)$-approximations of all the points of space with respect to X.

The proved statement allows to approximate a lot of geometric clustering problems reducible to finding optimal centers in high-dimensional Euclidean space. Actually, it gives a reduction of these problems to their discrete versions, where the desired centers must be selected from a finite input set. One of general problems which can be approximated with the proposed approach is selecting k points in space $\mathbb{R}^d$ (centers) to minimize an arbitrary objective function satisfying a natural continuity-type condition which requires that small

G. Nicosia et al. (Eds.): LOD 2019, LNCS 11943, pp. 284–295, 2019.
https://doi.org/10.1007/978-3-030-37599-7_24

relative changes of the distances from the input points to the selected centers give a small relative change of the objective function value.

Related Work. Perhaps the most studied geometric center-based clustering problems are k-Means and Euclidean k-Median. They consist of finding k centers in space $\mathbb{R}^d$ minimizing the sum of the squared or non-squared distances respectively from the input points to the nearest centers. In the case when the parameters k and d belong to the input, k-Means and the discrete version of Euclidean k-Median are APX-hard and can be approximated within a constant factor [2, 10, 14]. If the number of clusters is fixed, both problems admit fast randomized approximation schemes, most of which are based on constructing a set of candidate approximate centers by using coresets [7, 9] and random sampling techniques [5, 6, 11, 13]. However, k-Means remains NP-hard even if $k = 2$ [4], while the complexity of Euclidean 2-Median is still an open question.

In many cases, it is not required for the desired clusters to cover all the input points and the goal is to choose a subset of a given cardinality in the input set to minimize the cost of clustering on this subset. Such problems are called *partial clustering problems* or *problems with outliers*. Perhaps the simplest geometric center-based problems of this type are Smallest m-Enclosing Ball and m-Variance. The first consists of finding m input points minimizing the radius of the enclosing ball. The second problem consists of finding m input points with the minimum sum of squared distances from these points to their mean. In high dimensions, both problems are strongly NP-hard [12, 16, 17] but admit approximation schemes with running time $\mathcal{O}\big(n^{\lceil 1/\varepsilon \rceil} d\big)$ [1] and $\mathcal{O}\big(n^{\lceil 2/\varepsilon \rceil + 1} (9/\varepsilon)^{3/\varepsilon} d\big)$ [15] respectively. Both approximation schemes are based on constructing candidate approximate centers of the desired cluster.

Our Technique. Despite the approaches mentioned above give nice algorithms for classical clustering problems, one can face difficulties trying to extend these results to other similar problems. For example, random sampling for constructing candidate approximate centers works good for k-Means and Euclidean k-Median but seems to fail for partial clustering problems in which the total cardinality of the desired clusters may be arbitrarily small and, therefore, any constant number of random samples may "miss" good clusters. Also, it is not clear how to construct small coresets for variations of k-Means/Median in which the objective function has a more complicated dependence on the intra-cluster distances than the sum of these distances or of their squares.

Our technique is more universal. To construct a set of points which contains approximations of all the possible centers for a given input set, we do not use any information about the objective function. Unlike the known frameworks, we approximate optimal centers not by the objective function values but by the distances from the candidate centers to every input point. It allows to find an approximate minimum of any objective function which has a continuity-type dependence on these distances (Condition $\mathcal{C}$). As a result, we immediately get approximation algorithms for a wide range of geometric clustering problems in high-dimensional space (see Theorems 2–5), including approximation schemes in the case of a fixed number of clusters (see Corollaries 1 and 2).

One of features of our approach is its usability for partial clustering. For illustration, we propose an approximation algorithm for a problem which contains simultaneously m-Variance and both k-Means/Median problems with outliers. This problem consists of finding k clusters of a given total cardinality m in an n-element input set and selecting centers of these clusters in space $\mathbb{R}^d$ to minimize the weighted sum of the distances raised to different powers bounded by a fixed parameter $\alpha \geq 0$ between the elements of the clusters and their centers with different weight multipliers for each cluster. The proposed algorithm computes a $(1+\varepsilon)^\alpha$-approximate solution of the problem in time $\mathcal{O}\big((n/\varepsilon)^{\frac{2k}{\varepsilon}\log\frac{2}{\varepsilon}} nkd\big)$. So we have an approximation scheme PTAS in the case when k and α are constants. A similar result is obtained for the version of the problem in which all the desired clusters are required to have given cardinalities.

In fact, the proved statement describing a set of universal approximate centers (Theorem 1) gives a polynomial-time approximation-preserving reduction of general Euclidean center-based problems in high-dimensional space to their discrete versions, where all the centers must belong to a finite input set (see Theorem 6). In particular, we have a PTAS-reduction in the case of the k-Means/Median problems. It follows that the discrete version of k-Means is APX-hard and Euclidean k-Median can be approximated within a factor of $\beta+\varepsilon$, where β is the factor for its discrete version (currently, $\beta = 2.633$ [2]).

2 Constructing Universal Approximate Centers

Describe a polynomial-time algorithm which constructs a set of points which approximate all the points of space in terms of the distances to n given points.

Definition 1. *Given points $x, y \in \mathbb{R}^d$, a set $X \subseteq \mathbb{R}^d$, and a real number $\alpha \geq 1$, we say that y is an α-approximation of x with respect to X if $\|p-y\| \leq \alpha\|p-x\|$ for all $p \in X$, where $\|.\|$ denotes the values of the Euclidean norm.*

Definition 2. *Given sets $X, K \subseteq \mathbb{R}^d$ and a real number $\alpha \geq 1$, we say that K is an α-collection for X if K contains α-approximations of all the points of space with respect to X.*

Example. Any set $X \subseteq \mathbb{R}^d$ is a 2-collection for itself. Indeed, let x be an arbitrary point in $\mathbb{R}^d$ and y be a point of X nearest to x. Then, by the triangle inequality and the choice of y, we have $\|p-y\| \leq \|p-x\| + \|x-y\| \leq 2\|p-x\|$ for all $p \in X$, so y is a 2-approximation of x with respect to X.

Theorem 1. *For an arbitrary set X of n points in space $\mathbb{R}^d$ and for any $\varepsilon \in (0, 2]$, there exists a $(1+\varepsilon)$-collection for the set X which consists of $N(n, \varepsilon) = \mathcal{O}\big(n^{\frac{1}{\varepsilon}\log\frac{2}{\varepsilon}+1}(\frac{1}{\varepsilon})^{\frac{2}{\varepsilon}\log\frac{2}{\varepsilon}}\big)$ elements, where $\log$ is the logarithm to the base 2, and which can be constructed in time $\mathcal{O}\big(N(n,\varepsilon)\, d\big)$.*

Proof. As mentioned above, the set X is a 2-collection for itself. Therefore, if $\varepsilon \geq 1$, then a required $(1+\varepsilon)$-collection coincides with the input set and consists of n elements. Further, we will assume that $\varepsilon \in (0, 1)$.

First, make some geometric constructions and describe their properties underlying the future algorithm for computing a required collection. For any vectors $x, y \in \mathbb{R}^d$, denote by $x \cdot y$ their dot product and by $[x, y]$, the line segment between them. Given points $x_1, \ldots, x_t \in \mathbb{R}^d$, denote by $A(x_1, \ldots, x_t)$ the affine hull of the set $\{x_1, \ldots, x_t\}$. For any affine space A in $\mathbb{R}^d$ and a point $O \in \mathbb{R}^d$, denote by $Proj_A(O)$ the orthogonal projection of O into A.

Suppose that $\delta \in (0, \varepsilon)$ and O is an arbitrary point in $\mathbb{R}^d$. Construct the following sequences $(x_t)_{t \geq 1}$ and $(y_t)_{t \geq 1}$ depending on O and δ:

define $x_1 = y_1$ as a point of X nearest to O;

for $t \geq 2$, consider the ball B_t consisting of the points $x \in \mathbb{R}^d$ such that $\|x - y_{t-1}\| \geq (1 + \delta)\|x - O\|$;

if the set $X \cap B_t$ is empty, finish the sequences (x_t) and (y_t); otherwise, define x_t as any point from $X \cap B_t$ and put $y_t = Proj_{A(x_1, \ldots, x_t)}(O)$.

Denote by T the length of the constructed sequences (x_t) and (y_t).

Lemma 1. *If $2 \leq t \leq T$, then the vector $y_t - y_{t-1}$ is orthogonal to the affine space $A(x_1, \ldots, x_{t-1})$.*

Proof. Suppose that a and b are any points in the affine space $A(x_1, \ldots, x_{t-1})$. Since the point y_{t-1} is the orthogonal projection of O into this space, we have $(O - y_{t-1}) \cdot (a - b) = 0$. On the other hand, the point y_t is the orthogonal projection of O into the affine space $A(x_1, \ldots, x_t)$, which contains the affine space $A(x_1, \ldots, x_{t-1})$. It follows that $(O - y_t) \cdot (a - b) = 0$, so $(y_t - y_{t-1}) \cdot (a - b) = 0$. Lemma 1 is proved. □

Lemma 2. *If $2 \leq t \leq T$, then the vectors $x_2 - x_1, \ldots, x_t - x_1$ are linearly independent, $y_t \neq y_{t-1}$, and the vectors $e_{t-1} = \dfrac{y_t - y_{t-1}}{\|y_t - y_{t-1}\|}$ can be computed by the Gram-Schmidt process for orthonormalising the set $x_2 - x_1, \ldots, x_t - x_1$.*

Proof. Since $x_t \in B_t$, we have $(x_t - y_{t-1}) \cdot (O - y_{t-1}) > 0$. At the same time, the points y_{t-1} and x_1 belong to the affine space $A(x_1, \ldots, x_{t-1})$ and the point y_{t-1} is the orthogonal projection of O into this space, so $(y_{t-1} - x_1) \cdot (O - y_{t-1}) = 0$, which follows that $(x_t - x_1) \cdot (O - y_{t-1}) > 0$. On the other hand, the points x_t and x_1 belong to the affine space $A(x_1, \ldots, x_t)$ and the point y_t is the orthogonal projection of O into this space, so $(x_t - x_1) \cdot (O - y_t) = 0$, which implies the inequality $(x_t - x_1) \cdot (y_t - y_{t-1}) > 0$.

Then $y_t \neq y_{t-1}$ and the vector $x_t - x_1$ can not be a linear combination of the vectors $x_2 - x_1, \ldots, x_{t-1} - x_1$ since, by Lemma 1, the vector $y_t - y_{t-1}$ is orthogonal to the affine space $A(x_1, \ldots, x_{t-1})$. These observations hold for all $t \geq 2$, therefore, the vectors $x_2 - x_1, \ldots, x_t - x_1$ are linearly independent.

By Lemma 1 and by the above, the vector e_{t-1} is orthogonal to the vectors e_i, $i < t - 1$, and has a positive dot product with the vector $x_t - x_1$. But there is the only unit vector with such properties in the linear span of the vectors $x_2 - x_1, \ldots, x_t - x_1$, so e_{t-1} can be obtained by the normalization of the vector

$$x_t - x_1 - \sum \{e_i\left((x_t - x_1) \cdot e_i\right) \,|\, i < t - 1\}.$$

The latter holds for all $t \geq 2$, therefore, the vectors $e_1, \ldots, e_{t-1}$ can be computed by using the Gram-Schmidt process for orthonormalising the set of vectors $x_2 - x_1, \ldots, x_t - x_1$. Lemma 2 is proved. □

Lemma 3. *If* $2 \leq t \leq T$, *then* $\|y_t - y_{t-1}\| \leq \|y_{t-1} - O\| \leq \|y_1 - O\|$.

Proof. The right inequality follows from the choice of the point y_{t-1} and the fact that $y_1 \in A(x_1, \ldots, x_{t-1})$. To prove the left one, observe that the points y_{t-1} and y_t belong to the affine space $A(x_1, \ldots, x_t)$, so the vector $y_{t-1} - y_t$ is orthogonal to the vector $y_t - O$ by the definition of orthogonal projection. It follows that the line segment $[y_{t-1}, O]$ is the hypotenuse of the right triangle $\Delta y_{t-1} y_t O$, therefore, $\|y_{t-1} - y_t\| \leq \|y_{t-1} - O\|$. Lemma 3 is proved. □

Proposition 1. *If* $1 \leq t \leq T$, *then the point* y_t *belongs to the cube* $Cube_t(r_1)$, *where* $Cube_t(r) = \Big\{y_1 + \sum_{i=1}^{t-1} \alpha_i e_i \,\big|\, \alpha_1, \ldots, \alpha_{t-1} \in [0, r]\Big\}$ *and* $r_1 = \|x_1 - O\|$.

Proof. This statement directly follows from Lemmas 1 and 3. □

Lemma 4. *If* $T \geq 2$ *and* $dist_1 = \|x_2 - x_1\|/(1 + \delta)$, *then* $r_1 \in [dist_1 \delta, dist_1]$.

Proof. By the choice of the points x_1 and x_2, we have $r_1 \leq \|x_2 - O\|$ and $\|x_2 - x_1\| \geq (1 + \delta)\|x_2 - O\|$. Therefore, $r_1 \leq dist_1$.

Next, denote by Far the point of the ball B_2 farthest from x_1. Then we have $r_1 + \|Far - O\| = (1 + \delta)\|Far - O\|$, so $\|Far - O\| = r_1/\delta$. Since $x_2 \in B_2$, it follows that $\|x_2 - x_1\| \leq r_1 + \|Far - O\| = r_1(1 + \delta)/\delta$, so $r_1 \geq dist_1 \delta$. Lemma 4 is proved. □

Lemma 5. *The point* y_T *is a* $(1 + \delta)$*-approximation of the point* O *with respect to* X.

Proof. Since the set $X \cap B_{T+1}$ is empty, we have $\|p - y_T\| < (1 + \delta)\|p - O\|$ for all $p \in X$. Lemma 5 is proved. □

Lemma 6. *If* $2 \leq t \leq T$ *and* $r_t = \|y_t - O\|$, *then* $r_t \leq \left(\frac{1}{1+\delta}\right)^{t-1} r_1$.

Proof. First, state some properties of the ball B_t. Denote by $Near$ and Far the points of this ball nearest and farthest from y_{t-1} respectively. Then we have $r_{t-1} - \|Near - O\| = (1+\delta)\|Near - O\|$ and $r_{t-1} + \|Far - O\| = (1+\delta)\|Far - O\|$. So $\|Near - O\| = r_{t-1}/(2 + \delta)$ and $\|Far - O\| = r_{t-1}/\delta$. Therefore, the radius R of the ball B_t equals to the value of $\dfrac{r_{t-1}(1 + \delta)}{\delta(2 + \delta)}$.

It is easy to see that the distance from O to y_t is at most the distance from O to the line segment $[y_{t-1}, x_t]$. But the latter is at most the value of $r_{t-1} \sin \beta$, where β is the angle between the line segment $[y_{t-1}, O]$ and the tangent to the ball B_t dropped from the point y_{t-1}. On the other hand, $\sin \beta = R/(R + near)$, where $near = \|y_{t-1} - Near\| = r_{t-1} - \|Near - O\| = r_{t-1} - r_{t-1}/(2 + \delta) = \delta R$. It follows that $\sin \beta = \dfrac{1}{1 + \delta}$, so $\|y_t - O\| \leq \dfrac{r_{t-1}}{1 + \delta}$. By induction, we obtain that $r_t \leq \left(\frac{1}{1+\delta}\right)^{t-1} r_1$. Lemma 6 is proved. □

Proposition 2. *Let* $t = \min\{T, T(\delta)\}$*, where* $T(\delta) = \left\lceil \dfrac{\log \frac{1}{\delta}}{\log(1+\delta)} \right\rceil + 1$*. Then the point* y_t *is a* $(1+\delta)$*-approximation of the point* O *with respect to* X*.*

Proof. If $t = T$, the statement follows from Lemma 5. Suppose that $t = T(\delta)$.

By the triangle inequality and the choice of the point x_1, the point y_t is a $\left(1 + \dfrac{r_t}{r_1}\right)$-approximation of the point O with respect to X. On the other hand, by Lemma 6, we have $\dfrac{r_t}{r_1} \le \left(\dfrac{1}{1+\delta}\right)^{t-1}$. But $(1+\delta)^{T(\delta)-1} \ge \dfrac{1}{\delta}$ by the definition of $T(\delta)$. Therefore, $\dfrac{r_t}{r_1} \le \delta$. Proposition 2 is proved. □

Idea of the output collection construction. Propositions 1, 2 and Lemmas 2, 4 give an idea how to approximate any unknown point $O \in \mathbb{R}^d$. We can enumerate all the tuples $x_1, \ldots, x_t \in X$ for all $t \le T(\delta)$ and, for each of these tuples, construct the vectors $e_1, \ldots, e_{t-1}$ defined in Lemma 2. Next, we approximate the point y_t, which approximates O. For this aim, we construct a set R_1 containing approximations of the unknown value of r_1 by using the bounds established in Lemma 4. Finally, to approximate the point y_t, it remains to consider grids on the cubes $Cube_t(r)$, $r \in R_1$. Note that no information about the point O is used, so the constructed set will contain approximations of all the points of space.

Algorithm $\mathcal{A}$.
Select a real parameter $\delta \in (0, \varepsilon)$ and integer parameters $I, J \ge 1$.
`Step 1.` Include to the output set all the elements of X.
`Step 2.` Enumerate all the tuples $x_1, \ldots, x_t \in X$, $2 \le t \le T(\delta)$, such that the vectors $x_2 - x_1, \ldots, x_t - x_1$ are linearly independent and, for each of these tuples, perform Steps 3–5.
`Step 3.` Execute the Gram-Schmidt process for orthonormalising the set of vectors $x_2 - x_1, \ldots, x_t - x_1$ and obtain the orthonormal vectors $e_1, \ldots, e_{t-1}$.
`Step 4.` Construct the set R_1 of the numbers $dist_1 \delta^{i/I}$, $i = 0, \ldots, I$, where $dist_1 = \|x_2 - x_1\|/(1+\delta)$.
`Step 5.` For each value $r \in R_1$, include to the output set the nodes of the grid

$$Grid(x_1, \ldots, x_t; r, J) = \left\{ x_1 + \sum_{i=1}^{t-1} \frac{e_i r \alpha(i)}{J} \;\middle|\; \alpha \in \{0, \ldots, J\}^{t-1} \right\}.$$

Proposition 3. *The output set of Algorithm* $\mathcal{A}$ *is a* $(1+\delta+\delta_+)$*-collection for* X*, where* $\delta_+ = \dfrac{\sqrt{T(\delta)-1}}{2J\delta^{1/I}}$*.*

Proof. Suppose that O is an arbitrary point in space $\mathbb{R}^d$ and consider the sequences $(x_t)_{t\ge1}$ and $(y_t)_{t\ge1}$ defined for this point. By Proposition 2, there exists a number $t \le T(\delta)$ such that the point y_t is a $(1+\delta)$-approximation of the point O with respect to X. If $t = 1$, then Algorithm $\mathcal{A}$ outputs a required approximation at Step 1. Suppose that $t \ge 2$.

In this case, by Lemma 2, the vectors $x_2 - x_1, \ldots, x_t - x_1$ are linearly independent, so the tuple $x_1, \ldots, x_t$ is listed at Step 2. Lemma 4 implies that there exists a number $r \in R_1$ such that $r \geq r_1 \geq r\delta^{1/I}$, where $r_1 = \|x_1 - O\|$. Let z be a node of $Grid(x_1, \ldots, x_t; r, J)$ nearest to y_t. By Proposition 1, the point y_t belongs to the cube $Cube_t(r_1) \subseteq Cube_t(r)$, so the distance between z and y_t is at most $\frac{r\sqrt{t-1}}{2J}$. Then $\|p-z\| \leq \|p-y_t\| + \|y_t - z\| \leq (1+\delta)\|p-O\| + \frac{r\sqrt{t-1}}{2J}$ for any $p \in X$. But $r \leq r_1/\delta^{1/I} \leq \|p-O\|/\delta^{1/I}$ by the choice of x_1. Therefore, we have $\|p-z\| \leq \left(1+\delta+\frac{\sqrt{t-1}}{2J\delta^{1/I}}\right)\|p-O\|$. Proposition 3 is proved. □

Proposition 4. *The output set of Algorithm $\mathcal{A}$ consists of $N(n,\delta,I,J)$ points, where $N(n,\delta,I,J) = \mathcal{O}\big(n^{T(\delta)}(J+1)^{T(\delta)-1}I\big)$. The running time of the algorithm is $\mathcal{O}\big(N(n,\delta,I,J)\,d\big)$.*

Proof. The number of all the tuples $x_1, \ldots, x_t \in X$, where $t \leq T(\delta)$, is $\mathcal{O}(n^{T(\delta)})$. The set R_1 consists of $I+1$ elements. Each grid $Grid(x_1, \ldots, x_t; r, J)$ contains $(J+1)^{t-1}$ nodes. Thus, the total number of such nodes is $N(n,\delta,I,J)$.

Estimate the running time of Algorithm $\mathcal{A}$. For each tuple $x_1, \ldots, x_t \in X$, the vectors $e_1, \ldots, e_{t-1}$ can be constructed in time $\mathcal{O}(t^2 d)$. Vector operations in space $\mathbb{R}^d$ take time $\mathcal{O}(d)$. So the algorithm runs in time

$$\mathcal{O}\big(n^{T(\delta)}(T(\delta)^2 d + (J+1)^{T(\delta)-1} I d)\big) = \mathcal{O}\big(N(n,\delta,I,J)\,d\big).$$

Proposition 4 is proved. □

To prove Theorem 1, it remains to choose appropriate values of the parameters δ, I, J. Let $\delta = 0.9\,\varepsilon$, $J = \left\lceil \frac{\sqrt{T(\delta)-1}}{0.2\,\varepsilon\,\delta^{1/I}} \right\rceil$, and $I = 100$. Then $\delta_+ \leq 0.1\,\varepsilon$ and, by Proposition 3, Algorithm $\mathcal{A}$ outputs a $(1+\varepsilon)$-collection for X.

Estimate the cardinality of the constructed collection and the running time of the algorithm by using Proposition 4. Since $\delta = 0.9\,\varepsilon$, we have

$$T(\delta) = \zeta + 1, where \zeta = \left\lceil \frac{\log \frac{1}{0.9\,\varepsilon}}{\log(1+0.9\,\varepsilon)} \right\rceil,$$

and

$$(J+1)^{T(\delta)-1} = (J+1)^{\zeta} = \left(\left\lceil \frac{\sqrt{\zeta}}{0.2\,\varepsilon\,(0.9\,\varepsilon)^{0.01}} \right\rceil + 1\right)^{\zeta}.$$

The values of ζ and $(J+1)^{\zeta}$ can be estimated as follows. Consider the functions $a(\varepsilon) = \zeta\,\varepsilon/\log\frac{2}{\varepsilon}$ and $b(\varepsilon) = (J+1)^{\zeta}\,\varepsilon^{\frac{2}{\varepsilon}\log\frac{2}{\varepsilon}}$. By using asymptotic properties of these functions for small ε, estimates of their derivatives on the segments where ζ and $(J+1)^{\zeta}$ are constants, and computer calculations of the values of $a(.)$ and $b(.)$ on a grid with sufficiently small step, we obtain that $a(\varepsilon) \leq 1$ and $b(\varepsilon) < 44$ for all $\varepsilon \in (0,1)$. It follows that $\zeta \leq \frac{1}{\varepsilon}\log\frac{2}{\varepsilon}$ and $(J+1)^{\zeta} < 44\left(\frac{1}{\varepsilon}\right)^{\frac{2}{\varepsilon}\log\frac{2}{\varepsilon}}$. Thus, by Proposition 4, the number of points in the output of Algorithm $\mathcal{A}$ is

$$N(n,\varepsilon) = \mathcal{O}\left(n^{\frac{1}{\varepsilon}\log\frac{2}{\varepsilon}+1}\left(\tfrac{1}{\varepsilon}\right)^{\frac{2}{\varepsilon}\log\frac{2}{\varepsilon}}\right) = \mathcal{O}\left((n/\varepsilon)^{\frac{2}{\varepsilon}\log\frac{2}{\varepsilon}}\right)$$

and the algorithm runs in time $\mathcal{O}\big(N(n,\varepsilon)\,d\big)$. Theorem 1 is proved. □

3 Applications to Center-Based Clustering

Theorem 1 gives a way for constructing approximation algorithms for a lot of center-based clustering problems which can be reduced to finding centers of optimal clusters.

Let f be an arbitrary non-negative function defined for each finite set $X \subset \mathbb{R}^d$ and each tuple $c_1, \ldots, c_k \in \mathbb{R}^d$. Suppose that this function satisfies the following condition which expresses its dependence on the Euclidean distances from the points of X to the points $c_1, \ldots, c_k$:

Condition $\mathcal{C}$. There exists a function $\mu : [1, \infty) \to [1, \infty)$ such that, if $c_i, c_i' \in \mathbb{R}^d$, $\varepsilon > 0$, and $\|x - c_i'\| \le (1 + \varepsilon)\|x - c_i\|$ for all $x \in X$ and $i = 1, \ldots, k$, then

$$f(X; c_1', \ldots, c_k') \le \mu(1 + \varepsilon) f(X; c_1, \ldots, c_k).$$

Condition $\mathcal{C}$ means that, if we change the tuple $c_1, \ldots, c_k$ so that all the $n \times k$ distances from the points of X to the elements of the tuple increase at most in $1+\varepsilon$ times, then the value of the function f increases at most in $\mu(1+\varepsilon)$ times. It seems to be a very natural property for clustering. Examples of objective functions satisfying Condition $\mathcal{C}$ are those of many well-known geometric clustering problems such as k-Means/Median/Center, m-Variance, and Smallest m-Enclosing Ball. In the case of the k-Means and m-Variance, this condition holds with $\mu(1 + \varepsilon) = (1 + \varepsilon)^2$. In the case of k-Median, k-Center, and Smallest m-Enclosing Ball, we have $\mu(1 + \varepsilon) = (1 + \varepsilon)$.

3.1 k-Clustering

Consider a general geometric center-based k-clustering problem:

k-Clustering. Given a set X of n points in space $\mathbb{R}^d$ and an integer $k \ge 1$, find a tuple $c_1, \ldots, c_k \in \mathbb{R}^d$ to minimize the value of $f(X; c_1, \ldots, c_k)$.

Theorem 2. *For any non-negative function f satisfying Condition $\mathcal{C}$, there exists an algorithm which, given $\varepsilon \in (0, 2]$, computes a $\mu(1 + \varepsilon)$-approximate solution of k-Clustering with the function f in time $\mathcal{O}\big(N(n, \varepsilon)^k (T_f + d)\big)$, where T_f is the time complexity of calculating f.*

Proof. Suppose that a tuple $c_1^*, \ldots, c_k^*$ is an optimal solution of the problem. By using Algorithm $\mathcal{A}$, construct a $(1+\varepsilon)$-collection K for X. Then one of the tuples $c_1, \ldots, c_k \in K$ satisfies the property $\|x - c_i\| \le (1+\varepsilon)\|x - c_i^*\|$ for all $x \in X$ and $i = 1, \ldots, k$. By Condition $\mathcal{C}$, we have $f(X; c_1, \ldots, c_k) \le \mu(1+\varepsilon) f(X; c_1^*, \ldots, c_k^*)$, so the tuple $c_1, \ldots, c_k$ is a $\mu(1 + \varepsilon)$-approximate solution of the problem. It remains to note that, by Theorem 1, finding the set K, enumerating all the tuples $c_1, \ldots, c_k \in K$, and calculating the values of $f(X; c_1, \ldots, c_k)$ for all these tuples take time $\mathcal{O}\big(N(n, \varepsilon)^k (T_f + d)\big)$. Theorem 2 is proved. □

Corollary 1. *Suppose that the value of k is fixed, the function f is computable in polynomial time and satisfies Condition $\mathcal{C}$, where $\mu(1 + \varepsilon) \to 1$ as $\varepsilon \to 0$. Then k-Clustering with the function f admits an approximation scheme PTAS.*

3.2 Partial Clustering

It is easy to extend the obtained results to partial clustering. Consider a general geometric center-based partial clustering problem:

Partial Clustering. Given a set X of n points in space $\mathbb{R}^d$ and integers $k, m \geq 1$, find an m-element subset $S \subseteq X$ and select a tuple $c_1, \dots, c_k \in \mathbb{R}^d$ to minimize the value of $f(S; c_1, \dots, c_k)$.

Theorem 3. *Let f be any non-negative function satisfying Condition $\mathcal{C}$ and let F be an oracle which, given points $c_1, \dots, c_k \in \mathbb{R}^d$, returns an m-element subset of X with the minimum value of $f(\,.\,; c_1, \dots, c_k)$. Then there exists an algorithm which, given $\varepsilon \in (0, 2]$, computes a $\mu(1+\varepsilon)$-approximate solution of Partial Clustering with the function f in time $\mathcal{O}\big(N(n,\varepsilon)^k(T_F + d)\big)$, where T_F is the time complexity of the oracle F.*

Proof. Suppose that a set S^* and a tuple $c_1^*, \dots, c_k^*$ are an optimal solution of the problem. By using Algorithm $\mathcal{A}$, construct a $(1+\varepsilon)$-collection K for X. Then one of the tuples $c_1, \dots, c_k \in K$ satisfies the property $\|x - c_i\| \leq (1+\varepsilon)\|x - c_i^*\|$ for all $x \in X$ and $i = 1, \dots, k$. By Condition $\mathcal{C}$, it follows the inequality $f(S^*; c_1, \dots, c_k) \leq \mu(1+\varepsilon) f(S^*; c_1^*, \dots, c_k^*)$. But, by the choice of the oracle F, we have $f(S; c_1, \dots, c_k) \leq f(S^*; c_1, \dots, c_k)$, where $S = F(c_1, \dots, c_k)$. So the set S and the tuple $c_1, \dots, c_k$ form a $\mu(1+\varepsilon)$-approximate solution of the problem. It remains to note that, by Theorem 1, finding the set K, enumerating all the tuples $c_1, \dots, c_k \in K$, and executing the oracle F for all these tuples take time $\mathcal{O}\big(N(n,\varepsilon)^k(T_F + d)\big)$. Theorem 3 is proved. □

Corollary 2. *Suppose that the value of k is fixed, the function f satisfies Condition $\mathcal{C}$, where $\mu(1+\varepsilon) \to 1$ as $\varepsilon \to 0$, and the oracle F can be implemented in polynomial time. Then Partial Clustering with the function f admits an approximation scheme PTAS.*

3.3 Examples

To illustrate the obtained results, consider the following partial clustering problem. Let α be an arbitrary non-negative real parameter.

Problem 1. Given a set X of n points in space $\mathbb{R}^d$, integers $k, m \geq 1$, and real numbers $\alpha_i(x) \in [0, \alpha]$ and $w_i(x) \geq 0$, $i = 1, \dots, k$, $x \in X$, find disjoint subsets $S_1, \dots, S_k \subseteq X$ with the property $|S_1 \cup \dots \cup S_k| = m$ and select a tuple $c_1, \dots, c_k \in \mathbb{R}^d$ to minimize the value of

$$cost(S_1, \dots, S_k;\, c_1, \dots, c_k) = \sum_{i=1}^{k} \sum_{x \in S_i} w_i(x)\|x - c_i\|^{\alpha_i(x)}.$$

It is easy to see that Problem 1 contains m-Variance and both k-Means/Median problems with outliers. It may be actual for the case of non-uniform centers, e.g., when the cost of service of an input point x by a center c_i depends not only on the distance between them but also on some parameters specific for the pair (x, i), which is expressed in the values of $w_i(x)$ and $\alpha_i(x)$.

Theorem 4. *There exists an algorithm which, given* $\varepsilon \in (0,2]$, *computes a* $(1+\varepsilon)^\alpha$*-approximate solution of Problem* 1 *in time* $\mathcal{O}\big(N(n,\varepsilon)^k nkd\big)$.

Proof. Note that Problem 1 can be reduced to finding a tuple $c_1,\ldots,c_k \in \mathbb{R}^d$ and an m-element subset $S \subseteq X$ minimizing the function

$$f(S;c_1,\ldots,c_k) = \sum_{x\in S} D(x;c_1,\ldots,c_k),$$

where $D(x;c_1,\ldots,c_k) = \min\limits_{i=1,\ldots,k} w_i(x)\|x-c_i\|^{\alpha_i(x)}$. Clearly, the function f satisfies Condition $\mathcal{C}$ with $\mu(1+\varepsilon) = (1+\varepsilon)^\alpha$.

Next, given points $c_1,\ldots,c_k \in \mathbb{R}^d$, we can easily construct an m-element subset of X with the minimum value of $f(\,.\,;c_1,\ldots,c_k)$ by using the greedy algorithm which selects m points of X with the minimum values of $D(\,.\,;c_1,\ldots,c_k)$. Since choosing m minimum numbers in an n-element set takes linear time (e.g., see [3]), the oracle F mentioned in Theorem 3 runs in time $\mathcal{O}(nkd)$. Therefore, by Theorem 3, we can get a $(1+\varepsilon)^\alpha$-approximate solution of the considered problem in time $\mathcal{O}\big(N(n,\varepsilon)^k nkd\big)$. Theorem 4 is proved. □

Another example of center-based problems which can be approximated with the proposed approach is the version of Problem 1 in which every desired cluster is required to have a given cardinality.

Problem 2. Given a set X of n points in space $\mathbb{R}^d$, integers $m_1,\ldots,m_k \geq 1$, and real numbers $\alpha_i(x) \in [0,\alpha]$ and $w_i(x) \geq 0$, $i=1,\ldots,k$, $x \in X$, find disjoint subsets $S_1,\ldots,S_k \subseteq X$ with the property $|S_i| = m_i$, $i=1,\ldots,k$, and select a tuple $c_1,\ldots,c_k \in \mathbb{R}^d$ to minimize the value of $cost(S_1,\ldots,S_k;\ c_1,\ldots,c_k)$.

Theorem 5. *There exists an algorithm which, given* $\varepsilon \in (0,2]$, *computes a* $(1+\varepsilon)^\alpha$*-approximate solution of Problem* 2 *in time* $\mathcal{O}\big(N(n,\varepsilon)^k(n^3+nkd)\big)$.

Proof. Given a tuple $c_1,\ldots,c_k \in \mathbb{R}^d$, consider the following weighted bipartite graph $G = (V_1,V_2;E)$. The part V_1 consists of the points of X. The part V_2 consists of the vertices c_i^j, $i=1,\ldots,k$, $j=1,\ldots,m_i$, and also the vertices b^j, $j=1,\ldots,n-m$, where $m = m_1+\cdots+m_k$, if $n > m$. For each $x \in X$ and indices i and j, the weight of the edge $(x,\,c_i^j)$ is defined as the value of $w_i(x)\|x-c_i\|^{\alpha_i(x)}$ and the weight of every edge $(x,\,b^j)$ is zero.

It is easy to see that Problem 2 is reduced to finding a tuple $c_1,\ldots,c_k$ minimizing the weight of a perfect matching in the graph G. Denote this minimum weight by $f(X;c_1,\ldots,c_k)$. Then, clearly, the function f satisfies Condition $\mathcal{C}$ with $\mu(1+\varepsilon) = (1+\varepsilon)^\alpha$. On the other hand, given centers $c_1,\ldots,c_k$, constructing the graph G and calculating the value of $f(X;c_1,\ldots,c_k)$ take time $\mathcal{O}(n^3 + nkd)$ by using the known algorithms for matchings (i.e., see [8]). Therefore, by Theorem 2, we can get a $(1+\varepsilon)^\alpha$-approximate solution of Problem 2 in time $\mathcal{O}\big(N(n,\varepsilon)^k(n^3+nkd)\big)$. Theorem 5 is proved. □

Corollary 3. *If the values of k and α are fixed, Problems* 1 *and* 2 *admit approximation schemes PTAS.*

It is easy to prove that Theorems 4–5 and Corollary 3 remain true if we replace each term $w_i(x)\|x-c_i\|^{\alpha_i(x)}$ in the function *cost* by a more general value of $f_{i,x}\big(\,\|x-c_i\|\,\big)$, where each $f_{i,x}$ is any non-negative function satisfying the inequality $f_{i,x}(pq) \le p^{\alpha} f_{i,x}(q)$ for all $p>1$, $q \ge 0$.

3.4 Reduction to Discrete Euclidean Problems

In fact, Theorem 1 gives an approximation-preserving reduction of the described Euclidean center-based problems in high-dimensional space to their discrete versions, in which the desired centers must belong to a given finite set.

Discrete Version of k-Clustering. Given an n-element set $X \subset \mathbb{R}^d$, an ℓ-element set $Y \subset \mathbb{R}^d$, and integer $k \ge 1$, find a tuple $c_1,\dots,c_k \in Y$ to minimize the value of $f(X;c_1,\dots,c_k)$.

Discrete Version of Partial Clustering. Given an n-element set $X \subset \mathbb{R}^d$, an ℓ-element set $Y \subset \mathbb{R}^d$, and integers $k,m \ge 1$, find an m-element subset $S \subseteq X$ and select a tuple $c_1,\dots,c_k \in Y$ to minimize the value of $f(S;c_1,\dots,c_k)$.

Theorem 6. *Suppose that $\beta \ge 1$ and there exists an algorithm which computes a β-approximate solution of Discrete version of k-Clustering with a non-negative function f satisfying Condition $\mathcal{C}$ in time $T(n,\ell,k,d)$. Then there exists an algorithm which, given $\varepsilon \in (0,2]$, computes a $\beta\mu(1+\varepsilon)$-approximate solution of k-Clustering with the function f in time $\mathcal{O}\big(N(n,\varepsilon)\,d\big)+T\big(n,N(n,\varepsilon),k,d\big)$. The same holds for Partial Clustering.*

Proof. We prove the statement for Partial Clustering: the proof for k-Clustering is similar. Let a set S^* and a tuple $c_1^*,\dots,c_k^*$ be an optimal solution of the Partial Clustering problem on some input set X. Construct a $(1+\varepsilon)$-collection K for this set and find a set S and a tuple $c_1,\dots,c_k$ which form a β-approximate solution of the Discrete Partial Clustering problem on the input sets X and $Y=K$. By the definition of a $(1+\varepsilon)$-collection and by Condition $\mathcal{C}$, there exists a tuple $c_1',\dots,c_k' \in K$ such that $f(S^*,c_1',\dots,c_k') \le \mu(1+\varepsilon)f(S^*;c_1^*,\dots,c_k^*)$. So we have $f(S;c_1,\dots,c_k) \le \beta f(S^*;c_1',\dots,c_k') \le \beta\mu(1+\varepsilon)f(S^*;c_1^*,\dots,c_k^*)$.

It remains to note that, by Theorem 1, the set K can be constructed in time $\mathcal{O}\big(N(n,\varepsilon)\,d\big)$ and, since K consists of $N(n,\varepsilon)$ points, a β-approximate solution of the Discrete Partial Clustering problem on this set can be found in time $T\big(n,N(n,\varepsilon),k,d\big)$. Theorem 6 is proved. □

In particular, we have a PTAS-reduction of Euclidean k-Median and k-Means to their discrete versions. It implies the following approximability results. Let β be the approximation ratio of polynomial-time algorithms for Discrete version of Euclidean k-Median (currently, $\beta = 2.633$ [2]) and γ be the inapproximability bound for k-Means (currently, $\gamma = 1.0013$ [14]).

Corollary 4. *For any $\varepsilon > 0$, Euclidean k-Median can be approximated in polynomial time within a factor of $\beta+\varepsilon$.*

Corollary 5. *For any $\varepsilon > 0$, Discrete version of k-Means is NP-hard to approximate within a factor of $\gamma-\varepsilon$.*

4 Conclusion

We propose an algorithm constructing a polynomial-cardinality set of points which approximate all the points of space in terms of the distances to n given points. As a corollary, we obtain approximation algorithms for general clustering problems with arbitrary objective functions depending on the intra-cluster distances and satisfying a natural continuity-type condition.

Acknowledgments. The work was supported by the program of fundamental scientific researches of the SB RAS, project 0314-2019-0014.

References

1. Agarwal, P.K., Har-Peled, S., Varadarajan, K.R.: Geometric approximation via coresets combinatorial and computational geometry. MSRI **52**, 1–30 (2005)
2. Ahmadian S., Norouzi-Fard A., Svensson O., Ward J.: Better guarantees for k-means and Euclidean k-median by primal-dual algorithms. In: Proceedings of 58th Symposium on Foundations of Computer Science (FOCS 2017), pp. 61–72 (2017)
3. Aho, A.V., Hopcroft, J.E., Ullman, J.D.: The Design and Analysis of Computer Algorithms. Addison-Wesley, Boston (1974)
4. Aloise, D., Deshpande, A., Hansen, P., Popat, P.: NP-hardness of Euclidean sum-of-squares clustering. Mach. Learn. **75**(2), 245–248 (2009)
5. Bădoiu, M., Har-Peled, S., Indyk, P.: Approximate clustering via core-sets. In: Proceedings of 34th ACM Symposium on Theory of Computing (STOC 2002), pp. 250–257 (2002)
6. Bhattacharya, A., Jaiswal, R., Kumar, A.: Faster algorithms for the constrained k-means problem. Theory Comput. Syst. **62**(1), 93–115 (2018)
7. Chen, K.: On coresets for k-median and k-means clustering in metric and euclidean spaces and their applications. SIAM J. Comput. **39**(1), 923–947 (2009)
8. Cook, W., Rohe, A.: Computing minimum-weight perfect matchings. INFORMS J. Comput. **11**(2), 138–148 (1999)
9. Feldman, D., Monemizadeh, M., Sohler, C.: A PTAS for k-means clustering based on weak coresets. In: Proceedings of 23rd ACM Symposium on Computational Geometry, pp. 11–18 (2007)
10. Guruswami, V., Indyk, P.: Embeddings and non-approximability of geometric problems. In: Proceedings of 14th ACM-SIAM Symposium on Discrete Algorithms (SODA 2003), pp. 537–538 (2003)
11. Jaiswal, R., Kumar, A., Sen, S.: A simple D^2-sampling based PTAS for k-means and other clustering problems. Algorithmica **70**(1), 22–46 (2014)
12. Kel'manov, A.V., Pyatkin, A.V.: NP-completeness of some problems of choosing a vector subset. J. Appl. Industr. Math. **5**(3), 352–357 (2011)
13. Kumar, A., Sabharwal, Y., Sen, S.: Linear-time approximation schemes for clustering problems in any dimensions. J. ACM. **57**(2), 1–32 (2010)
14. Lee, E., Schmidt, M., Wright, J.: Improved and simplified inapproximability for k-means. Inf. Proc. Lett. **120**, 40–43 (2017)
15. Shenmaier, V.V.: An approximation scheme for a problem of search for a vector subset. J. Appl. Industr. Math. **6**(3), 381–386 (2012)
16. Shenmaier, V.V.: The problem of a minimal ball enclosing k points. J. Appl. Industr. Math. **7**(3), 444–448 (2013)
17. Shenmaier, V.V.: Complexity and approximation of the smallest k-enclosing ball problem. Eur. J. Comb. **48**, 81–87 (2015)

Modification of the k-MXT Algorithm and Its Application to the Geotagged Data Clustering

Anastasia Stepanova, Sergei V. Mironov, Sergei Sidorov(✉), and Alexey Faizliev

Saratov State University, Saratov 410012, Russian Federation
sidorovsp@info.sgu.ru

Abstract. The paper considers the problem of detection of the most attractive city sights using datasets with geotagged photographs. We form a graph on the basis of the geotagged spot coordinates and rewrite the problem as the problem of graph clustering. In this paper, we propose a modification of the k-MXT algorithm, which we called the k-MXT-W algorithm and which uses window functions. We compare the proposed algorithm with k-Means and k-MXT algorithms on simulated data using ARI, one of the most common metrics for assessing clustering quality. In this paper we also use the k-MXT-W algorithm to find the most popular places in St. Petersburg (Russia) and we compare the performance of the proposed algorithm with the k-MXT algorithm on real-world data using the modularity metric that does not require knowledge of true clustering.

Keywords: Graph clustering algorithms · Weighted graph · Geotagged data

1 Introduction

The k-MXT algorithm was proposed in [5] for solving the problem of clustering geotagged data and was applied to find the most tourist attractive city sights using the database with geotagged data of photographs. Currently, there is a large amount of geotagged data on the World Wide Web and their correct processing can be useful in many applications. Flickr (https://www.flickr.com/) contains a huge number of geotagged photographs and is one of such examples of dataset with geotagged data. Social networks also contain information with geotagged data which can be used in some applications. Clustering algorithms on geotagged records from Twitter is used in the paper [13] to make an assumption on geolocation of the tweets that do not contain geotagging.

When clustering geotagged data, it is necessary to divide the points lying on the two-dimensional surface into some groups (clusters). Currently, there

This work was supported by the Russian Science Foundation, project 19-18-00199.

G. Nicosia et al. (Eds.): LOD 2019, LNCS 11943, pp. 296–307, 2019.
https://doi.org/10.1007/978-3-030-37599-7_25

are many algorithms for clustering geotagged data. Each of them has its own peculiarities and drawbacks in different types of data sets. For example, the algorithms K-Means [2,9,10,21] and Mini Batch K-Means [1,19], BIRCH [24] perform poorly on data in which clusters have the form of a tape, i.e. data in which for any point at close distance there is another point belonging to the same cluster, and there are also points that are not located at close distance but belong to the same cluster. Examples of tape-shaped data are moon- and circle-shaped data sets, which we will use for a comparative analysis of clustering algorithms in this paper. On the contrary, the DBSCAN algorithm [7] works well on data with tape-shaped clusters, and at the same time, the DBSCAN algorithm performs incompetently on data with clusters that have a large difference in density.

In general, two types of metrics are used to evaluate the quality of clustering algorithms. The metrics of the first type imply the knowledge of true clusters. The examples of such clustering performance evaluation metrics are ARI metrics (Adjusted Rand Index) [8,16,23] and NMI (Normalized Mutual Information) [23]. In many cases these metrics work well regardless of the data type and are used for the evaluation of clustering algorithms. The second type is the metrics for assessing the quality of clustering, which do not require the knowledge of true clustering. The examples of such estimates are the modularity metric [3,4,6,12,14,15], the conductance metric [6,20], the silhouette coefficient [18], Kappas [11]. Clustering quality assessment metrics that do not imply the knowledge of true clustering are mainly used to assess the correctness of clustering in practical problems. In most cases, these metrics are based on the statement that clusters are some areas of the greatest density of the graph. Therefore, these metrics work worse on tape-shaped data.

In this paper, we propose a new modification of the k-MXT algorithm which we called the k-MXT-W algorithm and which uses window functions. We compare the proposed algorithm with K-Means and k-MXT algorithms on simulated data using ARI. In this paper, we also use the k-MXT-W algorithm to find the most popular places in St. Petersburg (Russia) and we compare the performance of the proposed algorithm with the k-MXT algorithm on real-world data using the modularity metric, quality assessment metric that does not require the knowledge of true clustering.

In this research, all algorithms were implemented in Python using the Jupyter Notebook environment. The simulated data sets were obtained using the functions of *scikit-learn* (Python library). To find the most popular places of St. Petersburg (Russia), geotagged data were obtained from Flickr. The data sets obtained for the city of St. Petersburg and the algorithm implementation can be found via the link https://github.com/stacy-s/Clustering.

2 Clustering Algorithms

2.1 k-MXT Algorithm

Often it is not known how many clusters you need to divide the points from the data set. Unlike the well-known K-Means algorithm, the k-MXT algorithm itself determines this number. In the k-MXT algorithm, the number of clusters depends on the parameters ε and k, which we describe later.

To apply the k-MXT algorithm to data with geotags, you need to build an undirected weighted graph $G = (V, E)$ (V is the set of vertices, E is the set of edges of the graph), where vertices are points from a dataset. The edges of G are the connections between the vertices located at a distance less than ε. The weight of each edge (q, r) is equal to the number of common neighbors of the vertex q and the vertex r.

According to the graph G you need to build an oriented graph $G' = (V', E')$ (V' is the set of vertices, E' is the set of arcs of the graph G'). Initially, the graph G' contains all vertices of the graph G, but does not contain any arcs $V' = V$, $E' = \varnothing$.

Let the set $E(p)$ contain all edges that incident to the vertex p. Then we add arcs to the directed graph G' as follows:

- If $E(p)$ contains less than k edges or exactly k edges, then all edges from $E(p)$ are added to E'.
- If $E(p)$ contains more than k edges, then k edges from $E(p)$ with the greatest weight are added to E'. If it is necessary to choose from several edges with the same weight, the selection is carried out randomly.

The clusters are the strongly connected components of the resulting graph G'.

2.2 k-MXT-W Algorithm

A cluster is a group of similar (in some sense) objects. In many geotagged data clustering problems, the measure of similarity of two objects is determined by the distance between these objects. The k-MXT algorithm does not explicitly use the distance between two vertices when calculating the weights of edges. Note that these distances can be easily calculated based on the known geotags. We modify the k-MXT algorithm so that the distances between the vertices could be used. Moreover, this modified k-MXT algorithm (further called k-MXT-W algorithm) will transform these distances with the help of a function that increases the edge weight between vertices with close distance.

We use window functions to take into account the distance between two vertices v, u to find the edge weight (v, u). The window functions must satisfy the following conditions:

- it must be symmetric about zero, i.e. $f(x) = f(-x)$ for any x.
- the function has its maximum at the point $x = 0$.
- if $x < -\varepsilon$ or $x > \varepsilon$, then $f(x) \to 0$.

Thus, the k-MXT-W algorithm differs from the k-MXT algorithm in the way that edge weights are calculated. The edge weight (q, r) is equal to the number of common neighbors of vertex q and vertex r multiplied by the value of the window function at x (the distance between the vertices q and r).

In our research, we obtained the results of the k-MXT-W algorithm for different window functions (1)–(3):

$$f(x) = \max\left((\varepsilon - |x|), 0\right) \tag{1}$$

$$f(x) = \max\left(c\left(\varepsilon^3 - |x|^3\right)^3, 0\right) \tag{2}$$

$$f(x) = \begin{cases} 0, & \text{if } |x| > \varepsilon, \\ \frac{c}{\sigma}\exp^{-\frac{x^2}{2\sigma^2}}, & \text{otherwise,} \end{cases} \tag{3}$$

where c is some constant, $\sigma = \frac{\varepsilon}{3}$.

It turned out that k-MXT-W algorithm with different choice of functions (1)–(3) performs similarly in terms of the quality of clusterization. The use of the considered functions when calculating the link weights gives a more correct clustering than k-MXT algorithm. Depending on the choice of the function, the values of the parameters k and ε at which the maximum value of the metric of clustering quality is achieved, are different. Among these functions, the best result was performed with use of function (3). Therefore, in this paper we present only the results for the function (3) as a window function. According to the three sigma rule, to fully cover all points located at a distance ε from a given vertex, it is sufficient to take $\sigma = \varepsilon/3$. k-MXT algorithm modification based on function (3) was considered in the paper [22]. This paper presents an extended analysis of k-MXT-W algorithm.

3 Comparison of Algorithms on Simulated Data

3.1 Methodology

For numeric experiments data was simulated using *scikit-learn*, a free software machine learning library for the Python programming language (https://scikit-learn.org/). Blobs-, circles- and moons-type data were obtained using functions `sklearn.datasets.make_blobs`, `sklearn.datasets.make_circles` and `sklearn.datasets.make_moons` respectively with the following parameters:

- `n_samples`, the total number of generated points,
- `cluster_std`, the standard deviations of clusters,
- `random_state`, random number generator parameter,
- `factor`, scaling factor between inner and outer circle,
- `noise`, the Gaussian noise with standard deviation.

In all data types, the unspecified parameters of the functions were taken by default, the number of vertices `n_samples` was 200.

We use the ARI (Adjusted Rand Index) metric to evaluate the quality of clustering. This metric compares how much two vertices splittings into clusters are similar. In this case, the resulting clusterings will be compared with the true ones. Let $U = \{u_1, u_2 \dots u_r\}$ represent the original partition of a dataset, where u_i denote a subset of the objects associated with cluster i. Equivalently, let $V = \{v_1, v_2 \dots v_r\}$ represent the partition found by a cluster algorithm. Then $n_{i,j}$ is the number of objects belonging to both subset u_i and v_j, $n_i = \sum_j n_{i,j}$. ARI is defined as follows [8]:

$$ARI = \frac{\sum_{i,j}\binom{n_{i,j}}{2} - [\sum_i \binom{n_i}{2} \sum_j \binom{n_j}{2}]/\binom{n}{2}}{\frac{1}{2}[\sum_i \binom{n_i}{2} + \sum_j \binom{n_j}{2}] - [\sum_i \binom{n_i}{2} \sum_j \binom{n_j}{2}]/\binom{n}{2}}. \quad (4)$$

A comparison of well-known clustering methods with use of ARI metric was performed in the paper [17].

To compare the k-MXT-W and k-MXT algorithms we conducted numerical experiments, the results of which are presented later in this section. We randomly generated 1000 datasets for each of the three data types (blobs-, circles- and moons-types). All data sets are generated with different values of the `random_state` parameter.

For each randomly generated dataset, we calculated ARI values for both k-MXT-W and k-MXT algorithms (with optimal values of ε). Next we found the average ARI values (over all 1000 datasets) as well as their sample standard deviations, and then we conducted the Aspin–Welch t-test, i.e. the test for comparing two independent sample means with unknown standard deviations. Note that it does not make sense to consider the results of clustering for $k \leq 2$, since the vertices are divided into a large number of small clusters, and the ARI metric is close to zero.

3.2 The Blobs Type Data

With a fixed parameter k, the clusterization quality for both k-MXT-W and k-MXT algorithms depends on the parameter ε. In order for the comparison of two algorithms with fixed k to be correct, we should first select the value of ε, at which the quality of clustering would be optimal. It should be noted that optimal values of ε for k-MXT-W and k-MXT algorithms (with the same k) are different. As an example, we present the results of the algorithms for $k = 12$. Figure 1,*a* plots the dependence of the ARI metric on the values of the parameters ε (values were taken from 0.1 to 3.9 with a step 0.1) on the blobs data type with `cluster_std` = 0.5 and `random_state` = 0. In the following figures the same color indicates the vertices belonging to the same cluster.

Table 1 (the left part) presents average values of ARI metric and their sample standard deviations obtained by k-MXT-W and k-MXT algorithms with $4 \leq k \leq 12$. Note that the average value of ARI metric of k-MXT algorithm is close to that of k-MXT-W algorithm only when $k \leq 3$. As a result of clustering with parameter $k = 1$, we obtain a large number of clusters consisting of one or two points, which does not represent practical value.

Table 1 (the left part) shows that k-MXT-W algorithm has a greater average value of ARI metric than k-MXT algorithm for all $k \geq 4$. The Aspin–Welch t-test validates that the means values of ARI metric for k-MXT-W algorithm are statistically greater than or equal to the means values of ARI metric for k-MXT

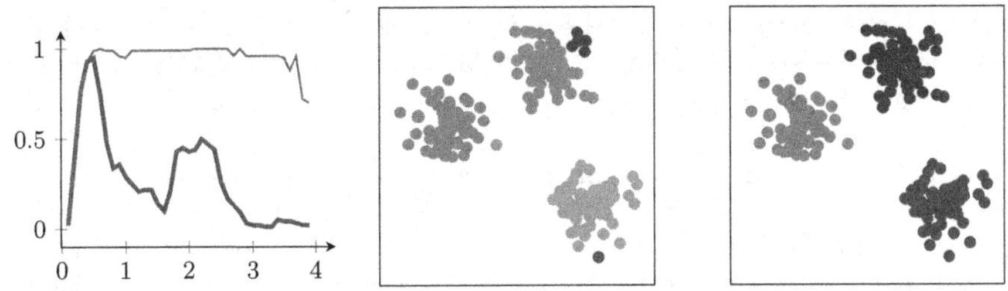

(a) (b) (c)

Fig. 1. The results of 12-MXT algorithm and 12-MXT-W algorithm on blobs data type with `cluster_std` = 0.5 and `random_state` = 0: (*a*) Dependence of ARI metric on the values of parameter ε for 12-MXT algorithm (blue line) and for 12-MXT-W algorithm (red line). The performance of the algorithms with the best value of ARI: (*b*) $ARI = 0.95$ for 12-MXT algorithm with $\varepsilon = 0.5$; (*c*) $ARI = 1.00$ for 12-MXT-W algorithm with $\varepsilon = 0.6$. (Color figure online)

Table 1. The results obtained by k-MXT and k-MXT-W algorithms for different k on blobs type datasets; the number of vertices is 200; μ denotes the average value of ARI and σ is the sample standard deviation.

k	Datasets with $cluster\_std = 0.5$						Datasets, $cluster\_std = [1.0, 1.5, 0.5]$					
	k-MXT			k-MXT-W			k-MXT			k-MXT-W		
	ε	μ	σ	ε	μ	σ	ε	μ	σ	ε	μ	σ
4	2.0	0.45	0.12	2.1	0.66	0.12	2.3	0.24	0.13	4.0	0.56	0.16
5	2.2	0.66	0.14	2.3	0.85	0.08	2.4	0.28	0.14	3.8	0.69	0.18
6	0.3	0.49	0.12	2.3	0.92	0.07	0.6	0.33	0.10	3.5	0.75	0.19
7	0.3	0.63	0.11	2.4	0.95	0.07	0.6	0.42	0.10	2.9	0.78	0.20
8	0.4	0.70	0.12	2.3	0.95	0.07	0.9	0.49	0.12	2.2	0.81	0.20
9	0.4	0.79	0.10	2.3	0.96	0.07	0.9	0.55	0.12	2.2	0.81	0.21
10	0.4	0.85	0.07	2.2	0.97	0.07	1.0	0.58	0.13	2.1	0.81	0.22
11	0.4	0.88	0.06	2.2	0.97	0.07	0.6	0.56	0.10	1.9	0.81	0.22
12	0.5	0.89	0.08	0.6	0.97	0.05	0.6	0.59	0.11	1.8	0.81	0.23

algorithm. When $k \geq 7$, the average value of ARI metric of k-MXT-W algorithm is close to 1, while k-MXT algorithm does not approach the best possible value of ARI metric. Starting at $k = 7$, k-MXT-W algorithm reaches the value of ARI metric close to 1 for many values of ε. This is not the case for k-MXT algorithm (see Fig. 1).

k-MXT-W algorithm with correctly selected parameters on this data type splits the points into clusters as well as the K-Means algorithm does.

We have considered how k-MXT and k-MXT-W algorithms work with datasets with the same density clusters. Now consider the case when datasets have different density clusters. To do this, we generate data of blobs type with parameters `cluster_std` = [1.0, 1.5, 0.5], `random_state` = 170. Table 1 (the right-hand part) shows that the average value of ARI metric of k-MXT-W algorithm is statistically greater than the average value of ARI metric of k-MXT algorithm for all $k \geq 3$.

When $k = 10$ k-MXT-W algorithm reaches value of ARI metric close to 1. For all the considered values of parameter k, k-MXT algorithm does not reach the maximum possible value of ARI metric. A comparison of the best clustering results obtained using the k-MXT algorithm revealed some solution disadvantages. In some cases, the greatest number of vertices assigned to the wrong cluster is in the area of the graph with the highest density. In other cases, the vertices from the graph areas with the lowest density are mistakenly divided into clusters. When using small values of parameter k, k-MXT-W algorithm correctly splits into clusters vertices belonging to the high-density graph area. As the value of parameter k increases, this algorithm begins to divide the areas of the graph that have a lower density into clusters properly.

For this data type, k-MXT-W algorithm with properly selected parameters produces a more correct result than the K-Means algorithm.

3.3 The Circles Data

In this subsection we consider how the k-MXT and k-MXT-W algorithms work on the circles data type. We have generated data of this type using *scikit-learn*, free software machine learning library for the Python programming language (available via link https://scikit-learn.org/), with the parameters `noise` = 0.05 and `factor` = 0.4. As before, we consider all values of ARI metric for values of k from 2 to 12 and values of ε from 0.1 to 3.9 with a step 0.1. As an example, Fig. 2,a plots the dependence of the ARI metric on the values of the parameters ε on the circles data type for $k = 6$.

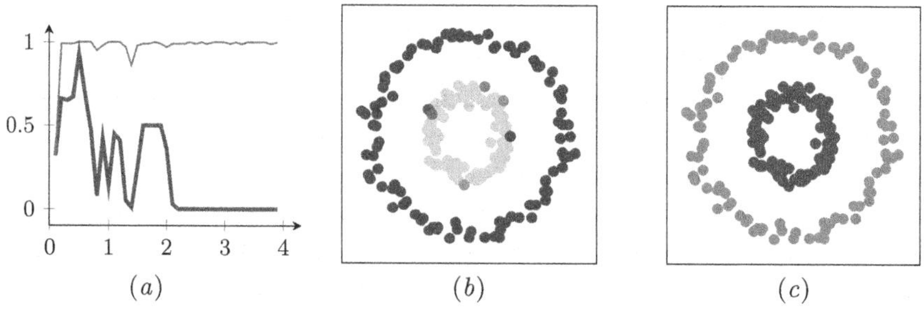

Fig. 2. The results of 6-MXT algorithm and 6-MXT-W algorithm on circles data type with `noise` = 0.05 and `factor` = 0.4: (a) Dependence of ARI metric on the values of parameter ε for 6-MXT algorithm (blue line) and for 6-MXT-W algorithm (red line). The performance of the algorithms with the best value of ARI: (b) $ARI = 0.93$ for 6-MXT algorithm with $\varepsilon = 0.5$; (c) $ARI = 1.00$ for 6-MXT-W algorithm with $\varepsilon = 0.5$. (Color figure online)

Starting from $k = 3$, the average value of the ARI metric for k-MXT-W algorithm is higher than that for k-MXT algorithm (Table 2, the left-hand side). The Aspin–Welch t-test validates that the means values of ARI metric for k-MXT-W algorithm are statistically greater than or equal to the means values of ARI metric for k-MXT algorithm. When the parameter $k \geq 5$, the algorithm k-MXT-W approaches 1, i.e. the best possible value of ARI metric, while k-MXT algorithm approaches 1 as $k \geq 11$.

If we consider the dependence of the value of ARI metric on the values of ε for k-MXT-W algorithm, for $5 \leq k \leq 12$ there are many values of the ε parameter, where ARI is close to 1.

Table 2. The results obtained by k-MXT and k-MXT-W algorithms for different k on circles type datasets with `noise` = 0.05, `factor` = 0.4 and on moons type datasets with `noise` = 0.05; the number of vertices is 200; μ denotes the average value of ARI and σ is the sample standard deviation.

k	Data sets on circles type						Data sets on moons type					
	k-MXT			k-MXT-W			k-MXT			k-MXT-W		
	ε	μ	σ	ε	μ	σ	ε	μ	σ	ε	μ	σ
4	1.6	0.46	0.03	3.0	0.83	0.15	0.1	0.10	0.03	3.6	0.93	0.10
5	0.4	0.60	0.12	2.1	0.87	0.14	0.2	0.30	0.11	3.4	0.91	0.13
6	0.5	0.76	0.13	0.5	0.95	0.09	0.3	0.49	0.16	0.3	0.79	0.17
7	0.2	0.87	0.14	0.2	0.94	0.10	0.3	0.76	0.15	0.2	0.90	0.13
8	0.2	0.90	0.13	0.2	0.95	0.09	0.3	0.89	0.10	0.2	0.97	0.08
9	0.2	0.92	0.12	0.2	0.95	0.09	0.2	0.88	0.15	0.2	0.99	0.04
10	0.2	0.93	0.10	0.2	0.95	0.09	0.2	0.98	0.06	0.2	1.00	0.02
11	0.2	0.95	0.10	0.2	0.95	0.09	0.2	1.00	0.02	0.2	1.00	0.01
12	0.2	0.95	0.09	0.2	0.95	0.09	0.2	1.00	0.01	0.2	1.00	0.01

Unlike K-Means algorithm, algorithms k-MXT and k-MXT-W with optimal parameter values produce the correct partitioning into clusters.

3.4 The Moons Data Type

The moons-type data also have a complex shape. Some algorithms distinguish clusters incorrectly on this data type (including K-Means algorithm). We obtained the results for k-MXT and k-MXT-W algorithms on the moons-type data produced by `sklearn.datasets.make_moons` with `noise` = 0.05, ε ranges from 0.1 to 3.9 with 0.1 steps; k is taken from 2 to 12.

Table 2 shows that if $k \geq 3$ then k-MXT-W algorithm performs better in terms of ARI metric than k-MXT algorithm. The Aspin–Welch t-test validates that the means values of ARI metric for k-MXT-W algorithm is statistically greater than or equal to the means values of ARI metric for k-MXT algorithm.

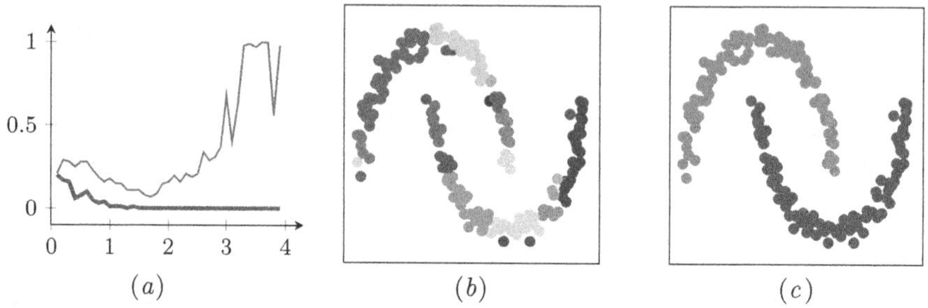

Fig. 3. The results of 4-MXT algorithm and 4-MXT-W algorithm on moons-shaped data with `noise` = 0.05: (*a*) Dependence of ARI metric on the values of parameter ε for 4-MXT algorithm (blue line) and for 4-MXT-W algorithm (red line). The performance of the algorithms with the best value of ARI: (*b*) $ARI = 0.20$ for 6-MXT algorithm with $\varepsilon = 0.1$; (*c*) $ARI = 1.00$ for 4-MXT-W algorithm with $\varepsilon = 3.6$. (Color figure online)

On the other hand, k-MXT algorithm has approximately the same ARI metric value only for $k \geq 9$. The empirical results show that the k-MXT-W algorithm reaches the best value of the ARI metric in most cases, which is not observed for the k-MXT algorithm. For example, if $k = 4$ then the best value of the ARI metric for the k-MXT algorithm is 5 times less than the best value of the ARI metric of the k-MXT-W algorithm on the same moons-shaped data sets (Fig. 3).

The k-MXT and k-MXT-W algorithms running with optimal values of their parameters perform on moons-shaped data much better than K-Means algorithm.

4 Comparison of the Results of the *k*-MXT and *k*-MXT-W Algorithms on Real Data Sets

In this section we compare the performance of the k-MXT and k-MXT-W algorithms on geotagged data sets to identify popular sights of St. Petersburg (Russia). In this task, the vertices of the graph will correspond to geotagged photographs taken from the Flickr. The edges will be added to this graph if the distance between the spots where the photos were taken is less than some fixed value ε. The selection of clusters of the graph allows you to identify locations in which people most often take photos, i.e. the most popular places in the city.

The data were collected using the Flickr API and consist of more than 230,000 photos. For more accurate clustering, only one entry record with the same longitude, latitude and the owner name was taken into account (more than 2,000 photos).

The results of the clustering performed by the k-MXT and k-MXT-W algorithms are shown in Fig. 4. In all subsequent figures, noise vertices are colored in red.

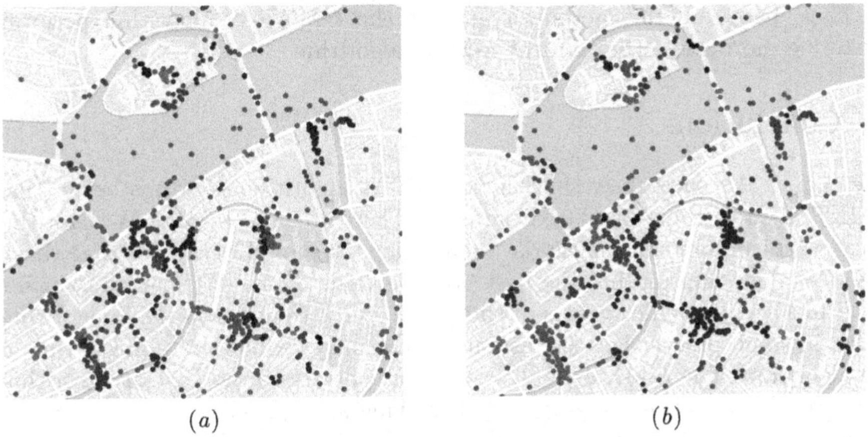

Fig. 4. Results of clustering on the geotagged data for the central part of St. Petersburg: (a) 7-MXT algorithm with $\varepsilon = 50$ and (b) 7-MXT-W with $\varepsilon = 50$ (Color figure online)

For the results of the k-MXT-W algorithm, the value of the modularity metric is equal to 0.17, while for k-MXT algorithm it is only 0.07. Such a difference in the values of the modularity metric is due to the fact that with a large accumulation of photos at one location point, the algorithms divide the photos into clusters in different ways. When comparing the clustering results presented in Fig. 5, it is noticeable that the k-MXT algorithm splits a set of vertices into a large number of clusters and marks many vertices as noise. On the contrary, the k-MXT-W algorithm generates a smaller number of clusters.

Based on the empirical analysis similar to the analysis from Sect. 3 we found that algorithms performs better with parameter values $k = 7$ and $\varepsilon = 50$. The splitting of vertices into clusters almost completely coincides with the actual popular sites of St. Petersburg.

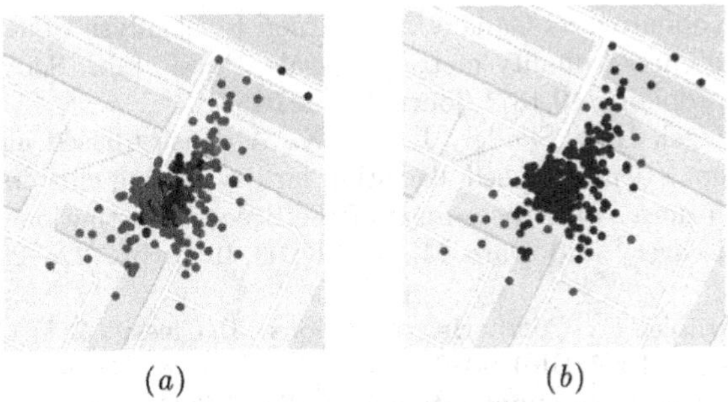

Fig. 5. Clustering results: (a) 7-MXT algorithm with $\varepsilon = 50$ and (b) 7-MXT-W with $\varepsilon = 50$.

Thus, based on the modularity metric the k-MXT-W algorithm performs clustering more correctly than the k-MXT algorithm.

5 Conclusions

The empirical results show that the k-MXT-W algorithm performs better than k-MXT algorithm on simulated datasets. Therefore, the k-MXT-W algorithm looks promising for automatic data clustering. It should be noted, that the k-MXT-W algorithm reaches the best possible value of the ARI metric (which is equal to 1) with smaller values of the parameter k than the k-MXT algorithm, which in some practical problems can significantly reduce the running time of the algorithm. The k-MXT-W algorithm, in contrast to the k-MXT algorithm, performs clusterization on graphs with varying spacial density much more correctly. These features can serve as the basis for the successful application of the k-MXT-W algorithm to solving problems of clustering geotagged data (including the analysis of diffusion and spillover effects on geotagged systems).

References

1. Béjar, J.: K-means vs mini batch k-means: a comparison. Technical report, Universitat Politècnica de Catalunya (2013)
2. Bottou, L., Bengio, Y.: Convergence properties of the k-means algorithms. In: Advances in Neural Information Processing System (NIPS 1994), pp. 585–592 (1994)
3. Brandes, U., et al.: On modularity clustering. IEEE Trans. Knowl. Data Eng. **20**(2), 172–188 (2008). https://doi.org/10.1109/TKDE.2007.190689
4. Clauset, A., Newman, M.E.J., Moore, C.: Finding community structure in very large networks. Phys. Rev. E **70**(6), 066111 (2004). https://doi.org/10.1103/PhysRevE.70.066111
5. Cooper, C., Vu, N.: An experimental study of the k-MXT algorithm with applications to clustering geo-tagged data. In: Bonato, A., Prałat, P., Raigorodskii, A. (eds.) WAW 2018. LNCS, vol. 10836, pp. 145–169. Springer, Cham (2018). https://doi.org/10.1007/978-3-319-92871-5_10
6. Emmons, S., Kobourov, S., Gallant, M., Börner, K.: Analysis of network clustering algorithms and cluster quality metrics at scale. Public Libr. Sci. **11**(7), e0159161 (2016). https://doi.org/10.1371/journal.pone.0159161
7. Ester, M., Kriegel, H.P., Sander, J., Xu, X.: A density-based algorithm for discovering clusters a density-based algorithm for discovering clusters in large spatial databases with noise. In: Proceedings of the Second International Conference on Knowledge Discovery and Data Mining, KDD 1996, pp. 226–231. AAAI Press (1996)
8. Hubert, L., Arabie, P.: Comparing partitions. J. Classif. **2**(1), 193–218 (1985). https://doi.org/10.1007/BF01908075
9. Lloyd, S.: Least squares quantization in PCM. IEEE Trans. Inf. Theory **28**(2), 129–137 (1982). https://doi.org/10.1109/tit.1982.1056489
10. Macqueen, J.: Some methods for classification and analysis of multivariate observations. In: 5th Berkeley Symposium on Mathematical Statistics and Probability, pp. 281–297 (1967)

11. Miasnikof, P., Shestopaloff, A.Y., Bonner, A.J., Lawryshyn, Y.: A statistical performance analysis of graph clustering algorithms. In: Bonato, A., Prałat, P., Raigorodskii, A. (eds.) WAW 2018. LNCS, vol. 10836, pp. 170–184. Springer, Cham (2018). https://doi.org/10.1007/978-3-319-92871-5_11
12. Newman, M.E.J., Girvan, M.: Finding and evaluating community structure in networks. Phys. Rev. E **69**(2), 026113 (2004). https://doi.org/10.1103/PhysRevE.69.026113
13. Oku, K., Hattori, F., Kawagoe, K.: Tweet-mapping method for tourist spots based on now-tweets and spot-photos. Procedia Comput. Sci. **60**, 1318–1327 (2015). https://doi.org/10.1016/j.procs.2015.08.202
14. Prokhorenkova, L.O., Prałat, P., Raigorodskii, A.: Modularity in several random graph models. Electron. Notes Discrete Math. **61**, 947–953 (2017). https://doi.org/10.1016/j.endm.2017.07.058
15. Prokhorenkova, L.O., Raigorodskii, A., Pralat, P.: Modularity of complex networks models. Internet Math. (2017). https://doi.org/10.24166/im.12.2017
16. Rand, W.M.: Objective criteria for the evaluation of clustering methods. J. Am. Stat. Assoc. **66**(336), 846–850 (1971). https://doi.org/10.2307/2284239
17. Rodriguez, M.Z., et al.: Clustering algorithms: a comparative approach. PLoS ONE **14**(1), e0210236 (2019). https://doi.org/10.1371/journal.pone.0210236
18. Rousseeuw, P.J.: Silhouettes: a graphical aid to the interpretation and validation of cluster analysis. J. Comput. Appl. Math. **20**, 53–65 (1987). https://doi.org/10.1016/0377-0427(87)90125-7
19. Sculley, D.: Web-scale k-means clustering. In: Proceedings of the 19th International Conference on World Wide Web - WWW 2010, pp. 1177–1178. ACM Press (2010). https://doi.org/10.1145/1772690.1772862
20. Spielman, D.A., Teng, S.H.: A local clustering algorithm for massive graphs and its application to nearly linear time graph partitioning. SIAM J. Comput. **42**, 1–26 (2013). https://doi.org/10.1137/080744888
21. Steinhaus, H.: Sur la division des corps materiels en parties. Bull. Acad. Polon. Sci. **4**(12), 801–804 (1956)
22. Stepanova, A., Mironov, S., Korobov, E., Sidorov, S.: The clusterization of geotagged data for finding city sights with use of a modification of k-MXT algorithm (2019). https://doi.org/10.2991/cmdm-18.2019.4
23. Vinh, N.X., Epps, J., Bailey, J.: Information theoretic measures for clusterings comparison: variants, properties, normalization and correction for chance. J. Mach. Learn. Res. **11**, 2837–2854 (2010)
24. Zhang, T., Ramakrishnan, R., Livny, M.: BIRCH: an efficient data clustering method for very large databases. SIGMOD Rec. **25**(2), 103–114 (1996). https://doi.org/10.1145/235968.233324

CoPASample: A Heuristics Based Covariance Preserving Data Augmentation

Rishabh Agrawal(✉) and Paridhi Kothari

Department of Mathematics, Indian Institute of Technology Guwahati, Guwahati 781039, India
rishabh.agrawal694@gmail.com, paridhi.pk7@gmail.com

Abstract. An efficient data augmentation algorithm generates samples that improves accuracy and robustness of training models. Augmentation with informative samples imparts meaning to the augmented data set. In this paper, we propose CoPASample (Covariance Preserving Algorithm for generating Samples), a data augmentation algorithm that generates samples which reflects the first and second order statistical information of the data set, thereby augmenting the data set in a manner that preserves the total covariance of the data. To address the issue of exponential computations in the generation of points for augmentation, we formulate an optimisation problem motivated by the approach used in ν-SVR to iteratively compute a heuristics based optimal set of points for augmentation in polynomial time. Experimental results for several data sets and comparisons with other data augmentation algorithms validate the potential of our proposed algorithm.

Keywords: Data augmentation · Covariance · Optimisation · Clustering · Insufficient data

1 Introduction

Machine learning models are increasingly being used to make sense of the available data and they perform efficiently by training on large data sets. Hence, availability of a sufficient amount of data is critical for their performance. When the data set is small, over-fitting of both the training data and the validation set becomes difficult to avoid. Moreover, outliers and noise have significant impact on the performance of the models. In practice, data corresponding to certain important events may not be sufficiently available. Data corresponding to diseases like particular forms of cancer is insufficient owing to their rare occurrence. Manufacturing related data on pilot runs and order acceptance procedures can only be obtained in small quantities.

R. Agrawal and P. Kothari—Contributed equally.

G. Nicosia et al. (Eds.): LOD 2019, LNCS 11943, pp. 308–320, 2019.
https://doi.org/10.1007/978-3-030-37599-7_26

To address this issue, many data augmentation techniques have been proposed. Data augmentation is a technique through which data is enlarged by adding new sample points. In literature, addition of noise to the data [1–3] has been proposed. Synthetic sampling techniques like SMOTE [4] and ADASYN [5] have been proposed to counter the problem of imbalanced data sets. SMOTE is an oversampling technique that generates synthetic data points in the minority class using k-nearest neighbours heuristic. ADASYN primarily generates synthetic data points in the neighbourhood of the minority samples that are harder to learn. Variants of SMOTE and ADASYN, namely, SMOTEBoost [6] and RAMOBoost [7] have also been proposed to address the issue of imbalance. Others [8,9] have discussed conditional and marginal data augmentation schemes. Bootstrapping [10] is a widely utilised method to generate virtual samples in case of data insufficiency [11,12]. Non-parametric Bootstrap selects random sample points with replacement from the available data for augmentation. EigenSample [13] is another algorithm that generates synthetic data points such that the cluster information in the data set is minimally distorted.

Covariance has multifaceted utilisations in engineering. Principal Component Analysis (PCA) [14], a dimensionality reduction method that depends extensively on the covariance of the data set, finds many applications in various domains [15–17]. COSVM [18] is an SVM (Support Vector Machine) algorithm that incorporates covariance into the involved optimisation problem. In signal processing, [19] utilises the covariance of the data as a tool for estimation techniques. Data Mining is yet another field that exploits covariance [20].

Motivated by the evident vitality of covariance and the reality of insufficient data, we have proposed CoPASample, a covariance preserving data augmentation algorithm. CoPASample first clusters the original data and obtains the mean of each cluster, the sample mean and the total covariance of the data. Samples on the line joining the mean of each cluster with the corresponding cluster data points, along with their reflections about the mean of respective cluster, are then computed. From this set of computed points, our proposed algorithm chooses a subset that would preserve the (1) sample mean and (2) total covariance after the chosen subset is augmented to the original data.

The rest of the paper is organised as follows. Section 2 provides a detailed description of CoPASample. In Sect. 3 experimental results and comparisons are presented. Finally, Sect. 4 concludes our research.

2 CoPASample Framework

The objective of CoPASample is twofold: to improve the learning of classification algorithms and to preserve the first and second order statistical information (sample mean and total covariance) of the data after augmentation. Consider a classification problem which trains on $D = \{x_1, x_2, \ldots, x_n\}$ using the class labels $y = \{y_1, y_2, \ldots, y_n\}$, where each sample point $x_i \in \mathbb{R}^d$ and each label $y_i \in Y$ (a finite label space). Let c be the number of clusters of data set, n_i be the number of points in the ith cluster and z_i be the mean of the ith cluster. The sample mean of the data is given by $\bar{x}$ and the total covariance matrix $S = [s_{pq}] \in \mathbb{R}^{d \times d}$ is such that s_{pq} is the covariance between the pth and q^{th} dimensions.

After clustering, let x_{ij} represents the jth data point in the ith cluster for $i \in \{1, 2, \ldots, c\}$ and for $j \in \{1, 2, \ldots, n_i\}$. Now, we compute k points on the line joining a sample point x_{ij} with the mean of the cluster it belongs to (z_i) and obtain $x'_{ij1}, x'_{ij2}, \ldots, x'_{ijk}$. Here, $x'_{ijl} = \lambda_l z_i + (1-\lambda_l)x_{ij}$ and $0 < \lambda_l < 1$, for all $l \in \{1, 2, \ldots, k\}$. Also obtain the reflections $r'_{ij1}, r'_{ij2}, \ldots, r'_{ijk}$ of $x'_{ij1}, x'_{ij2}, \ldots, x'_{ijk}$ about z_i given by $r'_{ijl} = 2z_i - x'_{ijl}$. Of all these computed points, only those points in the ith cluster are augmented to the original data set which on augmentation satisfy the following two criteria.

2.1 Sample Mean Preservation

Suppose we choose $x'_{i1}, x'_{i2}, \ldots, x'_{it_i}$ from $\{x'_{i11}, x'_{i12}, \ldots, x'_{i1k}, x'_{i21}, x'_{i22}, \ldots, x'_{i2k}, \ldots, x'_{in_i1}, x'_{in_i2}, \ldots, x'_{in_ik}\}$ in the ith cluster along with their reflection $r'_{i1}, r'_{i2}, \ldots, r'_{it_i}$ about z_i for augmentation. The mean (z'_i) of the chosen points along with the initial points in this cluster is given by:

$$z'_i = \frac{\sum_{l=1}^{n_i} x_{il} + \sum_{l=1}^{t_i} x'_{il} + \sum_{l=1}^{t_i} r'_{il}}{n_i + 2t_i} = z_i. \quad (1)$$

The second equality in Eq. 1 follows directly after substituting $r'_{il} = 2z_i - x'_{il}$. After augmentation, the new sample mean ($\bar{x}'$) is given by:

$$\bar{x}' = \frac{\sum_{i=1}^{c}(\sum_{l=1}^{n_i} x_{il} + \sum_{l=1}^{t_i} x'_{il} + \sum_{l=1}^{t_i} r'_{il})}{\sum_{i=1}^{c}(n_i + 2t_i)} = \frac{\sum_{i=1}^{c}(n_i + 2t_i)z_i}{\sum_{i=1}^{c}(n_i + 2t_i)}. \quad (2)$$

The second equality follows from Eq. 1. If $2t_i = \mu n_i$ for all $i \in \{1, 2, \ldots, c\}$, where μ is a is positive integer, then $\bar{x}' = \bar{x}$. In this paper, we have taken $2t_i = n_i$, for all $i \in \{1, 2, \ldots, c\}$. Hence, for sample mean preservation, for every point we choose, its corresponding reflected point is also chosen and $2t_i = n_i$.

2.2 Total Covariance Preservation

The covariance of the original data set between the pth and q^{th} dimensions ($1 \leq p, q \leq d$) is given by:

$$s_{pq} = \frac{\sum_{i=1}^{c}\sum_{l=1}^{n_i}((x_{il})_p - (\bar{x})_p)((x_{il})_q - (\bar{x})_q)}{\sum_{i=1}^{c} n_i - 1}, \quad (3)$$

where $(x_{il})_p$ denotes the pth dimension value of the sample point x_{il} and $(\bar{x})_p$ denotes the pth dimension value of $\bar{x}$. Upon augmenting the data set with the points chosen in Subsect. 2.1 and with $2t_i = n_i$, for all $i \in \{1, 2, \ldots, c\}$ so that $\bar{x}' = \bar{x}$, the covariance of the augmented data set between the pth and the qth dimension is given by:

$$s'_{pq} = \frac{1}{\sum_{i=1}^{c} 2n_i - 1}\Bigg(\Big(\sum_{i=1}^{c} n_i - 1\Big)s_{pq} + \sum_{i=1}^{c}\sum_{l=1}^{t_i}\Bigg(\Big((x'_{il})_p - (\bar{x})_p\Big)\Big((x'_{il})_q - (\bar{x})_q\Big) + \Big((r'_{il})_p - (\bar{x})_p\Big)\Big((r'_{il})_q - (\bar{x})_q\Big)\Bigg)\Bigg). \quad (4)$$

Now,

$$s'_{pq} = \frac{(\sum_{i=1}^{c}(n_i + 2t_i) - 1)s_{pq}}{\sum_{i=1}^{c}(n_i + 2t_i) - 1} = s_{pq}, \quad (5)$$

if, for all $i \in \{1, 2, \ldots, c\}$

$$2t_i s_{pq} = \sum_{l=1}^{t_i}\Bigg(\Big((x'_{il})_p - (\bar{x})_p\Big)\Big((x'_{il})_q - (\bar{x})_q\Big) + \Big((r'_{il})_p - (\bar{x})_p\Big)\Big((r'_{il})_q - (\bar{x})_q\Big)\Bigg). \quad (6)$$

If the points chosen in Subsect. 2.1 satisfy Eq. 6 for $1 \leq p, q \leq d$, in every cluster, then the total covariance of the augmented data is equal to the total covariance of the original data set. For the ith cluster, to choose an optimal set of n_i points that best approximates s_{pq}, for all $p, q \in \{1, 2, \ldots, d\}$, $\binom{kn_i}{t_i}$ sets of $2t_i$ points (points and their reflections about z_i) are required to be processed. To reduce this exponential complexity to polynomial time, a system of simultaneous linear equations is formulated to choose the points that should be augmented to the original data set such that the total covariance is preserved after augmentation. The procedure for the same is discussed below.

Formulation of Matrix Vector Equation. Of the computed $2kn_i$ points in the ith cluster, $2t_i(= n_i)$ points are to be chosen for covariance preservation. In order to choose these points corresponding to the ith cluster, we construct a matrix vector equation $Aw = b$, where $A = [a_{uv}] \in \mathbb{R}^{d^2 \times kn_i}$, $b^T = [s_{11}, s_{12}, \ldots, s_{1d}, \ldots, s_{d1}, s_{d2}, \ldots, s_{dd}]$ and $w^T = [w_1, w_2, \ldots, w_{kn_i}]$. The entries a_{uv} of A is given by:

$$a_{uv} = \frac{((x'_{ijl})_p - (\bar{x})_p)((x'_{ijl})_q - (\bar{x})_q) + ((r'_{ijl})_p - (\bar{x})_p)((r'_{ijl})_q - (\bar{x})_q)}{2t_i}, \quad (7)$$

where,

$$j = \lceil v/k \rceil, \qquad p = \lceil u/d \rceil,$$

$$l = \begin{cases} v\,mod(k), & \text{if } v\,mod(k) \neq 0 \\ k, & \text{otherwise,} \end{cases} \qquad q = \begin{cases} u\,mod(d), & \text{if } u\,mod(d) \neq 0 \\ d, & \text{otherwise.} \end{cases}$$

Solving Matrix Vector Equation. To solve the matrix vector equation $Aw = b$, we construct an optimisation problem given by the following objective function and constraints, as used in the least square version of ν-SVR [21]:

$$Minimize\ \tau(w, \xi, \xi^*, \epsilon) = \frac{1}{2}||w||^2 + C(\nu\epsilon + \frac{1}{d^2}\sum_{l=1}^{d^2}((\xi)_l^2 + (\xi^*)_l^2)), \tag{8}$$

$$Subject\ to\ Aw - \xi \leq b + \epsilon\mathbf{1}, \tag{9}$$

$$Aw + \xi^* \geq b - \epsilon\mathbf{1}, \tag{10}$$

$$\epsilon \geq 0. \tag{11}$$

An error of ϵ is allowed for each constraint and ξ, ξ^* are slack variables that capture the error above ϵ, C is a regularization constant and ν is another constant. The latter two are chosen a-priori. $\mathbf{1}$ is a column vector of size d^2 whose each entry is 1. Taking the Lagrangian of the entire problem, we obtain:

$$\begin{aligned} L(w, \alpha, \alpha^*, \xi, \xi^*, \beta, \epsilon) = &\frac{1}{2}||w||^2 + C\Big(\nu\epsilon + \frac{1}{d^2}\sum_{l=1}^{d^2}\left((\xi)_l^2 + (\xi^*)_l^2\right)\Big) - \beta\epsilon \\ &+ \sum_{l=1}^{d^2}(\alpha)_l((Aw)_l - (\xi)_l - (b)_l - \epsilon) \\ &- \sum_{l=1}^{d^2}(\alpha^*)_l((Aw)_l + (\xi^*)_l - (b)_l + \epsilon), \qquad 1 \leq l \leq d^2. \end{aligned} \tag{12}$$

Here, α, α^*, β are the Lagrangian coefficients. On setting the derivatives of L w.r.t. the primal variables equal to zero and on further calculations, we obtain:

$$w = A^T(\alpha^* - \alpha), \tag{13}$$

$$\begin{bmatrix} \alpha \\ \alpha^* \\ \beta \end{bmatrix} = pinv\left(\begin{bmatrix} -(AA^T + \frac{d^2 I}{2C}) & AA^T & \mathbf{0} \\ AA^T & -(AA^T + \frac{d^2 I}{2C}) & \mathbf{0} \\ \mathbf{1} & \mathbf{1} & 1 \end{bmatrix}\right)\begin{bmatrix} b + \epsilon \\ \epsilon - b \\ C\nu \end{bmatrix}, \tag{14}$$

where I is an identity matrix of order d^2, $\mathbf{0}$ is a column vector and $\mathbf{1}$ is a row vector of appropriate sizes. Solution of Eq. 14 is plugged into Eq. 13 to obtain w.

Interpretation of the Solution of Matrix Vector Equation. The vector b is the linear combination of columns of A and the lth component of w gives the coefficient of the lth column involved in this linear combination. A is a matrix of kn_i columns where the lth column corresponds to the contribution of the lth computed point and its reflection towards the covariance. We interpret the lth component of w to indicate the degree to which the point and its reflection represented by the lth column of A affects the covariance preservation. That is, the larger the value of the lth component of w, the more the corresponding point and its reflection contribute towards covariance preservation. Since we have to choose only $2t_i$ points from $2kn_i$ computed points, we select the points and their reflections corresponding to the t_i largest components of w for augmentation.

First and Second Order Statistical Information of the Sample Points Chosen for Augmentation. For $2t_i = n_i$, Eq. 2 implies that the sample mean of the points chosen for augmentation is equal to the sample mean of the original data. The covariance between the pth and q^{th} dimensions ($\hat{s}_{pq}$) of the chosen points is given by:

$$\hat{s}_{pq} = \frac{\sum_{i=1}^{c}\sum_{l=1}^{t_i}\left(\left((x'_{il})_p - (\bar{x})_p\right)\left((x'_{il})_q - (\bar{x})_q\right) + \left((r'_{il})_p - (\bar{x})_p\right)\left((r'_{il})_q - (\bar{x})_q\right)\right)}{\sum_{i=1}^{c} 2t_i - 1}. \tag{15}$$

Using Eq. 6,

$$\hat{s}_{pq} = \frac{\sum_{i=1}^{c} 2t_i s_{pq}}{\sum_{i=1}^{c} 2t_i - 1} \approx s_{pq}. \tag{16}$$

From Eq. 16, we note that the second order statistical information of the chosen points is approximately equal to that of the original data set. Therefore, CoPASample augments the data set with informative samples.

Optimal Choice of Points for Augmentation. For different values of k, we obtain a different set of points for augmentation. We choose the set that best approximates the covariance. For this, we calculate the error $\|e\|_2$ ($\|\cdot\|_2$ represents 2-norm) where e is the difference of sum of the columns of the matrix A corresponding to the chosen points and the vector b. The set of points for which this computed error $\|e\|_2$ is minimum is the optimal choice of points for augmentation.

Time Complexity of CoPASample. The pseudo-code of CoPASample is presented in Algorithm 1. Steps 9 and 10 translate to $O(kn_i)$ and $O(kn_id^2 + d^2)$, respectively, in terms of time complexity. Steps 11 through 13 form the core operations of the heuristics discussed. Step 11 includes computation of AA^T, pseudo inverse as given in Eq. 14 and product of the pseudo inverse with b which takes $O(kn_id^4)$, $O(d^6)$ and $O(d^4)$ time respectively. $O(kn_ilog(kn_i))$ accounts for the sorting algorithm involved in step 12 and step 13 takes $O(d^2log(n_i))$ to compute the 2-norm of e, wherein each element of the vector addition involves $O(log(n_i))$ time. Hence, the upper bound of the runtime can now be given by $\sum_{i=1}^{c} log(n_i)(log(n_i)n_id^4 + d^6 + log(n_i)n_ilog(log(n_i)n_i)))$. The entire algorithm allows for a polynomial time complexity of $O(nlog^2(n)d^4 + clog(n)d^6 + nlog^3n)$. The resultant complexity takes into account the observation that the time complexity of most clustering algorithms are bounded above by it.

Algorithm 1. CoPASample

Input: D, y, n, c, d
Compute the covariance matrix S, sample mean $\bar{x}$
Cluster the data set D with training labels y using clustering algorithm
for each cluster i **do**
  Get n_i, z_i, t_i
  Set $max\_iter = log_2(n_i)$.
  Set (min_error, ν, C)
  for each k in 1 to max_iter **do**
    Compute kn_i points and their reflection about z_i
    Compute the matrix A and vector b
    Solve $Aw = b$ using the constructed optimisation problem to obtain w
    Obtain t_i points and their reflection
    Compute the error $\|e\|_2$
    if $\|e\|_2 < min\_error$ **then**
      Set $min\_error = \|e\|_2$
      Choose these t_i points and their reflections
    end if
  end for
  Augment the chosen n_i points to D
end for
Obtain class labels $\hat{y}$ for the chosen points $\hat{D}$ using the trained classifier
Output: Augmented data $[D; \hat{D}]$, class labels $[y; \hat{y}]$

3 Experimental Results

To illustrate the effectiveness of CoPASample, we conducted experiments on data sets taken from the official repository of University of California at Irvine (UCI) [22]. The description of the data sets are given in Table 1. We compared classifier performance of the augmented data obtained from CoPASample with

Table 1. Description of the UCI data sets used in experimentation

Data set	Description of data set				
	#Datapoints	*#Features*	*#Points with labels +*	*#Points with labels −*	*Comment*
Breast Cancer	568	30	211	357	-
Ecoli	220	5	77	143	2 of the 8 classes (cp, im) considered
Ionosphere	351	34	126	225	-
Parkinsons	195	21	48	147	-
Planning Relax	182	12	52	130	-
Seed	140	7	70	70	2 of the 3 classes (Kama, Rosa) considered
Yeast	626	8	163	463	2 of the 10 classes (CYT, ME3) considered
Wine	178	13	59	119	Combined classes 2 and 3 considered as class '-'
PID	768	8	268	500	-
SONAR	208	60	97	111	-

Table 2. Comparison of overall accuracy, precision, recall, F-measure and g-mean of multiple data augmentation algorithms vs CoPASample on applying k-means algorithm on the data sets

Data set	Method	OA	Precision	Recall	F-measure	G-mean
Breast Cancer	CoPASample	**0.9242**	**0.9013**	**1.000**	**0.9481**	**0.9494**
	SMOTE	0.7121	0.5751	**1.000**	0.7303	0.7584
	ADASYN	0.8185	0.7340	0.9972	0.8456	0.8555
	Bootstrapping	0.8523	0.8109	0.9972	0.8944	0.8992
Ecoli	CoPASample	**0.9790**	**0.9632**	0.9912	**0.9769**	**0.9831**
	SMOTE	0.9339	0.8691	0.9928	0.9269	0.9289
	ADASYN	0.8334	0.7513	**0.9930**	0.8554	0.8637
	Bootstrapping	0.9500	0.9370	0.9895	0.9625	0.9629
Ionosphere	CoPASample	**0.9068**	**1.000**	**0.8311**	**0.9078**	**0.9116**
	SMOTE	0.7194	0.6495	0.6893	0.6688	0.6691
	ADASYN	0.6543	0.6713	0.6190	0.6441	0.6446
	Bootstrapping	0.6900	0.5766	0.7460	0.6505	0.6559
Parkinsons	CoPASample	0.9219	0.8524	0.8965	0.8739	0.8742
	SMOTE	**0.9471**	0.9351	**0.9512**	**0.9430**	**0.9499**
	ADASYN	0.9389	**0.9384**	0.9383	0.9383	0.9383
	Bootstrapping	0.7972	0.5809	0.6354	0.6069	0.6075
Wine	CoPASample	**0.9578**	**1.000**	0.9327	**0.9652**	**0.9657**
	SMOTE	0.8912	0.8102	**0.9910**	0.8915	0.9025
	ADASYN	0.8354	0.9072	0.7457	0.8186	0.8225
	Bootstrapping	0.8910	0.8474	0.8474	0.8474	0.8850

that obtained from implementation of SMOTE [23], ADASYN [24] and Bootstrapping.

Table 3. Comparison of overall accuracies (in %) of multiple clustering algorithms when performed on the original and the augmented data set using CoPASample. Here OD = Overall Accuracy of original data and AD = Overall Accuracy of augmented data set.

Data set	k-means		GT2 FCM		SVM		RF		DT	
	OD	*AD*	*OD*	*AD*	*OD*	*AD*	*OD*	*AD*	*OD*	*AD*
Breast Cancer	85.4	**92.4**	93.7	84.8	91.5	91.4	92.6	92.4	91.9	**92.3**
Ecoli	95.9	**97.9**	95.9	**97.8**	99	**99.1**	98.2	97.7	98.6	97
Ionosphere	71.2	**90.7**	71.8	**94.4**	94.3	**96.5**	92.6	**92.9**	88.6	**91.3**
Parkinsons	72.8	**92.2**	73.4	**92.9**	88.7	**93.1**	89.2	**94.6**	88.2	**92.5**
Planning Relax	51.6	**78.8**	52.2	**80.9**	73.6	**83.1**	69.2	**79.6**	63.7	**80.7**
Seed	92.1	**94.2**	93.6	**96.4**	95.7	94.6	95.7	95.3	97.1	94.2
Yeast	53.6	**86.6**	54.3	**87.8**	96.6	**97.2**	95.7	92.5	95.8	92.5
Wine	91.6	**95.8**	91.8	**95.8**	100	96.9	98.3	97.7	98.3	97.7
PID	70.1	**89.9**	70.8	**92.9**	77.7	**93.6**	75.7	**92.6**	74.2	**90.2**
Sonar	54.3	**93.2**	55.3	**93.1**	88.0	**94.2**	84.1	**93.5**	71.6	**85.6**

Table 4. Total covariance of ecoli data set before and after augmentation using CoPASample.

Ecoli data set: total covariance					
Before	0.0261	0.0040	0.0026	0.0130	0.0118
	0.0040	0.0096	0.0008	0.0099	0.0068
	0.0026	0.0008	0.0111	0.0076	0.0087
	0.0130	0.0099	0.0076	0.0553	0.0427
	0.0118	0.0068	0.0087	0.0427	0.0413
After	0.0272	0.0041	0.0025	0.0131	0.0120
	0.0041	0.0097	0.0007	0.0104	0.0067
	0.0025	0.0007	0.0111	0.0072	0.0087
	0.0131	0.0104	0.0072	0.0583	0.0438
	0.0120	0.0067	0.0087	0.0438	0.0433

Table 5. Total covariance of seed data set before and after augmentation using CoPASample.

Seed data set: total covariance							
Before	5.7908	2.6246	0.0101	0.9196	0.6491	0.9152	1.1869
	2.6246	1.2078	0.0026	0.4322	0.2851	0.4234	0.5558
	0.0101	0.0026	0.0003	−0.0002	0.0022	0.0006	0.0000
	0.9196	0.4322	−0.0002	0.1654	0.0936	0.1443	0.2083
	0.6491	0.2851	0.0022	0.0936	0.0799	0.1052	0.1215
	0.9152	0.4234	0.0006	0.1443	0.1052	1.6180	0.2159
	1.1869	0.5558	0.0000	0.2083	0.1215	0.2159	0.2859
After	6.0041	2.7247	0.0107	0.9561	0.6731	0.9680	1.2259
	2.7247	1.2541	0.0028	0.4481	0.2973	0.4461	0.5731
	0.0107	0.0028	0.0002	−0.0001	0.0022	0.0006	0.0000
	0.9561	0.4481	−0.0001	0.1708	0.0984	0.1475	0.2138
	0.6731	0.2973	0.0022	0.0984	0.0819	0.1144	0.1267
	0.9680	0.4461	0.0006	0.1475	0.1144	1.7572	0.2285
	1.2259	0.5731	0.0000	0.2138	0.1267	0.2285	0.2920

Given that the overall accuracy is not a complete representative of the performance of a classifier, we have used 5 evaluation metrics - overall accuracy, precision, recall, F-measure and g-mean - as the basis for our comparison. Overall Accuracy $(OA) = \frac{TP+TN}{TP+FN+TN+FP}$, precision $= \frac{TP}{TP+FP}$, recall $= \frac{TP}{TP+FN}$, F-measure $= \frac{(1+\beta^2)\times recall \times precision}{\beta^2 \times recall + precision}$ (we take $\beta = 1$) and g-mean $= \sqrt{\frac{TP}{TP+FN} \times \frac{TN}{TN+FP}}$, where $TP =$ True Positive (number of positive

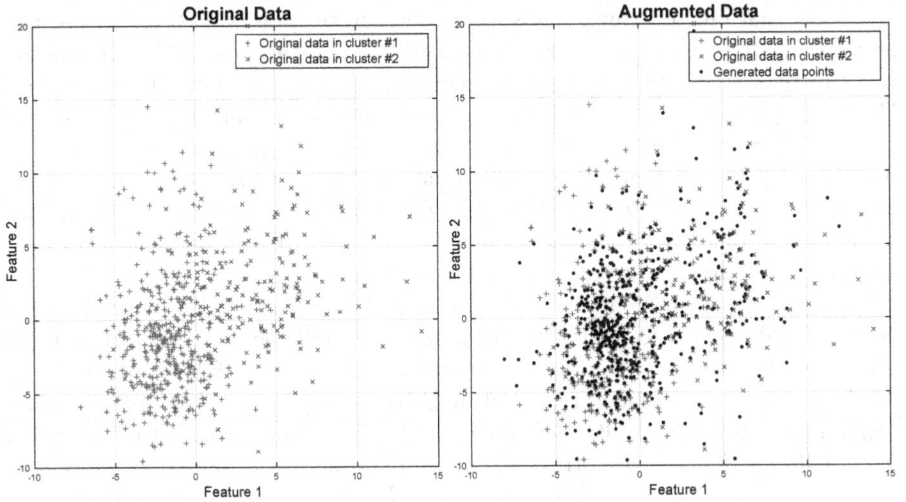

Fig. 1. Breast Cancer data set: Feature 1 and Feature 2 of the original data set and the augmented data set using CoPASample

samples correctly classified), FN = False Negative (number of positive samples incorrectly classified), TN = True Negative (number of negative samples correctly classified) and FP = False Positive (number of negative samples incorrectly classified). For this, we used k-means clustering algorithm for classification and have summarised the results on **5** data sets - Breast Cancer, Ecoli, Ionosphere, Parkinsons [25], and Wine in Table 2. SMOTE and ADASYN aim at improving learning from imbalanced data sets and therefore for a meaningful comparison with these algorithms, only imbalanced data sets have been used. On Breast Cancer and Ionosphere, CoPASample outperforms the other three algorithms. On Ecoli and Wine, CoPASample performs better on four of the five metrics. ADASYN and SMOTE give the best recall values for Ecoli and Wine respectively. In case of Parkinsons, SMOTE performs the best on all metrics except precision. We observe that CoPASample results in remarkable g-mean values except in case of Parkinsons. This implies that k-means clustering on the augmented data set obtained from CoPASample results in efficient classification of both the positive and negative samples.

A comparison of the classifier performances of k-means [26], GT2 FCM [27] (with the fuzzifier range for α-planes as proposed in [28]), SVM [29], Decision Tree [30] and Random Forest [31] classification algorithms on the original and augmented data sets has been reported in Table 3. 10 data sets - Breast Cancer, Ecoli, Ionosphere, Parkinsons, Planning Relax [32], Seed, Yeast, Wine, Pima Indian Diabetes (PID) and SONAR were used for this comparison. We observed that k-means clustering accuracy increases in every case. The enhanced performance of k-means clustering on the augmented data sets can be attributed to the fact that in case of k-means, the centre of each cluster is the mean of the samples in that cluster. Since our approach generates points with respect to the mean of the samples in a cluster, it is notable that the density of the sample points increases around the centre of each cluster given by k-means algorithm. GT2 FCM performs well on all data sets except Breast Cancer. Classifier performance of SVM, Decision Tree and Random Forest improves in most cases after augmentation. Even in cases where accuracies decrease, they remain comparable to the ones obtained before augmentation. Figure 1 shows the first and second feature of the original and the augmented Breast Cancer data set obtained through CoPASample clustered using k-means clustering.

Furthermore, CoPASample was successful in preserving the sample mean and the total covariance when implemented on multiple data sets. The total covariance obtained before and after augmentation using CoPASample on Ecoli and Seed data sets has been reported in Tables 4 and 5. We obtained the same sample means (0.4038, 0.4401, 0.4830, 0.4684, 0.5125) and (16.33, 15.21, 0.8817, 5.828, 3.461, 3.156, 5.553) corresponding to Ecoli and Seed data set before and after augmentation using CoPASample. We observed a relative change of $\approx$ 3% to 4% in most of the components of the total covariance matrix. The approximate preservation of the total covariance verifies our heuristics for the optimal selection of sets of sample points for augmentation.

4 Conclusion

The first and second order statistical information encode significant properties of a data set. To counter the problem of overfitting while training models on small data sets, we presented a heuristics based data augmentation algorithm that introduces a possible approach towards preservation of covariance post-augmentation. The choice of an optimal set of points for augmentation required exponential computation and therefore, a heuristics for a polynomial time solution has been proposed to determine an optimal set. Experimental results obtained on implementation of CoPASample validates our heuristics and demonstrates that an approximate preservation of covariance can be achieved. In addition, we augment the original data set with informative samples and a theoretical justification for the same has been provided. Simulation results across various evaluation metrics illustrate the effectiveness of CoPASample. In this paper, implementations on two class data sets have been carried out, however, CoPASample can also be implemented to augment multiclass data sets.

CoPASample can be effectively utilised for analysis of scarce data corresponding to phenomenons like rare diseases where augmentation allowing for preservation of information is vital. In future, to deal with high dimensional data sets, a suitable dimensionality reduction technique could be incorporated in CoPASample such that the task of preserving covariance is efficiently accomplished.

Acknowledgement. The authors would like to acknowledge Dr. Sriparna Bandopadhyay (Indian Institute of Technology Guwahati) and Dr. Ayon Ganguly (Indian Institute of Technology Guwahati) for their valuable feedbacks.

References

1. Wang, C., Principe, J.C.: Training neural networks with additive noise in the desired signal. In: The Proceedings of the 1998 IEEE International Joint Conference on Neural Networks, pp. 1084–1089 (1998)
2. Brown, W.M., Gedeon, T.D., Groves, D.I.: Use of noise to augment training data: a neural network method of mineral potential mapping in regions of limited known deposit examples. Nat. Resour. Res. **12**(2), 141–152 (2003)
3. Karystinos, G.N., Pados, D.A.: On overfitting, generalization, and randomly expanded training sets. IEEE Trans. Neural Netw. **11**(5), 1050–1057 (2000)
4. Chawla, N.V., Bowyer, K.W., Hall, L.O., Kegelmeyer, W.P.: SMOTE: synthetic minority over-sampling technique. J. Artif. Intell. Res. **16**, 321–357 (2002)
5. He, H., Bai, Y., Garcia, E.A., Li, S.: ADASYN: adaptive synthetic sampling approach for imbalanced learning. In: The Proceedings of the 2008 IEEE International Joint Conference on Neural Networks, pp. 1322–1328 (2008)
6. Chawla, N.V., Lazarevic, A., Hall, L.O., Bowyer, K.W.: SMOTEBoost: improving prediction of the minority class in boosting. In: Lavrač, N., Gamberger, D., Todorovski, L., Blockeel, H. (eds.) PKDD 2003. LNCS (LNAI), vol. 2838, pp. 107–119. Springer, Heidelberg (2003). https://doi.org/10.1007/978-3-540-39804-2_12
7. Chen, S., He, H., Garcia, E.A.: RAMOBoost: ranked minority oversampling in boosting. IEEE Trans. Neural Netw. **21**(10), 1624–1642 (2010)
8. Polson, N.G., Scott, S.L.: Data augmentation for support vector machines. Bayesian Anal. **6**(1), 1–23 (2011)
9. Meng, X.L., van Dyk, D.A.: Seeking efficient data augmentation schemes via conditional and marginal augmentation. Biometrika **86**(2), 301–320 (1999)
10. Efron, B., Tibshirani, R.J.: An Introduction to the Bootstrap, 1st edn. Chapman & Hall/CRC, Boca Raton (1993)
11. Ivănescu, V.C., Bertrand, J.W.M., Fransoo, J.C., Kleijnen, J.P.C.: Bootstrapping to solve the limited data problem in production control: an application in batch process industries. J. Oper. Res. Soc. **57**(1), 2–9 (2006)
12. Tsai, T.I., Li, D.C.: Utilize bootstrap in small data set learning for pilot run modeling of manufacturing systems. Expert Syst. Appl. **35**(3), 1293–1300 (2008)
13. Jayadeva, Soman, S., Saxena, S.: EigenSample: a non-iterative technique for adding samples to small datasets. Appl. Soft Comput. **70**, 1064–1077 (2018)
14. Pearson, K.: On lines and planes of closest fit to systems of points in space. London Edinb. Dublin Philos. Mag. J. Sci. **2**(11), 559–572 (1901)
15. David, C.C., Jacobs, D.J.: Principal component analysis: a method for determining the essential dynamics of proteins. Methods Mol. Biol. **1084**, 193–226 (2014)
16. van Nieuwenburg, E.P.L., Liu, Y.H., Huber, S.D.: Learning phase transitions by confusion. Nat. Phys. **13**, 435–439 (2017)
17. Yang, M.H., Kriegman, D.J., Ahuja, N.: Detecting faces in images: a survey. IEEE Trans. Pattern Anal. Mach. Intell. **24**(1), 34–58 (2002)
18. Khan, N.M., Ksantini, R., Ahmad, I.S., Guan, L.: Covariance-guided one-class support vector machine. Pattern Recogn. **47**(6), 2165–2177 (2014)
19. Ottersten, B., Stoica, P., Roy, R.: Covariance matching estimation techniques for array signal processing applications. Digit. Signal Proc. **8**(3), 185–210 (1998)
20. Alqallah, F.A., Konis, K.P., Martin, R.D., Zamar, R.H.: Scalable robust covariance and correlation estimates for data mining. In: The Proceedings of the 8th ACM SIGKDD International Conference on Knowledge Discovery and Data Mining, pp. 14–23 (2002)

21. Schölkopf, B., Smola, A.J., Williamson, R.C., Bartlett, P.L.: New support vector algorithms. Neural Comput. **12**(5), 1207–1245 (2000)
22. Dua, D., Graff, C.: UCI Machine Learning Repository. University of California, School of Information and Computer Science, Irvine, CA (2019). http://archive.ics.uci.edu/ml
23. SMOTE MATLAB Code. https://in.mathworks.com/matlabcentral/fileexchange/38830-smote-synthetic-minority-over-sampling-technique. Accessed 10 May 2019
24. ADASYN MATLAB Code. https://in.mathworks.com/matlabcentral/fileexchange/50541-adasyn-improves-class-balance-extension-of-smote. Accessed 10 May 2019
25. Little, M.A., McSharry, P.E., Roberts, S.J., Costello, D.A.E., Moroz, I.M.: Exploiting nonlinear recurrence and fractal scaling properties for voice disorder detection. Biomed. Eng. Online **6**(23) (2007)
26. Kanungo, T., Mount, D.M., Netanyahu, N.S., Piatko, C.D., Silverman, R., Wu, A.Y.: An efficient k-means clustering algorithm: analysis and implementation. IEEE Trans. Pattern Anal. Mach. Intell. **24**(7), 881–892 (2002)
27. Linda, O., Manic, M.: General type-2 fuzzy c-means algorithm for uncertain fuzzy clustering. IEEE Trans. Fuzzy Syst. **20**(5), 883–897 (2012)
28. Kulkarni, S., Agrawal, R., Rhee, F.C.H.: Determining the optimal fuzzifier range for alpha-planes of general type-2 fuzzy sets. In: The Proceedings of The 2018 IEEE International Conference on Fuzzy Systems (2018)
29. Cortes, C., Vapnik, V.: Support-vector networks. Mach. Learn. **20**, 273–297 (1995)
30. Breiman, L.: Classification and Regression Trees, 1st edn. Chapman & Hall/CRC, New York (1984)
31. Breiman, L.: Random forests. Mach. Learn. **45**(1), 5–32 (2001)
32. Bhatt, R.: Planning-Relax Dataset for Automatic Classification of EEG Signals. UCI Machine Learning Repository (2012)

Active Matrix Completion for Algorithm Selection

Mustafa Mısır[1,2(✉)]

[1] Department of Computer Engineering, Istinye University, Istanbul, Turkey
mustafa.misir@istinye.edu.tr
[2] College of Computer Science and Technology,
Nanjing University of Aeronautics and Astronautics, Nanjing, China

Abstract. The present work accommodates active matrix completion to generate cheap and informative incomplete algorithm selection datasets. Algorithm selection is being used to detect the best possible algorithm(s) for a given problem ($\sim$ instance). Although its success has been shown in varying problem domains, the performance of an algorithm selection technique heavily depends on the quality of the existing dataset. One critical and likely to be the most expensive part of an algorithm selection dataset is its performance data. Performance data involves the performance of a group of algorithms on a set of instance of a particular problem. Thus, matrix completion [1] has been studied to be able to perform algorithm selection when the performance data is incomplete. The focus of this study is to come up with a strategy to generate/sample low-cost, incomplete performance data that can lead to effective completion results. For this purpose, a number of matrix completion methods are utilized in the form of active matrix completion. The empirical analysis carried out on a set of algorithm selection datasets revealed significant gains in terms of the computation time, required to produce the relevant performance data.

1 Introduction

The experimental studies comparing different algorithms on a set of problem instances usually report that while a particular algorithm works well on a group of instances, it fails to outperform the competing algorithms on the other instances. In other words, there is no one algorithm that always perform the best, as also referred in the No Free Lunch (NFL) theorem [2]. Algorithm Selection [3] is an automated way of aiming at choosing the (near-) best algorithm(s) for solving a given problem instance. Thus, the target instances can be solved with the help of multiple algorithms rather than just one. In that respect, algorithm selection can offer an opportunity of defeating any given problem specific algorithm designed by the domain experts, as in the SAT competitions[1].

[1] http://www.satcompetition.org/.

G. Nicosia et al. (Eds.): LOD 2019, LNCS 11943, pp. 321–334, 2019.
https://doi.org/10.1007/978-3-030-37599-7_27

A traditional algorithm selection approach derives performance prediction models [4] that can tell the performance of a given set of algorithms on a new problem instance. For generating such models, the performance of these algorithms on a suite of training instances should be known. Besides, a set of features that can effectively characterize the target problem's instances is essential. Then, an algorithm selection model can simply map these instance features to the algorithms' performance. However, it can be challenging to generate the performance data in the first place. Finding representative instance features can also be complicated depending on the available problem domain expertise. Regarding the performance data, the main issue is the cost of generating it. Especially, if the computational resources are limited, it could take days, months, even years [5] to generate a single algorithm selection dataset. This drawback can restrict the use of algorithm selection by anyone who would like to perform algorithm selection on a new problem domain.

A collaborative filtering based algorithm selection technique, i.e. ALORS [1], was introduced to resolve this issue to a certain extent. Collaborative filtering [6] is a popular field of recommender systems to predict the interest of a user on an item such as a book or a movie, s/he haven't seen yet. The prediction process particularly relies on the user's preferences on multiple items, such as the scores given by her/him on these items. By taking all the other users' partial preferences into account, the existing partial preference information of this user can be utilized to determine whether s/he will like those items. From this perspective, if two users share similar preferences, their preferences on the unobserved items are also expected to be similar. The ALORS' collaborative filtering capability comes from the use of matrix completion, which is the performance data in this context. As its name suggests, matrix completion is about filling the unavailable entries of an incomplete matrix. In relation to the matrix completion task, ALORS showed that it can outperform the single best algorithm on varying problems with up to the 90% sparsity. Its matrix completion component also showed success in process mining [7].

The present study focuses on incorporating Active Matrix Completion (AMC) [8,9] into algorithm selection considering the data generation cost and quality. The AMC problem is defined as: for a given matrix $\mathcal{M}$ with unobserved entries, determine new entries to be queried, Q so that $\mathcal{M}' = \mathcal{M} \cup Q$ carries sufficient information for successful completion compared to $\mathcal{M}$. Minimizing $\mid Q \mid$ can also be targeted during querying. AMC is practical to determine the most profitable entries to add into a matrix for improved matrix completion. For this purpose, I apply various matrix completion techniques and together as ensembles (Ensemble Matrix Completion) in the form of AMC to specify the most suitable instance-algorithm entries to be sampled. In [10], a matrix completion method was used to perform both sampling and completion in ALORS. This study extends [10] by considering a set of existing matrix completion methods while examining their combined power as AMC ensembles. For analysis, a series of experiments is performed on the Algorithm Selection library (ASlib)[2] [5] datasets.

[2] http://aslib.net.

In the remainder of the paper, Sect. 2 gives background information both on algorithm selection and matrix completion. The proposed approach is detailed in Sect. 3. Section 4 provides an empirical analysis through the computational results. The paper is finalized with a summary and discussion in Sect. 5.

2 Background

2.1 Algorithm Selection

Algorithm Selection has been applied to various domains such as Boolean Satisfiability [11], Constraint Satisfaction [12], Combinatorial Optimization [13], Machine Learning [14] and Game Playing [15]. The majority of the methods called algorithm selection is considered working *Offline*. This means that the selection process is being performed before the target problem ($\sim$ instance) is being solved. This study is related to the Offline algorithm selection. Among the existing offline algorithm selection studies, SATZilla [11] is known as one of the premier works. It incorporates pre-solvers and backup solvers to address easy and hard instances, respectively. A rather traditional runtime prediction strategy is utilized to determine the best possible algorithm on the remaining instances. Instance features are used together with the performance data are used to generate the prediction models. SATZilla was further extended [16] with cost-sensitive models, as also in [17]. In [18], a collaborative filtering based recommender system, i.e. [19], was accommodated for AS. Hydra [20], inspired from ISAC [21], was developed to apply parameter tuning for generating algorithm portfolios/diverse algorithm sets, using a single algorithm. 3S [22] was studied particularly to offer algorithm schedules for determining which algorithm to run how long on each target problem instance. For this purpose, the resource constrained set covering problem with column generation was solved. In [23], deep learning was applied for algorithm selection when the instance features are unavailable. The corresponding instance files are converted into images for directly being used with the deep learning's feature extraction capabilities. As a high level strategy on algorithm selection, AutoFolio [24] was introduced to perform parameter tuning to come up with the best possible algorithm selection setting. Some of these AS systems as well as a variety of new designs were competed in the two AS competitions [25], took place in 2015 [26] and 2017[3] [27]. Algorithm scheduling was particularly critical for the successful AS designs like ASAP [28].

AS has been also studied as the *Online* operating methods for choosing algorithms on-the-fly while a problem instance is being solved. Thus, it is particularly suitable for optimization. Selection hyper-heuristics [29] mainly cover this field by combining selection with additional mechanisms. Adaptive Operator Selection (AOS) [30] has been additionally referred purely for Online selection.

[3] https://www.coseal.net/open-algorithm-selection-challenge-2017-oasc/.

2.2 Matrix Completion

Although data might be present in different ways, it is usually not perfect. The being perfect relates to quality and availability. Referring to the data quality, the data at hand might be enough in terms of size yet misleading or involving limited useful information. For availability, some data entries might be missing, requiring to have complete data. Thus, in reality, the systems focusing on data might be required to deal with both issues. In recommender systems, the latter issue is mainly approached as the matrix completion problem, for collaborative filtering [6]. The Netflix challenge[4] [31] was the primary venue made collaborative filtering popular. In the challenge, the goal was to predict missing 99% data from 1% observed entries of a large group of the user scores on many movies. The data was involving the scores regarding the preferences of the users on the target movies. In the literature, the methods addressing this problem are categorized as the *memory-based* (neighbourhood methods) and *model-based* (latent factor models) approaches. The memory-based approaches solely depend the available incomplete data to fill the missing entries. The model-based approaches look for ways to generate models that can perform matrix completion. The model-based approaches are known to be effective especially for the high incompleteness cases.

Matrix factorization [32] has been widely handled by the model-based approaches in collaborative filtering. It has been used to extract latent (hidden) factors that can characterize the given incomplete data such that the missing entries could be effectively predicted. A Probabilistic Matrix Factorization (PMF) method [33] that scales linearly with the number of observations was introduced. CofiRank [34] was devised as a matrix factorization method optimizing an upper-bound on the Normalized Discounted Cumulative Gain (NDCG) criterion for given rank matrices. ListRank-MF [35] was offered as an extension of matrix factorization with a list-wise learning-to-rank algorithm. In [36], matrix completion was studied to show its capabilities under noise. As a growing research direction in matrix factorization, non-negative matrix factorization [37,38] has been studied mainly to deliver meaningful and interpretable factors.

Related to the present work, active learning has been incorporated into matrix factorization [8]. An AMC framework was proposed and tested with various techniques in [8]. In [39], AMC was studied to address the completion problem for the low-rank matrices. An AMC method called `Order&Extend` was introduced to perform both sampling and completion together.

3 Method

A simple data sampling approach for algorithm selection, exemplified in Fig. 1, was investigated in [10]. The idea is to apply matrix completion to predict both cheap and informative unobserved entries. As detailed in Algorithm 1, a given completion method θ is first applied. Then, for each instance (matrix row), at most n number of entries to be sampled are determined. n is set for maintaining

[4] http://www.netflixprize.com.

balanced sampling between the instances. This step is basically performed by running the algorithms on the corresponding instances based on their expected performance. The top n, expectedly, the best missing entries are chosen for this purpose. Considering the target dataset with the performance metric of runtime, the best entries here also mean that the ones requiring the least computation effort for sampling. After running those selected algorithms for their target instances, a partial matrix is delivered, assuming that some entries are initially available.

$$\mathcal{M} = \begin{pmatrix} - & - & 7 & \cdots & - & - & 1 \\ 1 & - & - & \cdots & - & 6 & - \\ - & 6 & - & \cdots & 1 & - & 5 \\ \vdots & \vdots & \vdots & \vdots & \vdots & \vdots & \vdots \\ - & - & 3 & \cdots & - & 1 & 5 \\ - & 4 & - & \cdots & 5 & - & - \\ - & 1 & - & \cdots & - & - & 3 \end{pmatrix} \xrightarrow[\textbf{run algorithm } j \textbf{ on instance } i]{\textbf{sample } \mathcal{M}_{i,j} \textbf{ entries with } \theta} \begin{pmatrix} - & 5 & 7 & \cdots & - & - & 1 \\ 1 & - & 2 & \cdots & - & 6 & 4 \\ - & 6 & - & \cdots & 1 & - & 5 \\ \vdots & \vdots & \vdots & \vdots & \vdots & \vdots & \vdots \\ 8 & - & 3 & \cdots & - & 1 & 5 \\ - & 4 & - & \cdots & 5 & 2 & - \\ 5 & 1 & - & \cdots & - & 7 & 3 \end{pmatrix}$$

Fig. 1. An active matrix completion example

For evaluation, the matrix is needed to be completely filled, following this active matrix completion sampling step. As in [1,10], the completion is carried out using the basic cosine-similarity approach, commonly used in collaborative filtering. The completed matrix can then be utilized to determine the best/strong algorithm on each instance that is partially known through the incomplete, observed entries. Besides that, it can also be used to perform traditional algorithm selection which additionally requires instance features. The instance features are ignored as the focus here is solely on generating sufficiently informative algorithm selection datasets.

Algorithm 1. Active Matrix Completion for Algorithm Selection

Input: An incomplete performance matrix $\mathcal{M}$ of I instances and J algorithms, Matrix completion method θ, Number of samples to be added: N

Matrix Completion: $\theta(\mathcal{M}) \rightarrow \widehat{\mathcal{M}}$
Per Instance Sample Size: $n = N / \mid I \mid$
foreach $\widehat{\mathcal{M}}_i$ *where* $i \in I$ **do**
  $\arg\min_{J'} (\widehat{\mathcal{M}}_i, n)$ where $\mathcal{M}_{i,j} = \texttt{NaN}, \quad \mid J' \mid \leq n$
  foreach $j \in J'$ **do**
    $\mathcal{M}_{i,j} \leftarrow$ run algorithm j on instance i
  end
end

The same completion approach is followed yet with different matrix completion methods as the sampling strategies, as listed in Table 1. The first five algorithms[5] (ϑ) perform rather simple and very cheap completion. Referring to [1], more complex approaches, in particular the model-based ones or the ones performing heavy optimization, are avoided. Besides these methods, three simple ensemble based approaches, i.e. MeanE, MedianE and MinE using ϑ, are implemented. Finally, pure random sampling (Random) from [1] and the matrix completion method (MC) [10] used in the same study are utilized for comparison.

Table 1. Matrix completion algorithms, θ (all in ϑ with their default values from fancyimpute 0.1.0, $\vartheta = \{$SimpleFill, KNN, SoftImpute, IterativeSVD, MICE$\}$)

Name	Type
SimpleFill	Column mean
KNN	k-nearest neighbour ($k = 5$)
SoftImpute [40]	Iterative soft thresholding of SVD
IterativeSVD [41]	Iterative low-rank SVD
MICE [42]	Multiple imputation by chained equations
MeanE	$\sum_i^n \vartheta_i(\mathcal{M})/n$ for $n = \mid \vartheta \mid$
MedianE	$\mu_{1/2}\vartheta_i(\mathcal{M})$ for $\forall i$
MinE	$\min \vartheta_i(\mathcal{M})$ for $\forall i$
Random [1]	Random sampling
MC [10]	Row-based cosine similarity

4 Computational Results

The computational experiments are performed on 13 Algorithm Selection library (ASlib)[6] [5] (Table 2) datasets. The performance data of each ASlib dataset is converted into a rank form as in [1]. For each dataset, AMC is utilized to sample data for the decreasing incompleteness levels, from 90% incompleteness to 10% incompleteness. As the starting point, AMC is applied to each dataset after randomly picking the first 10% of its entries. AMC is tested by gradually sampling new 10% data. For the sake of randomness, initial 10% random sampling is repeated 10 times on each dataset.

Figure 2 reports the performance of each active matrix completion method in terms of time which is the data generation cost. Time (t_C) is normalized w.r.t. picking out the least costly ($t_{C_{best}}$) and the most costly ($t_{C_{worst}}$) samples for the given incompleteness levels, as in $\frac{t_C - t_{C_{best}}}{t_{C_{worst}} - t_{C_{best}}}$. For the SAT12 datasets, all the methods can significantly improve the performance of random sampling (Random) by choosing less costly instance-algorithm samples. For instance,

[5] From fancyimpute 0.1.0: https://pypi.python.org/pypi/fancyimpute.
[6] http://aslib.net.

adding the 10% performance data after the initially selected 10% for SAT12-RAND with MICE can save ~565 CPU hours (~24 days) compared to Random. The time gain can reach to ~1967 CPU hours (~82 days) (Table 3), on PROTEUS-2014. Significant performance difference is also achieved for the SAT11 and ASP-POTASSCO datasets.

Table 2. The ASlib benchmarks, with the number of CPU days required to generate

Dataset	#Instances	#Algorithms	#CPU
ASP-POTASSCO	1294	11	25
CSP-2010	2024	2	52
MAXSAT12-PMS	876	6	56
PRE.-ASTAR-2015	527	4	28
PROTEUS-2014	4021	22	596
QBF-2011	1368	5	163
SAT11-HAND	296	15	168
SAT11-INDU	300	18	128
SAT11-RAND	600	9	158
SAT12-ALL	1614	31	415
SAT12-HAND	767	31	234
SAT12-INDU	1167	31	284
SAT12-RAND	1362	31	447

Table 3. Average time gain achieved in hours by MICE compared to Random (0.9 is ignored since it is the initial random sampling for both methods)

Dataset	Incompleteness levels							
	0.1	0.2	0.3	0.4	0.5	0.6	0.7	0.8
ASP-POTASSCO	33	30	33	39	40	31	19	6
CSP-2010	−107	−21	−78	−83	−60	−31	−1	12
MAXSAT12-PMS	−21	37	28	60	72	53	32	−27
PRE.-ASTAR-2015	−95	−58	−26	−55	−55	−46	−67	−58
PROTEUS-2014	−516	242	1106	1702	1967	1932	1634	1095
QBF-2011	−169	−101	−74	−53	−41	−94	−118	−160
SAT11-HAND	74	108	135	127	125	122	67	6
SAT11-INDU	76	98	103	108	72	58	11	1
SAT11-RAND	47	134	156	169	170	113	80	1
SAT12-ALL	140	526	964	1156	1177	1096	817	403
SAT12-HAND	97	202	313	345	322	277	185	67
SAT12-INDU	138	507	910	1005	886	712	481	225
SAT12-RAND	137	313	560	801	1004	1160	1008	565

For PROTEUS-2014, MC, SimpleFill and KNN start choosing costly samples after the observed entries exceed 60% of the whole dataset. The remaining algorithms show similar behavior when the observed entries reaches 80%. For MAXSAT12-PMS, only SimpleFill, MICE and MeanE are able to detect cheap entries while the rest consistently picks the costly entries. On CSP-2010, PRE.-ASTAR-2015 and QBF-2011, the methods query the costly instance-algorithm pairs. The common characteristic between these three datasets is having a limited number of algorithms: 2, 4 and 5. The reason behind this issue is that for matrix completion, having high-dimensional matrices can increase the chance of high quality completion. For instance, in the aforementioned Netflix challenge, the dataset is composed of ∼480K users and ∼18K movies. Since both dimensions of the user-movie matrix are large, having only 1% of the complete data can be more practical in terms of matrix completion than a small scale matrix. Still, choosing costly entries can be helpful for the quality of the completion due to having misleading observed entries. Referring to the performance of the ensemble methods, despite the success of MeanE, MinE delivers poor performance while MedianE achieves average quality performance compared to the rest. The performance of MC, as the method of running the exact same completion method for both sampling and completion, it comes with the best performance on SAT12-INDU and SAT12-RAND together with the other matrix completion methods. However, it shows average performance on the remaining datasets except CSP-2010 where it delivers the worst performance.

Figure 3 presents the matrix completion performance in terms of average ranks, following each AMC application. The average rank calculation refers to choosing the lowest ranked algorithm for each instance w.r.t. the filled matrix, then evaluating its true rank performance. For SAT12-RAND, all the methods significantly outperform Random except KNN and MinE which provide similar performance to Random. SimpleFill, IterativeSVD, MC, MICE and MeanE particularly shows superior performance compared to Random. On SAT12-INDU, the results indicate that especially SimpleFill, MC, MICE and MeanE are able to provide significant performance. Yet, when the observed entries reach 60%, their corresponding rank performance starting to degrade. For SAT12-HAND, the significant improvement in terms of time doesn't help to reach to outperform Random. However, IterativeSVD, MICE, MeanE and MedianE are able to deliver similar performance to Random when the observed entries is below 50%. For the ASP-POTASSCO, PROTEUS-2014 and SAT11 datasets, similar rank performance can be seen for varying matrix completion methods. The majority of the methods is able to catch the performance of Random throughout all the tested incompleteness levels. On the CSP-2010 dataset, significantly better rank performance is achieved compared to Random yet as mentioned above more computationally costly samples are requested. Apart from preferring the expensive entries, the results indicate that these samples are able to elevate the quality of the matrix completion process. On the remaining datasets with the relatively small algorithm sets, i.e. PRE.-ASTAR-2015 and QBF-2011, similar performance is achieved against Random. However, it should be noted that this average rank performance is delivered

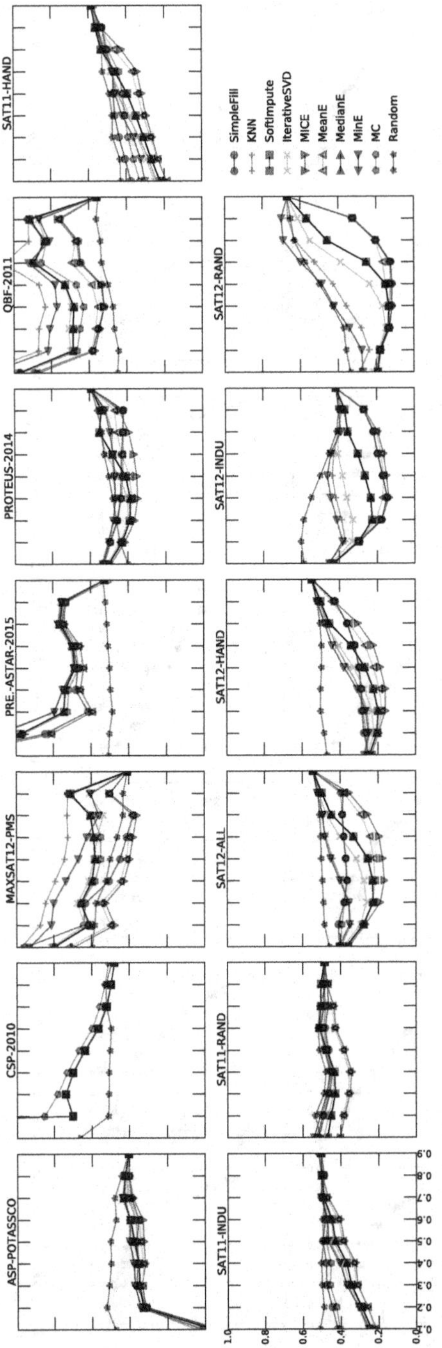

Fig. 2. Normalized time ratio (smaller is better) for varying incompleteness levels, 0.1 → 0.9 (0.9 refers to the initial 10% data case)

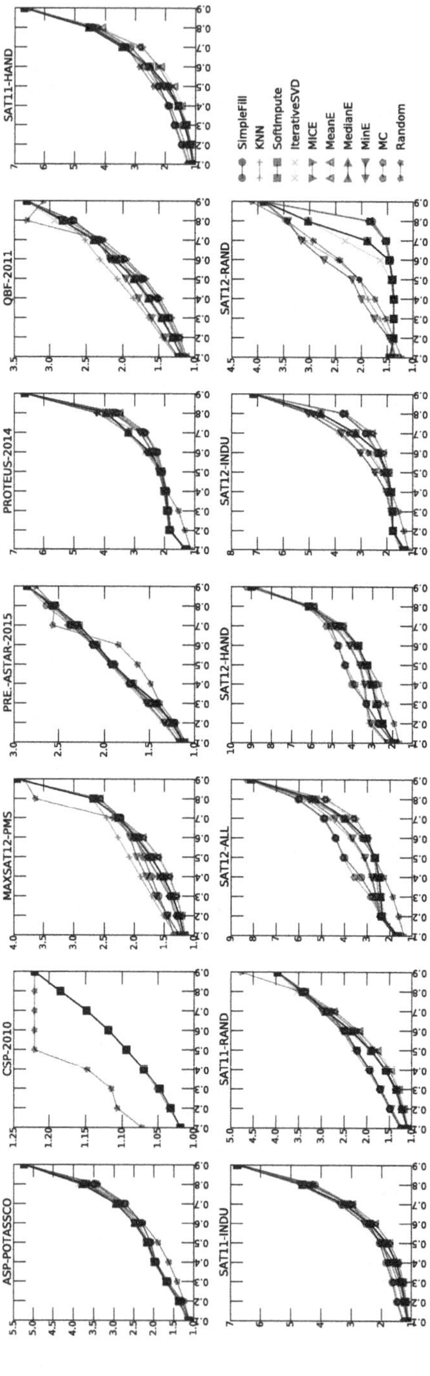

Fig. 3. Average rank (smaller is better) for varying incompleteness levels, 0.1 → 0.9 (0.9 refers to the initial 10% data case)

with more costly entries than Random. For the ensemble methods, their behavior on the cost of generating the instance-algorithm data is reflected to their average rank performance. Similar to the MC's average cost saving performance on the data generation/sampling, its rank performance is either being outperformed or matched with the other tested matrix completion techniques.

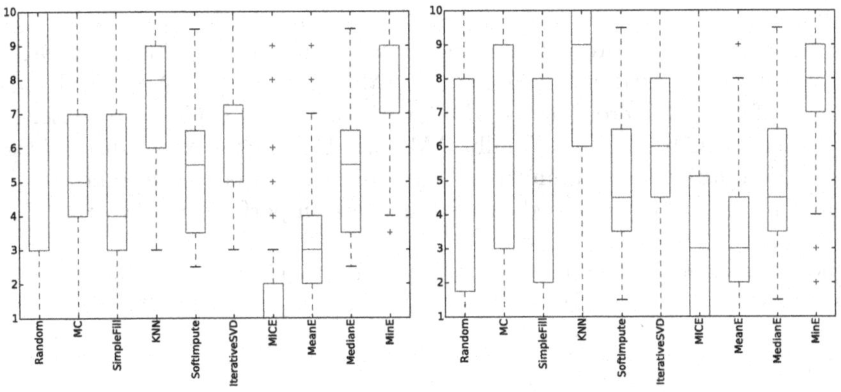

Fig. 4. Overall rank of the matrix completion methods as AMC, based on (a) normalized time ratio and (b) average rank

Figure 4 shows the overall performance of AMC across all the ASlib datasets with the aforementioned incompleteness levels. In terms of generating cheap performance data, MICE comes as the best approach, followed by an ensemble method, i.e. MeanE. Random delivers the worst performance. MinE, another ensemble method, also performs poorly together with KNN. For their average rank performance, MICE and MeanE come as the best performing methods. The overall rank performance of KNN and MinE are even worse than Random. Besides that, MC performs similarly to Random.

5 Conclusion

This study applies active matrix completion (AMC) to algorithm selection for providing high quality yet computationally cheap incomplete performance data. The idea of algorithm selection is about automatically choosing algorithm(s) for a given problem (∼instance). The selection process requires a set of algorithms and a group of problem instances. Performance data concerning the success of these algorithms on the instances plays a central role on the algorithm selection performance. However, generating this performance data could be quite expensive. ALORS [1] applies matrix completion to perform algorithm selection with limited/incomplete performance data. The goal of this work is deliver an informed data sampling approach to determine how the incomplete performance data is to be generated with the cheap to calculate entries while maintaining the data quality. For this purpose, a number of simple and fast matrix completion methods is utilized. The experimental results on the algorithm selection library (ASlib) benchmarks showed that AMC can provide substantial time gain on generating performance data for algorithm selection while delivering strong matrix completion performance, especially for the datasets with large algorithm sets.

The follow-up research plan covers investigating the effects of matrix completion on cold start which is the traditional algorithm selection task, i.e. choosing algorithms for unseen problem instances. Next, the problem instance features will be utilized for further improving the AMC performance. Afterwards, algorithm portfolios [43] across many AMC methods will be explored by extending the utilized AMC techniques. Finally, AMC will be targeted as a multi-objective optimization problem for minimizing the performance data generation time while maximizing the commonly used algorithm selection performance metrics.

Acknowledgement. This study was partially supported by an ITC Conference Grant from the COST Action CA15140.

References

1. Mısır, M., Sebag, M.: ALORS: an algorithm recommender system. Artif. Intell. **244**, 291–314 (2017)
2. Wolpert, D., Macready, W.: No free lunch theorems for optimization. IEEE Trans. Evol. Comput. **1**, 67–82 (1997)
3. Kerschke, P., Hoos, H.H., Neumann, F., Trautmann, H.: Automated algorithm selection: survey and perspectives. Evol. Comput. **27**(1), 3–45 (2019). MIT Press
4. Hutter, F., Xu, L., Hoos, H.H., Leyton-Brown, K.: Algorithm runtime prediction: methods & evaluation. Artif. Intell. **206**, 79–111 (2014)
5. Bischl, B., et al.: ASlib: a benchmark library for algorithm selection. Artif. Intell. **237**, 41–58 (2017)
6. Su, X., Khoshgoftaar, T.M.: A survey of collaborative filtering techniques. Adv. Artif. Intell. **2009**, 4 (2009)
7. Ribeiro, J., Carmona, J., Mısır, M., Sebag, M.: A recommender system for process discovery. In: Sadiq, S., Soffer, P., Völzer, H. (eds.) BPM 2014. LNCS, vol. 8659, pp. 67–83. Springer, Cham (2014). https://doi.org/10.1007/978-3-319-10172-9_5
8. Chakraborty, S., Zhou, J., Balasubramanian, V., Panchanathan, S., Davidson, I., Ye, J.: Active matrix completion. In: The 13th IEEE ICDM, pp. 81–90 (2013)
9. Ruchansky, N., Crovella, M., Terzi, E.: Matrix completion with queries. In: Proceedings of the 21th ACM SIGKDD KDD, pp. 1025–1034 (2015)
10. Mısır, M.: Data sampling through collaborative filtering for algorithm selection. In: the 16th IEEE CEC, pp. 2494–2501 (2017)
11. Xu, L., Hutter, F., Hoos, H., Leyton-Brown, K.: SATzilla: portfolio-based algorithm selection for SAT. J. Artif. Intell. Res. **32**(1), 565–606 (2008)
12. Yun, X., Epstein, S.L.: Learning algorithm portfolios for parallel execution. In: Hamadi, Y., Schoenauer, M. (eds.) LION 2012. LNCS, pp. 323–338. Springer, Heidelberg (2012). https://doi.org/10.1007/978-3-642-34413-8_23
13. Beham, A., Affenzeller, M., Wagner, S.: Instance-based algorithm selection on quadratic assignment problem landscapes. In: ACM GECCO Comp, pp. 1471–1478 (2017)

14. Vanschoren, J.: Meta-learning: a survey. arXiv preprint arXiv:1810.03548 (2018)
15. Stephenson, M., Renz, J.: Creating a hyper-agent for solving angry birds levels. In: AAAI AIIDE (2017)
16. Xu, L., Hutter, F., Shen, J., Hoos, H., Leyton-Brown, K.: SATzilla2012: improved algorithm selection based on cost-sensitive classification models. In: Proceedings of SAT Challenge 2012: Solver and Benchmark Descriptions, pp. 57–58 (2012)
17. Malitsky, Y., Sabharwal, A., Samulowitz, H., Sellmann, M.: Algorithm portfolios based on cost-sensitive hierarchical clustering. In: The 23rd IJCAI, pp. 608–614 (2013)
18. Stern, D., Herbrich, R., Graepel, T., Samulowitz, H., Pulina, L., Tacchella, A.: Collaborative expert portfolio management. In: The 24th AAAI, pp. 179–184 (2010)
19. Stern, D.H., Herbrich, R., Graepel, T.: Matchbox: large scale online Bayesian recommendations. In: The 18th ACM WWW, pp. 111–120 (2009)
20. Xu, L., Hoos, H., Leyton-Brown, K.: Hydra: automatically configuring algorithms for portfolio-based selection. In: The 24th AAAI, pp. 210–216 (2010)
21. Kadioglu, S., Malitsky, Y., Sellmann, M., Tierney, K.: ISAC-instance-specific algorithm configuration. In: The 19th ECAI, pp. 751–756 (2010)
22. Kadioglu, S., Malitsky, Y., Sabharwal, A., Samulowitz, H., Sellmann, M.: Algorithm selection and scheduling. In: Lee, J. (ed.) CP 2011. LNCS, vol. 6876, pp. 454–469. Springer, Heidelberg (2011). https://doi.org/10.1007/978-3-642-23786-7_35
23. Loreggia, A., Malitsky, Y., Samulowitz, H., Saraswat, V.A.: Deep learning for algorithm portfolios. In: The 13th AAAI, pp. 1280–1286 (2016)
24. Lindauer, M., Hoos, H.H., Hutter, F., Schaub, T.: AutoFolio: an automatically configured algorithm selector. JAIR **53**, 745–778 (2015)
25. Lindauer, M., van Rijn, J.N., Kotthoff, L.: The algorithm selection competitions 2015 and 2017. arXiv preprint arXiv:1805.01214 (2018)
26. Kotthoff, L.: ICON challenge on algorithm selection. arXiv preprint arXiv:1511.04326 (2015)
27. Lindauer, M., van Rijn, J.N., Kotthoff, L.: Open algorithm selection challenge 2017: setup and scenarios. In: OASC 2017, pp. 1–7 (2017)
28. Gonard, F., Schoenauer, M., Sebag, M.: Algorithm selector and prescheduler in the ICON challenge. In: Talbi, E.-G., Nakib, A. (eds.) Bioinspired Heuristics for Optimization. SCI, vol. 774, pp. 203–219. Springer, Cham (2019). https://doi.org/10.1007/978-3-319-95104-1_13
29. Pillay, N., Qu, R.: Hyper-Heuristics: Theory and Applications. Natural Computing Series. Springer, Cham (2018). https://doi.org/10.1007/978-3-319-96514-7
30. Da Costa, L., Fialho, A., Schoenauer, M., Sebag, M.: Adaptive operator selection with dynamic multi-armed bandits. In: GECCO, pp. 913–920 (2008)
31. Bennett, J., Lanning, S., et al.: The netflix prize. In: Proceedings of KDD Cup and Workshop, New York, NY, USA, vol. 2007, p. 35 (2007)
32. Koren, Y., Bell, R., Volinsky, C.: Matrix factorization techniques for recommender systems. Computer **42**(8), 30–37 (2009)
33. Salakhutdinov, R., Mnih, A.: Probabilistic matrix factorization. In: Platt, J.C., Koller, D., Singer, Y., Roweis, S.T. (eds.) The 21st NIPS, pp. 1257–1264 (2007)
34. Weimer, M., Karatzoglou, A., Le, Q.V., Smola, A.: CofiRank - maximum margin matrix factorization for collaborative ranking. In: The 21st NIPS, pp. 222–230 (2007)
35. Shi, Y., Larson, M., Hanjalic, A.: List-wise learning to rank with matrix factorization for collaborative filtering. In: The 4th ACM RecSys, pp. 269–272 (2010)

36. Candes, E.J., Plan, Y.: Matrix completion with noise. Proc. IEEE **98**(6), 925–936 (2010)
37. Wang, Y.X., Zhang, Y.J.: Nonnegative matrix factorization: a comprehensive review. IEEE TKDE **25**(6), 1336–1353 (2013)
38. Gillis, N., Vavasis, S.A.: Fast and robust recursive algorithms for separable non-negative matrix factorization. IEEE TPAMI **36**(4), 698–714 (2014)
39. Sutherland, D.J., Póczos, B., Schneider, J.: Active learning and search on low-rank matrices. In: Proceedings of the 19th ACM SIGKDD KDD, pp. 212–220 (2013)
40. Mazumder, R., Hastie, T., Tibshirani, R.: Spectral regularization algorithms for learning large incomplete matrices. JMLR **11**, 2287–2322 (2010)
41. Troyanskaya, O., et al.: Missing value estimation methods for DNA microarrays. Bioinformatics **17**(6), 520–525 (2001)
42. Azur, M.J., Stuart, E.A., Frangakis, C., Leaf, P.J.: Multiple imputation by chained equations: what is it and how does it work? Int. J. Methods Psych. Res. **20**(1), 40–49 (2011)
43. Gomes, C., Selman, B.: Algorithm portfolio design: theory vs. practice. In: The 13th UAI, pp. 190–197 (1997)

A Framework for Multi-fidelity Modeling in Global Optimization Approaches

Zelda B. Zabinsky[1(✉)], Giulia Pedrielli[2], and Hao Huang[3]

[1] University of Washington, Seattle, WA 98195, USA
zelda@u.washington.edu
[2] Arizona State University, Tempe, AZ, USA
Giulia.Pedrielli@asu.edu
[3] Yuan Ze University, Taoyuan City, Taiwan
haohuang@saturn.yzu.edu.tw

Abstract. Optimization of complex systems often involves running a detailed simulation model that requires large computational time per function evaluation. Many methods have been researched to use a few detailed, high-fidelity, function evaluations to construct a low-fidelity model, or surrogate, including Kriging, Gaussian processes, response surface approximation, and meta-modeling. We present a framework for global optimization of a high-fidelity model that takes advantage of low-fidelity models by iteratively evaluating the low-fidelity model and providing a mechanism to decide when and where to evaluate the high-fidelity model. This is achieved by sequentially refining the prediction of the computationally expensive high-fidelity model based on observed values in both high- and low-fidelity. The proposed multi-fidelity algorithm combines Probabilistic Branch and Bound, that uses a partitioning scheme to estimate subregions with near-optimal performance, with Gaussian processes, that provide predictive capability for the high-fidelity function. The output of the multi-fidelity algorithm is a set of subregions that approximates a target level set of best solutions in the feasible region. We present the algorithm for the first time and an analysis that characterizes the finite-time performance in terms of incorrect elimination of subregions of the solution space.

Keywords: Global optimization · Multi-fidelity models · Meta-models · Probabilistic Branch and Bound · Gaussian processes

1 Introduction

Complex systems are often represented and evaluated by means of a detailed, high-fidelity simulation model that requires significant computational time to execute. Since the high-fidelity model is time consuming to run, a low-fidelity

This work has been supported in part by the National Science Foundation, Grant CMMI-1632793.

G. Nicosia et al. (Eds.): LOD 2019, LNCS 11943, pp. 335–346, 2019.
https://doi.org/10.1007/978-3-030-37599-7_28

model is often constructed that takes far less computational time to run, but whose output is affected by error. For example, in engineering design, the high-fidelity model may involve a finite element analysis with a fine grid, whereas a low-fidelity version may use a coarse grid in the finite element analysis. The coarse grid is faster to execute, but provides less accurate performance metrics of the design. As another example, in manufacturing, a high-fidelity model may include a detailed discrete-event simulation with a complicated network of queues, whereas a low-fidelity version may be an analytical Markov chain model that is constructed by making simplifying assumptions.

An interesting and relevant question is how to make use of low-fidelity models to increase the likelihood of determining a solution, or set of solutions, that achieve good high-fidelity performance. The importance of the problem is well documented by the rich literature on the topic across different areas of engineering and computer science [4,7,11,14,16,20]. Most of the approaches can be brought back to the large category of Bayesian optimization methods.

Bayesian optimization is a well-established approach [12,13,17] to optimizing a complex system described by a potentially multi-modal, non-differentiable, black-box objective function such as the high-fidelity model we refer to. A Bayesian method starts with an a priori distribution, commonly a Gaussian distribution with a special covariance matrix, that represents the unknown objective function. Given several function evaluations of the objective function, the posterior (conditional) distribution of the objective function is updated. A tutorial on Bayesian optimization is given in [2]. We follow the basic procedure of updating the spatial covariance matrix using the observed function values.

There are several alternatives to Gaussian distributions to describe the objective function. In particular, radial basis functions have shown a remarkable success [19], additive Gaussian Processes that assume a dependency structure among the co-variates exists and can be learned [3,8]. Moreover, embeddings have been investigated in order to tackle the problem of scalability of model-based approaches [9,18].

Our approach is to use the statistical power of Gaussian distributions to relate the low-fidelity model to the high-fidelity model, and thus use fewer high-fidelity function evaluations. Several other papers have used a combination of low- and high-fidelity models, however, our approach is unique in that it embeds the Gaussian process into Probabilistic Branch and Bound [5,23], which provides a statistical confidence interval on how close the solutions obtained in the algorithm are to the global optimum.

In the literature relevant to this work, Xu et al. [21] proposed MO^{2}TOS (Multi-Fidelity Optimization with Ordinal Transformation and Optimal Sampling) that relies on the concept of Ordinal Transformation (OT). OT is a mapping $\mathbb{X} \rightarrow \mathcal{H}$, where $\mathbb{X}$ is a d-dimensional discrete space and $\mathcal{H}$ is a one-dimensional rank space constructed by associating to each point of $\mathbb{X}$, the rank computed according to the evaluation returned by the low-fidelity model. This mapping, as defined by the authors, can be applied to any finite space $\mathbb{X}$. Once the mapped space is computed, the solutions are grouped in subsets defined using

$\mathcal{H}$, and sampled according to the Optimal Sampling (OS) scheme. The theoretical analysis performed by the authors provides properties that a low-fidelity model should have to guarantee an improved performance of the proposed algorithm with respect to a benchmark version not using any low-fidelity information. Xu et al. [22] further extends the previous contribution by proposing an innovative optimal sampling methodology that maximizes the estimated probability of selecting the best solution. In [6], the authors extend the framework to continuous optimization by proposing a novel *additive model* that captures the relationship between the high- and low-fidelity functions and trying to sample mostly with the low-fidelity model. A potential drawback of this approach is that it assumes that a unique additive model can be used that fits the function across the entire solution space. In the direction to consider different behaviors, in [10], a multi-fidelity algorithm for global optimization was introduced that used Probabilistic Branch and Bound (PBnB) [5,23] to approximate a level set for the low-fidelity model, and under assumptions of consistency between the low-fidelity and high-fidelity models, the paper showed an increased probability of sampling high-quality solutions within a low-fidelity level set.

In this paper, we relax the assumption in [10] that the low-fidelity level set and high-fidelity level set need to intersect and the assumption of a unique model in [6], by combining the work in [10] with [6]. The new approach derives several predictive models of the original high-fidelity function using both high- and low-fidelity evaluations, in the subregions identified by PBnB. Specifically, when the predictive model(s) fails a certification test, it indicates that either more high-fidelity observations are needed, or that the subregion should be branched into smaller subregions. When the predictive model is good, we use it to decide which subregion should be explored more to discover global optima. A theoretical analysis of this new algorithm provides a probability of correctly focusing on good subregions on any iteration k, providing new finite-time results.

2 Framework

We consider an optimization problem with a high-fidelity black-box function f_H

$$\min_x \ f_H(x) \tag{1}$$

$$\text{subject to } x \in S$$

where $S \subset \Re^d$ is the feasible region, and $f_H : S \to \Re$. We also consider a low-fidelity model, $f_L : S \to \Re$, and assume that the computation time to evaluate $f_L(x)$ is much less than that to evaluate $f_H(x)$.

We are interested in determining near-optimal solutions in a target set that consists of the best δ-quantile of solutions, which can be defined as a level set bounded by a quantile $y_H(\delta, S)$,

$$y_H(\delta, S) = \arg\min_y \{P(f_H(X) \leq y | x \in S) \geq \delta\}, \text{ for } 0 < \delta < 1, \tag{2}$$

where X is uniformly distributed over S. Using $y_H(\delta, S)$, the target level set is defined as

$$L_H(\delta, S) = \{x \in S : f_H(x) \leq y_H(\delta, S)\}, \text{ for } 0 < \delta < 1. \tag{3}$$

Similarly, we define $y_L(\delta, S)$ and $L_L(\delta, S)$ as quantile and target set associated with the low-fidelity model, respectively. We note that for quantile level δ, $\delta = \frac{\nu(L_H(\delta,S))}{\nu(S)} = \frac{\nu(L_L(\delta,S))}{\nu(S)}$, where $\nu(\cdot)$ is the d-dimensional volume (i.e., Lebesgue measure) of a set.

The goal of the algorithm introduced in this paper is to approximate the target level set $L_H(\delta, S)$ using relatively few high-fidelity function evaluations and allow many more low-fidelity function evaluations.

2.1 Using Statistical Learning to Bridge High and Low Fidelity Models

As is common in Bayesian optimization [2], we use Gaussian processes as a framework to provide a statistical relationship between the low-fidelity function and the high-fidelity function, so that we can use the low-fidelity function evaluations to predict improving regions and focus the execution of the high-fidelity function. We next briefly summarize a Gaussian process modeling framework and introduce notation.

Within the Gaussian process modeling framework, a function $f(x)$ is interpreted as a realization from an infinite family of random functions, i.e., a stationary Gaussian process $Y(x)$. The statistical model estimating the behavior of the unknown function Y is characterized in terms of an optimal predictor $\hat{y}(x)$ and predictive error $s^2(x)$ based on a set of k observations, $x_i \in S$, with function evaluations $f(x_i)$, for $i = 1, \ldots, k$. We let $\mathbb{X}$ represent the set of observed locations $\mathbb{X} = \{x_1, \ldots, x_k\}$, and let $\bar{f}$ be a k-vector of the observed function evaluations $f(x_i)$, for $i = 1, \ldots, k$.

The statistical model of Y, conditional on the sampled observations, is:

$$Y(x) \sim \mathcal{N}\left(\hat{y}(x), s^2(x)\right) \tag{4}$$

for any $x \in S$ that has not yet been observed,[1] such that:

$$\hat{y}(x) := \hat{\mu} + \mathbf{c}^T\mathbf{K}^{-1}\left(\bar{f} - \hat{\mu}1\right) \quad \text{and} \quad \hat{\mu} := \frac{1^T\mathbf{K}^{-1}\bar{f}}{1^T\mathbf{K}^{-1}1} \tag{5}$$

and

$$s^2(x) := \tau^2\left(1 - \mathbf{c}^T\mathbf{K}^{-1}\mathbf{c} + \frac{\left(1 - 1^T\mathbf{K}^{-1}\mathbf{c}\right)^2}{1^T\mathbf{K}^{-1}1}\right). \tag{6}$$

Here, 1 is a k-vector having all elements equal to 1, $\mathbf{K}$ is a $k \times k$ spatial variance-covariance matrix parameterized over ϕ, and ϕ is a d-dimensional vector of weighting parameters, obtained upon the sampling of k points.

[1] For an observed point x_i, $\hat{y}(x_i) = f(x_i)$ and $s^2(x_i) = 0$.

The scale correlation coefficients in the d-dimensional vector ϕ are sensitivity parameters that control how fast the correlation decays with the *distance* between two observed points (x_i, x_j). Given k observations, the $d+1$ parameters (τ and $\phi_i, i = 1, \ldots, d$) are estimated using maximum likelihood [15].

The spatial variance-covariance matrix often appears in its exponential form, where the (i,j)th element of $\mathbf{K}$ is

$$\mathbf{K}(\mathbb{X}, \phi)_{i,j} = \exp\left[-\sum_{l=1}^{d} \phi_l \left(x_{i,l} - x_{j,l}\right)^t\right] \tag{7}$$

and $x_{i,l}$ is the lth element of x_i. The parameter t controls the smoothness of the response. When $t = 1$, Eq. (7) is known as the exponential correlation function, and when $t = 2$ it is known as the Gaussian correlation function [15]. In this paper, we adopt the Gaussian correlation function, with $t = 2$.

The k-vector $\mathbf{c}$ contains the spatial correlation between the prediction point $x \in S$ and the sampled locations x_j, $j = 1, \ldots, k$, where the jth element of $\mathbf{c}$ uses the lth element of the prediction point x_l and the sampled location $x_{j,l}$:

$$\mathbf{c}(x, \mathbb{X}, \phi)_j = \exp\left[-\sum_{l=1}^{d} \phi_l \left(x_l - x_{j,l}\right)^t\right] \tag{8}$$

where, again, we let $t = 2$.

Given k observations, it is straight-forward to build a Gaussian process model $Y(x)$ with predictor $\hat{y}(x)$ as in (5) and predictive error $s^2(x)$ as in (6) (Fig. 1).

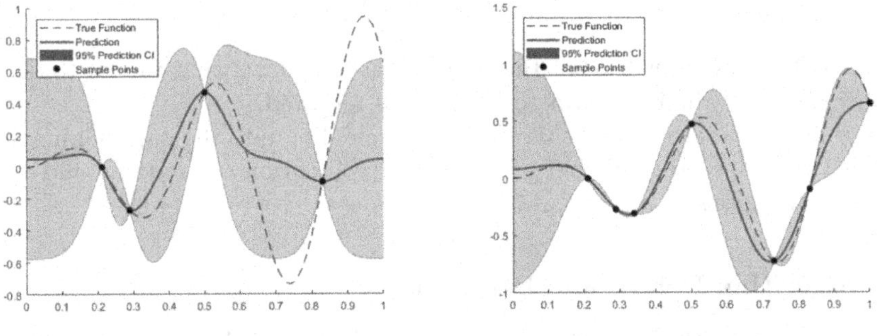

(a) Illustration using four observations. (b) Illustration using seven observations.

Fig. 1. Illustration of a one-dimensional function $f(x)$ with a predictor function $\hat{y}(x)$ and predictive error $s^2(x)$, using four observations in (a) and seven observations in (b).

Our approach to determine a relationship between the high-fidelity function f_H and the low-fidelity function f_L is to build several Gaussian process models based on high- and low-fidelity function observations. Following the approach in [6], we let $\mathbb{X}_H$ represent the set of observations that are sampled in S and only evaluated with high-fidelity (computationally expensive) function evaluations; let $\mathbb{X}_L$ represent the set of observations which are sampled and only evaluated with the low-fidelity (very fast computation) function; and let $\mathbb{X}_B$ be the set of points independently sampled that are evaluated with both high- and low-fidelity functions. The observations in $\mathbb{X}_B$ are used to estimate the bias using both cheap and expensive function evaluations, as a type of training set.

Under a Gaussian process multi-fidelity perspective, consider the following high-fidelity, low-fidelity, and bias statistical models used in the algorithm:

$$\begin{aligned} Y_{\cdot}(x) &\sim \mathcal{N}\left(\hat{y}_{\cdot}(x), s_{\cdot}^2(x)\right) \\ \hat{y}_{\cdot}(x) &:= \hat{\mu}_{\cdot} + \mathbf{c}_{\cdot}^T \mathbf{K}_{\cdot}^{-1}\left(\bar{f}_{\cdot} - \hat{\mu}_{\cdot}1\right) \\ s_{\cdot}^2(x) &:= \tau_{\cdot}^2\left(1 - \mathbf{c}_{\cdot}^T \mathbf{K}_{\cdot}^{-1}\mathbf{c}_{\cdot} + \frac{\left(1 - 1^T \mathbf{K}_{\cdot}^{-1}\mathbf{c}_{\cdot}\right)^2}{1^T \mathbf{K}_{\cdot}^{-1} 1}\right) \end{aligned} \tag{9}$$

where the subscript $\cdot$ is replaced by either H, L, or B indicating which function evaluations are used in the statistical model (high-fidelity, low- fidelity or bias, respectively). Specifically, the high-fidelity model uses $\bar{f}_H(x_i) = f_H(x_i)$ for $x_i \in \mathbb{X}_H$, the low-fidelity model uses $\bar{f}_L(x_i) = f_L(x_i)$ for $x_i \in \mathbb{X}_L$, and the bias model uses $\bar{f}_B(x_i) = f_H(x_i) - f_L(x_i)$ for $x_i \in \mathbb{X}_B$.

The statistical relationship between the high- and low-fidelity models is:

$$Y_{H|L}(x) = Y_L(x) + Y_B(x) \tag{10}$$

and use the set of observations $\mathbb{X}_B$ to reconstruct the prediction of the low-fidelity model conditional on the high-fidelity simulations performed, namely,

$$Y_{L|H}(x) = Y_H(x) - Y_B(x) \tag{11}$$

for $x \in S$. This model will be at the basis of the certificate that we use to decide whether or not evaluations in high-fidelity are required.

Our objective is to embed this statistical modeling into the partitioning logic of PBnB to reduce the number of high-fidelity function evaluations and focus them on subregions of interest.

2.2 Algorithm Details and Behavior

The main idea for the algorithm is to make a few high-fidelity function evaluations and more low-fidelity function evaluations, construct the statistical models, and use them to focus the location of more high-fidelity observations to eventually obtain subregions that are likely to contain the target set $L_H(\delta, S)$. We iteratively partition subregions of S, take additional high- and low-fidelity function evaluations, refine the statistical models, and subdivide the subregions maintaining statistical confidence that they contain the target set.

In the algorithm, we test whether $Y_{L|H}$ generates a good predictor $\hat{y}_L$ on a subregion of S. The hypothesis is that the model in (11) generates an accurate prediction of the low-fidelity response. Such a test is relevant to the search since the model in (11) is built using also the high-fidelity observations. The rationale is that, if $Y_{L|H}$ generates a good predictor for f_L, then $Y_{H|L}$ will also give good performance in predicting f_H, the function we are interested in optimizing. However, testing over the model $Y_{H|L}$ would require many high-fidelity function evaluations, which is undesirable due to the high evaluation cost.

When we have a good predictor over a subregion, we can use the predictor to decide if the subregion is likely or unlikely to contain the δ-quantile target set. If it is likely to contain the target set, we refine the partition of the subregion to focus the use of the expensive high-fidelity function evaluations. If the subregion is determined to be unlikely to contain the target set, we conclude that there is enough statistical evidence (at 1-α) to refrain from making more high-fidelity evaluations in that subregion. We use the framework of PBnB, integrated with the statistical models, to make decisions whether to evaluate using a high or low precision model and to guide the sampling locations.

Overview of Multi-fidelity Algorithm:

Step 0. *Initialize*: Set confidence level $\alpha, 0 < \alpha < 1$, target quantile $\delta, 0 < \delta < 1$, tolerated volume $\epsilon, \epsilon > 0$, and partitioning number B integer valued. Initialize the set of current subregions $\Sigma_1 = \{\sigma_1, \ldots, \sigma_B\}$, with $\cup_{j=1}^{B}\sigma_j = S$ and $\cap_{j=1}^{B}\sigma_j = \emptyset$, the number of subregions $J = B$, and the iteration counter $k = 1$.

Step 1. *Evaluate functions and build statistical models*:
Generate additional sample points in each subregion σ_j for all $j = 1, \ldots, J$, and $\sigma_j \in \Sigma_k$ so there is at least one point in each subregion for high-fidelity evaluations, for low-fidelity evaluations, and for both high- and low-fidelity function evaluations. Update the sets $\mathbb{X}_{H_k}$, $\mathbb{X}_{L_k}$, and $\mathbb{X}_{B_k}$ for those points with high-, low-, and both high- and low-fidelity function evaluations. While the number of samples is important for cross validating the statistical models, the idea is that the number of high-fidelity evaluations is much less than the number of low-fidelity evaluations, $|\mathbb{X}_H| \approx |\mathbb{X}_B| << |\mathbb{X}_L|$, where $|\cdot|$ represents the cardinality of a set. In the numerical results, we let $|\mathbb{X}_L| \geq 10Bd$ as is common when constructing Gaussian processes. However, this may be too large for computationally expensive high-fidelity function evaluations, so we just ensure that $|\mathbb{X}_H \cap \sigma_j| \geq 1$ and $|\mathbb{X}_B \cap \sigma_j| \geq 1$ for $j = 1, \ldots, J$.

Update $\bar{f}_H(x_i) = f_H(x_i)$ for $x_i \in \mathbb{X}_{H_k}$, $\bar{f}_L(x_i) = f_L(x_i)$ for $x_i \in \mathbb{X}_{L_k}$, and $\bar{f}_B(x_i) = f_H(x_i) - f_L(x_i)$ for $x_i \in \mathbb{X}_{B_k}$.

Given the function evaluations, build cross-validated models using the available observations for the two fidelities and the bias as well as the conditional densities: $Y_H, Y_L, Y_B, Y_{L|H}$, and $Y_{H|L}$.

Step 2. *Test predictive capability of low-fidelity model*:
As in PBnB, partition each subregion in the current set of subregions Σ_k into B new subregions, denoted $\sigma_1, \ldots, \sigma_J$, where $J = |\Sigma_k|B$ and $\cap_{j=1}^{J}\sigma_j = \emptyset$. Update Σ_k with the newly branched subregions.

Only using the low-fidelity samples, build J low-fidelity Gaussian process models, denoted $Y_{L,j}$, for each newly created subregion σ_j, $j = 1, \ldots, J$. Note that there may be no points in $\mathbb{X}_{L_k}$ that are also in a new σ_j, in which case, more points may be generated and evaluated with f_L. Also, if there are no points in $\mathbb{X}_{H_k}$ or $\mathbb{X}_{B_k}$ that are also in σ_j, use an extrapolation procedure to build the subregion-specific models $Y_{H,j}$ and $Y_{B,j}$.

For each subregion σ_j, without evaluating the high-fidelity function at any new points, use $Y_{L,j}$, $Y_{H,j}$ and $Y_{B,j}$ to build subregion-specific models $Y_{L|H,j}$ and $Y_{H|L,j}$ for $j = 1, \ldots, J$.

For each subregion σ_j, test the hypothesis that the Gaussian process $Y_{L|H,j}$ generates an accurate prediction of the low-fidelity response $Y_{L,j}$. We propose the following test to derive a low-fidelity certificate:

$$\mathrm{H}_0 : Y_{L|H,j}(x) \sim \mathcal{N}\left(\hat{y}_{L|H,j}(x), s^2_{L|H,j}(x)\right) \quad \text{and} \quad Q(x) = \frac{f_L(x) - \hat{y}_{L|H,j}(x)}{s_{L|H,j}(x)}$$

for $x \in \sigma_j$ and $j = 1, \ldots, J$. We compute the test statistic Q on a grid of points in σ_j to estimate the quantile and compare it to a standard normal value $z_{\alpha/2}$, assuming the normality assumption in the null hypothesis H_0.

If the test fails in a subregion, then, with $1 - \alpha$ confidence, we reject the null hypothesis and conclude that the predictor is providing poor performance in that subregion. In this case, we need to make more high-fidelity evaluations in the subregions that fail, so the algorithm goes to Step 1.

If the test does not fail, then proceed to Step 3 with $Y_{H|L,j}$ for each subregion σ_j, $j = 1, \ldots, J$.

Step 3. *Use $Y_{H|L,j}$ with additional samples to make pruning decision*:

For each subregion σ_j in Σ_k, $j = 1, \ldots, J$, uniformly and independently sample $N_k = \left\lceil \frac{\ln \alpha_k}{\ln\left(1 - \frac{\epsilon}{v(S)}\right)} \right\rceil$ points $\tilde{x}_{\sigma_j,n}$ for $n = 1, \ldots, N_k$, and evaluate them with $\hat{y}_{H|L,j}$. Notice that this is relatively fast to compute and does not require any high-fidelity function evaluations. Within each subregion σ_j, order these sampled points by their predicted values $\hat{y}_{H|L,j}$ and denote them $\tilde{x}_{\sigma_j,(1)}, \ldots, \tilde{x}_{\sigma_j,(N_k)}$, where

$$\hat{y}_{H|L,j}(\tilde{x}_{\sigma_j,(1)}) \leq \hat{y}_{H|L,j}(\tilde{x}_{\sigma_j,(2)}) \leq \cdots \leq \hat{y}_{H|L,j}(\tilde{x}_{\sigma_j,(N_k)}).$$

Also order *all* of the points $\tilde{x}_{\sigma_j,n}$ for $j = 1, \ldots, J$ and $n = 1, \ldots, N_k$ by their predicted values. Since the predicted value function is subregion specific, for notational purposes, let $\hat{y}^k_{H|L}(\tilde{x}) = \hat{y}_{H|L,j}(\tilde{x})$ for $\tilde{x} \in \sigma_j$. Then, we can use this function $\hat{y}^k_{H|L}$ to order the points $\tilde{x}_{\sigma_j,n}$ and denote them $\tilde{x}_{\Sigma_k,(1)}, \ldots, \tilde{x}_{\Sigma_k,(JN_k)}$, where

$$\hat{y}^k_{H|L}(\tilde{x}_{\Sigma_k,(1)}) \leq \hat{y}^k_{H|L}(\tilde{x}_{\Sigma_k,(2)}) \leq \cdots \leq \hat{y}^k_{H|L}(\tilde{x}_{\Sigma_k,(JN_k)}).$$

Perform the following comparison, the pruning test, for each subregion σ_j,

$$\hat{y}_{H|L,j}\left(\tilde{x}_{\sigma_j,(1)}\right) - z_{\frac{\alpha_k}{2}} s_{H|L,j}(\tilde{x}_{\sigma_j,(1)}) > \hat{y}^k_{H|L}\left(\tilde{x}_{\Sigma_k,(s)}\right) + z_{\frac{\alpha_k}{2}} s^k_{H|L}(x_{\Sigma_k,(s)}) \qquad (12)$$

where s satisfies,

$$\min s : \sum_{i=0}^{s-1} \binom{N_k}{i} (\delta_k)^i (1 - \delta_k)^{N_k - i} \geq 1 - \frac{\alpha_k}{2}. \qquad (13)$$

The test in (12) compares the best value in a subregion with the s-best value overall, accounting for an error term with confidence α_k.

If (12) is satisfied, prune σ_j. If (12) is not satisfied, keep σ_j. Update Σ_{k+1} with the subregions in Σ_k that have not been pruned.

Update $\alpha_{k+1} = \alpha_k / B$ and

$$\delta_{k+1} = \frac{\delta_k v(\Sigma_k)}{v(\Sigma_k) - v(\Sigma_k \setminus \Sigma_{k+1})}. \tag{14}$$

Step 4. *Continue?* Check a stopping criterion and either stop and return the current set of subregions in Σ_{k+1} to provide an approximation to the target level set, or increment the iteration counter $k \leftarrow k+1$ and go back to Step 1.

Illustrative Example. We showcase the proposed algorithm with a simple example where the true (unknown) function, with a discontinuity at $x = 0$, and the corresponding low fidelity are:

$$f_H = \begin{cases} 2 \cdot \sin(x) + 10 & \text{if } -5 \le x < 0 \\ 1.5 \cdot \sin(2x) - 5 & \text{if } \ 0 \le x \le 5. \end{cases}, \quad f_L = \begin{cases} 2.2 \cdot \sin(x) + 7 & \text{if } -5 \le x < 0 \\ 1.5 \cdot \sin(0.75x) - 5 & \text{if } \ 0 \le x \le 5. \end{cases}$$

Figure 2(a) shows the quality of the predictor $\hat{y}_{H|L}(x)$, when 4 points are evaluated in low-fidelity and only 2 in high-fidelity. In Fig. 2(b), the interval is partitioned into two subregions, and 4 more points are evaluated in low-fidelity, while only 2 more in high-fidelity. The partition is very effective in this example, and Step 3 is able to confidently prune the left half.

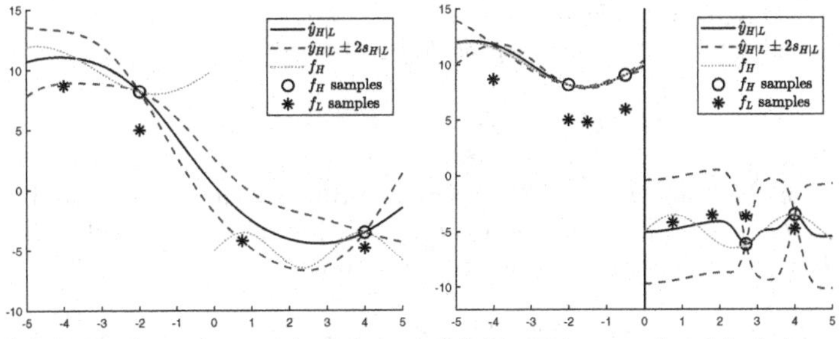

(a) First iteration with 4 low-fidelity and 2 high-fidelity evaluations.

(b) Partition and add 4 low-fidelity and 2 high-fidelity evaluations.

Fig. 2. Illustration of a one-dimensional function $f_H(x)$ with a predictor function $\hat{y}_{H|L}(x)$ and predictive error $s^2_{H|L}(x)$, across two iterations of the algorithm with increased simulation budget. Note that only high fidelity values are interpolated.

3 Algorithm Analysis

The analysis of the multi-fidelity algorithm provides a probabilistic guarantee that the approximation of the target set does not incorrectly prune a large volume. The main result in Theorem 1 says that, on any kth iteration, the volume of accumulated region that was pruned incorrectly is less than ϵ with probability $(1-\alpha)^2$.

Lemma 1. *For any iteration $k \geq 1$, suppose all previous pruning is correct, that is, the comparison in (12) is satisfied for all of the pruned subregions, $x \in S \setminus \Sigma_k$. Then, the δ_k updated according to (14) can be used to determine the quantile over the entire set S, i.e.,*

$$y^k_{H|L}(\delta, S) = y^k_{H|L}(\delta_k, \Sigma_k) \tag{15}$$

where the quantile notation, as in (2), is used with function $\hat{y}^k_{H|L}$.

Proof. Lemma 1 is a special case of Theorem 1 in Huang and Zabinsky [5] with no maintained subregions.

Lemma 2. *Suppose σ_p has been pruned on the kth iteration. Also, suppose $y^k_{H|L}(\delta, S) \leq \hat{y}^k_{H|L}\left(x_{\Sigma_k,(s)}\right)$. Then, the volume of the incorrectly pruned region, i.e., $v(L^k_{H|L}(\delta, S) \cap \sigma_p)$, is less than or equal to ϵ_k with probability at least $1-\alpha_k$:*

$$P\left(v(L^k_{H|L}(\delta, S) \cap \sigma_p) \leq \epsilon_k | y^k_{H|L}(\delta, S) \leq \hat{y}^k_{H|L}\left(x_{\Sigma_k,(s)}\right)\right) \geq (1-\alpha_k)^2, \epsilon_k = \frac{\epsilon}{B^k}$$

Proof. Lemma 2 is a special case of Theorem 2 in Huang and Zabinsky [5] applied to function $\hat{y}^k_{H|L}$.

Theorem 1. *The pruned subregions on the kth iteration contain at most ϵ volume of the target δ level set $L^k_{H|L}(\delta, S)$ with probability $(1-\alpha)^2$,*

$$P\left(v(L^k_{H|L}(\delta, S) \setminus \Sigma_k\right) \leq \epsilon) \geq (1-\alpha)^2. \tag{16}$$

Proof. Suppose there are $d_m < B^m$ subregions pruned at iteration m and $\epsilon_m = \frac{\epsilon}{B^m}$, then we have $d_m \epsilon_m < \epsilon$. Considering a subregion σ^m_p that was pruned at iteration m, $m = 1, \ldots, k$, then the incorrect pruned volume results

$$\begin{aligned} P_v &= P\left(v(L^m_{H|L}(\delta, S) \setminus \cup^{d_m}_{p=1} \sigma^m_p) \leq d_m \epsilon_m | y^m_{H|L}(\delta, S) \leq \hat{y}^m_{H|L}\left(x_{\Sigma_m,(s)}\right)\right) \\ &= \prod^{d_m}_{p=1} P\left(v(L^m_{H|L}(\delta, S) \setminus \sigma^m_p) \leq \epsilon_m | y^m_{H|L}(\delta, S) \leq \hat{y}^m_{H|L}\left(x_{\Sigma_m,(s)}\right)\right) \quad (17) \\ &\geq (1-\alpha_m)^{2d_m} \quad (18) \end{aligned}$$

where (17) holds since each subregion is pruned independently, and (18) holds due to Lemma 2. Assuming that the pruning decision in iteration k is made independently from the decisions on prior iterations, the following holds,

$$
\begin{aligned}
P(v(L^k_{H|L}(\delta,S)\setminus\Sigma_k)\le\epsilon) &\ge P\left(v(L^k_{H|L}(\delta,S)\setminus\cup^k_{m=1}\cup^{d_m}_{p=1}\sigma^m_p)\le d_m\epsilon_m\right)\\
&=\prod_{m=1}^k P\left(v(L^k_{H|L}(\delta,S)\setminus\cup_p\sigma^m_p)\le d_m\epsilon_m | y_{H|L}(\delta,S)\le\hat{y}_{H|L}\left(x_{\Sigma_m,(s)}\right)\right)\\
&\qquad\cdot P\left(y_{H|L}(\delta,S)\le\hat{y}_{H|L}\left(x_{\Sigma_m,(s)}\right)\right)\\
&\ge\prod_{m=1}^k(1-\alpha_m)^{2d_m}P\left(y_{H|L}(\delta_m,\Sigma_m)\le\hat{y}_{H|L}\left(x_{\Sigma_m,(s)}\right)\right) &(19)\\
&\ge\prod_{m=1}^k(1-\alpha_m)^{2d_m}(1-\alpha_m) &(20)\\
&\ge\prod_{m=1}^k(1-\alpha_m)^{2B}=\prod_{m=1}^k(1-\alpha_m)^B(1-\alpha_m)^B &(21)\\
&\ge(1-\alpha)(1-\alpha)=(1-\alpha)^2 &(22)
\end{aligned}
$$

where (19) holds based on Lemma 1 and (20) applies the interval quantile estimated from [1] with s calculated as in (13). The inequality in (21) holds since $d_m < B$. Finally, the inequality in (22) is obtained by repeatedly applying Bernoulli's inequality $(1-\frac{\alpha_k}{B})^B \ge (1-B\frac{\alpha_k}{B})$. □

4 Conclusions

Our multi-fidelity algorithm iteratively constructs a predictor function that is updated and specialized over subregions of the entire feasible region. Since the predictor is constructed to pass a test indicating that it is of good accuracy, we can use it to guide the placement of high-fidelity function evaluations. Leveraging our statistical model, Theorem 1 provides a finite time probabilistic guarantee for the quality of the resulting approximate target level set.

References

1. Conover, W.J.: Practical Nonparametric Statistics. Wiley, Hoboken (1980)
2. Frazier, P.: A tutorial on Bayesian optimization. arXiv:1807.02811v1 [stat.ML], 8 July 2018 (2018)
3. Gardner, J., Guo, C., Weinberger, K., Garnett, R., Grosse, R.: Discovering and exploiting additive structure for Bayesian optimization. In: Artificial Intelligence and Statistics, pp. 1311–1319 (2017)
4. Hoag, E., Doppa, J.R.: Bayesian optimization meets search based optimization: a hybrid approach for multi-fidelity optimization. In: Thirty-Second AAAI Conference on Artificial Intelligence (2018)
5. Huang, H., Zabinsky, Z.B.: Adaptive probabilistic branch and bound with confidence intervals for level set approximation. In: Proceedings of the 2013 Winter Simulation Conference: Simulation: Making Decisions in a Complex World, pp. 980–991. IEEE Press (2013)
6. Inanlouganji, A., Pedrielli, G., Fainekos, G., Pokutta, S.: Continuous simulation optimization with model mismatch using Gaussian process regression. In: 2018 Winter Simulation Conference (WSC), pp. 2131–2142. IEEE (2018)

7. Kandasamy, K., Dasarathy, G., Schneider, J., Póczos, B.: Multi-fidelity Bayesian optimisation with continuous approximations. In: Proceedings of the 34th International Conference on Machine Learning, vol. 70. pp. 1799–1808. JMLR.org (2017)
8. Kandasamy, K., Schneider, J., Póczos, B.: High dimensional Bayesian optimisation and bandits via additive models. In: International Conference on Machine Learning, pp. 295–304 (2015)
9. Li, C., Gupta, S., Rana, S., Nguyen, V., Venkatesh, S., Shilton, A.: High dimensional Bayesian optimization using dropout. arXiv:1802.05400 (2018)
10. Linz, D.D., Huang, H., Zabinsky, Z.B.: Multi-fidelity simulation optimization with level set approximation using probabilistic branch and bound. In: 2017 Winter Simulation Conference (WSC), pp. 2057–2068. IEEE (2017)
11. March, A., Willcox, K.: Provably convergent multifidelity optimization algorithm not requiring high-fidelity derivatives. AIAA J. **50**(5), 1079–1089 (2012)
12. Mockus, J.: Bayesian Approach to Global Optimization. Kluwer Academic Publishers, Dordrecht (1989)
13. Mockus, J.: Application of Bayesian approach to numerical methods of global and stochastic optimization. J. Global Optim. **4**, 347–365 (1994)
14. Poloczek, M., Wang, J., Frazier, P.: Multi-information source optimization. In: Advances in Neural Information Processing Systems, pp. 4288–4298 (2017)
15. Santner, T.J., Williams, B.J., Notz, W., Williams, B.J.: The Design and Analysis of Computer Experiments, vol. 1. Springer, New York (2003). https://doi.org/10.1007/978-1-4757-3799-8
16. Takeno, S., et al.: Multi-fidelity Bayesian optimization with max-value entropy search. arXiv preprint arXiv:1901.08275 (2019)
17. Törn, A., Zilinskas, A.: Global Optimization. Springer, Berlin (1989). https://doi.org/10.1007/3-540-50871-6
18. Wang, Z., Zoghi, M., Hutter, F., Matheson, D., De Freitas, N.: Bayesian optimization in high dimensions via random embeddings. In: Twenty-Third International Joint Conference on Artificial Intelligence (2013)
19. Wild, S.M., Regis, R.G., Shoemaker, C.A.: ORBIT: optimization by radial basis function interpolation in trust-regions. SIAM J. Sci. Comput. **30**(6), 3197–3219 (2008)
20. Wu, J., Toscano-Palmerin, S., Frazier, P.I., Wilson, A.G.: Practical multi-fidelity Bayesian optimization for hyperparameter tuning. arXiv:1903.04703 (2019)
21. Xu, J., Zhang, S., Huang, E., Chen, C.H., Lee, L.H., Celik, N.: An ordinal transformation framework for multi-fidelity simulation optimization. In: 2014 IEEE International Conference on Automation Science and Engineering (CASE), pp. 385–390. IEEE (2014)
22. Xu, J., Zhang, S., Huang, E., Chen, C.H., Lee, L.H., Celik, N.: MO2TOS: multi-fidelity optimization with ordinal transformation and optimal sampling. Asia-Pac. J. Oper. Res. **33**(03), 1650017 (2016)
23. Zabinsky, Z.B., Huang, H.: A partition-based optimization approach for level set approximation: probabilistic branch and bound. In: Smith, A.E. (ed.) Women in Industrial and Systems Engineering. WES, pp. 113–155. Springer, Cham (2020). https://doi.org/10.1007/978-3-030-11866-2_6

Performance Evaluation of Local Surrogate Models in Bilevel Optimization

Jaqueline S. Angelo[1(✉)], Eduardo Krempser[2], and Helio J. C. Barbosa[1,3]

[1] Laboratório Nacional de Computação Científica, Petrópolis, RJ, Brazil
{jsangelo,hcbm}@lncc.br
[2] Fundação Oswaldo Cruz, Rio de Janeiro, RJ, Brazil
eduardo.krempser@fiocruz.br
[3] Universidade Federal de Juiz de Fora, Juiz de Fora, MG, Brazil

Abstract. Bilevel problems (BLPs) involve solving two nested levels of optimization, namely the upper (leader) and the lower (follower) level, usually motivated by real-world situations involving a hierarchical structure. BLPs are known to be hard and computationally expensive. When the computation of the objective functions and/or constraints require an expensive computer simulation, as in "black-box" optimization, evolutionary algorithms (EAs) are often used. As EAs may become very expensive by requiring a large number of function evaluations, surrogate models can help overcome this drawback, either by replacing expensive evaluations or allowing for increased exploration of the search space. Here we apply different types of local surrogate models at the upper level optimization, in order to enhance the overall performance of the proposed method, which is studied by means of computational experiments conducted on the well-known SMD benchmark problems.

Keywords: Bilevel optimization · Surrogate models · Differential evolution

1 Introduction

Bilevel problems (BLPs) have gained significant attention due to their ability to model decentralized decision making in which two agents are hierarchically related. Since its first appearance in the area of game theory [24], the mathematical optimization community has applied great efforts to develop efficient methods to deal with the many challenges in BLP. Even the simplest case of BLPs with linear functions on both levels may be non-linear [4]. BLPs were shown to be strongly NP-hard [13], and just to identify if a solution is optimal or not, is itself NP-hard [27].

In BLPs, two decision makers are hierarchically related, the leader in the upper level (UL) and the follower in the lower level (LL). Each decision maker has control over a set of variables, seeking to optimize its own objective function subject to a set of constraints. Once the leader chooses the value of its variables,

G. Nicosia et al. (Eds.): LOD 2019, LNCS 11943, pp. 347–359, 2019.
https://doi.org/10.1007/978-3-030-37599-7_29

the follower reacts by choosing the optimal response in order to optimize its objective function. The class of BLPs considered here are:

$$\begin{aligned} &\min_{x \in X} F(x, y) \\ &\text{s.t.} \ \min_{y \in Y} f(x, y) \end{aligned} \tag{1}$$

where $x \in X \subset \mathbb{R}^n$ and $y \in Y \subset \mathbb{R}^m$ are the (bound constrained) control variables of the leader and the follower, respectively, with $F : \mathbb{R}^{n+m} \to \mathbb{R}$ and $f : \mathbb{R}^{n+m} \to \mathbb{R}$ being the objective function of the upper and lower level, respectively. The leader has control over x and makes his/her decision first, fixing x. The y variables are then set by the follower in response to the given x.

Compared with single-level optimization, the nested structure present in BLPs imposes additional challenges. The main one is that unless a solution is optimal for the LL problem, it cannot be feasible for the BLP. This indicates that approximate methods could not be used to solve the LL problem, as they do not guarantee that the optimal solution is actually found. Nevertheless, from the practical point of view, near-optimal solutions are commonly acceptable [10], especially when the LL problem is too costly to be exactly solved, therefore rendering the use of exact methods impractical. In addition, such approximate LL solutions may affect the optimization process, resulting in an objective value better than the true optimal at the UL [1].

A number of exact methods have been proposed involving the use of enumerative schemes, penalty function, gradient based methods, and single level reduction by replacing the LL problem by its KKT conditions. Those methods are mostly limited to handle BLPs assuming linearity, convexity, differentiability, etc, and in the case of single level reduction, it was shown that, in some cases, the transformed problem is not equivalent to the original bilevel formulation [11]. Due to the complexity of BLPs, using evolutionary algorithms (EAs) became a promising alternative.

A considerable amount of work has been published using EAs to solve different applications in bilevel optimization. Ant colony optimization, genetic algorithms, differential evolution, particle swarm optimization and coevolutionary methods are among the metaheuristics proposed for solving BLPs. Recent reviews of applications and solution methods for BLPs, both classical and evolutionary, can be found in [12,20]. Other derivative-free methods have also been investigated, such as trust-region based methods [8,9] and direct search [29].

In our first attempt to solve BLPs, we proposed a Differential Evolution (DE) algorithm which used two nested DE methods, one for optimizing each level of the BLP [2]. Although this procedure was considered computationally expensive, it was the first method to use DE on both levels of BLP, showing to be promising and useful for further developments. Thereby, to improve the previously developed method, in [3] we proposed the use of a Similarity-Based Surrogate Model (SBSM) to approximate the vector of the LL variables, by replacing the LL optimization, thus bypassing the DE in the LL problem, in order to save function evaluations at both levels.

Here, assumptions such as linearity, convexity, and differentiability are not included for the upper and lower level functions, and we focus our attention on the UL optimization, assuming that the UL objective function and constraints are much more computationally expensive than their LL counterparts, as can be found in structural optimization [5], robust optimization for circuit design [7,9], among other problems. In this way, we are going to analyze the performance of the proposed method when equipped with local SBSMs at the UL optimization. Note that, a different strategy to bypass the lower lever optimization is employed here. Instead of using a surrogate model to approximate the LL variables, as we have done in [3], the surrogate models will be used to approximate the UL objective function and constraints. The latter approach is much simpler and straightforward for applying metamodels in BLPs. To show the competitiveness of the proposed approach, computational experiments will be conducted using the SMD benchmark problems [23] and the results will be compared with other methods from the literature.

The main contribution of the paper is to present a new way of using well established methods in machine learning (k-Nearest Neighbors, Local Linear Regression and Radial Basis Function Networks) as surrogates for objective functions in bilevel programming, opening space for a more significant reduction of computational costs involved in solving a BLP, expanding its perspectives of application in new formulations and in higher dimensions.

The paper is organized as follows. Section 2 presents a description of the DE method and how it can be applied to solve BLPs. In Sect. 3 a literature review is conducted showing evolutionary methods that use surrogate models in bilevel optimization. The proposed approach in presented in Sect. 4, where the SBSMs applied are described, along with the description of the proposed algorithm. The computation experiments are analyzed and discussed in Sect. 5. Finally, Sect. 6 concludes the paper.

2 Differential Evolution for BLP

Differential Evolution (DE) [25,26] is a simple stochastic population based algorithm originally developed for continuous global optimization. It requires very few control parameters, basically the population size (pop), a scale factor (F), and the crossover rate (CR). DE starts with a set of candidate solutions, represented by vectors, randomly distributed in the search space. New candidate solutions are generated by basic operations. The simplest variant of DE, namely DE/rand/1/bin, performs the following mutation operation

$$v_{i,G+1} = x_{r_1,G} + \mathtt{F} \times (x_{r_2,G} - x_{r_3,G}) \tag{2}$$

where $i = 1, \ldots, \mathtt{pop}$, x_{r_1}, x_{r_2} and x_{r_3} are randomly selected individuals in the current population, with $r_1 \neq r2 \neq r3 \neq i$, F is the scale factor used to control the amplitude of the search in the direction of the applied difference and G is the current generation. In addition, a crossover operation is performed, using a user defined crossover rate parameter CR, in which the trial vector $u_{i,G+1}$ is created with elements from both the target vector $x_{i,G}$ and the donor vector $v_{i,G+1}$ as

$$u_{i,j,G+1} = \begin{cases} v_{i,j,G+1}, & \text{if } \mathtt{rand}(0,1) < \mathtt{CR} \text{ or } j = jRand \\ x_{i,j,G}, & \text{otherwise} \end{cases} \tag{3}$$

where $j = 1, \ldots, n$, $\mathtt{rand}(0,1)$ is a random number uniformly distributed in $[0,1]$, and $jRand$ is a randomly generated integer from 1 to n so that, $u_{i,G+1} \neq x_{i,G}$. The fitness or quality of the target vector $x_{i,G}$ is compared to that of the trial vector $u_{i,G+1}$, the one with the best solution value is selected to the next generation. This process continues until a stopping criterion is reached.

To apply DE in BLP, we proposed in [2] a nested procedure, called BLDE, in which two DE were used, each one responsible to optimize one level of the BLP. At the UL the solution method maintain two populations, $\mathtt{pop_L}$ which contains the UL members, and $\mathtt{pop_{LF}}$ containing the associated optimal LL solutions. The leader's DE is responsible to generate and evolve the UL variables in $\mathtt{pop_L}$ For each UL member, the follower's DE is executed, by generating and evolving the LL variables in $\mathtt{pop_F}$, and the optimal LL solution is returned to the leader. Given the follower's response, the UL solution can be evaluated with respect to the UL objective function and constraints. The target and the trial vectors are compared, and the one with the best solution value is selected to the next generation, placing x in $\mathtt{pop_L}$ and the optimal response y in $\mathtt{pop_{LF}}$. The BLDE algorithm returns the best UL solution with its best LL response.

3 Review of Methods Using Surrogate Models in BLP

In the domain of evolutionary bilevel optimization, few works can be found addressing the use of surrogate models. Most of them applied surrogate models at the LL, where the objective functions and/or constraints of the LL are approximated. We briefly outline some of them in cronological order.

The algorithm called BLEAQ, proposed in [22], constructed quadratic approximations to predict the LL optimal variables, suppressing the need to solve the LL optimization. The approximations are constructed based on a database of UL solutions and the corresponding LL optimum solutions. Further improvements for BLEAQ were proposed in [19] by incorporating archiving and local search procedure.

The surrogate assisted BLDE proposed by the authors in [3], used the k-Nearest Neighbors method (k-NN) as a surrogate model. The idea was to approximate the vector of the LL variables by replacing the LL optimization, in this way bypassing the execution of the LL DE algorithm. Based on a database with truly evaluated UL solutions, i.e. with the LL variables obtained via the execution of the follower's DE, the surrogate model approximates the values of the LL variables. A probability β was used to balance the choice of using a true evaluation or a surrogate model to approximate the vector of LL variables.

The SABLA algorithm proposed in [14] uses DE algorithm at the UL and multiple surrogate models at the LL. During the generations, LL solutions are approximated through a set of surrogate models, composed by Response Surface method and Kriging. The metamodel that best approximate the new solutions is chosen for prediction in subsequent generations of LL optimization. A periodically nested local search was also performed at both levels to improve the performance of the algorithm. In [15] the authors extended the idea used in SABLA by applying Efficient Global Optimization (EGO) as the surrogate model. Two methods were proposed, BLEGO_{V1} that used EGO only at the LL, and BLEGO_{V2} that used EGO on both levels.

In this paper, we are interested in efficiently solving BLPs where the UL objective function and constraints are much more computationally expensive than their lower-level counterparts. Knowing that surrogate models are capable to improve the performance of EAs in computationally expensive single level optimization problems [16], the idea is to extend the research done in [3] by analysing local SBSMs now at the UL optimization. It is important to highlight that in [3] the metamodel was applied at the LL to approximate the vector of LL variables, while in the proposed approach different surrogate models will be applied at the UL to approximate the (expensive) UL objective function value.

4 The Proposed Approach

In contrast to neural networks and polynomial response surfaces, which generate a model and then discard the inputs, SBSMs store the inputs and defer processing until a prediction of the fitness value of a new candidate solution is required. The evaluations are based on individuals stored in a archive, which were evaluated by the original model, and then they reply by combining their stored data (previous samples) using a similarity measure.

In [16], it was shown that the use of a local surrogate model to solve computationally expensive single level optimization problems is capable to improves the performance of DE algorithm. Hence, in order to improve the bilevel method previously proposed in [3], in this paper we analyse three different local SBSMs to (hopefully) obtain accurate solutions, at the same time reducing the number of UL function evaluations. The proposed method, named BLDES, uses two nested DE algorithms, one for each level, and incorporates a local surrogate model at the UL optimization.

4.1 Surrogate Models

For the local SBSMs used in this paper, the approximations are built based on a database $\mathcal{D}$, which stores η individuals (samples) with their respective original objective function value, in which

$$\mathcal{D} = \{x_i, F(x_i, y_i), y_i, f(x_i, y_i), i = 1, ..., \eta\} \quad (4)$$

where x_i and y_i are the variables of the leader and the corresponding follower (obtained by the follower's DE procedure) respectively, and $F(x_i, y_i)$ and

$f(x_i, y_i)$ are its respectively objective function values. Each time x_i and y_i are evaluated exactly, the set $\mathcal{D}$ is updated with those values. Notice that, constraints values can also be incorporated to the set $\mathcal{D}$ where the approximations could be performed the same way as done for the objective function.

k-Nearest Neighbors (k-NN). The k-nearest neighbors method (k-NN) [17] is a simple and intuitive surrogate model. Given a candidate solution x and the archive $\mathcal{D}$, the following approximation is considered:

$$F(x, y) \approx \widehat{F}(x, y) = \frac{\sum_j^{|\mathcal{N}|} s(x, x_j^{\mathcal{N}})^p F(x_j^{\mathcal{N}})}{\sum_j^{|\mathcal{N}|} s(x, x_j^{\mathcal{N}})^p} \tag{5}$$

where $|.|$ denotes the cardinality of a set, the $x_j^{\mathcal{N}} \in \mathcal{N}$ are the nearest neighbors of x, with $\mathcal{N}$ composed by the k elements in the set $\mathcal{D}$ most similar to x, $s(x, x_j^{\mathcal{N}})$ is a similarity measure between x and $x_j^{\mathcal{N}}$, and p is set to 2. Here,

$$s(x, x_j^{\mathcal{N}}) = 1 - \frac{d_E(x, x_j^{\mathcal{N}})}{d_E(x^u, x^l)}, \tag{6}$$

where $d_E(x, x_j^{\mathcal{N}})$ is the Euclidean distance between x and $x_j^{\mathcal{N}}$, and x^u and x^l are the vectors containing the upper and lower bounds of each variable, respectively. If $x = x_i$ for some $x_i \in \mathcal{D}$ then $\widehat{F}(x, y) = F(x_i, y_i)$. In this method, the only parameter to be set by the user is the number of the nearest neighbors k.

Local Linear Regression (LLR). The local linear regression [28] is another kind of metamodel technique analysed here. The general form of the linear regression is given by

$$\widehat{F}(x, y) = \theta_0 x_0 + \theta_1 x_1 + \cdots + \theta_n x_n, \quad \text{with } x_0 = 1. \tag{7}$$

where the parameters of the linear model $\theta \in R^{n+1}$ must be calculated. Given a matrix X with a set of candidate solutions in $\mathcal{D}$ and $\overrightarrow{F}$ their corresponding objective function values, the least squares estimation leads to

$$\theta = (X^T X)^{-1} X^T \overrightarrow{F}, \tag{8}$$

were the superscript T denotes transposition and $(X^T X)$ must be a non-singular matrix. In practice, a pseudo-inverse of $(X^T X)$ is computed in the singular case. Here, the singular value decomposition (SVD) from the Eigen library[1] was used.

We assume that, given a set of neighbors of a query point, it is possible to obtain a linear approximation $\widehat{F}(x, y) \approx F(x, y)$. The actual number of neighbors is the minimum number of data points for which the linear system can be solved. To obtain a linear system with the minimum number of points, the number of neighbors is initialized with the dimension of the problem plus 1. Until the linear system cannot be solved, the number of neighbors is increased by one in each trial, up to twice the dimension of the problem. After that, the number of neighbors is duplicated in each trial.

[1] http://eigen.tuxfamily.org.

Radial Basis Function (RBF) Networks. The Radial Basis Function Networks (RBF) [6] is a special neural network architecture with three layers. The units of the hidden layer are called neurons and each neuron applies a radial basis function. The model can be formulated as

$$\widehat{F}(x, y) = \sum_{i=1}^{k} \theta_i \phi(||x - c_i||) \tag{9}$$

where the nearest solutions previously calculated are the centers of the approximation, given by c_i, $i = 1, \ldots, k$. θ_i, $i = 1, \ldots, k$, are the weights connecting the hidden and output layers, and the radial basis function (ϕ) considered is the Gaussian function

$$\phi(r) = e^{-\frac{r^2}{2\sigma^2}} \tag{10}$$

where r is the distance between an individual and the basis i. The σ value is a real parameter, here defined by the distance between the bounds of the problem.

The learning process of the RBF is basically the selection of the centers and the definition of the weights. Here, the RBF was applied following the SBSM idea, i. e., for each new candidate solution a local approximation was performed selecting as centers the nearest solutions previously calculated.

4.2 Description of the Method

The pseudo-code of the algorithm is presented in Algorithm 1 (`BLDES`) and Algorithm 2 (`DEFollower`). The main steps are summarized as follows:

Step 1: Initialization. In the UL two populations are maintained, $\texttt{pop}_\text{L}$ containing individuals with $x \in R^n$ variables and $\texttt{pop}_\text{LF}$ containing individuals with the optimal LL response $y \in R^m$, both populations with $\texttt{size}_\text{L}$ members. In the initialization procedure (line 2 of `BLDES`), each UL member is initialized with random values and the corresponding LL variables are determined by calling the LL procedure `DEFollower`, which generates the optimal solution y of LL variables. The UL members are then evaluated based on UL objective function and constraints.

Step 2. Upper level procedure. For each UL member, w offsprings are generated using mutation and recombination operations (line 6 of `BLDES`), and the evaluations are done by the local surrogate model (line 7 of `BLDES`). Note that, when the surrogate model is used, the `DEFollower` is no longer executed. The best offspring solution, with respect to the UL objective function and constraints, is selected (line 8 of `BLDES`) and evaluated exactly (line 10 of `BLDES`), that is, the `DEFollower` is executed and the optimal LL solution is returned to the leader (line 9 of `BLDES`). This approach is important to not mislead the results with inaccurate LL solutions. After that, the UL solution x and its associated LL solution y are stored in the archive $\mathcal{D}$ (line 11 of `BLDES`). The fitness of the target vector (x_{i,G_L}, y_{i,G_L}) is then compared to that of the trial vector $(u_{i,G_L+1}, y_{i,G_L+1})$ (line 12 of `BLDES`), the one with the best solution value, with respect to the UL objective function and constraints, is selected to the next generation (line 13 of `BLDES`).

Step 2.1. Lower level procedure. For a given UL solution, the LL procedure is executed. The individuals of $\mathtt{pop_F}$, with $\mathtt{size_F}$ members, are initialized with random values and are evaluated based on LL objective function and constraints (line 2 of `DEFollower`). The LL individuals are generated using mutation and recombination operations (line 5 of `DEFollower`), and the solutions are evaluated based on LL objective function and constraints (line 6 of `DEFollower`). The fitness of the target vector (y_{i,G_F}) is then compared to that of the trial vector (u_{i,G_F+1}) (line 7 of `DEFollower`), and the one with the best solution value, based on LL objective function and constraints, is selected to the next generation (line 8 of `DEFollower`). The `DEFollower` returns the best value of the LL problem, given x. The LL procedure stops the execution when the function value of the best $\mathtt{pop_F}$ member reaches a desired objective function value accuracy or if exceeds a maximum number of generations.

Step 3. Final solution. The algorithm `BLDES` stops the execution when the function value of the best member of $\mathtt{pop_L}$ (x_{best}), with its associated y in $\mathtt{pop_{LF}}$ (y_{best}), reaches a desired objective function value accuracy or if exceeds a maximum number of generations. The algorithm returns the best UL solution x_{best} with its optimal LL solution y_{best}.

Algorithm 1. `Algorithm BLDES`

Input : npL (size of leader's pop.), npF (size of follower's pop.), genF (follower's generations), CR (crossover rate), F (mutation scaling), w (number of offsprings evaluated by the metamodel)

$G_L \leftarrow 0$;
GenerateRandomInitialPopulation($pop_L^{G_L}$, $pop_{LF}^{G_L}$);
while *termination criteria not met* **do**
  for $i \leftarrow 1$ **to** npL **do**
    for $l \leftarrow 1$ **to** w **do**
      $\vec{u}^{\,l}_{i,G_L+1} \leftarrow$ ApplyOperators($pop_{L,i}^{G_L}$, CR, F); /* Eqs.(2-3) */
      EvaluateWithLocalSurrogate $\widehat{F}(\vec{u}^{\,l}_{i,G_L+1})$;
    $\vec{u}_{i,G_L+1} \leftarrow$ SelectBestOffspring($\vec{u}^{\,l}_{i,G_L+1}$); /* With $l \in 1, \ldots, \mathsf{w}$ */
    $\vec{y}_{i,G_L+1} \leftarrow$ DEFollower($\vec{u}_{i,G_L+1}$);
    Evaluate $F(\vec{u}_{i,G_L+1}, \vec{y}_{i,G_L+1})$;
    InsertDatabase($\vec{u}_{i,G+1}$, $\vec{y}_{i,G_L+1}$);
    $(x_{i,G_L+1}, y_{i,G_L+1}) \leftarrow$ SelectBest((x_{i,G_L}, y_{i,G_L}), $(u_{i,G_L+1}, y_{i,G_L+1})$);
    UpdatePopulation($pop_{L,i}^{G_L+1} \leftarrow \vec{x}_{i,G_L+1}, pop_{LF,i}^{G_L+1} \leftarrow \vec{y}_{i,G_L+1}$);
  G_L ++;
return $(\vec{x}_{best}, \vec{y}_{best})$

5 Computational Experiments

The performance of the proposed method is analysed on the unconstrained SMD benchmark problems [23] and the results are compared with those in works [3] and [21]. The SMD test problems analyzed contain different levels of difficulty

Algorithm 2. `DEFollower`

input : $\vec{v}$ (leader's variables), npF (size of follower's population), genF (follower's generations), CR (crossover rate), F (mutation scaling)

$G_F \leftarrow 0$;
`CreateRandomInitialPopulation`($pop_F^{G_F}$);
while *termination criteria not met* **do**
  for $i \leftarrow 1$ **to** npF **do**
    $\vec{u}_{i,G_F+1} \leftarrow$ `ApplyOperators`($pop_F^{G_F}$, i, CR, F); /* Eqs.(2-3) */
    `Evaluate` $f(\vec{v}, \vec{u}_{i,G_F+1})$;
    $u_{i,G_F+1} \leftarrow$ `SelectBest`(u_{i,G_F+1}, y_{i,G_F});
    `UpdatePopulation`($pop_{F,i}^{G+1} \leftarrow u_{i,G_F+1}$);
  G_F ++;
return $\vec{y}_{best}$

in terms of convergence, complexity of interaction between the two levels and multimodality. Each problem was tested with 5 dimensions, 2 variables at the UL and 3 variables at the LL, using $p = 1$, $q = 2$ and $r = 1$ for all SMD problems except SMD6. For the SMD6 we used $p = 1$, $q = 1$, $r = 1$ and $s = 1$.

The parameters setting of BLDES algorithm are the following: the upper and lower level population size was set to 10 and 20, respectively; for a fair comparison, the stopping criterion is the same as that used in [21], that is, when the function value reaches an accuracy of 10^{-2} on both levels. In addition, for the LL procedure, a maximum number of 50 generation was also employed; the crossover rate CR was set to 0.8; the scaling factor F was set to 0.9; the DE variant used was the DE/rand/1/bin; and the number of offspring w generated by each UL member varies to 2, 5, 10 and 20.

For the SMD benchmark problems, the solution of both levels are known. In this way, the accuracy is calculated using an error metric given by $\varepsilon_{ul} = |F^* - F_t^*|$ for the UL, where F^* is the best value obtained by the algorithm at the end of execution, and F_t^* is the true optimum, and similarly $\varepsilon_{ll} = |f^* - f_t^*|$ for the LL procedure.

5.1 Results and Discussions

Table 1 shows the median of UL and LL function evaluations used by the methods when solving the test problems SMD1 to SMD4 and SMD5 to SMD8, respectively. The numbers in boldface indicate the lowest values of UL function evaluations. The results of the proposed BLDES algorithm are compared with those provided by algorithms BLDE [2] and BLEAQ-II [21]. Twelve configurations of BLDES were tested, where in Table 1 the numbers after BLDES indicates the number of offspring w each UL member generates, and the last three letters indicate the local surrogated model equipped, which are, KNN, LLR and RBF. The last two lines in Table 1 show the percentage of UL function evaluations savings from the best configuration of BLDES, indicated by the numbers in boldface.

Table 1. Median of function evaluations on SMD1 to SMD8.

Method	SMD1		SMD2		SMD3		SMD4		SMD5		SMD6		SMD7		SMD8	
	UL	LL	UL	LL	UL	LL	UL	LL	UL	LL	UL	LL	UL	LL	UL	LL
BLDES-2-KNN	160	88230	150	85070	170	114070	100	71480	170	122370	160	90160	190	111180	260	173210
BLDES-2-RBF	150	85460	130	73400	160	108970	510	218710	160	112690	150	83480	170	96470	250	166630
BLDES-2-LLR	190	105080	150	83210	170	114160	150	101530	170	125270	190	105420	260	147150	280	186570
BLDES-5-KNN	120	66900	80	46820	170	119480	130	86590	160	109960	130	72630	170	95680	400	191280
BLDES-5-RBF	100	56290	80	43740	130	91460	120	65780	110	73910	110	57660	130	67570	**190**	126680
BLDES-5-LLR	160	91290	110	64500	180	120650	100	66670	150	105280	180	99050	220	127180	260	170020
BLDES-10-KNN	90	49300	70	35790	120	77820	210	101160	120	76940	140	78560	90	52380	510	256170
BLDES-10-RBF	80	43860	70	37410	80	53380	60	42020	80	50530	**80**	45140	120	67560	290	154970
BLDES-10-LLR	160	89910	110	59940	180	121350	90	62440	170	121750	160	88430	190	101640	240	163950
BLDES-20-KNN	**70**	35420	70	34240	200	136160	60	37150	**70**	48540	140	77180	120	67300	510	241230
BLDES-20-RBF	**70**	37440	**60**	32330	**70**	46860	**50**	33650	**70**	46170	**80**	41440	**70**	35420	210	123800
BLDES-20-LLR	180	104880	110	63620	180	124510	130	86490	170	123820	170	94110	160	95150	250	163210
BLDE	190	109550	140	81800	200	138530	510	248440	180	128100	200	110710	210	123060	290	199780
BLEAQ-II	123	8462	114	7264	264	12452	272	8600	126	14490	259	914	180	8242	644	22866
BLDES savings of UL function evaluations (in %)																
BLDE	53		57		65		90		61		66		67		34	
BLEAQ-II	27		47		73		82		44		69		61		70	

Regarding the number of UL function evaluations, the proposed BLDES equipped with RBF and $w = 20$ (BLDES-20-RBF), provided the best performance, reaching the lowest values of UL function evaluations in all test problems, with exception of SMD8, in which BLDES-5-RBF reaches the lowest value. Particularly for SMD4, BLDES-20-RBF reduced the number of UL functions evaluations by 90% compared with BLDE, and more then 80% when compared with BLEAQ-II. It is also noticed that, most configurations of BLDES outperform BLDE in all test problems regarding the number of upper and lower level function evaluations. When compared with BLEAQ-II, configurations with $w = 5, 10, 20$ using KNN and RBF performs best regarding the number of UL function evaluations.

It is possible to observe that, in BLDES when the number of offspring w increases, in general, the number of function evaluations decrease in both levels. The opposite can be seen, e.g., in BLDES equipped with LLR on SMD4, where the number of upper and lower level function evaluations decrease until $w = 10$ and increase when $w = 20$. This phenomena was already observed in our previous work [3], where when the probability of using the metamodel increased, it caused a deterioration on the quality of the result.

It is important to highlight that when the application of the metamodel is increased, here represented by increasing the number of offspring evaluated by it, the impact of the approximation of the metamodel in the quality of the expected solution is also increased. If the individuals chosen to be part of the approximation are underrepresented for estimating the original function, this could lead to low quality approximations.

Regarding the number of LL function evaluations, BLEAQ-II outperforms BLDES in all test problems analysed. In BLEAQ-II a genetic algorithm is used and quadratic approximations are used if the LL problem is convex. In addition, local searches are constantly performed by meta-modeling the upper and lower level functions.

6 Conclusions

In this work, we considered bilevel problems where the UL objective function and/or constraints are much more computationally expensive than their lower-level counterparts. Hence, an evolutionary bilevel method equipped with local surrogate models at the UL optimization was proposed. In previous works [3,16], the use of surrogate models proved to be reliable and efficient, and in some cases, are the most straightforward alternative to reduce computationally expensive simulations.

The experiments showed the ability of the local surrogate models to improve the performance of the method, by allowing the algorithm to reach a desired accuracy of objective function value with less UL objective function evaluations. Although this work focused on the UL optimization, investigations are being conducted also concerned with reducing the number of LL function evaluations.

In future works the surrogate model will also be used to perform a pre-selection of the best offspring among those obtained by using different DE mutation variants, as previously done in [18]. This approach alleviates the user from the task of defining *a priori* which type of mutation formula to use in the DE. Another line of investigation is to explore the number of times the surrogate model is applied, analyzing the compromise relation between the number of times the individuals are evaluated exactly and by the metamodel, in order to find more accurate solutions at the same time preventing the UL optimization from being mislead due to inaccurate LL results.

Acknowledgement. The authors thank the reviewers for their comments and observations, and PNPD/CAPES and CNPq (grant 312337/2017-5), for the financial support.

References

1. Angelo, J.S., Barbosa, H.J.C.: A study on the use of heuristics to solve a bilevel programming problem. Int. Trans. Oper. Res. **22**(5), 861–882 (2015)
2. Angelo, J.S., Krempser, E., Barbosa, H.J.C.: Differential evolution for bilevel programming. In: IEEE Congress on Evolutionary Computation, pp. 470–477 (2013)
3. Angelo, J.S., Krempser, E., Barbosa, H.J.C.: Differential evolution assisted by a surrogate model for bilevel programming problems. In: IEEE Congress on Evolutionary Computation, pp. 1784–1791 (2014)
4. Bard, J.F.: Practical Bilevel Optimization. Kluwer Academic Publisher, Boston (1998)

5. Barjhoux, P.J., Diouane, Y., Grihon, S., Bettebghor, D., Morlier, J.: A bilevel methodology for solving a structural optimization problem with both continuous and categorical variables. In: 2018 Multidisciplinary Analysis and Optimization Conference, pp. 1–16 (2018)
6. Broomhead, D., Lowe, D.: Multivariable functional interpolation and adaptive networks. Complex Syst. **2**, 321–355 (1988)
7. Ciccazzo, A., Latorre, V., Liuzzi, G., Lucidi, S., Rinaldi, F.: Derivative-free robust optimization for circuit design. J. Optim. Theory Appl. **164**(3), 842–861 (2015)
8. Colson, B., Marcotte, P., Savard, G.: A trust-region method for nonlinear bilevel programming: algorithm and computational experience. Comput. Optim. Appl. **30**(3), 211–227 (2005)
9. Conn, A.R., Vicente, L.N.: Bilevel derivative-free optimization and its application to robust optimization. Optim. Methods Softw. **27**(3), 561–577 (2012)
10. Deb, K., Sinha, A.: Solving bilevel multi-objective optimization problems using evolutionary algorithms. In: Ehrgott, M., Fonseca, C.M., Gandibleux, X., Hao, J.-K., Sevaux, M. (eds.) EMO 2009. LNCS, vol. 5467, pp. 110–124. Springer, Heidelberg (2009). https://doi.org/10.1007/978-3-642-01020-0_13
11. Dempe, S.: Foundations of Bilivel Programming. Kluwer Academic Publisher, Dordrecht (2002)
12. Dempe, S.: Bilevel optimization: theory, algorithms and applications (2018). https://tu-freiberg.de/sites/default/files/media/fakultaet-fuer-mathematik-und-informatik-fakultaet-1-9277/prep/preprint_2018_11_dempe.pdf
13. Hansen, P., Jaumard, B., Savard, G.: New branch-and-bound rules for linear bilevel programming. SIAM J. Sci. Stat. Comput. **13**(5), 1194–1217 (1992)
14. Islam, M.M., Singh, H.K., Ray, T.: A surrogate assisted approach for single-objective bilevel optimization. IEEE Trans. Evol. Comput. **21**(5), 681–696 (2017)
15. Islam, M.M., Singh, H.K., Ray, T.: Efficient global optimization for solving computationally expensive bilevel optimization problems. In: 2018 IEEE Congress on Evolutionary Computation, pp. 1–8, July 2018
16. Krempser, E., Bernardino, H.S., Barbosa, H.J., Lemonge, A.C.: Performance evaluation of local surrogate models in differential evolution-based optimum design of truss structures. Eng. Comput. **34**(2), 499–547 (2017)
17. Shepard, D.: A two-dimensional interpolation function for irregularly-spaced data. In: Proceedings of the 23rd ACM National Conference, New York, NY, USA, pp. 517–524 (1968)
18. da Silva, E.K., Barbosa, H.J.C., Lemonge, A.C.C.: An adaptive constraint handling technique for differential evolution with dynamic use of variants in engineering optimization. Optim. Eng. **12**, 31–54 (2011)
19. Sinha, A., Malo, P., Deb, K.: An improved bilevel evolutionary algorithm based on quadratic approximations. In: 2014 IEEE Congress on Evolutionary Computation, pp. 1870–1877 (2014)
20. Sinha, A., Malo, P., Deb, K.: A review on bilevel optimization: from classical to evolutionary approaches and applications. IEEE Trans. Evol. Comput. **22**(2), 276–295 (2018)
21. Sinha, A., Lu, Z., Deb, K., Malo, P.: Bilevel optimization based on iterative approximation of multiple mappings. arXiv preprint arXiv:1702.03394 (2017)
22. Sinha, A., Malo, P., Deb, K.: Efficient evolutionary algorithm for single-objective bilevel optimization. CoRR abs/1303.3901 (2013)
23. Sinha, A., Malo, P., Deb, K.: Test problem construction for single-objective bilevel optimization. Evol. Comput. **22**(3), 439–477 (2014). pMID: 24364674

24. Stackelberg, H.V.: Marktform und Gleichgewicht. Springer, Berlin (1934). English translation: The Theory of the Market Economy. Oxford University Press, Oxford (1952)
25. Storn, R., Price, K.V.: Differential evolution: a simple and efficient adaptive scheme for global optimization over continuous spaces (1995). iCSI, USA, Technical report, TR-95-012 (1995). http://icsi.berkeley.edu/~storn/litera.html
26. Storn, R., Price, K.V.: Differential evolution - a simple and efficient heuristic for global optimization over continuous spaces. J. Global Optim. **11**, 341–359 (1997)
27. Vicente, L.N., Savard, G., Júdice, J.J.: Descent approaches for quadratic bilevel programming. J. Optim. Theory Appl. **81**(2), 379–399 (1994)
28. Yan, X., Su, X.G.: Linear Regression Analysis: Theory and Computing. World Scientific Publishing Company, Singapore (2009)
29. Zhang, D., Lin, G.H.: Bilevel direct search method for leader-follower problems and application in health insurance. Comput. Oper. Res. **41**, 359–373 (2014)

BowTie - A Deep Learning Feedforward Neural Network for Sentiment Analysis

Apostol Vassilev(✉)

National Institute of Standards and Technology, 100 Bureau Drive, Gaithersburg, MD 20899, USA
apostol.vassilev@nist.gov

Abstract. How to model and encode the semantics of human-written text and select the type of neural network to process it are not settled issues in sentiment analysis. Accuracy and transferability are critical issues in machine learning in general. These properties are closely related to the loss estimates for the trained model. I present a computationally-efficient and accurate feedforward neural network for sentiment prediction capable of maintaining low losses. When coupled with an effective semantics model of the text, it provides highly accurate models with low losses. Experimental results on representative benchmark datasets and comparisons to other methods (DISCLAIMER: This paper is not subject to copyright in the United States. Commercial products are identified in order to adequately specify certain procedures. In no case does such identification imply recommendation or endorsement by the National Institute of Standards and Technology, nor does it imply that the identified products are necessarily the best available for the purpose.) show the advantages of the new approach.

Keywords: Deep learning · Sentiment analysis · Natural Language Processing

1 Introduction

When approaching the problem of applying deep learning to sentiment analysis one faces at least five classes of issues to resolve. First, what is the best way to encode the semantics in natural language text so that the resulting digital representation captures well the semantics in their entirety and in a way that can be processed reliably and efficiently by a neural network and result in a highly accurate model? This is a critically important question in machine learning because it directly impacts the viability of the chosen approach. There are multiple ways to encode sentences or text using neural networks, ranging from a simple encoding based on treating words as atomic units represented by their rank in a vocabulary [1], to using word embeddings or distributed representation of words [2], to using sentence embeddings. Each of these encoding types

Supported by NIST Information Technology Laboratory Grant #7735282-000.

G. Nicosia et al. (Eds.): LOD 2019, LNCS 11943, pp. 360–371, 2019.
https://doi.org/10.1007/978-3-030-37599-7_30

has different complexity and rate of success when applied to a variety of tasks. The simple encoding offers simplicity and robustness. The usefulness of word embeddings has been established in several application domains, but it is still an open question how much better it is than the simple encoding in capturing the entire semantics of the text in natural language processing (NLP) to provide higher prediction accuracy in sentiment analysis. Although intuitively one may think that because word embeddings do capture some of the semantics contained in the text this should help, the available empirical test evidence is inconclusive. Attempts to utilize sentence embeddings have been even less successful [3].

Second, given an encoding, what kind of neural network should be used? Some specific areas of applications of machine learning have an established leading network type. For example, convolutional neural networks are preferred in computer vision. However, because of the several different types of word and sentence encoding in natural language processing (NLP), there are multiple choices for neural network architectures, ranging from feedforward to convolutional to recurrent neural networks.

Third, what dataset should be used for training? In all cases the size of the training dataset is very important for the quality of training but the way the dataset is constructed and the amount of meta-data it includes also play a role. The Keras IMDB Movie reviews Dataset [4] (KID) for sentiment classification contains human-written movie reviews. A larger dataset of similar type is the Stanford Large Movie Review Dataset (SLMRD) [5]. I consider KID and SLMRD in Sects. 2.3 and 2.4 respectively. Generally, simpler encodings and models trained on large amounts of data tend to outperform complex systems trained on smaller datasets [2].

Fourth, what kind of training procedure should be employed - supervised or unsupervised? Traditionally, NLP systems are trained on large unsupervised corpora and then applied on new data. However, researchers have been able to leverage the advantages of supervised learning and transfer trained models to new data by retaining the transfer accuracy [3].

Fifth, when training a model for transfer to other datasets, what are the model characterizing features that guarantee maintaining high/comparable transfer accuracy on the new dataset? Certainly, training and validation accuracy are important but so are the training and validation losses. Some researchers argue that the gradient descent method has an implicit bias that is not yet fully understood, especially in cases where there are multiple solutions that properly classify a given dataset [6]. Thus, it is important to have a neural network with low loss estimates for a trained model to hope for a good and reliable transfer accuracy.

The primary goal of this paper is to shed light on how to address these issues in practice. To do this, I introduce a new feedforward neural network for sentiment analysis and draw on the experiences from using it with two different types of word encoding: a simple one based on the word ranking in the dataset vocabulary; the other judiciously enhanced with meta-data related to word polarity. The main contribution of this paper is the design of the BowTie neural network in Sect. 3.

2 Data Encoding and Datasets

As discussed above, there are many different types of encodings of text with different complexity and degree of effectiveness. Since there is no convincing positive correlation established in the literature between complexity of the encoding and higher prediction accuracy, it is important to investigate the extent to which simple data encodings can be used for sentiment analysis. Simpler encodings have been shown to be robust and efficient [1]. But can they provide high prediction accuracy in sentiment analysis?

I investigate this open question by evaluating the accuracy one may attain using two types of text encoding on representative benchmark datasets.

2.1 Multi-hot Encoding

The first encoding is the well-known *multi-hot* encoding of text [7]. The multi-hot encoding represents a very simple model of the semantics in text.

2.2 Polarity-Weighted Multi-hot Encoding

The second encoding I consider is the polarity-weighted multi-hot encoding of text.

Let π be a linguistic type (e.g. morpheme, word) and let Π_D be the set of all such linguistic types in a dataset D. Let $M = |\Pi_D|$ be the cardinality of Π_D. Let ψ be a linguistic text type (e.g., a movie review) and let Ψ_D be the set of all texts in D. Let $N = |\Psi_D|$ be the cardinality of Ψ_D. Let Π_D and Ψ_D be finite sets such that the elements in each set are enumerated by $\{0, ..., M\}$ and $\{0, ..., N\}$ respectively. Let T^{NxM} be a tensor of real numbers of dimensions N by M, whose elements are weighted by the cumulative effect of the polarity of each word present in a given text π, as computed by [8]. Let $c_{\pi,\psi}$ be the number of tokens of the linguistic type π in a text ψ. Let ξ_π be the polarity rating of the token $\pi \in \Pi_D$. Naturally, I assume that if Ξ_D is the set of all polarity ratings for tokens $\pi \in \Pi_D$, then $|\Xi_D| = |\Pi_D|$. Let $\omega_{\xi\pi\psi} = \xi_\pi * c_{\pi,\psi}$ be the cumulative polarity of π in the text ψ. Let $\Omega_D = \{\omega_i\}_{i=0}^{M}$ and $C_D = \{c_i\}_{i=0}^{M}$. Let Θ^{NxM} be a tensor of real numbers of dimensions N by M, whose elements are set as follows:

$$\{\theta_{jk}\} = \begin{cases} \omega_{\xi_k \pi_k \psi_j}, & \text{if } \pi_k \in \psi_j; \\ 0, & \text{otherwise.} \end{cases} \tag{1}$$

The polarity-weighted multi-hot encoding (1) represents a more comprehensive model of the semantics in ψ, $\forall \psi \in \Psi_D$, that captures more information about ψ. I will attempt to investigate if and how much this additional information helps to improve the sentiment predictions in Sect. 4. An example of a polarity-weighted multi-hot encoded text is shown in Fig. 1.

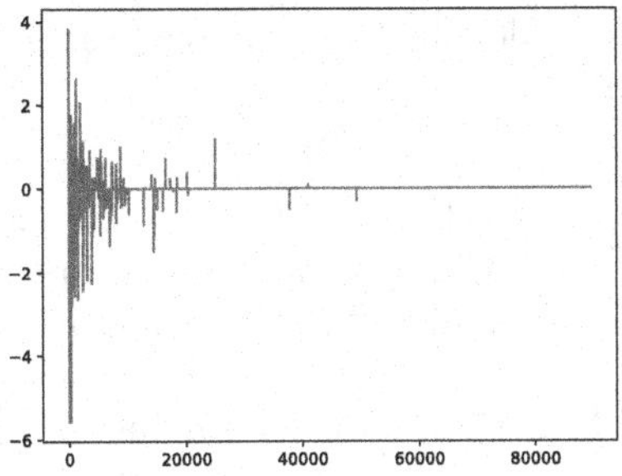

Fig. 1. A polarity-weighted multi-hot encoded text in D, Π_D with $M = 89527$.

2.3 The Keras IMDB Dataset (KID)

The KID [4] contains 50 000 human-written movie reviews that are split in equal subsets of 25 000 for training and testing and further into equal categories labeled as positive or negative. For convenience, the reviews have been pre-processed and each review is encoded as a sequence of integers, representing the ranking of the corresponding word in Π_D with $|\Pi_D| = 88\,587$. Hence, it can be easily encoded by the multi-hot encoding [7].

2.4 The Stanford Large Movie Review Dataset (SLMRD)

SLMRD contains 50 000 movie reviews, 25 000 of them for training and the rest for testing. The dataset comes also with a processed bag of words and a word polarity index [5,10]. SLMRD contains also 50 000 unlabeled reviews intended for unsupervised learning. It comes with a Π_D, polarity ratings Ω_D, and word counts C_D with $|\Omega_D| = |C_D| = |\Pi_D| = 89\,527$.

3 The BowTie Feedforward Neural Network

The ability of a neural network to provide accurate predictions and maintain low losses is very important for transferability to other datasets with the same or higher prediction accuracy as on the training dataset.

I now introduce a feedforward neural network with that criteria in mind. By way of background [11], logistic regression computes the probability of a binary output $\hat{y}_i$ given an input x_i as follows:

$$P(\hat{\mathbf{y}}|\mathbf{X}, \mathbf{w}) = \prod_{i=1}^{n} \mathbf{Ber}[\hat{y}_i|\mathbf{sigm}(x_i\mathbf{w})], \quad (2)$$

where $\mathbf{Ber}[\;]$ is the Bernouli distribution, $\mathbf{sigm}()$ is the sigmoid function, $\mathbf{w}$ is a vector of weights. The cost function to minimize is $\mathbf{C}(\mathbf{w}) = -\log P(\hat{\mathbf{y}}|\mathbf{X}, \mathbf{w})$. This method is particularly suitable for sentiment prediction. One critical observation is that logistic regression can be seen as a special case of the generalized linear model. Hence, it is analogous to linear regression. In matrix form, linear regression can be written as

$$\hat{\mathbf{y}} = \mathbf{X}\mathbf{w} + \epsilon, \tag{3}$$

where $\hat{\mathbf{y}}$ is a vector of predicted values $\hat{y}_i$ that the model predicts for $\mathbf{y}$, $\mathbf{X}$ is a matrix of row vectors x_i called regressors, $\mathbf{w}$ are the regression weights, and ϵ is an error that captures all factors that may influence $\hat{\mathbf{y}}$ other than the regressors $\mathbf{X}$.

The gradient descent algorithm used for solving such problems [11] may be written as

$$\mathbf{w}^{(k+1)} = \mathbf{w}^{(k)} - \rho^{(k)}\mathbf{g}^{(k)} + \epsilon^{(k)}, \tag{4}$$

where $\mathbf{g}^{(k)}$ is the gradient of the cost function $\mathbf{C}(\mathbf{w})$, $\rho^{(k)}$ is the learning rate or step size, and $\epsilon^{(k)}$ is the error at step k of the iterative process. One error-introducing factor in particular is the numerical model itself and the errors generated and propagated by the gradient descent iterations with poorly conditioned matrices running on a computer with limited-precision floating-point numbers. Even if regularization is used, the specific parameters used to weigh them in the equation, e.g. the L_2-term weight or the dropout rate, may not be optimal in practice thus leading to potentially higher numerical error. This is why it is important to look for numerical techniques that can reduce the numerical error effectively. This observation inspires searching for techniques similar to multigrid from numerical analysis that are very effective at reducing the numerical error [12].

The Neural Network Design

The feedforward neural network at the bottom of Fig. 2 consists of one encoding layer, a cascade of dense linear layers with L_2-regularizers and of appropriate output size ($Z_i@K$, where K is the layer output size) followed by a dropout regularizer and a sigmoid. The encoder takes care of encoding the input data for processing by the neural network. In this paper I experiment with the two encodings defined in Sect. 2: the simple multi-hot encoding and the polarity-weighted multi-hot encoding.

The sigmoid produces the estimated output probability $P(\hat{\mathbf{y}}^{(i)}|\mathbf{x}^{(i)}, \mathbf{w}^{(i)})$, which may be used to compute the negative log-loss or binary cross-entropy [11] as

$$-\left[\mathbf{y}\log(P(\hat{\mathbf{y}}^{(i)}|\mathbf{x}^{(i)}, \mathbf{w}^{(i)})) + (1-\mathbf{y})\log(1 - P(\hat{\mathbf{y}}^{(i)}|\mathbf{x}^{(i)}, \mathbf{w}^{(i)}))\right]. \tag{5}$$

The binary cross-entropy provides a measure for quality and robustness of the computed model. If the model predicts correct results with higher probability then the binary cross-entropy tends to be lower. If, however, the model predicts correct results with probability close to the discriminator value or predicts an incorrect category, the binary cross-entropy tends to be high. Naturally, it is desirable to have models that confidently predict correct results. It is also important to have models that maintain low binary cross-entropy for many training epochs because, depending on the training dataset, the iterative process (4) may need several steps to reach a desired validation accuracy. Models that quickly accumulate high cross-entropy estimates tend to overfit the training data and do poorly on the validation data and on new datasets.

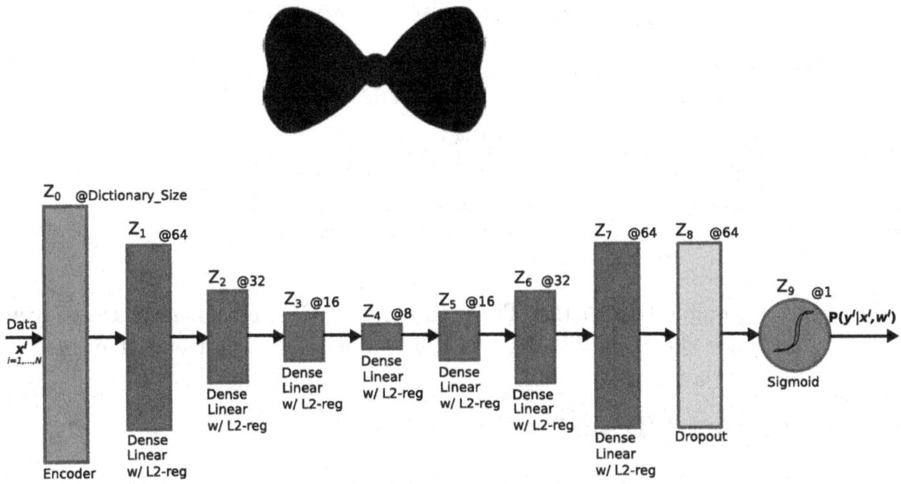

Fig. 2. The BowTie neural network. The classic bow tie (shown on top) originated among Croatian mercenaries during the Thirty Years' War of the 17th century. It was soon adopted by the upper classes in France, then a leader in fashion, and flourished in the 18th and 19th centuries. (Wikipedia). The estimated probability $P(\hat{\mathbf{y}}^{(i)}|\mathbf{x}^{(i)}, \mathbf{w}^{(i)})$ may be fed into a post-processing discriminator component to assign a category (pos/neg) for the input $\mathbf{x}^{(i)}$ with respect to a discriminator value $\delta \in [0, 1]$. All experiments presented in this paper use $\delta = 0.5$.

Hyperparameters. There are several hyperparameters that influence the behavior of BowTie, see Table 1. For optimal performance the choice of the dense layer activation should be coordinated with the choice for the L_2-regularization weight. This recommendation is based on the computational experience with BowTie and is in line with the findings in [14] about the impact of the activation layer on the training of neural networks in general. The optimal L_2-regularization weight for the linear network (no dense layer activation) is 0.019 whereas rectified linear unit (RELU) activation favors L_2-regularization weights close to 0.01. The network can tolerate a range of dropout rates but good stability and accuracy is attained with a dropout rate of 0.2. It is interesting to note that Root Mean Square Propagation (RMSProp) tends to converge faster to a solution and sometimes with a higher validation accuracy than Nesterov adaptive momentum (NADAM) but the transfer accuracy of the models computed with NADAM tends to be higher than for models computed with RMSPRop. For example, a model trained on SLMRD with validation accuracy of 89.24%, higher than any of the data in Table 4 below, yielded 91.04% transfer accuracy over KID, which is lower than the results in Table 4. This experimental finding is consistent over many tests with the two optimizers and needs further investigation in future research to explore the theoretical basis for it.

Table 1. BowTie hyperparameters.

Hyperparameters	
Name	Values/range
L_2-regularization weight	0.01 to 0.02
Dropout rate	0.2 to 0.5
Optimizer	NADAM or RMSProp
Dense layer activation	None (Linear network), RELU

It may be possible to run BowTie with other optimizers or activation layer choices than those shown in Table 1 but I have not tested their effectiveness.

4 Training and Transfer Scenarios

This section defines the objectives for the testing of the BowTie neural network shown in Fig. 2 in terms of four training and transfer scenarios.

But first it is important to decide on the type of training to employ - supervised or unsupervised. I embark on supervised training based on the findings in [3] about the advantages of supervised training and the availability of corpora of large labeled benchmark datasets [4] and [5]. The scenarios to explore are:

- **Scenario 1 (Train and validate):** Explore the accuracy and robustness of the BowTie neural network with the simple multi-hot encoding by training and validating on KID.
- **Scenario 2 (Train and validate):** Explore the accuracy and robustness of the BowTie neural network with the simple multi-hot encoding by training and validating on SLMRD.
- **Scenario 3 (Train and validate):** Explore the accuracy and robustness of the BowTie neural network with the polarity-weighted multi-hot encoding by training and validating on SLMRD.
- **Scenario 4 (Train, validate, and transfer):** Explore the transfer accuracy of the BowTie neural network with polarity-weighted multi-hot encoding by training on SLMRD and predicting on KID.

The primary goal of this exploration is to establish some baseline ratings of the properties of the BowTie neural network with the different encodings and compare against similar results for other neural networks with other types of encoding. This provides a quantitative criteria for comparative judging.

Results

In this section I report the results from executing Scenarios 1 to 4 from Sect. 4 using TensorFlow [7], version 1.12, on a 2017 MacBook Pro, *without* GPU acceleration. The modeling code is written in Python 3. The speed of computation improved on a platform *with* eight Tesla V100-SXM2 GPUs: speedup of nearly a factor of two but in training the acceleration plateaued if more than two GPUs were used.

Scenario 1. In this test, the BowTie neural network is tested with encoding [7]. The results in Table 2 show high accuracy and low binary cross-entropy estimates.

Table 2. Scenario 1 results. Data from experiments with training a model on KID until it attains some validation accuracy grater than 88%. Note that each time the data is loaded for training, it is shuffled randomly, hence the small variation in computational results.

Training and validating on KID	
Validation accuracy (%)	Validation binary cross-entropy
88.08	0.2955
88.18	0.2887
88.21	0.2945

To assess the relative computational efficiency of the BowTie neural network, I compared it to the convolutional neural network in [13] with a 10 000 word dictionary. The network reached accuracy of 88.92% at Epoch 4 with binary cross-entropy of 0.2682. However, it took 91 seconds/Epoch, which matched the numbers reported by the authors for the CPU-only computational platform [13]. In addition, after Epoch 4, the binary cross-entropy started to increase steadily while the accuracy started to decline. For example, the binary cross-entropy reached 0.4325 at Epoch 10 with validation accuracy of 87.76% and 0.5536 and 87.40% correspondingly at Epoch 15. In comparison, BowTie takes only 3 seconds/Epoch for the same dictionary size and attains accuracy of 88.20% with binary cross-entropy of 0.2898. The binary cross-entropy stays below 0.38 for a large number of Epochs.

Scenario 2. SLMRD is more challenging than KID for reasons that are visible in the test results for Scenarios 3 and 4, hence the slightly lower validation accuracy attained by BowTie using the simple multihot encoding [7] - it easily meets or exceeds the threshold accuracy of 87.95% but could not surpass the 88% level in several experiments.

Table 3. Scenario 2 results. Data from experiments with training a model on KID until it attains validation accuracy of at least 87.95%. Note that each time the data is loaded for training, it is shuffled randomly, hence the small variation in computational results.

Training and validating on SLMRD	
Validation accuracy (%)	Validation binary cross-entropy
87.98	0.2959
87.95	0.2996
87.96	0.3001

Scenarios 3 and 4. I combine the reporting for Scenarios 3 and 4 because once the model is trained under Scenario 3 it is then transferred to compute predictions on KID. To perform the transfer testing on KID one needs to reconcile the difference in $|\Pi_{SLMRD}|$ and $|\Pi_{KID}|$. As I noted in Sect. 2, $|\Pi_{SLMRD}| = 89\,527$ and $|\Pi_{KID}| = 88\,587$. Moreover, $\Pi_{KID} \not\subset \Pi_{SLMRD}$. Let $\Pi_{\Delta} = \Pi_{KID} \setminus (\Pi_{KID} \cap \Pi_{SLMRD})$. It turns out that

$$\Pi_{\Delta} = \left\{ \begin{array}{l} 0's, 1990s, 5, 18th, 80s, 90s, 2006, 2008, \\ 85, 86, 0, 5, 10, tri, 25, 4, 40's, 70's, 1975, \\ 1981, 1984, 1995, 2007, dah, walmington, \\ 19, 40s, 12, 1938, 1998, 2, 1940's, 3, 000, 15, 50. \end{array} \right\}.$$

Clearly, $|\Pi_{\Delta}|$ is small and of the words in Π_{Δ}, only 'walmington' looks like a plausible English word. This is the name of a fictional town in a British TV series from the 1970's. As such it has no computable polarity index and is semantically negligible. Based on this, I dropped all these words during the mapping of $\pi \in \Pi_{\Delta}$ into the corresponding $\pi\prime \in \Pi_{SLRMD}$. Note that the mapping $\pi \rightarrow \pi\prime$ enables the encoding of the semantics of the texts in KID according to (1).

Some Simple but Revealing Statistics About the Data. With encoding (1), the elements of the matrix T_{SLMRD} for the training set in SLMRD show cumulative polarity values in the range $[-50.072\,837, 58.753\,546]$ and the cumulative polarity of the elements of the matrix T_{SLMRD} for the test set is in the

Table 4. Scenarios 3 and 4 results. Data from experiments with training a model on SLMRD until it attains validation accuracy grater than 89% and using that model to predict the category for each labeled review in KID, computing over the entire set of 50 000 reviews in KID.

Training on SLMRD		Predicting on KID
Validation accuracy (%)	Val. binary cross-entropy	Tansfer accuracy (%)
89.02	0.2791	91.56
89.12	0.2815	91.63
89.17	0.2772	91.76

range [−48.960 107, 63.346 164]. This suggests that SLMRD is pretty neutral and an excellent dataset for training models. In contrast, the elements of the matrix T_{KID} are in the range [−49.500 000, 197.842 862]. Table 4 contains the results from executing Scenarios 3 and 4. The validation and transfer accuracy results in Table 4 are better than those shown in Tables 2 and 3. This demonstrates the value word polarity brings in capturing the semantics of the text. The transfer accuracy over KID is also higher than the results obtained by using the convolutional neural network [13] on KID. The results in Table 4 are higher than the results reported for sentiment prediction of movie reviews in [3] but in agreement with the reported experience by these authors about consistently improved accuracy from supervised learning on a large representative corpus before transferring the trained model to a corpus of interest.

Note also that the validation accuracy of BowTie with the encoding (1) on SLMRD is higher than the results in [10] for the same dataset. The observation in [10] that even small percentage improvements on a large corpus result in a significant number of correctly classified reviews applies to the data in Table 4: that is, there are between 172 and 210 more correctly classified reviews by BowTie.

5 Discussion and Next Steps

The experimental results from sentiment prediction presented above show the great potential deep learning has to enable automation in areas previously considered impossible. At the same time, we are witnessing troubling trends of deterioration in cybersecurity that have permeated the business and home environments: people often cannot access the tools they need to work or lose the data that is important to them [15]. Traditionally, governments and industry groups have approached this problem by establishing security testing and validation programs whose purpose is to identify and eliminate security flaws in IT products before adopting them for use. One area of specific concern in cybersecurity is cryptography. Society recognizes cryptography's fundamental role in protecting sensitive information from unauthorized disclosure or modification. The cybersecurity recommendations in [15] list relying on cryptography as a means of data protection as one of the top recommendations to the business community for the past several years. However, the validation programs to this day remain heavily based on human activities involving reading and assessing human-written documents in the form of technical essays. This validation model worked well for the level of the technology available at the time when the programs were created more than two decades ago. As technology has advanced, however, this model no longer satisfies current day industry and government operational needs in the context of increased number and intensity of cybersecurity breaches [15].

There are several factors for this. First, current cybersecurity recommendations [15] require patching promptly, including applying patches to cryptographic modules. Technology products are very complex and the cost of testing them fully to guarantee trouble-free use is prohibitively high. As a result, products contain vulnerabilities that hackers and technology producers are competing to discover first: the companies to fix, the hackers to exploit. Patching products changes the game for hackers and slows down their progress. However, patching changes also the environment in which a cryptographic module runs and may also change the module itself, thus invalidating the previously validated configuration. Users who depend on validated cryptography face a dilemma when frequent updates and patches are important for staying ahead of the attackers, but the existing validation process does not permit rapid implementation of these updates while maintaining a validated status because of the slow human-based validation activities.

The second factor hindering the effectiveness of the traditional validation model is the demand for speed in the context of the cognitive abilities of the human brain. Recent scientific research points out that humans are limited in their ability to process quickly and objectively large amounts of complex data [16].

Going back to the results on sentiment analysis with deep learning from above and in spite of that success, people may always question the ability of machines to replace humans in solving such cognitive and analytical tasks. They will always ask why the accuracy is not one hundred percent? Or, notwithstanding the available scientific evidence [16], say that if a human was reviewing the text she would have never made a mistake.

Changing public opinion may be a slow process. Besides, it seems that cybersecurity and machine learning/artificial intelligence (AI) will always be joined at the hip because the more we rely on machines to solve ever more complex tasks, the higher the risk those machines may be attacked to subvert their operation. Can AI fight back though? The results presented in this paper suggest that computer-based validation of cryptographic test evidence may be the only viable alternative that would allow objective and accurate assessment of large volumes of data at the speed needed to respond to the present day cybersecurity challenge [15]. This paper demonstrates that deep learning neural networks are capable of tackling the core tasks in security validation and thereby automating existing programs [9]. If this effort is indeed successful one may reason that by helping to improve the process of validation and thereby increase cybersecurity, albeit indirectly, AI will in fact be defending itself from cybersecurity threats. Over time, this may lead to societal acceptance of AI into sensitive domains such as the validation of critical components for the IT infrastructure.

Next Steps. The importance of incorporating word polarity into the model is illustrated clearly by the results presented in this paper. However, the type of language used in the validation test reports tends to be different than the colloquial English used in movie reviews. The technical jargon in test reports uses many common words whose meaning changes in this context. Moreover, the assessments for the test requirements are written in a way different from movie reviews. Here the author provides arguments that justify her conclusion about compliance. The challenge is to distinguish weak/faulty arguments from solid ones and develop an appropriate polarity model, along with assembling a representative corpus of labeled validation test report data for training and validation.

References

1. Brants, T., Popat, A., Xu, P., Och, F., Dean, J.: Large language models in machine translation. In: Proceedings of the 2007 Joint Conference on Empirical Methods in Natural Language Processing and Computational Natural Language Learning (Prague), pp. 858–867 (2007)
2. Mikolov, T., Sutskever, I., Chen, K., Corrado, G., Dean, J.: Distributed representations of words and phrases and their compositionality. In: Advances in Neural Information Processing Systems, pp. 3111–3119 (2013)
3. Conneau, A., Kiela, D., Schwenk, H., Barraul, L., Bordes, A.: Supervised learning of universal sentence representations from natural language inference data. In: Proceedings of the 2017 Conference on Empirical Methods in Natural Language Processing (Copenhagen), pp. 670–680. Association of Computational Linguistics (2017)
4. IMDB Movie reviews sentiment classification, Keras Documentation. https://keras.io/datasets/. Accessed 1 Mar 2019
5. Large Movie Review Dataset, Stanford University. http://ai.stanford.edu/~amaas/data/sentiment/. Accessed 1 Mar 2019
6. Nacson, M., Lee. J., Gunasekar, S., Savarese, P., Srebro, N., Soudry, D.: Convergence of Gradient Descent on Separable Data. https://arxiv.org/abs/1803.01905v2. Accessed 15 Mar 2019
7. TensorFlow: an open source machine learning framework for everyone, Google LLC. https://www.tensorflow.org/. Accessed 1 Mar 2019
8. Potts, C.: On the negativity of negation. In: Proceedings of Semantics and Linguistic Theory, vol. 20, pp. 636–659 (2011)
9. Automated Cryptographic Validation Testing, NIST. https://csrc.nist.gov/projects/acvt/. Accessed 1 Mar 2019
10. Maas, A.L., Daly, R.E., Pham, P.T., Huang, D., Ng, A.Y., Potts, C.: Learning word vectors for sentiment analysis. In: Proceedings of the 49th Annual Meeting of the Association for Computational Linguistics: Human Language Technologies, 2011, Portland, Oregon, USA, pp. 142–150. Association for Computational Linguistics (2011)
11. Goodfellow, I., Bengio, Y., Courville, A.: Deep Learning. MIT Press, Cambridge (2016)
12. Bramble, J.: Multigrid Methods. Pitman Research Notes in Mathematics, vol. 294. Wiley (1993). ISBN: 0-470233524
13. Demonstration of the use of Convolution1D for text classification, Keras. https://github.com/keras-team/keras/blob/master/examples/imdb_cnn.py. Accessed 1 Feb 2019
14. Hayou, S., Doucet, A., Rousseau, J.: On the Impact of the Activation Function on Deep Neural Networks Training. https://arxiv.org/abs/1902.06853. Accessed 1 Mar 2019
15. Data Breach Investigations Report, Verizon. https://enterprise.verizon.com/resources/reports/dbir/. Accessed 1 Feb 2019
16. Kahneman, D.: Thinking, Fast and Slow. Farrar, Straus and Giroux, New York (2011). ISBN: 9780374275631 0374275637

To What Extent Can Text Classification Help with Making Inferences About Students' Understanding

A. J. Beaumont(✉) and T. Al-Shaghdari

Aston University, Birmingham, UK
a.j.beaumont@aston.ac.uk

Abstract. In this paper we apply supervised machine learning algorithms to automatically classify the text of students' reflective learning journals from an introductory Java programming module with the aim of identifying students who need help with their understanding of the topic they are reflecting on. Such a system could alert teaching staff to students who may need an intervention to support their learning.

Several different classifier algorithms have been validated on the training data set to find the best model in two situations; with equal cost for a positive or negative classification and with cost sensitive classification. Methods were used to identify those individual parameters which maximise the performance of each algorithm. Precision, recall and F1-score, as well as confusion matrices were used to understand the behaviour of each classifier and choose the one with the best performance.

The classifiers that obtained the best results from the validation were then evaluated on a testing data set containing different data to that used for training.

We believe that although the results could be improved with further work, our initial results show that machine learning could be applied to students' reflective writing to assist staff in identifying those students who are struggling to understand the topic.

1 Introduction

"Text classification is the automated grouping of textual or partially texted entities" [20]. It is a machine learning problem that can be solved by training classifiers to learn the distinctions between the classes and build a trained model that automatically assigns documents to the correct classes. This investigation aims to discover the effectiveness of training a supervised learning algorithm on labelled[1] examples of student's writing from reflective learning journals to see if it is possible to identify students who are struggling with their learning.

During the teaching of Java Programming at Aston University, first year students are tasked with submitting weekly reflective learning journals in which they reflect on their learning during that week. That is based on the premise that

[1] The labels correspond to the final exam results.

G. Nicosia et al. (Eds.): LOD 2019, LNCS 11943, pp. 372–383, 2019.
https://doi.org/10.1007/978-3-030-37599-7_31

writing about new information or ideas enables students to better understand and remember them, and that articulating connections between new information or ideas and existing knowledge secures and extends learning [24,26]. Each learning journal entry is structured into a number of sections and our data combines each section into a single entry. Teaching staff do not impose any constraints on the length of each entry.

Providing individual feedback on students' journal entries every week is a labour intensive process because over two hundred journal entries of varying quality are submitted each week.

The motivation for our research is the importance of providing feedback to students to guide them on the right track. The question proposed is, Can machine learning help in automatic classification of learning journals to predict whether students need support in their learning or not?

Our aim is to select a machine learning algorithm that has a high accuracy performance that can be used as a plug-in to our reflective learning journal system. As a plug-in, it would classify student journal submissions and provide initial feedback to the teaching staff who may follow up with the student where required. Specifically in this part of the research we aim to:

- Conduct a comprehensive study on the analysis of different classifiers and understand their weaknesses and strengths in the context of the research question.
- Define the areas of development for classification; feature selection and data preprocessing.

The question proposed makes this research project unique for several reasons, these are:

- Although extensive research exists on text categorisation and the domains that it can be used in, no study has been found on whether this method can be used to identify students who need support with their learning based on reflective writing they complete throughout a taught module.
- Despite best efforts, a comparison study of different classifiers in any scenario that is similar to the scenario in the research question could not be found.
- Research exists on using classification to predict essay scoring [19,21,28] but not for identifying students who need help with their learning. Additionally, researchers mainly used one or two approaches without making a strong case for their methodology.
- Despite there being research in feature selection algorithms, there's remains a gap between the progress made in this research area and classifiers algorithms although it's as important, or perhaps more so.

2 Text Classification

"Text categorisation research has a long history starting in the early 1960s" [15]. It has been used for information retrieval, data mining, and machine learning. More recently it has become one of the main techniques for data analysis and has been used in a variety of domains [1–3,38]. In the field of education, text classification has been used to predict essay scoring [19,21,28] and also in auto-marking students' work [31,35]. The classification of text involves labelling some training documents according to their class. For example, when addressing classification for student assessment, training documents can be labelled as either pass (positive class) or fail (negative class). Each document is processed to create a numerical vector representing features of the text that will be used to distinguish the relevant classes. Thereafter, a machine learning algorithm will train a classifier that will classify the training documents accurately. The learning method will try to improve on its accuracy depending on the validation set. Once the classifier is trained, it is used to predict the class of other documents outside of the training data to evaluate its performance. "Algorithm selection is not the only factor that effects classification result" [37]. The choice of text representation models and text preprocessing options also influence the classification performance.

2.1 Representation Models

A string of character documents must be transformed into a suitable representation for the learning algorithm. Multiple representation methodologies have been employed in previous research with different classifiers, those methodologies can be categorised under three models;

- Bag of Words model (BOW): This representation "collects the words as a set (or bag), ignoring any syntactic structure or word order information but counting the occurrences of each word" [22]. The occurrences could be represented as the Total Frequency (TF) of each word, or by factoring in an Inverse Document Frequency (TF-IDF) factor to diminish the weight of frequently used terms and increase the weight of terms that occur rarely [16]. A drawback of the BOW is that there is no information on syntactic structure or semantic content.
- Structured Representation model: the text will be converted to syntactic relational semi structured text. One way is to extend the bag of words model to use a fixed length n-gram (bi-gram, tri-gram or four-gram). Another way is to produce structural model by a Natural Language Processor, e.g. MedLEE [8] which takes input text and uses clinical vocabulary to map words and phrases to standards terms.
- Word Embedding Model: where words are mapped to vectors represented with real numbers in the vector space as in [23]. This representation puts words of similar meaning closer to each other in the vector space e.g.(doing, do, done).

2.2 Preprocessing and Feature Selection

To reduce the space dimension of the data, The data needs to be preprocessed before it can be used to train the classifier. There is some research that has investigated the impact of preprocessing particularly in text classification [4,34]. The preprocessing tasks for texts usually are filtering, tokenization, lemmatization and stemming. Furthermore, there is also feature selection that could be applied to text before modelling the data. Feature selection helps in understanding data, reducing computation cost, decreasing the effect of the high dimensional space and improving the predictor performance. The focus of feature selection is to select a subset of variables from the input which can efficiently describe the input data while reducing effects from noise or irrelevant variables and still provide good prediction results.

In this research, exploration will be performed to identify its characteristics leading to the preprocessing. Further more, combining multiple feature selection methods with different classifier algorithms will also be applied to get the best results.

2.3 Methods of Text Classification

There are many different classifiers that are used to categorise texts. There have been various studies arguing which of these classifiers are better for specified problems; Naive Bayes classifier is particularly suited when the dimensionality of the inputs is high [25]. Passive-Aggressive is also good with classification problems [5]. Although K-Nearest Neighbour has no model other than storing the entire data set, results in [33] show that KNN with TF-IDF as a feature selection method performs pretty well at text classification regardless of the K factor value. Logistic and Ridge regression have been used for classification tasks [7,30]. Ridge has achieved satisfactory results with high dimensional data. "The Ridge regression estimator has been shown to give more precise estimate of the regression perimeters because its variants and mean square error are generally smaller than that of the least square estimates" [11]. In [10], results show that Random Forest algorithm with high dimensional data is better than a decision tree and it performs well with text data [12,36].

Convolutional Neural Networks (CNN) [9] are also known to be good with natural language processing (NLP). CNNs have achieved superior results in visual tasks as well as voice recognition and natural language processing. There is evidence that CNNs using the SST data set are "more suitable for constructing the semantic representation of texts" [18], they claim that the reason for this is that the pooling layers in CNN can select more discriminative features and the convolutional layers capture contextual information. Another advantage of CNNs is that it runs a lot faster compared to others [17].

In this project, various algorithms will be chosen based on the findings from the research cited above. CNN will be the only Neural Network to be used as it has very positive feedback when applied to similar problems to this project.

We followed a knowledge discovery from data (KDD) process to collect and explore the data. Then we cleaned the data, did feature selection, transformation, data mining (creating model), and evaluation. The language used with all the steps in this process was Python v3.5.4. The main two libraries used are Scikit-Learn [27] and Tensorflow [32].

3 Data Collection and Exploration

Our training data-set was organised into two CSV files, one containing 3597 records where each record represented a whole journal entry with all sections merged together (title, description, reflection, concepts and what's next), while the other file contained the label corresponding to the record on the same line in the first file. The label is either neg or pos based on the exam mark that the student who wrote the entry achieved in the Java Programming exam. Hence, neg means that the student who wrote the entry obtained a fail mark (less than 40%) and pos means that the student obtained a pass mark. We have excluded journal entries from students who got a zero mark on the exam.

We started by exploring the feature selection and preprocessing that would improve the performance of the information extraction. Some characteristics of the text that were found are: the entries are not purely English, they have some pseudo or Java program code, there are some duplicate entries where multiple students have submitted the same text, entries are of different lengths and the data is unbalanced. There are 1078 neg entries and 2315 pos entries. The visualisation shown in Fig. 1 uses yellow to represent the neg samples and dark blue to represent the pos samples. The samples have been vectorised using TF-IDF (TfidfVectorizer in Scikit-Learn) and principle component analysis (PCA) to reduce the data dimensionality. There are some outliers that can be seen that don't seem to belong to the rest of the group. Some of the vectors that are associated with the pos class are on top of vectors from the neg class, or very close to them, which shows that the data is quite hard to separate in the vector space.

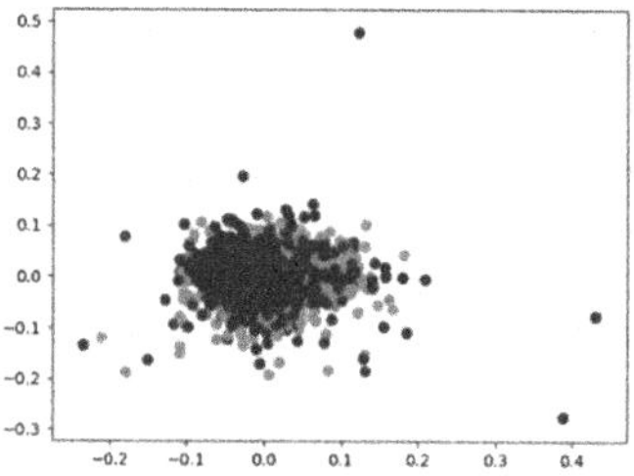

Fig. 1. Data representation with PCA and TfidfVectorizer (Color figure online)

4 Data Cleaning and Transformation

It is essential to preprocess the data and prepare it for the classification algorithms. First, we *preprocess* the data to deal with missing values and noise by removing empty entries or samples that have only one sentence because they might represent errors or cause unreliable results.

The classes are not represented equally so under-sampling was applied to delete instances from the over-represented class. Hence half of the pos class samples have been deleted. Tokenization, filtering (removing stop words), converting to lowercase, stemming and lemmatization are all applied using the Natural Language Toolkit (NLTK) package.

The next step is to *prepare* the data for the classification algorithms. There are splitting methods available in Scikit-Learn to split the data for training and validation. Cross-validation is applied where data is partitioned to k-folds of independent subsets. A model will be built using (k − 1) folds and the remaining fold will be used for testing. This is done using *StratifiedKFold* which preserves the percentage of samples in both classes.

The next step is *feature selection* and visualisation of the data features is helpful in selecting those features that are helpful to the classifier:

- We used the CountVectorizer function in Scikit-Learn to convert the text into token counts matrix and vectors. 1000 Positive samples and 1000 negative samples were converted separately and we used Plotly [14] to visualise the results of that. Comparing Figs. 2 and 3, we can notice that although the word counts matrix for the pos samples is more distributed in the two dimensional space, the scaling in the neg class is larger because of the outliers. Both pos and neg matrix vectors are concentrated in an overlapping area, indicating that both classes contain similar words with similar frequency.
  To explore that hypotheses, further analysis is needed to understand how both classes are using the words. By extracting the feature names from CountVectorizer (`get_feature_names()`), we plot the first twenty most frequently used words in each class. Figure 4 shows the most twenty frequently used words in the pos samples while Fig. 5 shows that regarding the neg samples.
  We can also see that the frequency scale in Fig. 4 is larger that in Fig. 5, which shows that more words are used in the pos samples.
- Another feature selection that should be considered is Term Frequency times Inverse Document Frequency (TF-IDF). If we simply use an occurrence count, shorter documents will have lower average word counts than longer documents, even though they might be speaking about the same topics. A solution is to divide the number of occurrences of each word in a document by the total number of words in the document which gives us the TF (Term Frequency) then downscale weights for words that occur in many documents because they are less informative than the words occurring less frequently in the data.
- We explored the use of Chi Square (χ^2) to identify the most relevant terms in the pos and neg samples in the hope that it could provide some meaningful information about the data and improve computational efficiency and classification performances. However we found that it reduced the accuracy of some classifiers and so we decided not to use it.

After our feature selection investigation, countVectoriser and TF-IDF were selected and used.

After these preprocessing steps, the data is now ready for classification and for exploring different classifier algorithms with different parameters and different n-gram representation.

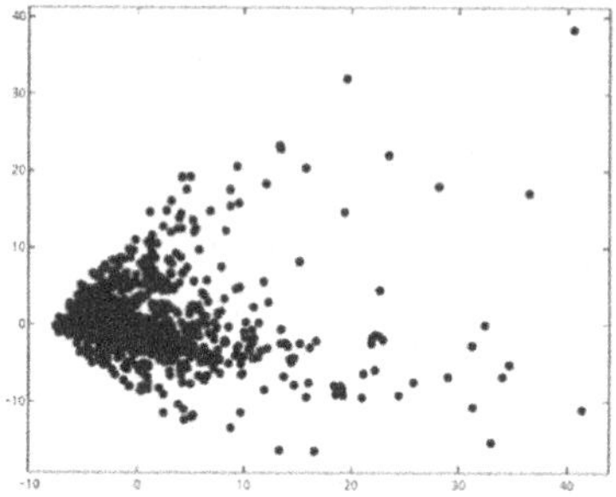

Fig. 2. Pos samples

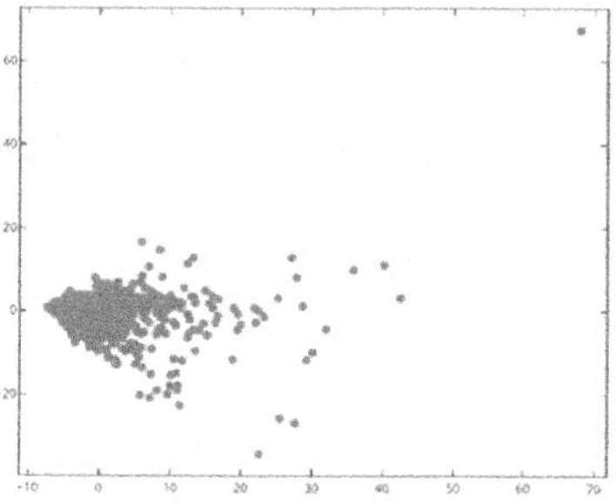

Fig. 3. Neg samples

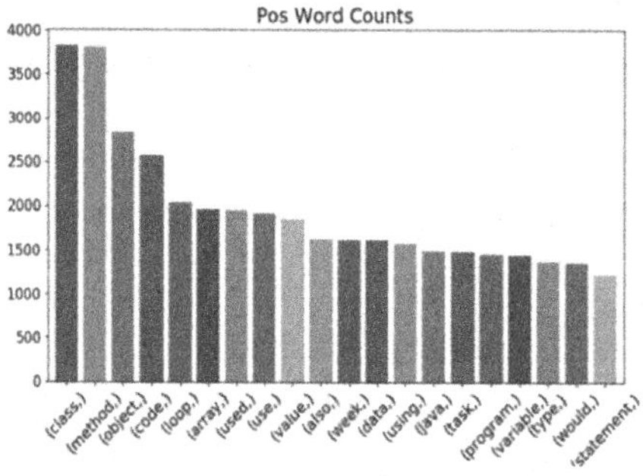

Fig. 4. Pos

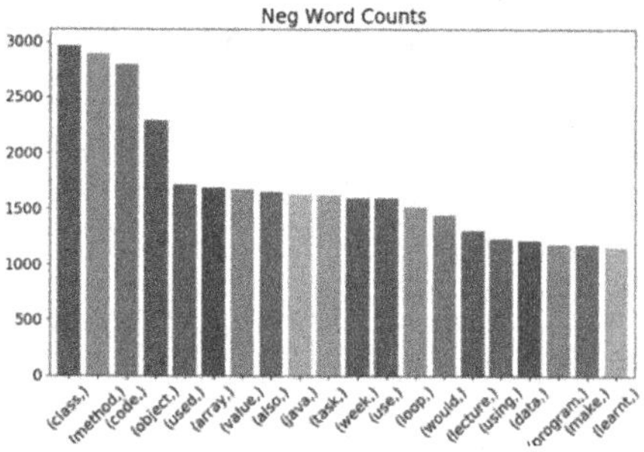

Fig. 5. Neg

4.1 Model Development

"Classification is a form of data analysis that extracts models describing important data classes. Such models, called classifiers, predict categorical (discrete, unordered) class labels" [13].

The classifiers we used from Scikit-Learn are; the Ridge Classifier, Naive Bayes, Passive Aggressive, K Nearest Neighbours, and Random Forest. We also used the Logistic Regression algorithm which was applied with the Stochastic Gradient Descent (SGD) classifier.

We have chosen various classifiers based on our background research and applied those with different representations of the text data (bag of words, 2-gram and 3-gram) and different feature selections (CountVectorizer and TF-IDF).

We also carried out parameter tuning for the classifiers, such as the learning rate, random state and others parameters that Scikit-Learn provides. Tuning was done using GridSearchCV which generates candidates from a grid of parameter values that we specified.

The validation is done with the cross-validation methodology as mentioned in the data preparation using the StratifiedKFold method with k = 10.

The Python Pipeline library was also used to apply multiple transforms (such as the preprocessing and feature selection) sequentially then apply a final estimator which is the classifier. Changing the feature selection method makes a significant change on the algorithm behaviour.

We implemented Convolutional Neural Network (CNN) using TensorFlow which is used for numerical computation and large-scale machine learning. An attached visualisation tool called TensorBoard was used to show the learning curve to help understand the performance of the algorithm. The CNN algorithm contained an embedding layer to build a word embedding representation of the data, a convolutional layer followed by pooling layer for 128 filters, drop out layer as a regularisation method and early stopping methodology to avoid over-fitting.

5 Experimental Results

The performance of the various classifier algorithms will be analysed using precision, recall and also a confusion matrix.

Precision, also known as specificity, provides a measure of the performance on correctly identifying positive samples, taking into account the number of false positive classifications. The formula for calculating precision is:

$$Precision = \frac{TruePositive}{TruePositive + FalsePositive}$$

Recall, also known as sensitivity, provides a measure of the performance on correctly identifying positive samples, taking into account the number of false negative classifications. The formula for calculating recall is:

$$Recall = \frac{TruePositive}{TruePositive + FalseNegative}$$

These two measures can be combined into one measure known as the F1-score, and F1 is high when both precision and recall are high. The formula to calculate the F1-score is:

$$F1 = \frac{2}{\frac{1}{Precision} + \frac{1}{Recall}}$$

Table 1 shows the precision, recall and F1-score for the various different classifier algorithms we have explored.

We can see that none of the classifiers seem to be biased towards positive or negative classification.

Cost function to the algorithms is applied in which the negative class has been given a higher weighting so the cost of misclassifying a negative journal sample is more expensive than misclassifying a positive journal. The rationale for this is that our aim is to identify students who are struggling with their understanding, so it is more important to correctly identify samples from those students. If we apply an intervention, initially talking to the students whose

Table 1. Precision, recall and F1 for the classification algorithms

Algorithm classifier	Precision	Recall	F1-score
Ridge regression	71	70	69
Multinomial Naive Bayes	71	69	68
K Nearest Neighbours	68	68	67
Random Forest	64	64	64
SGD classifier	66	64	62
Passive-aggressive	62	62	62

journal entries are classified as negative, we should be able to easily identify those false negatives. We would prefer not to provide any intervention when journal entries are classified as positive so long as false positives are minimised.

The cost function for the SGD Classifier weights the positive class at 30% and so the negative class has 70% weighting. The result was the SGD classifier performed with an accuracy of 70.59% on the validation.

The Ridge classifier with a cost function classified all test data as neg and so we present the figures for the Ridge classifier without cost modelling.

More information on the performance in correctly classifying the samples is shown as positive or negative using a confusion matrix. Figure 6 shows the format of a confusion matrix.

		Predicted Value	
		Neg	**Pos**
Actual Value	**Neg**	True Negative	False Positive
	Pos	False Negative	True Positive

Fig. 6. Confusion Matrix (CM)

$$\begin{pmatrix} 65 & 32 \\ 28 & 79 \end{pmatrix}$$

Fig. 7. CM for SGD

The confusion matrix for the SGD classifier which produced the best overall performance is shown in Fig. 7.

The training phase included some validation using the k-fold technique and the accuracy on this phase was typically higher than the accuracy when we tested the algorithms with unseen data of 200 entries. The figures for accuracy in both validation and training can be seen in Table 2.

When we tested the classifiers with a new data set (testing data-set) of 200 samples, the CNN algorithm had an accuracy of 68%, the Ridge classifier performed with an accuracy of 65.5%, and the SGD classifier an accuracy of 66%.

The SGD and CNN classifiers produce good performance on the classification in terms of accuracy and cost. Either could be used as the basis of a plug-in to our reflective learning journal system to provide feedback to the teaching staff. Their accuracy is not good enough to provide feedback directly to the students

Table 2. Accuracy the classification algorithms

Algorithm classifier	Accuracy percentage	
	Validation	Testing
CNN	80	68
SGD classifier	70.59	66
Ridge regression	70.09	65.5

but is good enough to identify a smaller subset of students who would potentially benefit from intervention by the teaching staff.

In general, labelling the documents according to student's exam results could be an area to consider as to why classification is not performing very well. As some students do very well when writing with open resources yet under-perform when it comes to the exam. The opposite can also be observed whereby students may seem to be under performing through the year, yet during exams they will achieve exceptional results. Although some of the challenges occurred during the development of the project such as lack of the data set and some missing methods when using Scikit-Learn, the best classifier has been chosen depending on the results and the evaluation. However the results returned were not very satisfying, most likely because the data samples (learning journals) were labelled according to a different assessment criteria (exams) hence why clustering [6,29] needs to be explored as an alternative approach for future work.

6 Conclusion

As shown in Table 2, the CNN classifier performed better than the other methods in term of its accuracy on both the validation and test data. The SGD classifier had comparable accuracy on the test data. We believe that the accuracy of the CNN classifier is good enough for us to trial it as an aid in identifying students who are struggling to understand programming from their written reflections.

Analysing the data and understanding the feature selection is important before applying any classifier. One challenge is the relatively small size of the data set, but with more journals being written every year, we will be able to gather a larger data set to work with in future. The journal entries were labelled according to their mark on a closed book exam. There is an interesting open question as to the extent the evaluation of students reflective writing is related to the grade they obtain on a closed book exam.

For future development, we would like to explore the use of clustering and topic modelling to help categories the data according to their content and bring more knowledge of the topics in the text [29].

We believe that our current results can be improved with further work and that the results reported here show that machine learning could be applied to students' reflective writing to identify those students who are struggling to understand the topic.

References

1. Aphinyanaphongs, Y., Tsamardinos, I., Statnikov, A., Hardin, D., Aliferis, C.F.: Text categorization models for high-quality article retrieval in internal medicine. J. Am. Med. Inform. Assoc. **12**(2), 207–216 (2005). https://doi.org/10.1197/jamia.M1641
2. Argamon, S., Koppel, M., Pennebaker, J., Schler, J.: Automatically profiling the author of an anonymous text. Commun. ACM **52**, 119–123 (2009)
3. Carreras, X., Marquez, L.: Boosting trees for anti-spam email filtering (2001). https://arxiv.org/abs/cs/0109015. Accessed 12 Jun 2018
4. Chawla, N.V.: C4.5 and imbalanced data sets: investigating the effect of sampling method, probabilistic estimate, and decision tree structure. In: Proceedings of the ICML, vol. 3, p. 66 (2003)
5. Crammer, K., Dekel, O., Keshet, J., Shalev-Shwartz, S., Singer, Y.: Online passive-aggressive algorithms. J. Mach. Learn. Res. **7**, 551–585 (2006)
6. Elankavi, R., Kalaiprasath, R., Udayakumar, D.R.: A fast clustering algorithm for high-dimensional data. Int. J. Civ. Eng. Technol. (IJCIET) **8**(5), 1220–1227 (2017)
7. Endelman, J.B.: Ridge regression and other kernels for genomic selection with R package rrBLUP. Plant Genome **4**(3), 250–255 (2011)
8. Friedman, C.: A broad-coverage natural language processing system. In: Proceedings of the AMIA Symposium, pp. 270–274 (2000)
9. Géron, A.: Hands-On Machine Learning with Scikit-Learn and TensorFlow: Concepts, Tools, and Techniques to Build Intelligent Systems. O'Reilly Media, Inc. (2017)
10. Gokgoz, E., Subasi, A.: Comparison of decision tree algorithms for EMG signal classification using DWT. Biomed. Signal Process. Control **18**, 138–144 (2015)
11. Gruber, M.: Improving Efficiency by Shrinkage: The James–Stein and RidgeRegression Estimators. Routledge (2017)
12. Ham, J., Chen, Y., Crawford, M.M., Ghosh, J.: Investigation of the random forest framework for classification of hyperspectral data. IEEE Trans. Geosci. Remote Sens. **43**(3), 492–501 (2005)
13. Han, J., Pei, J., Kamber, M.: Data Mining: Concepts and Techniques. Elsevier, Amsterdam (2011)
14. Plotly Technologies Inc.: Collaborative data science (2015). https://plot.ly
15. Joachims, T.: Learning to Classify Text Using Support Vector Machines: Methods, Theory and Algorithms. Kluwer Academic Publishers, Norwell (2002)
16. Jones, K.S.: A statistical interpretation of term specificity and its application in retrieval. J. Documentation **28**(1), 11–21 (1972). https://doi.org/10.1108/eb026526
17. Joulin, A., Grave, E., Bojanowski, P., Mikolov, T.: Bag of tricks for efficient text classification. arXiv preprint arXiv:1607.01759 (2016)
18. Lai, S., Xu, L., Liu, K., Zhao, J.: Recurrent convolutional neural networks for text classification. In: AAAI, vol. 333, pp. 2267–2273 (2015)
19. Larkey, L.: Automatic essay grading using text categorization techniques. In: Proceedings of the 21st Annual International ACM SIGIR Conference on Research and Development in Information Retrieval, pp. 90–95. ACM, August 1998
20. Lewis, D., Gale, W.: A sequential algorithm for training text classifiers. In: ACM SIGIR Conference on Research and Development in Information Retrieval, pp. 3–12. Springer, New York (1994). https://doi.org/10.1007/978-1-4471-2099-5_1

21. McNamara, D., Crossley, S., Roscoe, R., Allen, L., Dai, J.: A hierarchical classification approach to automated essay scoring. Assessing Writ. **23**, 35–59 (2015)
22. McTear, M., Callejas, Z., Griol, D.: The Conversational Interface: Talking to Smart Devices, 1st edn. Springer, Cham (2016). https://doi.org/10.1007/978-3-319-32967-3
23. Mikolov, T., Sutskever, I., Chen, K., Corrado, G., Dean, J.: Distributed representations of words and phrases and their compositionality. CoRR abs/1310.4546 (2013). http://arxiv.org/abs/1310.4546
24. Moon, J.: Reflection in Learning and Professional Development. Routledge, London (1999)
25. Murphy, K.P., et al.: Naive Bayes Classifiers, p. 18. University of British Columbia (2006)
26. O'Rourke, R.: The learning journal: from chaos to coherence. Assessment Eval. High. Educ. **23**(4), 403–413 (1998)
27. Pedregosa, F., et al.: Scikit-learn: machine learning in Python. J. Mach. Learn. Res. **12**, 2825–2830 (2011)
28. Rudner, L., Liang, T.: Automated essay scoring using Bayes' theorem. J. Technol. Learn. Assessment **1**(2) (2002)
29. Silge, J., Robinson, D.: Text Mining with R: A Tidy Approach. O'Reilly Media, Inc. (2017)
30. Staal, J., Abràmoff, M.D., Niemeijer, M., Viergever, M.A., Van Ginneken, B.: Ridge-based vessel segmentation in color images of the retina. IEEE Trans. Med. Imaging **23**(4), 501–509 (2004)
31. Sukkarieh, J.Z., Pulman, S.G., Raikes, N.: Auto-marking: using computational linguistics to score short, free text responses. In: 29th Annual Conference of the International Association for Educational Assessment (IAEA), Manchester, UK (2003)
32. TensorFlow: large-scale machine learning on heterogeneous systems (2015). https://www.tensorflow.org/, software available from tensorflow.org
33. Trstenjak, B., Mikac, S., Donko, D.: KNN with TF-IDF based framework for text categorization. Procedia Eng. **69**, 1356–1364 (2014)
34. Uysal, A.K., Gunal, S.: The impact of preprocessing on text classification. Inf. Process. Manage. **50**(1), 104–112 (2014)
35. Valenti, S., Neri, F., Cucchiarelli, A.: An overview of current research on automated essay grading. J. Inf. Technol. Educ. Res. **2**, 319–330 (2003)
36. Wu, Q., Ye, Y., Zhang, H., Ng, M.K., Ho, S.S.: ForesTexter: an efficient random forest algorithm for imbalanced text categorization. Knowl. Based Syst. **67**, 105–116 (2014)
37. Yu, B.: An evaluation of text classification methods for literary study. Literary Linguist. Comput. **23**(3), 327–343 (2008)
38. Zhou, B., Yao, Y., Luo, J.: Cost-sensitive three-way email spam filtering. J. Intell. Inf. Syst. **42**(1), 19–45 (2014)

Combinatorial Learning in Traffic Management

Giorgio Sartor(✉), Carlo Mannino, and Lukas Bach

SINTEF Digital, Oslo, Norway
{giorgio.sartor,carlo.mannino,lukas.bach}@sintef.no

Abstract. We describe an exact *combinatorial learning* approach to solve dynamic job-shop scheduling problems arising in traffic management. When a set of vehicles has to be controlled in real-time, a new schedule must be computed whenever a deviation from the current plan is detected, or periodically after a short amount of time. This suggests that each two (or more) consecutive instances will be very similar. We exploit a recently introduced MILP formulation for job-shop scheduling (called *path&cycle*) to develop an effective solution algorithm based on delayed row generation. In our re-optimization framework, the algorithm maintains a pool of combinatorial cuts separated during the solution of previous instances, and adapts them to warm start each new instance. In our experiments, this adaptive approach led to a 4-time average speedup over the static approach (where each instance is solved independently) for a critical application in air traffic management.

Keywords: Job-shop scheduling · Re-optimization · Mixed Integer Linear Programming

1 Introduction

In many real-life applications one needs to solve a sequence of instances of an optimization problem while the underlying system evolves over time. Each new instance is only "slightly" different from the previous one, and it would only be natural to exploit any knowledge acquired by solving a previous instance (e.g., solution, generated cuts) to help solving a new one. Re-optimization has been the focus of several recent studies. A formal, unifying framework has recently been introduced in [11], where it is observed that solving a re-optimization problem involves two challenges: (*i*) computing a (close to) optimal solution for the new instance; and (*ii*) efficiently converting the current solution to the new one. The authors also observe that these challenges give rise to many theoretical and practical questions. Here we focus on exact methods that efficiently exploit previously acquired knowledge to tackle a combinatorial re-optimization problem in Air Traffic Management (ATM).

Despite the fact that several applications embed dynamic aspects, there are not many examples of exact re-optimization algorithms applied to relevant real-life problems. The great majority of approaches are heuristic (see, e.g. [8] for

G. Nicosia et al. (Eds.): LOD 2019, LNCS 11943, pp. 384–395, 2019.
https://doi.org/10.1007/978-3-030-37599-7_32

dynamic vehicle routing). An interesting exception is provided by a recent paper for vehicle routing [7], where columns generated at the previous iterations are massaged and adapted to subsequent iterations.

The approach to re-optimization presented in this paper exploits the properties of a non-compact Mixed Integer Linear Programming (MILP) formulation for vehicle scheduling recently introduced in [4,10]. There exists a vast body of literature on vehicle scheduling, even when limited to train/air traffic management. In both cases, the problem amounts to schedule (and often route) a set of vehicles on capacited resources, minimizing some measure of the deviation from a desired schedule. This is a (blocking, no-wait) job-shop scheduling problem [6], often successfully attacked by means of MILP. The great majority of MILP approaches presented in the literature are based on two contending formulations, so called big-M and time-indexed[1]. The best approach depends on the application and the input data. Despite their weak dual bounds, big-M based approaches usually prevail on real-time applications, because the relaxations can be solved very efficiently (see experiments and discussion in [5]).

The *path&cycle* formulation that we recently developed, originally applied to train scheduling [4] and then to flight scheduling [10], is derived in [4] by strengthening (and lifting) the standard Benders' reformulation of a natural big-M formulation of the original vehicle scheduling problem. Despite the fact that its rows may grow exponentially with the number of vehicles, it outperformed the original big-M formulation in both real-life applications.

Like many other Benders-like decomposition techniques, the path&cycle formulation is very well suited for a delayed row and column generation solution approach (see Algorithm 2 in [10]). We start with a small subset of constraints (and variables), and solve the current master problem to integer optimality. We then check for constraints that are violated by the current solution, add them to the formulation, and iterate. The process terminates when the current solution does not violate any (missing) constraint. As also discussed in [2], this basically corresponds to an iterative approximation of the objective function through the optimality cuts of the Benders' reformulation. It would only be natural to assume that knowing the correct approximation in advance (i.e., all the missing constraints) would lead to a reduction of the solution time. This observation is indeed the key of our re-optimization algorithm.

In traffic management applications one needs to continuously compute and return new solutions while the system evolves. In train dispatching, for example, a new solution must be presented to human controllers every few seconds. Similarly, in the ATM hotspot problem addressed in this paper, a new solution is computed every 60 s. It goes without saying that, in almost all cases, not much can have happened, and the new instance is indeed very similar to the last. This implies that many of the cuts separated while solving a previous instance will still be valid at the next one, prior to some small and quick adjustment. This

[1] For theoretical and computational discussions, and comparisons of these formulations for the single machine scheduling problem see [3,9].

would also suggest that the new instance may be solved to optimality much faster than if solved from scratch.

Our re-optimization framework consists of maintaining a pool of cuts generated while solving a sequence of instances of the path&cycle formulation. Every time a cut is separated, it is automatically added to the pool. Whenever a new instance becomes available, we check if the cuts contained in the pool are feasible for the new instance. Some cuts will be dropped permanently from the pool (e.g., a vehicle exits the system), some will be adjusted (e.g., a vehicle slightly changes its schedule), and some others will be kept untouched. Then, all the cuts contained in the pool are added to the initial master problem of the new instance, which is solved through an exact delayed row (and column) generation approach. This way, each new instance is warm started with some of the knowledge acquired while solving previous similar instances. Even better, all the cuts generated by the path&cycle formulation are combinatorial and interpretable, making the pool easy to maintain and to understand. For example, a typical cut can be of the kind "If vehicle a meets vehicle b in resource s, then vehicle a will be delayed (in respect to its original schedule) by at least x seconds".

It is interesting to observe that storing previously generated constraints is very different from simply storing the previous incumbent solution. Indeed, it is as if we were storing information on how to approximate the behaviour of the system or, in other words, it is as if we were continuously learning the dynamics of the system. We would like to stress the fact that, while this approach can be seen as an iterative approximation of a dynamic system, it is also part of an exact framework that guarantees the convergence towards an optimal solution.

2 A Natural Formulation for the Hotspot Problem

A big-M formulation and a path&cycle formulation to solve the hotspot problem have been recently presented in [10]. Here, we reintroduce some of the key ideas, starting with a more natural disjunctive (not necessarily big-M) formulation, and then moving towards the full path&cycle formulation. Only this time, we also focus on describing the new underlying re-optimization framework.

The hotpost problem arises from strict regulations on the management of airplanes flying within a certain region of the airspace. In fact, we can imagine the airspace partitioned in a set S of volumes called sectors, each assigned to one or more air traffic controllers. Each controller, for safety reasons, can monitor simultaneously only a limited number of airplanes. For each sector $s \in S$ we let c_s be the capacity of s, namely the maximum number of aircraft allowed to be in the sector at the same time. Due to unforeseen circumstances (e.g., unexpected delays, bad weather conditions), there may be too many aircraft scheduled to simultaneously cross a certain sector at a certain time in the future. Given all the scheduled flights F, we say that a set of flights $\{f_1, \ldots, f_n\} \subseteq F$ produces a *hotspot* in sector $s \in S$ if the schedule of $f_1, \ldots, f_n$ has a point in time in which they are simultaneously in s and $n > c_s$. If we have a hotspot in sector s, then there exists a set $Q \subseteq F$ of $c_s + 1$ flights simultaneously crossing the sector, i.e., "meeting" in s. Note that the number of pairs of flights from Q that meet in s is precisely $\binom{c_s+1}{2}$. In order to prevent a hotspot, one or more airplanes can be temporary held on ground, delaying their departure. In some cases, flights can also be rerouted, but adding this possibility to problem is left for future research. In its basic version, the Hotspot Problem consists of computing a hotspot-free schedule for the flights in F such that the total delay (in respect the original schedule) is minimized. This boils down to choosing which flights to delay and for how long. Figure 1 shows a snapshot (in time) of the original and optimal schedule of an instance.

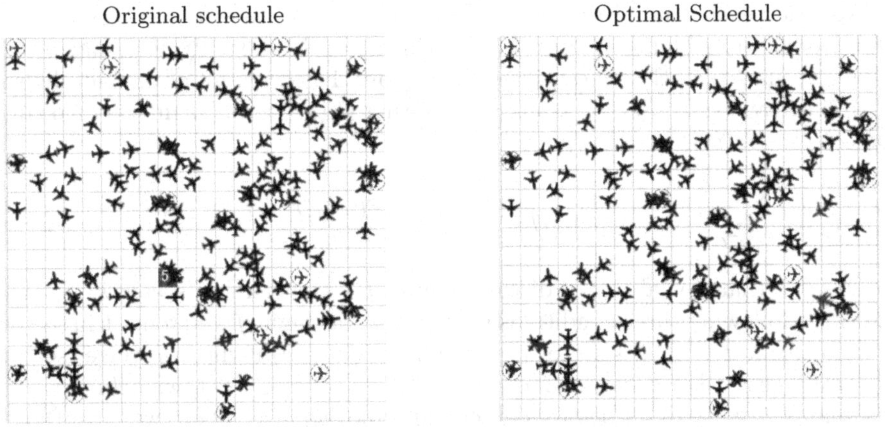

Fig. 1. Two snapshots of the first instance used in the tests. The figure to the left shows a particular point in time of the original schedule. The red square highlights a hotspot with 5 airplanes (in this case, the capacity c_s is equal to 4 for all sectors). The figure to the right depicts the situation at the same point in time if the airplanes followed the hotspot-free optimal schedule. Highlighted in green are the flights that required to be delayed in order to prevent *all* the hotspots. (Color figure online)

In the Hotpost Problem, each flight $f \in F$ is assigned a route, i.e., an ordered sequence $n^f_{s_1}, n^f_{s_2}, \ldots, n^f_{s_q}$ of *route nodes*, where $s_i \in S, i = 1, \ldots, q$ is a sector and s_1, s_q are the sectors in which the departure and arrival airports are located, respectively. With some abuse of notation, we denote by n^f_{s+1} the route node that immediately follows n^f_s in the flight route of f.

Let N be the set of all route nodes for all flights in F, $D \subset N$ be the set of all departure nodes, and $A \subset N$ the set of all arrival nodes. We denote by $\Lambda_{f,s}$ the time necessary for flight f to traverse sector s. Flights are usually not delayed when already airborne (at least not for hotspot resolving), thus we assume $\Lambda_{f,s}$ to be fixed for all flights and sectors. Also, a flight $f \in F$ cannot depart earlier than the scheduled departure time Γ_f.

Next, we describe a (natural) disjunctive formulation for the hotspot problem. We associate a scheduling variable $t_{f,s} \in \mathbb{R}$ with each route node $n^f_s \in N$, representing the time flight f enters sector s. We also introduce a fictitious variable $t_o \in \mathbb{R}$, which serves as a reference time for all airplanes. Thus, we have

$$t_{f,s} - t_o \geq \Gamma_f, \quad n_s^f \in D. \tag{1}$$

Now let $n_s^f, n_{s+1}^f \in N$ be two consecutive route nodes in a particular flight route. Note that the time a flight exits a particular sector is precisely the time the flight enters the subsequent sector in its route. Since the traversing time is fixed, the following *precedence constraints* holds:

$$t_{f,s+1} - t_{f,s} = \Lambda_{f,s}. \tag{2}$$

Now, for each pair of distinct flights $f, g \in F$ meeting in a sector $s \in S$, we introduce the binary quantity x_{fg}^s, which is 1 if and only if f and g are simultaneously (i.e., they *meet*) in sector s. Consider now a set of distinct flights $F(s) \subseteq F$ traversing a sector s, and assume that $|F(s)| > c_s$. Then, the following *hotspot constraints* must hold:

$$\sum_{\{f,g\} \subseteq Q} x_{fg}^s \leq \binom{c_s + 1}{2} - 1, \quad Q \subseteq F(s), |Q| = c_s + 1, s \in S. \tag{3}$$

Observe that, for a pair of distinct flights f, g traversing a sector s, exactly one of the following three conditions must occur: (a) flight f and g meet in sector s, or (b) flight f traverses sector s before flight g, or (c) flight g traverses sector s before flight f. For each ordered pair of flights $(f, g) \in F(s)$, we define y_{fg}^s to be equal to 1 if f exits s before g enters, and 0 otherwise. Then, we have that

$$y_{fg}^s + y_{gf}^s + x_{fg}^s = 1, \quad \{f, g\} \subseteq F(s), s \in S. \tag{4}$$

Accordingly, for every $\{f, g\} \subseteq F(s), s \in S$, the schedule t will satisfy a family of (indicator) *disjunctive constraints* as follows:

$$\begin{array}{lll} (i) \ y_{fg}^s = 1 & \Longrightarrow & t_{g,s} - t_{f,s+1} \geq 0, \\ (ii) \ y_{gf}^s = 1 & \Longrightarrow & t_{f,s} - t_{g,s+1} \geq 0, \\ (iii) \ x_{fg}^s = 1 & \Longrightarrow & \begin{cases} t_{g,s+1} - t_{f,s} \geq 0 \\ t_{f,s+1} - t_{g,s} \geq 0 \end{cases}, \\ & y_{fg}^s, y_{gf}^s, x_{fg}^s \in \{0, 1\}. & \end{array} \tag{5}$$

Indeed, $y_{fg}^s = 1$ implies that f exits s before g enters s and, similarly, $y_{gf}^s = 1$ implies that g exits s before f enters. Similarly, when $x_{fg}^s = 1$, then both f and g exit the sector s after the other flight enters it (i.e., they meet in s).

In conclusion, a complete disjunctive formulation for the Hotspot Problem can be obtained by considering the minimization of the delay at destination of all flights subject to constraints (1)–(5). In principle, constraints (5) could be easily linearized using the infamous big-M method. However, it is well known that big-M formulations tend to produce weak bounds and large search trees. Moreover, in a re-optimization context, one cannot hope to use information from previous iterations (apart from the previous incumbent solution), and the problem must be solved form scratch whenever even a small change occurs.

3 The path&cycle Formulation and a Re-optimization Framework

Although this paper focuses on a specific application, the formulation and re-optimization framework that we present in this section can be applied not only to other problems in traffic management, but also to a large variety of more generic job-shop scheduling problems. More importantly, being able to store information collected in the search tree at one iteration and possibly re-use it in the following iterations helps producing a formulation that evolves over time together with the sequence of instances. This computational aspect is not well-investigated in the literature yet, but presents interesting challenges and promising results.

Our approach is based on the methodology first developed in [4, 10]. In particular, we apply a Benders-like decomposition to obtain a (master) problem only in the binary variables, plus a few continuous variables to represent the objective function. In [4] we show how to strengthen and lift the constraints of the Benders' reformulation of (the big-M representation of) constraints (1) to (5). Remarkably, the constraints of the reformulated master correspond to basic graph structures in the disjunctive graph (see [1,6]) associated with constraints (1), (2) and (5), such as paths and cycles. The key observation behind our re-optimization framework is that this disjunctive graph will not change drastically from one iteration to another, and most of the constraints generated in one iteration will remain (almost) valid at the next one.

We sketch here how to construct the disjunctive graph, and how to derive the path&cycle formulation. The disjunctive graph is a directed graph $G = (V, E)$ obtained by considering a vertex for every route node $n \in N$, plus an extra node: the origin o. A directed edge (u, v) of length l_{uv} in the disjunctive graph represents an inequality $t_v - t_u \geq l_{uv}$, indicating that the minimum travel time from route node u to route node v is l_{uv}. Using this property, we can encode constraints (1), (2), and (5) into the disjunctive graph just by adding the associated edges (see [10] for a detailed description on how to construct the disjunctive graph). The edges corresponding to constraints (1) and (2) are called *precedence edges.* They encode the schedule and the running times of each flight. The edges associated with constraints (5.i)–(5.iii) are called *conflict edges.* They encode the "meeting decisions", that is whether or not a pair of flights meets in a certain sector. We denote by K the set of all conflict edges. Note that while the precedence edges make up for the static structure of the graph, the conflict edges should be binding only when the associated variable (either y^s_{fg} or x^s_{fg}, $f, g \in F$, $s \in S$ in (5)) is equal to 1. Let $G(\bar{y}, \bar{x})$ be the graph obtained from the disjunctive graph by removing all the conflict edges whose associated variables are equal to 0. Then the following lemma holds.

Lemma 1. *Consider the problem of finding a solution $(\bar{y}, \bar{x}, \bar{t})$ to (1)–(5) which minimizes a non-decreasing function of the schedule $c(\bar{t})$. (i.) If $(\bar{y}, \bar{x}, \bar{t})$ is feasible to (1)–(5), then $G(\bar{y}, \bar{x})$ does not contain strictly positive directed cycles. (ii.) If $(y^*, x^*, \bar{t})$ is an optimal solution, then there exists an optimal solution (y^*, x^*, t^*)*

where t_n^* is the the length of the longest path from the origin o to route node $n \in N$ in $G(y^*, x^*)$.

See [10] for the proof. As a straightforward consequence of the lemma, when $c(t)$ is the sum of delays over all flights, then the Hotspot Problem (HP) can be formulated as: *find a feasible incidence vector $(y, x) \in \{0,1\}^K$ of conflict edges such that $G(y, x)$ does not contain a strictly positive directed cycle, the sum of the lengths of the longest paths from the origin o to the arrival nodes $u \in A$ is minimum, and the resulting schedule is hotspot-free.*

Let us denote by $\mathcal{C}$ the set of strictly positive length directed cycles of $G(y, x)$, and by $L^*(y, x, u)$ the length of the longest path from o to u in $G(y, x)$. Then path&cycle formulation for the Hotspot Problem can be written as follows:

$$
\begin{array}{lll}
\min \sum\limits_{u \in A} \mu_u & & \\
s.t. & & \\
(i) \;\; y^s_{fg} + y^s_{gf} + x^s_{fg} = 1, & \{f,g\} \in F, s \in S, & \\
(ii) \;\; \sum\limits_{e \in C \cap K_y} y_e + \sum\limits_{e \in C \cap K_x} x_e \leq |C \cap K| - 1, & C \in \mathcal{C}, & \quad (6)\\
(iii) \;\; \sum\limits_{\{f,g\} \subseteq Q} x^s_{fg} \leq \binom{|Q|}{2} - 1, & s \in S, Q \subseteq F(s), |Q| = c_s + 1, & \\
(iv) \;\; \mu_u \geq L^*(u, y, x), & u \in A, & \\
\qquad y \in \{0,1\}^{|K_y|}, x \in \{0,1\}^{|K_x|}, \mu \in \mathbb{R}^{|A|}. & &
\end{array}
$$

Constraints (6.*ii*) called *cycle inequalities* or *feasibility cuts* ensure that one does not select all the conflict edges contained in any strictly positive directed cycle. Constraints (6.*iii*) called *hotspot constraints* prevent the hotspots. Constraints (6.*iv*) called *path inequalities* or *optimality cuts* allow to correctly evaluate the objective value. Indeed, for any feasible choice of y and x, the arrival time $t_{f,s}$ of the flight corresponding to $u = n^f_s \in A$ equals the longest path length from o to u in $G(y, x)$ (equality being granted from the sign of the objective function).

We can finally rewrite constraints (6.*iv*) in a way that can be immediately exploited in a delayed row generation algorithm. We denote by $\mathcal{T}$ the set of all $G(y, x)$ for (y, x) satisfying (6.*i*), (6.*ii*), (6.*iii*). If $T \in \mathcal{T}$, then we denote by $P_u(T)$ the (set of edges of a) longest path from o to u in T, and by $L_u(T)$ the length of $P_u(T)$. The final reformulation can be obtained from (6) by replacing (6.iv) with the following:

$$
(iv.b) \quad \sum_{e \in K_y \cap P_u(T)} y_e + \sum_{e \in K_x \cap P_u(T)} x_e - |K \cap P_u(T)| + 1 \leq \frac{\mu_u}{L_u(T)}, \qquad u \in A, T \in \mathcal{T}
$$

Indeed, consider a feasible solution $(\bar{y}, \bar{x}, \bar{\mu})$ to $(iv.b)$. Let $\bar{T} = G(\bar{y}, \bar{x})$, and let $\bar{P}_u(\bar{T})$ be a longest path from o to u in $\bar{T}$. Then we have:

$$
\sum_{e \in K_y \cap \bar{P}_u(\bar{T})} \bar{y}_e + \sum_{e \in K_x \cap \bar{P}_u(\bar{T})} \bar{x}_e - |K \cap \bar{P}_u(\bar{T})| + 1 = 1
$$

which in turn implies

$$\bar{\mu}_u \geq L_u(\bar{T}) = L^*(\bar{y}, \bar{x}, u).$$

However, when one or more edges in $\bar{P}_u(\bar{T})$ are not selected in the current solution, then the constraint is dominated.

We call problem (6) with (iv) replaced by $(iv.b)$ the *path&cycle* formulation (PC) of the Hotpot Problem. Note that all constraints are combinatorial and easily interpretable. In fact, constraints $(6.i)$ read as "Either a pair of flights meets in a certain sector or one of the two flights precedes the other"; constraints $(6.ii)$ as "This set of flight meetings/precedences (represented by the x and y variables, respectively) is not feasible"; constraints $(6.iii)$ as "At least one pair of flights in a certain set should not meet in a certain sector"; finally, constraints $(6.iv.b)$ read as "If a certain set of flight meetings/precedences take place, then a certain flight will delayed by at least a certain amount of time".

In [10], we present a delayed row generation algorithm to solve one instance of the Hotspot Problem. The algorithm starts by setting up a MILP that contains only constraints $(6.i)$, while the rows corresponding to $(6.ii)$, $(6.iii)$, and $(6.iv.b)$ are generated on the fly only when violated by the current incumbent solution. When the algorithm terminates (i.e., an optimal solution has been found), only a small fraction of the constraints of PC will be present in the final MILP.

When managing air traffic in real-time, one needs to address a sequence of hotspot problems over time, say $HP_1, HP_2, \ldots$. Let PC_j be the path&cycle formulation for problem HP_j, and let R_j be the rows generated while solving PC_j. Our dynamic re-optimization framework consists of trying to add the rows stored in R_j to the initial MILP of PC_{j+1}. If the underlying problems HP_j, HP_{j+1} are similar, then many of the constraints in R_j are still valid for PC_{j+1}. In particular, let us call $\mathcal{C}_j$, $\mathcal{P}_j$, and $\mathcal{H}_j$ the cycle, path and hotspot constraints stored in R_j, respectively. It is easy to see that a cycle $C \in \mathcal{C}_j$ is valid for PC_{j+1} if and only if C is a strictly positive cycle in the disjunctive graph associated with HP_{j+1}. Similarly, a hotspot constraint $H \in \mathcal{H}_j$ is valid for PC_{j+1} if and only if the corresponding hotspot still exists in problem HP_{j+1}. Instead, paths may require some additional work. For each path constraint $P \in \mathcal{P}_j$, one must first check that the corresponding path still exists in the disjunctive graph associated with problem HP_{j+1}. Next, the length of the path should be recomputed (since flights have moved from one iteration to another), and the constraint updated accordingly.

This procedure can be seen as a warm start for the path&cycle formulation where, instead of providing an initial solution to the subsequent iteration (to which the term "warm start" usually refers to), we provide a set of valid constraints. The pool of these valid combinatorial constraints grows, contracts, and adapts over time, following the dynamics of the system. This is what the term *combinatorial learning* refers to. To the best of our knowledge, this is the first time a generic re-optimization framework that exploits previously generated constraints has been studied.

4 Computational Results

Testing an algorithm in a dynamic environment is usually not an easy task. The most challenging issue with dynamic scheduling is that the underlying problem evolves not only based on the original schedule, but also based on the decisions taken by the scheduling algorithm. For example, in our context, an algorithm could decide (during a particular iteration) to delay a certain flight in order to prevent a hotspot, but another algorithm could decide to keep that flight on schedule and delay another one (because there are multiple optima or because it is heuristic). This may create completely different scenarios at subsequent iterations.

Rather than using sophisticated simulation tools that would be difficult to distribute or recreate, we generated a "static" set of problems (similar to those used in [10]) that, by following simple rules, can be easily integrated in a dynamic simulated environment[2]. This would make future comparisons extremely easy. The airspace is composed of 400 sectors, 20 airports (randomly positioned across the sectors), and around 250 flights. The flight route of each airplane is randomly generated by choosing a pair of airports and a departure time within a time window of 3 h (for simplicity and without loss of generality, we assume that all flights travel at the same speed). The average travel time is around 3 h. While we have access to real instances, we are not allowed to share them publicly. However, these random instances have been generated to resemble realistic ones. Part of the dynamic behaviour is given by the fact that the schedule of each airplane is only available two hours before the departure time. This means that at the beginning of the 3 h time window only a fraction of the total number of airplanes will be available to be scheduled. As time goes by, more flights will become available, some will depart, and some will land. For each of the 10 ATM problems, the test is performed by solving one instance every minute for the entire 180 min time window. After one instance has been solved, we move the time one minute forward, assuming that all airplanes will simply follow the optimal schedule given by the solution found at the previous instance. Moreover, flights that are already departed cannot be rescheduled, but flights that are still waiting for departure can always be rescheduled.

Therefore, starting from an initial "static" schedule, one could construct a dynamic simulation by following these rules:

- At any time t, only the flights bound to depart in the next two hours are available to be scheduled.
- Flights that have already departed cannot be rescheduled.
- If the interval of time between two consecutive iterations is δ, then in the interval of time $[t, t + \delta]$ all flights will follow the optimal schedule computed at time t.

[2] The problems are available from the authors upon request.

All the tests have been performed on a single thread of an Intel(R) Core(TM) i7-7700HQ CPU @ 2.80 GHz with 32 GB of memory, using CPLEX 12.8 through the C♯ API.

Table 1 reports the results of the path&cycle algorithm in its basic implementation (*PC*) and in its augmented version that make use of constraints generated while solving previous instances of the same problem (*PC+*). Note how both the average and maximum computation time and the average and maximum number of visited nodes in the branching tree drops substantially when using *PC+*.

Figure 2 shows the typical behaviour of some key measures across the 180 instances of the 3 h time window. It also shows the total number of rows added to the master problem when the optimal solution was found. It is interesting to note that the MILPs of *PC+* contained many more rows than the ones of *PC*. This has to do with the fact that *PC+* also makes use of all the constraints contained in the pool that is maintained across the instances. As the experiments shows, this usually produces a significant advantage.

Table 1. Results on 10 randomly generated problems. The table shows the number of scheduled *flights*, how many *hotspot* constraints have been generated by each algorithm, the total number of *nodes* visited by the branch and bound algorithm, and the total *time* required to find the optimal solution. Each of these numbers is the average across 180 instances (the maximum in parenthesis), moving each time one minute forward in the schedules.

ID	$\|F\|$	Hotspots		Visited nodes		Time (s)	
		PC	*PC+*	*PC*	*PC+*	*PC*	*PC+*
ATM1+	323	18	38	14161 (83610)	441 (15178)	3.08 (18.16)	0.51 (5.38)
ATM2+	236	8	12	46 (237)	6 (266)	0.28 (1.33)	0.11 (0.84)
ATM3+	250	17	30	6955 (119877)	297 (22231)	3.58 (56.39)	0.67 (11.6)
ATM4+	215	15	29	9128 (127705)	721 (42626)	1.69 (17.6)	0.44 (9.19)
ATM5+	257	9	20	1956 (42712)	570 (42526)	0.88 (15.79)	0.58 (14.5)
ATM6+	220	6	6	127 (312)	10 (76)	0.58 (57.03)	0.35 (24.34)
ATM7+	218	10	22	453 (2756)	66 (1937)	0.42 (1.28)	0.13 (1.2)
ATM8+	236	8	14	685 (5762)	79 (774)	0.71 (8.81)	0.32 (4.97)
ATM9+	238	10	15	218 (1812)	32 (782)	0.37 (1.13)	0.14 (0.79)
ATM10+	250	12	22	4490 (39357)	647 (18075)	1.71 (19.59)	1.19 (33.24)

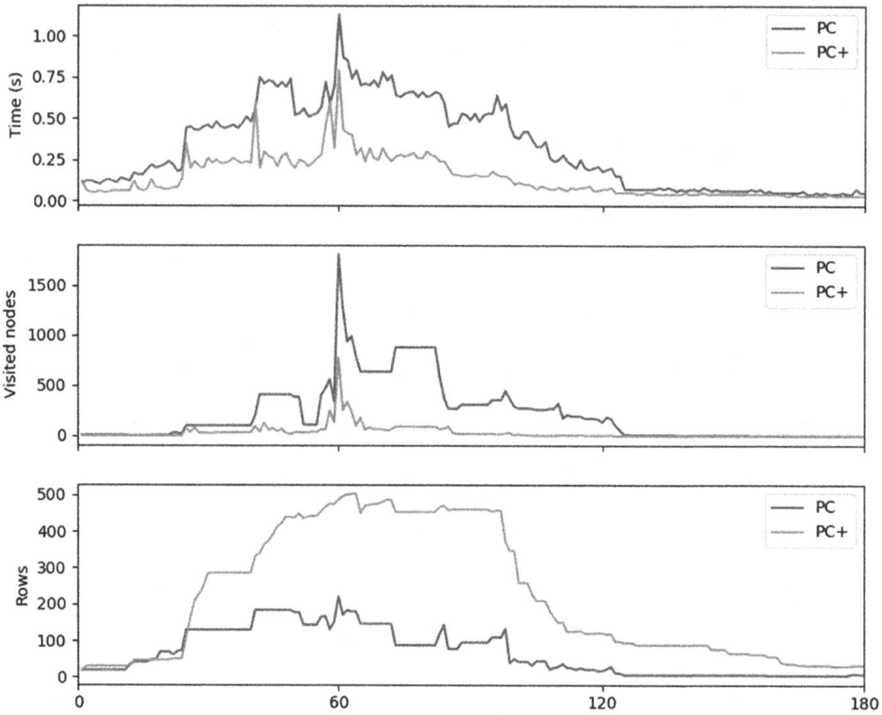

Fig. 2. Computation *time*, visited *nodes* of the branching tree, and total number of *rows* in the MILP of each of the 180 instances generated when solving the problem ATM9+ every minute (of schedule time) for 3 h.

5 Conclusions

In this work we showed how a lazy MILP formulation can be used in a re-optimization framework (that we call *combinatorial learning*) to speed-up the computation of consecutive iterations of a mathematical model representing a dynamic system.

While this algorithmic idea could be applied to a generic lazy MILP formulation (i.e., a formulation where the constraints are added only when needed), it is particularly suited for our *path&cycle* formulation. In fact, the constraints of the *path&cycle* are almost purely combinatorial and, most importantly, do not contain any scheduling variable t. Instead, the values of the scheduling variables (which are necessary to determine the actual schedules of the flights) can be computed a posteriori from the values of the binary variables of the solution of the *path&cycle*. Having a formulation where the constraints do not explicitly contain time-dependent variables is crucial, because they can possibly be reused without changes when the dynamic system evolves with time.

This paper describes an application of *combinatorial learning* to a relevant problem that arises in Air Traffic Management, and shows that it can lead to a significant improvement in computation time when compared to the classic "static" approach. This approach is suitable for other traffic management problems as well. In particular, the *path&cycle* formulation can be used for any scheduling problem that can be described via a disjunctive graph (e.g., production scheduling).

Finally, exact re-optimization techniques are still not well-studied, but they will certainly play an important in the future. The results obtained by the novel approach presented in this paper show that there is still a lot to contribute to this branch of research.

References

1. Balas, E.: Machine sequencing via disjunctive graphs: an implicit enumeration algorithm. Oper. Res. **17**(6), 941–957 (1969)
2. Fischetti, M., Ljubić, I., Sinnl, M.: Redesigning benders decomposition for large-scale facility location. Manage. Sci. **63**(7), 2146–2162 (2016)
3. Keha, A.B., Khowala, K., Fowler, J.W.: Mixed integer programming formulations for single machine scheduling problems. Comput. Ind. Eng. **56**(1), 357–367 (2009)
4. Lamorgese, L., Mannino, C.: A non-compact formulation for job-shop scheduling problems in transportation. Oper. Res. (to appear)
5. Mannino, C., Mascis, A.: Optimal real-time traffic control in metro stations. Oper. Res. **57**(4), 1026–1039 (2009)
6. Mascis, A., Pacciarelli, D.: Job-shop scheduling with blocking and no-wait constraints. Eur. J. Oper. Res. **143**(3), 498–517 (2002)
7. Ozbaygin, G., Savelsbergh, M.: An iterative re-optimization framework for the dynamic vehicle routing problem with roaming delivery locations. http://www.optimization-online.org/DB_FILE/2018/08/6784.pdf (2018)
8. Psaraftis, H.N., Wen, M., Kontovas, C.A.: Dynamic vehicle routing problems: three decades and counting. Networks **67**(1), 3–31 (2016)
9. Queyranne, M., Schulz, A.S.: Polyhedral approaches to machine scheduling. TU, Fachbereich 3, Berlin (1994)
10. Sartor, G., Mannino, C.: The path&cycle formulation for the hotspot problem in air traffic management. In: 18th Workshop on Algorithmic Approaches for Transportation Modelling, Optimization, and Systems, ATMOS 2018, Helsinki, Finland, 23–24 August 2018 (2018)
11. Schieber, B., Shachnai, H., Tamir, G., Tamir, T.: A theory and algorithms for combinatorial reoptimization. Algorithmica **80**(2), 576–607 (2018)

Cartesian Genetic Programming with Guided and Single Active Mutations for Designing Combinational Logic Circuits

José Eduardo H. da Silva, Lucas A. M. de Souza, and Heder S. Bernardino(✉)

Universidade Federal de Juiz de Fora, Juiz de Fora, MG, Brazil
{jehenriques,heder}@ice.ufjf.br, lucas.mullers@gmail.com

Abstract. The design of digital circuits using Cartesian Genetic Programming (CGP) has been widely investigated but the evolution of complex combinational logic circuits is a hard task for CGP. We introduce here a new mutation operator for CGP that aims to reduce the number of evaluations needed to find a feasible solution by modifying the subgraph of the worst output of the candidate circuits. Also, we propose a variant of the standard evolutionary strategy commonly adopted in CGP, where (i) the Single Active Mutation (SAM) and (ii) the proposed mutation operator is used in order to improve the capacity of CGP in generating feasible circuits. The proposals are applied to a benchmark of combinational logic circuits with multiple outputs and the results obtained are compared to those found by a CGP with SAM. The main advantages observed when both mutation operators are combined are the reduction of the number of objective function evaluations required to find a feasible solution and the improvement in the success rate.

Keywords: Cartesian genetic programming · Combinational logic circuit · Mutation

1 Introduction

The design of digital electronic circuits is a task that requires time and knowledge of specific rules. These circuits design's complexity increases when the number of inputs and outputs become greater. The traditional design of electronic circuits is aided by several methods, such as the Karnaugh Maps and ESPRESSO [1]. On the other hand, evolutionary computation proved to be a powerful tool in this task. The so-called Evolvable Hardware (EH) consists of the use of evolutionary computation techniques in the design and implementation of hardware in general.

The first efforts for evolving electronic circuits were made by Koza [7], using Genetic Programming. Later, EH methods were also developed by Coello and his collaborators [2–4] and Miller and his collaborators [11–13]. In particular, Miller proposed Cartesian Genetic Programming (CGP), in which programs are

G. Nicosia et al. (Eds.): LOD 2019, LNCS 11943, pp. 396–408, 2019.
https://doi.org/10.1007/978-3-030-37599-7_33

represented as directed acyclic graphs (DAGs), encoded by a matrix of nodes. In addition, CGP is pointed out as the most efficient technique for the evolution of digital circuits [18]. Most of these circuits are combinational logic circuits (CLCs), i.e., digital circuits whose outputs are represented only by a combination of its inputs. There are no memory elements.

Despite the great success of CGP's application in this task, several works recently published in [16] point out that the EH researches still works with the same problems of 15 years ago and are walking through a very challenging scenario for balancing the real needs of the industry with the capacity of designing circuits via CGP, where the most complex circuit already evolved is 28-input benchmark [19]. The evolution of more complex circuits using CGP can be found only minimizing the number of gates of known models [18].

Although CGP has difficulties in evolving complex circuits, this metaheuristic is still used as it proved to be efficient obtaining circuits that escape the traditionalism of human design, especially when the goal is to find more compact circuits in terms of reducing the number of logic gates. One of the reasons that CGP finds it difficult to evolve larger circuits is the problem of scalability, that is, the complexity of the circuit design increases exponentially with the number of inputs [12]. This is directly linked to the very definition of CLCs. As a consequence, several works aim to increase and to improve the CGPs capability of exploring the search space, such as [5,6,9,10,14,15]. This exploitation capacity is directly linked to the search space and to the genetic operators involved. CGP normally only uses mutation to generate new individuals, and point mutation and SAM (Single Active Mutation) are the most adopted approaches.

We propose a new mutation operator for designing CLCs, where the solutions created in each generation aim to improve the subgraphs whose outputs present the minimum amount of hits with respect to the truth table, increasing the assurance that the mutations will not be harmful to the good subgraphs already obtained by the evolutionary process. The Evolution Strategy (ES) commonly used in CGP is also modified. In this way, we propose (i) a new mutation operator to CGP when applied to CLCs and, (ii) the combination of the proposed mutation with SAM modifying the ES $1 + \lambda$ commonly used in CGP. The results of the proposals are compared to those obtained by a baseline CGP using SAM.

2 Cartesian Genetic Programming

Cartesian Genetic Programming (CGP) is a type of Genetic Programming proposed by Miller [12]. In CGP, the programs are DAGs represented by a matrix of processing nodes and each node contains genes that describe its function and connections to the other nodes. Each node of this matrix is represented by its inputs and one function. Here, the functions represent logic gates. Also, each node connects to other ones on its left columns in the matrix. Figure 1 illustrates a CGP individual with a 3×3 matrix when solving a problem with 3 inputs and 3 outputs. CGP has three user-defined parameters: number of columns (n_c) and rows (n_r) of the matrix that encodes the graph, and levels-back (lb). Levels-back is the number of columns (on the left side) in which each node can be connected.

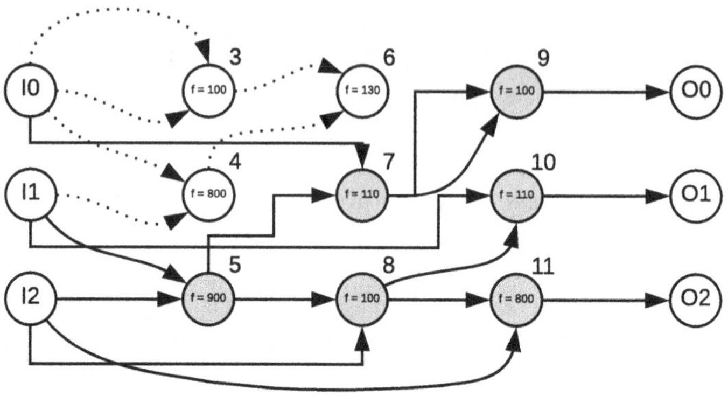

Fig. 1. Example of a CGP individual with *levels-back* = *number of columns*. The continuous lines connect the active nodes (in gray) which form the phenotype. The value into each node represents its function which is applied to its inputs.

Algorithm 1. CGP's (1 + λ)-ES general procedure [12].

Randomly generate the individual population
Select the fittest individual as parent
while the maximum number of evaluations allowed is not reached **do**
    Generate λ new individuals (by mutating the parent individual)
    Evaluate the λ new individuals
    Select as parent the best individual considering the $(1 + \lambda)$ current individuals
end while

The most common search technique used in CGP is an Evolutionary Strategy (ES) $(1+\lambda)$ [12], where λ is the number of new solutions generated at each iteration, as presented in Algorithm 1. In this case, the best individual between the parent and the λ new generated solutions is selected to the next generation.

In CGP, the most used mutation is the point mutation [12]. However, when using the point mutation, the changes may occur in nodes which do not interfere in the output of the solutions *(inactive nodes)* leading to a lack of modifications in the phenotype. These nodes are the white ones in Fig. 1. The single active mutation (SAM) [5] is another common mutation and was proposed for reducing the wasted evaluations of objective functions, where (i) one node and one of its elements (inputs or function) are selected at random, and (ii) the value is changed to another valid value. The steps (i) and (ii) are repeated until an active node is modified. So SAM ensures that one active node is changed.

Here, the evolutionary process is divided into two steps where the objectives are (i) to maximize the number of correct outputs when compared to the truth table (to obtain a feasible solution) and (ii) to minimize the number of logic gates. In the first step, the best individual is the one with the most matches with respect to the truth table. After a feasible solution is found, the best is the feasible individual with the smaller number of logic gates. The best individual is promoted as father to the subsequent generation.

2.1 Binary Decision Diagrams

Binary Decision Diagrams (BDDs) can be used to evaluate the individuals [19]. BDD is a directed acyclic graph with one root and two terminal nodes that are referred to as 0 and 1. All other nodes are called non-terminal nodes. Each non-terminal node (v) is labeled with a boolean variable (var), and have two child nodes, representing high (var(v) = 1) and low (var(v) = 0) levels. An Ordered Binary Decision Diagram (OBDD) is a BDD with a defined variable ordering, a tuple $T(z_1, ..., z_m)$. A node labeled as z_i has two child nodes z_j and z_k. The node labeled as z_j has a higher index in T than its father $(i < j)$. The same occurs for the other child node. This ordering allows for the simplification of OBDDs.

A Reduced Ordered Binary Decision Diagram (ROBDD) is an OBDD where each node represents a unique logic function [19], and can be reduced using two rules: elimination and isomorphism. In elimination, the nodes which children point to the same node can be removed, keeping only the children. In isomorphism, given nodes v and w, v can be removed and all its incoming edges can be redirected to w when: (i) v and w are terminals and have the same value, and (ii) v and w are inner nodes, they are labeled by the same boolean variable, and their children are the same. ROBDDs can be used to evaluate the candidate circuits in CGP once it is canonical for a particular function and variable order.

Given a XOR gate with the outputs of two CLCs (C_A and C_B with outputs $y_1, ..., y_m$ and $y'_1, ..., y'_m$ respectively) as its inputs, these two circuits, represented by ROBDDs, are functionally equivalent when for all output i, the result of $k_i = y_i$ XOR y'_i is an empty set in the set(k_i). It is possible to use the Sat-Count function on all k_i outputs and count all the results. The resulting value represents the Hamming distance between C_A and C_B. When the result of the Sat-Count is equal to zero, the circuits C_A and C_B are functionally equivalent.

The desired circuit and the candidate solution are represented by independent ROBDDs. Thus, the fitness function adopted is to minimize the Sat-Count between their outputs. The Sat-Count function used is provided by the Buddy Package [8]. The use of ROBDD for the individuals' representation and the Sat-Count as fitness function used in this work are the same presented in [19].

3 The Proposed Approaches

Here two proposals are presented. First, a new mutation operator focused on mutating the worst individuals' subgraph (GAM). Then, a proposal combining both mutation operators (SAM and GAM) are presented.

3.1 Guided Active Mutation

The proposed mutation operator, labeled here as GAM (Guided Active Mutation) consists of modifying active(s) node(s) on the subgraph from the inputs to the outputs with the smallest number of correct values when compared to the truth table. Thus, GAM does not mutate nodes in subgraphs which already match their corresponding values in the truth table. This feature is important due to the reduction of wasted evaluations. Modifications in subgraphs which generate correct outputs will be worst or equal to the preliminary candidate circuit, as it already matches its full truth table.

GAM is presented in Algorithm 2 and the procedure is as follows: (i) select the output of the circuit with the smallest number of correct values with respect to the truth table (called here worst output), (ii) choose one of the worst outputs randomly when a tie occurs, (iii) select the active nodes in the subgraph from the inputs to this selected output (even if the nodes are in the path of other outputs), and (iv) mutate one (or more) randomly selected active node this subgraph. This mutation is made by changing one of the node's inputs or function. In (i), there is an array that stores the fitness value of each individual's outputs. In (ii), the active nodes of each individual's outputs are also stored in arrays.

Algorithm 2. The GAM operator.

```
1: worst_out ← output with the worst values when compared to truth table
2: active_nodes ← active nodes in the subgraph from the inputs to worst_out
3: while the number of active nodes mutated is not reached do
4:     modify a randomly selected active node
5: end while
```

Here we modify only one active node in each new candidate solution, but this can be modified (`while` instruction in Algorithm 2). One can notice that GAM tries to improve the worst subgraph of the candidate solution. Also, GAM always modifies an active node only, differently of SAM, in which inactive nodes are modified up to an active note is changed.

Figure 2 illustrates the GAM operator with an individual with three inputs (I0, I1, and I2) and three outputs (O0, O1, and O2). The numbered circles represent the nodes of the individual (gates of the circuit). The nodes in gray are the active ones, and those in white are the inactive ones. The arrows represent the directed connections between the nodes and the continuous lines connect the active nodes. The phenotype (circuit) is composed of inputs, active nodes and the connections between them, and outputs. In this way, the path to the output O2 of the first individual is formed by I2, and nodes 5, 8 and 11.

For this illustrative example, the number of correct values of the outputs of the individuals of iterations 1 and 2 is presented in Table 1. In generation 1, output O1 is the worst one, as it matches 3 of 8 possible bits of the truth table. The other outputs of this circuit hit 8 and 6 of the values of the truth table. Then, GAM modifies only the active nodes in the subgraph of output O1, which are the red nodes highlighted in Fig. 2(b).

Considering that the offspring created during generation 1 is the parent of generation 2, generated by a mutation in node 10, as presented in Fig. 2(c), such modification improved output O1: 6 values are correct now. In this case, the worst output of this individual can be O1 or O2. So, the GAM procedure

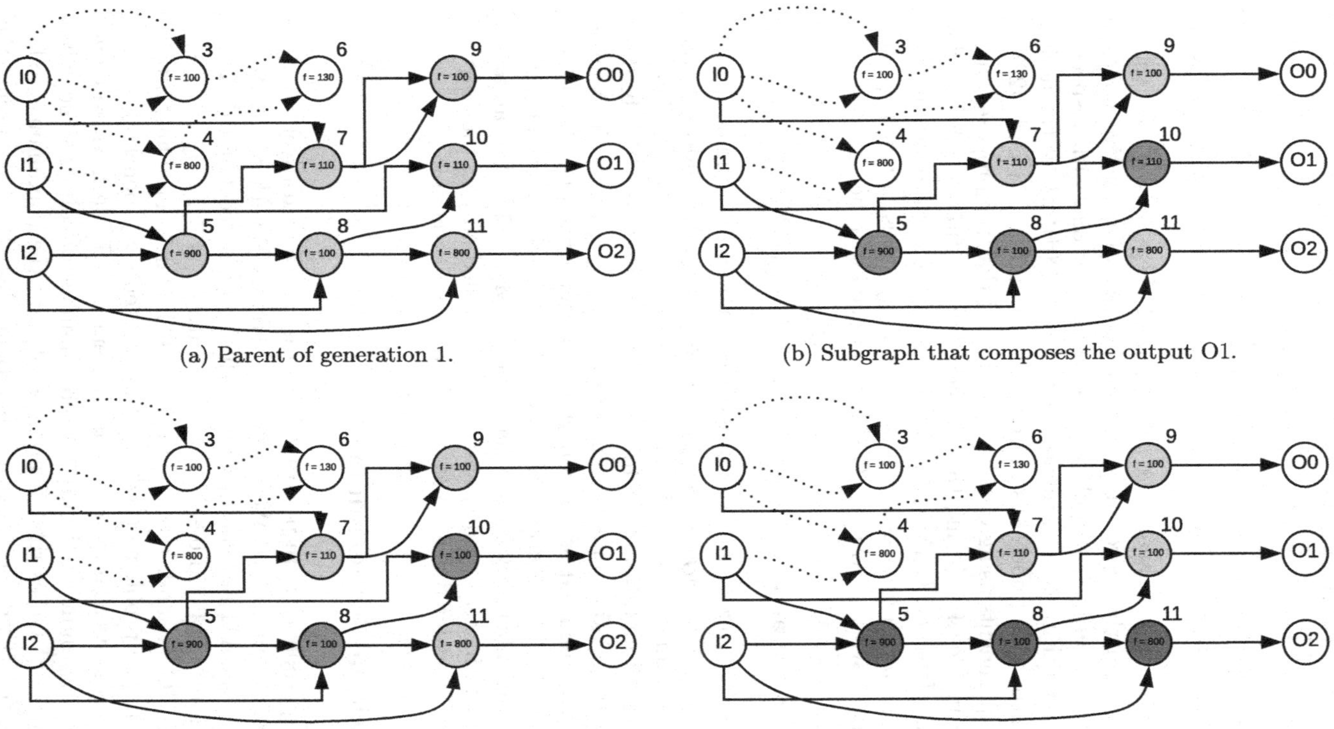

(a) Parent of generation 1.

(b) Subgraph that composes the output O1.

(c) Parent of generation 2.

(d) Subgraph that composes the output O2.

Fig. 2. (a) Are the parent of generation 1 and its subgraph that composes the output O1 is showed in (b), in red. (c) is the parent of generation 2 created by a mutation in node 10 from the previous parent with its respective subgraph in red. Both subgraphs in red of (c) and in blue of (d) match 6 of 8 bits with respect to the truth table. One of them will be randomly chosen to receive the GAM operator to generate the subsequent offspring. (Color figure online)

Table 1. Matches with respect to the truth table per generation.

Generation	O0	O1	O2	Fitness
1	8	3	6	17
2	8	6	6	20

will randomly choose one of them. The subgraphs of these outputs are shown in Fig. 2(c) and (d). Furthermore, in cases that outputs share active nodes, a mutation can improve an output and deteriorate another one with respect to the matches to the truth table, as shown in Fig. 2(c) and (d), where the node 5 is shared by all outputs and node 8 is shared between outputs O1 and O2.

Preliminary experiments using GAM demonstrated that the proposed mutation operator is capable of reducing the number of evaluations needed to find a feasible solution. However, GAM did not reach high success rates (number of independent runs in which a feasible solution is found). Thus, a new approach is also proposed here, where SAM and GAM are applied together.

3.2 CGP with SAM and GAM: $(1 + \lambda_{SAM} + \lambda_{GAM})$-ES

In order to increase the success rate, we proposed to modify the commonly used $(1+\lambda)$-ES to a $(1+\lambda_{SAM}+\lambda_{GAM})$-ES. In this proposal, λ_{SAM} individuals are generated using SAM and λ_{GAM} individuals are created using GAM. Thus, we take advantage of both mutation operators into a single approach (SAMGAM). This is justified as the SAM can mutate inactive nodes, which favors the Neutral Genetic Drift (NGD) reported as an important characteristic of CGP [12,17]. So, even if SAM and GAM are applied independently, each operator will be responsible for acting differently on the genotype. SAM is applied to an active node and one or more inactive nodes, and GAM is applied only to the outputs' subgraphs that are considered the worst, trying to improve them.

4 Computational Experiments

Computational experiments were performed in order to comparatively analyze the performance of the proposals. We considered here: CGP with Single Active Mutation (labeled as SAM and 4+0), CGP with the proposed Guided Active Mutation (labeled as GAM and 0+4), and CGP with an $(1 + \lambda_{SAM} + \lambda_{GAM})$-ES (labeled as SAMGAM and, its variants 1+3, 2+2 and 3+1). Thus, 5 techniques were tested. The problems used in all experiments are benchmarks and arithmetic circuits, and its characteristics are presented in Table 2, where name, number of inputs (ni), number of outputs (no), number of columns (nc) used by CGP, the number of Gates related in RevLib[1], and the maximum number of evaluations allowed (Eval.) are shown. Also, Balancing (Bal.) represents the

[1] http://revlib.org/functions.php.

ratio between the number of 0s observed in the outputs of the truth table and the total the number of outputs of the truth table (number of outputs × number of lines), and Simplification (Simp.) is the ratio between the number of gates related in column Gates and the number of gates required to form the circuit through the sum of products. Previous studies have shown that CGP has ease in solving problems whose the boolean expressions are simplifiable (smaller values of Simp. imply simplifiable circuits). In addition, balancing of the number of 0s and 1s in the outputs of the truth tables also can affect the performance of the methods. This is similar to the unbalancing problem in machine learning. Thus, the number of objective function evaluations allowed to the search methods are based on the number of inputs of the circuit, and the rates of balancing and simplification. The number of function evaluations is ni $\times$ $(obj_{\text{balancing}} + obj_{\text{simplification}})$, where $obj_{\text{balancing}} = 800,000$ if the balancing ratio is in $[0.5, 0.75]$, and $1,600,000$ otherwise, and $obj_{\text{simplification}} = 1,600,000$ if the simplification ratio is in $[0, 0.0138]$, and $800,000$ otherwise. These values were chosen empirically.

Table 2. Problems used in the computational experiments.

Name	Inputs	Outputs	nc	Gates	Eval.	Bal.	Simp.
dc1	4	7	100	39	6, 400, 000	0.598	0.129
dc2	8	7	100	75	19, 200, 000	0.753	0.014
alu4	14	8	1000	1063	33, 600, 000	0.565	0.001
z4ml	7	4	100	48	11, 200, 000	0.500	0.018
cm85a	11	3	1000	69	26, 400, 000	0.734	0.002
C17	5	2	100	6	12, 000, 000	0.938	0.300
sao2	10	4	1000	88	32, 000, 000	0.818	0.007
cu	14	11	1000	40	44, 800, 000	0.924	0.001
apla	10	12	1000	80	24, 000, 000	0.987	0.029
inc	7	9	100	93	16, 800, 000	0.756	0.031
f51m	8	8	100	43	19, 200, 000	0.500	0.004
6x6-Adder	12	12	1000	229	28, 800, 000	0.709	0.001

The search process is (i) to find a feasible solution and (ii) to minimize the number of gates. The parameters adopted here are $\lambda = 4$, $n_r = 1$ and $\text{lb} = n_c/2$, as lower values of lb tends to reduce the number of active nodes [10]. The number of independent runs is 30 and the function set is $\Gamma = \{$AND, OR, NOT, NAND, NOR, XOR, XNOR$\}$ for all the tested problems. Also, the algorithm runs until the maximum number of evaluations of the objective function is reached. The source-codes of the proposals and supplementary material are available[2].

The median of the number of objective function evaluations observed to find a feasible solution and the percentage of the number of independent runs in which

[2] https://github.com/ciml.

Table 3. Number of objective function evaluations to find a feasible solution and p-values of the Kruskal-Wallis (p_{kw}).

Method	Median	SR (%)	p_{kw}	Method	Median	SR (%)	p_{kw}	Method	Median	SR (%)	p_{kw}
dc1				dc2				alu4			
0+4	82327	93.33	1.06E-1	0+4	**2916409**	3.33	3.44E-1	0+4	14958613	6.66	5.34E-1
1+3	88343	**100**		1+3	7370571	93.33		1+3	15297191	60	
2+2	**61383**	**100**		2+2	6880624	90		2+2	14487059	80	
3+1	71383	**100**		3+1	6437425	93.33		3+1	11570917	**93.33**	
4+0	92649	**100**		4+0	5386677	**96.67**		4+0	**11510149**	**93.33**	
z4ml				cm85a				C17			
0+4	238629	30	9.63E-2	0+4	503249	96.67	2.56E-2	0+4	3063	**100**	9.42E-1
1+3	137857	**100**		1+3	**371347**	**100**		1+3	2713	**100**	
2+2	169565	**100**		2+2	446549	**100**		2+2	2827	**100**	
3+1	**103043**	**100**		3+1	502865	**100**		3+1	2617	**100**	
4+0	199311	**100**		4+0	550441	**100**		4+0	**2449**	**100**	
sao2				cu				apla			
0+4	**2523281**	96.67	9.27E-1	0+4	2825729	**100**	1.64E-5	0+4	4801839	**100**	7.47E-5
1+3	2532073	**100**		1+3	2391943	**100**		1+3	3700897	**100**	
2+2	2990237	**100**		2+2	**1847761**	**100**		2+2	**3166145**	**100**	
3+1	3197411	**100**		3+1	2736021	**100**		3+1	3510565	**100**	
4+0	3070581	**100**		4+0	3816927	**100**		4+0	5822959	**100**	
inc				f51m				6x6-Adder			
0+4	–	0	8.44E-1	0+4	–	0	2.24E-1	0+4	2572245	66.66	2.16E-2
1+3	**98988.53**	**30**		1+3	**1857245**	96.67		1+3	**1080389**	**100**	
2+2	10912945	23.33		2+2	1987763	**100**		2+2	1649199	**100**	
3+1	11863591	20		3+1	2208043	**100**		3+1	1335817	**100**	
4+0	11806739	26.66		4+0	3292863	**100**		4+0	1957119	**100**	

Table 4. Number of gates for the first (First) and final (Final) solution and reduction (Red. %) observed for SAM, GAM and SAMGAM with variable λ.

Method	Number of gates		Red. (%)	Method	Number of gates		Red. (%)	Method	Number of gates		Red. (%)
	First	Final			First	Final			First	Final	
dc1				dc2				alu4			
0+4	51.93	29.64	42.92	0+4	63	52	17.46	0+4	239	129	**46.03**
1+3	53.7	25.5	52.51	1+3	64.04	53.86	15.90	1+3	187.05	118.29	36.76
2+2	50.9	24.5	51.87	2+2	64.88	52.22	19.51	2+2	185.33	112.5	39.30
3+1	51.6	24.13	**53.24**	3+1	61.79	49.11	**20.52**	3+1	167.43	110.56	33.97
4+0	**49.97**	**23.97**	52.03	4+0	**59.86**	**48.97**	18.19	4+0	**159.14**	**97.82**	38.53
z4ml				cm85a				C17			
0+4	46	29.11	36.72	0+4	119.59	31.79	73.42	0+4	25.9	5.83	77.49
1+3	39.03	18.4	**52.86**	1+3	102.8	27.1	**73.64**	1+3	24.97	4.57	81.70
2+2	**36**	**18**	50	2+2	96.03	26.6	72.30	2+2	26.43	4.73	**82.10**
3+1	37.77	18.67	50.57	3+1	83.97	27.5	67.25	3+1	25.4	4.63	81.77
4+0	36.77	17.77	51.67	4+0	**72.4**	**26.47**	63.44	4+0	**24.23**	**4.53**	81.30
sao2				cu				apla			
0+4	139.34	70.83	49.17	0+4	203.6	43.43	**78.67**	0+4	254.07	90.5	**64.38**
1+3	141.03	67.03	52.47	1+3	167.9	38.97	76.79	1+3	234.73	85.8	63.45
2+2	**132**	**61.87**	53.13	2+2	146.07	39.1	73.23	2+2	232.53	85.87	63.07
3+1	141.23	65.33	53.74	3+1	146.9	**38.77**	73.61	3+1	224.77	83.1	63.03
4+0	139.13	64.1	**53.93**	4+0	**121**	**38.77**	67.96	4+0	**204.9**	**80.13**	60.89
inc				f51m				6x6-Adder			
0+4	–	–	–	0+4	–	–	–	0+4	175.85	66.75	62.04
1+3	77.78	72.11	7.29	1+3	53.9	39.34	27.01	1+3	147.97	47.73	67.74
2+2	81.43	73.71	9.48	2+2	53.7	40.73	24.15	2+2	143.07	49.6	65.33
3+1	77.83	**70.33**	**9.64**	3+1	49.37	36.33	26.41	3+1	134.57	42.73	**68.27**
4+0	**76.13**	71	6.74	4+0	**48.13**	**34.47**	**28.38**	4+0	**116.07**	**42.23**	63.62

a circuit that produces the same outputs observed in the truth table (success rate: SR) are presented in Table 3. Kruskal-Wallis test was performed in order to compare the results statistically. The best results are highlighted in boldface. SAM was capable of achieving the highest success rate on all problems, except for inc. When compared to SAM (4+0), GAM (0+4) is able to reduce the number of objective function evaluations needed to find a feasible solution in 6 out of 12 problems in median but obtained the worst success rates. SAMGAM approaches (1+3, 2+2 and 3+1) obtained the best results in median (8 of 12 problems). Also, approaches in which GAM generates more than one offspring, the best median results are obtained in 7 of 12 problems. This means that the major influence of SAM when using SAMGAM is perceptible only in problem z4ml. Regarding the statistical tests, only 4 of the 12 problems presented statistical differences between the proposals.

In Table 4, the mean of the total number of gates are presented for the first feasible solution and for the best solution found. The column Red. (Reduction) presents the relative difference between the values observed in the first feasible solution and the best solution. When SAM (4+0) is used, its first feasible solutions (10 of 12 problems) and its best solutions (9 of 12 problems) have fewer logic gates. Also, the circuits found by GAM are not compact. With respect to the reduction capability between the first feasible solution and the best solution obtained, approaches using SAM and GAM were able to find the best values for reducing the number of logic gates in 7 of the 12 problems. Based on the information of the problems in which each of the methods was better, it is possible to conclude that GAM shows a good performance for circuits (i) with smaller values of Simp., (ii) with higher values of Bal., and (iii) with a larger number of inputs. SAMGAM presented good results for circuits that (i) are not simplifiable, (ii) have high unbalance rate, and (iii) fewer inputs. No common features were observed in the problems in which SAM achieved the best performance.

5 Concluding Remarks and Future Work

A new mutation operator for Cartesian Genetic Programming (CGP) was introduced here in order to design combinational logic circuits. The proposed mutation operator was compared to a baseline CGP with Single Active Mutation (SAM). Also, the proposed Guided Active Mutation (GAM) was combined with SAM to increase the capability of GAM in finding feasible solutions and performed better than the other approaches considered here.

GAM modifies the subgraphs of the individuals which can still be improved, reducing wasted mutations on full match subgraphs. Thus, GAM was able to find feasible solutions with a reduced number of objective function evaluations when compared to the baseline CGP with SAM in most of the problems.

It is possible to conclude that SAM (i) obtained the highest success rates for most problems, (ii) obtains first feasible solution with lower number of logic gates for most problems, and (iii) obtains the best solution with the least number of logic gates in 9 of the 12 problems. The proposed GAM operator (i) requires less number of objective function evaluations to find a feasible solution when compared to SAM, (ii) obtains low success rates for all problems, (iii) does not have the capacity to find compact circuits (fewer logic gates) and, (iv) works well for circuits that are simplifiable, unbalanced and with a larger number of inputs. In order to obtain an approach that reaches high success rates (SAM) and low number of objective functions for finding a feasible solution (GAM), it is possible to conclude that the SAMGAM approaches (i) obtained the best results for most problems in the median, (ii) obtained high success rates, and (iii) works well for the circuits that are not simplifiable, have high unbalance rate, and fewer inputs.

It is important to notice that when the solutions share active nodes, GAM can improve one output but deteriorating the other one. The crossover operator proposed in [14] can take advantage of this specific improvement and discard the deterioration in the other output. Thus, the exploration of using this crossover into SAMGAM is an interesting research avenue. Finally, we also intend to apply the proposed SAMGAM to more complex problems and conduct a study regarding the mutation of more than one active node when using GAM.

Acknowledgments. We thank the reviewers for your suggestions and the support provided by CNPq (grant 312682/2018-2), FAPEMIG (grant APQ-00337-18), Capes, PPGCC/UFJF, PPGMC/UFJF.

References

1. Brayton, R.K., Hachtel, G.D., McMullen, C., Sangiovanni-Vincentelli, A.: Logic Minimization Algorithms for VLSI Synthesis, vol. 2. Springer, Heidelberg (1984). https://doi.org/10.1007/978-1-4613-2821-6
2. Coello, C.A.C., Aguirre, A.H.: Design of combinational logic circuits through an evolutionary multiobjective optimization approach. AI EDAM **16**(1), 39–53 (2002)
3. Coello, C.A.C., Alba, E., Luque, G.: Comparing different serial and parallel heuristics to design combinational logic circuits. In: Proceedings of the NASA/DoD Conference on Evolvable Hardware, pp. 3–12 (2003)
4. Coello, C.A.C., Luna, E.H., Aguirre, A.H.: Use of particle swarm optimization to design combinational logic circuits. In: Tyrrell, A.A.M., Haddow, P.C., Torresen, J. (eds.) ICES 2003. LNCS, vol. 2606, pp. 398–409. Springer, Heidelberg (2003). https://doi.org/10.1007/3-540-36553-2_36
5. Goldman, B.W., Punch, W.F.: Reducing wasted evaluations in cartesian genetic programming. In: Krawiec, K., Moraglio, A., Hu, T., Etaner-Uyar, A.Ş., Hu, B. (eds.) EuroGP 2013. LNCS, vol. 7831, pp. 61–72. Springer, Heidelberg (2013). https://doi.org/10.1007/978-3-642-37207-0_6
6. Husa, J., Kalkreuth, R.: A comparative study on crossover in cartesian genetic programming. In: Castelli, M., Sekanina, L., Zhang, M., Cagnoni, S., García-Sánchez, P. (eds.) EuroGP 2018. LNCS, vol. 10781, pp. 203–219. Springer, Cham (2018). https://doi.org/10.1007/978-3-319-77553-1_13
7. Koza, J.R.: Genetic Programming II, Automatic Discovery of Reusable Subprograms. MIT Press, Cambridge (1992)
8. Lind-Nielsen, J., Cohen, H.: Buddy - a binary decision diagram package (2014). https://sourceforge.net/projects/buddy/

9. Manfrini, F.A.L., Bernardino, H.S., Barbosa, H.J.C.: A novel efficient mutation for evolutionary design of combinational logic circuits. In: Handl, J., Hart, E., Lewis, P.R., López-Ibáñez, M., Ochoa, G., Paechter, B. (eds.) PPSN 2016. LNCS, vol. 9921, pp. 665–674. Springer, Cham (2016). https://doi.org/10.1007/978-3-319-45823-6_62
10. Manfrini, F.A.L., Bernardino, H.S., Barbosa, H.J.C.: On heuristics for seeding the initial population of cartesian genetic programming applied to combinational logic circuits. In: Proceedings of GECCO, pp. 105–106 (2016)
11. Miller, J.F.: An empirical study of the efficiency of learning boolean functions using a cartesian genetic programming approach. In: Proceedings of the 1st Annual Conference on Genetic and Evolutionary Computation, vol. 2, pp. 1135–1142. Morgan Kaufmann Pub. Inc. (1999)
12. Miller, J.F.: Cartesian genetic programming. In: Miller, J. (ed.) Cartesian Genetic Programming, pp. 17–34. Springer, Heidelberg (2011). https://doi.org/10.1007/978-3-642-17310-3_2
13. Miller, J.F., Job, D., Vassilev, V.K.: Principles in the evolutionary design of digital circuits - Part I. Genet. Program Evolvable Mach. **1**(1–2), 7–35 (2000)
14. da Silva, J.E., Bernardino, H.: Cartesian genetic programming with crossover for designing combinational logic circuits. In: Proceedings of the 7th Brazilian Conference on Intelligent Systems (BRACIS), pp. 145–150. IEEE (2018)
15. da Silva, J.E.H., Manfrini, F.A., Bernardino, H.S., Barbosa, H.J.: Biased mutation and tournament selection approaches for designing combinational logic circuits via cartesian genetic programming. In: ENIAC, pp. 835–846 (2018)
16. Stepney, S., Adamatzky, A. (eds.): Inspired by Nature: Essays Presented to Julian F. Miller on the Occasion of his 60th Birthday. ECC, vol. 28. Springer, Cham (2018). https://doi.org/10.1007/978-3-319-67997-6
17. Turner, A.J., Miller, J.F.: Neutral genetic drift: an investigation using cartesian genetic programming. GP Evolvable Mach. **16**(4), 531–558 (2015)
18. Vasicek, Z.: Cartesian GP in optimization of combinational circuits with hundreds of inputs and thousands of gates. In: Machado, P., et al. (eds.) EuroGP 2015. LNCS, vol. 9025, pp. 139–150. Springer, Cham (2015). https://doi.org/10.1007/978-3-319-16501-1_12
19. Vasicek, Z., Sekanina, L.: How to evolve complex combinational circuits from scratch? In: Proceedings of the Conference on Evolvable Systems (ICES), pp. 133–140. IEEE (2014)

Designing an Optimal and Resilient iBGP Overlay with Extended ORRTD

Cristina Mayr(✉), Claudio Risso, and Eduardo Grampín

Instituto de Computación, Universidad de la República, Montevideo, Uruguay
{mayr,crisso,grampin}@fing.edu.uy

Abstract. The Internet results from interconnecting several thousands of Autonomous Systems (ASes), which are networks under a single administrative domain such as: corporations, service providers, universities, and content providers, among others. To ensure global communication between end users, it is necessary that routers of every AS get to learn all IP addresses in this immense and extremely decentralized network. The Border Gateway Protocol (BGP) is the responsible of learning and distributing that reachability information among ASes in the form of groups of addresses (a.k.a. prefixes). Unlike other routing protocols, BGP routers communicate through administratively set point-to-point BGP sessions over TCP. BGP sessions are either external (eBGP, between routers of different ASes, a.k.a. Border Routers or ASBRs) or internal (iBGP, between routers whit to the same AS). While eBGP is needed to exchange reachability information among ASes, iBGP makes it possible for internal routers to learn prefixes necessary to forward IP packets to the appropriate ASBRs. To make sure that the whole information is learnt and no traffic deflection occur, a full-mesh of iBGP sessions among routers within each AS can be used, which causes scalability issues. Route Reflectors (RR) is a mechanism to improve performance, but designing a: correct, reliable and consistent iBGP overlay of sessions whith RRs is a delicate, far from easy task for ASes engineers, even though several popular heuristics are common practice. In previous works we proposed combinatorial optimization models to design consistent and resilient BGP overlays, when only non-Border-Routers are eligible for RRs. The present work extends previous models to allow any router (including Border Routers) to be Route Reflectors, which matches better to some application contexts.

Keywords: Network overlay design · Route Reflection · BGP · Internet routing · Combinatorial optimization · BGP resilience · Network resilience · Internet prefix classes · Border Routers

1 Introduction

Autonomous Systems (ASes) are networks or sets of networks under a single and clearly defined external routing policy. The Internet is composed of the interconnection of several thousands ASes, which use the Border Gateway Protocol

G. Nicosia et al. (Eds.): LOD 2019, LNCS 11943, pp. 409–421, 2019.
https://doi.org/10.1007/978-3-030-37599-7_34

(BGP, [1]) to exchange network prefixes (aggregations of IP addresses) reachability advertisements. BGP advertisements (or updates) are sent over BGP sessions administratively set between pairs of routers. Those sessions have two variants: internal BGP (iBGP) is used between routers belonging to the same AS, and external BGP (eBGP) when the routers belong to different ASes. In the last case, BGP routers are called Autonomous System Border Routers (ASBRs), while those running only iBGP sessions are referred to as Internal Routers (IRs).

Global IP reachability information is acquired using BGP, but each AS also needs to deploy an Internal Gateway Protocol (IGP) intra-domain, so as to know the internal topology, and guarantee the delivery of IP packets within the domain. BGP is tightly tied to the IGP; indeed, when there are more than one next-hop option for a given IP prefix, the BGP decision process compares different attributes to break the tie, eventually reaching the IGP metric to the next-hop ("hot potato routing").

Traditional iBGP implementations require a full-mesh of sessions among routers of each AS. This is due to the *split horizon* rule, under which iBGP routers do not re-advertise routes learned via iBGP to other iBGP peers. As a result, a number of $\frac{n \times (n-1)}{2}$ iBGP sessions is needed for an AS with n routers. *Route Reflection* [2] is used as an alternative to reduce BGP sessions and gain efficiency in CPU and memory usage. With Reflection, one or more routers within the AS are designated as Route Reflectors (RRs) and they are allowed to re-advertise routes learned from an internal peer to other internal peers. The rest of the routers are *clients* of some RRs. A *client* is an iBGP router that the RR will "reflect" routes to. Note that RRs re-advertise only best routes after running their own decision process, and also note that re-advertisements are biased by the placement of RRs within AS's topology, because as it was previously mentioned, prefixes selection considers IGP metric during the BGP path-selection. Routing is called *FM-optimal* whenever selected gateways for prefixes are those routers would have chosen under a full-mesh overlay. The *iBGP overlay design problem* consists in deciding which routers are to be route reflectors, and what sessions are to be established between clients and those RRs. Route Reflection has been extensively studied [3–5] with respect to reliability. There are also previous researches about how to locate the RRs [6–8].

Our approach for the optimization problem consist in minimizing the number of RRs in such a way that reliability is preserved and no sub-optimal route is chosen. We have studied the problem not only in the nominal case (refer to [9]) but also the resilient case (see [10,11]) when any single link or one router might fail, and we have proposed a technique called Optimal Route Reflector Topology Design (ORRTD) to minimize the number of route reflectors that could be chosen among the IRs. In our previous models, Internal Routers were the only candidates to become Route Reflectors, which is a realistic constraint for many Internet Service Providers (ISP). However, in ASes whose goal is providing connectivity to other ASes (i.e. transit networks), Border Routers proportionally are too many to disregard of them, turning previous premises impractical. This work expands previous models to allow ASBRs to become RRs. The new technique is called Extended ORRTD, and we have obtained promising results in experimental environment: as networks become larger, the number of RRs obtained can be significantly lower.

This document is organized as follows. Section 2 presents iBGP overlay based on RRs. Section 3 describes the impact of allowing border routers to be eligible as RRs. Section 4 presents experimental results over real-world network topologies, and Sect. 5 summarizes our main conclusions and lines for further research.

2 The iBGP Overlay

2.1 Protocol Concepts

ISPs and most large networks use some dynamic routing protocol intra-domain, called Internal Gateway Protocols (IGP). The most popular are Open Shortest Path First (OSPF) and Intermediate System - Intermediate System (IS-IS), which are efficient and safe, and fall in the link-state protocol category, since they build a complete network state database, using flooding of link state information among the routers in the domain; link costs may take into account actual links parameters, such as bandwidth. On the other hand, inter-domain routing (i.e., routing among ASes) is a job for BGP, which is a *path-vector* protocol, meaning that without administrative policies, the metric that determines the shortest path for a given prefix is the number of ASes that a routing announcement have crossed (the *AS_PATH*), used as a *hop* metric. Each BGP announcement contains a number of *attributes* (either mandatory or optional) which characterizes the routing information contained in the announcement (called Network Layer Reachability Information - NLRI, or simply "a route"). Some of the well-known, mandatory attributes are the aforementioned *AS_PATH*, and *NEXT_HOP*, which is the IP address of the router from a neighbour AS that has sent a particular route announcement to the ASBR (the exit point of the current AS). BGP routers usually receive multiple route alternatives for a given destination prefix, and therefore they need to run the BGP decision process to select the best path. As mentioned above, this decision process eventualy reach the point where the route with the lowest IGP metric (the IGP cost) towards the BGP next hop is chosen. When using RRs, a hierarchy among the iBGP speakers is created by clustering a subset of iBGP speakers with each RR. RRs must form a full mesh among themselves in order to make all announcements reachable, while clients peer with one or more of those RRs. To ensure reliability and loop-less [12,13], it is important to carefully design this overlay.

2.2 Optimal Route Reflector Topology Design (ORRTD)

An optimization model can be formulated as an integer programming problem to minimize the number of RRs and BGP sessions under certain conditions: constraints are introduced to avoid problems described in [9], not only for steady/non-faulty state, but also in the case of each possible single node or link failure [10,11], and they have also been chosen to always select IGP optimal routes. The technique is called *Optimal Route Reflector Topology Design* (ORRTD) and is described in detail in [9,10]. To formulate the problem we need the following definitions:

The *internal-to-border (IR-ASBR)* graph represents the preferred border router from each IR, based on the IGP costs (the one with the lowest cost path).

The *internal-to-internal (IR-IR)* graph represents the affinity among IRs: IRs that share a common preferred ASBR for a common *class* of prefixes, could serve as the reflector of each other for that class.

Prefix classes are aggregations of IP prefixes whose prefixes are indifferent at the time of choosing one ASBR over another, i.e. they take the same decisions in the previous steps of the BGP best path selection algorithm.

For this problem, we assume that the prefixes classes are built and known in advance. The basic idea is to construct an optimal IR-ASBR and an IR-IR graph based on the network topology represented by an undirected graph where the weight of each link is the IGP cost, and derive the resulting connections constraints. Then we introduce a set of control variables to help decide which routers are to be chosen as route reflectors and how to connect clients to those route reflectors. ORRTD also takes into account the *prefixes classes* that arrive through the ASBRs. In Fig. 1 we have an example of a graph transformation where there are two prefixes classes A (received by routers A and AB) and B (received by routers B and AB). Nodes 1 through 6 represent the internal routers. The leftmost graph is the original weighted graph and at the right part we have the corresponding IR-ASBR (for prefixes classes A and B, the ASBRs with the lowest cost path for each IR) and IR-IR graphs for those prefixes classes. For example, the preferred border router from internal router 5 for prefixes class A are both A and AB, as the path cost is 6. The preferred ASBR for prefixes class B from IR 5 is AB. So in both IR-ASBR sub-graphs for prefixes class A and B there is an edge from 5 to AB, and besides, for prefixes class A there is also an edge from 5 to A. Based on the information provided by the new graphs, we construct an optimization model where the adjacencies are represented in the constraints.

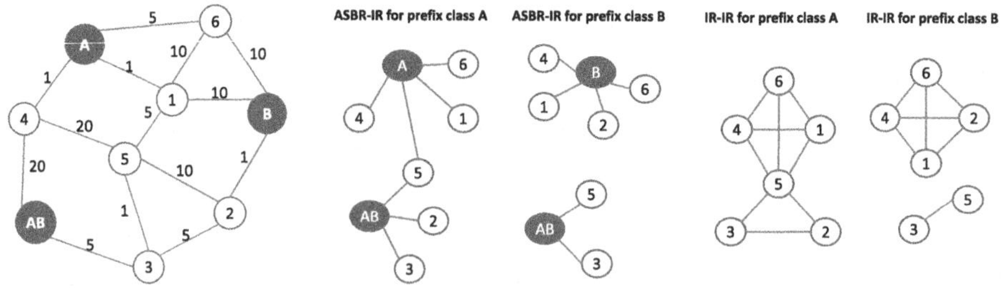

Fig. 1. Original graph and its IR-ASBR and IR-IR Adjacency graphs

3 Border Routers and Route Reflectors

In the previous version (see [9,10]), only IRs could be chosen as RRs. We represent a network topology as an undirected weighted graph. So an optimal internal-to-border router graph is built [9,10] for each class of prefixes; IRs that share a common ASBR for a common prefixes class, could serve as the reflector of each other for that class. But the fact is that there are ASes that have most of the routers working as ASBRs, so introducing this relaxation has a practical interest.

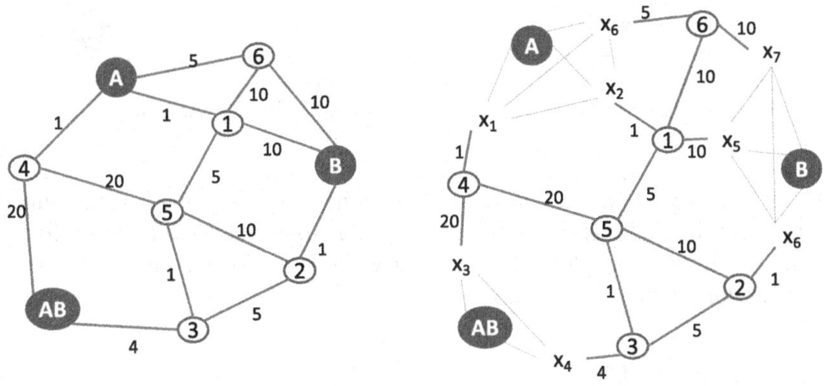

Fig. 2. Graph transformation with fictitious nodes and links

A first option consists in applying the optimization model described in [10], ORRTD, to the graph representing the AS network but with the following adaptation: as border routers cannot be RRs, for each link from an ASBR to some other router, introduce a fictitious internal node in such a way that the cost from each ASBR to the fictitious node is 0, while the cost from the fictitious node to the original adjacent node is the real weight. To illustrate the situation, let's take the weighted graph shown in the left part of Fig. 2. Instead of the link from ASBR A to IR 4 with cost 1, we have a link from ASBR A to fictitious IR x_1 with cost 0, and another link from IR x_1 to IR 4 with cost 1. The dotted lines represent the fictitious links. Then connect all the fictitious nodes associated to the same ASBR in a full mesh. Once the new graph is obtained, we can find the IR-to-ASBR graph affinity by applying Dijkstra's algorithm, and continuing with the reasoning as described in [9,10]. This simple transformation works fine for small networks, and a reduced number of ASBRs and prefixes classes, but, as we have demonstrated in [10], this is a NP-hard problem, which means that when increasing the input as in the previous transformation, and when considering the resilient version of the problem where fictitious prefixes classes are introduced for every alternative path to a different ASBR (even though failure of fictitious nodes is not considered, but failure of the links do apply), it becomes more difficult to solve. We can think of small improvements to the original implementation, like modifying Dijkstra's algorithm to get, among the shortest paths (by introducing the fictitious links with 0 cost there will be many of the same cost), the one with the lower quantity of edges, but anyway, the problem is still harder than the original one. So how many fictitious edges and nodes do we add? It seems reasonable to keep this quantity as low as possible.

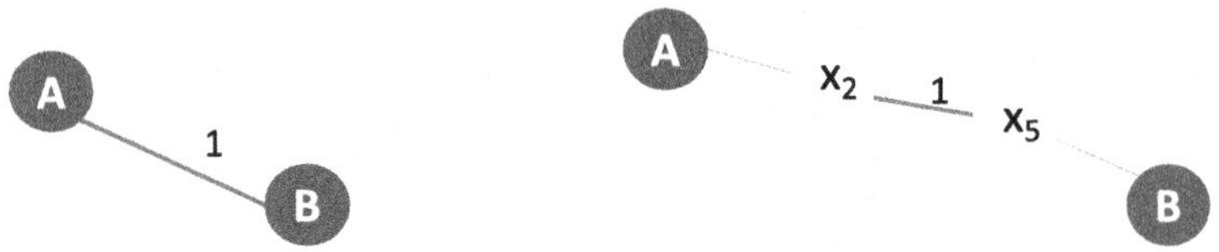

Fig. 3. Fictitious nodes and links for 2 adjacent ASBRs

In Fig. 3 we show that we have one path with two fictitious routers and two fictitious links between the ASBRs. But when applying to the case shown in Fig. 4, many fictitious nodes (the x_s) and links (the dotted lines) are needed, introducing complexity.

3.1 Extended ORRTD

Given the previous considerations regarding the problem size growth, the most convenient alternative is to adapt the problem formulation to the new condition: border routers can also be designated as route reflectors, which leads to a new set of constraints shown in (1). In this research we assume that any border router (ASBR) or internal router (IR) can be a RR. The objective function in (1) pushes down to get the minimum number of RRs, where every router in the network could potentially be designated as route reflector. But this objective has several constraints, stated in equation groups (i) to (x). Equation groups (i) to (vii) ensure optimality, while those in groups $(viii)$ to (x) ensure resilience.

Equations (1) have the following **input sets**:

V : set of all routers, $V = \{IR \cup BR\}$
$\mathcal{C}$: set of prefixes classes
$\{S^k\}$: set of router-to-border BGP affinity matrices $S^k_{ij} = 1$ if and only if $j \in$ ASBR-to-Router for prefixes class k, with $k \in \mathcal{C}$, $i \in$ BR, $j \in$ IR $\cup$ BR
$\{T^k\}$: set of router-to-router BGP affinity matrices
FC : set of fictitious prefixes classes
$\{P^l\}$: set of new BGP best path nodes from ASBR-to-Router affinity matrices
$\{Q^l\}$: set of new BGP best path Router-to-Router affinity matrices

Equations (1) have the following **parameters** to support resilience:

BR : set of all Autonomous System Border Routers
IR : set of all Internal Routers
S' : $\{S^k\} \cup \{FC^l\}$, set of router-to- border router affinity for every prefixes class, including the fictitious prefixes classes
C' : $C \cup FC$
T' : $\{T^k\} \cup \{Q^l\}$

and the following boolean **variables**:

x_i : 1 if router $i \in BR \cup IR$ is to be a RR and 0 otherwise;
y_{ij}^k : 1 if ASBR i is to be iBGP adjacent to router $j \in BR \cup IR$ for prefixes class k and 0 otherwise;
z_{ij}^k : 1 if router $i \in BR \cup IR$ is to be iBGP adjacent to router $j \in BR \cup IR$ for prefixes class k and 0 otherwise;
w_{gh}^l : 1 if nodes $g, h \in P^l$, i.e., the alternative best path

$$\begin{cases} \min \sum\limits_{i \in IR} x_i \\ \text{Subject to :} \\ \sum\limits_{(ij) \in S'^k} y_{ij}^k \geq 1, & \forall i \in BR, k \in C', S'^k \neq \emptyset \quad (i) \\ x_i + x_j - y_{ij}^k \geq 0, & \forall i \in BR, k \in C', (ij) \in S'^k \; (ii) \\ x_i + x_j - y_{ij}^k \leq 1, & \forall i \in BR, k \in C', \quad (iii) \\ & (ij) \in S'^k \\ x_j + \sum\limits_{(ij) \in T'^k} z_{ij}^k \geq 1, & \forall j \in IR \cup BR, k \in C' \quad (iv) \\ x_i + x_j - z_{ij}^k \geq 0, & \forall i \in IR \cup BR, k \in C' \quad (v) \\ & (ij) \in T'^k \\ x_i + x_j + z_{ij}^k \leq 2, & \forall i \in IR \cup BR, k \in C' \quad (vi) \\ & (ij) \in T'^k \\ \sum\limits_{(jh) \in S'^k} y_{jh}^k - z_{ih}^k \geq 0, & \forall j \in IR, k \in C' \quad (vii) \\ & (ih) \in T'^k \\ \sum\limits_{i \in IR \cup BR} x_i \geq 2, & \forall i \in IR \cup BR \quad (viii) \\ w_{gh}^l \geq y_{ij}^k, & \forall i \in BR, j \in IR \cup BR, \\ & k \in C', g, h \in FC^l \quad (ix) \\ \sum\limits_{(ij) \in P^l} y_{ij}^l \geq 1, & \forall i \in BR, l \in \mathcal{FC} \quad (x) \\ x_i, y_{ij}^k, z_{ij}^k, w_{gh}^l \in \{0,1\}, \forall i,j \in V, k \in C', l \in FC \end{cases} \tag{1}$$

4 Experimental Results

In this section we analyze the results of applying the proposed theoretical model to a selection of network topologies. Some of the topologies were taken from "The Internet Topology Zoo" repository [16] and slightly adapted to ensure 2-node connectivity (by introducing the minimum number of additional edges) to make finding a resilient topology design viable; other topologies are theoretical cases. They can be found in [17], where we show, for each topology, nodes, edges and the IGP costs. The models were solved with CPLEX Optimization Studio V12.6.3, running on an Intel Core I7 2.3 GHz and 8 Gb RAM. For the purpose of this test we assume there are: four, ten, fifty and one hundred prefixes classes (PC), and that we know in advance which ASBRs advertise each prefixes class.

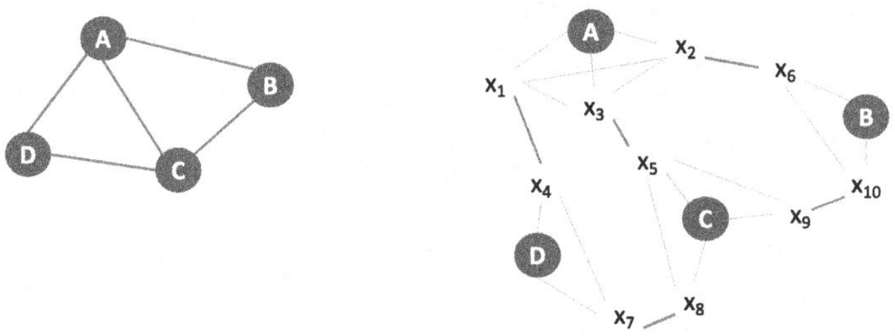

Fig. 4. Fictitious nodes and links for 4 adjacent ASBRs

Table 1 shows the results of applying the original model, ORRTD, with the addition of fictitious nodes and links, and the new model or extended ORRTD (eORRTD for short), varying from four to one hundred prefixes classes to the AB topology of Fig. 2. The leftmost columns of Table 1 show the network name, the number of border routers and internal routers and the number of prefixes classes. The rest of the columns are the results obtained: number of RRs, number of BGP sessions, and number of equations (constraints) both of the extended ORRTD technique and the original one (ORRTD) with the addition of fictitious nodes. The number of resulting RRs is the same (three route reflectors) in all

Table 1. Comparison of models

Network	# BRs	#IRs	#PCs	#RRs	BGP sessions	Eqs. eORRTD	Eqs. ORRTD with fictitious nodes
AB5	3	5	4	3	18	331	1324
AB5	3	5	10	3	18	666	2066
AB5	3	5	50	3	19	2742	8073
AB5	3	5	100	3	19	5404	16141

Table 2. eORRTD: RRs, BGP sessions and equations - different topologies

Network	# BRs	#IRs	#PCs	#RRs	BGP sessions	BGP sessions FM	#Eqs.
AB5	3	5	4	3	18	28	331
AB10	3	10	4	2	23	78	1361
Abilene2	3	8	4	2	16	55	856
Cernet2	4	37	4	3	112	820	15765
Garr2	7	48	4	7	246	1485	11233
Jgn2Plus2	6	11	4	5	45	136	2598
UniC	3	24	4	3	68	351	8811
Uran	5	18	4	4	48	253	4448
WideJpn	11	19	4	5	77	435	5452
TtNew	4	96	4	6	532	4950	91499
AB5	3	5	10	3	18	28	666
AB10	3	10	10	2	24	78	2418
Abilene	3	8	10	3	25	55	1560
Cernet2	4	37	10	3	112	820	30784
Garr2	7	48	10	7	272	1485	26634
Jgn2Plus2	6	11	10	5	47	136	4635
UniC	3	24	10	3	69	351	13974
Uran	5	18	10	4	67	253	8700
WideJpn	11	19	10	5	89	435	10107
TtnNew	4	96	10	6	561	4950	173009
AB5	3	5	50	3	19	28	2742
AB10	3	10	50	2	24	78	9015
Abilene	3	8	50	3	26	55	5868
Cernet2	4	37	50	3	112	820	142748
Garr2	7	48	50	7	265	1485	103263
Jgn2Plus2	6	11	50	6	79	136	18041
UniC	3	24	50	3	68	351	46740
Uran	5	18	50	5	78	253	35354
WideJpn	11	19	50	5	113	435	36686
TtnNew	6	94	50	6	573	4950	789153

cases and the number of BGP sessions is clearly better than in a full mesh (that would have fifty six sessions), but the quantity of equations grows faster when adding the fictitious nodes and links.

Table 2 shows the resulting number of RRs and BGP sessions for different network topologies, when having four, ten and fifty prefixes classes, and allowing ASBRs to be RRs. It can be appreciated that the number of BGP sessions is

Table 3. ORRTD vs eORRTD

Network	# BRs	#IRs	#PCs	#RRs eORRTD	#RRs ORRTD
TtNew	6	94	10	6	18
SwitchL3	12	30	10	3	3
Garr2	7	48	10	7	28
Uran	5	18	10	4	5

significantly lower than the necessary BGP session in full mesh. The last column also shows the evolution of the number of equations in the model, which grows, depending on the size of the network and the quantity of prefixes classes.

Table 3 shows that for the bigger networks studied, and supposing all ASBRs are eligible, the number of RRs can be reduced with eORRTD. The problem can be solved with any popular solver like GLPK or CPLEX. Figure 5 shows an example of the execution time of the solver for different number of prefixes classes, setting y-axis as log10 for better visualization. For more complex problems, with hundreds of classes, a heuristic approach should be used.

We also present results obtained in the emulation environment proposed by [18] which is based on Quagga[1], MiniNExT[2] and ExaBGP[3] for injecting BGP messages. After certain period of injection, we wait for the BGP network to become stable, i.e., best routes are selected after applying routing policies in each router, and then analyze the BGP tables. We study the content of the LOC_RIB table of each router in the network. LOC_RIB table contains the best route out of all those available for each distinct destination, and the NEXT_HOP attribute (the preferred exit router for each prefixes class). Whenever a routing prefix is received from a neighbor, BGP process decides if any of the neighbor's new routes are preferred to routes already installed in the LOC_RIB and it replaces it as required. For the purpose of the experiment, we consider AB topology shown in Fig. 2 described in [10] and one hundred prefixes classes, each one represented by one prefix, e.g. 195.66.4.0/24. For each router in that network and every prefixes class, we compare the LOC_RIB table content in the emulation environment when applying the BGP overlay design resulting from the new model eORRTD, and ORRTD with fictitious nodes and links.

In Table 4 we show an extract of the LOC_RIB table, which is completely coinciding for both overlay designs.

[1] Quagga Routing Suite. Available at: https://www.quagga.net/. Accessed: 2018-09-01.

[2] MiniNExT (Mininet ExTended). Available at: https://www.quagga.net/. Accessed: 2019-03-01.

[3] https://github.com/Exa-Networks/exabgp.

Table 4. LOC_RIB table for topology AB

Prefix class	Next_Hop	Router	Id
195.66.1.0/24	192.168.0.8	2	1
195.66.2.0/24	192.168.0.7	2	2
195.66.3.0/24	192.168.0.7	2	3
195.66.4.0/24	192.168.0.7	2	4
195.66.5.0/24	192.168.0.7	2	5
.....	.....	..	..
195.66.100.0/24	192.168.0.7	1	354
195.66.1.0/24	172.16.3.2	AB	382
195.66.2.0/24	172.16.3.2	AB	383
195.66.3.0/24	172.16.3.2	AB	384

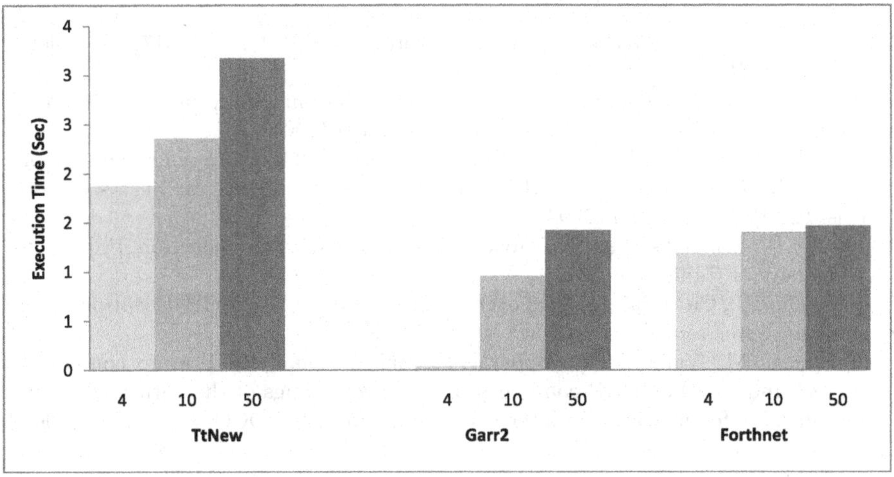

Fig. 5. Execution time for 4, 10 and 50 prefix classes

5 Conclusion

In this article we focus on the efficient use of BGP, particularly in the intra-domain scope with Route Reflection. We describe an Integer Programming Problem to select the route reflectors, allowing both, internal and border routers to be eligible as RRs. We also consider the impact of Internet prefix categorization in the proposed solution and show experimental results with different number of prefixes classes. The technique has been called *extended Optimal Route Reflector Topology Design*, or eORRTD for short. Among other advantages, with eORRTD there is no need to modify or augment existing BGP standards. Experimental results in emulation environments demonstrate the theoretical consistency of eORRTD, in the nominal case and in the event of fails over single nodes or links. In previous works [9,10] we have shown that ORRTD outperforms other heuristic approaches, both in the number of route reflectors and the BGP sessions needed. In the present paper we show eORRTD outperforms full-mesh and ORRTD. Besides, we analyze the impact of increasing number of PCs arriving to different border routers in the design of the best solution. Finally, an important line of future research work is the construction of the prefixes classes using the routing information, eventually considering the dynamic classification of prefixes. We are actually working on real data from an international provider; as a preliminary result, if we take the AS_PATH as in [14,15], the number of prefix can be reduced in one order of magnitude. This may constitute a whole line of research, for example, using machine learning or other techniques to build the prefixes classes based on the BGP updates.

References

1. Rekhter, Y., Li, T.: A Border Gateway Protocol 4 (BGP-4), RFC1771 Obsoleted by RFC 4271 (1995)
2. Bates, T., Chen, E., Chandra, R.: BGP route reflection: an alternative to full mesh internal BGP (iBGP), RFC4456 (Draft Standard) (2006)
3. Xiao, L., Wang, J., Nahrstedt, K.: Reliability-aware iBGP route reflection topology design. In: Proceedings - 11th IEEE International Conference on Network Protocols, ICNP, pp. 180–189 (2003)
4. Park, J.H.: Understanding the impact of internal BGP route reflection, PhD thesis, University of California (2011)
5. Vissicchio, S., Cittadini, L., Vanbever, L., Bonaventure, O.: iBGP deceptions: more sessions, fewer routes. In: INFOCOM, IEEE (2012)
6. Vutukuru, M., Valiant, P., Kopparty, S., Balakrishnan, H.: How to construct a correct and scalable iBGP configuration. In: Proceedings of the 25th IEEE International Conference on Computer Communications, INFOCOM, pp. 1–12 (2006)
7. Zhao, F., Lu, X., Zhu, P., Zhao, J.: BGPSep_D: An Improved Algorithm for Constructing Correct and Scalable IBGP Configurations Based on Vertexes Degree. In: Gerndt, M., Kranzlmüller, D. (eds.) HPCC 2006. LNCS, vol. 4208, pp. 406–415. Springer, Heidelberg (2006). https://doi.org/10.1007/11847366_42
8. Zhang, R., Bartell, M.: BGP Design and Implementation, pp. 264–266. Cisco Press, Indianapolis (2003)
9. Mayr, C., Grampín, E., Risso, C.: Optimal route reflection topology design. In: 10th Latinamerican Networking Conference (LANC 2018), pp. 65–72. ACM, New York (2018)
10. Mayr, C., Grampín, E., Risso, C.: A combinatorial optimization framework for the design of resilient iBGP overlays. In 15th International Conference on the Design of Reliable Communication Networks (DRCN 2019) (2019, to be published)
11. Mayr, C., Grampín, E., Risso, C.: A Combinatorial Optimization Framework for the Design of Resilient iBGP Overlays, Technical Report (2019). http://www.fing.edu.uy/~crisso/PID5740833.pdf
12. Flavel, A.: BGP, not as easy as 1-2-3. Ph.D. Dissertation, University of Adelaide, Australia (2009)

13. Pelsser, C., Masuda, A., Shiomoto, K.: A novel internal BGP route distribution architecture. In: IEICE General Conference (2010)
14. Broido, A., Claffy, K.C.: Analysis of RouteViews BGP data: policy atoms, Cooperative Association for Internet Data Analysis - CAIDA, San Diego Supercomputer Center, University of California, San Diego. In: Proceedings of NRDM workshop Santa Barbara (2001)
15. Afek, Y., Ben-Shalom, O., Bremler-Barr, A.: On the structure and application of BGP 'Policy Atoms'. In: Proceedings of the 2nd ACM SIGCOMM Workshop on Internet Measurment (IMW 2002), pp. 209–214. ACM, New York (2002). https://doi.org/10.1145/637201.637234
16. Knight, S., Nguyen, H.X., Falkner, N., Bowden, R., Roughan, M.: The internet topology zoo, selected areas in communications. IEEE J. **29**(9), 1765–1775 (2011)
17. Mayr, C., Grampín, E., Risso, C.: Examples of Internet Topologies (2019). https://www.fing.edu.uy/~crisso/Topologies.pdf
18. Solla, V., Jambrina, G., Grampín, E.: Route reflection topology planning in service provider networks. In: 2017 IEEE URUCON, pp. 1–4 (2017)

GRASP Heuristics for the Stochastic Weighted Graph Fragmentation Problem

Nicole Rosenstock(✉), Juan Piccini(✉), Guillermo Rela, Franco Robledo, and Pablo Romero

Instituto de Matemática y Estadística, IMERL, Facultad de Ingeniería - Universidad de la República, Montevideo, Uruguay
{nicole.rosenstock,piccini,grela,frobledo,promero}@fing.edu.uy

Abstract. Vulnerability metrics play a key role nowadays in order to cope with natural disasters and deliberate attacks. The scientific literature is vast, sometimes focused on connectivity aspects in networks or cascading failures, among others. The Graph Fragmentation Problem (GFP) has its origin in epidemiology, and it represents the worst case analysis of a random attack.

The contributions of this paper are three-fold. First, a valuable interplay between the GFP and a connectivity-based metric, called the Critical Node Detection (CND) problem, is here explored. Second, a novel combinatorial optimization problem is introduced. It is called Stochastic Weighted Graph Fragmentation Problem (SWGFP), and it is a generalization of the GFP. The node-importance is included in the model, as well as a level of network awareness from the attacker viewpoint. Finally, a novel GRASP (Greedy Randomized Adaptive Search Procedure) heuristic is proposed for the SWGFP. A faithful comparison with a previous heuristic shows its competitiveness.

Keywords: Vulnerability metrics · Combinatorial optimization · Computational complexity · Metaheuristics

1 Motivation

Every society should be prepared to cope with natural disasters such as pandemics, fires or lightning shocks. Nevertheless, the history reveals severe losses of human lives and/or economic wealth in catastrophic environments. The Spanish flu from 1918 cause global fatalities for more than 20 million (WWII more than 61 million), and half the population of the world has been exposed to the virus [18].

A lightning shock in 1977 produced a struck to the Edison electrical transmission line causing one of the major blackouts in New York City [1]. An infernal fire in October 1871 ravaged part of Chicago, leaving more than 90.000 homeless and 300 deaths [14].

The contributions of this paper can be summarized in the following items:

G. Nicosia et al. (Eds.): LOD 2019, LNCS 11943, pp. 422–433, 2019.
https://doi.org/10.1007/978-3-030-37599-7_35

1. The Stochastic Weighted Graph Fragmentation Problem (SWGFP) is here introduced. This combinatorial optimization problem is an extension of the previously studied GFP [2,15,16].
2. We formally prove that the GFP is completely equivalent to the Critical Node Detection or CND [12].
3. We derive a universal inapproximability result for the CND as a corollary of the previous coincidence.
4. A novel GRASP heuristic is developed to address the SWGFP, which belongs to the class of $\mathcal{NP}$-Hard problems. This GRASP is connectivity-based, supported by the coincidence with the CND, and exploits properties of independent sets. It will be called GRASP-IS (Independent Sets).
5. A slight variation of a previous GRASP version for the GFP is proposed in order to address the SWGFP. Here, it is called GRASP-GFP.

This paper is organized as follows. Section 2 formally defines the GFP, CND and the SWGFP respectively. Section 3 states the hardness and inapproximability results for the three problems. Additionally, the coincidence of the CND and GFP as combinatorial optimization problems is established. GRASP-IS and GRASP-GFP heuristics are detailed in Sect. 4. An experimental analysis is carried out in Sect. 5. Finally, Sect. 6 contains concluding remarks and trends for future work.

2 Models

Here we describe the three combinatorial optimization problems under study, to know, GFP, CND and SWGFP. The main focus of this paper is the SWGFP. However, an interplay between different research lines is available after the coincidence of both combinatorial problems, GFP and CND, expressed in Theorem 1. As a consequence, the SWGFP is an extension of both models, and the inapproximability for the CND holds as a corollary of the inapproximability for the GFP (see Theorem 3). Due to margin constraints, we could not include the graph-theoretic terms. We refer the reader to Frank Harary Graph Theory book [10].

2.1 GFP

We are given a population represented by a graph $G = (V, E)$, and a budget constraint B, which is a natural number B such that $0 \leq B \leq |V|$. We can choose B nodes and immunize them: we delete the nodes from G obtaining a subgraph G', so that the chosen nodes cannot be affected by the disaster. The nature picks a node v uniformly at random from G'. The disaster kills all the members of the same connected component as v.

The goal is to minimize the expected number of deaths. Mathematically, if the subgraph G' has $V' = n$ nodes and k connected components with orders $n_1, \ldots, n_k$, the probability to choose component i is n_i/n. Therefore, the expected number of deaths is $E(G') = \sum_{i=1}^{k} n_i p_i$, with $p_i = n_i/n$. The goal of the Graph Fragmentation Problem (GFP) is to choose the node-immunization set in order to minimize the expected number of deaths:

$$\min_{U \subseteq V} f(U) = \sum_{i=1}^{k} \frac{n_i^2}{n}$$
$$s.t. |U| = B.$$

2.2 CND

The instance is identical to that of the GFP, but the objective is slightly different [12]. The goal is to minimize the connectivity of G'. Specifically, we count all pairs of the node-set V' that belong to the same component. There are $\binom{n_i}{2}$ pairs that belong to the component with size n_i. Therefore, the CND is formally defined as the following combinatorial optimization problem:

$$\min_{U \subseteq V} g(U) = \sum_{i=1}^{k} \frac{n_i(n_i - 1)}{2}$$
$$s.t. |U| = B.$$

2.3 SWGFP

Two additional elements are considered in the new model:

1. Non-negative weights are assigned to the nodes (i.e., a notion of importance).
2. The attacker selects some probability law in the nodes to determine the attack.

Therefore, we consider a weighted graph $G = (V, E, w)$ instead, where $w : V \rightarrow \mathbb{R}^+$ is the weight function (i.e., a priority assignment). We must pick $B = |U|$ nodes for protection such that the resulting graph $G^* = G - U$ is resilient against the attack. Assume that G^* is partitioned into k connected components $V_1, \ldots, V_k$ with respective cardinalities $n_1, \ldots, n_k$, $n = \sum_{i=1}^{k} n_i$ and $W_i = \sum_{v_{ij} \in V_i} w(v_{ij})$ is the sum-weight of component V_i. The attack is governed by a probability law $p_V(x)$ in the nodes $x \in V$. The goal in the SWGFP is to minimize the expected losses after the attack:

$$\min_{U} L(U) = \sum_{i=1}^{k} W_i \sum_{x \in V_i} p_V(x_i),$$
$$s.t. |U| = B,$$

Observe that p_V induces a probability law $(p_i)_{i \in V_i}$, being $p_i = \sum_{x \in V_i} p_V(x_i)$. The SWGFP can be simplified:

$$\min_{U} L(U) = \sum_{i=1}^{k} W_i p_i,$$
$$s.t. |U| = B,$$

In this paper, we will consider three outstanding probability-laws for the attacker:

1. *Random Attack*: $p_i = |V_i|/n$.
2. *Weighted Attack*: $p_i = W_i/W$.
3. *Attack under full knowledge*: $p_i = 1$ iff V_i has the largest weight.

3 Analysis

In this section we study the complexity and the interplay between the different combinatorial problems. First, we show that the GFP and the CND are equivalent combinatorial problems. The hardness of both problems are established, and, as corollary, we know that the SWGFP is also hard (by inclusion). Furthermore, we offer strong inapproximability results of the GFP and CND. Since the SWGFP is an extension of the GFP, the hardness and inapproximability also holds. This fact promotes the development of heuristics.

3.1 Equivalence Between CND and GFP

In this paragraph we see that both combinatorial problems are identical in terms of inputs and globally optimum solutions.

Theorem 1. *GFP and CND are equivalent problems.*

Proof. The inputs of both problems are identical. Furthermore, an elementary calculation shows that $nf(U) = 2g(U) + n$, being $n = |V'|$ a constant. Therefore, the globally optimum solution for both problems are identical as well.

A full proof is included in [2].

3.2 Complexity Analysis

Stephen Cook formally proved that the joint satisfiability of an input set of clauses in disjunctive form is the first $\mathcal{NP}$-Complete decision problem [3]. Furthermore, he provided a systematic procedure to prove that a certain problem is $\mathcal{NP}$-Complete. Specifically, it suffices to prove that the decision problem belongs to set $\mathcal{NP}$, and that it is at least as hard as an $\mathcal{NP}$-Complete problem. Richard Karp followed this hint, and presented the first 21 combinatorial problems that belong to this class [11]. In particular, Vertex Cover belongs to this list. The reader is invited to consult an authoritative book in Complexity Theory, which has a larger list of $\mathcal{NP}$-Complete problems and a rich number of bibliographic references [7].

Definition 1 (Vertex Cover). *Instance: simple graph $G = (V, E)$ and positive integer k.*
Does there exist a node-set U such that $|U| \leq k$ and every link is incident to some node that belongs to U?

The hardness of the GFP has theoretical importance. Indeed a large set of immunization problems are at least as hard as the GFP. Therefore, this proves that immunization is a hard task, and the intuition from epidemiologist is correct. Here, we prove the hardness of the GFP in a more simple way than in [15].

Theorem 2. *The GFP belongs to the $\mathcal{NP}$-Hard class.*

Proof. The resulting graph $G^* = G - U$ has isolated nodes if and only if U is a vertex cover, where $|U| \leq B$. Thus, the GFP is at least as hard as Vertex Cover.

The following problem will be considered in order to prove a stronger inapproximability result for the GFP.

Definition 2 (Multiway k-cut). *Instance: simple graph $G = (V, E)$, terminal set $K \subseteq V$ with $|K| = k$, positive integer B.*
Does there exist a separator set $U \subseteq V - K$ with $|U| \leq B$ such that each terminal node belongs to different components in $G - U$?

We know that Multiway 2-cut is in $\mathcal{P}$. A polynomial time algorithm is provided by Ford and Fulkerson [6]. However, Multiway k-cut is $\mathcal{NP}$-Complete for every fixed $k \geq 3$ [8].

Theorem 3. *It is $\mathcal{NP}$-Hard to approximate GFP within any factor $\alpha < \frac{5}{3}$.*

Proof. Consider an instance of Multiway 3-cut with ground graph $G = (V, E)$, distinguished nodes $K = \{v_1, v_2, v_3\}$ and positive integer B. Replace those nodes by large cliques $\{K_N, K_N, K_N\}$, where $N >> |V|$. The order of the new graph G^* is roughly $3N$. If the instance admits a 3-cut, i.e. a subset whose removal separates v_1, v_2, v_3, then the optimal cost in the GFP with instance (G^*, B) is roughly N. Otherwise, the expected number of dead nodes is never lower than $\frac{(2N)^2+N^2}{3N} = \frac{5N}{3}$. Therefore, an approximation algorithm with factor strictly less than $5/3$ would determine whether (G, K) admits a 3-cut using B nodes or not, thus solving Multiway 3-cut.

The CND has an extensive study from the scientific literature. Its hardness has been already established. However, to the best of our knowledge, there are no efficient approximation algorithms available, neither inapproximability results.

Corollary 1. *It is $\mathcal{NP}$-Hard to approximate CND within any factor $\alpha\frac{n}{3}$ for some $\alpha < 1$.*

Proof. Assume for a moment that there exists an algorithm $\mathcal{A}$ for CND such that $U = g(\mathcal{A}(G, B))$ for every instance (G, B), and $g(U) \leq \alpha\frac{n}{3}OPT$ for some $\alpha < 1$. But then:

$$\begin{aligned} f(U) &= \frac{2}{n}g(U) + 1 \leq \frac{2\alpha}{3}OPT + 1 \\ &\leq \frac{2\alpha}{3}OPT + OPT = \frac{2\alpha + 3}{3}OPT \\ &= \frac{5}{3}\alpha' OPT, \end{aligned}$$

being $\alpha' = (2\alpha + 3)/5 < 1$. It was used the fact that $OPT \geq 1$ by definition of the GFP. Therefore, we could find an approximation algorithm of factor $\frac{5}{3}\alpha' < \frac{5}{3}$ for the GFP whenever there exists an algorithm $\mathcal{A}$, and the statement holds.

4 Heuristics

GRASP is a multi-start metaheuristic for combinatorial optimization problems, in which each iteration consists basically of two phases [4,5,17]: construction and local search. The construction phase builds a feasible solution, whose neighborhood is investigated until a local minimum is found during the local search phase. The best overall solution is kept as the result.

Algorithm 1. $G_{out} = Alg\_GRASP(G, B, \alpha, w)$

for $i = 1$ **to** B **do**
  $SC_l \leftarrow$ *LowestReduction(G)*
  $SC_h \leftarrow$ *HighestReduction(G)*
  $RCL \leftarrow \{v : S_c(G - v) < SC_l + \alpha(SC_h - SC_l)\}$
  $v \leftarrow$ *ChooseRandom(RCL)*
  $G \leftarrow G - \{v\}$
  endfor
while *Improve(G)* = *True* **do**
  $(G, LocalImprove) \leftarrow Swap(G)$
  endwhile
return G_{out}

An intensification strategy based on path-relinking [13] is frequently used to improve solution quality and to reduce computation times by exploring elite solutions previously found along the search.

The reader is referred to Resende and Ribeiro [17] for a complete description of GRASP heuristic. We also invite the reader to consult the comprehensive Handbook of Metaheuristics for further information [9].

We develop two GRASP heuristics for the SWGFP. Our proposal, GRASP-IS exploits properties of independent sets. A previous GRASP implementation for the GFP was adapted in order to compare the effectiveness of our proposal.

GRASP for GFP in Piccini et al. [2,15,16] is inspired by GRASP methodology. We also invite the reader to consult the reference.

4.1 GRASP-GFP

4.1.1 GRASP-GFP Construction and Local Search Phases

Essentially, weighted-nodes are considered, and the score is found as a function of the attack. In this GRASP implementation, a construction phase is first applied and then a Local Search phase takes place. The best idea is to immunize exactly

B nodes so the **for** loop has precisely B iterations. During each for loop, a single node is picked and removed. The lowest and highest feasible score reductions are found; a classical Restricted Candidate List (RCL) is built, regarding the elements with the largest score reduction.

Algorithm 2. $GRASP(G, B, w, MaxIter, \alpha)$

```
1: eliteset ← ∅
2: for i = 1 to MaxIter do
3:     ŝ ←Construction_Phase(G, B, w, α);
4:     s ←Local_Search(ŝ);
5:     solutions ← solutions ∪ {s}
6:     endfor
7: eliteset ←build_elite_set(solutions)
8: best_sol ←PathRelinking(eliteset)
9: return best_sol
```

Parameter α represents the fraction of all feasible candidates for node-immunization. The reader can appreciate that the parameter α trades greediness for randomization. A singleton $\{v\}$ is chosen uniformly at random among the nodes from the RCL. The node v is deleted from G. A local search phase is performed composed by an elementary *Swap* operation, until a local optima is met. *Swap* picks iteratively a single (protected, non-protected) pair, and their roles are exchanged only if the solution is better. The result is a local optima, which is returned.

4.1.2 GRASP-GFP with Path-Relinking

Once the Construction Phase together with the Local Search find $Maxiter$ solutions, the solutions with the lowest scores are chosen to integrate an elite set. In order to improve, Path-relinking is applied to every pair of solutions in the elite set. A detailed explanation was avoided for available space reasons, as well as the explanation of GRASP-IS with Path-relinking.

4.2 GRASP-IS

In order to disrupt graph connectivity and minimize SWGFP score, we propose a GRASP combined with Path-relinking procedure based on the construction of maximal independent sets as a starting point. Graph connectivity can be disrupted through the removal of a vertex cover of that graph, leaving only an independent set. We started by finding a maximal independent set using a greedy randomized approach and then, some nodes in the vertex cover were added back to the graph in a greedy fashion, until the vertex cover had exactly B nodes. Unlike previous works, we considered a graph with weighted nodes.

An adaptive, randomized and greedy construction phase was used to generate $Maxiter$ solutions and to perform a local search in each one of them. From this

Algorithm 3. $GRASP - ISConstructionPhase(G(V, E), w, B, \alpha)$

```
IS ← ∅
while 𝓑 ≠ ∅ do
    minweight ←min{weight(v) : v ∈ 𝓑}
    𝓡𝓒𝓛 = {v ∈ 𝓑 : weight(v) ≤ min_weight + α × (max_weight − min_weight)}
    Select v* ∈ 𝓡𝓒𝓛 at random
    IS ← IS ∪ {v*}
    𝓑 = 𝓑\{v*}\{u ∈ 𝓑 : (u, v*) ∈ E}
    endwhile
S0 ← V\IS
while |S0| > B do
    x ←argmin(∑ Wi²/W : i are components in MIS ∪ chosen)
    chosen ← chosen ∪ {x}

    endwhile
return S0
```

set, the best solutions (the solutions with lower score) were selected to build an elite set. Then, Path-relinking was applied to elite set to obtain the best solution.

GRASP-IS with Path-relinking for SWGRP is analogous to GRASP with Path-relinking for SWGRP.

4.2.1 GRASP-IS Construction Phase for SWGFP

In Algorithm 3, this phase starts by finding a maximal independent set (IS) for a given graph G. This is done in a greedy randomized fashion. The minimum weight (*min_weight*) of the available nodes in each step is computed.

Algorithm 4. $GRASP - ISLocalSearch(G(V, E), w, S_0, iter)$

```
improve ← TRUE
while (improve) do
    Ŝ0 ← S0\{v}
    nodes ← △(S0, V)
    improve ← FALSE
    while (not improve) and (nodes ≠ ∅) do
        Ŝ0 ← S0 ∪ {u ∈ nodes}
        if f(Ŝ0) < f(S0) then
            S0 ← Ŝ0
            improve ← TRUE
        else
            nodes ← nodes \ {u}
            endif
        endwhile
    endwhile
return S0
```

A Restricted Candidate List (RCL) is built with every node whose weight (w) does not exceed $min\_weight + \alpha(max\_weight - min\_weight)$ and is not adjacent to the previously chosen nodes. A node is chosen randomly from the RCL and it is included in the IS. The set of available nodes is updated by deleting the chosen node and its adjacent nodes. These steps are repeated until it cannot be found other nodes to be included ($\mathcal{B} = \emptyset$) and the maximal independent set is built. Nodes in the vertex cover obtained after building the IS, are the initial solution S_0. In general, S_0 has more than B nodes so it represents an infeasible solution to the SWGFP.

However, a feasible solution can be built by removing nodes from S_0 and greedily adding them back to the IS, until S_0 remains with exactly B nodes. The node chosen at each step (x) is the one that minimizes the SWGFP function

Gr.	B.	Random Attack			Weighted Attack			Best Attack		
		Rand.	Grasp	GspIS	Rand.	Grasp	GspIS	Rand.	Grasp	GspIS
Arp.	1	46 (3)	46 (0)	46 (0)	46 (3)	46 (0)	46 (0)	46 (4)	46 (0)	46 (0)
	2	41 (101)	22 (0)	22 (0)	41 (106)	22 (0)	22 (0)	27 (78)	24 (0)	24 (0)
	3	35 (234)	12 (0)	12 (0)	23 (198)	13 (0)	13 (0)	32 (119)	19 (0)	19 (0)
	4	15 (306)	8 (0)	8 (0)	11 (274)	8 (0)	8 (0)	15 (190)	11 (0)	11 (0)
	5	16 (366)	6 (0)	6 (0)	8 (342)	6 (0)	6 (0)	15 (210)	9 (0)	9 (0)
	6	11 (350)	4 (0)	4 (0)	11 (395)	4 (0)	4 (0)	14 (317)	6 (6)	6 (0)
Dod.	1	57 (0)	57 (0)	57 (0)	57 (0)	57 (0)	57 (0)	57 (0)	57 (0)	57 (0)
	2	54 (0)	54 (0)	54 (0)	54 (0)	54 (0)	54 (0)	54 (0)	54 (0)	54 (0)
	3	45 (12)	45 (0)	45 (0)	51 (12)	45 (0)	45 (0)	48 (6)	48 (0)	48 (0)
	4	42 (27)	38 (0)	38 (0)	42 (24)	38 (0)	38 (0)	48 (14)	42 (0)	42 (0)
	5	30 (71)	25 (0)	25 (0)	35 (76)	25 (0)	25 (0)	42 (48)	30 (0)	30 (0)
	6	21 (93)	19 (0)	19 (0)	27 (99)	19 (0)	19 (0)	30 (86)	21 (0)	21 (0)
EON	1	66 (3)	66 (0)	66 (0)	66 (4)	66 (0)	66 (0)	66 (3)	66 (0)	66 (0)
	2	61 (44)	45 (0)	46 (2)	61 (36)	48 (0)	48 (0)	63 (21)	54 (0)	54 (0)
	3	52 (66)	35 (0)	35 (0)	46 (64)	36 (0)	36 (0)	54 (40)	43 (0)	43 (0)
	4	39 (157)	20 (0)	20 (0)	41 (143)	23 (0)	23 (0)	52 (106)	27 (0)	27 (0)
	5	31 (253)	14 (0)	14 (0)	38 (224)	15 (0)	15 (0)	42 (202)	17 (0)	17 (0)
	6	36 (363)	9 (0)	9 (0)	20 (367)	10 (0)	10 (0)	41 (292)	12 (0)	12 (0)
FON	1	245 (3)	241 (0)	241 (0)	244 (1)	244 (0)	244 (0)	245 (1)	244 (0)	244 (0)
	2	239 (6)	232 (0)	231 (0)	238 (4)	236 (0)	235 (0)	240 (3)	237 (0)	237 (0)
	3	232 (11)	215 (0)	215(0)	231 (7)	227 (2)	222 (1)	235 (5)	229 (0)	229 (0)
	4	227 (14)	206 (0)	206 (0)	226 (9)	219 (2)	214 (0)	231 (7)	221 (0)	221 (0)
	5	216 (21)	198 (4)	191 (0)	225 (16)	207 (4)	199 (1)	230 (10)	213 (0)	212 (0)
	6	217 (27)	182 (2)	178 (0)	215 (22)	186 (0)	186 (0)	218 (14)	204 (2)	200 (1)
IEEE	1	733 (15)	646 (0)	646 (0)	734 (15)	647 (0)	647 (0)	737 (8)	690 (0)	690 (0)
	2	728 (75)	424 (0)	424 (0)	724 (74)	424 (0)	424 (0)	735 (44)	513 (0)	513 (0)
	3	647 (87)	388 (0)	388 (0)	722 (88)	392 (0)	392 (0)	728 (52)	483 (0)	483 (0)
	4	711 (122)	355 (8)	328 (0)	633 (118)	359 (8)	333 (0)	728 (82)	403 (0)	403 (0)
	5	635 (160)	342 (23)	279 (2)	719 (158)	343 (21)	283 (0)	702 (98)	473 (2)	367 (0)
	6	624 (207)	246 (8)	232 (0)	631 (204)	273 (16)	235 (1)	701 (203)	239 (15)	239 (1)

Fig. 1. Average percent deviation - between brackets - and optimal score for Arpanet, Dodecahedron, EON, FON and IEEE300 graphs under Random, Weigted and Best Attack

when added to the subgraph formed by the IS and previously chosen nodes. The chosen node x is deleted from S_0, and S_0 is returned.

4.2.2 GRASP-IS Local Search for SWGFP

Once a solution S_0 is built in the construction phase, a local search phase is performed on this solution (Algorithm 4). A first-improvement local search strategy was chosen. In a first-improvement strategy the algorithm chooses any neighbour with a better value for objective function. During the neighbourhood search, the first-improving step is performed. At each search position s, the procedure evaluates the neighbouring candidate solutions $s' \in N(s)$ and the first s_0 that presents a lower value for the objective function is selected. A $2 - node - exchange$ (or *swap*) neighbourhood is used in this algorithm: a node $v \in S_0$ is exchanged with a node $u \in V \setminus S_0$ and a new score is computed. If the score is lower, the new solution is kept as the best solution; the local search is restarted from this new solution, continuing until as long as improvements exists. A node v is removed from S_0, the symmetric difference between S_0 and V is computed and stored in *nodes* and the *improve* flag is set as $FALSE$.

A node (from the *nodes* set) is added to S_0; computed the score, if it is lower than the original solution, is replaced by the new solution. The *improve* flag is set to $TRUE$. If it does not improve, the inspected node is removed from *nodes*. The search continues through the neighbourhood until it has been completely inspected and no improvement can be done, returning the best solution S_0.

5 Experimental Results

We tuned the parameters of both algorithms finding the optimality in small-sized graphs: *Arpanet* (20 nodes - 30 links), *Dodecahedron* (20 nodes - 30 links), EON (European Optical Network, 19 nodes - 36 links), FON (Fiber Optic Network, 62 nodes - 126 links), $IEEE300$ (Power Electric Grid 265 - 373 links) in Random, Weighted and Best Attack (Fig. 1).

The Random Algorithm (picks uniformly at random a solution of B nodes from the graph and calculates the score when the selected set is removed from the graph), $GRASP$ and $GRASP - IS$ was run 10 times for the five graphs under the three types of attack, varying parameter B from 1 to 6, obtaining better score reduction, as B evolves, in both GRASP algorithms than in Random algorithm, as well as lower scores an deviation values [15].

6 Conclusions

Operational researchers are engaged with the analysis of natural disasters, its prediction and protection mechanisms. Usually, the way to understand natural disasters is either using simulation of different scenarios, or mathematical models.

The Graph Fragmentation Problem (GFP) is an epidemic-inspired model that considers the worst case analysis of a disaster that entirely affects the connected population (i.e., the same connected component of a network). In this model we assume that we can protect some individuals subject to a fixed budget constraint. This model is suitable for decision-making in electrical networks (the location of electromagnetic protectors to cope with electric shocks). Potential applications include fire-fighting and epidemics as well.

In this paper, a more realistic model, called Stochastic Weighted GFP (SWGFP) is introduced. The model includes random, deliberated and priority-based attacks, as well as node-importance. GRASP-GFP is an adaptation of a previous heuristic for the GFP, while GRASP-IS is a novel proposal that exploits properties of independent sets. The experimental analysis shows that the novel proposal, GRASP-IS, outperforms GRASP-GFP in all the networks under study. Furthermore, the gap with respect to the globally optimum solution is small, or null, in most scenarios. As future work, we would like to interconnect previous works in the literature that deal with the CND, and the more recent GFP, that are incidentally the same problems, coming from different research interests (to know, epidemiology and network connectivity). Furthermore, we would like to find a game-theoretic framework to analyze attacker-defender strategies of a network subject to different attacks or failures.

Acknowledgements. This work is partially supported by Projects 395 CSIC I+D *Sistemas Binarios Estocásticos Dinámicos*, COST Action 15127 and Math-AMSUD Raredep *Rare events analysis in multi-component systems with dependent components.*

References

1. American Association for the Advancement of Science: Tales from the blackout. Science **301**(5636), 1029 (2003)
2. Aprile, M., Castro, N., Ferreira, G., Piccini, J., Robledo, F., Romero, P.: Graph fragmentation problem: analysis and synthesis. Int. Trans. Oper. Res. **26**(1), 41–53 (2019)
3. Cook, S.A.: The complexity of theorem-proving procedures. In: Proceedings of the Third Annual ACM Symposium on Theory of Computing, STOC 1971, pp. 151–158. ACM, New York (1971)
4. Festa, P., Resende, M.G.C.: An annotated bibliography of grasp - part l: algorithms. Int. Trans. Oper. Res. **16**(1), 1–24 (2009)
5. Festa, P., Resende, M.G.C.: An annotated bibliography of grasp - part ll: aplications. Int. Trans. Oper. Res. **16**(2), 131–172 (2009)
6. Ford, L.R., Fulkerson, D.R.: Maximal flow through a network. Can. J. Math. **8**, 399–404 (1956)
7. Garey, M.R., Johnson, D.S.: Computers and Intractability: A Guide to the Theory of NP-Completeness. W. H. Freeman and Company, New York (1979)
8. Garg, N., Vazirani, V.V., Yannakakis, M.: Multiway cuts in node weighted graphs. J. Algorithms **50**(1), 49–61 (2004)
9. Gendreau, M., Potvin, J.-Y.: Handbook of Metaheuristics, 2nd edn. Springer, Boston (2010). https://doi.org/10.1007/978-1-4419-1665-5
10. Harary, F.: Graph Theory. Addison-Wesley Series in Mathematics. Addison Wesley, Boston (1969)

11. Karp, R.M.: Reducibility among combinatorial problems. In: Miller, R.E., Thatcher, J.W. (eds.) Complexity of Computer Computations, pp. 85–103. Plenum Press (1972)
12. Lalou, M., Tahraoui, M.A., Kheddouci, H.: The critical node detection problem in networks: a survey. Comput. Sci. Rev. **28**, 92–117 (2018)
13. Frinhani, R.M.D., Silva, R.M.A., Mateus, G.R., Festa, P., Resende, M.G.C.: GRASP with path-relinking for data clustering: a case study for biological data. In: Pardalos, P.M., Rebennack, S. (eds.) SEA 2011. LNCS, vol. 6630, pp. 410–420. Springer, Heidelberg (2011). https://doi.org/10.1007/978-3-642-20662-7_35
14. Pauly, J.J.: The great chicago fire as a national event. Am. Q. **36**(5), 668–683 (1984)
15. Piccini, J., Robledo, F., Romero, P.: Complexity among combinatorial problems from epidemics. Int. Trans. Oper. Res. **25**(1), 295–318 (2018)
16. Piccini, J., Robledo, F., Romero, P.: Node-immunization strategies in a stochastic epidemic model. In: Pardalos, P., Pavone, M., Farinella, G.M., Cutello, V. (eds.) MOD 2015. LNCS, vol. 9432, pp. 222–232. Springer, Cham (2015). https://doi.org/10.1007/978-3-319-27926-8_19
17. Resende, M.G.C., Ribeiro, C.C.: Optimization by GRASP. Springer, New York (2016). https://doi.org/10.1007/978-1-4939-6530-4
18. Taubenberger, J.K., Reid, A.H., Lourens, R.M., Wang, R., Jin, G., Fanning, T.G.: Molecular virology: was the 1918 pandemic caused by a bird flu? Was the 1918 flu avian in origin? (Reply). Nature **440**, 9–10 (2006)

Uniformly Most-Reliable Graphs and Antiholes

Guillermo Rela, Franco Robledo, and Pablo Romero(✉)

Instituto de Matemática y Estadística, IMERL, Facultad de Ingeniería - Universidad de la República, Montevideo, Uruguay
{grela,frobledo,promero}@fing.edu.uy

Abstract. In network design, the all-terminal reliability maximization is of paramount importance. In this classical setting, we assume a simple graph with perfect nodes but independent link failures with identical probability ρ. The goal is to communicate p terminals using q links, in such a way that the connectedness probability of the resulting random graph is maximum. A graph with p nodes and q links that meets the maximum reliability property is called uniformly most-reliable (p, q)-graph (UMRG). The discovery of these graphs is a challenging problem that involves an interplay between extremal graph theory and computational optimization. Recent works confirm the existence of special cubic UMRG, such as Wagner, Petersen and Yutsis graphs. To the best of our knowledge, there is no prior works from the literature that find 4-regular UMRG. In this paper, we revisit the breakthroughs in the theory of UMRG. Finally, we mathematically prove that the complement of a cycle with seven nodes, $\overline{C_7}$, is a 4-regular UMRG. This graph is also identified as an odd antihole using terms from perfect graph theory.

Keywords: Network reliability · Uniformly most-reliable graphs · Antiholes

1 Motivation

Extremal graph theory is inspirational for network design [7]. Berge, in the second book ever written in graph theory, challenged the readers to find a graph with maximum connectivity among all graphs with a fixed number of nodes and links. Later, Frank Harary provided a full answer, finding also connected graphs with minimum and maximum diameter [12]. The building blocks of communication design are credited to Kirchhoff [15]. In this pioneer work he introduced the Matrix-Tree theorem, where he counts the number of spanning trees of a connected graph or *tree-number* as any cofactor of the Laplacian matrix. Curiously enough, this is an intermediate result to develop his theory of linear-time resistive circuits or *Kirchhoff's laws*, and this is probably the first result of Algebraic Graph Theory [3].

Network reliability analysis deals with probabilistic-based models, where the goal is to determine the probability of correct operation of a system [11].

G. Nicosia et al. (Eds.): LOD 2019, LNCS 11943, pp. 434–444, 2019.
https://doi.org/10.1007/978-3-030-37599-7_36

In its most elementary setting, we are given a simple graph G with perfect nodes but random link failures with identical and independent probability ρ. The all-terminal reliability is the probability that the resulting random graph remains connected.

Our motivation is to understand the interplay between network reliability analysis and inspirational problems from connectivity theory. This paper is organized as follows. Section 2 presents a formal definition of UMRG and classical terminology from graph theory. The related work is presented in Sect. 3. The main contribution is presented in Sect. 4, where it is formally proved that $\overline{C_7}$ is the only uniformly most-reliable $(7, 14)$-graph. Section 5 presents concluding remarks and trends for future work.

2 Definitions and Terminology

Given a simple graph $G = (V, E)$ with perfect nodes and unreliable links that fail independently with identical failure probability ρ. The all-terminal reliability $R_G(\rho)$ measures the probability that the resulting random graph remains connected. For convenience, we work with the unreliability $U_G(\rho) = 1 - R_G(\rho)$. Let us denote $p = |V|$ and $q = |E|$ the respective order and size of the graph G. In this paper, a *cutset* is a link-set U such that $G - U$ is not connected. Denote by $m_k(G)$ the set of all cutsets with cardinality k. By sum-rule, the unreliability polynomial can be expressed as follows:

$$U_G(\rho) = \sum_{k=0}^{q} m_k(G)\rho^k(1-\rho)^{q-k}. \tag{1}$$

A (p, q)-graph is a graph with p nodes and q links. Clearly, if we consider a fixed $\rho \in [0, 1]$, there is at least one graph H that attains the minimum unreliability, i.e., $U_H(\rho) \leq U_G(\rho)$ for all (p, q) graph G. Further, if the previous condition holds for all $\rho \in [0, 1]$ and all (p, q)-graphs G, then H is a UMRG.

The link-connectivity $\lambda(G)$ is the smallest λ such that $m_\lambda > 0$. A *trivial cutset* is a cutset that includes all the links adjacent to a fixed node. The *degree* of a node $v \in V$ is the number of links that are incident to v. A graph is *regular* if all the nodes have identical degrees. The *minimum degree* of a graph G is denoted by $\delta(G)$. Using trivial cutsets, $\lambda(G) \leq \delta(G)$. A graph is super-$\lambda$, or *superconnected*, if it is λ-regular and further, it has only trivial cutsets: $m_\lambda = p$. A *bridge* is a single link e such that $G - e$ is not connected. A *cut-point* is a node v such that $G - v$ is not connected. A graph G is said to be *biconnected* if it has no cut-points. A *matching* is a disjoint (or non-adjacent) set of links. A *perfect matching* is a matching that meets all the nodes of a graph. The elementary cycle with p nodes is denoted C_p, and K_p represents the complete graph with p nodes, where all the nodes are adjacent. The complementary graph of G is denoted by $\overline{G}$. A *hole* in a graph G is an induced subgraph of G isomorphic to a cycle of length at least 4. An *antihole* is an induced subgraph H of G, such that $\overline{H}$ is hole of G. A hole (resp. antihole) is odd or even according to the number of its nodes.

3 Related Work

In order to simplify the readability, this section presents, in order, the etymology and elements on UMRG. Then, we outline the findings of sparse and dense (almost complete) UMRG. It is worth to remark that our graph under study, $\overline{C_7}$, is a special case of K_n minus n links, and falls within the analysis of almost complete graphs.

3.1 Etymology

The most-reliable graphs for all values of ρ were called *uniformly optimally reliable* networks by Boesch in his seminal work [4]. Later, Wendy Myrvold offered a nice survey up to 1986 [18]. She explains that the term uniformly most-reliable graph is adopted to avoid a tongue-twister.

3.2 Elements

Expression (1) suggests a sufficient criterion for a graph H to be UMRG, this is, $m_k(H) \leq m_k(G)$ for all k and all (p, q)-graph G. In 1986, Boesch conjectured that the converse holds, and this is yet one of the major open problems in the field [4]. As far as we know, the most recent survey in this topic was written a decade ago, and it summarizes different trends in network reliability analysis [6]. In the last decade, the progress in this field shows to be slow, and few novel results are available. Still today, the search of UMRG lays on the minimization of all the coefficients m_k. This approach is promoted by the following result, which can be proved using elementary calculus [1]:

Proposition 1.

(i) If there exists some integer k such that $m_i(H) = m_i(G)$ for all $i < k$ but $m_k(H) < m_k(G)$, then there exists $\rho_0 > 0$ such that $U_H(\rho) < U_G(\rho)$ for all $\rho \in (0, \rho_0)$.

(ii) If there exists some integer k such that $m_i(H) = m_i(G)$ for all $i > k$ but $m_k(H) < m_k(G)$, then there exists $\rho_1 < 1$ such that $U_H(\rho) < U_G(\rho)$ for all $\rho \in (\rho_1, 1)$.

By definition, there are no cutsets with lower cardinality than the link connectivity λ. Therefore, $m_i(G) = 0$ for all $i < \lambda$, and by Proposition 1-(i) UMRG must have the maximum link-connectivity λ. Furthermore, the number of cutsets m_λ must be minimized. On the other hand, $m_i(G) = \binom{q}{i}$ for all $i > q - p + 1$, since trees are minimally connected with $q = p - 1$ links. The number of connected sets with $q - p + 1$ links is precisely the tree-number $\tau(G)$, so $m_{q-p+1}(G) = \binom{q}{q-p+1} - \tau(G)$. Using Proposition 1-(ii), the maximization of the tree-number is a necessary condition. Prior observations directly connect this network design problem with extremal graph theory:

Corollary 1. *If H is UMRG, it must have the maximum tree-number $\tau(H)$, maximum connectivity $\lambda(H)$, and the minimum number of cutsets $m_\lambda(H)$ among all (p,q)-graphs with maximum connectivity.*

For convenience we say that a (p,q)-graph, H, is t-optimal if $\tau(H) \geq \tau(G)$ for every (p,q) graph G. Briefly, Corollary 1 claims that UMRG must be t-optimal and max-λ min-m_λ.

It is worth to remark that there are (p,q)-pairs where a UMRG does not exist. Indeed, Wendy Myrvold proposed an infinite family of counterexamples [19]. The historical conjecture of Leggett and Bedrosian asserts that t-optimal graphs must be almost regular, that is, the degrees differ at most by one [17]. Even though closed formulas are available for the tree-number of specific graphs, the progress on t-optimality is effective under special regularity conditions [10], almost-complete graphs or other special graphs with few links [20].

Frank Harary found the maximum connectivity of a (p,q) graph. By Handshaking, the average-degree of every (p,q)-graph is $\frac{2q}{p}$. If $\delta(G)$ denotes the minimum degree, we immediately get that $\lambda(G) \leq \delta(G) \leq \lfloor \frac{2q}{p} \rfloor$. The candidate connectivity is $\lambda_{max} = \lfloor \frac{2q}{p} \rfloor$. It suffices to find a (p,q)-graph with connectivity λ_{max} whenever $p \geq q-1$ (otherwise, the graph is not connected). Harary provided graphs with maximum connectivity $\lambda_{\max} = \lfloor \frac{2q}{p} \rfloor$, so, they are max-$\lambda$. The number of cutsets should be minimized as well; max-λ graphs that minimize the cutsets with λ nodes are called max-λ min-m_λ graphs.

Prior works from Bauer et al. fully characterize max-λ min-m_λ graphs [2]. The key is to observe that in a max-λ graph, the number of cutsets m_λ is at least the number of nodes with degree λ. If this bound is achieved, a max-λ min-m_λ graph is retrieved. For that purpose, they define generalized Harary graphs, which are just an augmentation of the original Harary graphs with random matchings (this is, non-adjacent links). In that way, the number of nodes with degree λ is minimized, and the authors show that no other cutsets with that size exists.

3.3 Finding Sparse UMRG

By Corollary 1, Bauer et al. provide a family of graphs that contains all UMRG. Later works try to find uniformly $(p,p+i)$-most-reliable graphs for i small (i.e., sparse graphs), by a simultaneous minimization of all the coefficients m_k. When $i=0$ we have $q=p$, and the elementary cycle C_p is t-optimal. All the other graphs with $p=q$ are not 2-connected, and by direct inspection we can see that C_p achieves the minimum coefficients m_k.

Perhaps the first non-trivial UMRG were found by Boesch et al. in 1991 [5]. A new reading of Bauer et al. construction lead them to find that Monma graphs are $(n,n+1)$ UMRG, when the path lengths are as even as possible.

A more challenging problem is to find $(n,n+2)$ UMRG. Boesch et al. minimize the four effective terms m_0, m_1, m_2 and m_3 from Expression (1).

An $(n,n+2)$ max-λ min-m_λ graph already minimizes the first three terms. If in addition the tree-number is minimized all the coefficients are simultaneously minimized, and the result must be a UMRG. The merit of the paper [5] is to adequately select the feasible graphs from Bauer et al. that minimizes the tree-number. Observe that K_4 can be partitioned into three perfect matchings, PM_1, PM_2 and PM_3.

The result is that we should insert $n-4$ points in the six links of K_4 in such a way that:

(i) the number of inserted nodes in all the links differ by at most one, and
(ii) if we insert the same number of nodes in two different matchings $PM_i \neq PM_j$, then the number of nodes in the four links from $PM_i \cup PM_j$ are identical.

The resulting $(n, n+2)$-graph defines, for every $n \geq 4$, a single graph up to isomorphism. The authors formally prove that the resulting graph is uniformly most-reliable $(n, n+2)$-graph. Furthermore, inspired by a previous research on t-optimality in multipartite graphs authored by Cheng [10], they conjecture that all uniformly most-reliable $(n, n+3)$-graphs with more than 6 nodes are elementary subdivisions of $K_{(3,3)}$. This conjecture is correct, and it was proved by Wang [22].

For every even natural n, Möbius graph M_n is constructed from the cycle C_{2n} adding n new links joining every pair of opposite nodes. It is worth to note that $M_2 = K_4$ and $M_3 = K_{(3,3)}$ are Mobius graphs. More recently, Romero formally proved using iterative augmentation that M_4, known as Wagner graph, is uniformly most-reliable as well. In this sense, Möbius graphs apparently generalize the particular result for $K_{(3,3)}$ and K_4. The author proposed the following conjecture:

Conjecture 1. All $(n, n+i)$ UMRG are elementary subdivisions of Möbius graph M_n., whenever $i < n$.

Conjecture 1 is not true. Indeed, the same authors formally proved in a previous work that Petersen graph is UMRG [21]. This work reinforces Donald Knuth's statement that Petersen graph serves as a counterexample to several optimistic predictions in graph theory [16]. A less ambitious conjecture is still open:

Conjecture 2. All UMRG $(n, n+4)$-graphs with $n \geq 8$ are elementary subdivisions of Wagner graph.

3.4 Finding Almost-Complete UMRG

For a large number of nodes n, it is known the shape of UMRG when we can remove few links to the complete graph K_n. In fact, Kelmans and Chelnokov formally proved that if we must remove $n/2$ links or less, then a UMRG is retrieved whenever we remove a matching [14]. Further research tries to characterize UMRG when we must remove e links meeting the inequality $n/2 \leq e \leq n$. Kelmans again found a characterization of those graphs having up to n links removed from K_n [13]. Petingi et al. formally proved the t-optimality of the graphs suggested by Kelmans, for special cases where $n = 3k$, being k some positive integer [20].

3.5 Recent Progress

For practical reasons, the most recent works try to find highly reliable (p, q)-graphs instead. In [8], the authors develop a GRASP/VND heuristic in order to find highly reliable cubic graphs. They found new candidates of 3-regular UMRG, such as Yutsis and Heawood-Kantor graphs. In [9], it is mathematically proved that Yutsis is UMRG. Furthermore, a reliability-improving graph transformation is proposed, where non-biconnected graphs can be always transformed into biconnected graphs whenever $q \geq p$. In Sect. 4 we use this fact in order to discard the graphs with $\delta(G) = 1$ for comparison. Indeed, a stronger result holds:

Theorem 1. *Let $q \geq p \geq 3$. For any connected (p, q)-graph G with a cut-point, there is some biconnected (p, q)-graph G' such that $m_k(G') \leq m_k(G)$ for all k.*

The reader is invited to consult [9] for a formal proof. Note that if G is biconnected, its minimum-degree must be $\delta(G) \geq 2$. The reliability-improving transformation from Theorem 1 is schematically illustrated in two steps. In Step 1, we can observe that (v, w) is not a bridge any more in G^* (see Fig. 1). Furthermore, if G^* has some cut-point, it can be removed and find a biconnected graph G' (see Fig. 2).

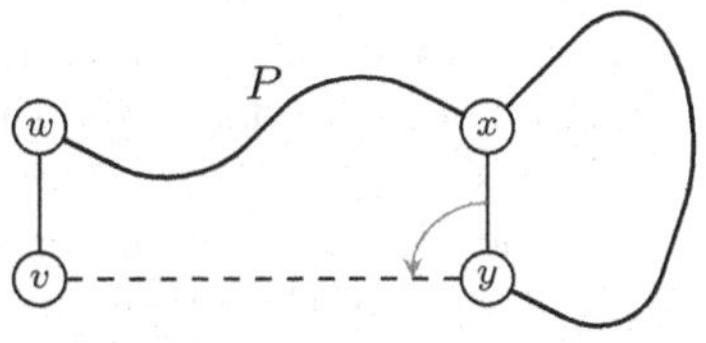

Fig. 1. Building a bridgeless graph G^*.

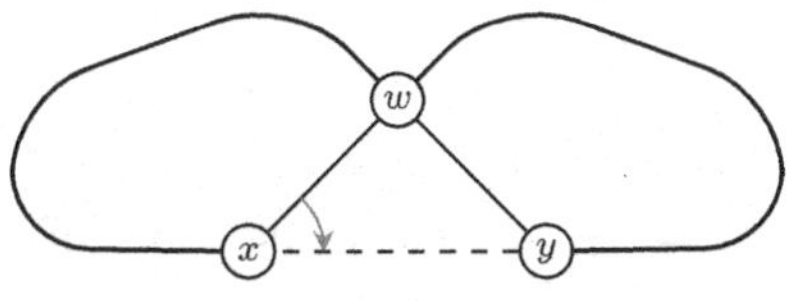

Fig. 2. Building a biconnected graph G'.

4 The Graph $H = \overline{C_7}$ is UMRG

In order to simplify the understanding of the main result of this paper, first we show a toy sample, where we formally prove that the elementary cycle is UMRG among all (n, n)-graphs. This example will serve as a smaller example, but the rationale behind the proof is identical.

Proposition 2. *The elementary cycle C_n is UMR (n,n)-graph.*

Proof. Disconnected graphs are immediately discarded, since they have null reliability. Observe that connected (n,n)-graphs are either the elementary cycle C_n, or a tree with the addition of a sigle link. Let us denote T_e to a graph that consists of a tree with a link addition, that is not the elementary cycle C_n. Let us compare the number of cutsets $m_k(T_e)$ and $m_k(C_n)$:

- Both graphs are connected, so $m_0(C_n) = m_0(T_e) = 0$.
- By construction, T_e must have a bridge, but C_n is bridgeless. Therefore: $m_1(C_n) < m_1(T_e)$.
- The deletion of 2 or more links disconnect both graphs: $m_k(C_n) = m_k(T_e) = \binom{n}{k}, \forall k \geq 2$.

As a consequence, $m_k(C_n) \leq m_k(G)$ for all (n,n)-graph G, and the unreliability $U_{C_n}(\rho)$ is dominated by $U_G(\rho)$. This means that C_n is UMRG, as desired.

The antihole $H = \overline{C_7}$ is depicted in Fig. 3. The reader can appreciate that $\overline{C_7}$ falls in the case of an almost-complete graph K_7 removing exactly 7 links. The intuition of Kelmans [13] is correct, where he formally proved that $H = \overline{C_7}$ is t-optimal, and he suggested that $H = \overline{C_7}$ could be UMRG. Following the necessary criterion from Corollary 1, the candidate from the set of $(7,14)$-graphs must be 4-regular. Consider an arbitrary 4-regular $(7,14)$ graph G. Observe that the complement $\overline{G}$ is a 2-regular simple graph, and it has 7 links. The only possibilities are C_7 and $C_3 \cup C_4$. Therefore, the only 4-regular graphs are their complements: $H = \overline{C_7}$ and $H' = \overline{C_3 \cup C_4}$. Recall that the candidate must be t-optimal as well. The odd-antihole H is a circulant graph (i.e. each row-vector of the adjacency matrix is a right-shift of the preceding row-vector). Therefore, we can find a closed formula for its tree-number using algebraic graph theory [3]. Let us denote $p(G;\lambda)$ to the characteristic polynomial of the adjacency matrix of a graph G. The tree-number $\kappa(H)$ of a k-regular graph H with an odd number of nodes n can be expressed in terms of its complementary graph as follows [3]:

$$\kappa(H) = \frac{1}{n^2} p(H^C; -1-k) \tag{2}$$

In the case under study, $\overline{H} = C_7$, and the eigenvalues of C_7 are $\lambda_0 = 2$ (simple) and $\lambda_i = 2cos(\frac{2i\pi}{n}), i = 1, \ldots, \frac{(n-1)}{2}$ (double roots). Therefore, we can replace in Equation (2) using $k = n-3 = 4$ to find the tree-number of the odd-antihole $H = \overline{C_7}$:

$$\kappa(H) = \frac{1}{7} \prod_{i=1}^{3} [4sin^2\left(\frac{i\pi}{n}\right) - n]^2 = 1183. \tag{3}$$

Analogously, we find the tree-number of $H' = \overline{C_3 \cup C_4}$ using Equation (2):

$$\kappa(H') = \frac{1}{49} P(C_3, -5) p(C_4; -5) = 15 \times (5 + 2cos\left(\frac{2\pi}{3}\right)^2 = 540. \tag{4}$$

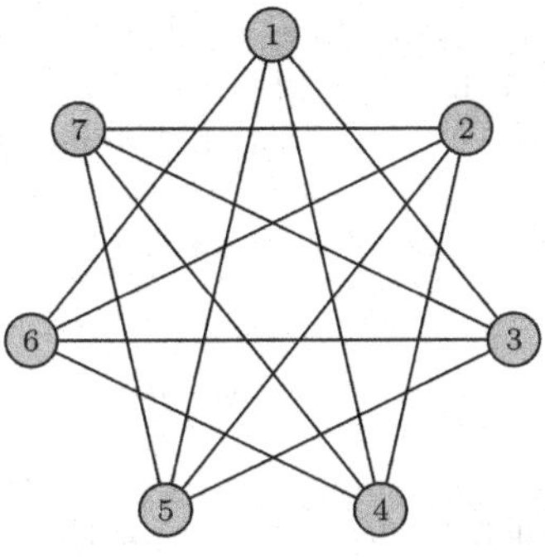

Fig. 3. Antihole Graph $H = \overline{C_7}$-

Since $\kappa(H) > \kappa(H')$, H has the largest tree-number among all 4-regular graphs. By the necessary criterion from Corollary 1, the odd-antihole is the only candidate to become a UMRG.

Since, H is circulant, it has super-connectivity $\lambda = 4$, and $m_4(H) = 7$ is minimum among all $(7, 14)$-graphs. All its 5-cutsets are trivial, so $m_5(H) = 7 \times 10 = 70$. In order to count $m_6(H)$ we observe that the 5-cutsets are either trivial or isolate links, so, $m_6 = 7 \times \binom{10}{2} + 14 = 329$. In order to count $m_7(H)$ we observe that the 7-cutsets can isolate both nodes and links, so, $m_7(H) = 7 \times \binom{10}{3} + 14 \times \binom{7}{1} = 938$. We will count $m_8(H)$ using two alternative ways. The first way is to observe that $m_8(H) + \kappa(H) = \binom{14}{8}$, since the graphs are either connected or not. Therefore, $m_8(H) = 3003 - 1183 = 1820$. The second way is to note that 8-cutset either isolate nodes, links, 2-paths or 3-paths, so, $m_8(H) = 7 \times \binom{10}{4} - 21 + 14 \times \binom{7}{2} + 42 + 35 = 1820$.

In the following, we will prove that H is UMRG. Our strategy is to partition the family of $(7, 14)$-graphs using the minimum degree. By Handshaking, if G is a $(7, 14)$-graph necessarily $\delta(G) \in \{1, 2, 3, 4\}$. By Theorem 1, if $\delta(G) = 1$ there exists a biconnected $(12, 18)$-graph, G', such that $m_k(G') \leq m_k(G)$ for all k. Since G' is biconnected, in particular we know that $\delta(G') \geq 2$. As a consequence, the case $\delta(G) = 1$ is established once we prove the result for the remaining cases. There are only 2 graphs with $\delta(G) = 4$, to know, $H = \overline{C_7}$ and $H' = \overline{C_3 \cup C_4}$. By the way, H' has as much cutsets as H for all feasible sizes, but $m_8(H) < m_8(H')$, since the inequality $\kappa(H) > \kappa(H')$ has been already established for the respective tree-numbers. From this analysis, we conclude that it suffices to prove that $m_k(H) \leq m_k(G)$ for all $(7, 14)$-graphs with $\delta(G) \in \{2, 3\}$. The following four lemmas study the different size for the cutsets in a step-by-step fashion. Under otherwise stated, we only count trivial cutsets, using inclusion-exclusion principle.

Lemma 1. *The coefficient m_5 is minimized in the odd antihole H.*

Proof. If $\delta(G) = 2$ it is sufficient to count trivial cutsets. Indeed, the presence of a single node v with $deg(v) = 2$ implies that $m_5(G) \geq 1 \times \binom{12}{3} > m_5(H)$.

If $\delta(G) = 3$ and there is a single node v such that $deg(v) = 3$, the degree-sequence must be $(3,4,4,4,4,4,5)$. In this case, counting trivial cutsets we know that $m_5(G) \geq \binom{11}{2} + 5 \times \binom{10}{1} > m_5(H)$. If there exists two nodes with this degree, the degree-sequence must be either $(3,3,4,4,4,5,5)$ or $(3,3,4,4,4,4,6)$, and $m_5(G) \geq 2 \times \binom{11}{2} > m_5(H)$. The number of cutsets is greater under the presence of more nodes with degree 3.

Lemma 2. *The coefficient m_6 is minimized in the odd antihole H.*

Proof. If $\delta(G) = 2$, we get that $m_6(G) \geq \binom{12}{4} > m_6(H)$. If $\delta(G) = 3$ and there is a single node v such that $deg(v) = 3$, the degree-sequence must be $(3,4,4,4,4,4,5)$, and $m_6(G) \geq \binom{11}{3} + 5 \times \binom{10}{2} > m_6(H)$. If there exists two nodes with this degree, the degree-sequence must be either $(3,3,4,4,4,5,5)$ or $(3,3,4,4,4,4,6)$, and $m_6(G) \geq 2 \times \binom{11}{3} > m_6(H)$. The number of cutsets is greater under the presence of more nodes with degree 3.

Lemma 3. *The coefficient m_7 is minimized in the odd antihole H.*

Proof. If $\delta(G) = 2$, $m_7(G) \geq \binom{12}{5} + 5 \times \binom{10}{3} > m_7(H)$. If $\delta(G) = 3$ and there is a single node v such that $deg(v) = 3$, the degree-sequence must be $(3,4,4,4,4,4,5)$. In this case, counting trivial cutsets we know that $m_7(G) \geq 1 \times \binom{11}{4} + 5 \times \binom{10}{3} - 5 + 1 \times \binom{9}{2} > m_7(H)$. If there exists two nodes with this degree, the degree-sequence must be either $(3,3,4,4,4,5,5)$ or $(3,3,4,4,4,4,6)$, and $m_7(G) \geq 2 \times \binom{11}{4} - 8 + 3 \times \binom{10}{3} - 6 > m_7(H)$. The number of cutsets is greater under the presence of more nodes with degree 3.

Lemma 4. *The coefficient m_8 is minimized in the odd antihole H.*

Proof. If $\delta(G) = 2$, let us discuss according to the degree-sequence d:

1. If $d = (2,3,4,4,4,5,6)$, it suffices to count trivial cutsets: $m_8(G) \geq \binom{12}{6} + \binom{11}{5} - \binom{9}{3} + 3 \times (\binom{10}{4} - \binom{8}{2} - \binom{7}{1}) - 5 + \binom{9}{3} - \binom{7}{1} + \binom{8}{2} > m_8(H)$. The case $d = (2,3,4,4,5,5,5)$ is analogous but link separators are added to the trivial cutsets.
2. If $d = (2,3,3,4,5,5,6)$, it suffices to count trivial cutsets: $m_8(G) \geq \binom{12}{6} + 2 \times (\binom{11}{5} - \binom{9}{3}) - 2 \times \binom{8}{2} + \binom{10}{4} - 2 \times \binom{7}{1} + 2 \times 81 > m_8(H)$. The cases $d = (2,3,3,4,4,6,6)$ and $d = (2,3,3,5,5,5,5)$ are covered analogously.
3. If $d = (2,3,3,3,5,6,6)$, it suffices to count trivial cutsets: $m_8(G) \geq 1 \times \binom{12}{6} + 3 \times (\binom{11}{5} - \binom{9}{3} - \binom{8}{2} - 1) > m_8$.
4. If $d = (2,4,4,4,4,5,5)$, counting link separators and trivial cutsets: $m_8(G) \geq 1 \times \binom{12}{6} + 4 \times (\binom{10}{4} - \binom{8}{2}) - 6 + 2 \times (\binom{9}{3} - \binom{7}{1}) + 2 \times \binom{9}{3} > m_8$. The case $d = (2,4,4,4,4,4,6)$ is analogous.
5. If $d = (2,2,4,4,4,6,6)$, $(2,2,4,4,5,5,6)$ or $(2,2,4,5,5,5,5)$, the gaps in the inequalities are even greater than in the previous cases.

If $\delta(G) = 3$, let us discuss again according to the degree sequence d:

1. If $d = (3,4,4,4,4,4,5)$, we count 2-path, link-separator and trivial ones: $m_8(G) \geq \binom{11}{5} + 5 \times (\binom{10}{4} - \binom{7}{1}) - 11 + \binom{9}{3} + 2 \times \binom{8}{3} + 8 \times \binom{7}{2} + 4 \times 6 + \binom{14}{2} > m_8$.
2. The remaining cases have more degree-3 nodes, and the gap in the inequality is even greater than the previous case.

Theorem 2. *Antihole $H = \overline{C_7}$ is uniformly most-reliable.*

Proof. The graph $H = \overline{C_7}$ is circulant. Therefore, it is superconnected, and it minimizes m_4. Clearly, $m_i = 0$ for $i \in \{0,1,2,3\}$, and $m_i = \binom{14}{i}$ for all $(7,14)$-graphs, when $i \geq 9$. By Lemmas 1–4, H minimizes m_5, m_6, m_7 and m_8. Therefore, H simultaneously minimizes all the coefficients m_i among $(7,14)$-graphs, and it is uniformly most-reliable.

5 Conclusions and Trends for Future Work

Uniformly most-reliable graphs (UMRG) represent a synthesis in network reliability analysis. Finding them is a hard task not well understood. An exhaustive comparison is computationally prohibitive for most cases. Prior works in the field try to globally minimize the cutsets. This methodology provides uniformly most-reliable $(n, n+i)$ graphs for $i \leq 6$. In this paper, we formally proved that the odd Antihole $H = \overline{C_7}$ is UMRG. To the best of our knowledge, this is the first non-trivial 4-regular UMRG in the literature (since K_5 is 4-regular).

There are several trends for future work. Even though all UMRG share special properties such as the greatest girth and minimum diameter, there are no proofs available for these conjectures. Boesch conjecture is open since 1986. A powerful methodology to find UMRG is not known. The problem of t-optimality is still not well understood. Conjecture 2 could be studied with a similar reasoning as in Boesch [5] and Wang [22]. A more general result of Theorem 2 considering an infinite family of odd-antiholes is also challenging.

Acknowledgements. This work is partially supported by Projects 395 CSIC I+D *Sistemas Binarios Estocásticos Dinámicos*, COST Action 15127 *Resilient communication services protecting end-user applications from disaster-based failures*, Math-AMSUD Raredep *Rare events analysis in multi-component systems with dependent components* and STIC-AMSUD ACCON, *Algorithms for the capacity crunch problem in optical networks*.

References

1. Bauer, D., Boesch, F., Suffel, C., Van Slyke, R.: On the validity of a reduction of reliable network design to a graph extremal problem. IEEE Trans. Circuits Syst. **34**(12), 1579–1581 (1987)
2. Bauer, D., Boesch, F., Suffel, C., Tindell, R.: Combinatorial optimization problems in the analysis and design of probabilistic networks. Networks **15**(2), 257–271 (1985)

3. Biggs, N.: Algebraic Graph Theory. Cambridge Mathematical Library, Cambridge University Press, Cambridge (1993)
4. Boesch, F.T.: On unreliability polynomials and graph connectivity in reliable network synthesis. J. Graph Theory **10**(3), 339–352 (1986)
5. Boesch, F.T., Li, X., Suffel, C.: On the existence of uniformly optimally reliable networks. Networks **21**(2), 181–194 (1991)
6. Boesch, F.T., Satyanarayana, A., Suffel, C.L.: A survey of some network reliability analysis and synthesis results. Networks **54**(2), 99–107 (2009)
7. Bollobás, B.: Extremal Graph Theory. Dover Books on Mathematics, Dover Publications (2004)
8. Bourel, M., Canale, E., Robledo, F., Romero, P., Stábile, L.: A Hybrid GRASP/VND heuristic for the design of highly reliable networks. In: Blesa Aguilera, M.J., Blum, C., Gambini Santos, H., Pinacho-Davidson, P., Godoy del Campo, J. (eds.) HM 2019. LNCS, vol. 11299, pp. 78–92. Springer, Cham (2019). https://doi.org/10.1007/978-3-030-05983-5_6
9. Canale, E., Robledo, F., Romero, P., Viera, J.: Building reliability-improving network transformations. In: 2019 15th International Conference on Design of Reliable Communication Networks (DRCN 2019), Coimbra, Portugal, March 2019
10. Cheng, C.-S.: Maximizing the total number of spanning trees in a graph: two related problems in graph theory and optimum design theory. J. Comb. Theory Ser. B **31**(2), 240–248 (1981)
11. Colbourn, C.J.: The Combinatorics of Network Reliability. Oxford University Press Inc., New York (1987)
12. Harary, F.: The maximum connectivity of a graph. Proc. Nat. Acad. Sci. U.S.A. **48**(7), 1142–1146 (1962)
13. Kelmans, A.: On graphs with the maximum number of spanning trees. Random Struct. Algorithms **9**(1–2), 177–192 (1996)
14. Kelmans, A.K., Chelnokov, V.M.: A certain polynomial of a graph and graphs with an extremal number of trees. J. Comb. Theory Ser. B **16**(3), 197–214 (1974)
15. Kirchhoff, G.: Über die auflösung der gleichungen, auf welche man bei der untersuchung der linearen verteilung galvanischer ströme geführt wird. Ann. Phys. Chem. **72**, 497–508 (1847)
16. Knuth, D.E.: The art of computer programming: introduction to combinatiorial algorithms and Boolean functions. In: Addison-Wesley Series in Computer Science and Information Proceedings, Addison-Wesley (2008)
17. Leggett, J.D., Bedrosian, S.D.: On networks with the maximum numbers of trees. In: Proceedings of Eighth Midwest Symposium on Circuit Theory, pp. 1–8, June 1965
18. Myrvold, W.: Reliable network synthesis: some recent developments. In: Proceedings of the Eighth International Conference on Graph Theory, Combinatorics, Algorithms and Applications, vol. II, pp. 650–660 (1996)
19. Myrvold, W., Cheung, K.H., Page, L.B., Perry, J.E.: Uniformly-most reliable networks do not always exist. Networks **21**(4), 417–419 (1991)
20. Petingi, L., Boesch, F., Suffel, C.: On the characterization of graphs with maximum number of spanning trees. Discrete Math. **179**(1), 155–166 (1998)
21. Rela, G., Robledo, F., Romero, P.: Petersen graph is uniformly most-reliable. In: Nicosia, G., Pardalos, P., Giuffrida, G., Umeton, R. (eds.) MOD 2017. LNCS, vol. 10710, pp. 426–435. Springer, Cham (2018). https://doi.org/10.1007/978-3-319-72926-8_35
22. Wang, G.: A proof of Boesch's conjecture. Networks **24**(5), 277–284 (1994)

Merging Quality Estimation for Binary Decision Diagrams with Binary Classifiers

Nikolaus Frohner(✉) and Günther R. Raidl

Institute of Logic and Computation, TU Wien, Vienna, Austria
{nfrohner,raidl}@ac.tuwien.ac.at

Abstract. Relaxed binary decision diagrams (BDDs) are used in combinatorial optimization as a compact representation of a relaxed solution space. They are directed acyclic multigraphs which are derived from the state space of a recursive dynamic programming formulation of the considered optimization problem. The compactness of a relaxed BDD is achieved by superimposing states, which corresponds to merging BDD nodes in the classical layer-wise top-down BDD construction. Selecting which nodes to merge crucially determines the quality of the resulting BDD and is the task of a merging heuristic, for which the *minimum longest path value* (minLP) heuristic has turned out to be highly effective for a number of problems. This heuristic sorts the nodes in a layer by decreasing current longest path value and merges the necessary number of worst ranked nodes into one. There are, however, also other merging heuristics available and usually it is not easy to decide which one is more promising to use in which situation. In this work we propose a prediction mechanism to evaluate a set of different merging mechanisms at each layer during the construction of a relaxed BDD, in order to always select and apply the most promising heuristic. This prediction is implemented by either a perfect or by a k-layers lookahead construction of the BDD, gathering feature vectors for two competing merging heuristics which are then fed into a binary classifier. Models based on statistical tests and a feed-forward neural network are considered for the classifier. We study this approach for the maximum weighted independent set problem and in conjunction with a parameterized merging heuristic that takes also the similarity between states into account. We train and validate the binary classifiers on random graphs and finally test on weighted DIMACS instances. Results indicate that relaxed BDDs can be obtained whose upper bounds are on average up to ≈16% better than those of BDDs constructed with the sole use of minLP.

Keywords: Binary decision diagrams · Merging heuristics · Lookahead construction · Binary classifiers

1 Introduction

Binary decision diagrams (BDDs) were introduced in the 1950s by Lee [7] as a compact representation for boolean functions. In the last decade, they have

G. Nicosia et al. (Eds.): LOD 2019, LNCS 11943, pp. 445–457, 2019.
https://doi.org/10.1007/978-3-030-37599-7_37

gained increasing popularity in the field of combinatorial optimization, where BDDs are used as a graphical representation of the solution space of a given optimization problem, constructed from a dynamic-programming-like recursive formulation of the solution space. Formally, a BDD is a directed acyclic multi-graph $B = (U, A)$ with node set U and arc set A. Each $u \in U$ is associated with a state $s(u)$ of the respective recursive formulation. Paths from a root node $\mathbf{r}$ through the BDD correspond to (partial) solutions and carry a length, corresponding to the solution costs; a longest path to a designated target node $\mathbf{t}$ then corresponds to an optimal solution for a maximization problem[1]. For a thorough introduction, we recommend the book by Bergman, Cire, van Hoeve, and Hooker [3].

Throughout this paper, we focus specifically on *relaxed limited-width* BDDs, which are constructed layer-by-layer in a breadth-first-search fashion. While exact BDDs model the solution space exactly but typically have exponential size for hard combinatorial optimization problems, relaxed BDDs represent a discrete relaxation and are kept compact by limiting the width, i.e., the number of nodes, at each layer. This width limitation is achieved by layer-wise *merging* of nodes: Whenever a layer is about to become too large, nodes are selected and merged, which means that their states are superimposed in a way that guarantees not to lose any feasible solutions. These merging operations, however, in general introduce new paths that do not represent feasible solutions. Therefore, the relaxed BDD represents a discrete relaxation of the original problem and the length of the longest path corresponds to an upper bound of the optimal solution value.

The method to select the nodes to be merged, called *merging heuristic*, is crucial for the quality of the resulting bound. We propose a method to evaluate multiple available merging heuristics at a given layer and to choose the believed-to-be locally best one.

The next section recaps the well-known *minimum longest path value* (minLP) merging heuristic and the parameterized state-similarity based merging heuristic we introduced in [5]. In Sect. 3, we introduce a method to estimate the quality relative to minLP of any merging heuristic applied at a given layer by conducting either a perfect or a k-layers lookahead and predicting the resulting bound at the final layer. This allows us to select different merging heuristics at each layer and gracefully deviate from the minLP merging heuristic. We always compare merging heuristics pairwise using a binary classifier, either based on a simple feature comparison test, linear regression, a Wilcoxon signed rank sum test or an artificial neural network, see Sect. 4, which gives us a probabilistic estimate whether one is better than the other. We present the data preparation for our binary classifier training and validation together with our computational study in Sect. 5, where we consider the maximum weighted independent set problem

[1] We consider only maximization throughout this paper. The methods are, however, equally applicable to minimization by changing the sign of the objective function.

(MWISP) with training on random weighted graphs and final tests on weighted DIMACS[2] instances. We conclude and give indications for future work in Sect. 6.

In the MWISP, we are given a graph $G = (V, E)$ and costs $c_j \in \mathbb{R}$ for each node $j \in V$. We seek to find a set of nodes $S \subset V$ with maximum costs $\sum_{j \in S} c_j$ for which no two nodes are adjacent in G. In a recursive formulation, we assign to each node of the graph a binary decision variable x_i and impose an ordering π_i on these. A state is the set of nodes that can still be added to the current independent set. When at step i of the recursion, we decide either to add the node $\pi_i = j$ to the independent set, setting $x_i = 1$, which removes the node and its neighborhood $N(j)$ from the set, or to leave it, setting $x_i = 0$, which only removes the node itself; this is encoded by a corresponding state transition function τ:

$$\tau : \{0, 1\} \times 2^V \to 2^V \tag{1}$$

$$(0, s_i) \mapsto s_{i+1} = \tau(0, s_i) = s_i - \{j\} \tag{2}$$

$$(1, s_i) \mapsto s_{i+1} = \tau(1, s_i) = s_i - \{j\} - N(j) \tag{3}$$

We denote with D_{s_i} the admissible values for x_i in state s_i, which is $\{0, 1\}$ when $j \in s_i$, otherwise $\{0\}$. The maximization problem is then given by the following Bellman equations:

$$z^*(s_i) = \max_{d \in D_{s_i}} \{c_j d + z^*(\tau(d, s_i))\} \tag{4}$$

$$z^*(s_n) = 0 \tag{5}$$

$z^*(s_0)$, where the initial state is given by $s_0 = V$, yields the cost of a maximum weighted independent set.

2 Related Work

For the layer-wise construction of limited-width BDDs for MWISP instances, we follow the algorithm described by Bergman et al. [2], which employs zero-suppressing long-arcs, the minimal state (minState) variable ordering heuristic, and the minimum longest path length (minLP) merging heuristic. For the MWISP, the state $s(u)$ associated with a node u of the BDD is the set of nodes that can still be included in the independent set. Successors of a state are obtained by setting a still open variable to either one or zero, representing the decision that the corresponding node is either included or excluded, respectively. Arc lengths correspond to the gains in the objective function, i.e., the weight of the respective node if included or zero otherwise. The used minState variable ordering heuristic selects at each layer always a variable that appears in the fewest number of states associated with the BDD nodes that were generated by previous layers and still need to be placed on a layer. The minLP merging heuristic sorts the nodes u at a current layer in decreasing order of their currently longest path length $z^{\mathrm{lp}}(u)$ from the root to them and merges the necessary

[2] https://github.com/jamestrimble/max-weight-clique-instances/tree/master/DIMACS.

number of nodes from the back into one node, so that the maximum width is kept. Merging is done by applying the set union over all affected states.

In [5], we identified the similarity between states as worth considering for the merging decision and introduced a parameterized merging algorithm that also begins by bulk merging nodes with smallest longest path value into one node but not enough to reach the maximum width. Instead, the method then applies pairwise merging of the remaining nodes that have longest path values below some threshold by iteratively selecting two nodes with minimal dissimilarity. This dissimilarity between two nodes $u, v \in U$ can be defined in different ways but in [5] we found that considering an upper bound on the costs-to-go from the state $s(w)$ we would obtain when merging the two nodes into one new node w is particularly useful. In case of the MWISP, we choose the weighted sum over the remaining graph vertices that can still be selected after merging u and v:

$$d_{\mathrm{ub}}(u, v) = z^{\mathrm{ub}}_{\mathrm{MWISP}}(w) = \sum_{j \in s(w)} c_j \tag{6}$$

The merging boundary is a contiguous set of nodes in a layer that have the same rank when sorted by the longest path length to them and that could participate in the classical minLP merging. The parameters that control which nodes take part in the iterative pairwise similarity based merging are $(\delta_{\mathrm{l}}, \delta_{\mathrm{r}}) \in [0, 1]^2$. They define as a relative measure depending on the longest path values of the nodes how many nodes left (determined by δ_{l}) and how many nodes right (determined by δ_{r}) of the merging boundary should be taken, see Fig. 1 for the conceptual differences between the merging heuristics. The extreme cases are $(1.0, 1.0)$, where all nodes are potential merging candidates, and $(0.0, 0.0)$, which corresponds to minLP with an additional tie breaking when nodes at the merging boundary have the same longest path value and the pure rank-based merging would not be unique—for them the similarity based merging is applied.

Lookahead approaches [9] are very common in deterministic games, where different possible moves are compared by conducting a limited playout and evaluating the resulting configurations by an approximating evaluation function. The idea of a k-layers lookahead approach for BDDs has been presented by Bergman et al. [1] in the context of a dynamic variable ordering for the maximum independent set problem.

3 Merging Quality Estimation

So far, when constructing a limited-width BDD layer-wise, a predefined merging heuristic is repeatedly applied at every layer that would exceed the maximum width. We aim at higher flexibility by having a set of merging heuristics $\mathcal{H}$ at our disposal, and hope that a careful application of different heuristics improves the resulting quality of the BDD. At a current layer l of the BDD construction, the central question then is which heuristic to select to conduct the actual merging. Unfortunately, a reliable free-standing way of judging the potential of each of the heuristics in dependence of the current situation is not obvious. We therefore

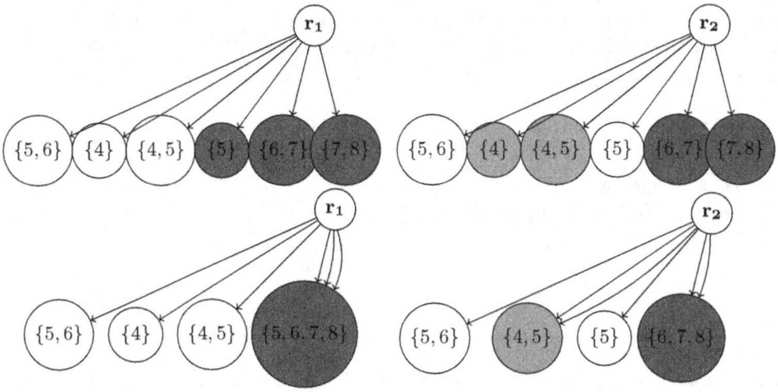

Fig. 1. An exemplary layer with set-based states where merging has to take place to reduce the width to 4. Left: the classical minLP merging heuristic which bulk-merges the necessary number of nodes with currently shortest path length (dark), potentially leading to large states; right: the merging heuristic from [5] which combines bulk-merging nodes with currently shortest path length (dark) with pairwise merging of nodes under consideration of their similarity (bright), distributing the state sizes more evenly.

suggest a different, amenable approach that always keeps the established minLP heuristic in $\mathcal{H}$ and uses it as a baseline to define a relative quality measure for each of the considered merging heuristics.

Definition 1 (Merging Quality). *Given are a set of merging heuristics* $\mathcal{H}$*, where minLP* $\in \mathcal{H}$ *and a BDD in construction facing a layer* l *to be merged. For each* $H \in \mathcal{H}$ *we create a shallow copy of the BDD, conduct the merging determined by* H *and finish the construction of the BDD afterwards by only applying minLP. The resulting upper bounds on the objective value act as measure of quality for* H *at layer* l*: We write* $H \prec H'$ *when* H *yields a strictly tighter bound than* H'*, and* H *is then considered the locally better choice than* H'*.*

Overall, after evaluating each merging heuristic in the described way, we actually select and apply a dominating one and continue at the next layer in the same way; ties are broken randomly. Clearly, this complete lookahead for each considered merging heuristic at each level is computationally expensive, but what we obtain is the possibility to measure the impact of the different heuristics and finally we can also study how often each method has been applied in the construction of BDDs. Thus, we can see whether this combination of multiple merging heuristics may in principle improve the bounds of the resulting BDDs for our given problem instances. A pseudo-code for this lookahead algorithm is shown in Algorithm 1.

The approach also allows to generate the ground truth for a corresponding classification problem: given heuristics H and H', let f be a binary function returning 1, iff $H \prec H'$, i.e., H provides a tighter bound than H', and 0 otherwise. To make the approach viable in practice, we move from the

Input: BDD B under construction, current layer l, maximum layer $l_{\max}$, set of competing merging heuristics $\mathcal{H}$ including minLP
Output: Winning merging heuristic $H \in \mathcal{H}$

Function *perfect-lookahead*$(B, l, l_{\max}, \mathcal{H})$
  for $H \in \mathcal{H}$ **do**
    $B' \leftarrow$ shallow copy of B;
    apply H to B' at layer l;
    continue B' with minLP until reaching final layer $l_{\max}$;
    $z_H^{\mathrm{lp}} \leftarrow z^{\mathrm{lp}}(\mathbf{t})$ in B';
  end
  $H^* \leftarrow \operatorname{argmin}_{H \in \mathcal{H}} z_H^{\mathrm{lp}}$;
  return H^*;

Algorithm 1: Perfect lookahead algorithm for deciding which merging heuristic $H \in \mathcal{H}$ to use at layer l.

evaluation by *perfect lookahead* to a statistically estimated variant that only considers the next k layers, i.e., a *k-layers lookahead.* During construction with alternative heuristics, when at layer l, we apply H and continue the construction of a shallow copy of the BDD for $k-1$ more layers using minLP, yielding a feature matrix $\mathbf{Y_H} \in \mathbb{R}^{p \times k}$ for the looked-ahead layers $\{l, \ldots, l+k-1\}$, where p is the number of features. The distinguished baseline feature matrix is $\mathbf{Y_{minLP}}$, when only minLP was applied for the layers $\{l, \ldots, l+k-1\}$ in a shallow copy of the BDD. The learning goal now is to find a classifier function $h \colon \mathbb{R}^{p \times k} \times \mathbb{R}^{p \times k} \to \{0,1\}$ for which wrong classifications in the sense of $h(\mathbf{Y_H}, \mathbf{Y_{H'}}) \neq f(\mathbf{Y_H}, \mathbf{Y_{H'}})$ are unlikely. Every layer provides a fixed number of features, for which we have to aggregate information from the variable number of nodes per layer. By taking the maximum, mean value, and minimum of the longest path values $z^{\mathrm{lp}}(u)$ over all nodes u at the layer and likewise of the upper bound values $z^{\mathrm{ub}}_{\mathrm{MWISP}}(u)$, we identified six natural options to be used as features per layer. In the following sections, we consider different types of binary classifiers for the statistical lookahead problem based on a linear regression model, on the Wilcoxon signed rank sum test, and on training an artificial neural network on random weighted graphs.

4 Binary Classifiers

We consider parameterized binary classifiers $h_\alpha \in \mathcal{B}$ that are constructed by taking a function $\tilde{h}$ that provides an estimation of the probability that $H \prec H'$ and apply a threshold $\alpha \in (0,1)$:

$$\tilde{h} \colon \mathbb{R}^{p \times k} \times \mathbb{R}^{p \times k} \to [0,1]. \tag{7}$$

$$h(\mathbf{Y}, \mathbf{Y}') = \begin{cases} 0, & \tilde{h}(\mathbf{Y}, \mathbf{Y}') < \alpha \\ 1, & \tilde{h}(\mathbf{Y}, \mathbf{Y}') \geq \alpha \end{cases} \tag{8}$$

Input: BDD under construction B, current layer l, number of layers to look ahead k, maximum layer $l_{\max}$, acceptance threshold α, set of merging heuristics to test $\mathcal{H}$, probabilistic binary classifier $\tilde{h}$

Output: Winning merging heuristic $H \in \mathcal{H}$

Function *k-layers-lookahead*(B, l, $l_{\max}$, $\mathcal{H}$, $\tilde{h}$, α)
    if $l + k \geq l_{\max}$ **then**
        return *perfect-lookahead*$(B, l, l_{\max}, \mathcal{H})$;
    end
    apply minLP to shallow copy B' at layer l, continue for $k-1$ layers with minLP;
    $\mathbf{Y}_{\text{minLP}} \leftarrow$ feature vectors for layers $\{l, \ldots, l+k-1\}$;
    for $H \in \mathcal{H} \setminus \{\text{minLP}\}$ **do**
        apply H to shallow copy B' at layer l, continue for $k-1$ layers with minLP;
        $\mathbf{Y}_H \leftarrow$ feature vectors for layers $\{l, \ldots, l+k-1\}$;
        $p_H \leftarrow \tilde{h}(\mathbf{Y}_H, \mathbf{Y}_{\text{minLP}})$;
    end
    $H^* \leftarrow \text{argmax}_{H \in \mathcal{H} \setminus \{\text{minLP}\}}\, p_H$;
    if $p_{H^*} \geq \alpha$ **then**
        return H^*;
    return minLP;

Algorithm 2: k-layers lookahead algorithm for deciding which merging heuristic $H \in \mathcal{H}$ to use at layer l by means of a probabilistic binary classifier $\tilde{h}$ parameterized by threshold α.

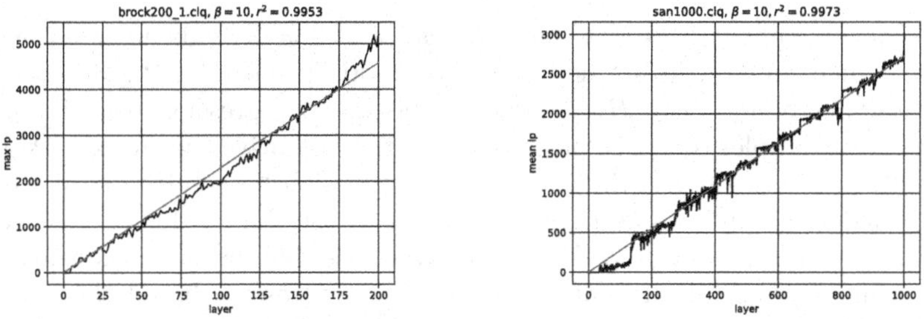

Fig. 2. Evolution of the maximum and mean longest path lengths over the layers with a linear regression line for different weighted DIMACS instances with maximum layer width $\beta = 10$.

Equipped with an $\tilde{h}$, we can formulate the k-layers-lookahead merging heuristic selection algorithm as listed in Algorithm 2. It compares every merging heuristic with minLP by feeding their feature matrices to the probabilistic binary classifier and saves for each H the resulting probability estimate p_H. If the largest p_H is greater than or equal to the acceptance threshold α, the corresponding winning heuristic H^* is returned, otherwise minLP.

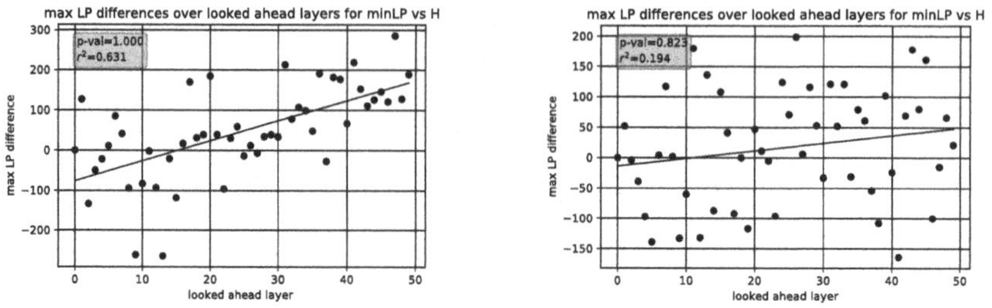

Fig. 3. Examples for a $k = 50$ layers lookahead with minLP versus a competing heuristic H, regressing the differences of the maximum longest path length (maxLP) values on the layers. Left: a true positive case (minLP worse than H), right: a true negative case (minLP not worse than H).

We now consider different possibilities for the probabilistic binary classifier $\tilde{h}$. In some preliminary experiments we observed that as a first approximation a linear dependence between layers and the maximum and mean longest path values is reasonable to assume, see Fig. 2. Considering minLP as default merging strategy, we want to test whether the growth trend for another merging heuristic H is significantly smaller, i.e., $H \prec \text{minLP}$. To do so, we restrict ourselves to one feature, for example the maximum longest path length for a layer (maxLP), calculate $\Delta\mathbf{Y}_H = \mathbf{Y}_{\text{minLP}} - \mathbf{Y}_H$, where $\Delta\mathbf{Y}_H \in \mathbb{R}^{1\times k}$, and solve the linear regression (LR) model

$$\Delta\hat{Y}_H(l;\theta,d) = \theta\, l + d. \tag{9}$$

A subsequent student t-test for the significance of $\hat{\theta} \neq 0$ yields a corresponding p-value, which can be transformed to a belief in $[0,1]$ that minLP grows towards a worse upper bound than H. We denote this linear regression based classifier as $\tilde{h}^{\text{LR}}$. For examples of a true positive and a true negative case for a lookahead length of $k = 50$, see Fig. 3. Since the linear regression based classifier considers only the slope, a non-parametric alternative is to use the p-value of a Wilcoxon signed rank sum test over the $\Delta\mathbf{Y}_H$, the layer-wise differences of the features.

As a more powerful alternative to the linear regression based classifier, we consider a feed-forward neural network (NN) $\tilde{h}^{\text{NN}}(\mathbf{Y}_H, \mathbf{Y}_{\text{minLP}})$ that yields a score in $[0,1]$ which we interpret as probability for $H \prec H'$. As features for the k layers created by the two different merging heuristics applied at layer l, we consider two from the possibilities namely the differences of the maximum longest path values and the difference of the maxima of the upper bounds over the given nodes by layer. Furthermore, we provide the graph density and the layer progress $l/l_{\max}$ as input, resulting in an input layer of the NN consisting of $2(k+1)$ neurons. These differences are normalized by dividing them by their maximum absolute value to feed values from $[-1,1]$ into the network in order to facilitate the training. Two hidden layers with twice the neurons of the input layer, i.e., $4(k+1)$, follow with a single neuron in the output layer. Each non-final neuron is configured with a reLU activation function, the final one with a sigmoid activation function to obtain a value in $[0, 1]$. We use binary cross-entropy loss and train with minibatch gradient descent with a batch size of 64. We start with a learning rate of 10^{-3} and decrease by factors of 10 after reaching a plateau. To prevent overfitting, we use a weight decay factor of 10^{-3} and use early stopping by monitoring the accuracy on a validation set consisting of 20% of the original training samples. We trained the neural network using Keras[3] 2.0.8 with TensorFlow 1.2.1 backend.

To evaluate the performance of the binary classifiers, we calculate precision-recall curves by varying the acceptance threshold α of the binary classifier and gather the corresponding precision-recall data points. The precision is the number of true positives divided by the number of classified as positives and the recall the number of true positives divided by the overall positives. This allows to tune the classifier to the required behavior. Furthermore, a single area-under-curve (AUC) value can be calculated from this curve and compared for different classifiers. As a simple baseline classifier that is not based on a probability score, given one-dimensional feature vectors for both merging heuristics, we compare the maxima of the feature vectors and return 1 if $\max \mathbf{Y_H} < \max \mathbf{Y_{H'}}$ and otherwise zero. Its performance is then evaluated by a single precision-recall point. When we use the maxLP values as features, we call this max-maxLP, since we compare the maxima of the maxima.

5 Computational Study

For training and validation data, we created 1000 $G(n, p)$ random weighted graphs with parameters $n \in [100, 2000]$ and $p \in [0.05, 0.95]$ drawn uniformly at random. Weights are assigned from $\{1, \ldots, 200\}$ in dependence of the index of a vertex j via $j \bmod 200 + 1$. For each graph G_i, $i = 1, \ldots, 1000$, we construct a binary decision diagram with maximum width $\beta = 10$ and merging parameters $\delta_\mathrm{l} = \delta_\mathrm{r} = 0$ (i.e., minLP with tie breaking according to [5]) and save the resulting upper bound u_i and the features by layer $\mathbf{Y}^i_{\mathrm{minLP}}$. Furthermore, we sample a layer $l' \in \{1, \ldots, n_i\}$ uniformly at random, construct the corresponding binary decision diagram up to layer l' in the same fashion as before. If no merging is needed, we restart; otherwise we sample merging parameters $\delta_\mathrm{l}, \delta_\mathrm{r}$ uniformly at random and likely apply a different merging. After this, we continue the construction as before with $\delta_\mathrm{l} = \delta_\mathrm{r} = 0$ and save the resulting upper bounds $\tilde{u}^m_i$ and feature vectors by layer $\mathbf{Y}^i_m$ from l' to the last layer, where $m \in \{1, \ldots, 20\}$, resulting into 20 training samples per graph. This creates the ground truth for $f(\mathbf{Y}^i_m, \mathbf{Y}^i_{\mathrm{minLP}})$ which is one if the resulting upper bound $\tilde{u}^m_i$ of the alternative merging is strictly smaller than the pure minLP upper bound u_i, otherwise zero. When we train or test our binary classifiers, we set the fixed lookahead length k and extract only the corresponding subpart of the saved feature matrices in a preprocessing step.

[3] https://keras.io.

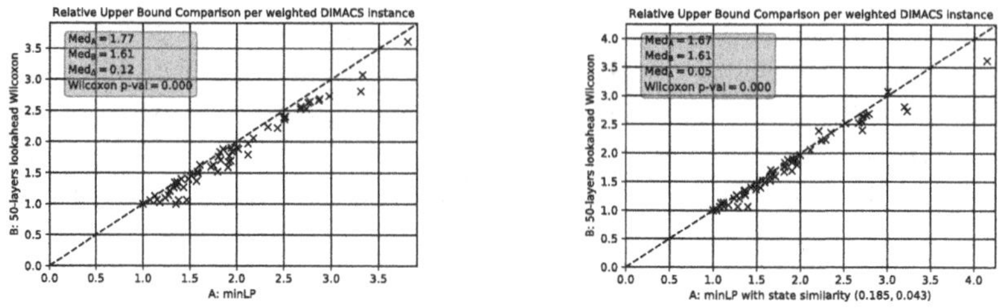

Fig. 4. Comparison of pure minLP (left) vs. minLP with state similarity and raced parameters (right) vs a combination of both on weighted DIMACS instances with statistical lookahead of 50 layers, using the Wilcoxon test classifier and an acceptance threshold of $\alpha = 0.95$.

We conducted the final tests on a weighted DIMACS graph set[4], from which we consider $N = 64$ instances $\mathcal{I}_{\text{WDIMACS}}$ that we could solve to optimality. This allows us to calculate for each graph instance $I \in \mathcal{I}_{\text{WDIMACS}}$ a relative bound u_I^{rel} from the absolute bound $u_I^{\text{rel}} = u_I / z_I^*$ derived from the construction of a relaxed BDD with one of the described approaches. As figures of merit when comparing two approaches, we consider the median and the mean of the pairwise differences $\widetilde{\Delta} := \text{Med}_{I \in \mathcal{I}_{\text{WDIMACS}}}[\Delta u_I^{\text{rel}}]$, $\overline{\Delta}$ of the relative upper bounds. First, we applied the perfect lookahead approach with the parameterized similarity based merging heuristic as introduced in [5]. The baseline merging heuristic is minLP with tie breaking, corresponding to parameters $\delta_{\text{l}} = \delta_{\text{r}} = 0$ and four further competing parameter sets $(\delta_{\text{l}}, \delta_{\text{r}}) \in \{(0.185, 0.043), (0.2, 0.2), (0.3, 0.3), (0.4, 0.4)\}$. The competing merging heuristics are used in only 4% of the layers with merging, still resulting in a median relative bound improvement of 0.16 over using only minLP alone. In [5], we tuned the parameter set $(0.185, 0.043)$ using irace [8], which gave a median improvement of 0.05, when applied alone. For the k-layers lookahead, we evaluate the linear regression, the Wilcoxon signed rank sum based variant, the NN, and the simple max-maxLP binary classifiers on the validation set of the random weighted graph instances described in Sect. 5. The NN classifier outperforms the other classifiers beginning from a lookahead length of $k = 30$ as can be seen in the precision-recall plot in Fig. 5 for $k = 50$ and for the precision-recall area-under-curve values calculated for $k \in \{10, 20, \ldots, 90\}$. For example, when we tune to a modest recall of 0.1, we get only 0.58 precision for the LR and Wilcoxon classifier but approximately 0.68 for the NN, which is not surprising since the latter considers more features, including upper bound information and graph density, whereas the former only relies on the maximum longest path values as feature. In general, the precision of the classifiers is relatively weak indicating a difficult classification problem.

The final test is conducted on the weighted DIMACS instances, and the median and mean of differences of relative upper bounds are used again as figures of merit and summarized in Table 1 for three different lookahead-lengths

[4] https://github.com/jamestrimble/max-weight-clique-instances/tree/master/DIMACS.

$\{30, 50, 70\}$. We compare with the pure application of minLP and pure application of minLP with state similarity merging parameterized by $(0.185, 0.043)$. We see that the naive, completely parameter-less max-maxLP classifier that considers one heuristic to be worse than the other when the maximum of the maximum longest path values over the looked ahead layers is strictly greater yields results comparable to using the linear regression with a median relative bound improvement between 0.09 and 0.10. The also rather simple Wilcoxon test performed second best in our final test, yielding figures between 0.09 and 0.12 (Fig. 4), slightly worse than the NN classifier with figures between 0.11 and 0.16.

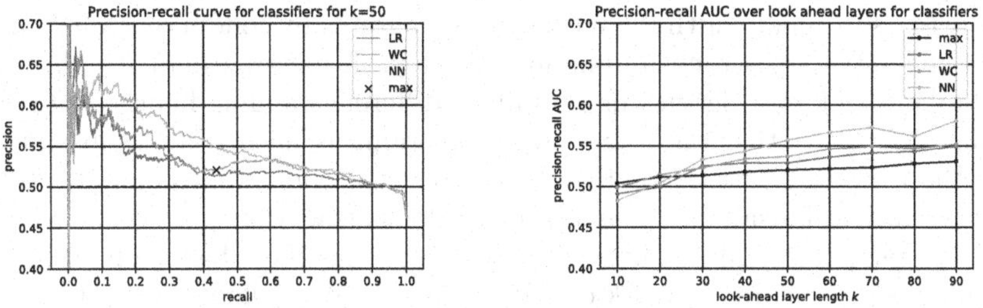

Fig. 5. Left: precision-recall curves for different classifiers with k-layers lookahead with $k = 50$ on random weighted graph instances; right: corresponding precision-recall AUC over lookahead length $k \in \{10, \ldots, 90\}$.

Table 1. Median $\widetilde{\Delta}$ and mean $\overline{\Delta}$ of pairwise differences of relative bounds for perfect lookahead and $\{30, 50, 70\}$-layers lookahead with different classifiers vs pure minLP and vs minLP with state similarity based merging with parameters $\delta_l = 0.185, \delta_r = 0.043$, on weighted DIMACS instances. (PLA = perfect lookahead, max = max-maxLP, LR = linear regression, WC = Wilcoxon, NN = neural network).

Comparing approach	PLA		k	max		LR		WC		NN	
	$\widetilde{\Delta}$	$\overline{\Delta}$		$\widetilde{\Delta}$	$\overline{\Delta}$	$\widetilde{\Delta}$	$\overline{\Delta}$	$\widetilde{\Delta}$	$\overline{\Delta}$	$\widetilde{\Delta}$	$\overline{\Delta}$
pure minLP	0.16	0.17	30	0.09	0.11	0.07	0.08	0.09	0.11	0.11	0.11
minLP with state similarity	0.09	0.11		0.04	0.06	0.02	0.03	0.04	0.06	0.04	0.06
pure minLP	0.16	0.17	50	0.09	0.11	0.09	0.11	0.12	0.13	0.12	0.13
minLP with state similarity	0.09	0.11		0.03	0.06	0.03	0.06	0.05	0.08	0.08	0.08
pure minLP	0.16	0.17	70	0.10	0.12	0.10	0.12	0.12	0.14	0.15	0.16
minLP with state similarity	0.09	0.11		0.04	0.06	0.03	0.07	0.05	0.09	0.08	0.11

To do a runtime comparison, we conducted all experiments on an Intel Xeon E5-2640 processor with 2.40 GHz in single-threaded mode and a memory limit of 8 GB using Python3.6. We measured runtimes when constructing the BDDs with 30-, 50-, and 70-layers-lookahead for two competing merging heuristics and relative to plain minLP, which gave us a median factor of $\approx 2k\times$, for example for a 30-layers-lookahead with the NN classifier $\approx 70\times$. This is what we expected given this computationally demanding approach. Still, if the resulting decision diagram is more compact than another one with the same bound and is traversed many times afterwards, then this initial construction overhead may pay off. See for instance [6], where arcs in an already constructed relaxed decision diagram are repeatedly filtered to remove infeasible paths. An idea under investigation for further runtime reduction is to conduct the lookahead with a smaller width.

6 Conclusion and Future Work

In this paper, we have shown a method to locally evaluate the quality of merging heuristics in small-width binary decision diagrams by conducting a lookahead using the simple yet strong minimum longest path merging heuristic. We used this method to devise algorithms that allow different merging heuristics to compete against each other on a layer where merging is needed and subsequently apply the winning heuristic. The evaluation is either done by a computationally intensive perfect lookahead or by a k-layers lookahead where we try to approximate the perfect lookahead by means of binary classifiers based on either statistical tests or a neural network classifier. We trained, validated, and tuned the classifiers on random weighted graph instances and finally tested on a weighted DIMACS graph set where we could show significant bound improvements by combining minLP and our parameterized state-similarity based merging heuristic over only using either one alone.

Further research is needed to validate this lookahead approach on other problems, compare with other classification approaches such as logistic regression or support vector machines, and to possibly find computationally less demanding local quality estimation methods. In the context of branching heuristics of mixed-integer programming solvers, the "Dynamic approach for switching heuristics" [4] creates clusters of sub-problems in a feature space during an off-line training phase where a different heuristic works best for each cluster and then dynamically switches between these heuristics during traversal of the branch-and-bound tree. This could potentially also be an interesting approach for selecting different merging heuristics during the construction of a BDD.

References

1. Bergman, D., Cire, A.A., van Hoeve, W.-J., Hooker, J.N.: Variable ordering for the application of BDDs to the maximum independent set problem. In: Beldiceanu, N., Jussien, N., Pinson, É. (eds.) CPAIOR 2012. LNCS, vol. 7298, pp. 34–49. Springer, Heidelberg (2012). https://doi.org/10.1007/978-3-642-29828-8_3
2. Bergman, D., Cire, A.A., van Hoeve, W.J., Hooker, J.N.: Optimization bounds from binary decision diagrams. INFORMS J. Comput. **26**(2), 253–268 (2013)
3. Bergman, D., Cire, A.A., van Hoeve, W.J., Hooker, J.N.: Decision Diagrams for Optimization. Artificial Intelligence: Foundations, Theory, and Algorithms. Springer, Cham (2016). https://doi.org/10.1007/978-3-319-42849-9
4. Di Liberto, G., Kadioglu, S., Leo, K., Malitsky, Y.: DASH: dynamic approach for switching heuristics. Eur. J. Oper. Res. **248**(3), 943–953 (2016)

5. Frohner, N., Raidl, G.R.: Towards improving merging heuristics for binary decision diagrams. In: Proceedings of LION 13–13th International Conference on Learning and Intelligent Optimization. Lecture Notes in Computer Science, Springer (2019, to appear)
6. Horn, M., Raidl, G.R.: Decision diagram based limited discrepancy search for a job sequencing problem. In: Computer Aided Systems Theory - EUROCAST 2019. Lecture Notes in Computer Science, Springer (2019, to appear)
7. Lee, C.Y.: Representation of switching circuits by binary-decision programs. Bell Syst. Tech. J. **38**(4), 985–999 (1959)
8. López-Ibáñez, M., Dubois-Lacoste, J., Cáceres, L.P., Birattari, M., Stützle, T.: The irace package: Iterated racing for automatic algorithm configuration. Oper. Res. Persp. **3**, 43–58 (2016)
9. Pearl, J.: Heuristics: Intelligent Search Strategies for Computer Problem Solving. Addison-Wesley Publishing Co., Inc., Reading (1984)

Directed Acyclic Graph Reconstruction Leveraging Prior Partial Ordering Information

Pei-Li Wang[1] and George Michailidis[1,2(✉)]

[1] Informatics Institute, University of Florida, Gainesville, USA
gmichail@ufl.edu
[2] Department of Statistics, University of Florida, Gainesville, USA

Abstract. Reconstructing directed acyclic graphs (DAGs) from observed data constitutes an important machine learning task. It has important applications in systems biology and functional genomics. However, it is a challenging learning problem, due to the need to consider all possible orderings of the network nodes and score the resulting network structures based on the available data. The resulting computational complexity for enumerating all possible orderings is exponential in the size of the network. A large number of methods based on various modeling formalisms have been developed to address this problem, primarily focusing on developing fast algorithms to reduce computational time. On many instances, partial topological information may be available for subsets of the nodes; for example, in biology one has information about transcription factors that regulate (precede) other genes, or such information can be obtained from perturbation/silencing experiments for subsets of DAG nodes (genes).

We develop a framework for estimating DAGs from observational data under the assumption that the nodes are partitioned into sets and a *complete topological ordering* exists amongst them. The proposed approach combines (i) (penalized) regression to estimate edges between nodes across different sets, with (ii) the popular PC-algorithm that identifies the skeleton of the graph within each set. In the final step, we combine the results from the previous two steps to screen out redundant edges. We illustrate the performance of the proposed approach on topologies extracted from the DREAM3 competition. The numerical results showcase the usefulness of the additional partial topological ordering information for this challenging learning problem.

1 Introduction

Regulatory networks control the growth and development of organisms and their ability to respond to environmental conditions, through mechanisms that span multiple molecular levels [2]. The structure of regulatory networks reflects the interactions between the regulatory elements in biological systems, such as genes

Supported by grants NIH 1U01CA23548701 and 5R01GM11402904 to G. Michailidis.

G. Nicosia et al. (Eds.): LOD 2019, LNCS 11943, pp. 458–471, 2019.
https://doi.org/10.1007/978-3-030-37599-7_38

and proteins [12,19]. Hence, their reconstruction from observational Omics data is a key, but challenging task, even when the underlying network is assumed to have a directed acyclic structure. Nevertheless, such reconstructed networks can provide insights into biological processes and mechanisms and help generate new hypotheses. For this reason, there has been a growing number of computational approaches based on different formalisms for addressing this problem.

Certain methods require data obtained from experiments involving gene knockouts/knockdowns of their expression level (see, e.g, [1,15] and references therein). Other methods require time series expression data [18]. Approaches that integrate multiple sources of data have also been recently examined [13].

Another class of methods, statistical in nature, require as input observational steady state expression data. It also uses the framework of statistical graphical models described by directed acyclic graphs (DAGs). However, determining directed graphs from observational data is computationally NP-hard [24].

One popular approach (the PC-algorithm) constructs an undirected graph from the data and then employs a series of conditional independence tests to orient the edges [10,11], thus recovering an equivalence class that contains the true DAG (see details in the next Section). The problem becomes computationally significantly simpler, if a complete topological ordering of the nodes in the DAG is available. To that end, [22] provided a lasso penalized regression based method for learning sparse DAGs. A Bayesian variant of the penalized regression method with complete topological ordering information was developed by [16] and subsequently applied to gene regulatory networks. Other penalties were explored in a series of papers [3,5,8,9], but avoided making the assumption of knowing a complete topological ordering of the nodes. However, the resulting optimization problem is non-convex and hence computational more challenging.

Other approaches used the framework of Structural Equations Models coupled with the assumption of non-Gaussian data [4], or non-linearities in the model [20]) to aid the reconstruction of the DAG. Finally, some approaches combined observational data with data from perturbation experiments that led to improved DAG reconstructions [21].

In this study, we propose a framework that assumes that partial ordering information amongst sets of nodes of the DAG is available. In that sense, this is similar to the setting in [14]. However, the graph structure within each set is another DAG. Hence, we propose to combine a penalized regression approach to obtain estimates of the direct edges between sets and apply the PC-algorithm to learn edges within the sets. The resulting graph will be a superset of the underlying DAG. Subsequently, we refine the edge set through an additional step involving regressions and finally use ideas from the PC-algorithm methodology to obtain the direction of any remaining non-oriented edges. Extensive numerical work demonstrates the competitive nature of the method, but also points out the importance of having good quality partial ordering information.

Availability of such partial ordering information heavily depends on the application context. In biology, such information can be obtained from databases (e.g.

the Kyoto Encyclopedia of Genes and genomes) and the literature (for a relevant discussion, see [21]).

The remainder of the paper is organized as follows: important concepts on DAGs and linear structural equations models, together with a step-by-step description of the proposed approach is presented in the next Section. Selected numerical results based on data from the DREAM3 competition are provided in the performance evaluation Section, followed by some concluding remarks.

2 Methods and Algorithms

As discussed in the introductory Section, we restrict our attention to inferring regulatory networks based on DAGs. Next, we provide a detailed description of our technical framework, emphasizing the case of partitioning the set of nodes into two sets for conveying the main ideas of the proposed framework and present the key algorithmic steps for inferring the DAG. We then extend the approach to the case of partitioning the node set into multiple subsets for which a complete topological ordering is available. We start in Sects. 2.1 and 2.2 by introducing several technical concepts from the literature, necessary for the technical developments in the proposed framework.

2.1 DAGs and Linear Structural Equation Models

Consider a DAG $\mathcal{G} := (V, E)$ with node set V and edge set $E \subseteq V \times V$. A DAG is characterized by the property that there are no directed paths starting and ending at the same node. For each node $j \in V$, let $\mathrm{Pa}(j) \equiv \{k \in V : (k, j) \in E\}$ denote the parent set of j. A DAG $\mathcal{G}$ represents a probability distribution $F(X_1, \cdots, X_p)$ over the random vector $\boldsymbol{X} = (X_1, \cdots, X_p)$ with p being the size of V, if F factorizes as $F(x_1, \cdots, x_p) \propto \prod_{j=1}^{p} F\big(x_j | X_{\mathrm{Pa}(j)}\big)$. Further, a permutation π of the vertex set V is a *topological ordering* for $\mathcal{G}$ if $\pi(j) < \pi(k)$ whenever $(j, k) \in E$. Such a topological ordering exists for any DAG [7], although it may not be unique. The above factorization implies that node X_j is *conditionally independent* of all nodes that are non-descendent of j, conditioning on $\mathrm{Pa}(j)$. There are important connections between DAGs and linear structural equation models (SEM) of the form:

$$\boldsymbol{X} = \mathbf{B}'\boldsymbol{X} + \boldsymbol{\varepsilon}, \quad \boldsymbol{\varepsilon} \sim N_p(\mathbf{0}, \mathbf{D}),$$

where $\mathbf{B}$ is a strictly upper triangular matrix and further assume that $\mathbb{E}(\boldsymbol{X}) = \mathbb{E}(\boldsymbol{\varepsilon}) = 0$ and $\varepsilon_j \perp\!\!\!\perp (X_1, \cdots, X_{j-1})$ for all j. Note that DAG $\mathcal{G}$ with vertex set V and edge set $E = \{(j, k) : \mathbf{B}(j, k) \neq 0\}$ represents the joint distribution F on $\boldsymbol{X}$. The definition of a linear SEM implies that $F(X_j | X_1, \cdots, X_{j-1}) = F(X_j | \boldsymbol{X}_{\mathrm{Pa}(j)})$, which leads to the factorization $F(\boldsymbol{X}) = \prod_{j=1}^{p} F\big(x_j | \boldsymbol{X}_{\mathrm{Pa}(j)}\big)$.

Then, the learning task becomes to learn/estimate the unknown matrix $\mathbf{B}$, from which $\mathcal{G}$ can be recovered (for further connections between DAGs and linear SEMs, see [17]). However, recovery of the true data generating DAG from

$\mathbf{B}$ under the posited assumption of normality of the error term is not unique. Note that $\boldsymbol{X}$ is normally distributed, with covariance $\boldsymbol{\Sigma}$ given by:

$$\boldsymbol{X} \sim \mathcal{N}_p(\mathbf{0}, \boldsymbol{\Sigma}), \quad \text{where} \quad \boldsymbol{\Sigma} = (\mathbf{I} - \mathbf{B})^{-T}\mathbf{D}(\mathbf{I} - \mathbf{B})^{-1}.$$

For notation convenience, we use throughout the paper $A^{-T} \equiv (A^T)^{-1}$ for any square matrix A. Since ex-hypothesis, $\mathbf{B}$ is a strictly upper triangular matrix $\mathbf{I} - \mathbf{B}$ is nonsingular. However, there could exist multiple DAGs and SEMs with different parameters $(\mathbf{B}, \mathbf{D})$ that give rise to the same $\boldsymbol{\Sigma}$. To see this, consider three random variables, where the skeleton of the underlying DAG (no direction of edges) is given by $(X_1 - X_2 - X_3)$. Further, consider the following two DAGs with the same skeleton:

$$\mathcal{G}_1 : X_1 \to X_2 \to X_3. \qquad \boldsymbol{X} = \mathbf{B}_1^T \boldsymbol{X} + \varepsilon_1, \quad \varepsilon_1 \sim N_p(\mathbf{0}, \mathbf{D}_1),$$

$$\mathcal{G}_2 : X_1 \leftarrow X_2 \to X_3. \qquad \boldsymbol{X} = \mathbf{B}_2^T \boldsymbol{X} + \varepsilon_2, \quad \varepsilon_2 \sim N_p(\mathbf{0}, \mathbf{D}_2),$$

where

$$\mathbf{B}_1 = \begin{pmatrix} 0\,1\,0 \\ 0\,0\,1 \\ 0\,0\,0 \end{pmatrix}, \mathbf{D}_1 = \begin{pmatrix} 1\,0\,0 \\ 0\,1\,0 \\ 0\,0\,1 \end{pmatrix},$$

$$\mathbf{B}_2 = \begin{pmatrix} 0 & 0\,0 \\ 0.5 & 0\,1 \\ 0 & 0\,0 \end{pmatrix}, \text{ and } \mathbf{D}_2 = \begin{pmatrix} 0.5 & 0\,0 \\ 0 & 2\,0 \\ 0 & 0\,1 \end{pmatrix}.$$

Then, the covariance matrix of both models is given by

$$\boldsymbol{\Sigma} = (\mathbf{I} - \mathbf{B}_1)^{-T}\mathbf{D}(\mathbf{I} - \mathbf{B}_1)^{-1} = \begin{pmatrix} 1\,1\,1 \\ 1\,2\,2 \\ 1\,2\,3 \end{pmatrix} = (\mathbf{I} - \mathbf{B}_2)^{-T}\mathbf{D}_1(\mathbf{I} - \mathbf{B}_2)^{-1}$$

The upshot is that both DAGs share the same covariance matrix and the same *conditional independence* relationships. The latter fact implies that from observational data, only the *Markov equivalence class* (explicitly defined below) of the underlying DAG can be unambiguously estimated (see [24]), as explained next.

2.2 Completed Partially DAG and the PC Algorithm

We start by further examining and enhancing the example previously presented on DAGs and SEMs, where both DAGs $\mathcal{G}_1$ (chain structure) and $\mathcal{G}_2$ (fork structure) encoded the same conditional independence information. Considering the skeleton $X_1 - X_2 - X_3$ as before, the Table below presents additional DAGs that share the same skeleton, but some of them share the same conditional independence relationships, while one of them does not.

structure	DAG	conditional independence relationships
v-structure	$X_1 \to X_2 \leftarrow X_3$	$X_1 \perp\!\!\!\perp X_3; X_1 \not\perp\!\!\!\perp X_3 \mid X_2$
chain	$X_1 \to X_2 \to X_3$	$X_1 \not\perp\!\!\!\perp X_3; X_1 \perp\!\!\!\perp X_3 \mid X_2$
chain	$X_1 \leftarrow X_2 \leftarrow X_3$	$X_1 \not\perp\!\!\!\perp X_3; X_1 \perp\!\!\!\perp X_3 \mid X_2$
fork	$X_1 \leftarrow X_2 \to X_3$	$X_1 \not\perp\!\!\!\perp X_3; X_1 \perp\!\!\!\perp X_3 \mid X_2$

It can be seen that another chain (row 3) shares the same conditional independence relationships as $\mathcal{G}_1$ and $\mathcal{G}_2$, whereas a DAG exhibiting a so-called *v-structure* (row 1) does not, and its conditional independence relationships are unique. The four DAGs presented in the Table are called *Markov equivalent* and form a *Markov equivalence class*, comprising of DAGs with both the same skeleton and v-structures [24].

A compact representation of the Markov equivalence class is given by the *completed partially directed acyclic graph* (CPDAG). A CPDAG is a partially directed acyclic graph with the following properties [24]: (1) A Markov equivalence class can be described uniquely by a CPDAG. (2) X_i is a direct cause of X_j, only if $X_i \rightarrow X_j$; X_i is a possibly indirect cause of X_j if there is no directed path from X_i to X_j. (3) Every directed edge in the CPDAG implies that there are directed edges in all DAGs in the Markov equivalence class. (4) Every undirected edge in the CPDAG implies that there are directed edges of the form $X_i \rightarrow X_j$ in some DAGs, and $X_i \leftarrow X_j$ in other DAGs in the Markov equivalence class. The CPDAG of a DAG G is denoted by $\mathcal{C}(G)$.

An illustration of a CPDAG is given in Fig. 1. The left three panels show DAGs in the same Markov equivalence class, because they share the same v-structure and the same skeleton (panel four). This Markov equivalence class is represented by the CPDAG in the rightmost panel. On the other hand, an examination of the CPDAG in the right panel, shows that no other v-structure exists. Although the direction for edges $X_2 - X_1 - X_3$ is unspecified, DAGs with the following v-structure $X_2 \rightarrow X_1 \leftarrow X_3$ are not members of the Markov equivalence class under consideration.

A popular algorithm for estimating the CPDAG from observational data is the PC-algorithm [23,24], under the additional *faithfulness* assumption. The latter says that a probability distribution F is faithful with respect to a graph $\mathcal{G}$, if its conditional independence relationships can be inferred from so-called d-separation in the graph $\mathcal{G}$, and vice-versa. It was established in [10] that the PC-algorithm can reconstruct consistently the CPDAG of sparse large DAGs. However, the output of the original version of the algorithm depends on the order in which relationships between nodes are examined; an order-free variant was developed in [6].

The input to the PC-algorithm comprises of conditional independence information amongst all nodes, while its main steps are. (1) Identification of the graph skeleton, as well as its separation sets (i.e. sets of nodes that determine conditional independence between pairs of nodes). (2) Assign direction to triples of nodes in the skeleton exhibiting a v-structure based on the separation sets. (3) Assign direction to as many of the remaining undirected edges as possible, to ensure that no new v-structure and no directed cycle in the estimated CPDAG is present.

In summary, the SEM framework is dependent on the ordering of the nodes/variables, which is not available in most applications. The CPDAG framework is readily applicable, but in practice works well for very sparse DAGs. Next, we discuss our proposed framework that assumes that partial ordering information on subsets of the nodes and employs both SEMs and estimation of CPDAGs for each node subset. To convey the main ideas, we start with the case of a two-set partition of the node set V.

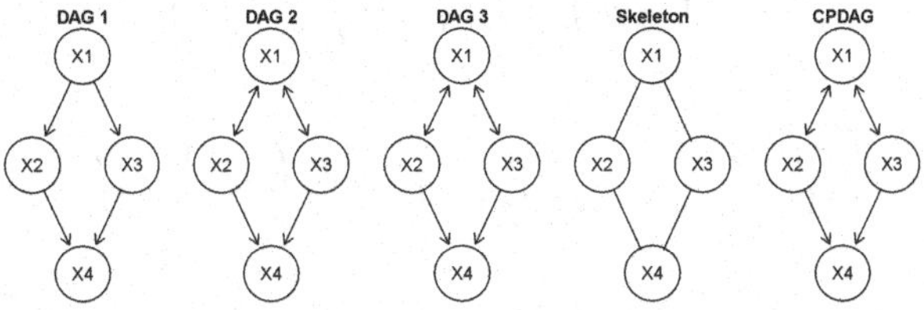

Fig. 1. Members of the Markov equivalence class of 4-node DAGs (panels 1–3) with the skeleton given in panel 4 and its CPDAG representation in panel 5.

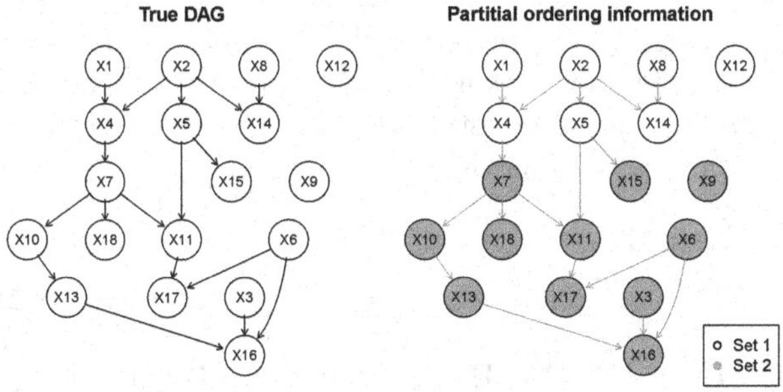

Fig. 2. Illustration of partial ordering information on a DAG: the left panel depicts the true underlying DAG, while the right panel partitions the nodes into sets S_1 (open circles) and S_2 (filled circles), that correspond to DAGs as well.

2.3 Linear SEMs for Two-Set Partitions with Partial Ordering Information

Let $S_1, S_2 \subset V$ and $V = S_1 \cup S_2, S_1 \cap S_2 = \emptyset$. Further, assume that in any topological ordering of V, nodes in S_1 precede those in S_2 and denoted by $S_1 \prec S_2$, which defines a *partial ordering* on the node set V. An illustration is given in Fig. 2, wherein the left panel depicts the underlying DAG comprising of 18 nodes. In the right panel, the two sets S_1 (open circles) and S_2 (filled circles) are shown. This partial prior knowledge implies that members of S_1 can be potential parents of members in S_2, but not the other way around. Note that [22] assume that the node set is partitioned into p singleton sets with $S_1 \prec \ldots \prec S_p$, which induces a complete topological ordering. As is well known [7], there exist many such complete orderings; e.g. for the DAG in Fig. 2, both $X_1 \prec X_2 \prec X_3 \prec \ldots X_{18}$ and $X_1 \prec X_2 \prec X_8 \prec X_{12} \prec X_4 \ldots \prec X_3 \prec X_{16}$ are valid. In the presence of more than two subsets $S_k, k \geq 2$ the same may occur for partial orderings. This issue is further examined in our numerical work in Sect. 3. Using the framework of linear SEMs and incorporating the partial ordering information, we get:

$$\boldsymbol{X} = \begin{pmatrix} \boldsymbol{X}^1 \\ \boldsymbol{X}^2 \end{pmatrix} = \mathbf{B}' \begin{pmatrix} \boldsymbol{X}^1 \\ \boldsymbol{X}^2 \end{pmatrix} + \begin{pmatrix} \boldsymbol{\varepsilon}^1 \\ \boldsymbol{\varepsilon}^2 \end{pmatrix}$$

where

$$\mathbf{B} = \begin{bmatrix} \mathbf{B}_{11} & \mathbf{B}_{12} \\ \mathbf{0} & \mathbf{B}_{22} \end{bmatrix}, \quad \boldsymbol{\varepsilon} = \begin{pmatrix} \boldsymbol{\varepsilon}^1 \\ \boldsymbol{\varepsilon}^2 \end{pmatrix} \sim \mathcal{N}_p \left(\mathbf{0}, \begin{bmatrix} \mathbf{D}_1 & \mathbf{0} \\ \mathbf{0} & \mathbf{D}_2 \end{bmatrix} \right).$$

That is

$$\begin{aligned} \boldsymbol{X}^1 &= \mathbf{B}_{11}' \boldsymbol{X}^1 + \boldsymbol{\varepsilon}^1, && \text{with } \boldsymbol{\varepsilon}^1 \sim \mathcal{N}_{p_1}(\mathbf{0}, \mathbf{D}_1) \\ \boldsymbol{X}^2 &= \mathbf{B}_{12}' \boldsymbol{X}^1 + \mathbf{B}_{22}' \boldsymbol{X}^2 + \boldsymbol{\varepsilon}^2 && \text{with } \boldsymbol{\varepsilon}^2 \sim \mathcal{N}_{p_2}(\mathbf{0}, \mathbf{D}_2) \end{aligned}$$

and diagonal covariance matrices $\mathbf{D}_i, i = 1, 2$. The upper block diagonal nature of the coefficient matrix $\mathbf{B}$ reflects the information that $S_1 \prec S_2$.

Note that block matrices $\mathbf{B}_{ii}$ represent the DAGs $\mathcal{G}_i, i = 1, 2$ for the partitions S_1, S_2, respectively, while the block matrix $\mathbf{B}_{12}$ captures the connections between S_1 to S_2. Hence, their consistent estimation would enable us to recover the full DAG $\mathcal{G}$. An equivalent way of writing the model is:

$$\begin{aligned} \boldsymbol{X}^1 &\sim \mathcal{N}_{p_1}(\mathbf{0}, \boldsymbol{\Sigma}_1) \\ \boldsymbol{X}^2 &= \mathbf{W}' \boldsymbol{X}^1 + \boldsymbol{U}, \text{ and } \quad \boldsymbol{U} = (\mathbf{I} - \mathbf{B}_{22})^{-T} \boldsymbol{\varepsilon}^2 \sim \mathcal{N}_{p_2}(\mathbf{0}, \boldsymbol{\Sigma}_u) \end{aligned}$$

where $\boldsymbol{\Sigma}_1 = (\mathbf{I} - \mathbf{B}_{11})^{-T} \mathbf{D}_1 (\mathbf{I} - \mathbf{B}_{11})^{-1}$ and $\mathbf{W} = \mathbf{B}_{12}(\mathbf{I} - \mathbf{B}_{22})^{-1}$. It can be seen that the structure of DAG $\mathcal{G}_1$ can be recovered by estimating the CPDAG encoded by the information in the covariance matrix $\boldsymbol{\Sigma}_1$. This can be accomplished by employing the PC-algorithm, under appropriate assumptions on sparsity and faithfulness of the corresponding distribution. Edges between S_1 and S_2 can be recovered through the multivariate regression of the $\boldsymbol{X}^2$ block on the $\boldsymbol{X}^1$ block, assuming that the coefficient matrix $\mathbf{W}$ and the covariance matrix $\boldsymbol{\Sigma}_u = (I - \mathbf{B}_{22})^{-T} \mathbf{D}_2 (I - \mathbf{B}_{22})^{-1}$ are also adequately sparse. The procedure used corresponds to a penalized regression, as outlined in [14]. Finally, note that

$$\boldsymbol{U} = \mathbf{B}_{22} \boldsymbol{U} + \boldsymbol{\varepsilon}^2, \text{ where } \boldsymbol{\varepsilon}^2 \sim \mathcal{N}_{p_2}(\mathbf{0}, \mathbf{D}_2)$$

so the structure of DAG $\mathcal{G}_2$ can be recovered by estimating the CPDAG encoded by the information in the covariance matrix $\boldsymbol{\Sigma}_u$, employing again the PC-algorithm.

Further, note that the influence of parent nodes $\boldsymbol{X}^1$ on child nodes $\boldsymbol{X}^2$ is described in the coefficient matrix

$$\mathbf{W} = \mathbf{B}_{12}(\mathbf{I} - \mathbf{B}_{22})^{-1} = \mathbf{B}_{12} + \mathbf{B}_{12}(\mathbf{B}_{22} + \ldots + \mathbf{B}_{22}^{p_2 - 1}).$$

It can be seen that the coefficient matrix $\mathbf{W}$ is composed of two parts: $\mathbf{B}_{12}$ is the direct effect of members of S_1 on those in S_2, while $\mathbf{B}_{12}(\mathbf{B}_{22} + \ldots + \mathbf{B}_{22}^{p_2 - 1})$

is the indirect effect from nodes in S_1 to those in S_2. As established in Sect. 2.3, the support of $\mathbf{W}$ is a superset of the support of $\mathbf{B}_{12}$. Since we are interested in the support of $\mathbf{B}_{12}$ (edges from S_1 to S_2), an adjustment step is introduced to correct the estimate $\widehat{\mathbf{W}}$, as detailed next.

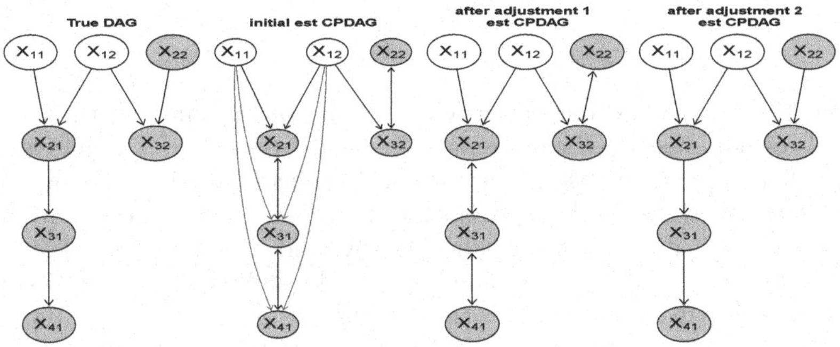

Fig. 3. Illustration of adjustment steps: (a) True DAG, and partitions S_1 (open circles) and S_2 (filled circles). (b) An estimate of CPDAG based on linear SEMs and application of the PC algorithm; the additional blue arrows represent the indirect effect from nodes in S_1 to those in S_2. (c) Updated CPDAG, after first adjustment step. (d) Updated CPDAG, after second adjustment step. (Color figure online)

Adjustment Steps: There are two issues when estimating the CPDAG using the linear SEM framework, coupled with application of the PC-algorithm. To illustrate them, we focus on the example depicted in Fig. 3. Panel 3(a) depicts the true DAG, with the two subsets S_1, S_2 shown by the open and filled circles, respectively. We start by focusing on the chain $X_{11} \to X_{21} \to X_{31} \to X_{41}$ in the true DAG, with $X_{11} \in S_1$, and $X_{21}, X_{31}, X_{41} \in S_2$. Since we regress X_{31} on nodes in S_1, without considering its true parent node X_{21}, there will be an edge from nodes X_{11}, X_{12}, as indicated in the estimated CPDAG depicted in panel 3(b); analogously, for the case of X_{41}. Hence, edges $X_{11} \to X_{31}, X_{12} \to X_{31}, X_{11} \to X_{41}, X_{12} \to X_{41}$ are present in $\mathbf{W}$, but not in $\mathbf{B}_{12}$. Subsequently, when the PC-algorithm is applied to the nodes in set S_2, and assuming strong enough signal, it will identify the following edges $X_{21} \leftrightarrow X_{31} \leftrightarrow X_{41}$ in CPDAG $\mathcal{C}(\mathcal{G}_2)$. Hence, when taking the union of edges discovered in $\mathcal{C}(\mathcal{G}_1), \mathcal{C}(\mathcal{G}_2)$ and through regression, we obtain the estimated CPDAG depicted in panel 3(b).

The first adjustment step considered corresponds to regressing nodes in S_2 to the union of parent nodes identified by the regression procedure and in $\mathcal{C}(\mathcal{G}_2)$. As an example, consider node X_{31}, which is regressed on predictors $X_{11}, X_{12}, X_{21}, X_{41}$; the first two specified through the partial ordering and the last two identified as candidates in $\mathcal{C}(\mathcal{G}_2)$. Assuming again adequate strength of signal, since X_{21} is the parent of X_{31} the edge from ancestors X_{11} and X_{12} will be eliminated. An analogous regression procedure, should eliminate edges between nodes in S_1 and X_{41}. In summary, the first adjustment step comprises of fitting regression models where the response are the nodes in S_2 and the predictors their *connected* parent nodes in $adj(\widehat{\mathcal{C}_w}(\mathcal{G}), X_j)$. The estimated adjusted "CPDAG" is shown in panel 3(c).

However, the result may not be a proper CPDAG and the second adjustment step addresses this issue. As an illustration, note that in panels 3(b) and (c), the triplets (X_{12}, X_{32}, X_{22}) and (X_{11}, X_{21}, X_{31}) have the following relationships:

$$X_{12} \rightarrow X_{32} \leftrightarrow X_{22} \text{ and } X_{11} \rightarrow X_{21} \leftrightarrow X_{31}.$$

However, they should exhibit either a v-structure or a chain structure. This is the result of not checking for existence of v-structures between members in S_1 and S_2. Hence, to resolve the structure for bi-directional edges X_j and X_k in S_2, we check whether a node X_i exists, such that $X_i \rightarrow X_j$ in $\mathbf{B}_{12}$, but at the same time there is not edge present between X_i and X_k. If such an edge exists, we test whether X_i and X_k are conditionally independent given X_j. If they are not conditionally independent, we conclude there is a v-structure. After determining the v-structures, the remaining undirected edges (if any) are updated, by repeated application of the rules in the PC algorithm to ensure absence of any remaining v-structure. In the illustrative example, after the second adjustment step, we will get $X_{12} \rightarrow X_{32} \leftarrow X_{22}$ and $X_{11} \rightarrow X_{21} \rightarrow X_{31}$, when checking for v-structures. Then, upon updating the remaining undirected edges to ensure absence of any new v-structures in S_2, we obtain $X_{21} \rightarrow X_{31} \rightarrow X_{41}$. Thus, the final result is shown in panel 3(d) that recovers the true DAG.

The full procedure is summarized in the Algorithm.

2.4 Extension to *K*-set Partitions

We assume that there exists a partition of V into K non-overlapping sets $S_1, \cdots, S_K$ and in addition, a complete topological ordering on them is available (which translates to partial ordering information on the nodes); specifically, $S_1 \prec S_2 \prec \ldots. \prec S_K$. The edges (interactions) between sets can be estimated by the following multivariate regressions:

$$\begin{aligned} \boldsymbol{X}_1 &= \boldsymbol{U}_1, \qquad \text{where } \boldsymbol{U}_1 \sim \mathcal{N}_2(\mathbf{0}, \boldsymbol{\Sigma}_1) \\ \boldsymbol{X}_2|\boldsymbol{X}_1 &= \mathbf{B}'_{12}\boldsymbol{X}_1 + \boldsymbol{U}_2, \qquad \text{where } \boldsymbol{U}_2 \sim \mathcal{N}_2(\mathbf{0}, \boldsymbol{\Sigma}_2) \\ &\cdots \\ \boldsymbol{X}_K|\boldsymbol{X}_1, \ldots, \boldsymbol{X}_{K-1} &= \mathbf{B}'_{1K}\boldsymbol{X}_1 + \ldots + \mathbf{B}'_{K-1,K}\boldsymbol{X}_{K-1} + \boldsymbol{U}_K, \end{aligned}$$

where $\boldsymbol{U}_K \sim \mathcal{N}_2(\mathbf{0}, \boldsymbol{\Sigma}_K)$.

After this step, the remaining quantities can be obtained as follows:

- $\boldsymbol{\Sigma}_1$ reflects the DAG within S_1 and an application of the PC-algorithm will yield the corresponding CPDAG $\mathcal{C}(\mathcal{G}_1)$.
- $\mathbf{B}_{ik}, \forall 1 < i < k \leq K$ contain directed edges from nodes in S_i to nodes in S_k.
- $\boldsymbol{\Sigma}_k, k = 1, \cdots, K$ reflect the DAG within S_k and an application of the PC-algorithm will yield the corresponding CPDAGs $\mathcal{C}(\mathcal{G}_k)$.

Algorithm: for a Two-Set Partition

```
Input : Data from predictors X^1 ∈ S_1 and responses X^2 ∈ S_2.
Estimate sub-DAG in parent layer:
    Apply PC-algorithm to the conditional independence information among
    variables in S_1 . Obtain Ĉ(G_1).
Obtain W and Ĉ(G_2):
    Regress X^2 on X^1 using procedure in [14] to get an estimate Ŵ and
    residuals data set U.
    Apply PC-algorithm to Σ_U and obtain the estimator of adjacency
    matrix in response set, Ĉ(G_2)
Adjustment step
  Regression step
    Let Ĉ_W(G) be the union of Ĉ(G_1), Ŵ, Ĉ(G_2).
    for j = 1,...,p do
        Obtain all parents of X_j according to Ĉ_W(G), X_j^pa,
        Regress X_j on X_j^pa, obtain the corresponding vector of p-values,
        and update the j-th column of Ĉ_W(G)
    end
  Updating CPDAG
    Let Ĉ_o(G) = Ĉ_w(G)
    Find new v-structure and update Ĉ_w(G): Check whether any bi-
    directional edges in Ĉ_w(G) could be determined.
    Let Ĉ_1(G) = Ĉ_w(G).
    while Ĉ_o(G) does not equal to Ĉ_1(G) do
        Ĉ_o(G) = Ĉ_1(G).
        For bi-directional edges, check that no new v-structure are present
        and update Ĉ_1
    end
Output: Final estimate of CPDAG: Ĉ_1(G)
```

As before, the $\mathbf{B}_{ik}$ will be estimated based on penalized lasso regressions outlined in [14] and obtain the corresponding residuals $\widehat{U}_k$. An application of the PC-algorithm on $\boldsymbol{\Sigma}_{U_k}$ will obtain initial estimates of $\mathcal{C}(\mathcal{G}_k)$. As previously discussed, we need to undertake the two adjustment steps for obtaining the final estimate of the DAG. They include: (i) running regressions, recursively for nodes in set S_k to all the nodes that are connected to them in sets $S_\ell, \ell \leq k$ in reverse order starting from set S_K and proceeding backwards to set S_2; and (ii) applying the CPDAG rules for finding all remaining v-structures.

3 Performance Evaluation

Next, we evaluate the performance of the proposed framework (denoted in the Tables as *Partial*) based on topologies extracted from the DREAM3 competition. We consider the following competing methods: (I) PC-algorithm with prior information denoted in the Tables as *PC-Prior* and (II) penalized regression based on a complete topological ordering (see [22]) denoted as *Complete.* Note that application of the PC-algorithm, yielded non-competitive results and hence the PC-Prior variant, described next, was used.

(I) PC-Prior: An application of the original PC-algorithm yields an estimate of the CPDAG. We then use the prior information contained in the partial orderings, together with the rules consistent with a CPDAG to orient bi-directional edges.
(II) Complete: In this case, we assume that a complete topological ordering on the nodes $X_1 \prec X_2 \prec \cdots \prec S_p$ is available and use penalized regression obtain an estimate of the DAG.

The performance metrics used include precision (true positive edges/(true positive + false positive edges), recall (true positive edges/(true positive + false negative edges) and the Matthews Correlation Coefficient (MCC) that takes values between +1 and −1 (higher values signify better performance) and is given by:

$$\text{MCC} = \frac{TP \times FP - TN \times FN}{\sqrt{(TP + FP)(TP + FN)(TN + FP)(TN + FN)}},$$

where TP = true positive edges, FP = false positive, TN = true negative, FN = false negative, respectively.

The DREAM (Dialogue on Reverse Engineering Assessments and Methods) project is a community-wide effort to assess the relative strengths and weaknesses of different computational methods for a set of core problems in systems biology. Figure 4 depicts four DAGs comprising of 50 nodes from the third annual DREAM competition (DREAM3). Two of them were extracted from the Ecoli regulatory network (Ecoli 1 and Ecoli2), while the remaining three from the Yeast regulatory network (Yeast 1 and Yeast 2). As can be seen in the Figure, they differ significantly in terms of their density and complexity. For example, the Ecoli DAGs have a clear layered structure that enables us to get an informative 2-set partition, which is not the case for Yeast. Further, the Yeast DAGs have much longer chains. On the other hand, there are a number of nodes with many descendants in Ecoli, which if assigned to the S_1 subset, will be very informative to the Partial method. The data used correspond to the one labeled "Null-mutants" and contain one sample for steady state levels of the genes for wild type cells and one sample where each of the genes has been knocked out, for a total of 51 samples.

Table 1 shows the results and the Partial method outperforms its competitors for the Ecoli DAG topologies, while it exhibits a better performance by a very thin margin on the Yeast topologies. On the other hand, the PC-Prior method lags behind across all topologies. Once again, the plain PC-algorithm does not prove competitive and thus its results are not included.

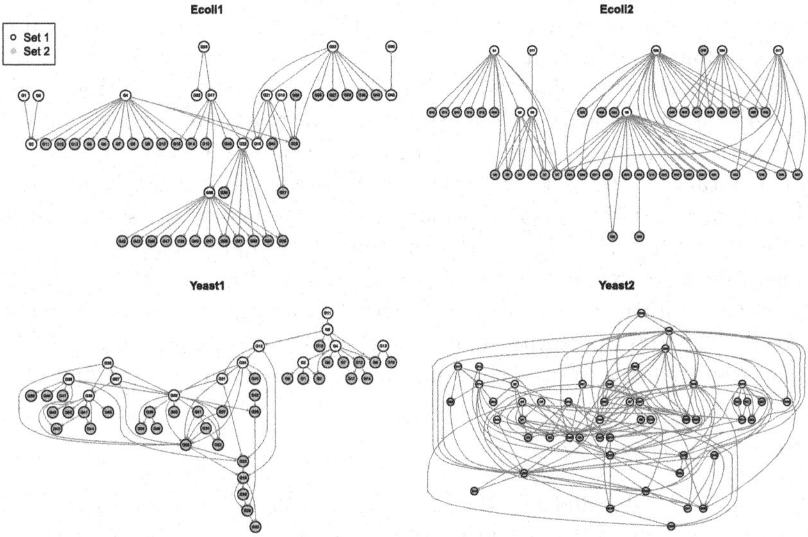

Fig. 4. DAGs from the DREAM3 competition, including the partial ordering partition information used in the numerical work; S_1 shown as open circles and S_2 as filled circles.

Table 1. Results based on DREAM3 data obtained from 51 null-mutant samples.

DREAM3 networks with $p = 50$ nodes and $n = 51$ null-mutant samples						
Method	Ecoli1			Ecoli2		
	2-set partition: 15, 35			2-set partition: 8, 42		
	Precision	Recall	MCC	Precision	Recall	MCC
Partial	0.743	0.419	0.542	0.788	0.500	0.608
PC-Prior	0.636	0.339	0.445	0.553	0.256	0.348
Complete	0.410	0.516	0.428	0.565	0.317	0.394
Method	Yeast1			Yeast2		
	2-set partition: 13, 37			2-set partition: 23, 27		
	Precision	Recall	MCC	Precision	Recall	MCC
Partial	0.619	0.338	0.432	0.730	0.169	0.314
PC-Prior	0.550	0.286	0.369	0.429	0.244	0.250
Complete	0.565	0.338	0.409	0.590	0.225	0.312

4 Concluding Remarks

We introduced computational methodology that leverages prior information about partial orderings of the nodes in the DAG. The extensive numerical work undertaken shows that such information is in general beneficial for reconstructing DAGs from observational data, since both our proposed method and our modification of the popular PC-algorithm demonstrate. Nevertheless, the Partial method leverages the partial ordering information in a more systematic way and hence obtains better performance that the modified PC-algorithm. As mentioned in the introduction, such information is available in many settings. Extensive additional numerical work both on simulated random network topologies and simulated measurements and on the DREAM3 topologies with simulated measurements confirm the results of the brief evaluation section; namely, that the Partial method outperforms its competitors by a wide margin for topologies with clear hierarchies and by a thin margin for those with more complex ones.

References

1. Anchang, B., et al.: Modeling the temporal interplay of molecular signaling and gene expression by using dynamic nested effects models. Proc. Nat. Acad. Sci. **106**(16), 6447–6452 (2009)
2. Andersson, S.A., Madigan, D., Perlman, M.D.: Alternative markov properties for chain graphs. Scand. J. Stat. **28**(1), 33–85 (2001)
3. Aragam, B., Zhou, Q.: Concave penalized estimation of sparse Gaussian Bayesian networks. J. Mach. Learn. Res. **16**, 2273–2328 (2015)
4. Bühlmann, P., Peters, J., Ernest, J., et al.: CAM: causal additive models, high-dimensional order search and penalized regression. Ann. Stat. **42**(6), 2526–2556 (2014)
5. Champion, M., Picheny, V., Vignes, M.: Inferring large graphs using ℓ_1-penalized likelihood. Stat. Comput. **28**(4), 905–921 (2018)
6. Colombo, D., Maathuis, M.H.: Order-independent constraint-based causal structure learning. J. Mach. Learn. Res. **15**(1), 3741–3782 (2014)
7. Cormen, T.H., Leiserson, C.E., Rivest, R.L., Stein, C.: Introduction to Algorithms. MIT Press, Cambridge (2009)
8. Fu, F., Zhou, Q.: Learning sparse causal Gaussian networks with experimental intervention: regularization and coordinate descent. J. Am. Stat. Assoc. **108**(501), 288–300 (2013)
9. Gu, J., Fu, F., Zhou, Q.: Penalized estimation of directed acyclic graphs from discrete data, March 2014. https://arxiv.org/abs/1403.2310
10. Kalisch, M., Bühlmann, P.: Estimating high-dimensional directed acyclic graphs with the PC-algorithm. J. Mach. Learn. Res. **8**(Mar), 613–636 (2007)
11. Kalisch, M., Mächler, M., Colombo, D., Maathuis, M.H., Bühlmann, P., et al.: Causal inference using graphical models with the R package pcalg. J. Stat. Softw. **47**(11), 1–26 (2012)
12. Lee, T.I., et al.: Transcriptional regulatory networks in Saccharomyces cerevisiae. Science **298**(5594), 799–804 (2002)
13. Liang, X., Young, W.C., Hung, L.H., Raftery, A.E., Yeung, K.Y.: Integration of multiple data sources for gene network inference using genetic perturbation data. J. Comput. Biol. **26**, 1113–1129 (2019)

14. Lin, J., Basu, S., Banerjee, M., Michailidis, G.: Penalized maximum likelihood estimation of multi-layered Gaussian graphical models. J. Mach. Learn. Res. **17**(146), 1–51 (2016)
15. Markowetz, F., Kostka, D., Troyanskaya, O.G., Spang, R.: Nested effects models for high-dimensional phenotyping screens. Bioinformatics **23**(13), i305–i312 (2007)
16. Ni, Y., Stingo, F.C., Baladandayuthapani, V.: Bayesian nonlinear model selection for gene regulatory networks. Biometrics **71**(3), 585–595 (2015)
17. Peters, J.M.: Restricted structural equation models for causal inference. Ph.D. thesis, ETH Zurich (2012)
18. Pinna, A., Soranzo, N., De La Fuente, A.: From knockouts to networks: establishing direct cause-effect relationships through graph analysis. PLoS ONE **5**(10), e12912 (2010)
19. Qin, S., Ma, F., Chen, L.: Gene regulatory networks by transcription factors and micrornas in breast cancer. Bioinformatics **31**(1), 76–83 (2014)
20. Shimizu, S., Hoyer, P.O., Hyvärinen, A., Kerminen, A.: A linear non-Gaussian acyclic model for causal discovery. J. Mach. Learn. Res. **7**(Oct), 2003–2030 (2006)
21. Shojaie, A., Jauhiainen, A., Kallitsis, M., Michailidis, G.: Inferring regulatory networks by combining perturbation screens and steady state gene expression profiles. PLoS ONE **9**(2), e82393 (2014)
22. Shojaie, A., Michailidis, G.: Penalized likelihood methods for estimation of sparse high-dimensional directed acyclic graphs. Biometrika **97**(3), 519–538 (2010)
23. Spirtes, P., Glymour, C., Scheines, R.: Causation, Prediction, and Search. Lecture Notes in Statistics. Springer, New York (1993). https://doi.org/10.1007/978-1-4612-2748-9
24. Spirtes, P., Glymour, C.N., Scheines, R.: Causation, Prediction, and Search. MIT Press, Cambridge (2000)

Learning Scale and Shift-Invariant Dictionary for Sparse Representation

Toshimitsu Aritake(✉) and Noboru Murata

Graduate School of Advanced Science and Engineering, Waseda University, Tokyo, Japan
toshimitsu.aritake@ruri.waseda.jp, noboru.murata@eb.waseda.ac.jp

Abstract. Sparse representation is a signal model to represent signals with a linear combination of a small number of prototype signals called atoms, and a set of atoms is called a dictionary. The design of the dictionary is a fundamental problem for sparse representation. However, when there are scaled or translated features in the signals, unstructured dictionary models cannot extract such features. In this paper, we propose a structured dictionary model which is scale and shift-invariant to extract features which commonly appear in several scales and locations. To achieve both scale and shift invariance, we assume that atoms of a dictionary are generated from vectors called ancestral atoms by scaling and shift operations, and an algorithm to learn these ancestral atoms is proposed.

Keywords: Sparse coding · Dictionary learning · Scale-invariance · Shift-invariance

1 Introduction

The past decade has witnessed a growing interest in the sparse representations of signals. Sparse representation is a signal model to represent signals with a linear combination of a small number of prototype signals called *atoms*, and a set of atoms is called a *dictionary*. Sparse representation of signals has been successfully applied to numerous problems such as denoising, super-resolution, and compressed sensing.

Sparse coding is a widely used method to obtain sparse representation and is formulated as follows. Let $\boldsymbol{y} \in \mathbb{R}^n$ be a signal whose size is n and $\boldsymbol{D} = (\boldsymbol{d}_1, \boldsymbol{d}_2, \cdots, \boldsymbol{d}_k) \in \mathbb{R}^{n \times k}$ be an over-complete dictionary which consists of k different atoms, where k is much larger than n. Given a signal $\boldsymbol{y}$ and the dictionary $\boldsymbol{D}$, the problem of sparse coding is to find a sparse linear combination of atoms that best represents the signal. Formally, the problem is to find a coefficients vector $\boldsymbol{x} = (x_1, x_2, \ldots, x_k)$ which best represents the signal $\boldsymbol{y}$ as $\boldsymbol{y} \simeq \sum_{i=1}^{k} x_i \boldsymbol{d}_i$. In this paper, we solve the following problem for sparse coding,

$$\underset{\boldsymbol{x}}{\text{minimize}} \, \|\boldsymbol{y} - \boldsymbol{D}\boldsymbol{x}\|_2^2 + \lambda \|\boldsymbol{x}\|_1, \tag{1}$$

G. Nicosia et al. (Eds.): LOD 2019, LNCS 11943, pp. 472–483, 2019.
https://doi.org/10.1007/978-3-030-37599-7_39

where $\|\boldsymbol{x}\|_1 = \sum_{j=1}^{k} |x_j|$ and $\lambda > 0$ is a regularization parameter to control sparsity of $\boldsymbol{x}$. Equation (1) is an ℓ_1 regularized least square problem and is known as lasso (least absolute shrinkage and selection operator) [1] or basis pursuit [2]. A number of algorithms have been proposed to solve this problem efficiently (see [3] and references therein).

An important issue of sparse coding is the choice of the dictionary $\boldsymbol{D}$ because it significantly affects the quality of overall signal processing. A learned dictionary is more widely used than prespecified dictionaries such as wavelets in the wide range of applications because of its better performance. The problem to learn a dictionary from a set of signals is called *dictionary learning*, and various algorithms such as gradient-based algorithm [4], method of optimal directions (MOD) [5], K-SVD [6] and online dictionary learning [7] have been proposed in the literature. One reason for the success of the dictionary learning is that the learned atoms can be considered as the proper features which characterize given signals.

However, the model of the dictionary which has no specific structure does not consider the scaling and shift of features explicitly. Therefore, when unique features appear at various scales and locations, such features cannot be learned by unstructured dictionary learning.

To learn such features, a scale and shift-invariant structured dictionary model should be considered. So far, several scale or shift-invariant dictionary models have been proposed to extract such features.

Shift-invariant dictionary [8] considers the shift structure of the dictionary. The atoms of the dictionary are translated template atoms. Therefore, learned features are invariant to a small shift of signals.

The multiscale K-SVD [9] learns several scale specific dictionaries where the scale of the dictionary corresponds to the size of signal patches. Namely, scaled features have a smaller size than the signals, hence the scaled features have limited non-zero elements in the n-dimensional signal space. The model of the dictionary assumes that there are different features in the different scales. The hierarchical multiscale dictionary [10] is an extension of the multiscale K-SVD. They considered the additional constraint for sparse coding so that the learned atoms have hierarchical relationships implicitly.

Another approach is to optimize filters of an orthogonal discrete wavelet transformation [11]. The dictionary follows the pyramid structure of multiresolution analysis and the atoms of the dictionary are generated by dilation-shift and recursive upsampling and interpolation along each dimension. A similar approach proposed earlier by [12] learns multiple wavelet filters instead of learning an orthogonal wavelet dictionary. However, features learned by these methods are less flexible, since the atoms are generated by upsampling and these models have a small number of learnable parameters.

Our previous work also presented a scale and shift-invariant dictionary [13]. We assumed that the atoms of a dictionary are generated from a single vector, which we call an *ancestral atom* or an *ancestor* for short, using downsampling and sliding window, and an algorithm for learning the ancestor was proposed.

We also assumed that the size of an ancestor is larger than the size of the training signals, hence features of the given signals are learned as the parts of the ancestor. The dictionary generation by downsampling allows us to have more learnable parameters than the optimized wavelets, thus more flexible features can be learned. However, the support of the atoms was not changed for each scale as the multiscale K-SVD or wavelets.

In this paper, we propose another model of scale and shift-invariant dictionary whose atoms are generated by downsampling and shifting multiple ancestors. The atom generation process is similar to our prior work, however, we use multiple ancestors whose dimension is the same as the signal instead of using a single large ancestor. Therefore, scale and shift-invariant features of signals can be obtained as ancestors. Atoms of a dictionary have limited non-zero elements in the signal space as the multiscale K-SVD while considering relationships among different scales explicitly. In addition, our model allows the overlap of the shifted atoms, hence our model is invariant to a smaller shift of features. Therefore, scale and shift-invariant features can be learned as ancestors flexibly.

Extracting scale and shift-invariant features are important when certain objects or features appear at several scales and positions in the signals. A possible application of learned scale and shift-invariant features is classification of time-series data or images [14]. Since the unstructured dictionary model is sensitive to small translation or scaling, using scale and shift-invariant features is beneficial to extract features from signals.

The rest of the paper is organized as follows. In Sect. 2, we briefly review classical dictionary learning models. Then, an atom generation process from ancestors is explained and a simple gradient-based optimization algorithm for learning ancestors is presented. In Sect. 3, the experimental results are shown, and we conclude and discuss further investigations of our problem in Sect. 4.

2 Problem Formulation

2.1 Dictionary Learning

Let $\boldsymbol{Y} = (\boldsymbol{y}_1, \boldsymbol{y}_2, \dots, \boldsymbol{y}_m) \in \mathbb{R}^{n\times m}$ be a finite set of signals where we denote the j-th column of $\boldsymbol{Y}$ as $\boldsymbol{y}_j$ $(j = 1, 2, \dots m)$. Given a finite set of signals $\boldsymbol{Y}$, dictionary learning finds the over-complete dictionary $\boldsymbol{D} \in \mathbb{R}^{n\times k}$ and the sparse coefficients $\boldsymbol{X} = (\boldsymbol{x}_1, \boldsymbol{x}_2, \dots, \boldsymbol{x}_m) \in \mathbb{R}^{k\times m}$ which best represents $\boldsymbol{Y}$ where $\boldsymbol{x}_j$ is the coefficient vector corresponds to the signal $\boldsymbol{y}_j$. The problem of dictionary learning is formulated as follows,

$$\text{minimize}_{\boldsymbol{D},\boldsymbol{X}} \ \|\boldsymbol{Y} - \boldsymbol{D}\boldsymbol{X}\|_F^2 + \lambda\|\boldsymbol{X}\|_1 \tag{2}$$

where $\|\boldsymbol{Y} - \boldsymbol{D}\boldsymbol{X}\|_F^2 = \sum_{j=1}^m \|\boldsymbol{y}_j - \boldsymbol{D}\boldsymbol{x}_j\|_2^2$ is a Frobenius norm of a matrix and $\|\boldsymbol{X}\|_1 = \sum_{j=1}^m \|\boldsymbol{x}_j\|_1$. However, Eq. (2) is not a jointly convex problem with respect to two variables $\boldsymbol{D}$ and $\boldsymbol{X}$. Therefore, most of the dictionary learning algorithms solve this problem by alternating two procedures, sparse coding and dictionary update.

In the sparse coding step, the coefficients $\boldsymbol{X}$ are updated by solving Eq. (1) for each pair of $\boldsymbol{y}_j$ and $\boldsymbol{x}_j$ while the temporal dictionary $\boldsymbol{D}$ is fixed. In the dictionary update step, the dictionary $\boldsymbol{D}$ is optimized while the coefficient matrix $\boldsymbol{X}$ is fixed, by solving

$$\underset{\boldsymbol{D}}{\text{minimize}} \, \|\boldsymbol{Y} - \boldsymbol{D}\boldsymbol{X}\|_F^2. \tag{3}$$

This problem is a convex optimization problem and various methods have been proposed to solve this problem.

2.2 Atom Generation from Ancestral Atoms

We explain our model of the dictionary: how atoms of a dictionary are generated from ancestral atoms to construct a scale and shift-invariant dictionary. We consider L ancestors whose size is n, $\boldsymbol{a}_l \in \mathbb{R}^n$ $(l = 1, 2, \ldots, L)$. First of all, the ancestors $\boldsymbol{a}_l$ themselves are used as the atoms of the dictionary. Specifically, the rest of the atoms of the dictionary are generated from the ancestors $\{\boldsymbol{a}_l\}_{l=1,2,\ldots L}$ by downsampling and shift operations. Therefore, the atoms generated from an ancestor are used to express similar patterns of the different scales and locations.

Unlike the scaling of the multiscale K-SVD or wavelets, we consider the non-dyadic downsampling for more flexible scaling. Namely, we consider downsampling of an ancestor whose size is n to any smaller size $n' = n-1, n-2, \ldots, 2$. Downsampling of an ancestor is done as follows. First, the ancestor whose size is n is expanded to $\texttt{lcm}(n, n')$ by repeating each element $\texttt{lcm}(n, n')/n$ times, where $\texttt{lcm}(n, n')$ is the least common multiple of n and n'. Then, the expanded ancestor is downsampled to size n' by taking averages of each adjacent $\texttt{lcm}(n, n')/n'$ elements. These expansion and averaging operations are both linear operations and therefore whole downsampling operation is linear.

Another important difference between our model and the multiscale K-SVD is the modeling of the relationship between atoms of different scales. The multiscale K-SVD just considers the relationships between atoms implicitly by sparse coding, hence the atoms of different scales do not necessarily represent similar patterns of different scales. On the other hand, our model considers the relationships explicitly by downsampling the ancestors. Therefore atoms generated from an ancestor represent similar patterns of different scales. In addition, unlike the multiscale K-SVD and wavelets, our model assumes that the amounts of shifts are not scale-dependent and more flexible shift structure can be considered, hence the generated dictionary is invariant to a translation of signals.

The atom generation process described above can be easily extended to the arbitrary dimension. For example, images are typical 2D signals and movies are 3D signals. When the signals are N-dimensional, we assume that the ancestors are also N-dimensional, and we apply downsampling and shift operations along each dimensional axis of the ancestor to generate atoms. For example, when the signals are 2D signals, we use ancestors whose size is $n \times n$, and these ancestors are downsampled to the size $n' \times n'$. Figure 1 shows the scaling of the 2D signal(image).

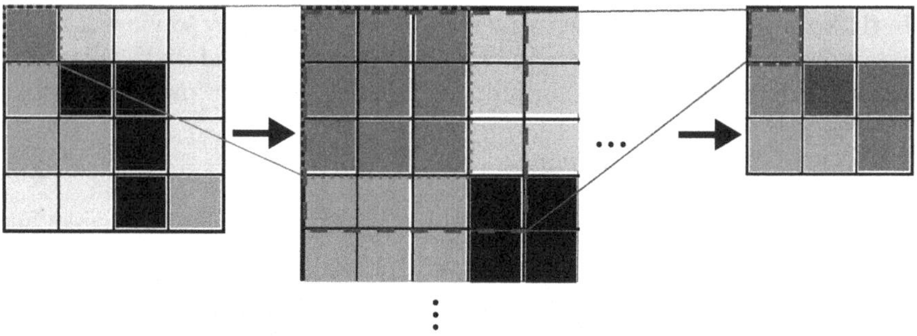

Fig. 1. A conceptual diagram of the scaling of a 4×4 image to a 3×3 image. First, the 4×4 image is expanded to a 12×12 image by repeating each pixel 3×3 times. Then, by taking an average of 4×4 pixels, the 3×3 image is calculated.

However, the downsampled ancestors have the size $n' < n$, hence the downsampled ancestors cannot be used as atoms of the dictionary for sparse coding directly. Therefore, we use zero-padding so that the downsampled ancestor can be used with the original ancestors. Zero-padding is also used for shifting the downsampled ancestors as illustrated in Fig. 2 by simply changing the position of non-zero elements of the zero-padded ancestors.

In this paper, we use the linear downsampling and shift operations, therefore the atom generation operator can be written as a matrix. Here, let $\boldsymbol{F}_{p,q} \in \mathbb{R}^{n \times n}$ be an operator which represents p-th downsampling and q-th shift, where $p = 0, 1, \ldots, P$ is the index of downsampling and $q = 0, 1, \ldots, Q$ is the index of the shift operation. For example, the downsampling of the 1D ancestors whose size is n to the size $n' = \frac{2}{3}n$ and shift of the single element construct the following atom generation operator,

$$\boldsymbol{F}_{1,1} = \begin{pmatrix} 0 & 0 & 0 & 0 \cdots 0 & 0 & 0 & 0 \\ 2/3 & 1/3 & & & & & & \\ & 1/3 & 2/3 & & & & \text{\Large 0} & \\ & & & \ddots & & & & \\ & & & & & 2/3 & 1/3 & \\ & \text{\Large 0} & & & & & 1/3 & 2/3 \\ & & & \text{\Large 0} & & & & \end{pmatrix} \in \mathbb{R}^{n \times n}. \tag{4}$$

Then, the atom generated from the ancestor $\boldsymbol{a}_l$ by $\boldsymbol{F}_{p,q}$ can be written as

$$\boldsymbol{d}_{p,q,l} = \boldsymbol{F}_{p,q}\boldsymbol{a}_l. \tag{5}$$

The set of possible pairs (p, q) is denoted by Λ and the cardinality of Λ is denoted by $|\Lambda|$. Then, the dictionary consists of atoms $\boldsymbol{d}_{p,q,l}$ $((p, q) \in \Lambda)$ can be written as

$$\boldsymbol{D}(\boldsymbol{a}_l) = (\boldsymbol{F}_{0,0}\boldsymbol{a}_l, \ldots, \boldsymbol{F}_{p,q}\boldsymbol{a}_l, \ldots, \boldsymbol{F}_{P,Q}\boldsymbol{a}_l) \in \mathbb{R}^{n \times |\Lambda|} \quad (l = 1, 2, \ldots, L). \tag{6}$$

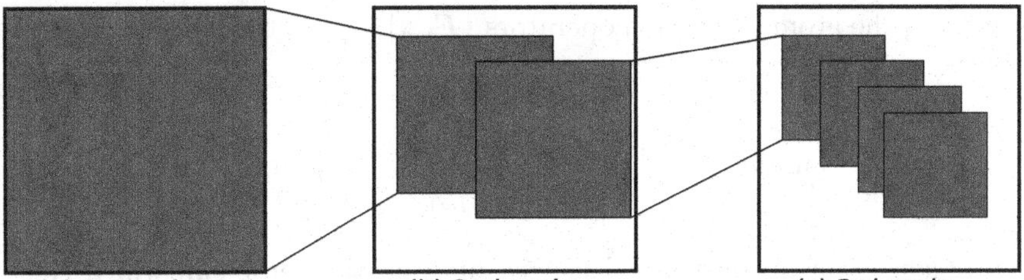

Fig. 2. A conceptual diagram of atom generation from a 2D ancestor. In the 1st scale, the ancestor itself is used as an atom. In the 2nd and 3rd scales, the scaled atom is placed in the empty image and used as an atom, where the white area is zero-padded. The scaled atom is shifted by changing the zero-padded area.

We generate L different dictionaries from each ancestor and the whole dictionary $\boldsymbol{D}(\boldsymbol{a}_1, \boldsymbol{a}_2, \ldots, \boldsymbol{a}_L)$ is constructed by concatenating these matrices, namely,

$$\boldsymbol{D}(\boldsymbol{a}_1, \boldsymbol{a}_2, \ldots, \boldsymbol{a}_L) = (\boldsymbol{D}(\boldsymbol{a}_1) \mid \boldsymbol{D}(\boldsymbol{a}_2) \mid \ldots \mid \boldsymbol{D}(\boldsymbol{a}_L)) \in \mathbb{R}^{n \times L|\Lambda|}. \tag{7}$$

When we use multidimensional signals and ancestors, these signals and atoms vectorized by stacking their columns. Then the atom generation operators described above also need to be modified to process these vectorized signals and ancestors.

2.3 Optimization of the Ancestors

In this subsection, we explain the algorithm to learn ancestors from a given set of signals $\boldsymbol{Y} \in \mathbb{R}^{n \times m}$. The problem of learning ancestors is formulated as follows,

$$\underset{\boldsymbol{a}_1, \ldots, \boldsymbol{a}_L, \boldsymbol{X}}{\text{minimize}} \ \|\boldsymbol{Y} - \boldsymbol{D}(\boldsymbol{a}_1, \ldots, \boldsymbol{a}_L)\boldsymbol{X}\|_F^2 + \|\boldsymbol{X}\|_1, \tag{8}$$

where a column of $\boldsymbol{X}$, $\boldsymbol{x}_i$ consists of coefficients of atoms $\boldsymbol{d}_{p,q,l}$ to represent $\boldsymbol{y}_i$, i.e.

$$\boldsymbol{x}_i = (x_i^{p,q,l})_{(p,q) \in \Lambda,\ l=1,2,\ldots,L} \in \mathbb{R}^{L|\Lambda|}. \tag{9}$$

The problem (8) is not jointly convex with respect to both coefficients $\boldsymbol{X}$ and ancestors $\boldsymbol{a}_l$ ($l = 1, 2, \ldots, L$). Therefore, we used the same strategy as dictionary learning, namely, our algorithm alternates sparse coding and ancestor update steps. We note that each column of the dictionary is normalized to unit length before sparse coding step. Then, each column of the dictionary and corresponding coefficients are rescaled after sparse coding step so that $\tilde{\boldsymbol{D}}\tilde{\boldsymbol{X}} = \boldsymbol{D}\boldsymbol{X}$ where $\tilde{\boldsymbol{D}}$ is the normalized dictionary and $\tilde{\boldsymbol{X}}$ is a coefficient matrix corresponds to $\tilde{\boldsymbol{D}}$. In the following ancestor update stage, $\boldsymbol{X}$ is used to update the ancestors. We can use any sparse coding algorithm for the sparse coding step, therefore we focus on the update of ancestors given $\boldsymbol{Y}$ and $\boldsymbol{X}$ in this subsection.

By using the atom extraction operators $\{\boldsymbol{F}_{p,q}\}_{(p,q)\in\Lambda}$, the problem of updating ancestors can be written as,

$$\underset{\boldsymbol{a}_1,\ldots,\boldsymbol{a}_L}{\text{minimize}} \sum_{i=1}^{m} \left\| \boldsymbol{y}_i - \sum_{(p,q)\in\Lambda} \sum_{l=1}^{L} x_i^{p,q,l} \boldsymbol{F}_{p,q} \boldsymbol{a}_l \right\|_2^2. \tag{10}$$

The problem is an unconstrained convex optimization problem, which can be efficiently solved. Although we can derive a closed-form solution, it needs matrix inversion which requires a high computational cost. Therefore, we used the stochastic gradient descent (SGD) instead to optimize the ancestors. The update rule of ancestors $\boldsymbol{a}_l$ $(l = 1, 2, \ldots, L)$ can be derived as follows,

$$\boldsymbol{a}_l^{(t+1)} = \boldsymbol{a}_l^{(t)} + \eta_j \left(\sum_{(p,q)\in\Lambda} x_i^{p,q,l} \boldsymbol{F}_{p,q} \right)^{\top} \left(\boldsymbol{y}_i - \sum_{(p,q)\in\Lambda} \sum_{l'=1}^{L} x_i^{p,q,l'} \boldsymbol{F}_{p,q} \boldsymbol{a}_{l'} \right), \tag{11}$$

where η_j is a step size at j-th iteration and an ancestor $\boldsymbol{a}_l$ after the t-th update is denoted by $\boldsymbol{a}_l^{(t)}$. We applied the above update rule for each pair of data and coefficient $\boldsymbol{y}_j$, $\boldsymbol{x}_j$ $(j = 1, 2, \ldots, m)$ sequentially. Our algorithm for learning ancestors is summarized in Algorithm 1. We repeat the algorithm several times and the order of the sample $\boldsymbol{y}_i$ is shuffled every time to obtain a better solution. Then, we alternate sparse coding and ancestor update for an appropriate number of iterations.

3 Experimental Results

We show experimental results of learning scale and shift-invariant features with 2D artificial and natural images. Our code used in the following experiments is available at https://github.com/t-aritake/ancestral_atom_learning.

Algorithm 1: update ancestors using stochastic gradient descent

Data: finite set of signals $\boldsymbol{Y}$, coefficients $\boldsymbol{X}$, atom extraction operators $\{\boldsymbol{F}_{p,q}\}_{(p,q)\in\Lambda}$, initial ancestors $\boldsymbol{a}_1^{(0)}, \boldsymbol{a}_2^{(0)}, \ldots, \boldsymbol{a}_L^{(0)}$, initial step size η_0
Result: updated ancestors $\boldsymbol{a}_1, \boldsymbol{a}_2, \ldots, \boldsymbol{a}_L$
for $i = 0$ *to* $m - 1$ **do**
  draw a sample $\boldsymbol{y}_i$ and corresponding coefficient $\boldsymbol{x}_i$;
  for $l = 1$ *to* L **do**
    update ancestor $\boldsymbol{a}_l^{(i)}$ to $\boldsymbol{a}_l^{(i+1)}$ by Eq. (11) with step size η_i;
  end
  update the step size η_i to η_{i+1};
end
return updated ancestors $\boldsymbol{a}_1^{(m)}, \boldsymbol{a}_2^{(m)}, \ldots, \boldsymbol{a}_L^{(m)}$

3.1 Experiments with Artificial Signals

We generated a set of signals $\boldsymbol{Y}$ using the 2D Gabor dictionary $\boldsymbol{D}$ [15]. The signals are 16×16 artificial images which are generated as linear combinations of randomly chosen atoms from the Gabor dictionary. The subset of the Gabor dictionary and the generated signals are shown in Fig. 3. We consider the three scales and three orientations of Gabor atoms. Then, these atoms and their shifted atoms were used to generate the artificial signals. However, these features do not easily be observed from the dataset.

In this experiment, we learned the features from the generated signals by the pyramid feature learning [12], the multiscale K-SVD [9] and our proposed method which we call ancestral atom learning. We used the lasso for the sparse coding for all three methods, and regularization parameter was set to $\lambda = 1.0 \times 10^{-2}$ and we repeated sparse coding and update of the dictionary 50 times for each method.

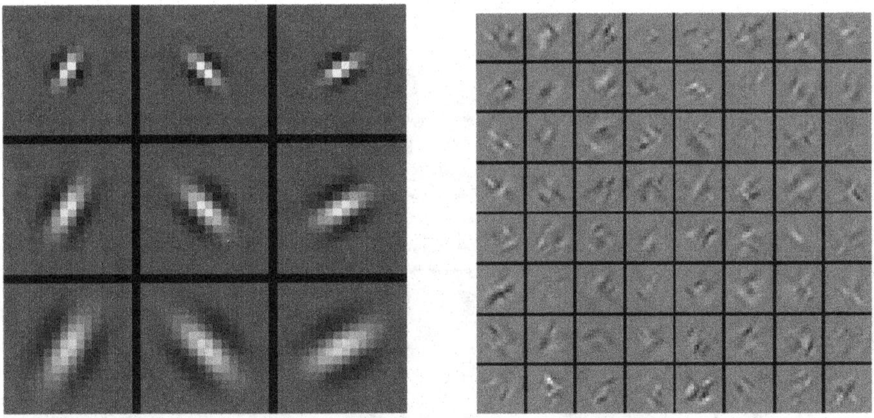

(a) subset of the Gabor dictionary (b) subset of the generated artificial images

Fig. 3. The subset of a Gabor dictionary and the subset of generated signals.

We assume that the number of orientations of the ground-truth dictionary is known in this experiment. Then, three 4×4 wavelet filters were used for the pyramid feature learning, and the number of atoms of the multiscale K-SVD for each scale is set to 10. We consider three scales for the pyramid feature learning and the multiscale K-SVD. We used three ancestors and five scales for our method. The sizes of the downsampled ancestors are 16×16, 14×14, 12×12, 10×10, 8×8 in this experiment. In the ancestor update step, we iterated Algorithm 1 for 20 times for each iteration. We put the initial step size for ancestor update $\eta_0 = 1.0 \times 10^{-4}$, and η_j is updated as $\eta_{j+1} = 0.95\eta_j$ for each step j.

The subset of learned atoms by each method is shown in Fig. 4. Figure 4(a) shows the learned pyramid features. We see that the wavelet filters (1st column) and their upsampled atoms (2nd, 3rd columns) are Gabor-like features. However, unlike the Gabor dictionary, the orientation of the features is limited to a single orientation. Figure 4(b) shows the learned features by the multiscale K-SVD. We observe that each feature represent one of the three orientations. However, both the 2nd scale (3rd, 4th columns) and the 3rd scale (5th, 6th columns) contain the features of the same size. Namely, some features are not learned as the atoms of the appropriate scale. Because the scale of the multiscale K-SVD is changed in a dyadic manner as 16×16, 8×8, and 4×4, some features of the multiscale K-SVD are learned as the features of different scales. Figure 4(c) shows the learned features by our proposed method. We see that the features have three different orientations as the Gabor dictionary. Moreover, the size of the learned features is the closest to the original Gabor dictionary. This result suggests that our proposed method can learn features which are invariant to both the small shift and the non-dyadic scaling.

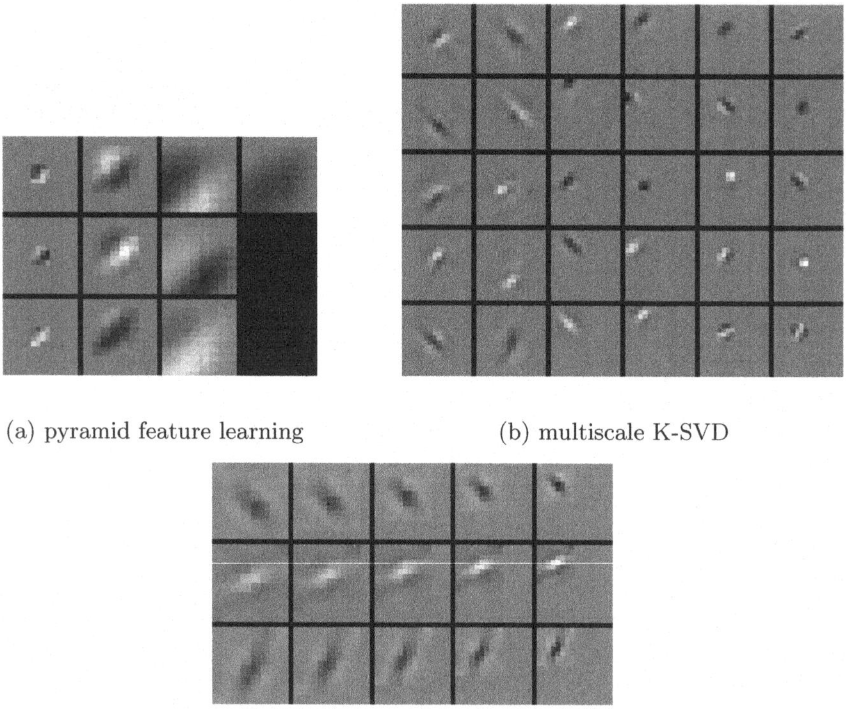

(a) pyramid feature learning

(b) multiscale K-SVD

(c) ancestral atom learning (proposed)

Fig. 4. Learned features from artificial images by each method

3.2 Experiments with Natural Images

Next, we show experimental results with 2D natural images. We used widely used images, Lena, Barbara, Boat, Mountain to create a set of training signals. We generated the training signals from these images by extracting 16×16 patches.

We used the lasso with regularization parameter $\lambda = 1.0 \times 10^{-3}$ for sparse coding, and the number of scales and the size of the features are set to the same values as the previous subsection. However, unlike the previous subsection, the true number of ancestors or wavelets is unknown, therefore, we used six 4×4 wavelet filters for the pyramid feature learning and 9 ancestors for our proposed

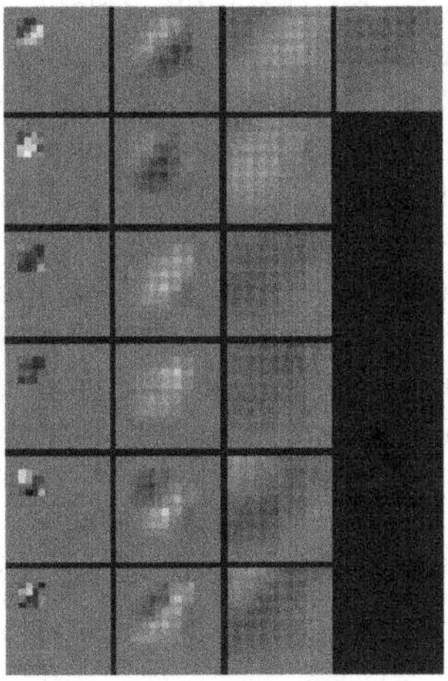

(a) pyramid feature learning

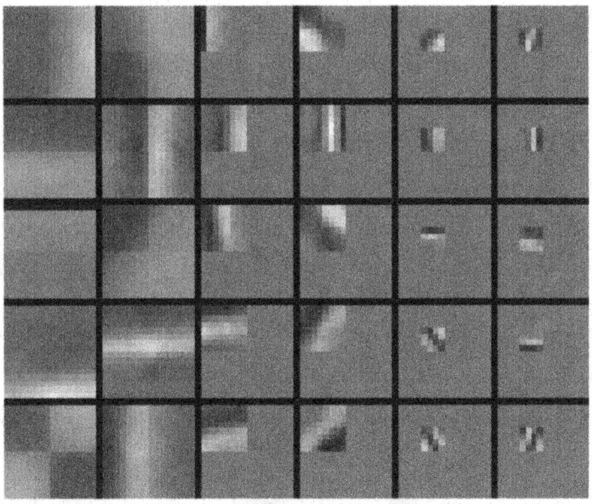

(b) multiscale K-SVD

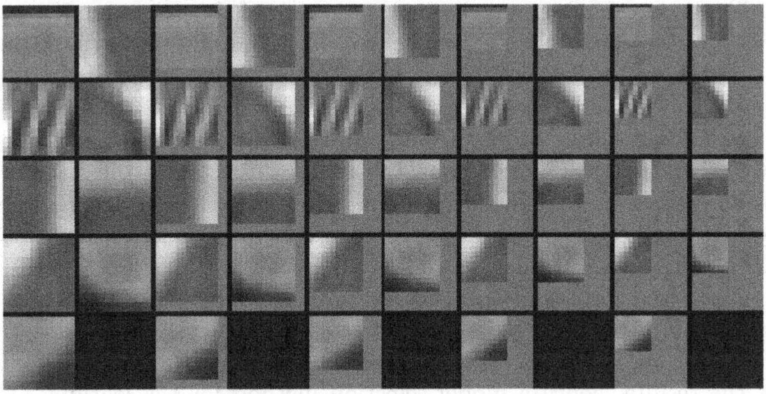

(c) ancestral atom learning (proposed)

Fig. 5. Learned features from natural images by each method

method. Currently, the number of ancestors is experimentally determined so that the generated atoms do not stuck at their initial values. We used the same number of atoms for the multiscale K-SVD as the previous subsection.

The subset of learned features by each method is shown in Fig. 5. We see that the learned pyramid features show Gabor-like features as the previous subsection. Still, the orientation of the features is limited to a single orientation and complex data-specific features were not learned. Because the pyramid features are generated from small wavelet filters by recursive upsampling and interpolation, it is difficult to generate data-specific features.

The Haar-like features and several edge-like features are obtained for the features of the multiscale K-SVD of each scale. Comparing with the pyramid features, more data-specific features are obtained by the multiscale K-SVD. However, as we can see, similar patterns commonly appear in different scales. Moreover, in the multiscale K-SVD dictionary, the texture-like features which appear in large patches tend to be represented by the linear combination of features of small patches. Therefore, appropriate post-processing is needed to extract complex features of large patches.

On the other hand, not only Haar-like and edge-like features, but also texture-like features are obtained by our proposed method. Because our model considers the relations between the features of different scales, as we can see in Fig. 5(c), the similar features commonly appear in the five scales. Namely, no post-processing is needed to extract common features from each scales.

4 Conclusion and Future Work

In this paper, we proposed a scale and shift-invariant model for dictionary learning. We assume that the atoms of a dictionary are generated from multiple ancestors by scaling and shift operations. The scaling operation of ancestors is realized by downsampling and the shift operation of the downsampled ancestors is realized by zero-padding. Our model considers more flexible scaling and shift of features than the prior methods. The learned ancestors can be regarded as features of the signals which are scale and shift invariant. Moreover, by considering multiple ancestors, other variant features such as rotations can be learned as different ancestors. The experimental results suggest that the features which appear at various scales and positions can be obtained by our proposed method. To apply learned scale and shift-invariant features to other tasks such as classification of time-series or images is our future work.

The proposed algorithm still leaves room for improvement. The parameters such as the number of ancestors, or a set of atom extraction operators affect the quality of the learned ancestors. Especially, the number of ancestors significantly affects the results, hence a model selection method for the learning ancestors should be developed. In addition, other atom generation operators should be considered. For example, if we include rotations of the features into the atom generation operators, it is expected that rotation-invariant features are learned by our proposed method.

Acknowledgement. This work was supported by JST CREST Grant Number JPMJCR1761 and JSPS KAKENHI Grant Numbers JP17H01793, JP18H03291.

References

1. Tibshirani, R.: Regression shrinkage and selection via the lasso. J. Roy. Stat. Soc. B **58**, 267–288 (1994)
2. Chen, S.S., Donoho, D.L., Saunders, M.A.: Atomic decomposition by basis pursuit. SIAM Rev. **43**(1), 129–159 (2001)
3. Hastie, T., Tibshirani, R., Wainwright, M.: Statistical Learning with Sparsity: The Lasso and Generalizations. Chapman & Hall/CRC, Boca Raton (2015)
4. Olshausen, B.A., et al.: Emergence of simple-cell receptive field properties by learning a sparse code for natural images. Nature **381**(6583), 607–609 (1996)
5. Engan, K., Aase, S.O., Husoy, J.H.: Method of optimal directions for frame design. In: 1999 Proceedings of the IEEE International Conference on Acoustics, Speech, and Signal Processing, vol. 5, pp. 2443–2446 (1999)
6. Aharon, M., Elad, M., Bruckstein, A.: K-SVD: an algorithm for designing overcomplete dictionaries for sparse representation. IEEE Trans. Sig. Process. **54**(11), 4311–4322 (2006)
7. Mairal, J., Bach, F., Ponce, J., Sapiro, G.: Online dictionary learning for sparse coding. In: Proceedings of the 26th Annual International Conference on Machine Learning, ICML 2009, pp. 689–696. ACM, New York (2009)
8. Barzideh, F., Skretting, K., Engan, K.: Imposing shift-invariance using flexible structure dictionary learning (FSDL). Digit. Sig. Proc. **69**, 162–173 (2017)
9. Mairal, J., Sapiro, G., Elad, M.: Learning multiscale sparse representations for image and video restoration. Multiscale Model. Simul. **7**(1), 214–241 (2008)
10. Shen, Y., Xiong, H., Dai, W.: Multiscale dictionary learning for hierarchical sparse representation. In: 2017 IEEE International Conference on Multimedia and Expo (ICME), pp. 1332–1337, July 2017
11. Grandits, T., Pock, T.: Optimizing wavelet bases for sparser representations. In: Pelillo, M., Hancock, E. (eds.) EMMCVPR 2017. LNCS, vol. 10746, pp. 249–262. Springer, Cham (2018). https://doi.org/10.1007/978-3-319-78199-0_17
12. Olshausen, B.A., Sallee, P., Lewicki, M.S.: Learning sparse image codes using a wavelet pyramid architecture. In: Advances in Neural Information Processing Systems, NIPS 2000, Denver, CO, USA, vol. 13, pp. 887–893 (2000)
13. Aritake, T., Hino, H., Murata, N.: Learning ancestral atom via sparse coding. IEEE J. Sel. Top. Sig. Process. **7**(4), 586–594 (2013)
14. Zheng, G., Yang, Y., Carbonell, J.: Efficient shift-invariant dictionary learning. In: Proceedings of the 22nd ACM SIGKDD International Conference on Knowledge Discovery and Data Mining, KDD 2016, pp. 2095–2104. ACM, New York (2016)
15. Daugman, J.G.: Two-dimensional spectral analysis of cortical receptive field profiles. Vis. Res. **20**(10), 847–856 (1980)

Robust Kernelized Bayesian Matrix Factorization for Video Background/Foreground Separation

Hong-Bo Xie[1(✉)], Caoyuan Li[2,3], Richard Yi Da Xu[3], and Kerrie Mengersen[1]

[1] The ARC Centre of Excellence for Mathematical and Statistical Frontiers, Queensland University of Technology, Brisbane, QLD 4001, Australia
hongbo.xie@qut.edu.au

[2] School of Computer Science and Technology, Beijing Institute of Technology (BIT), Beijing 100081, China

[3] Faculty of Engineering and Information Technology, University of Technology Sydney (UTS), Ultimo, NSW 2007, Australia

Abstract. Development of effective and efficient techniques for video analysis is an important research area in machine learning and computer vision. Matrix factorization (MF) is a powerful tool to perform such tasks. In this contribution, we present a hierarchical robust kernelized Bayesian matrix factorization (RKBMF) model to decompose a data set into low rank and sparse components. The RKBMF model automatically infers the parameters and latent variables including the reduced rank using variational Bayesian inference. Moreover, the model integrates the side information of similarity between frames to improve information extraction from the video. We employ RKBMF to extract background and foreground information from a traffic video. Experimental results demonstrate that RKBMF outperforms state-of-the-art approaches for background/foreground separation, particularly where the video is contaminated.

Keywords: Matrix factorization · Variational Bayesian inference · Sparse Bayesian learning · Background/foreground separation

1 Introduction

Using machine learning methods to find the low-rank and/or sparse approximation of a given data matrix is a fundamental problem in many computer vision applications, for example, background/foreground separation. By casting the problem into the penalization of the regularization term, a number of efforts have been devoted to applying convex or non-convex optimization methods to obtain the low rank and sparse components [8,9,16,17]. For most of these convex or non-convex methods, one has to manually choose some regularization

G. Nicosia et al. (Eds.): LOD 2019, LNCS 11943, pp. 484–495, 2019.
https://doi.org/10.1007/978-3-030-37599-7_40

parameters to properly control the trade-off between the data fitting error and the matrix rank when noise is involved. However, due to the lack of noise variance and rank, it is often unrealistic to determine the optimal regularization parameters.

Bayesian inference under probabilistic frameworks provides another essential principle to perform matrix factorization. Ding et al. [4] proposed a Bayesian robust principal component analysis (BRPCA) framework which infers an approximate representation for the noise statistics while simultaneously inferring the low rank and sparse components. However, this model is relatively complex, and the intractable posteriors are inferred by Gibbs sampling. Aicher [1] later improved the parameter inference in [4] by using the factorized variational Bayesian (VB) principle. Wang et al. [18] proposed a Bayesian robust matrix factorization model for image and video analysis. The Gaussian noise model is replaced by a Laplace mixture in [18] to enhance model robustness. Similarly, a Bayesian formulation of hierarchical L_1 norm low-rank matrix factorization is presented in [19]. In addition, Zhao et al. [20] presented a generative robust PCA model under the Bayesian framework with data noise modeled as a mixture of Gaussians (MoG). A common issue of the above models is that the optimal rank of the low rank component has to be manually pre-determined, which potentially either over-fits or under-fits the data. Babacan et al. [2] proposed to employ the automatic relevance determination principle in sparse Bayesian learning to determine the optimal rank of the low rank component.

Although these methods are successful in many areas including video processing, most of them simply ignore side information, or intrinsically, are not capable of exploiting it. On the other hand, many studies have indicated that kernelized matrix factorization to integrate side information, i.e., prior knowledge or data attributes for specific data, can significantly improve the performance of information extraction or prediction [5,10,14,21]. However, the inference of kernelized matrix factorization models using VB is still quite limited. Park et al. [13] placed Gaussian-Wishart priors on mean vectors and precision matrices of Gaussian user and item factor matrices, such that the mean of each prior distribution is regressed on corresponding side information. They developed a VB algorithm to approximate the posterior distributions over user and item factor matrices with a Bayesian Cramer-Rao bound. Very recently, Gönen and Kaski [6,7] extended the kernelized matrix factorization with a full VB treatment and with an ability to work with multiple side information sources expressed as different kernels. However, this model focused specifically on binary output matrices for multi-label classification. Moreover, both models in [6,7], and [13] lack of robustness, which is required to handle the sparse component or outliers in many real-world applications.

Along the line of this research, we present a generative model for robust kernelized Bayesian matrix factorization (RKBMF) which can integrate side information into inference. Our model adopts a different graphical model and priors as in [13]. A significant difference between our model and [6,7] is that the variance of a number of latent variables in [6,7] is set as constant, which is unacceptable in the case of video analysis with an unknown noise variance. The variance of each latent factor matrix is explicitly assigned as a latent variable with a specified prior in our model. The similarity information between frames is also integrated into RKBMF to improve the performance of video analysis. We test the performance of the model on simulated datasets and then apply this algorithm to perform the video background and foreground separation. The results demonstrated that RKBMF can accurately recover both low rank and sparse components in simulation, and generate background and foreground images with better visual effects than other three state-of-the-art robust matrix factorization approaches.

2 Robust Kernelized Bayesian Matrix Factorization

In this section, we first elaborate on the model specification of robust kernelized Bayesian matrix factorization. We then present the variational Bayesian method to infer all parameters and latent variables of this model in detail.

2.1 Model Specification

Considering the observation data as an $M \times N$ matrix $\mathbf{Y}$, the problem is to recover the original low-rank matrix $\mathbf{X}$ and sparse term $\mathbf{S}$, that is:

$$\mathbf{Y} = \mathbf{X} + \mathbf{S} + \mathbf{E} = \mathbf{U}\mathbf{V}^{\top} + \mathbf{S} + \mathbf{E}, \tag{1}$$

where $\mathbf{Y} \in \mathbb{R}^{M \times N}$, $\mathbf{U} \in \mathbb{R}^{M \times r}$, $\mathbf{V} \in \mathbb{R}^{N \times r}$, $\mathbf{S} \in \mathbb{R}^{M \times N}$, $\mathbf{E} \in \mathbb{R}^{M \times N}$, and r the rank or order of the low-rank term.

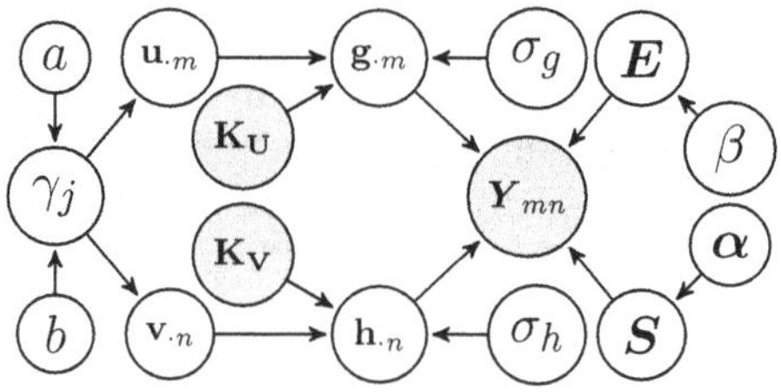

Fig. 1. Directed graphical representation of RKBMF model.

Figure 1 shows the graphical model of the proposed hierarchical robust kernelized Bayesian matrix factorization with latent variables and their corresponding priors. To automatically infer the rank of the low rank component, we impose sparsity into the low rank approximation model, assigning Gaussian priors to the columns of $\mathbf{U}$ and $\mathbf{V}$ with precisions (inverse variances) γ_j, namely,

$$p(\mathbf{U}|\gamma) = \prod_{j=1}^{r} \mathcal{N}(\mathbf{u}_{\cdot j}|\mathbf{0}, \gamma_j^{-1}\mathbf{I}_M), \tag{2}$$

$$p(\mathbf{V}|\gamma) = \prod_{j=1}^{r} \mathcal{N}(\mathbf{v}_{\cdot j}|\mathbf{0}, \gamma_j^{-1}\mathbf{I}_N), \tag{3}$$

where $\mathbf{I}_J \in \mathbb{R}^{J\times J}$ denotes an identity matrix. Therefore, the columns of $\mathbf{U}$ and $\mathbf{V}$ possess the same sparsity since they are enforced by the same precision γ_j. With such a constraint, most of the precision γ_j will be iteratively updated to very large values. The corresponding columns of $\mathbf{U}$ and $\mathbf{V}$ are removed since they make little contribution to the approximation $\mathbf{X}$, and hence the sparsity of latent factors $\mathbf{U}$ and $\mathbf{V}$ and low-rank of $\mathbf{X}$ are jointly satisfied. This sparse Bayesian learning formulation has been applied in compressive sensing and sparse coding [2,11,12].

To achieve the joint sparsity of $\mathbf{U}$ and $\mathbf{V}$, we further assign the conjugate Gamma hyper-prior to the precision γ_j:

$$p(\gamma_j) = \text{Gamma}(a, \frac{1}{b}) \propto \gamma_j^{a-1} exp(-b\gamma_j), \tag{4}$$

where very small values are assigned to the parameters a and b to achieve a diffuse hyper-prior. We also let $\mathbf{U}$ couple with the kernel matrx $\mathbf{K_U}$ resulting in a latent matrix $\mathbf{G}$, and assume that each entry of $\mathbf{G}$ follows Gaussian prior with precision σ_g, that is,

$$p(\mathbf{G}|\mathbf{U}, \mathbf{K_U}, \sigma_g) = \prod_{j=1}^{r} \mathcal{N}(\mathbf{g}_{\cdot j}|\mathbf{K_U}^\top \cdot \mathbf{u}_{\cdot j}, \sigma_g^{-1}\mathbf{I}_M). \tag{5}$$

Similarly, the prior of $\mathbf{H}$ is defined over the latent variable $\mathbf{V}$, kernel function $\mathbf{K_V}$, and precision σ_h:

$$p(\mathbf{H}|\mathbf{V}, \mathbf{K_V}, \sigma_h) = \prod_{j=1}^{r} \mathcal{N}(\mathbf{h}_{\cdot j}|\mathbf{K_V}^\top \cdot \mathbf{v}_{\cdot j}, \sigma_h^{-1}\mathbf{I}_N). \tag{6}$$

Here, the precisions σ_g and σ_h of the Gaussian distribution obey the Jeffery's prior:

$$p(\sigma_g) = \sigma_g^{-1}, \tag{7}$$

$$p(\sigma_h) = \sigma_h^{-1}. \tag{8}$$

The sparse component $\mathbf{S}$ is modeled with independent priors on each of the entries $\mathbf{S}_{ij}$ of the matrix $\mathbf{S}$, that is

$$p(\mathbf{S}|\boldsymbol{\alpha}) = \prod_{i=1}^{M}\prod_{j=1}^{N} \mathcal{N}(s_{ij}|0, \alpha_{ij}^{-1}). \tag{9}$$

In Eq. (1), we assume that the noise $\mathbf{E}$ obeys a Gaussian distribution with zero mean and unknown precision β. Hence, $\mathbf{E}$ is modeled as:

$$p(\mathbf{E}|\beta) = \prod_{i=1}^{M}\prod_{j=1}^{N} \mathcal{N}(e_{mn}|0, \beta^{-1}), \tag{10}$$

$$p(\beta) = \beta^{-1}, \tag{11}$$

where β also adopts the noninformative Jeffery's prior. Given the priors defined above, the conditional distribution for the observation model is as follows:

$$p(\mathbf{Y}|\mathbf{G}, \mathbf{H}, \mathbf{S}, \beta) = \mathcal{N}(\mathbf{Y}|\mathbf{G}\mathbf{H}^\top + \mathbf{S}, \beta^{-1}\mathbf{I}_{MN}). \tag{12}$$

With the conditional probability and all priors in hand, the joint distribution is given by:

$$\begin{aligned} &p(\mathbf{Y}, \mathbf{U}, \mathbf{V}, \mathbf{S}, \mathbf{G}, \mathbf{H}, \sigma_g, \sigma_h, \boldsymbol{\gamma}, \boldsymbol{\alpha}, \beta) \\ &= p(\mathbf{Y}|\mathbf{G}, \mathbf{H}, \mathbf{S}, \beta) p(\mathbf{G}|\mathbf{U}, \mathbf{K_U}, \sigma_g) p(\mathbf{H}|\mathbf{V}, \mathbf{K_V}, \sigma_h) \\ &\cdot p(\mathbf{U}|\boldsymbol{\gamma}) p(\mathbf{V}|\boldsymbol{\gamma}) p(\mathbf{S}|\boldsymbol{\alpha}) p(\sigma_g) p(\sigma_h) p(\boldsymbol{\gamma}) p(\boldsymbol{\alpha}) p(\beta). \end{aligned} \tag{13}$$

2.2 Model Inference of RKBMF

We use $\mathbf{Z}$ to represent the vector of all latent variables such that

$$\mathbf{Z} = (\mathbf{U}, \mathbf{V}, \mathbf{G}, \mathbf{H}, \mathbf{S}, \sigma_g, \sigma_h, \boldsymbol{\gamma}, \boldsymbol{\alpha}, \beta). \tag{14}$$

The approximate posterior distribution is therefore denoted by $q(\mathbf{Z})$.

$$q(\mathbf{Z}) = \prod_k q(\mathbf{Z}_k). \tag{15}$$

Within the VB framework, the expression of the optimal posterior approximation $q(\mathbf{Z}_k)$ can be denoted as

$$ln\ q(\mathbf{Z}_k) = \langle ln\ p(\mathbf{Y}, \mathbf{Z}) \rangle_{\mathbf{Z} \backslash \mathbf{Z}_k} + const, \tag{16}$$

where $\langle \cdot \rangle$ denotes the expectation and *const* denotes a constant which is not dependent on the current variable. $\mathbf{Z} \backslash \mathbf{Z}_k$ means the set of $\mathbf{Z}$ with $\mathbf{Z}_k$ removed. Each variable is updated in turn while holding others fixed. We detail the iteration rules for all unknown variables in Eq. (15).

Estimation of Latent Factors U and V. Combining the respective priors of $\mathbf{U}$ and $\mathbf{G}$ in Eqs. (2) and (5), the posterior approximation $ln\ q(\mathbf{U})$ is derived from Eq. (16) as:

$$\begin{aligned} ln\ q(\mathbf{U}) = \sum_j -\frac{1}{2}(&\mathbf{u}_{\cdot j}^\top (\langle \sigma_g \rangle \mathbf{K_U} \mathbf{K_U}^\top + \boldsymbol{\Gamma}_{\mathbf{u}_{\cdot j}}) \mathbf{u}_{\cdot j} \\ &- 2 \langle \sigma_g \rangle \mathbf{u}_{\cdot j}^\top \mathbf{K_U} \langle \mathbf{g}_{\cdot j} \rangle) + const, \end{aligned} \tag{17}$$

where $\boldsymbol{\Gamma}_{\mathbf{u}_{\cdot j}} = \langle \gamma_j \rangle \mathbf{I}_M$.

From Eq. (17) it is found that the posterior density of the jth column $\mathbf{u}_{\cdot j}$ of $\mathbf{U}$ obeys the multivariate Gaussian distribution:

$$q(\mathbf{u}_{\cdot j}) = \mathcal{N}(\mathbf{u}_{\cdot j} | \langle \mathbf{u}_{\cdot j} \rangle, \Sigma^{\mathbf{u}_{\cdot j}}), \tag{18}$$

with mean and covariance

$$\Sigma^{\mathbf{u}_{\cdot j}} = (\langle\sigma_g\rangle \cdot \mathbf{K_U}\mathbf{K_U}^\top + \boldsymbol{\Gamma}_{\mathbf{u}_{\cdot j}})^{-1}, \tag{19}$$

$$\langle\mathbf{u}_{\cdot j}\rangle = \langle\sigma_g\rangle \cdot \Sigma^{\mathbf{u}_{\cdot j}}\mathbf{K_U}\langle\mathbf{g}_{\cdot j}\rangle. \tag{20}$$

Similarly, the posterior approximation of $\mathbf{v}_{\cdot j}$ also obeys the multivariate Gaussian distribution with the density denoted by

$$q(\mathbf{v}_{\cdot j}) = \mathcal{N}(\mathbf{v}_{\cdot j}|\langle\mathbf{v}_{\cdot j}\rangle, \Sigma^{\mathbf{v}_{\cdot j}}), \tag{21}$$

and the mean and covariance are given by

$$\Sigma^{\mathbf{v}_{\cdot j}} = (\langle\sigma_h\rangle \cdot \mathbf{K_V}\mathbf{K_V}^\top + \boldsymbol{\Gamma}_{\mathbf{v}_{\cdot j}})^{-1}, \tag{22}$$

$$\langle\mathbf{v}_{\cdot j}\rangle = \langle\sigma_h\rangle \cdot \Sigma^{\mathbf{v}_{\cdot j}}\mathbf{K_V}\langle\mathbf{h}_{\cdot j}\rangle, \tag{23}$$

where $\boldsymbol{\Gamma}_{\mathbf{v}_{\cdot j}} = \langle\gamma_j\rangle\mathbf{I}_N$.

Estimation of Sparse Component $\boldsymbol{S}$. The posterior distribution of $\boldsymbol{S}$ is decomposed on its each entries s_{ij} which is a Gaussian distribution

$$q(s_{ij}) = \mathcal{N}(s_{ij}|\langle s_{ij}\rangle, \Sigma_{ij}^{\boldsymbol{S}}). \tag{24}$$

The covariance and mean are denoted as

$$\Sigma_{ij}^{\boldsymbol{S}} = \frac{1}{\langle\beta\rangle + \langle\alpha_{ij}\rangle}, \tag{25}$$

$$\langle s_{ij}\rangle = \langle\beta\rangle\Sigma_{ij}^{\boldsymbol{S}}(y_{ij} - \langle\boldsymbol{g}_{i\cdot}\rangle\langle\boldsymbol{h}_{i\cdot}^\top\rangle). \tag{26}$$

Estimation of $\boldsymbol{\gamma}$. Applying the priors of $\mathbf{U}, \mathbf{V}$ and $\boldsymbol{\gamma}$ in the same manner to Eq. (16), the posterior approximation of $ln\ q(\boldsymbol{\gamma})$ is given by

$$ln\ q(\boldsymbol{\gamma}) = ln(\gamma_j^{a-1+\frac{m+n}{2}} exp(-\frac{1}{2}\gamma_j(\langle\mathbf{u}_{\cdot j}^T\mathbf{u}_{\cdot j}\rangle + \langle\mathbf{v}_{\cdot j}^\top\mathbf{v}_{\cdot j}\rangle + 2b))) + const. \tag{27}$$

Hence the posterior distribution of γ_j is a Gamma distribution with mean

$$\langle\gamma_j\rangle = \frac{2a + M + N}{2b + \langle\mathbf{u}_{\cdot j}^\top\mathbf{u}_{\cdot j}\rangle + \langle\mathbf{v}_{\cdot j}^\top\mathbf{v}_{\cdot j}\rangle}. \tag{28}$$

The required expectations are found as

$$\langle\mathbf{u}_{\cdot j}^\top\mathbf{u}_{\cdot j}\rangle = \langle\mathbf{u}_{\cdot j}\rangle^\top\langle\mathbf{u}_{\cdot j}\rangle + \mathrm{tr}(\Sigma^{\mathbf{u}_{\cdot j}}), \tag{29}$$

$$\langle\mathbf{v}_{\cdot j}^\top\mathbf{v}_{\cdot j}\rangle = \langle\mathbf{v}_{\cdot j}\rangle^\top\langle\mathbf{v}_{\cdot j}\rangle + \mathrm{tr}(\Sigma^{\mathbf{v}_{\cdot j}}). \tag{30}$$

Estimation of G and H. Similar to the estimation of $\mathbf{U}$ and $\mathbf{V}$, the posterior approximation of $\mathbf{G}$ is given by

$$\begin{aligned} ln\ q(\mathbf{G}) =& \sum_i [-\frac{1}{2}(\mathbf{g}_{i\cdot}(\langle\beta\rangle\langle\mathbf{H}^\top\mathbf{H}\rangle + \langle\sigma_g\rangle\mathbf{I}_r)\mathbf{g}_{i\cdot}^T \\ &- 2\mathbf{g}_{i\cdot}(\langle\mathbf{H}\rangle^\top(\mathbf{y}_{i\cdot} - \mathbf{s}_{i\cdot})^\top + \langle\sigma_g\rangle\langle\mathbf{U}\rangle^\top\mathbf{K}_{\mathbf{U}_{\cdot i}}))] + const, \end{aligned} \tag{31}$$

which indicates that the ith row of $\mathbf{G}$ obeys the multivariate Gaussian distribution

$$q(\mathbf{g}_{i\cdot}) = \mathcal{N}(\mathbf{g}_{i\cdot}|\langle\mathbf{g}_{i\cdot}\rangle, \Sigma^{\mathbf{G}}). \tag{32}$$

The corresponding covariance and mean are denoted as

$$\Sigma^{\mathbf{G}} = (\langle\beta\rangle\langle\mathbf{H}^\top\mathbf{H}\rangle + \langle\sigma_g\rangle\mathbf{I}_r)^{-1}, \tag{33}$$

$$\langle\mathbf{g}_{i\cdot}\rangle^\top = \Sigma^{\mathbf{G}}(\langle\sigma_g\rangle\langle\mathbf{U}\rangle^\top\mathbf{K}_{\mathbf{u}_{\cdot i}} + \langle\beta\rangle\langle\mathbf{H}\rangle^\top(\mathbf{y}_{i\cdot} - \mathbf{s}_{i\cdot})^\top). \tag{34}$$

The jth row of $\mathbf{H}$ obeys another multivariate Gaussian distribution

$$q(\mathbf{h}_{j\cdot}) = \mathcal{N}(\mathbf{h}_{j\cdot}|\langle\mathbf{h}_{j\cdot}\rangle, \Sigma^{\mathbf{H}}), \tag{35}$$

with covariance and mean

$$\Sigma^{\mathbf{H}} = (\langle\beta\rangle\langle\mathbf{G}^\top\mathbf{G}\rangle + \langle\sigma_h\rangle\mathbf{I}_r)^{-1}, \tag{36}$$

$$\langle\mathbf{h}_{j\cdot}\rangle^\top = \Sigma^{\mathbf{H}}(\langle\sigma_h\rangle\langle\mathbf{V}\rangle^\top\mathbf{K}_{\mathbf{V}_{\cdot j}} + \langle\beta\rangle\langle\mathbf{G}\rangle^\top(\mathbf{y}_{\cdot j} - \mathbf{s}_{\cdot j}). \tag{37}$$

The required expectations are expressed as

$$\langle\mathbf{G}^\top\mathbf{G}\rangle = \langle\mathbf{G}\rangle^\top\langle\mathbf{G}\rangle + m\Sigma^{\mathbf{G}}, \tag{38}$$

$$\langle\mathbf{H}^\top\mathbf{H}\rangle = \langle\mathbf{H}\rangle^\top\langle\mathbf{H}\rangle + n\Sigma^{\mathbf{H}}. \tag{39}$$

Estimation of $\boldsymbol{\beta}$, $\boldsymbol{\sigma_g}$ and $\boldsymbol{\sigma_h}$. The posterior probability densities of β, σ_g and σ_h are all found to be Gamma distributed. For the noise precision β, we have

$$q(\beta) \propto \beta^{\frac{MN}{2}-1} exp(-\frac{1}{2}\beta\langle\|\mathbf{Y} - \mathbf{GH}^\top - \mathbf{S}\|_F^2\rangle), \tag{40}$$

with its expectation

$$\langle\beta\rangle = \frac{MN}{\langle\|\mathbf{Y} - \mathbf{GH}^\top - \mathbf{S}\|_F^2\rangle}. \tag{41}$$

The required expectation to estimate $\langle\beta\rangle$ is denoted as

$$\begin{aligned} \langle\|\mathbf{Y} - \mathbf{GH}^\top - \mathbf{S}\|_F^2\rangle =& \|\mathbf{Y} - \langle\mathbf{G}\rangle\langle\mathbf{H}\rangle^\top - \langle\mathbf{S}\rangle\|_F^2 + tr(N\langle\mathbf{G}\rangle^\top\langle\mathbf{G}\rangle\Sigma^{\mathbf{H}}) \\ &+ tr(M\langle\mathbf{H}\rangle^\top\langle\mathbf{H}\rangle\Sigma^{\mathbf{G}}) + tr(MN\Sigma^{\mathbf{G}}\Sigma^{\mathbf{H}}) + tr(\sum_{i=1}^{M}\sum_{j=1}^{N}\Sigma_{ij}^{S}). \end{aligned} \tag{42}$$

The updating rules for σ_g and σ_h are derived in the same manner:

$$\langle \sigma_g \rangle = \frac{Mr}{\langle \|\mathbf{G} - \mathbf{K_U}^\top \mathbf{U}\|_F^2 \rangle}, \tag{43}$$

$$\langle \sigma_h \rangle = \frac{Nr}{\langle \|\mathbf{H} - \mathbf{K_V}^\top \mathbf{V}\|_F^2 \rangle}, \tag{44}$$

with required expectations:

$$\begin{aligned} \langle \|\mathbf{G} - \mathbf{K_U}^\top \mathbf{U}\|_F^2 \rangle = \|\langle \mathbf{G} \rangle - \mathbf{K_U}^\top \langle \mathbf{U} \rangle\|_F^2 \\ + \operatorname{tr}(M\mathbf{K_U}\mathbf{K_U}^\top \Sigma^{\mathbf{U}}) + \operatorname{tr}(M\Sigma^{\mathbf{G}}), \end{aligned} \tag{45}$$

$$\begin{aligned} \langle \|\mathbf{H} - \mathbf{K_V}^\top \mathbf{V}\|_F^2 \rangle = \|\langle \mathbf{H} \rangle - \mathbf{K_V}^\top \langle \mathbf{V} \rangle\|_F^2 \\ + \operatorname{tr}(N\mathbf{K_V}\mathbf{K_V}^\top \Sigma^{\mathbf{V}}) + \operatorname{tr}(N\Sigma^{\mathbf{H}}). \end{aligned} \tag{46}$$

Estimation of $\boldsymbol{\alpha}$. Similar to β, σ_g and σ_h, the posterior probability density of α_{ij} is also found to be a Gamma distribution with

$$\langle \alpha_{ij} \rangle = \frac{1}{\langle s_{ij} \rangle^2 + \Sigma_{ij}^{\boldsymbol{S}}} \tag{47}$$

We update each parameter in turn while holding others fixed. By the properties of VB, convergence to a local minimum of the algorithm can be guaranteed after iterations [3].

2.3 Construction of the Kernel

Construction of an effective kernel plays an essential role in guaranteeing a good performance of kernelized matrix factorization. However, to the best of our knowledge, few studies have exploited the side information of similarity between video frames in such a model. In a similar scenario in image processing, the methods to take advantage of the side information, i.e., the similarity between patches, have resulted in superior performance scores in image restoration [21]. In this study, we aim to apply the RKBMF model with integrated side information to improve background subtraction and foreground detection. Here we present a new kernel which incorporates the similarity information between video frames into RKBMF. Denoting the Euclidean distance between a pair of frames (i, j) by $d_E^{i,j}$, we define the similarity between them, i.e., entry of $\mathbf{K_u}$ or $\mathbf{K_v}$ as

$$k_{ij} = \sqrt[4]{\frac{1}{1 + d_E^{i,j}/M}}, \tag{48}$$

where M is the total number of pixels in a frame of the video.

In the proposed method, the target frame is first vectorized as a column vector. The $M \times N$ matrix $\mathbf{Y}$ is constructed by grouping other $N-1$ frames with similar local spatial structures to the underlying one. Since each column shares similar underlying image structures, the noise-free low-rank matrix $\mathbf{X}$ corresponds to the background, while the sparse component corresponds to the foreground. With the kernel defined in Eq. (48), a similar frame with larger similarity value has a more substantial contribution in the RKBMF model to separate the background and foreground.

3 Results

3.1 Numerical Simulation

We first test the performance of the proposed algorithm on simulated matrices. We consider four square matrices with size $M = N = 500$, 1000, 1500 and 2000, respectively. The low-rank component $\mathbf{X}$ is simulated by the product of two matrices whose entries are independently drawn from $\mathcal{N}(0, 1/M)$. The sparse component $\mathbf{S}$ is simulated by the non-zero entries located uniformly at random with amplitudes obeyed uniform distribution within the range of $[-1, 1]$. The observation is generated as $\mathbf{Y} = \mathbf{X}+\mathbf{S}+\mathbf{E}$ with entries of $\mathbf{E}$ independently drawn from $\mathcal{N}(0, 10^{-4})$. The hyperparameters related to $\boldsymbol{\alpha}$, β, σ_g, σ_h are specified with a relatively small value, i.e., 10^{-6}. Three metrics, i.e., rank($\hat{\mathbf{X}}$), $||\hat{\mathbf{S}}||_0$ and $||\hat{\mathbf{S}} - \mathbf{S}||_F/||\mathbf{S}||_F$ are used to evaluate the performance of the algorithm. In this simulation, since no side information is available, it is reasonable to set $\boldsymbol{K}_u$ and $\boldsymbol{K}_v$ as identity matrices. From Table 1, it is clear that RKBMF can accurately approximate the rank of the low-rank component and $||\hat{\mathbf{S}}||_0$ with a very small reconstruction error of $||\hat{\mathbf{S}} - \mathbf{S}||_F/||\mathbf{S}||_F$.

Table 1. Comparison of reconstruction accuracy for noisy observation, with noise standard deviation $\sigma = 10^{-4}$. The true rank of the matrix $\mathbf{X}$ is $5\%N$, and the number of nonzero sparse elements is $5\%MN$.

N	rank($\mathbf{X}$)	$\|\|\mathbf{S}\|\|_0$	rank($\hat{\mathbf{X}}$)	$\frac{\|\|\hat{\mathbf{S}}-\mathbf{S}\|\|_F}{\|\|\mathbf{S}\|\|_F}$	$\|\|\hat{\mathbf{S}}\|\|_0$
500	25	12500	25	3.1×10^{-5}	12498
1000	50	50000	50	2.5×10^{-5}	50003
1500	75	112500	75	3.3×10^{-5}	112500
2000	100	200000	100	2.9×10^{-5}	199990

3.2 Video Example

In this section, we evaluate the performance of RKBMF to reconstruct the static background and moving foreground from a video sequence in traffic surveillance with a fixed camera[1]. Experiments are also conducted using Bayesian robust

[1] http://jacarini.dinf.usherbrooke.ca/dataset2012/.

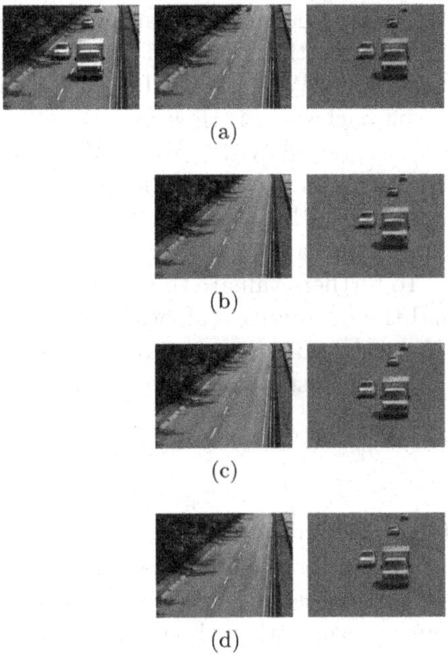

Fig. 2. Reconstruction of the background and the foreground. The video sequence contains 520 frames of size 320×240 pixels, and the results for frame 260 are shown. Left column: original image; middle: reconstruction of the low-rank component (background); and right: reconstruction of the sparse component (foreground). (a) Bayesian Robust PCA, (b) Mixture of Gaussians RPCA, (c) Online stochastic tensor decomposition and (d) RKBMF.

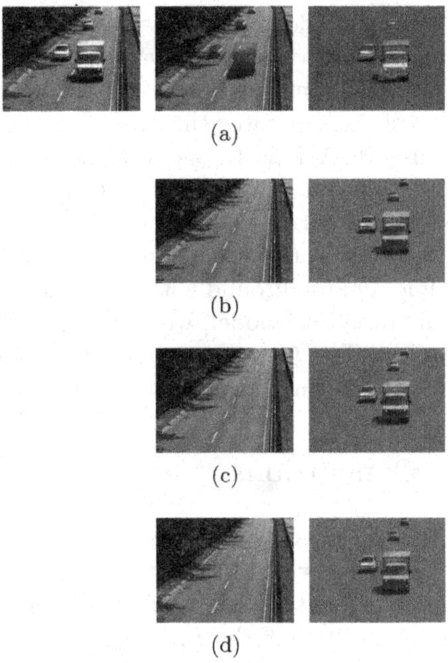

Fig. 3. Reconstruction of the background and the foreground under noisy observation. The additive white Gaussian noise has a standard deviation $\sigma = 10$. Left column: original noisy image; middle: background reconstruction; and right: foreground reconstruction. (a) Bayesian Robust PCA, (b) Mixture of Gaussians RPCA, (c) Online stochastic tensor decomposition and (d) RKBMF.

PCA [4], mixture of Gaussians RPCA [20], and online stochastic tensor decomposition [15], for comparison. The data are organized such that column of is constructed by concatenating all pixels of the frame from a grayscale video sequence. The background is then modeled as the low-rank component, and the moving foreground is modeled as the sparse component. The hyperparameters are the same as in Sect. 3.1. The kernel function for the latent factor matrix $\boldsymbol{K}_v$ is estimated using Eq. (48) while the kernel function for $\boldsymbol{K}_u$ is set as an identity matrix. The video sequence contains 520 frames of 320×240 pixels. Figure 2 shows the reconstruction of the background and the foreground for Frame 260 over four methods. It is observed that the four models produce good reconstructed background/foreground in this situation since the observation is relatively noise-free.

We then add Gaussian white noise with standard deviation 20 into the video sequence. Such noisy observations are common in practical applications. Figure 3 shows the reconstruction results of all four methods. We find that RKBMF still successfully separates the foreground from the background. However, Bayesian Robust PCA fails to separate background/foreground with part of the foreground existing in the low-rank component. Artifacts can still be found on the foreground extracted by the mixture of Gaussians RPCA and online stochastic tensor decomposition. In contrast, RKBMF generates the overall best background and foreground with fewer artifacts. To further evaluate the performance of the proposed model, we test RKBMF and the competitive algorithms on the CAVIAR Test Case Scenarios[2]. The experimental results are similar to the case of the traffic surveillance with RKBMF yielding the best separation performance.

4 Conclusions

In this paper, we have proposed a novel full Bayesian model for robust matrix factorization which integrates the side information for the low rank and sparse component extraction. Using both synthetic and real datasets, experimental results show that the proposed method outperforms other three state-of-the-art robust matrix factorization approaches. In particular, the proposed method can recover the background and slow-moving foreground even under high noise level. RKBMF can be further improved to accommodate streaming video and to integrate multiple side information.

References

1. Aicher, C.: A variational Bayes approach to robust principal component analysis. REU 2013 (2013)
2. Babacan, S.D., Luessi, M., Molina, R., Katsaggelos, A.K.: Sparse Bayesian methods for low-rank matrix estimation. IEEE Trans. Sig. Process. **60**(8), 3964–3977 (2012)
3. Bishop, C.M.: Pattern Recognition and Machine Learning. Springer, New York (2006)
4. Ding, X., He, L., Carin, L.: Bayesian robust principal component analysis. IEEE Trans. Image Process. **20**(12), 3419–3430 (2011)
5. Forsati, R., Mahdavi, M., Shamsfard, M., Sarwat, M.: Matrix factorization with explicit trust and distrust side information for improved social recommendation. ACM Trans. Inf. Syst. (TOIS) **32**(4), 17 (2014)
6. Gönen, M., Kaski, S.: Kernelized Bayesian matrix factorization. IEEE Trans. Pattern Anal. Mach. Intell. **36**(10), 2047 (2014)
7. Gönen, M., Khan, S., Kaski, S.: Kernelized Bayesian matrix factorization. In: International Conference on Machine Learning, pp. 864–872 (2013)
8. Hage, C., Kleinsteuber, M.: Robust PCA and subspace tracking from incomplete observations using ℓ_0-surrogates. Comput. Stat. **29**(3), 467–487 (2014)
9. Ji, H., Huang, S., Shen, Z., Xu, Y.: Robust video restoration by joint sparse and low rank matrix approximation. SIAM J. Imaging Sci. **4**(4), 1122–1142 (2011)

[2] http://homepages.inf.ed.ac.uk/rbf/CAVIARDATA1/.

10. Lan, L., et al.: Low-rank decomposition meets kernel learning: a generalized Nyström method. Artif. Intell. **250**, 1–15 (2017)
11. Lu, P., Gao, B., Woo, W.L., Li, X., Tian, G.Y.: Automatic relevance determination of adaptive variational Bayes sparse decomposition for micro-cracks detection in thermal sensing. IEEE Sens. J. **17**(16), 5220–5230 (2017)
12. Neal, R.M.: Bayesian Learning for Neural Networks, vol. 118. Springer, New York (2012). https://doi.org/10.1007/978-1-4612-0745-0
13. Park, S., Kim, Y.D., Choi, S.: Hierarchical Bayesian matrix factorization with side information. In: IJCAI, pp. 1593–1599 (2013)
14. Shah, V., Rao, N., Ding, W.: Matrix factorization with side and higher order information. STAT **1050**, 4 (2017)
15. Sobral, A., Javed, S., Ki Jung, S., Bouwmans, T., Zahzah, E.H.: Online stochastic tensor decomposition for background subtraction in multispectral video sequences. In: Proceedings of the IEEE International Conference on Computer Vision Workshops, pp. 106–113 (2015)
16. Sun, Q., Xiang, S., Ye, J.: Robust principal component analysis via capped norms. In: Proceedings of the 19th ACM SIGKDD International Conference on Knowledge Discovery and Data Mining, pp. 311–319. ACM (2013)
17. Tao, M., Yuan, X.: Recovering low-rank and sparse components of matrices from incomplete and noisy observations. SIAM J. Optim. **21**(1), 57–81 (2011)
18. Wang, N., Yeung, D.Y.: Bayesian robust matrix factorization for image and video processing. In: Proceedings of the IEEE International Conference on Computer Vision, pp. 1785–1792 (2013)
19. Zhao, Q., Meng, D., Xu, Z., Zuo, W., Yan, Y.: l_1-norm low-rank matrix factorization by variational Bayesian method. IEEE Trans. Neural Netw. Learn. Syst. **26**(4), 825–839 (2015)
20. Zhao, Q., Meng, D., Xu, Z., Zuo, W., Zhang, L.: Robust principal component analysis with complex noise. In: International Conference on Machine Learning, pp. 55–63 (2014)
21. Zhou, T., Shan, H., Banerjee, A., Sapiro, G.: Kernelized probabilistic matrix factorization: exploiting graphs and side information. In: Proceedings of the 2012 SIAM International Conference on Data mining, pp. 403–414. SIAM (2012)

Parameter Optimization of Polynomial Kernel SVM from miniCV

Li-Chia Yeh(✉) and Chung-Chin Lu

Department of Electrical Engineering, National Tsing Hua University, Hsinchu, Taiwan
lichiayeh@gmail.com, cclu@ee.nthu.edu.tw

Abstract. Polynomial kernel support vector machine (SVM) is one of the most computational efficient kernel-based SVM. Implementing an iterative optimization method, sequential minimal optimization (SMO) makes it more hardware independent. However, the test accuracy is sensitive to the values of hyperparameters. Moreover, polynomial kernel SVM has four hyperparameters which complicate cross-validation in parameter optimization. In this research, we transform polynomial kernels to have bounded values and analyze the relations between hyperparameters and the test error rate. Based on our discoveries, we propose mini core validation (miniCV) to fast screen out an optimized hyperparameter combination especially for large datasets. The proposed miniCV is a parameter optimization approach completely built on the distribution of the data generated via the iterative SMO training process. Since miniCV depends on the kernel matrix directly, it saves miniCV from cross-validation to optimize hyperparameters in kernel-based SVM.

Keywords: Support vector machine · Polynomial kernel · Multi-classification · Sequential minimal optimization · Parameter optimization

1 Introduction

Support vector machine (SVM) has been one of the most efficient binary classifiers since Vapnik [14] proposed it in 1995. SVM positions a partition hyperplane by solving a quadratic optimization problem. Generally, the instances are mapped to a higher dimensional space via a positive definite symmetric (PDS) kernel to improve the performance. Polynomial kernel $(\alpha + \gamma\langle x_r, x_{r'}\rangle)^d$ is one of PDS kernels to train a model with simple and fast computation, where d, γ and α stand for *degree*, *scalar* and *constant* respectively. However, a binary classifier is able to benefit from kernel-based SVM only under proper setting to all parameters.

The conventional method of parameter optimization is to utilize cross validation. With cross validation, researchers apply various algorithms to intelligently speed up parameter optimization within a confined region. Because these complicated parameter optimization methods, including various fitness functions, are

G. Nicosia et al. (Eds.): LOD 2019, LNCS 11943, pp. 496–507, 2019.
https://doi.org/10.1007/978-3-030-37599-7_41

under cross validation, it is time-consuming and requires a lot of computation. Based on our knowledge, most of current researches focus on RBF kernel SVM which only has two hyperparameters and implement algorithms on several small datasets from UCI [3]. These methods are able to be generalized to polynomial kernel SVM, but the four hyperparameters, d, γ, α and regulator C in SVM, indeed increase the complexity in implementation. Although hyperparameters in polynomial kernel SVM are optimized in [13], the duration of hyperparameter optimization is quite long, even on small datasets.

The main contribution of this research is proposing a light-weighted validation method via sequential minimal optimization (SMO) [6,11] to quickly screen out a more robust hyperparameter combination for polynomial kernel SVM with respect to a dataset itself. Unlike previous methods, there are no cross validation, fitness functions or confined hyperparameter regions in our method. The hyperparameters are optimized via the data extracted from the iterations of SMO directly, instead of applying another algorithm on top of kernel-based SVM. We call the presenting method as *mini Core Validation* (miniCV). Based on our previous exploration of 'miniCV on RBF kernel SVM' [16], we analyze and explore the relations among hyperparameters in polynomial kernel SVM and error rate to develop miniCV for polynomial kernel SVM.

The rest of this paper is organized as follows. Kernel-based SVM and SMO are briefly described in Sect. 3. Section 4 specifies the ideas and a key factor for analysis. All the experimental results are illustrated in Sect. 5. The discoveries in Sect. 5 motivate us to devise miniCV to optimize hyperparameters in polynomial kernel SVM. The implementation of miniCV and related performance is elaborated and demonstrated in Sect. 6. Section 7 gives a conclusion and future scope.

2 Related Work

The key issue confronted in applying kernel-based SVM is what a proper combination is to all hyperparameters, including the parameters in a kernel function and regulator C, for kernel-based SVM. An inappropriate hyperparameter combination results in a trained model with a bad test performance.

Many researchers spend effort on designing an approach or implementing an algorithm on top of SVM to optimize hyperparameters. Cross validation [2] is a statistical approach to target good combinations for these unknown parameters within a preset searching region. Especially, ν–fold cross-validation is usually the foundation of current hyperparameter optimization in machine learning. It divides the training set into ν folds and selects the hyperparameter combination with the best cross validation result among all grid searching combinations [4] in a bounded region. Later, instead of checking through all possible grid points, researchers start to utilize bio-inspired search algorithms as meta-strategies in optimizing hyperparameters for kernel-based SVM.

Genetic algorithms (GA) [8] encode each parameter into sequential gene bases and these gene bases together represent a phenotype encoded into a value of this parameter. Different parameters have their own gene bases and these gene bases are composed into one chromosome as one combination of hyperparameters. GA integrates the nature of genetic engineering, selection, mutation, duplication and cross-over, along with a fitness function to increase the variety of hyperparameter combination under cross validation. Eventually, all chromosomes converge to the same type representing optimized hyperparameters.

Similarly, particle swarm optimization (PSO) [1] denotes each possible hyperparameter combination as a particle. The position and velocity of a particle in a bounded hyperparameter region are adjusted by its current position, current velocity, the previous best position and the best position found by other particles. Similar to GA, the goodness of each particle is estimated by a fitness function via cross validation.

There are several GA or PSO related parameter optimization algorithms, either combining with different fitness functions or altering the approaches in cross validation or the procedure in GA or PSO, such as the bat algorithm in [13].

3 Review of Support Vector Machine

Let $\mathscr{X}$ be an input space and $\mathscr{Y} = \{1, -1\}$ a label space. In order to get a better training result, the data points in $\mathscr{X}$ are mapped to a Hilbert space, which is named as a feature space. A commonly used feature space for kernel-based SVM is the reproducing kernel Hilbert space (RKHS) $\mathbb{H}$ associated to a positive definite symmetric (PDS) kernel $K(\cdot,\cdot)$. Let Φ be the feature mapping from $\mathscr{X}$ to $\mathbb{H}$, where $K(x, x') = \langle \Phi(x), \Phi(x') \rangle$, $\forall x, x' \in \mathscr{X}$. Since $\mathbb{H}$ is the RKHS w.r.t. K, $\Phi(x) = K(x, \cdot)$, $\forall x \in \mathscr{X}$. A labeled sample $S = \{(x_1, y_1), \ldots, (x_m, y_m)\} \in (\mathscr{X} \times \mathscr{Y})^m$ of size m is given to train SVM. Due to the Representer theorem, the primal problem of a kernel-based SVM in $\mathbb{H}$ is equivalent to the one formed on the finite dimensional Hilbert space $\mathbb{H}_S \triangleq \text{Span}\{K(x_r, \cdot), x_r \in S, r \in [1, m]\}$. Then with regulator C and slack variables $\xi_1, \ldots, \xi_m$, the primal problem becomes

$$\begin{aligned} \text{Minimize: } & F(\boldsymbol{w}, b, \boldsymbol{\xi}) = \frac{1}{2}\|w\|_{\mathbb{H}_S}^2 + C\sum_{r=1}^{m} \xi_r \qquad (1) \\ \text{Subject to: } & 1 - \xi_r - y_r(\langle \boldsymbol{w}, \Phi(x_r) \rangle + b) \leqslant 0, -\xi_r \leqslant 0, \\ & r \in [1, m], (\boldsymbol{w}, b, \boldsymbol{\xi}) \in \mathbb{H}_S \times \mathbb{R} \times \mathbb{R}^m. \end{aligned}$$

The primal problem (1) can be optimized by solving its Lagrangian dual problem

$$\begin{aligned} \text{Minimize: } & \theta(\boldsymbol{\lambda}) = \frac{1}{2}\lambda_r \lambda_{r'} y_r y_{r'} K(x_r, x_{r'}) - \sum_{r=1}^{m} \lambda_r \qquad (2) \\ \text{Subject to: } & 0 \leqslant \lambda_r \leqslant C, \ \sum_{r=1}^{m} y_r \lambda_r = 0, \boldsymbol{\lambda} \in \mathbb{R}^m, \end{aligned}$$

where $\lambda_1, \lambda_2, \ldots, \lambda_m$ are effective Lagrangian multipliers. When the accompanied Karush-Kuhn-Tucher (KKT) necessary conditions are satisfied, this feasible solution is the optimal solution to both the Lagrangian dual problem (2) and the primal problem (1). Define $F_r \triangleq \langle \boldsymbol{w}, \Phi_{x_r} \rangle_{\mathbb{H}_S} - y_r$. Based on the value of label y_r and the range $\mathcal{R}(\lambda_r)$ of λ_r, the instances in S are partitioned into 5 groups, $I_0 \triangleq \{i \in [1,m] | 0 < \lambda_i < C\}$, $I_1 \triangleq \{i \in [1,m] | y_i = 1, \lambda_i = 0\}$, $I_2 \triangleq \{i \in [1,m] | y_i = -1, \lambda_i = 0\}$, $I_3 \triangleq \{i \in [1,m] | y_i = 1, \lambda_i = C\}$ and $I_4 \triangleq \{i \in [1,m] | y_i = -1, \lambda_i = C\}$. Platt [11] and Keerthi *et al.* [6] proposed an iterative algorithm called sequential minimal optimization (SMO), which only chooses a pair $(i,j) \in (I_0 \cup I_2 \cup I_3) \times (I_0 \cup I_1 \cup I_4)$ to trigger the updating, from $\boldsymbol{\lambda}^{(k-1)}, \boldsymbol{F}^{(k-1)}$ to $\boldsymbol{\lambda}^{(k)}, \boldsymbol{F}^{(k)}$ respectively, for the kth iteration, $k \in \mathbb{N}$ when $\Delta F^{(k-1)} \triangleq F_i^{(k-1)} - F_j^{(k-1)} > 0$. Let $K_{rr'} \triangleq K(x_r, x_{r'})$ and $\eta_{ij} \triangleq K_{ii} + K_{jj} - 2K_{ij}$. The simplified Lagrangian dual problem under SMO is

$$\min_{\lambda_j} \ \tilde{\Theta}^{(k)}(\lambda_j) = \frac{1}{2}(\lambda_j)^2 \eta_{ij} - \lambda_j \lambda_j^{(k-1)} \eta_{ij} - y_j \lambda_j \Delta F^{(k-1)} \tag{3}$$

$$\text{subject to: } 0 \leqslant \lambda_j, \tilde{R}_{ij} - y_i y_j \lambda_j \leqslant C, \tag{4}$$

where $\tilde{R}_{ij} = \lambda_i^{(k-1)} + y_i y_j \lambda_j^{(k-1)}$. A tolerance $\tau > 0$ is applied to loosen the termination criterion as $\Delta F^{(k-1)} \leqslant \tau$. The selected (i,j) in the kth iteration is named as a τ–violating pair [7]. In this paper, we call SVM implemented with SMO algorithm as SMO-SVM. $\tilde{\Theta}^{(k)}(\lambda_j^{(k-1)} + t^{(k)})$ in (3) is minimized at $t^{(k)} = y_j \frac{\Delta F^{(k-1)}}{\eta_{ij}}$ without considering the constraint (4). Once $\lambda_i^{(k-1)} - y_i y_j t^{(k)}, \lambda_j^{(k-1)} + t^{(k)} \notin [0, C]$, $t^{(k)}$ is clipped to let $\lambda_i, \lambda_j \in [0, C]$. Denote the update step as $t'^{(k)}$. If $t^{(k)}$ is not clipped, $t'^{(k)} = t^{(k)}$; otherwise $|t'^{(k)}| \leqslant |t^{(k)}|$. Then

$$F_r^{(k)} = F_r^{(k-1)} - y_j t'^{(k)} (K_{ir} - K_{jr}), \ r \in [1, m] \tag{5}$$

are updated for next iteration.

The initialization of SMO-SVM is to set $\boldsymbol{\lambda}^{(0)} = \boldsymbol{0}$, $\boldsymbol{F}^{(0)} = -\boldsymbol{y}$. With preset C and τ, SMO-SVM keeps grouping instances into I_0, I_1, I_2, I_3, I_4 and updating $\boldsymbol{\lambda}^{(k)}$ and $\boldsymbol{F}^{(k)}$ in each iteration until $\Delta F^{(k-1)} \leqslant \tau$ for all $i \in I_0 \cup I_2 \cup I_3$ and $j \in I_0 \cup I_1 \cup I_4$.

4 Experiments on Parameters

The original polynomial kernel function is $(\alpha + \gamma \langle x_r, x_{r'} \rangle)^d$. However, when the degree d becomes higher or the value of x larger, the overflow problem happens easily with inappropriate kernel parameters, α and γ. Unlike the other two conventional kernels, RBF kernel and sigmoidal kernel, have the restricted value between 0 and 1, polynomial kernel rises exponentially when d is increasing. In order to have a convinced computation results, we transform the polynomial kernel into a function which is bounded between -1 and 1, that is,

$$\left(\tilde{\alpha} + \frac{(1 - \tilde{\alpha})}{\max\limits_{i,j \in [1,m]} |\langle x_i, x_j \rangle|} \langle x_r, x_{r'} \rangle \right)^d, \ \tilde{\alpha} \in [0, 1), \tag{6}$$

where $\frac{\langle x_r, x_{r'} \rangle}{\max\limits_{i,j \in [1,m]} |\langle x_i, x_j \rangle|} \in [-1, 1]$. The relation between α and γ is replaced with $\tilde{\alpha}$ and $\frac{(1-\tilde{\alpha})}{\max\limits_{i,j \in [1,m]} |\langle x_i, x_j \rangle|}$. Note that transformed polynomial kernel function in (6) not only has a confined range $[-1, 1]$ but also reduces the kernel parameters from three (d, γ, α) to two $(d, \tilde{\alpha})$. In this research, we call kernel-based SVM with PDS kernel in (6) as transformed polynomial kernel SVM.

In the following, Subsect. 4.1 illustrates the experiment design in finding the influence of hyperparameters in transformed polynomial kernel SVM w.r.t. (6). Subsection 4.2 describes an intermediate variable extracted from the elements of the kernel matrix $\boldsymbol{K}$ during the iterations of SMO-SVM. This intermediate variable will be a key variable in miniCV.

4.1 Hyperparameters d, $\tilde{\alpha}$ and C

Equation (5) and the definitions of $t^{(k)}$ and $t^{'(k)}$ reveal the iterative mutual relations between t or t' and $\boldsymbol{F}^{(k)}$. In order to emphasize the role of the kernel matrix $\boldsymbol{K}$, we minimize the influence of C and τ by setting them to infinity and 0 respectively. Hence $t^{'(k)} = t^{(k)}$ for all k. Since $t^{(k)}$ is not clipped by C, we assume $t^{(k)}$ is fully characterized by $\boldsymbol{K}$. The distribution of $|t^{(k)}|$ will be utilized to estimate the value of C for each $\boldsymbol{K}$ corresponding to hyperparameters $(d, \tilde{\alpha})$. Next, the difference of the cost function in (3) between before and after updating

$$\Delta\tilde{\Theta}^{(k)} \triangleq \tilde{\Theta}^{(k)}(\lambda_j^{(k)}) - \tilde{\Theta}^{(k)}(\lambda_j^{(k-1)}) = -y_j t^{'(k)} \Delta F^{(k-1)} + \frac{1}{2}|t^{'(k)}|^2 \eta_{ij} \quad (7)$$

is recorded. Because the stopping criterion $\Delta F^{(k-1)} \leqslant \tau = 0$ is too tight, we terminate SMO when the cost function is almost converged and replace the stopping criterion to $|\Delta\tilde{\Theta}^{(k)}| < 10^{-3}$ in experiments. Let $\mathcal{D}_1 = \{2, \ldots, 11\}$ and $\mathcal{A}_1 = \{0.02, 0.04, \ldots, 0.98\}$. Transformed polynomial kernel SVM will be executed for each $(d, \tilde{\alpha}) \in \mathcal{D}_1 \times \mathcal{A}_1$. In addition to $t^{(k)}$ and $\Delta\tilde{\Theta}^{(k)}$, the values of an intermediate variable described in Subsect. 4.2 will also be collected.

4.2 An Intermediate Variable

Since, for a selected τ-violating pair (i, j) in the kth iteration, the updating of $\lambda_i^{(k-1)}, \lambda_j^{(k-1)}$ and $\boldsymbol{F}^{(k-1)}$ are related to η_{ij}, η_{ij} in each iteration will be recorded. Denote the collection of all η_{ij}'s before the nth iteration as $\mathfrak{E}_{n,d,\tilde{\alpha}}$ for each $(d, \tilde{\alpha}) \in \mathcal{D}_1 \times \mathcal{A}_1$ in Subsect. 4.1. Let $(r^{(k)}, r^{'(k)})$ be the τ-violating pair selected in the kth iteration. Then $\mathfrak{E}_{n,d,\tilde{\alpha}} = \{\eta_{r^{(k)} r^{'(k)}} | k < n\}$. Similar to [16], the variance $\mathrm{var}(\mathfrak{E}_{n,d,\tilde{\alpha}})$ of η_{ij}'s in $\mathfrak{E}_{n,d,\tilde{\alpha}}$ will be evaluated at the nth iteration to see the existence of any causal relation between $\mathrm{var}(\mathfrak{E}_{n,d,\tilde{\alpha}})$ and error rate.

5 Key Findings

This section demonstrates several key findings related to the variable $\mathrm{var}(\mathfrak{E}_{n,d,\tilde{\alpha}})$ introduced in Subsect. 4.2. The training and test datasets are selected from

MNIST [15] with labels 2 or 3 only, which together are named as *MNIST-23*. The sizes of the training and test datasets are 12089 and 2042 respectively.

5.1 Kernel Parameters d and $\tilde{\alpha}$

Following the experiments in Subsect. 4.1, we collect the relevant output data to verify whether there exists any causal relation between the variable $\mathrm{var}(\mathfrak{E}_{n,d,\tilde{\alpha}})$ in Subsect. 4.2 and error rate to facilitate parameter optimization in transformed polynomial kernel SVM. In addition, since we are interested in the possibility of fast screening out good hyperparameter combinations, n in $\mathrm{var}(\mathfrak{E}_{n,d,\tilde{\alpha}})$ is set between 30 and 200, depending on the size of the training dataset. The $\mathrm{var}(\mathfrak{E}_{n,d,\tilde{\alpha}})$'s for all trails $(d, \tilde{\alpha}) \in \mathcal{D}_1 \times \mathcal{A}_1$ in Subsect. 4.1 are all estimated at $n = \lfloor m/100 \rfloor = 120$.

Figure 1(a) shows the error rate curves versus $\tilde{\alpha}$ for various $d \in \mathcal{D}_1$. These curves look almost convex. Moreover, except $d = 2$, the lowest error rates of these curves are less than 0.25%. Figure 1(b) demonstrates the distribution of $\mathrm{var}(\mathfrak{E}_{n,6,\tilde{\alpha}})$ and error rate for $d = 6$ versus $\tilde{\alpha}$ in the left and right plots respectively. We observe that the distribution of $\mathrm{var}(\mathfrak{E}_{n,d,\tilde{\alpha}})$ is concave-like around the highest of $\mathrm{var}(\mathfrak{E}_{n,d,\tilde{\alpha}})$; while that of error rate in convex shape, which is also observed in the trials for other d than 6, except $d = 2$. Moreover, the two $\tilde{\alpha}$'s corresponding to the highest of $\mathrm{var}(\mathfrak{E}_{n,d=6,\tilde{\alpha}})$ and the lowest of error rate almost coincide with each other. It inspires us that $\mathrm{var}(\mathfrak{E}_{n,d,\tilde{\alpha}})$ may be utilized to trace the trend of error rate.

5.2 A Fast Estimation of an Appropriate $\tilde{\alpha}$

Furthermore, in order to find a method of speeding up the process in searching for an appropriate $\tilde{\alpha}$, we compute $\mathrm{var}(\mathfrak{E}_{n,d,\tilde{\alpha}})$ for $\tilde{\alpha} \in \mathcal{A}_1$ and apply least square estimation (LSE) regression to fit a polynomial curve $\widehat{\mathrm{var}(\mathfrak{E}_{n,d})}$ as a function of $\tilde{\alpha}$ for each d. Among all possible fitting polynomial formulas, $(a_2\tilde{\alpha}^2 + a_1\tilde{\alpha}^1 + a_0)$ shows the lowest rms error for all $d \in \mathcal{D}_1$ and the one for $d = 6$ is depicted as a solid curve in the left plot of Fig. 1(b).

Figure 2(a) demonstrates all the quadratic LSE fitted curves of $\widehat{\mathrm{var}(\mathfrak{E}_{n,d})}$ for $d \in \mathcal{D}_1$. The fitted formulas are shown in the bottom of Fig. 2(a) with a common magnitude 0.01 for (a_2, a_1, a_0) outside the parentheses.

Next, we compare Figs. 1(a) and 2(a). $\widehat{\mathrm{var}(\mathfrak{E}_{n,d=2})}$ in Fig. 2(a) is close to a straight line with negative slope maximized at $\tilde{\alpha} = 0$; while the error rate curve for $d = 2$ in Fig. 1(a) is approximated to a line with positive slope which has the minimum at $\tilde{\alpha} = 0$ too. Except $d = 2$, $\widehat{\mathrm{var}(\mathfrak{E}_{n,d})}$ for $d \in \mathcal{D}_1 \backslash \{2\}$ are all concave while error rate curves in Fig. 1(a) are almost convex for $d \in \mathcal{D}_1 \backslash \{2\}$. Moreover, since a fitted quadratic curve is maximized at $\frac{-a_1}{2a_2}$, we know $\frac{-a_1}{2a_2} < 0$ for $\widehat{\mathrm{var}(\mathfrak{E}_{n,d=2})}$, $\frac{-a_1}{2a_2} \in [0, 1)$ for $d \in \{3, 4, \ldots, 9\}$ and $\frac{-a_1}{2a_2} > 1$ for $d \in \{10, 11\}$.

In addition, we are interested in the robustness of transformed polynomial kernel SVM over $\tilde{\alpha}$ for each $d \in \mathcal{D}_1$. The robustness here will be represented by a largest interval of $\tilde{\alpha}$ where error rates of trials are not larger than a preset value ϵ. Therefore, since the top three lowest error rates among all trials shown in Fig. 1(a) are 0.147%, 0.196% and 0.245% and the error rate curves in Fig. 1(a) are almost convex, Fig. 2(b) demonstrates the robustness of $\tilde{\alpha}$ to error rate when $\epsilon = 0.15\%, 0.20\%, 0.25\%$ shown in solid, dotted and dash-dotted lines, respectively, for each $d \in \mathcal{D}_1$. These robustness intervals are moved from left to right progressively when d is increasing. In addition, we order the quality of $\widehat{\mathrm{var}(\mathfrak{C}_{n,d})}$ based on the distance from $\frac{-a_1}{2a_2}$ to the robustness interval with the lowest ϵ. The shorter the distance is, the better $\widehat{\mathrm{var}(\mathfrak{C}_{n,d})}$ is. Due to the constraint $\tilde{\alpha} \in [0, 1)$, $\arg\max_{\tilde{\alpha}} \widehat{\mathrm{var}(\mathfrak{C}_{n,d})} = \min\{\max\{\frac{-a_1}{2a_2}, 0\}, 0.99\}$ is pointed-marked in Fig. 2(b). The value of $\frac{-a_1}{2a_2}$ increases when d is raised. If $\frac{-a_1}{2a_2} \notin [0, 1)$, the corresponding d is regarded as bad and is marked with a dotted-line cup on the right side of Fig. 2(b). On the other hand, when $\frac{-a_1}{2a_2} \in [0, 1)$, d is qualified and is marked with a solid-line cup. We call these qualified ones as survival degrees. The lengths of robustness intervals w.r.t. a preset error rate $\epsilon \in \{0.15\%, 0.20\%, 0.25\%\}$ for these survival degrees are almost the same. The median $d = 6$ of the set $\{3, \dots, 9\}$ of survival degrees has the shortest distance between $\frac{-a_1}{2a_2}$ and the robustness interval w.r.t. $\epsilon = 0.15\%$. Note that the integers on the rightest side of Fig. 2(b) stand for degree d, and ($\tilde{\alpha}^*$,error rate) is evaluated from the error rate curves in Fig. 1(a) at $\tilde{\alpha}^* = \arg\max_{\tilde{\alpha}} \widehat{\mathrm{var}(\mathfrak{C}_{n,d})}$.

Therefore, we conclude our observations: when a quadratic curve $\widehat{\mathrm{var}(\mathfrak{C}_{n,d})}$ meets the two requirements $a_2 < 0$ and $\frac{-a_1}{2a_2} \in [0, 1)$, the corresponding degree d is a survival candidate for the best degree to transformed polynomial kernel SVM. In addition, the median of survival degrees can be the best and most robust one. These observations are utilized to develop an approach to kernel parameter $(d, \tilde{\alpha})$ optimization in transformed polynomial kernel SVM in Subsect. 6.1.

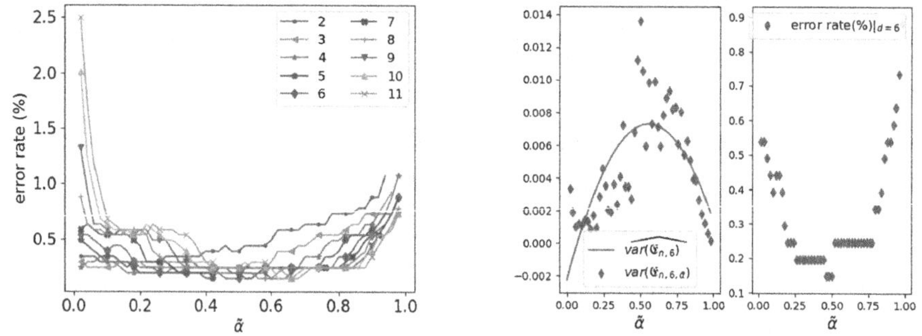

(a) error rate (%) (b) $\mathrm{var}(\mathfrak{C}_{n,6,\tilde{\alpha}})$ and error rate (%) @ $d = 6$

Fig. 1. The relations among $(d, \tilde{\alpha})$, error rate and $\mathrm{var}(\mathfrak{C}_n)$ on MNIST-23.

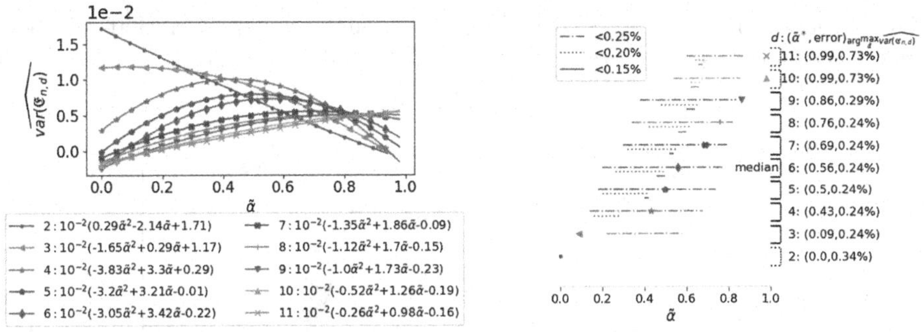

(a) $\widehat{\mathrm{var}(\mathfrak{E}_{n,d})}$ (b) robustness intervals

Fig. 2. The quadratic LSE fitted curves of $\widehat{\mathrm{var}(\mathfrak{E}_{n,d})}$'s and the corresponding robustness intervals on MNIST-23.

5.3 Regulator C

With decided $(d, \hat{\alpha}) = (6, 0.56)$, this subsection discusses the estimation of C w.r.t the change of the Lagrangian dual function value. The dash-dotted line in Fig. 3(a) marks the first iteration k' that meets $|\Delta\tilde{\Theta}^{(k)}| < 1$; while the dotted line represents the $\lfloor m/100 \rfloor$th iteration. Figure 3(a) shows that when $k < k'$, the Lagrangian dual function value $\theta(\boldsymbol{\lambda})$ in (2) is still not converging. Therefore, in order to get the distribution of non-clipped update step size $|t^{(k)}|$ before the Lagrangian dual function value starts to converge, keep collecting $|t^{(k)}|$ till $|\Delta\tilde{\Theta}^{(k)}| < 1$. Figure 3(b) shows a density histogram of $\mathcal{T}_{k'} \triangleq \{|t^{(k)}|, k < k'\}$, which looks like a gamma distribution. Hence, we set a null hypothesis that $|t^{(k)}|$ has a gamma distribution $\frac{1}{\Gamma(a)\beta^a}(z-c)^{a-1}e^{-\frac{z-c}{\beta}}, z \in (c, \infty)$, with parameter (a, β, c). Since the Chi-square test gives p-value≈ 1, it is hard to reject the null hypotheses that the variable $|t^{(k)}|$ has a gamma distribution with $(4.37, 7.74, 0.29)$ shown as the dash-dotted curve in Fig. 3(b). Because the multipliers λ_i, λ_j will be restricted in $[0, C]$ after updated by $t'^{(k)}$, the value of C will be selected to let most of τ–violating pairs have non-clipped update step $t'^{(k)} = t^{(k)}$. Therefore, we evaluate C based on the assumed gamma distribution of $|t^{(k)}|$.

Since we are interested in screening out hyperparameters in an early stage, $\mathcal{T}_{\hat{k}} \triangleq \{|t^{(k)}|, k < \hat{k} = \min\{k', \lfloor m/100 \rfloor\}\}$ will be used to fit a gamma distribution of $|t^{(k)}|$. Let $\hat{\mu} = E[\mathcal{T}_{\hat{k}}]$ and $\hat{\sigma^2} = \mathrm{var}(\mathcal{T}_{\hat{k}})$. The estimators of parameters a, β, c of a gamma distribution are $\hat{a} = \frac{\hat{\mu}^2}{\hat{\sigma^2}}$, $\hat{\beta} = \frac{\hat{\sigma^2}}{\hat{\mu}}$ and $\hat{c} = \frac{1}{2}\min_{t \in \mathcal{T}_{\hat{k}}} t$. Furthermore, denote $z_{0.99}$ as 99%-quantile of a gamma distribution with parameters $(\hat{a}, \hat{\beta}, \hat{c})$. The approximated gamma distribution w.r.t. $\mathcal{T}_{\hat{k}}$ is with $(\hat{a}, \hat{\beta}, \hat{c}) = (4.23, 8.07, 0.002)$ shown as the dotted curve in Fig. 3(b). The dash and solid lines in Fig. 3(b) represent $\hat{z}_{0.99}$ and $\hat{z}_{0.99} + \sqrt{\hat{\sigma^2}}$ respectively. Only three $|t^{(k)}|$'s locate between these two lines and none is above the dashed line. Since multipliers, λ_i, λ_j, are initially set to zeros, $C = \hat{z}_{0.99} + \sqrt{\hat{\sigma^2}}$ is large enough to make sure that selected τ-violating pairs in early iterations have non-clipped update step.

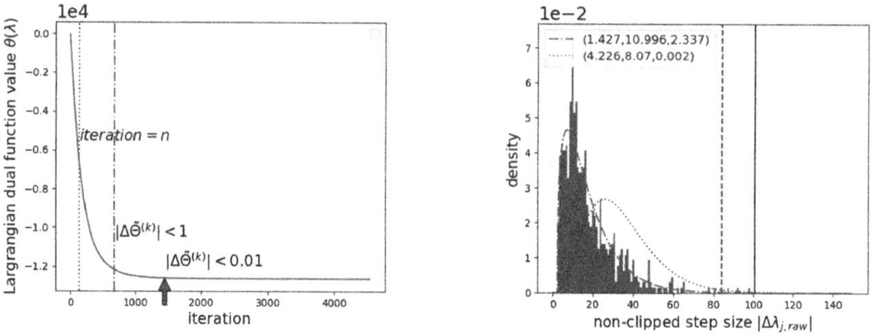

(a) the Lagrangian dual function value $\theta(\boldsymbol{\lambda})$ (b) density histogram of $|t|$

Fig. 3. The distribution of non-clipped step size $|t|$ when the Lagrangian dual function value is under fast decreasing, where $(d, \tilde{\alpha}) = (6, 0.56)$ on MNIST-23.

6 Devised Algorithm and Performance Evaluation

The observations in Sect. 5 motivate us to screen out d, $\tilde{\alpha}$ and C for transformed polynomial kernel SVM from the first few iterations of SMO-SVM. With the selected hyperparameters, a polynomial kernel SVM can be applied to complete the training. When SVM training is implemented with SMO, τ can be made larger when the cost-function in (2) is going to converge for earlier termination. Our approach in the parameter optimization of transformed polynomial kernel SVM will be described in details in Subsect. 6.1. Subsection 6.2 illustrates an estimator for an appropriate altered tolerance to terminate SMO-SVM training process earlier, where the tolerance is initialized with 0. At the end of this section, the performance of miniCV is demonstrated in Subsect. 6.3.

6.1 Devised Algorithm

We compile our key findings in Sect. 5 into a mini core validation via light-weighted SMO-SVM. The light-weighted SMO-SVM only iterates $n = \min\{\max\{30, \min\{\lfloor \frac{m}{100} \rfloor, k'\}\}, 200\}$ times in a trial, where k' is defined in Subsect. 5.3. The goal of mini core validation is to determine kernel parameters $\tilde{\alpha}$ and d for a robust transformed polynomial kernel SVM and then evaluate an appropriate C w.r.t. the decided $\tilde{\alpha}$ and d. We call this parameter optimization method as *miniCV*.

Following Subsects. 4.1 and 5.1, regulator C and tolerance τ are initially set to infinity and 0 respectively to ensure that $|t^{(k)}|$, $\Delta\tilde{\Theta}(\lambda_j^{(k)})$ and η_{ij} extracted from the kth iteration are not influenced by C and τ for all $k < n$. Each light-weighted SMO-SVM trial outputs $\text{var}(\mathfrak{E}_{n,d,\tilde{\alpha}})$ along with the corresponding kernel parameters d and $\tilde{\alpha}$. Denote $\mathcal{D}_2 = \{2, 3, \ldots, 10\}$ and $\mathcal{A}_2 = \{0.1, 0.2, \ldots, 0.9\}$. First light-weighted SMO-SVM is executed based on the grid search $(d, \tilde{\alpha}) \in \mathcal{D}_2 \times \mathcal{A}_2$. For each $d \in \mathcal{D}_2$, fit a quadratic curve for the set $\{\text{var}(\mathfrak{E}_{n,d,\tilde{\alpha}}) | \tilde{\alpha} \in \mathcal{A}_2\}$ by LSE regression. When the fitted curve for the largest $d \in \mathcal{D}_2$ still satisfies the two requirements (a) $a_2 < 0$ and (b) $0 < \frac{-a_1}{2a_2} < 1$, keep extending $\mathcal{D}_2$ with larger d until one of the requirements (a) or (b) is dissatisfied. Then collect the survival degrees which fulfill requirements (a) and (b) and choose the best degree d^* as the median of these survival degrees. Then set $\tilde{\alpha}^*$ to $\frac{-a_1}{2a_2}$ which is the point $\tilde{\alpha}$ at which $\widehat{\text{var}(\mathfrak{E}_{n,d^*})}$ reaches it maximum. Next with the decided $(d^*, \tilde{\alpha}^*)$, run light-weighted SMO-SVM one more time with n iterations to determine C^* by the approach in Subsect. 5.3.

Therefore the hyperparameters in transformed polynomial kernel SVM are optimized. These approaches in estimating $d, \tilde{\alpha}$ and C together define miniCV. For multiclassification polynomial kernel SVM with classes in Z where $|Z| > 2$, apply one-versus-all for all $z \in Z$ in miniCV and let D_z be the resulted set of survival degrees, i.e., $\widehat{\text{var}(\mathfrak{E}_{n,d_z})}$ satisfies requirements (a) and (b) $\forall d_z \in \mathcal{D}_z$. The common best degree d^* is selected as the median in the set $\bigcup_{z \in Z} \mathcal{D}_z$ of all survival degrees. Let Z' be the set of all classes $z \in Z$ such that $d^* \in \mathcal{D}_z$. The common best $\tilde{\alpha}^*$ is selected as the median in the set of all $\frac{-a_1}{2a_2}$, corresponding to the survival quadratic curve $\widehat{\text{var}(\mathfrak{E}_{n,d_z=d^*})}$ obtained in the one-versus-all miniCV for $z \in Z'$. Later, execute light-weighted SMO-SVM one more time for each $z \in Z$ to determine C_z^* individually with the common pair $(d^*, \tilde{\alpha}^*)$. The remaining is to train a polynomial kernel SVM with the determined hyperparameters.

6.2 An Alterable Tolerance

We use the experiment in Subsect. 5.3 to describe an estimator for an altered tolerance in order to terminate training process with SVM-SMO. When the first iteration k'' reaches the criterion $|\Delta\tilde{\Theta}^{(k)}| < 0.01$, marked by an arrow in Fig. 3(a), the value of the Lagrangian dual function is close to its converged value. Let $F_{I_{0,+}} = \{F_r^{(k'')}, r \in I_0 | y_r > 0\}$ $F_{I_{0,-}} = \{F_r^{(k'')}, r \in I_0 | y_r < 0\}$. Denote $Q(F, q)$, $q \in [0, 1]$ as the q-quantile of a set F and $IQR_F = Q(F, 0.75) - Q(F, 0.25)$. Then a range $\mathcal{R}(F_{I_0})$ for $F_{I_{0,+}} \cup F_{I_{0,-}}$ based on box plot concept, is defined as

$$\begin{aligned}\mathcal{R}(F_{I_0}) \triangleq [&\min\{Q(F_{I_{0,+}}, 0.25) - 1.5IQR_{F_{I_{0,+}}}, Q(F_{I_{0,-}}, 0.25) - 1.5IQR_{F_{I_{0,-}}}\},\\ &\max\{Q(F_{I_{0,-}}, 0.75) + 1.5IQR_{F_{I_{0,-}}}, Q(F_{I_{0,+}}, 0.75) + 1.5IQR_{F_{I_{0,+}}}\}]. \quad (8)\end{aligned}$$

By assuming $F_{I_{0,-}}$ and $F_{I_{0,+}}$ to have normal distributions, the possibility that an element in $F_{I_{0,-}}$ or $F_{I_{0,+}}$ is not in $\mathcal{R}(F_{I_0})$ is less than 0.7%. Due to the distribution of $\boldsymbol{F}$ at a converged stage, not shown, only F's related to I_0 are considered. Since $\theta(\boldsymbol{\lambda})$ is converging, we alter tolerance τ' into the length of the interval $\mathcal{R}(F_{I_0})$, which is tight enough to form the stopping criterion $\Delta F^{(k-1)} \leqslant \tau'$ w.r.t. the given sample S and transformed polynomial kernel SVM.

6.3 Performance Evaluation

The test performance of miniCV via all 10 handwritten digits in the MNIST dataset [15] is 1.76 % with $(d^*, \tilde{\alpha}^*) = (5, 0.604)$ for non-deskewing and 1.06% with $(d^*, \tilde{\alpha}^*) = (6, 0.557)$ for deskewing. The later is comparable to the MNIST benchmark in [15] with a polynomial kernel SVM ($d = 4$, error rate (%) = 1.1%). We also select three datasets, ionosphere, wilt [5] and HTRU2 [9], from UCI [3] to test miniCV and demonstrate the performance in Table 1. The first column is the name of the datasets and the number of classes are shown in the parentheses. The size of a dataset is listed in the second column. The selected UCI datasets are repeatedly randomly partitioned into training (70%) and test (30%) sets and used to train with transformed polynomial kernel SVMs using miniCV for 30 times. Denote $a \pm b = mean \pm std$ in Table 1. Then, the third to sixth columns stand for the statistics of d, $\tilde{\alpha}$, error rate (%) and the duration of miniCV respectively when the dataset is trained by transformed polynomial SVM with miniCV. The rest two columns with subscript 'o' represent that the current best other models and the related error rates (%). Table 1 shows when the size of dataset becomes larger, the performance of miniCV in d, $\tilde{\alpha}$ and error rate (%) is more stable with smaller *std* and more comparable to other best models. Furthermore, the duration of parameter optimization is more acceptable when miniCV is applied, compared to those in [13]. Note that the miniCV shown in Table 1 is implemented with python in sequential programming.

Table 1. Performance evaluation of miniCV implemented with Python.

Dataset	Size	d^*	$\tilde{\alpha}^*$	Error (%)	Duration (s)	model$_o$	error$_o$(%)
ionosphere (2)	351	4.6 ± 1.3	0.49 ± 0.1	11.5 ± 2.8	0.71 ± 0.23	BA-SVM	**3.7**[13]
wilt (2) [5]	4839	4.7 ± 1.8	0.45 ± 0.18	$\mathbf{1.9 \pm 0.7}$	9.1 ± 4.2	EIG-GA	9.6 [12]
HTRU2 (2) [9]	17898	4.5 ± 0.9	0.52 ± 0.03	2.1 ± 0.18	120.4 ± 30	fuzzy knn	2.2 [10]

7 Conclusion and Future Scope

In this research, we propose miniCV to optimize hyperparameters in transformed polynomial kernel SVMs. It speeds up the optimization via curve fitting of $\{\mathrm{var}(\mathfrak{C}_{n,d,\tilde{\alpha}}),\ \tilde{\alpha} \in \mathcal{A}_2\}$. Selecting medians of d and $\tilde{\alpha}$ among survival degrees maintains the robustness of the trained machine. In addition, since $|t^{(k)}|$ has a gamma distribution, regulator C is able to be estimated. When the dataset is large enough to account for the differences among classes, the hyperparameters screened out by miniCV are more robust to train kernel-based SVM. On the other hand, the results are usually fluctuating for small datasets. In the future, we will continuously explore our research on the usage of the variable $\mathrm{var}(\mathfrak{C}_{n,d,\tilde{\alpha}})$ and make miniCV more stable to both large and small datasets, and then apply miniCV to other machine learning topics and algorithms.

References

1. De Souza, B.F., De Carvalho, A.C.P.L.F., Calvo, R., Ishii, R.P.: Multiclass SVM model selection using particle swarm optimization. In: 2006 Sixth International Conference on Hybrid Intelligent Systems (HIS 2006), p. 31, December 2006. https://doi.org/10.1109/HIS.2006.264914
2. Devijver, P., Kittler, J.: Pattern Recognition: A Statistical Approach. Prentice-Hall, London (1982)
3. Dua, D., Graff, C.: UCI machine learning repository (2017). http://archive.ics.uci.edu/ml
4. Hsu, C.W., Lin, C.J.: A comparison of methods for multiclass support vector machines. Trans. Neur. Netw. **13**(2), 415–425 (2002). https://doi.org/10.1109/72.991427
5. Johnson, B., Tateishi, R., Hoan, N.: A hybrid pansharpening approach and multiscale object-based image analysis for mapping diseased pine and oak trees. Int. J. Remote Sens. **34**, 6969–6982 (2013). https://doi.org/10.1080/01431161.2013.810825
6. Keerthi, S.S., Shevade, S.K., Bhattacharyya, C., Murthy, K.R.K.: Improvements to Platt's SMO algorithm for SVM classifier design. Neural Comput. **13**(3), 637–649 (2001). https://doi.org/10.1162/089976601300014493
7. Keerthi, S., Gilbert, E.: Convergence of a generalized SMO algorithm for SVM classifier design. Mach. Learn. **46**(1), 351–360 (2002)
8. Lessmann, S., Stahlbock, R., Crone, S.F.: Genetic algorithms for support vector machine model selection. In: 2006 IEEE International Joint Conference on Neural Network Proceedings, pp. 3063–3069, July 2006. https://doi.org/10.1109/IJCNN.2006.247266
9. Lyon, R.J., Stappers, B.W., Cooper, S., Brooke, J.M., Knowles, J.D.: Fifty years of pulsar candidate selection: from simple filters to a new principled real-time classification approach. Mon. Not. Roy. Astron. Soc. **459**(1), 1104–1123 (2016). https://doi.org/10.1093/mnras/stw656
10. Mohamed, T.M.: Pulsar selection using fuzzy KNN classifier. Futur. Comput. Inform. J. **3**(1), 1–6 (2018)
11. Platt, J.C.: A fast algorithm for training support vector machines. In: Advances in Kernel Methods-Support Vector Learning, vol. 208, July 1998
12. Tahir, M.A.U.H., Asghar, S., Manzoor, A., Noor, M.A.: A classification model for class imbalance dataset using genetic programming. IEEE Access **7**, 71013–71037 (2019). https://doi.org/10.1109/ACCESS.2019.2915611
13. Tharwat, A., Hassanien, A.E., Elnaghi, B.E.: A BA-based algorithm for parameter optimization of support vector machine. Pattern Recognit. Lett. **93**, 13–22 (2017). Pattern Recognition Techniques in Data Mining
14. Vapnik, V.N.: The Nature of Statistical Learning Theory. Springer, Heidelberg (1995). https://doi.org/10.1007/978-1-4757-2440-0
15. LeCun, Y., Cortes, C., Burges, C.J.: The MNIST database of handwritten digits. http://yann.lecun.com/exdb/mnist/. Accessed 30 Nov 2018
16. Yeh, L.C., Lu, C.C.: Parameter optimization of RBF kernel SVM from miniCV. In: Proceedings of the 15th International Conference on Data Science (ICDATA 2019), July 2019

Analysing the Overfit of the Auto-sklearn Automated Machine Learning Tool

Fabio Fabris(✉) and Alex A. Freitas

School of Computing, University of Kent, Canterbury, Kent CT2 7NF, UK
{F.Fabris,A.A.Freitas}@kent.ac.uk

Abstract. With the ever-increasing number of pre-processing and classification algorithms, manually selecting the best algorithm and their best hyper-parameter settings (i.e. the best classification workflow) is a daunting task. Automated Machine Learning (Auto-ML) methods have been recently proposed to tackle this issue. Auto-ML tools aim to automatically choose the best classification workflow for a given dataset. In this work we analyse the predictive accuracy and overfit of the state-of-the-art auto-sklearn tool, which iteratively builds a classification ensemble optimised for the user's dataset. This work has 3 contributions. First, we measure 3 types of auto-sklearn's overfit, involving the differences of predictive accuracies measured on different data subsets: two parts of the training set (for learning and internal validation of the model) and the hold-out test set used for final evaluation. Second, we analyse the distribution of types of classification models selected by auto-sklearn across all 17 datasets. Third, we measure correlations between predictive accuracies on different data subsets and different types of overfitting. Overall, substantial degrees of overfitting were found in several datasets, and decision tree ensembles were the most frequently selected types of models.

Keywords: Automated Machine Learning · Overfit · Classification

1 Introduction

With the growing popularity and number of Machine Learning (ML) techniques, it is increasingly difficult for users to find the 'best' classification workflow (the combination of pre-processing methods, classification algorithms, and their hyper-parameter settings) to be applied to their data. This task becomes even more difficult when one considers the use of ensemble techniques, which may combine several classification workflows to make the final prediction.

Automated Machine Learning (Auto-ML) techniques were devised to solve the problem of how to automatically choose the best classification workflow for a given user's dataset. Typically, Auto-ML methods perform a search that works by using a dataset with instances with known class labels (the training dataset) and returning a fully-parameterised model to be used to predict the class labels of new unlabelled instances.

G. Nicosia et al. (Eds.): LOD 2019, LNCS 11943, pp. 508–520, 2019.
https://doi.org/10.1007/978-3-030-37599-7_42

The state-of-the-art method for Auto-ML is the auto-sklearn tool [3,8], which was the overall winner of the first ChaLearn Auto-ML challenge [4], and a variant of auto-sklearn also won the second, latest AutoML challenge [1]. Auto-sklearn uses meta-learning and the Sequential Model-based Algorithm Configuration (SMAC) method to build an ensemble of classification workflows given a training dataset. Meta-learning is used to initialise the SMAC search, suggesting a 'reasonable' classification model for the user's dataset, given the estimated predictive accuracy of the model in other datasets. Next, the iterative SMAC search uses a Bayesian approach to explore the huge space of possible classification workflows, training several candidate models per iteration and returning the best model found by the search.

SMAC methods normally return only the best classification workflow found by the search procedure as the final model. However, auto-sklearn exploits the fact the SMAC search procedure produces several 'good' candidate classification workflows that would normally be discarded. These classification workflows are used to build an ensemble instead of being discarded. By default, this ensemble contains at most 50 classification workflows at each iteration. Each workflow has a weight which is proportional to the workflow's relevance for the final prediction.

By default, auto-sklearn works by randomly dividing the training set into two disjoint sets, a learning set and a validation set. The learning set is used during the SMAC search to build the classification workflows. The validation set is used to estimate the accuracy of the workflows. One key aspect of Auto-ML tools like auto-sklearn, which has been to a large extent neglected in the literature thus far, is the degree of overfitting resulting from repeatedly using a fixed validation set across the search. Even though the training set has been properly divided into learning and validation sets, the fact that there are several iterations, each using the accuracy estimated in the validation set to guide the search, may lead to a high degree of overfitting to the validation set. That is, the search may select algorithms and their settings that classify the instances in the validation set very well (since it had several iterations to fine-tune its parameters) but fail to classify the test instances properly (due to model overfitting to the validation set). Besides the just defined overfit (between the validation and test sets), we also analyse two other types of overfit: (1) between the learning and validation sets and (2) between the learning and test sets.

This work has three contributions, all related to experimental analyses of auto-sklearn, as follows. First, we estimate the degree of overfitting of the tool using 3 measures of overfit. This analysis can be useful to ascertain to what extent SMAC's iterative learning procedure is actually hindering predictive accuracy due to overfit. Second, we identify the base classification algorithms most frequently selected by the SMAC method. This can be useful to find classification algorithms that are 'good' across several application domains and could be used as a principled 'first approach' to tackle classification problems. Third, we measure the correlations between several experimental results, aiming to uncover non-obvious relationships between the experimental variables that may lead to further insights about auto-sklearn.

We know of only one AutoML work [6] that measures and briefly analyses overfit, however, there is no work performing a comprehensive overfit analysis (considering 3 types of overfit) comparable to the one presented here. Actually, a very recent and comprehensive survey of Auto-ML studies [9] does not even mention the issue of overfitting.

The rest of this paper is organised as follows: Sect. 2 presents our experimental methodology. Section 3 presents the analysis of our results. Section 4 presents our conclusions and directions for future work.

2 Experimental Methodology

To measure the predictive accuracy of auto-sklearn we used 17 datasets which are pre-divided into training and test sets, taken from [7]. The training sets are further divided into a 'learning' set (which the SMAC method will use to build the ensemble) and a 'validation' set, which will be used to estimate the predictive accuracy of the models in each iteration of the SMAC search. The test set is never shown to the SMAC method, being reserved to estimate the predictive accuracy of the ensemble classifier created by each iteration of auto-sklearn. Note that, in a normal experimental scenario, the test set would be used only to evaluate the predictive accuracy of the ensemble returned after the last iteration of the SMAC search. However, since this study is interested in the overfit behaviour across iterations, we report results where the test set is also used to evaluate the ensemble at each iteration of the SMAC search. We emphasise that this procedure does not influence the SMAC search in any way.

Table 1 shows basic characteristics of the used datasets [7]. These datasets are very diverse in terms of application domain and dataset characteristics, varying from small datasets with 6 features and 1210 instances (car) to relatively large datasets with 3072 features (CIFAR-10-Small) or 43,500 instances (shuttle).

Auto-sklearn was run for 30 h on each dataset, using default settings for the other parameters, except that it optimized the AUROC measure (Sect. 2.1). We ran Auto-sklearn in a computing cluster comprising 20 8-core Intel Haswell machines, with a clock speed of 2.6 GHz and 16 GB of RAM memory.

2.1 Predictive Accuracy Estimation

We use the popular Area Under the Receiver Operating Characteristic curve (AUROC) to measure the predictive accuracy of auto-sklearn [5]. An AUROC of 1.0 indicates that the model correctly ranked all positive instances after the negative ones. An AUROC of 0.5 indicates that the classifier achieved an accuracy equivalent to randomly ranking the instances. For datasets with more than two class labels, the AUROC is calculated individually per class label and then averaged, weighted by the number of instances annotated with each class label.

Table 1. Dataset statistics.

Dataset name	Number of features	Number of training instances	Number of test instances	Number of class labels
gisette	5000	4900	2100	2
shuttle	9	43500	14500	7
kr-vs-kp	36	2238	958	2
car	6	1210	518	4
semeion	256	1116	477	10
abalone	8	2924	1253	28
amazon	10000	1050	450	50
convex	784	8000	50000	2
madelon	500	1820	780	2
waveform	40	3500	1500	3
CIFAR-10-Small	3072	10000	10000	10
dexter	20000	420	180	2
winequalitywhite	11	3425	1468	7
yeast	8	1034	445	9
german_credit	20	700	300	2
dorothea	100000	805	345	2
secom	590	1097	470	2

2.2 Estimation of Three Types of Overfitting

We measure three types of overfitting, which can be used to analyse different aspects of auto-sklearn's training procedure, as follows.

1. The *learning-validation* overfit – defined as the difference between the predictive accuracy in the learning and validation sets. This overfit can measure if auto-sklearn is successfully controlling the overfit of its training procedure by using the accuracy estimated in the validation set.
2. The *learning-test* overfit – defined as the difference between the predictive accuracy in the learning and test sets. This overfit measures the conventional overfit in standard classification, i.e., the difference between the accuracy in the learning set versus the expectedly smaller accuracy in the test set.
3. The *validation-test* overfit – defined as the difference between the predictive accuracy in the validation and test sets. This overfit can be interpreted as a measure of the effectiveness of using an internal validation set (part of the training set) to estimate the predictive accuracy on the test set (not used during training). That is, if the predictive accuracy in the validation set reflects the expected accuracy in the test set, this overfit should be close to zero. Note that even though the instances in the validation set are not

directly used to train the models, the accuracy of the classification models is repeatedly estimated across iterations using the validation set. Therefore, the model choice can overfit the validation set across iterations and the ensemble can perform badly in the final test set while achieving good accuracy in the validation set. Arguably, this is the most interesting type of overfitting from an Auto-ML perspective, and it is not normally investigated in the literature.

2.3 Analysis of the Selected Classification Models

To analyse the classification models present at the final iteration of auto-sklearn we measure the frequency each classification algorithm is chosen and the total relevance weights associated with each classification workflow. Note that a classification workflow may be selected several times to be present in the ensemble, but its total weight may be lower than a workflow that is selected only once.

3 Results

3.1 Predictive Accuracy and Overfit Results

Table 2 shows the main experimental results of our analysis, ordered by increasing degree of validation-test overfit. The columns show, respectively: the dataset name; the final learning set AUROC (the AUROC on the learning set at the last iteration of the SMAC search); the final validation set AUROC; the final test set AUROC; the learning-validation overfit (the final learning set AUROC minus the final validation AUROC); the learning-test overfit (the final learning set AUROC minus the final test AUROC); the training-test overfit (the final training set AUROC minus the final test AUROC); and the total number of iterations. The last row of this table shows the mean overfits across datasets. Note that we do not average the AUROCs as they are not directly comparable, easier problems will naturally have greater AUROCs than harder ones.

We can see in Table 2 that the learning AUROC (second column) in almost all datasets is 1.0 or very close to 1.0. Just the dataset "abalone" had an AUROC smaller than 0.97. This shows that the SMAC method is building models with high predictive accuracy in the learning set, as expected.

By analysing the column "Learning-Val. overfit" (fifth column) we can see that almost all datasets (except shuttle) exhibit this kind of overfit, the validation AUROC is almost always smaller than the learning AUROC. This is also expected, as SMAC did not have access to the validation instances during the training of each model. The degree of learning-validation overfit was smaller than 1% in 8 of the 17 datasets. However, a large degree of learning-validation overfit was observed in 6 datasets: 0.160 in german_credit, 0.149 in winequalitywhite, 0.131 in CIFAR-10-Small, 0.129 in yeast, 0.097 in secom, and 0.096 in abalone.

Also, by analysing the column "Val.-Test overfit" (sixth column) we can see that, with the exception of the first 4 datasets, the validation set AUROC is always over-optimistically estimated when compared to the test set AUROC, suggesting that the models are indeed overfitting in the validation set.

Table 2. Results ordered by increasing degree of validation-test overfit.

Dataset	Learning AUROC	Val. AUROC	Test AUROC	Learning-Val. overfit	Val.-Test overfit	Learning-Test overfit	Its
gisette	1.000	0.998	0.998	0.003	−0.001	0.002	210
shuttle	1.000	1.000	1.000	0.000	0.000	0.000	8
kr-vs-kp	1.000	0.999	1.000	0.001	0.000	0.000	34
car	1.000	0.999	0.999	0.001	0.000	0.001	324
semeion	1.000	0.999	0.998	0.001	0.001	0.003	174
abalone	0.883	0.788	0.785	0.096	0.003	0.099	161
amazon	1.000	0.995	0.991	0.005	0.003	0.009	207
convex	1.000	0.933	0.927	0.067	0.006	0.073	253
madelon	1.000	0.966	0.959	0.034	0.007	0.041	296
waveform	0.986	0.979	0.971	0.007	0.008	0.015	160
CIFAR-10-Small	1.000	0.869	0.861	0.131	0.008	0.139	193
dexter	1.000	0.994	0.982	0.006	0.012	0.018	428
winequalitywhite	1.000	0.851	0.829	0.149	0.023	0.171	186
yeast	0.997	0.868	0.840	0.129	0.028	0.157	74
german_credit	1.000	0.840	0.765	0.160	0.075	0.236	206
dorothea	0.979	0.966	0.878	0.013	0.088	0.101	241
secom	0.974	0.877	0.702	0.097	0.174	0.272	332
Mean overfit				0.053	0.026	0.079	

Finally, by analysing the column "Learning-Test overfit" (seventh column), we can see that, overall, it presents the largest overfit values across datasets. This is expected, as the SMAC method never had access to the test set to train the ensemble, but it did have direct access to the learning set (to train the classification models) and indirect access to the validation dataset (for predictive accuracy estimation).

Analysing the mean overfits (last row) we can see that the average learning-validation overfit is smaller than the learning-test overfit. This is expected, as the AUROC in the test set is usually smaller than in the validation set while the learning AUROC is the same for these two measures of overfitting. Also, the average validation-test overfit is smaller than the learning-test overfit, this is also expected, as the validation AUROC is naturally smaller than the learning AUROC, which drives the validation-test overfit value down.

Figures 1 and 2 show the evolution of the accuracy of the models across iterations by calculating the AUROC in the learning, validation and test sets. The datasets in the figures are ordered in the same sequence as in Table 2. To save space, we do not show figures associated with the first 5 datasets in Table 2, as those results are trivial (a horizontal line at AUROC = 1.0). Note that several plots appear to show only two lines, this is because the test and validation lines are overlapping.

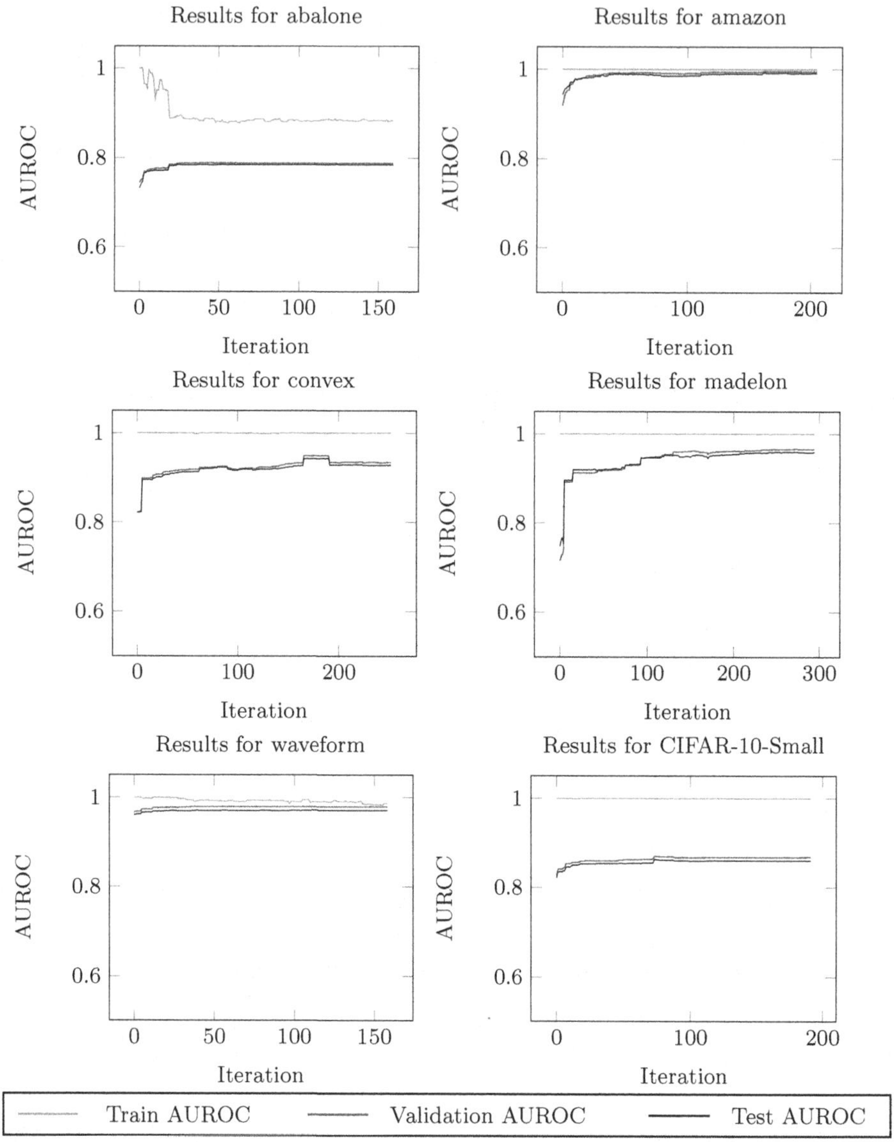

Fig. 1. Training, validation, and test AUROC variation across iterations. Overall, these plots show a 'good' convergence profile: the test AUROC increases with the iteration number.

Figure 1 and the first three plots in Fig. 2 show that the validation AUROC and the test AUROC are tracking very closely. Hence, for these datasets, the validation AUROC is a good estimator for the test AUROC, although there is little improvement in the validation and test AUROC along the search iterations for some datasets.

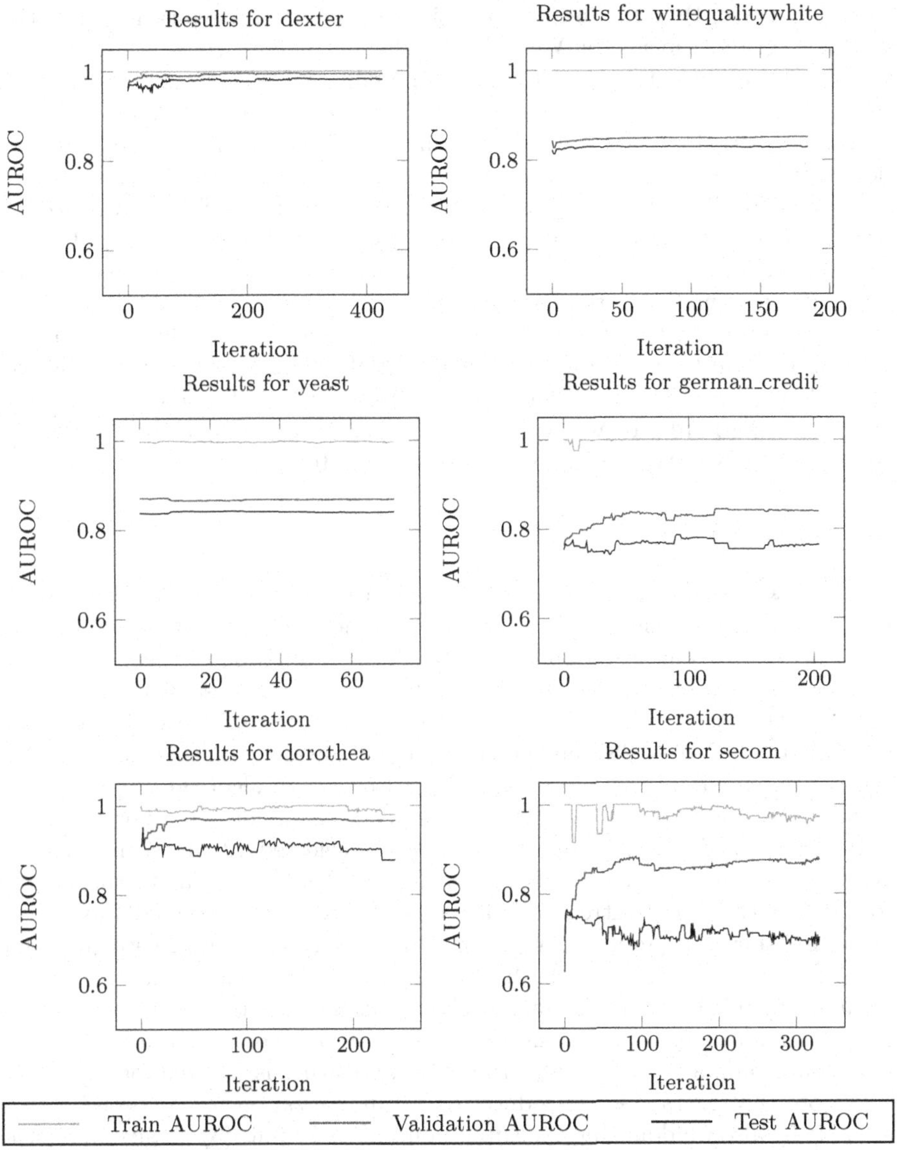

Fig. 2. Training, validation, and test AUROC values across iterations. The plots in this figure show a poor convergence profile: as the iteration number increases, the test AUROC remains close to its initial value, with no sign of clear improvement.

The last three plots in Fig. 2, however, show that auto-sklearn is clearly overfitting to the validation set. Note that, in these plots, the validation AUROC increases with the iterations, while the test AUROC decreases. The results for the dorothea dataset are especially interesting, since there was a test AUROC *decrease* between the first and last iterations of the SMAC algorithm while the validation AUROC increases. We attribute this behaviour, which is also observable to a lesser extent in datasets german_credit and secom, to the fact that the validation set, which is never directly used to build the model, is being constantly queried to estimate the performance of the models, which leads the selected models to overfit the validation set. That is, since the performance in the same validation set is being repeatedly used to guide the SMAC algorithm's search, SMAC is selecting parameter settings and algorithms that are, by some degree of random chance, good at classifying instances in the validation set, but have poor performance in the testing set.

On another note, 11 of the 17 datasets evaluated in this work had a difference between the final test AUROC and the test AUROC after one iteration of just 0.01 or less. Out of those 11 datasets, the first five (gisette, shuttle, kr-vs-kp, car and semeion) are probably too "easy" for classification algorithms, having a test AUROC after just one iteration already close to 1.

3.2 Selected Models

Table 3 shows the classification algorithms selected to be part of the final ensemble, with their total weight and the number of times each algorithm was selected to be part of the final iteration's ensemble. The summation of these frequency numbers is the number of members in the final iteration's ensemble.

Note that the algorithm frequency varies from just one final ensemble member in the shuttle dataset to 28 members in the dexter dataset; which is much smaller than the maximum allowed number of 50. Also, a high algorithm frequency in the final ensemble does not necessarily imply high importance. For instance, for the kr-vs-kp dataset, gradient boosting was selected 10 times, many more than the other algorithms. However, its weight (0.50) is similar to the weight associated with the extra_trees algorithm (0.46), which was selected only twice.

Table 4 shows a summary of Table 3, presenting the average weights and selection frequencies of each classification algorithm across all datasets.

Tables 3 and 4 contain 8 unique classification algorithms: (1) extra_trees and (2) random_forests, both ensembles of decision trees that randomly sample instances and a feature subset from which a feature is selected for each data split. Note that extra_trees introduce extra randomization to the decision tree ensemble by also randomizing the cutoff point of each split. (3) passive_agressive is an online version of SVM. (4) gradient_boosting induces a regression tree based on the negative gradient of the loss function. (5) lda is a Linear Discriminant Analysis classifier. (6) libsvm_svc and (7) liblinear_scv are both versions of a SVM classifier. (8) qda is the Quadratic Discriminant Analysis classifier.

Interestingly, the two classification algorithms with the highest average weight and average frequency across datasets are based on decision tree ensembles.

Table 3. Distribution of selected classification algorithms per dataset. The second column shows the final composition of the SMAC-generated ensemble, displaying the name of the base classification algorithm, followed by two numbers in parenthesis: the total weight of the algorithm in the interval [0, 1] and the number of times the algorithm was selected to be in the final ensemble for a given dataset.

Dataset	Selected algorithm (weight, frequency)
gisette	extra_trees (0.18, 1), gradient_boosting (0.30, 8), passive_aggressive (0.52, 1)
shuttle	random_forest (1.00, 1)
kr-vs-kp	liblinear_svc (0.04, 1), extra_trees (0.46, 2), gradient_boosting (0.50, 10)
car	passive_aggressive (1.00, 16)
semeion	passive_aggressive (0.42, 8), extra_trees (0.58, 5)
abalone	random_forest (0.42, 7), liblinear_svc (0.58, 16)
amazon	extra_tree (0.12, 4), random_forest (0.30, 12), passive_aggressive (0.58, 3)
convex	gradient_boosting (1.00, 15)
madelon	extra_trees (0.36, 7), libsvm_svc (0.64, 6)
waveform	liblinear_svc (0.02, 1), random_forest (0.22, 6), passive_aggressive (0.24, 3), lda (0.52, 5)
CIFAR-10-Small	random_forest (0.40, 10), qda (0.60, 1)
dexter	lda (1.00, 28)
winequalitywhite	extra_trees (1.00, 18)
yeast	random_forest (0.26, 1) extra_trees (0.74, 15)
german_credit	libsvm_svc (0.02, 1), random_forest (0.18, 3), extra_trees (0.80, 12)
dorothea	random_forest (1.00, 1)
secom	extra_trees (1.00, 12)

Table 4. Average weight and selection frequency of each algorithm across all datasets.

Classification algorithm	Avg. weight across datasets	Avg. freq. across datasets
extra_trees	0.31	4.47
random_forest	0.22	2.41
passive_aggressive	0.16	1.82
gradient_boosting	0.11	1.94
lda	0.09	1.94
libsvm_svc	0.04	0.41
liblinear_svc	0.04	1.06
qda	0.04	0.06

3.3 Statistical Analysis

The statistical analyses presented below are based on the Pearson's correlation coefficient (r) under the null hypothesis of $r = 0$ using a t-test. We have tested in total 12 hypotheses, one hypothesis for each unordered column pair in Table 2, excluding the hypotheses involving correlations between the variable pairs that are deterministically correlated (e.g.: 'Test AUROC' and 'Learning-test overfit', since Learning-test overfit is equal to Learning AUROC minus Test AUROC). We used the well-known Bonferroni correction for multiple hypotheses testing [5], so the adjusted threshold of statistical significance (the α value) we consider below is $\alpha = 0.05/12 \approx 0.004$. We show the results of these analyses in Table 5.

Table 5. Correlations between pairs of measures in Table 2. The top three rows show the statistically significant correlations ($\alpha = 0.004$) among the measures in Table 2 using Pearson's correlation coefficient (r), ordered by absolute r value. The remaining rows show the non-statistically significant correlations, also ordered by absolute r value. The first and second columns show the measures being tested, the third column shows the r value and the last column shows the p-value associated with that r value.

Measure 1	Measure 2	r	p-value
Validation AUROC	Test AUROC	0.90	1.21×10^{-6}
Test AUROC	Learning-validation overfit	-0.84	2.30×10^{-5}
Validation AUROC	Learning-test overfit	-0.81	7.61×10^{-5}
Learning AUROC	Validation AUROC	0.55	2.23×10^{-2}
Learning AUROC	Test AUROC	0.45	6.81×10^{-2}
Validation-test overfit	Iterations	0.31	2.33×10^{-1}
Learning-test overfit	Iterations	0.13	6.30×10^{-1}
Learning AUROC	Validation-test overfit	-0.11	6.80×10^{-1}
Test AUROC	Iterations	-0.10	6.97×10^{-1}
Validation AUROC	Iterations	0.06	8.14×10^{-1}
Learn.val.overfit	Iterations	-0.05	8.42×10^{-1}
Learning AUROC	Iterations	0.04	8.68×10^{-1}

As expected, there is a strong statistically significant correlation between the validation AUROC and the test AUROC ($r = 0.90$). Hence, the model's AUROC in the validation set is a good predictor for the AUROC in the test set.

There is a strong, highly statistically significant, negative correlation between the test AUROC and the learning-validation overfit ($r = -0.84$). That is, the greater the learning-validation overfit, the lower the test AUROC. This is expected, as models with a high overfit tend to perform worst in the testing set. Similarly, there is a strong, highly statistically significant correlation between the validation AUROC and the learning-test overfit ($r = -0.81$).

Somewhat surprisingly, there is no statistically significant correlation between the test AUROC and the learning AUROC nor between the validation AUROC and the learning AUROC. This reinforces the need for using validation sets to properly estimate the accuracy of the SMAC method.

Also, unexpectedly, there is no statistically significant correlation between the number of SMAC iterations and any measure of predictive accuracy or overfit. We were expecting that the more the validation set is used to estimate SMAC's performance, the greater would be the potential for overfit, but our analysis did not support this notion.

4 Conclusions and Future Work

In this work we have analysed the following two important aspects of the auto-sklearn tool using 17 datasets: (1) the degree of overfitting of the tool in terms of 3 types of overfitting, and (2) the diversity of the base classification algorithms most selected by the tool. The three overfits are defined as follows. The *learning-validation* overfit is the difference between the predictive accuracy in the learning and validation sets. The *learning-test* overfit is the difference between the predictive accuracy in the learning and test sets The *validation-test* overfit is the difference between the predictive accuracy in the validation and test sets.

We have concluded that there is a strong statistically significant correlation between the AUROC in the validation and testing sets, which suggests that, overall, the AUROC in the validation set is a useful proxy for the AUROC in the test set. We have also detected a strong significant negative correlation between the test AUROC and the learning-validation overfit, which suggests that reducing learning-validation overfit could be an effective approach to increase test AUROC. This is an intuitive conclusion since overfitting to the validations set (part of the training set) should reduce the AUROC on the test set. This conclusion is also actionable, since approaches can be developed to control learning-validation overfit during training, such as re-sampling the learning and validation sets or using cross-validation across SMAC's iterations [2]. Finally, we have also detected a statistically significant negative correlation between the validation AUROC and the learning-test overfit, which suggests that improving the validation AUROC (which is accessible during training) can lead to reduced learning-test overfit.

Regarding the base classification algorithms selected by auto-sklearn across all 17 datasets, the 2 most selected algorithms (with higher average weights and average selection frequency) were ensembles of decision trees.

Future work includes comparing the results obtained using auto-sklearn with other AutoML tools (such as Auto-Weka [7]), as well as investigating the scalability of Auto-sklearn to much larger datasets.

References

1. Feurer, M., Eggensperger, K., Falkner, S., Lindauer, M., Hutter, F.: Practical automated machine learning for the automl challenge 2018. In: International Workshop on Automatic Machine Learning, ICML 2018, pp. 1–12 (2018)

2. Feurer, M., Hutter, F.: Towards further automation in AutoML. In: ICML AutoML Workshop, p. 13 (2018)
3. Feurer, M., Klein, A., Eggensperger, K., Springenberg, J., Blum, M., Hutter, F.: Efficient and robust automated machine learning. In: Advances in Neural Information Processing Systems, vol. 28, pp. 2962–2970. Curran Associates, Inc. (2015)
4. Guyon, I., et al.: A brief review of the ChaLearn AutoML challenge: any-time any-dataset learning without human intervention. In: Workshop on Automatic Machine Learning, pp. 21–30 (2016)
5. Japkowicz, N., Shah, M.: Evaluating Learning Algorithms A Classification Perspective. Cambridge University Press, Cambridge (2011)
6. Kordík, P., Černý, J., Frýda, T.: Discovering predictive ensembles for transfer learning and meta-learning. Mach. Learn. **107**(1), 177–207 (2018)
7. Kotthoff, L., Thornton, C., Hoos, H.H., Hutter, F., Leyton-Brown, K.: Auto-WEKA 2.0: automatic model selection and hyperparameter optimization in WEKA. J. Mach. Learn. Res. **18**(1), 826–830 (2017)
8. Mohr, F., Wever, M., Hüllermeier, E.: Ml-Plan: automated machine learning via hierarchical planning. Mach. Learn. **107**(8–10), 1495–1515 (2018)
9. Yao, Q., et al.: Taking human out of learning applications: a survey on automated machine learning. CoRR abs/1810.13306 (2018)

A New Baseline for Automated Hyper-Parameter Optimization

Marius Geitle(✉) and Roland Olsson

Faculty of Computer Science, Østfold University College, Halden, Norway
{mariusge,rolando}@hiof.no

Abstract. Finding the optimal hyper-parameters values for a given problem is essential for most machine learning algorithms. In this paper, we propose a novel hyper-parameter optimization algorithm that is very simple to implement and still competitive with the state-of-the-art L-SHADE variant of Differential Evolution. While the most common method for hyper-parameter optimization is a combination of grid and manual search, random search has recently shown itself to be more effective and has been proposed as a baseline against which to measure other methods. In this paper, we compare three optimization algorithms, namely, the state-of-the-art L-SHADE algorithm, the random search algorithm, and our novel and simple adaptive random search algorithm. We find that our simple adaptive random search strategy is capable of finding parameters that achieve results comparable to the state-of-the-art L-SHADE algorithm, both of which achieve significantly better performance than random search when optimizing the hyper-parameters of the state-of-the-art XGBoost algorithm for 11 datasets. Because of the significant performance increase of our simple algorithm when compared to random search, we propose this as the new go-to method for tuning the hyper-parameters of machine learning algorithms when desiring a simple-to-implement algorithm, and also propose to use this algorithm as a new baseline against which other strategies should be measured.

Keywords: Hyper-parameter tuning · XGBoost · Random search · L-SHADE · Differential Evolution · Adaptive random search

1 Introduction

Most machine learning methods attempt to train a model whose performance minimizes the expected loss on a set of withheld samples i.i.d to the samples used during training. This is commonly achieved by using optimization strategies that attempt to efficiently estimate the structure and/or parameters of a model from the data.

However, most learning algorithms also have internal mechanics that affect how well the learned models will generalize to unseen data, whose parameter values cannot be estimated directly. These parameters are called hyper-parameters

G. Nicosia et al. (Eds.): LOD 2019, LNCS 11943, pp. 521–530, 2019.
https://doi.org/10.1007/978-3-030-37599-7_43

and finding the best values for these parameters is essential for the machine learning algorithms to build the best possible models.

Since it is not possible to calculate the optimal hyper-parameter values directly, the only viable method of finding them is blind testing of multiple configurations.

While extensive research has been conducted on optimization algorithms throughout the decades, the most common method for finding these hyper-parameters is still a combination of grid and manual searches. However, in 2012, Bergstra and Bengio [2] showed that simply generating and testing a large collection of random points will outperform that approach as the number of hyper-parameters grows. This observation has led to the interest in more advanced algorithms like Bayesian optimization [11].

While Bayesian optimization is likely to be the best algorithm at finding the best parameters with the lowest number of tests, this is an algorithm of significant complexity that requires considerable computation time while simultaneously being difficult to parallelize which makes it unsuitable for many tasks. In this paper, we therefore compare three optimization strategies that all have negligible runtime. Specifically, we compare the random search algorithm [2], the state-of-the-art L-SHADE algorithm [12], and a simple adaptive random search algorithm we introduce in this paper.

The algorithms were tested by optimizing eight of the 10 hyper-parameters of the XGBoost algorithm for 11 datasets, with the algorithms limited to using only 400 evaluations [2]. What we found is that our simple adaptive random search algorithm achieves significantly better results than the basic random search algorithm, and produces results comparable with the much more complicated state-of-the-art L-SHADE algorithm. We therefore propose this adaptive random algorithm as the new default hyper-parameter tuning method.

The remainder of this paper is structured as follows: In Sect. 2, we review related work on hyper-parameter optimization, Sect. 3 describes the optimization algorithms that have been tested in this paper, including the adaptive random algorithm, Sect. 4 describes the methodology used to test the algorithms, and Sect. 5 presents the results and compares the algorithms. Finally, a conclusion is given at the end.

2 Related Work

For finding the hyper-parameters in neural networks, methods based on Bayesian optimization have recently attracted considerable attention. These methods are often based on Gaussian processes and attempt to learn the hyper-parameter distribution. While these approaches have been shown to work well for low-dimensional problems [11], they are also difficult to parallelize, and require significant amount of computation time to calculate the next trial, though there has recently been progress on solving this problem [6,9].

Our work in this paper, however, is focused on optimization algorithms where the runtime requirements for the algorithm itself is negligible. There have been many such algorithms proposed for hyper-parameter optimization, including CMA-ES [8], particle swarm optimization [7], and AUTO-WEKA [13]. Because the published results do not typically allow for a direct empirical comparison of the algorithms between papers, we have relied on the results from the CEC competitions to identify an algorithm to use as a representative of the state-of-the-art in this paper. We have chosen the L-SHADE algorithm, as variations have won many CEC competitions in recent years, and has been top contenders in several more [1,4,10,12].

We should also note that Jaderberg et al. [5] has recently proposed a method for finding both the parameters and hyper-parameters simultaneously for neural networks, thereby avoiding the need for an outer optimization algorithm. Because the gradient boosting method tested in this paper works similarly to the gradient descent method used to train neural networks, it is possible that a similar approach could be used for gradient tree boosting. However, this would require a specialized algorithm.

3 Optimization Algorithms

Thoroughly testing even a small portion of the derivative-free optimization algorithms that have been developed over the years would be impractical. Therefore, we have opted to test only the L-SHADE algorithm as the state-of-the-art representative. As far as we know, this is the first time that L-SHADE has been employed for hyper-parameter optimization. Additionally, we present the basic random search algorithm, which is among the most simple optimization algorithms possible, and our novel adaptive random search algorithm.

3.1 Random Search

Random search is a simple algorithm that involve testing random points independent drawn from a uniform distribution spanning the entire search space, and selecting the best performing configuration as measured by an objective function.

3.2 L-SHADE

L-SHADE is a state-of-the-art variant of the Differential Evolution algorithm that combines the mechanism introduced by the earlier SHADE algorithm for adapting the crossover rate CR and scaling factor F, with a linear reduction in population size. Unlike the other two algorithms tested in this paper, this is a significantly more complicated algorithm, and will therefore not be fully described here. For a complete description, we refer the reader to [12].

The adaptation mechanism works by storing the weighted Lehmer mean of the successful parameters used during one generation into a memory slot. There are M memory slots that are accessed by an index $k(1 \leq k \leq H)$ which is cyclically incremented after every generation in which a new individual is accepted into the population. The value of the current memory slot is then used as the mean of the distributions from which the crossover and scaling factors are drawn.

When the number of function evaluations is sufficiently high, the default number of memory cells, $M = 5$, is likely to work well. However, when the number of function evaluations is severely limited, as in this paper, the adaptation mechanism will be unable to work properly. Similarly, the size of the archive in which discarded individuals are stored for a time and the initial population size might also need to be adjusted.

Therefore, the parameters used for the comparison in this paper were $M = 4$, archive rate $r^{arc} = 1.4$ and initial population size $N_{init} = 32$. These were found by tuning the algorithm on the Soybean dataset.

3.3 Adaptive Random Search

Because the random search algorithm samples the points in the search space independently, it is unable to exploit any of the information about the search space given by previous evaluations. A basic necessity in improving the search performance must therefore be to ensure that subsequent points are in some way drawn from a distribution that is biased towards points that are more likely to be optimal.

We therefore propose to use a new and simple population-based algorithm that is able to achieve results comparable to the state of the art, and much more complicated, L-SHADE algorithm. It is very simple to implement, and embarrassingly easy to parallelize.

procedure ADAPTIVERANDOM(N, $t_{max} \in \mathbb{N}^*$, $\alpha \in [1, N]$, $\zeta \in \mathbb{R}_{>0}$)
  $P_1 \leftarrow N$ samples drawn from a uniform distribution
  $t \leftarrow 1$
  while $t \leq t_{max}$ **do**
    $S \leftarrow$ best α individuals from P_t
    $\boldsymbol{\mu} \leftarrow$ mean(S)
    $\boldsymbol{\sigma} \leftarrow$ std(S)
    $P_{t+1} \leftarrow$ N samples drawn from $U(\boldsymbol{\mu} - \zeta\boldsymbol{\sigma}, \boldsymbol{\mu} + \zeta\boldsymbol{\sigma})$
    $t \leftarrow t + 1$
  end while
  return Best individual from P_{max}
end procedure

Fig. 1. Pseudocode for the adaptive random search algorithm.

The entire algorithm is contained in Fig. 1. It starts by initializing a population P to N individuals drawn independently from a uniform distribution spanning the entire search space. For each generation at time t, the mean μ and standard deviation σ of the α best individuals in P_t is calculated for each dimension. The next generation is then built by independently sampling from a uniform distribution spanning the subspace defined by $\boldsymbol{\mu} \pm \zeta \boldsymbol{\sigma}$.

Values for N, t_{max}, α and ζ were found by tuning the algorithm on the Soybean dataset while ensuring that $Nt_{max} = 400$. These parameters were then used across all datasets. The values found were $\zeta = 2.0$, $N = 50$, and $\alpha = 5$.

4 The XGBoost Case Study

To compare the effectiveness of the optimization algorithms, we optimize the hyper-parameters for the well known gradient tree boosting algorithm XGBoost. This algorithm serves as a good benchmark because it features a large number of hyper-parameters that require the optimization algorithm to handle many common challenges simultaneously.

These challenges include dependency between sets of parameters, such as the case when a larger number of trees must be matched by a smaller learning rate, and how a larger change is needed to cause a measurable shift in the performance when the number of trees becomes large.

4.1 The XGBoost Algorithm

XGBoost is a popular state-of-the-art, gradient boosting decision tree algorithm that is capable of both regression and classification [3]. The algorithm works by using gradient tree boosting to incrementally optimize a model f for a regularized learning objective:

$$L(f) = \sum_{i=1}^{n} L(\hat{y}_i, y_i) + \sum_{m=1}^{M} \Omega(\delta_m). \tag{1}$$

Here, n is the number of training instances, $L(\cdot)$ is the loss function, M is the number of trees, δ_m is the m-th tree, and $\hat{y}$ and y are the predicted and correct targets respectively. Finally, $\Omega(\delta_m)$ is the regularization term:

$$\Omega(\delta) = \alpha|\delta| + \frac{1}{2}\lambda||w||^2, \tag{2}$$

where α and λ are hyper-parameters, and the last term is ℓ^2 regularization.

In total, XGBoost features 10 hyper-parameters that affect multiple different aspects of the learned model; including the model complexity which is controlled by the number of the trees and the maximum depth of each tree, and parameters that affect the behavior of the final model, including the learning rate, and minimum loss reduction needed to further split a leaf.

4.2 The Search Space

We have chosen to optimize only eight of the 10 hyper-parameters of the XGBoost algorithm, as optimizing all 10 is unlikely to result in significantly improved performance of the resulting models; and optimizing eight parameters using only the 400 function evaluations allowed is already a significant challenge with the large parameter ranges that this space contains.

For ease of implementation, the algorithms all optimize the search space bounded by the hypercube $[0,1]^D$. The points are then mapped to the parameter ranges used by the XGBoost by the equations listed in Table 1. These mappings have been designed to cover larger domains than what is typical when tuning a gradient boosting algorithm using a automated hyper-parameter search. Our motivation for this is to challenge the hyper-parameter optimization algorithms.

Table 1. Definition of the mappings between the search space used by the optimization algorithms and the parameter values of XGBoost. In the mappings, $|A|$ is the number of attributes in the dataset.

Parameter	Mapping
learning rate	x_1
min split loss	$x_2 * 10.0$
max depth	$2.0^{x_3 * \log_2(\|A\|)}$
colsample bytree	x_4
min child weight	$x_5 * 20.0$
subsample	x_6
num round	$x_7 * (1000.0 - 25.0) + 25.0$
reg alpha	$x_8 * 5.0$

Depending on the dataset, this search space has low effective dimensionality, and reasonable performance can often be obtained by optimizing only five of these parameters using the simple grid search defined by the caret R package. Grid search however, is not tested in this paper, as the 400 function evaluations we allow would make it impractical to use a grid search to optimize all eight hyper-parameters.

4.3 Datasets

The algorithms are used to optimize XGBoost for the 11 classification datasets listed in Table 2. These represent a large variety in the number and type of attributes, in addition to a spread in the number of instances. Additionally, none of the datasets are trivially solvable.

The size of the datasets have a large effect on how challenging the hyper-parameter optimization problem will be. When optimizing for a small dataset, there will be a significant amount of noise in the estimated performance of the classifier, which is a problem that will be reduced for the larger datasets.

The datasets are largely used as-is without any preprocessing and having the missing values kept intact to be handled by XGBoosts internal mechanisms. In the cases where the datasets were published as separate training and tests sets, the two sets were combined into a single large dataset.

Table 2. Overview of the datasets used to test the optimization algorithms.

Dataset	Attributes	Instances
Biodeg	42	1055
Contraceptive	10	1473
Landsat	37	6435
Mammographic	6	961
Phishing websites	31	11055
Phoneme	6	5404
Soybean	36	683
Splice	61	3190
Vehicle	19	846
Wilt	6	4839
Winequality red	12	1599

5 Results

In this section we describe how we measure the performance of the algorithms. Additionally, because it is desirable for the algorithms to consistently produce the best possible models, with little variance between independent runs, we also discuss the variance of the algorithms.

5.1 Measuring Performance

The optimization algorithms were allotted a budget of 50 function evaluations per dimension resulting in a total of $50 * 8 = 400$ evaluations. Each evaluation consists of a 5-fold stratified cross-validation that is repeated three times. In order to evaluate the parameters that were found, we then employ a 5-fold stratified cross-validation repeated 50 times with different seeds to produce the result of one optimization trial.

As it is well known that the randomness involved in the initialization and operation of the algorithms will result in some degree of uncertainty of one experiment, each experiment is repeated 20 times for each dataset/algorithm combination.

The results of these trials are then used to calculate a score for the algorithm in terms of the median and mean calculated as

$$\text{Score} = \sum_{a \in \text{datasets}} mean(f_a) + median(f_a). \quad (3)$$

The best algorithm will obtain the lowest score. This score model is based the model used to rank algorithms for the CEC 2015 - real-parameter single objective computationally expensive optimization competition.

5.2 Comparison

As shown in Table 3, L-SHADE has the best overall performance according to our score model, closely followed by our adaptive random algorithm. Both are significantly better than random search. While the dominance of the state-of-the-art L-SHADE algorithm was expected, it was surprising how well the simple adaptive random algorithm performed. As shown in Table 4, the random search is unable to achieve similar results on any of the tested datasets.

Table 3. Ranking of the algorithms according to the score calculated by Eq. (3).

Algorithm	Score	Rank
L-SHADE	3.120579	1
Adaptive random search	3.145194	2
Random search	3.305767	3

Table 4. The results of each optimization algorithm when optimizing the XGBoost algorithm on 11 datasets. This table shows the mean and standard deviation based on 20 trials for each algorithm/dataset pairs.

Dataset	L-SHADE		Adaptive random		Random	
	Mean	Std.	Mean	Std.	Mean	Std.
Biodeg	0.128027	0.001586	0.128161	0.001300	0.133940	0.003705
Contraceptive	0.443900	0.001277	0.442992	0.001472	0.445326	0.001959
Landsat	0.077260	0.000951	0.078412	0.001712	0.085575	0.002923
Mammographic	0.165784	0.001204	0.165289	0.000905	0.167276	0.001587
Phishing websites	0.026456	0.000357	0.027040	0.000873	0.031693	0.002770
Phoneme	0.092643	0.001057	0.095974	0.003864	0.108249	0.004596
Soybean	0.051668	0.001686	0.052995	0.002748	0.064248	0.006355
Splice	0.030576	0.000780	0.031627	0.001353	0.034066	0.001655
Vehicle	0.220953	0.003816	0.222866	0.004917	0.235851	0.004252
Wilt	0.014503	0.000291	0.014744	0.000385	0.016033	0.000770
Winequality red	0.309346	0.003222	0.315816	0.008922	0.331308	0.009269

5.3 Variance

Our concern with variance is motivated by how a set of hyper-parameters can be very expensive to evaluate, often making multiple optimization runs impractical. Therefore, we need the algorithm to consistently find a good set of hyper-parameters from a single run.

As we can see in Table 4, the random search algorithm has a significantly larger standard deviation than the other two, with the L-SHADE algorithm having much lower variance than the other two on most datasets.

6 Conclusion

The experiments of Bergstra and Bengio [2] showed that random search typically outperformed the commonly used combination of a grid search and manually controlled sequential search. This is in large part because it is much more effective at exploiting the low effective dimensionality of many hyper-parameter optimization tasks. However, the effectiveness of a random search is still poor compared to most other methods because it is unable to utilize any of the information that is contained in the previously evaluated points.

In our analysis, we have empirically compared three different algorithms, namely, the random search algorithm, the state-of-the-art L-SHADE algorithm, and our adaptive random search. We found that both the adaptive random search and L-SHADE found significantly better models than a pure random search when optimizing the hyper-parameters of XGBoost for 11 datasets. To our knowledge, neither a state-of-the-art differential evolution variant like the L-SHADE algorithm, or an algorithm like our adaptive random search algorithm have been thoroughly tested for hyper-parameter optimization before.

Because the differences in performance are so small between the adaptive random search, and the L-SHADE algorithm, we recommend using this adaptive random algorithm as the go-to method for hyper-parameter optimization tasks from a practical perspective. When compared to L-SHADE, the algorithm has multiple properties that make it more attractive as the default approach:

- Much simpler - the algorithm can be implemented in only a few lines of code using popular numerical libraries available in most languages.
- Trivial to parallelize - unlike L-SHADE, the algorithm has a constant population size and will therefore be easier to parallelize.

Our novel algorithm is almost as simple as random search, but shows significantly better performance. It should therefore be the new baseline for hyper-parameter optimization.

Acknowledgement. This research was supported in part with computational resources at UIT provided by NOTUR, http://www.sigma2.no.

References

1. Awad, N.H., Ali, M.Z., Suganthan, P.N., Reynolds, R.G.: An ensemble sinusoidal parameter adaptation incorporated with L-SHADE for solving CEC2014 benchmark problems. In: 2016 IEEE Congress on Evolutionary Computation (CEC), pp. 2958–2965. IEEE (2016)
2. Bergstra, J., Bengio, Y.: Random search for hyper-parameter optimization. J. Mach. Learn. Res. **13**(Feb), 281–305 (2012)
3. Chen, T., Guestrin, C.: XGBoost: a scalable tree boosting system. In: Proceedings of the 22nd ACM SIGKDD International Conference on Knowledge Discovery and Data Mining, KDD 2016, pp. 785–794. ACM, New York (2016)
4. Guo, S.M., Tsai, J.S.H., Yang, C.C., Hsu, P.H.: A self-optimization approach for L-SHADE incorporated with eigenvector-based crossover and successful-parent-selecting framework on CEC 2015 benchmark set. In: 2015 IEEE Congress on Evolutionary Computation (CEC), pp. 1003–1010. IEEE (2015)
5. Jaderberg, M., et al.: Population based training of neural networks. arXiv preprint arXiv:1711.09846 (2017)
6. Klein, A., Falkner, S., Bartels, S., Hennig, P., Hutter, F.: Fast Bayesian optimization of machine learning hyperparameters on large datasets. arXiv preprint arXiv:1605.07079 (2016)
7. Lorenzo, P.R., Nalepa, J., Ramos, L.S., Pastor, J.R.: Hyper-parameter selection in deep neural networks using parallel particle swarm optimization. In: Proceedings of the Genetic and Evolutionary Computation Conference Companion, pp. 1864–1871. ACM (2017)
8. Loshchilov, I., Hutter, F.: CMA-ES for hyperparameter optimization of deep neural networks. arXiv preprint arXiv:1604.07269 (2016)
9. Martinez-Cantin, R.: BayesOpt: a Bayesian optimization library for nonlinear optimization, experimental design and bandits. J. Mach. Learn. Res. **15**(1), 3735–3739 (2014)
10. Mohamed, A.W., Hadi, A.A., Fattouh, A.M., Jambi, K.M.: LSHADE with semi-parameter adaptation hybrid with CMA-ES for solving CEC 2017 benchmark problems. In: 2017 IEEE Congress on Evolutionary Computation (CEC), pp. 145–152. IEEE (2017)
11. Snoek, J., Larochelle, H., Adams, R.P.: Practical Bayesian optimization of machine learning algorithms. In: Advances in Neural Information Processing Systems, pp. 2951–2959 (2012)
12. Tanabe, R., Fukunaga, A.S.: Improving the search performance of SHADE using linear population size reduction. In: 2014 IEEE Congress on Evolutionary Computation (CEC), pp. 1658–1665. IEEE (2014)
13. Thornton, C., Hutter, F., Hoos, H.H., Leyton-Brown, K.: Auto-WEKA: automated selection and hyper-parameter optimization of classification algorithms. CoRR, abs/1208.3719 (2012)

Optimal Trade-Off Between Sample Size and Precision of Supervision for the Fixed Effects Panel Data Model

Giorgio Gnecco(✉) and Federico Nutarelli

IMT School for Advanced Studies, Lucca, Italy
{giorgio.gnecco,federico.nutarelli}@imtlucca.it

Abstract. We investigate a modification of the classical fixed effects panel data model (a linear regression model able to represent unobserved heterogeneity in the data), in which one has the additional possibility of controlling the conditional variance of the output given the input, by varying the cost associated with the supervision of each training example. Assuming an upper bound on the total supervision cost, we analyze and optimize the trade-off between the sample size and the precision of supervision (the reciprocal of the conditional variance of the output), by formulating and solving a suitable optimization problem, based on a large-sample approximation of the output of the classical algorithm used to estimate the parameters of the fixed effects panel data model. Considering a specific functional form for that precision, we prove that, depending on the "returns to scale" of the precision with respect to the supervision cost per example, in some cases "many but bad" examples provide a smaller generalization error than "few but good" ones, whereas in other cases the opposite occurs. The results extend to the fixed effects panel data model the ones we obtained in recent works for a simpler linear regression model. We conclude discussing possible applications of our results, and extensions of the proposed optimization framework to other panel data models.

Keywords: Fixed effects panel data model · Generalization error · Large-sample approximation · Optimal sample size

1 Introduction

In several applications in economics, engineering, and many other fields, one has to approximate a function from a finite set of input-output noisy examples. This belongs to the typical class of problems studied by supervised machine learning [11]. In some cases, the noise variance of the output can be reduced to some extent by the researcher, by increasing the cost of each supervision. For instance, measurement devices with larger precision (hence, larger cost) could be employed. Similarly, the supervision could be provided by teachers (e.g., doctors) with a higher level of expertise (hence, higher cost). In all such cases, the

G. Nicosia et al. (Eds.): LOD 2019, LNCS 11943, pp. 531–542, 2019.
https://doi.org/10.1007/978-3-030-37599-7_44

investigation of an optimal trade-off between the sample size and the precision of supervision is needed. In [6], this analysis was conducted employing a modification of the classical linear regression model, in which one is additionally given the possibility of controlling the conditional variance of the output given the input, by varying the time (hence, the cost) dedicated to the supervision of each training example, and fixing an upper bound on the total available supervision time. Based on a large-sample approximation of the output of the ordinary least squares regression algorithm, it was shown therein that the optimal choice of the supervision time per example is highly dependent on the noise model.

In this work, we extend the analysis of the optimal trade-off between sample size and precision of supervision performed in [6], by considering a more general linear model of the input-output relationship, which is the fixed effects panel data model. In this model, observations associated with different units (individuals) are associated with different constants, which are able to represent unobserved heterogeneity in the data. Moreover, the same unit is observed along another dimension, which is typically time. The model is commonly applied in the analysis of microeconomics and macroeconomics data [13], where each unit may represent, e.g., a firm, or a country. It is also applied in biostatistics [5], psychology, political science, sociology, and life sciences [1]. Consistently with the results of the analysis performed in [6], for the classical linear regression model, we show that, also for the fixed effects panel data model, the following holds: when the precision of the supervision (the reciprocal of the conditional variance of the output) increases less than proportionally with respect to the supervision cost per example, the minimum (large-sample approximation of the) generalization error (conditioned on the training input data) is obtained in correspondence of the smallest supervision cost per example (hence, of the largest number of examples); when the precision increases more than proportionally with respect to the supervision cost per example, the optimal supervision cost per example is the largest one (which is associated with the smallest number of examples). In summary, the results of the theoretical analyses performed in [6] and, for a different regression model, in this paper highlight that an increase of the sample size is not always beneficial, if it is feasible to collect a smaller number of more reliable data. Hence, not only their number, but also their quality matters. This looks particularly relevant when one has the possibility of designing the data collection process. A similar conclusion is expected to hold also for other regression models, such as the ones mentioned at the end of this work.

The paper is structured as follows. Section 2 provides a background on the fixed effects panel data model. Section 3 presents the analysis of its conditional generalization error, and of the large-sample approximation of the latter with respect to time. Section 4 formulates and solves the optimization problem modeling the trade-off between sample size and precision of supervision for the fixed effects panel data model, using the large-sample approximation above. Finally, Sect. 5 discusses some possible applications and extensions of the theoretical results obtained in the work.

2 Background

We recall some basic facts about the following (static) fixed effects panel data model (see, e.g., [13, Chap. 10])[1]:

$$y_{n,t} := \eta_n + \boldsymbol{\beta}' \boldsymbol{x}_{n,t}, n = 1, \ldots, N, t = 1, \ldots, T. \tag{1}$$

Here, the outputs $y_{n,t}$'s are scalar, whereas the inputs $\boldsymbol{x}_{n,t}$'s ($n = 1, \ldots, N, t = 1, \ldots, T$) are column vectors in $\mathbb{R}^p$, which are modeled as random vectors. The parameters of the model are the individual constants η_n ($n = 1, \ldots, N$), one for each unit, and the column vector $\boldsymbol{\beta} \in \mathbb{R}^p$. Equation (1) represents a balanced panel data model, in which each unit n is associated with the same number T of outputs, each one at a different time t.

Noisy measures $\tilde{y}_{n,t}$'s of the outputs $y_{n,t}$'s are available. They are generated according to the following additive noise model:

$$\tilde{y}_{n,t} := y_{n,t} + \varepsilon_{n,t}, \tag{2}$$

where the $\varepsilon_{n,t}$'s are mutually independent and identically distributed random variables, having mean 0 and the same variance σ^2. Moreover, they are independent also from all the $\boldsymbol{x}_{n,t}$'s.

The training input-output pairs $(\boldsymbol{x}_{n,t}, \tilde{y}_{n,t})$ ($n = 1, \ldots, N, t = 1, \ldots, T$) are used to estimate the parameters of the model. Assuming the invertibility of the matrix $\sum_{n=1}^{N} \boldsymbol{X}_n' \boldsymbol{Q} \boldsymbol{X}_n$, the fixed effects estimate of $\boldsymbol{\beta}$ is

$$\begin{aligned}\hat{\boldsymbol{\beta}}_{FE} &:= \left(\sum_{n=1}^{N} \boldsymbol{X}_n' \boldsymbol{Q} \boldsymbol{X}_n \right)^{-1} \left(\sum_{n=1}^{N} \boldsymbol{X}_n' \boldsymbol{Q} \tilde{\boldsymbol{y}}_n \right) \\ &= \left(\sum_{n=1}^{N} \boldsymbol{X}_n' \boldsymbol{Q}' \boldsymbol{Q} \boldsymbol{X}_n \right)^{-1} \left(\sum_{n=1}^{N} \boldsymbol{X}_n' \boldsymbol{Q}' \boldsymbol{Q} \tilde{\boldsymbol{y}}_n \right),\end{aligned} \tag{3}$$

where $\boldsymbol{X}_n \in \mathbb{R}^{T,p}$ is a matrix whose rows are the transposes of the $\boldsymbol{x}_{n,t}$'s, $\tilde{\boldsymbol{y}}_n$ is a column vector which collects the noisy measures $\tilde{y}_{n,t}$'s,

$$\boldsymbol{Q} := \boldsymbol{I}_T - \frac{1}{T} \boldsymbol{1}_T \boldsymbol{1}_T' \tag{4}$$

(being $\boldsymbol{I}_T \in \mathbb{R}^{T \times T}$ the identity matrix, and $\boldsymbol{1}_T \in \mathbb{R}^T$ a column vector whose elements are all equal to 1) is a symmetric and idempotent matrix (i.e., $\boldsymbol{Q}' = \boldsymbol{Q} = \boldsymbol{Q}^2$). Hence, for each unit n,

$$\boldsymbol{Q} \boldsymbol{X}_n = \begin{bmatrix} \boldsymbol{x}_{n,1} - \frac{1}{T} \sum_{t=1}^{T} \boldsymbol{x}_{n,t} \\ \boldsymbol{x}_{n,2} - \frac{1}{T} \sum_{t=1}^{T} \boldsymbol{x}_{n,t} \\ \cdots \\ \boldsymbol{x}_{n,T} - \frac{1}{T} \sum_{t=1}^{T} \boldsymbol{x}_{n,t} \end{bmatrix}, \tag{5}$$

[1] For simplicity of exposition, here the model is not presented in its most general form (e.g., the disturbances $\varepsilon_{n,t}$'s are simply assumed to be mutually independent).

and

$$\boldsymbol{Q}\tilde{\boldsymbol{y}}_n = \begin{bmatrix} \tilde{y}_{n,1} - \frac{1}{T}\sum_{t=1}^{T}\tilde{y}_{n,t} \\ \tilde{y}_{n,2} - \frac{1}{T}\sum_{t=1}^{T}\tilde{y}_{n,t} \\ \cdots \\ \tilde{y}_{n,T} - \frac{1}{T}\sum_{t=1}^{T}\tilde{y}_{n,t} \end{bmatrix} \tag{6}$$

represent, respectively, the matrix of time de-meaned training inputs, and the vector of time de-meaned corrupted training outputs.

The fixed effects estimates of the η_n's are

$$\hat{\eta}_{n,FE} := \frac{1}{T}\sum_{t=1}^{T}\left(\tilde{y}_{n,t} - \hat{\boldsymbol{\beta}}'_{FE}\boldsymbol{x}_{n,t}\right). \tag{7}$$

The estimates (3) and (7) are unbiased, i.e.,

$$\mathbb{E}\left\{\hat{\boldsymbol{\beta}}_{FE} - \boldsymbol{\beta}\right\} = \mathbf{0}_p \tag{8}$$

(being $\mathbf{0}_p \in \mathbb{R}^p$ a column vector whose elements are all equal to 0), and

$$\mathbb{E}\left\{\hat{\eta}_{n,FE} - \eta_n\right\} = 0. \tag{9}$$

Finally, the covariance matrix of $\hat{\boldsymbol{\beta}}_{FE}$, conditioned on the training input data $\{\boldsymbol{x}_{n,t}\}_{n=1,\ldots,N}^{t=1,\ldots,T}$, is

$$\begin{aligned}\mathrm{Var}\left(\hat{\boldsymbol{\beta}}_{FE}|\{\boldsymbol{x}_{n,t}\}_{n=1,\ldots,N}^{t=1,\ldots,T}\right) &= \sigma^2\left(\sum_{n=1}^{N}\boldsymbol{X}'_n\boldsymbol{Q}\boldsymbol{X}_n\right)^{-1} \\ &= \sigma^2\left(\sum_{n=1}^{N}\boldsymbol{X}'_n\boldsymbol{Q}'\boldsymbol{Q}\boldsymbol{X}_n\right)^{-1}.\end{aligned} \tag{10}$$

3 Conditional Generalization Error and Its Large-Sample Approximation

We define the generalization error for the i-th unit ($i = 1,\ldots,N$), conditioned on the training input data, as follows:

$$\mathbb{E}\left\{\left(\hat{\eta}_{i,FE} + \hat{\boldsymbol{\beta}}'_{FE}\boldsymbol{x}_i^{test} - \eta_i - \boldsymbol{\beta}'\boldsymbol{x}_i^{test}\right)^2 |\{\boldsymbol{x}_{n,t}\}_{n=1,\ldots,N}^{t=1,\ldots,T}\right\}, \tag{11}$$

where $\boldsymbol{x}_i^{test} \in \mathbb{R}^p$ is independent from the training data. It is the expected mean squared error of the prediction of the output associated with a test input, conditioned on the training input data.

For $n = 1,\ldots,N$, let $\boldsymbol{\varepsilon}_n \in \mathbb{R}^T$ be the column vector whose elements are the $\varepsilon_{n,t}$'s, and let $\boldsymbol{\eta}_n \in \mathbb{R}^T$ be the column vector whose elements are all equal to η_n. Using

$$\mathbb{E}\left\{\boldsymbol{\varepsilon}_n\boldsymbol{\varepsilon}'_m\right\} = 0 \tag{12}$$

for $n \neq m$,

$$\mathbb{E}\left\{\boldsymbol{\varepsilon}_n \boldsymbol{\varepsilon}_n'\right\} = \sigma^2 \boldsymbol{I}_T, \tag{13}$$

$$\boldsymbol{Q}'\boldsymbol{Q} = \boldsymbol{Q}, \tag{14}$$

$$\boldsymbol{Q}\boldsymbol{Q}'\boldsymbol{Q}'\boldsymbol{Q} = \boldsymbol{Q}'\boldsymbol{Q}, \tag{15}$$

$$\boldsymbol{Q}\boldsymbol{\eta}_n = \boldsymbol{Q}'\boldsymbol{\eta}_n = \boldsymbol{0}_T, \tag{16}$$

and

$$\boldsymbol{Q}\boldsymbol{1}_T = \boldsymbol{Q}'\boldsymbol{1}_T = \boldsymbol{0}_T, \tag{17}$$

we can simplify the expression (11) of the conditional generalization error as follows, highlighting its dependence on σ^2 and T (see the Appendix for the details):

$$\begin{aligned}(11) = {} & \frac{\sigma^2}{T^2}\boldsymbol{1}_T'\boldsymbol{X}_i\left(\sum_{n=1}^N \boldsymbol{X}_n'\boldsymbol{Q}'\boldsymbol{Q}\boldsymbol{X}_n\right)^{-1}\boldsymbol{X}_i'\boldsymbol{1}_T + \frac{\sigma^2}{T} \\ & + \mathbb{E}\left\{\sigma^2\left(\boldsymbol{x}_i^{test}\right)'\left(\sum_{n=1}^N \boldsymbol{X}_n'\boldsymbol{Q}'\boldsymbol{Q}\boldsymbol{X}_n\right)^{-1}\boldsymbol{x}_i^{test}\Big|\{\boldsymbol{x}_{n,t}\}_{n=1,\dots,N}^{t=1,\dots,T}\right\} \\ & - 2\mathbb{E}\left\{\frac{\sigma^2}{T}\boldsymbol{1}_T'\boldsymbol{X}_i\left(\sum_{n=1}^N \boldsymbol{X}_n'\boldsymbol{Q}'\boldsymbol{Q}\boldsymbol{X}_n\right)^{-1}\boldsymbol{x}_i^{test}\Big|\{\boldsymbol{x}_{n,t}\}_{n=1,\dots,N}^{t=1,\dots,T}\right\}. \end{aligned} \tag{18}$$

Next, we obtain a large-sample approximation of the conditional generalization error (18) with respect to T, for a fixed number of units N. Such an approximation is useful, e.g., in the application of the model to macroeconomics data[2], for which it is common to investigate the case of a large horizon T.

Under mild conditions[3], the following convergences in probability[4] hold, which follow from Chebyshev's law of large numbers [10, Sect. 13.4.2]:

$$\text{plim}_{T\to+\infty}\frac{1}{T}\boldsymbol{1}_T'\boldsymbol{X}_i = \left(\mathbb{E}\left\{\boldsymbol{x}_{i,1}\right\}\right)', \tag{19}$$

and

$$\text{plim}_{T\to+\infty}\frac{1}{T}\sum_{n=1}^N \boldsymbol{X}_n'\boldsymbol{Q}'\boldsymbol{Q}\boldsymbol{X}_n = \boldsymbol{A}_N, \tag{20}$$

[2] The case of finite T and large N is of more interest for microeconometrics, and will be investigated in future research.

[3] E.g., if the $\boldsymbol{x}_{n,t}$'s are independent, identically distributed, and have finite moments up to the order 4.

[4] We recall that a sequence of random real matrices $\boldsymbol{M}_T$, $T = 1, \dots, +\infty$ converges in probability to the real matrix $\boldsymbol{M}$ if, for every $\varepsilon > 0$, Prob $(\|\boldsymbol{M}_T - \boldsymbol{M}\| > \varepsilon)$ (where $\|\cdot\|$ is an arbitrary matrix norm) tends to 0 as T tends to $+\infty$. In this case, one writes $\text{plim}_{T\to+\infty}\boldsymbol{M}_T = \boldsymbol{M}$.

where

$$\boldsymbol{A}_N = \boldsymbol{A}_N' := \sum_{n=1}^{N} \mathbb{E}\left\{(\boldsymbol{x}_{n,1} - \mathbb{E}\{\boldsymbol{x}_{n,1}\})' (\boldsymbol{x}_{n,1} - \mathbb{E}\{\boldsymbol{x}_{n,1}\})\right\} \tag{21}$$

is a symmetric and positive semi-definite matrix. In the following, its positive definiteness (hence, its invertibility) is also assumed[5].

When (19) and (20) hold, the conditional generalization error (18) has the following large-sample approximation with respect to T[6]:

$$\begin{aligned}(18) &\simeq \frac{\sigma^2}{T} (\mathbb{E}\{\boldsymbol{x}_{i,1}\})' \boldsymbol{A}_N^{-1} \mathbb{E}\{\boldsymbol{x}_{i,1}\} + \frac{\sigma^2}{T} \\ &+ \frac{\sigma^2}{T} \mathbb{E}\left\{\left(\boldsymbol{x}_i^{test}\right)' \boldsymbol{A}_N^{-1} \boldsymbol{x}_i^{test}\right\} - 2\frac{\sigma^2}{T} (\mathbb{E}\{\boldsymbol{x}_{i,1}\})' \boldsymbol{A}_N^{-1} \mathbb{E}\left\{\boldsymbol{x}_i^{test}\right\} \\ &= \frac{\sigma^2}{T}\left(1 + \mathbb{E}\left\{\left\|\boldsymbol{A}_N^{-\frac{1}{2}} \left(\mathbb{E}\{\boldsymbol{x}_{i,1}\} - \boldsymbol{x}_i^{test}\right)\right\|_2^2\right\}\right),\end{aligned} \tag{22}$$

where $\|\cdot\|_2$ denotes the l_2-norm, and $\boldsymbol{A}_N^{-\frac{1}{2}}$ is the principal square root (i.e., the symmetric and positive definite square root) of the symmetric and positive definite matrix $\boldsymbol{A}_N^{-1}$.

Interestingly, the large-sample approximation (22) has the form $\frac{\sigma^2}{T} K_i$, where

$$K_i := \left(1 + \mathbb{E}\left\{\left\|\boldsymbol{A}_N^{-\frac{1}{2}} \left(\mathbb{E}\{\boldsymbol{x}_{i,1}\} - \boldsymbol{x}_i^{test}\right)\right\|_2^2\right\}\right) \tag{23}$$

is a positive constant. This simplifies the analysis of the trade-off between sample size and precision of supervision performed in the next section, since one dos not need to compute the exact expression of K_i to find the optimal trade-off.

4 Optimal Trade-Off Between Sample Size and Precision of Supervision for the Fixed Effects Panel Data Model Under the Large-Sample Approximation

In this section, we are interested in optimizing the large-sample approximation (22) of the conditional generalization error when the variance σ^2 is modeled as a decreasing function of the supervision cost per example c, and there is an upper bound C on the total supervision cost NTc associated with the whole training set. In the analysis, N is fixed, and T is chosen as $\left\lfloor \frac{C}{Nc} \right\rfloor$. Moreover, the supervision cost per example c is allowed to take values on the interval $[c_{\min}, c_{\max}]$, where

[5] The existence of the probability limit (20) and the assumed positive definiteness of the matrix $\boldsymbol{A}_N$ guarantee that the invertibility of the matrix $\sum_{n=1}^N \boldsymbol{X}_n' \boldsymbol{Q} \boldsymbol{X}_n = \sum_{n=1}^N \boldsymbol{X}_n' \boldsymbol{Q}' \boldsymbol{Q} \boldsymbol{X}_n$ (see Sect. 2) holds with probability near 1 for large T.

[6] This is obtained taking also into account that, as a consequence of the Continuous Mapping Theorem [4, Theorem 7.33], the probability limit of the product of two random variables equals the product of their probability limits, when the latter two exist.

$0 < c_{\min} < c_{\max}$, so that the resulting T belongs to $\left\{\left\lfloor \frac{C}{Nc_{\max}} \right\rfloor, \ldots, \left\lfloor \frac{C}{Nc_{\min}} \right\rfloor\right\}$. In the following, C is supposed to be sufficiently large, so that the large-sample approximation (22) can be assumed to hold for every $c \in [c_{\min}, c_{\max}]$.

Consistently with [6], we adopt the following model for the variance σ^2, as a function of the supervision cost per example c:

$$\sigma^2(c) = kc^{-\alpha}, \tag{24}$$

where $k, \alpha > 0$. For $0 < \alpha < 1$, the precision of each supervision is characterized by "decreasing returns of scale" with respect to its cost because, if one doubles the supervision cost per example c, then the precision $1/\sigma^2(c)$ becomes less than two times its initial value (or equivalently, the variance $\sigma^2(c)$ becomes more than one half its initial value). Conversely, for $\alpha > 1$, there are "increasing returns of scale" because, if one doubles the supervision cost per example c, then the precision $1/\sigma^2(c)$ becomes more than two times its initial value (or equivalently, the variance $\sigma^2(c)$ becomes less than one half its initial value). The case $\alpha = 1$ is intermediate and refers to "constant returns of scale". In all the cases above, the precision of each supervision increases by increasing the supervision cost per example c.

Concluding, under the assumptions above, the optimal trade-off between the sample size and the precision of supervision for the fixed effects panel data model is modeled by the following optimization problem:

$$\text{minimize}_{c \in [c_{\min}, c_{\max}]} K_i k \frac{c^{-\alpha}}{\left\lfloor \frac{C}{Nc} \right\rfloor}. \tag{25}$$

When C is sufficiently large, the objective function $CK_i k \frac{c^{-\alpha}}{\left\lfloor \frac{C}{Nc} \right\rfloor}$ of the optimization problem (25), rescaled by the multiplicative factor C, can be approximated, with a negligible error[7] in the maximum norm on $[c_{\min}, c_{\max}]$, by $NK_i kc^{1-\alpha}$. In order to illustrate this issue, Fig. 1 shows the behavior of the rescaled objective functions $CK_i k \frac{c^{-\alpha}}{\left\lfloor \frac{C}{Nc} \right\rfloor}$ and $NK_i kc^{1-\alpha}$ for the three cases $0 < \alpha = 0.5 < 1$, $\alpha = 1.5 > 1$, and $\alpha = 1$ (the values of the other parameters are $k = 0.5$, $K_i = 2$, $N = 10$, $C = 100$, $c_{\min} = 0.4$, and $c_{\max} = 0.8$). One can show that, for $C \to +\infty$, the number of discontinuity points of the rescaled objective function $CK_i k \frac{c^{-\alpha}}{\left\lfloor \frac{C}{Nc} \right\rfloor}$ tends to infinity, whereas the amplitude of its oscillations above the lower envelope $NK_i kc^{1-\alpha}$ tends to 0 uniformly with respect to $c \in [c_{\min}, c_{\max}]$.

Concluding, under the approximation above, one can replace the optimization problem (25) with

$$\text{minimize}_{c \in [c_{\min}, c_{\max}]} NK_i kc^{1-\alpha}, \tag{26}$$

whose optimal solutions c° have the following expressions:

[7] By an argument similar to that used in [6], one can show that the approximation is exact, at optimality, when C is a multiple of both $Nc_{\min}$ and $Nc_{\max}$.

1. if $0 < \alpha < 1$ ("decreasing returns of scale"): $c^\circ = c_{\min}$;
2. if $\alpha > 1$ ("increasing returns of scale"): $c^\circ = c_{\max}$;
3. if $\alpha = 1$ ("constant returns of scale"): c° = any cost c in the interval $[c_{\min}, c_{\max}]$.

In summary, the results of the analysis show that, in the case of "decreasing returns of scale", "many but bad" examples are associated with a smaller generalization error than "few but good" ones. The opposite occurs for "increasing returns of scale", whereas the case of "constant returns of scale" is intermediate. These results are qualitatively in line with the ones obtained in [6] for a simpler linear regression problem, to which the ordinary least squares algorithm was applied. This depends on the fact that, in both cases, the conditional generalization error has the functional form $\frac{\sigma^2}{T} K_i$ (although two different positive constants K_i are involved in the two different cases).

One can observe that, in order to discriminate among the three cases of the analysis reported above, one does not need to know the exact values of the constants k, K_i, and N. Moreover, to discriminate between the first two cases, it is not necessary to know the exact value of the positive constant α (indeed, it suffices to know if α belongs, respectively, to the interval $(0, 1)$ or the one $(1, +\infty)$). Finally, no precise knowledge of the probability distributions of the input examples (one for each unit) is needed. In particular, different such probability distributions may be associated with different units, without affecting the results of the analysis.

5 Discussion

Up to our knowledge, the analysis and the optimization of the trade-off between sample size and precision of supervision in regression is investigated only seldomly in the machine-learning literature. Nevertheless, the approach exploited in the paper is similar to the one used in the optimization of sample survey design, in which some parameters of the design are optimized to minimize the sampling variance (see, e.g., the case of Neyman allocation in stratified sampling [8], for a given size of the dataset). It is also similar to the one used in [9] - in a context, however, in which linear regression is marginally involved, since only arithmetic averages of measurement results are considered - for the optimization of the design of measurement devices. The search for optimal sample designs can be also addressed by the Optimal Computing Budget Allocation (OCBA) method [3]. Differently from that approach, however, our analysis provides - for the specific noise model - the optimal design (the optimal choice of the supervision cost per example) *a priori* (i.e., before actually seeing the data). Our work can also be related to a recent strand of literature, which combines methods from both machine learning and econometrics [2,12] (e.g., the generalization error - which is typically considered in machine learning - is not investigated in the classical analysis of the fixed effects panel data model [7, Chap. 13]).

For what concerns practical applications, the theoretical results obtained in this work could be applied to the acquisition design of fixed effects panel data in

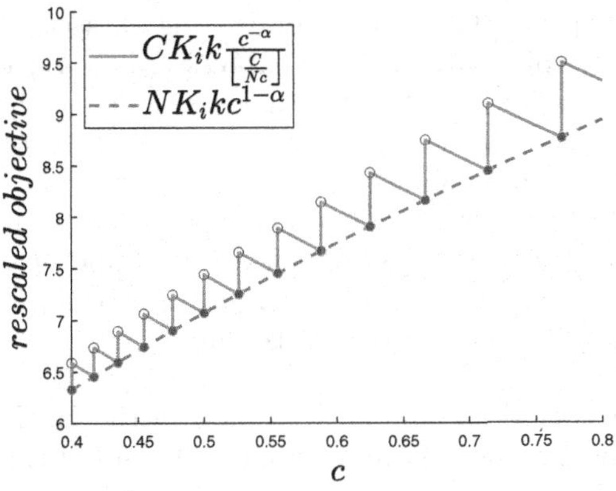

(a)

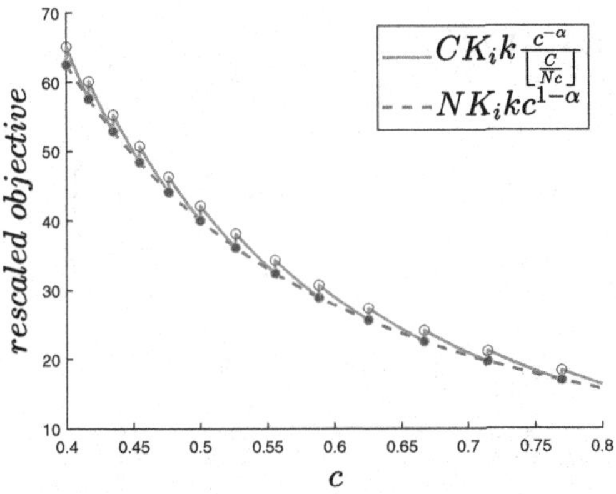

(b)

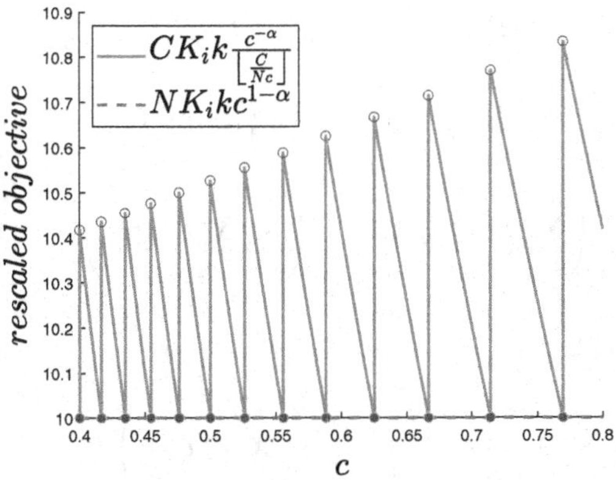

(c)

Fig. 1. Plots of the rescaled objective functions $CK_ik\frac{c^{-\alpha}}{\lfloor\frac{C}{Nc}\rfloor}$ and $NK_ikc^{1-\alpha}$ for $\alpha = 0.5$ (a), $\alpha = 1.5$ (b), and $\alpha = 1$ (c). See the main text for the values of the other parameters.

econometrics [7, Chap. 13]. The analysis of existing datasets could be performed by inserting artificial noise with variance expressed as in Eq. (24), possibly with the addition of a constant term in the variance modeling the original dataset. Moreover, the analysis of the large-sample case could be extended to deal with large N, or with both large N and T. As mentioned in the paper, the former case would be of interest for its potential applications in microeconometrics. As another possible extension, one could investigate and optimize the trade-off between sample size and precision of supervision for the random effects panel data model [7, Chap. 13], which is also commonly applied in the analysis of economic data. This differs from the fixed effects panel data model in that its parameters are considered as random variables. In the fixed effects panel data model, however, one can also obtain estimates of the individual constants η_n's (see Eq. (7)), which appear in the expression (11) of the conditional generalization error.

Appendix: Derivation of Eq. (18)

$$
\begin{aligned}
&\mathbb{E}\left\{\left(\hat{\eta}_{i,FE} + \hat{\boldsymbol{\beta}}_{FE}'\boldsymbol{x}_i^{test} - \eta_i - \boldsymbol{\beta}'\boldsymbol{x}_i^{test}\right)^2 \Big| \{\boldsymbol{x}_{n,t}\}_{n=1,\dots,N}^{t=1,\dots,T}\right\} \\
&= \mathbb{E}\left\{\left((\hat{\eta}_{i,FE} - \eta_i) + \left(\hat{\boldsymbol{\beta}}_{FE} - \boldsymbol{\beta}\right)'\boldsymbol{x}_i^{test}\right)^2 \Big| \{\boldsymbol{x}_{n,t}\}_{n=1,\dots,N}^{t=1,\dots,T}\right\} \\
&= \mathrm{Var}\left(\hat{\eta}_{i,FE} \big| \{\boldsymbol{x}_{n,t}\}_{n=1,\dots,N}^{t=1,\dots,T}\right) + \mathbb{E}\left\{\left(\left(\hat{\boldsymbol{\beta}}_{FE} - \boldsymbol{\beta}\right)'\boldsymbol{x}_i^{test}\right)^2 \Big| \{\boldsymbol{x}_{n,t}\}_{n=1,\dots,N}^{t=1,\dots,T}\right\} \\
&\quad + 2\mathbb{E}\left\{(\hat{\eta}_{i,FE} - \eta_i)\left(\hat{\boldsymbol{\beta}}_{FE} - \boldsymbol{\beta}\right)'\boldsymbol{x}_i^{test} \Big| \{\boldsymbol{x}_{n,t}\}_{n=1,\dots,N}^{t=1,\dots,T}\right\} \\
&= \mathbb{E}\left\{\left(\frac{1}{T}\sum_{t=1}^{T}\left(\left(\boldsymbol{\beta} - \hat{\boldsymbol{\beta}}_{FE}\right)'\boldsymbol{x}_{i,t} + \varepsilon_{i,t}\right)\right)^2 \Big| \{\boldsymbol{x}_{n,t}\}_{n=1,\dots,N}^{t=1,\dots,T}\right\} \\
&\quad + \mathbb{E}\left\{(\boldsymbol{x}_i^{test})'\left(\hat{\boldsymbol{\beta}}_{FE} - \boldsymbol{\beta}\right)\left(\hat{\boldsymbol{\beta}}_{FE} - \boldsymbol{\beta}\right)'\boldsymbol{x}_i^{test} \Big| \{\boldsymbol{x}_{n,t}\}_{n=1,\dots,N}^{t=1,\dots,T}\right\} \\
&\quad + 2\mathbb{E}\left\{\left(\frac{1}{T}\sum_{t=1}^{T}\left(\left(\boldsymbol{\beta} - \hat{\boldsymbol{\beta}}_{FE}\right)'\boldsymbol{x}_{i,t} + \varepsilon_{i,t}\right)\right)\left(\hat{\boldsymbol{\beta}}_{FE} - \boldsymbol{\beta}\right)'\boldsymbol{x}_i^{test} \Big| \{\boldsymbol{x}_{n,t}\}_{n=1,\dots,N}^{t=1,\dots,T}\right\} \\
&= \mathbb{E}\left\{\left(\frac{\boldsymbol{1}_T'}{T}\left(-\boldsymbol{X}_i\left(\sum_{n=1}^{N}\boldsymbol{X}_n'\boldsymbol{Q}'\boldsymbol{Q}\boldsymbol{X}_n\right)^{-1}\left(\sum_{n=1}^{N}\boldsymbol{X}_n'\boldsymbol{Q}'\boldsymbol{Q}\boldsymbol{\varepsilon}_n\right) + \boldsymbol{\varepsilon}_i\right)\right)^2\right. \\
&\qquad \left. \Big| \{\boldsymbol{x}_{n,t}\}_{n=1,\dots,N}^{t=1,\dots,T}\right\} \\
&\quad + \mathbb{E}\left\{(\boldsymbol{x}_i^{test})'\left(\sum_{n=1}^{N}\boldsymbol{X}_n'\boldsymbol{Q}'\boldsymbol{Q}\boldsymbol{X}_n\right)^{-1}\left(\sum_{n=1}^{N}\boldsymbol{X}_n'\boldsymbol{Q}'\boldsymbol{Q}\boldsymbol{\varepsilon}_n\right)\right. \\
&\qquad \left. \cdot\left(\left(\sum_{n=1}^{N}\boldsymbol{X}_n'\boldsymbol{Q}'\boldsymbol{Q}\boldsymbol{X}_n\right)^{-1}\left(\sum_{n=1}^{N}\boldsymbol{X}_n'\boldsymbol{Q}'\boldsymbol{Q}\boldsymbol{\varepsilon}_n\right)\right)'\boldsymbol{x}_i^{test} \Big| \{\boldsymbol{x}_{n,t}\}_{n=1,\dots,N}^{t=1,\dots,T}\right\}
\end{aligned}
$$

$$
\begin{aligned}
&+ 2\mathbb{E}\Bigg\{ \left(\frac{\mathbf{1}_T'}{T} \left(-\boldsymbol{X}_i \left(\sum_{n=1}^{N} \boldsymbol{X}_n' \boldsymbol{Q}' \boldsymbol{Q} \boldsymbol{X}_n \right)^{-1} \left(\sum_{n=1}^{N} \boldsymbol{X}_n' \boldsymbol{Q}' \boldsymbol{Q} \boldsymbol{\varepsilon}_n \right) + \boldsymbol{\varepsilon}_i \right) \right) \\
&\quad \cdot \left(\left(\sum_{n=1}^{N} \boldsymbol{X}_n' \boldsymbol{Q}' \boldsymbol{Q} \boldsymbol{X}_n \right)^{-1} \left(\sum_{n=1}^{N} \boldsymbol{X}_n' \boldsymbol{Q}' \boldsymbol{Q} \boldsymbol{\varepsilon}_n \right) \right)' \boldsymbol{x}_i^{test} \Big| \{\boldsymbol{x}_{n,t}\}_{n=1,\ldots,N}^{t=1,\ldots,T} \Bigg\} \\
&= \mathbb{E}\Bigg\{ \frac{\mathbf{1}_T' \boldsymbol{X}_i}{T^2} \left(\sum_{n=1}^{N} \boldsymbol{X}_n' \boldsymbol{Q}' \boldsymbol{Q} \boldsymbol{X}_n \right)^{-1} \left(\sum_{n=1}^{N} \boldsymbol{X}_n' \boldsymbol{Q}' \boldsymbol{Q} \boldsymbol{\varepsilon}_n \right) \\
&\quad \cdot \left(\sum_{n=1}^{N} \boldsymbol{\varepsilon}_n' \boldsymbol{Q}' \boldsymbol{Q} \boldsymbol{X}_n \right) \left(\left(\sum_{n=1}^{N} \boldsymbol{X}_n' \boldsymbol{Q}' \boldsymbol{Q} \boldsymbol{X}_n \right)^{-1} \right)' \boldsymbol{X}_i' \mathbf{1}_T \Big| \{\boldsymbol{x}_{n,t}\}_{n=1,\ldots,N}^{t=1,\ldots,T} \Bigg\} \\
&+ \mathbb{E}\left\{ \frac{\mathbf{1}_T' \boldsymbol{\varepsilon}_i \boldsymbol{\varepsilon}_i' \mathbf{1}_T}{T^2} \Big| \{\boldsymbol{x}_{n,t}\}_{n=1,\ldots,N}^{t=1,\ldots,T} \right\} \\
&- 2\mathbb{E}\left\{ \frac{\mathbf{1}_T' \boldsymbol{X}_i}{T^2} \left(\sum_{n=1}^{N} \boldsymbol{X}_n' \boldsymbol{Q}' \boldsymbol{Q} \boldsymbol{X}_n \right)^{-1} \left(\sum_{n=1}^{N} \boldsymbol{X}_n' \boldsymbol{Q}' \boldsymbol{Q} \boldsymbol{\varepsilon}_n \right) \boldsymbol{\varepsilon}_i' \mathbf{1}_T \Big| \{\boldsymbol{x}_{n,t}\}_{n=1,\ldots,N}^{t=1,\ldots,T} \right\} \\
&+ \mathbb{E}\Bigg\{ (\boldsymbol{x}_i^{test})' \left(\sum_{n=1}^{N} \boldsymbol{X}_n' \boldsymbol{Q}' \boldsymbol{Q} \boldsymbol{X}_n \right)^{-1} \left(\sum_{n=1}^{N} \boldsymbol{X}_n' \boldsymbol{Q}' \boldsymbol{Q} \boldsymbol{\varepsilon}_n \right) \\
&\quad \cdot \left(\sum_{n=1}^{N} \boldsymbol{\varepsilon}_n' \boldsymbol{Q}' \boldsymbol{Q} \boldsymbol{X}_n \right) \left(\left(\sum_{n=1}^{N} \boldsymbol{X}_n' \boldsymbol{Q}' \boldsymbol{Q} \boldsymbol{X}_n \right)^{-1} \right)' \boldsymbol{x}_i^{test} \Big| \{\boldsymbol{x}_{n,t}\}_{n=1,\ldots,N}^{t=1,\ldots,T} \Bigg\} \\
&- 2\mathbb{E}\Bigg\{ \frac{\mathbf{1}_T' \boldsymbol{X}_i}{T} \left(\sum_{n=1}^{N} \boldsymbol{X}_n' \boldsymbol{Q}' \boldsymbol{Q} \boldsymbol{X}_n \right)^{-1} \left(\sum_{n=1}^{N} \boldsymbol{X}_n' \boldsymbol{Q}' \boldsymbol{Q} \boldsymbol{\varepsilon}_n \right) \\
&\quad \cdot \left(\sum_{n=1}^{N} \boldsymbol{\varepsilon}_n' \boldsymbol{Q}' \boldsymbol{Q} \boldsymbol{X}_n \right) \left(\left(\sum_{n=1}^{N} \boldsymbol{X}_n' \boldsymbol{Q}' \boldsymbol{Q} \boldsymbol{X}_n \right)^{-1} \right)' \boldsymbol{x}_i^{test} \Big| \{\boldsymbol{x}_{n,t}\}_{n=1,\ldots,N}^{t=1,\ldots,T} \Bigg\} \\
&+ 2\mathbb{E}\Bigg\{ \frac{\mathbf{1}_T' \boldsymbol{\varepsilon}_i}{T} \left(\sum_{n=1}^{N} \boldsymbol{\varepsilon}_n' \boldsymbol{Q}' \boldsymbol{Q} \boldsymbol{X}_n \right) \left(\left(\sum_{n=1}^{N} \boldsymbol{X}_n' \boldsymbol{Q}' \boldsymbol{Q} \boldsymbol{X}_n \right)^{-1} \right)' \boldsymbol{x}_i^{test} \Big| \{\boldsymbol{x}_{n,t}\}_{n=1,\ldots,N}^{t=1,\ldots,T} \Bigg\} \\
&= \frac{\sigma^2 \mathbf{1}_T' \boldsymbol{X}_i}{T^2} \left(\sum_{n=1}^{N} \boldsymbol{X}_n' \boldsymbol{Q}' \boldsymbol{Q} \boldsymbol{X}_n \right)^{-1} \boldsymbol{X}_i' \mathbf{1}_T + \frac{\sigma^2}{T} \\
&- 2 \frac{\sigma^2 \mathbf{1}_T' \boldsymbol{X}_i}{T^2} \left(\sum_{n=1}^{N} \boldsymbol{X}_n' \boldsymbol{Q}' \boldsymbol{Q} \boldsymbol{X}_n \right)^{-1} \boldsymbol{X}_i' \boldsymbol{Q}' \boldsymbol{Q} \mathbf{1}_T \\
&+ \mathbb{E}\left\{ \sigma^2 (\boldsymbol{x}_i^{test})' \left(\sum_{n=1}^{N} \boldsymbol{X}_n' \boldsymbol{Q}' \boldsymbol{Q} \boldsymbol{X}_n \right)^{-1} \boldsymbol{x}_i^{test} \Big| \{\boldsymbol{x}_{n,t}\}_{n=1,\ldots,N}^{t=1,\ldots,T} \right\} \\
&- 2\mathbb{E}\left\{ \frac{\sigma^2 \mathbf{1}_T' \boldsymbol{X}_i}{T} \left(\sum_{n=1}^{N} \boldsymbol{X}_n' \boldsymbol{Q}' \boldsymbol{Q} \boldsymbol{X}_n \right)^{-1} \boldsymbol{x}_i^{test} \Big| \{\boldsymbol{x}_{n,t}\}_{n=1,\ldots,N}^{t=1,\ldots,T} \right\} \\
&+ 2\mathbb{E}\left\{ \frac{\sigma^2 \mathbf{1}_T' \boldsymbol{Q} \boldsymbol{X}_i}{T} \left(\sum_{n=1}^{N} \boldsymbol{X}_n' \boldsymbol{Q}' \boldsymbol{Q} \boldsymbol{X}_n \right)^{-1} \boldsymbol{x}_i^{test} \Big| \{\boldsymbol{x}_{n,t}\}_{n=1,\ldots,N}^{t=1,\ldots,T} \right\}
\end{aligned}
$$

$$= \frac{\sigma^2 \mathbf{1}_T' \boldsymbol{X}_i}{T^2} \left(\sum_{n=1}^{N} \boldsymbol{X}_n' \boldsymbol{Q}' \boldsymbol{Q} \boldsymbol{X}_n \right)^{-1} \boldsymbol{X}_i' \mathbf{1}_T + \frac{\sigma^2}{T}$$

$$+ \mathbb{E}\left\{ \sigma^2 \left(\boldsymbol{x}_i^{test}\right)' \left(\sum_{n=1}^{N} \boldsymbol{X}_n' \boldsymbol{Q}' \boldsymbol{Q} \boldsymbol{X}_n \right)^{-1} \boldsymbol{x}_i^{test} \middle| \{\boldsymbol{x}_{n,t}\}_{n=1,\ldots,N}^{t=1,\ldots,T} \right\}$$

$$- 2\mathbb{E}\left\{ \frac{\sigma^2 \mathbf{1}_T' \boldsymbol{X}_i}{T} \left(\sum_{n=1}^{N} \boldsymbol{X}_n' \boldsymbol{Q}' \boldsymbol{Q} \boldsymbol{X}_n \right)^{-1} \boldsymbol{x}_i^{test} \middle| \{\boldsymbol{x}_{n,t}\}_{n=1,\ldots,N}^{t=1,\ldots,T} \right\}. \tag{27}$$

References

1. Andreß, H.-J., Golsch, K., Schmidt, A.W.: Applied Panel Data Analysis for Economic and Social Surveys. Springer, Heidelberg (2013). https://doi.org/10.1007/978-3-642-32914-2
2. Athey, S., Imbens, G.: Recursive partitioning for heterogeneous causal effects. Proc. Nat. Acad. Sci. **113**, 7353–7360 (2016)
3. Chen, C.-H., Lee, L.H.: Stochastic Simulation Optimization: An Optimal Computing Budget Allocation. World Scientific, Singapore (2010)
4. Florescu, I.: Probability and Stochastic Processes. Wiley, Hoboken (2015)
5. Frees, E.W.: Longitudinal and Panel Data: Analysis and Applications in the Social Sciences. Cambridge University Press, Cambridge (2004)
6. Gnecco, G., Nutarelli, F.: On the trade-off between number of examples and precision of supervision in regression. In: Oneto, L., Navarin, N., Sperduti, A., Anguita, D. (eds.) INNSBDDL 2019. PINNS, vol. 1, pp. 1–6. Springer, Cham (2020). https://doi.org/10.1007/978-3-030-16841-4_1
7. Greene, W.H.: Econometric Analysis. Pearson Education, London (2003)
8. Groves, R.M., Fowler Jr., F.J., Couper, M.P., Lepkowski, J.M., Singer, E., Tourangeau, R.: Survey Methodology. Wiley, Hoboken (2004)
9. Nguyen, H.T., Kosheleva, O., Kreinovich, V., Ferson, S.: Trade-off between sample size and accuracy: case of measurements under interval uncertainty. Int. J. Approx. Reason. **50**, 1164–1176 (2009)
10. Ruud, P.A.: An Introduction to Classical Econometric Theory. Oxford University Press, Oxford (2000)
11. Vapnik, V.N.: Statistical Learning Theory. Wiley, Hoboken (1998)
12. Varian, H.R.: Big Data: new tricks for econometrics. J. Econ. Perspect. **28**, 3–38 (2014)
13. Wooldridge, J.M.: Econometric Analysis of Cross Section and Panel Data. MIT Press, Cambridge (2002)

Restaurant Health Inspections and Crime Statistics Predict the Real Estate Market in New York City

Rafael M. Moraes(✉), Anasse Bari, and Jiachen Zhu

Courant Institute of Mathematical Sciences, Computer Science Department, New York University, New York City, USA
{rafael.moraes,abari,jiachen.zhu}@nyu.edu

Abstract. Predictions of apartments prices in New York City (NYC) have always been of interest to new homeowners, investors, Wall Street funds managers, and inhabitants of the city. In recent years, average prices have risen to the highest ever recorded rebounding after the 2008 economic recession. Although prices are trending up, not all apartments are. Different regions of the city have appreciated differently over time; knowing where to buy or sell is essential for all stakeholders. In this project, we propose a predictive analytics framework that analyzes new alternative data sources to extract predictive features of the NYC real estate market. Our experiments indicated that restaurant health inspection data and crime statistics can help predict apartments prices in NYC. The framework we introduce in this work uses an artificial recurrent neural network with Long Short-Term Memory (LSTM) units and incorporates the two latter predictive features to predict future prices of apartments. Empirical results show that feeding predictive features from (1) restaurant inspections data and (2) crime statistics to a neural network with LSTM units results in smaller errors than the traditional Autoregressive Integrated Moving Average (ARIMA) model, which is normally used for this type of regression. Predictive analytics based on non-linear models with features from alternative data sources can capture hidden relationships that linear models are not able not discover. The framework presented in this study has the potential to serve as a supplement to the traditional forecasting tools of real estate markets.

Keywords: Artificial Intelligence · Predictive analytics · Supervised learning · Recurrent neural networks · Open Data · Alternative data · Wall Street · Real estate markets

1 Introduction

The price of apartments in New York City is a popular topic among homeowners, investors, real-estate agencies, city government and general inhabitants of the city. Average prices have risen in the last decades to the highest ever recorded in

G. Nicosia et al. (Eds.): LOD 2019, LNCS 11943, pp. 543–552, 2019.
https://doi.org/10.1007/978-3-030-37599-7_45

recent years, despite the economic crisis in 2008–10 [18]. However, not all apartments are created equal: different regions of the city appreciated very differently over time, which compounded to very disparate prices of apartments that may have seemed to be similar in the past [11].

Apartments, as other real estate properties, are valued not based on some existing standard price or inflation; instead, market supply and demand determine how valuable each apartment in each region of the city is. Even in the same building, it is possible to have apartments whose prices varied differently when undergoing the same changes, for example: a new commercial building in front may boost the price of apartments in lower floors, but decrease the prices of those in higher floors (e.g. because they lost their nice view of the park). Therefore, predicting these prices is inherently hard, given the specificity of each apartment, building and location.

On the other hand, housing is a large part of a person's expenditures in life [20], so any help in forecasting can become a decisive factor for buyers and sellers, both to decide how valuable apartments are and the right time to act.

The contributions in this study can be summarized as: (i) a predictive analytics framework for valuation of apartments in New York City given historical buy and sell prices and two alternative data sources: restaurant health inspections data and crime statistics; (ii) an artificial neural network model for predicting future prices with new predictive features for real estate markets; (iii) we experimentally show that crime statistics and restaurant health inspections, when used as predictive features in an LSTM model, provide lower prediction error than a traditional forecasting model; and (iv) we show a use case where linear predictive algorithms can fail to model reality whereas non-linear models based on recurrent neural networks are able discover hidden complex relationships.

In the next section we introduce the preprocessing work we did on the data sets used in this study.

2 Datasets

2.1 Description

Since 2012, New York City has laws that require government agencies to make much of their data accessible to the public [10], improving transparency and enabling other to put these data to good use. With this legislation, the NYC Open Data was created, where thousands of different datasets related in some way to the city can be found and used by anyone. We describe three of those, which were used in this work:

Rolling Sales Data [19]. Supported by the NYC's Dept. of Finance. Every time an apartment is sold in New York City, the operation needs to be registered with the city's Department of Finance (DoF), in order to generate the correct sales tax and proper documentation approval. Once the sale is registered, the DoF includes it as a record in its database and shares it with the public in the

Rolling Sales Data, which is updated frequently. This dataset contains tabular information about each apartment sold in the city since 2003, with fields such as: zip code, street address, apartment number, sale price, sale date, and tax class. We have chosen to use the data from 2008–2017, which contained around 197,000 sales records, of around 110,000 different apartments.

NYC Restaurant Inspection Results [16]. Supported by the NYC's Dept. of Health and Mental Hygiene. Every year, each of NYC's 24,000 restaurants go through at least one unannounced public health inspection, which looks for hygiene violations, adherence to regulations and assigns a score based on its compliance to health standards; the commonly known restaurant grades A, B and C derive from the score obtained in this inspection. This dataset contains fields such as: restaurant ID (i.e. CAMIS), address, zip code, and each specific violation code and textual description by the inspector. We have chosen to use the data from 2014–2017, which contained around 380,000 records, where each is one violation registered in an inspection, amounting to multiple violations per inspection, on average.

Citywide Crime Statistics [17]. Supported by NYC's Police Department (NYPD). They include all the arrests, shootings, complaints and crimes registered by the police around the city with details, such as: incident time and date, offense type (i.e. misdemeanor or felony), and location. We decided to use the Incident-level Complaint data, from 2006 to 2017, which contained more than 5 million records of misdemeanors and felonies committed across the city over these years. There are many details available for each record, such as a textual description and a categorization of the crime, details about the location and time it happened, and details about the suspect, among others.

2.2 Data Preprocessing

We performed several data preprocessing and data cleaning steps to the datasets used including feature extraction and feature selection.

In the Rolling Sales Data, in order to be able to match apartments that appear multiple times in the dataset, it was essential to standardize all the addresses. However, we have noticed that the addresses seemed to be typed manually at the source, leading to equal addresses being represented by different texts. For example: "10 West 15TH Street" and "10 W 15th ST". To homogenize them, we have used the Google Geocoding API [12], which receives an address and returns its "standardized" form (i.e. following Google's standardization).

A second problem faced was that many buy/sell operations had their prices listed too low when compared to other nearby apartments. For example, many operations had prices below $20,000. Therefore we discarded any record whose price was below $200,000. As a last step, zip codes of areas considered too small were removed from the analysis - such as 10118, which points exclusively to the Empire State Building - resulting in a total of 36 zip codes analyzed.

Regarding the two other datasets, the score of inspections was averaged and the total number of crimes (i.e. summing misdemeanors and felonies) was consolidated: in both cases it was done per month and per zip code. Details about the crimes or suspects were not considered, only the total number of records per month and per zip code. A similar approach was used for the restaurant inspections: additional details of the records were not considered, only the final scores of the inspections, which were averaged per month and per zip code. No other preprocessing step was necessary in these datasets.

3 Methodology

3.1 Predicting Present Apartments' Prices

To create the final dataset we first filtered only apartments that appeared at least twice in the data and stored their prices and dates of sale. Then these were used to calculate the average price growth per month between dates that, when compounded, would yield the price difference observed. This was calculated for all matching apartments. We assumed that this price growth per month could be extended a little from before the first sale date to after the last one, totaling an additional 20% extension, 10% for each side. This seemed reasonable since the price of an apartment does not start or stop growing when it is sold, instead it can be seen as a smoother process over time. For example, if the difference between sales is 5 years, we assume the growth rate is defined and constant in the 6 months prior to the first sale and 6 months after the last sale. With these calculations done for each apartment, we clustered them by zip codes and calculated a corresponding average price growth of each zip code per month. We note that clustering based solely on zip code has a low granularity and may not achieve optimal results, but this was not necessary for the analysis shown here, so a more precise clustering is left as future work.

3.2 Predicting Future Apartments' Prices

We have reasoned that even the current price of an apartment is unknown and we have explained a simple method to obtain an estimate for it based on nearby apartments. However, stakeholders would benefit further if there were a reasonably reliable system that could predict where the prices are going. The first idea one can imagine is simply doing a time-series regression with the data we already have, but this, as we show, is not optimal. We propose using all three datasets (i.e. apartment prices, crime, restaurant violations) to predict future apartment prices. The idea is that the two other datasets are much noisier but they still contain a faint signal that can improve our predictions. This is arguably plausible given the complexity of big cities, where many measurable factors are interconnected and can reinforce each other with certain time delays.

As baseline, we use an ARIMA (Autoregressive Integrated Moving Average) model for comparison, which is normally used for time-series analysis [21]. Initially we propose a simple linear model that tries to combine the time-series but, as shown below, this does not yield good results. Our final proposed model consists of a Long Short-Term Memory (LSTM) [15] neural network model that is able to combine the three datasets in a non-linear fashion and reaches the best performance. To validate these results, we ran thousands of experiments by varying hyperparameters and comparing the loss metric.

The analyzed time series have data for a period of 48 months for each chosen zip code in Manhattan. We use the first 42 months as train data and the last 6 months as validation. The goal is to compare the predictions generated by both the ARIMA and LSTM (with and without additional data) models with the ground-truth values. We use the mean squared error (MSE) metric to make this comparison.

3.3 Data Linearity and Stationarity

When dealing with time series, it is a common assumption that the autoregressive nature of the stochastic process that generated it can be reasonably approximated by linear models. This is a mere simplification and certainly not the best approach, but it is a widely used procedure. Based on this assumption, the Wold Decomposition theorem [1] states that any time-series that is weakly (covariance) stationary [9] (i.e. mean and autocovariance do not change over time) can be approximated by a sum of a deterministic and a stochastic time series. We will first apply this idea to the datasets presented here and compare this linear approach with a more general non-linear model.

The first step is verifying if the data is stationary, which can be done by using the Augmented Dickey-Fuller test [14]. This will allow the use of models such as ARMA (AutoRegressive Moving Average), which assumes the points in the data are a weighted combination of the previous p points and some random noise. However, data that shows some signs of non-stationarity can be better approximated by the ARIMA (AutoRegressive Integrated Moving Average), which has an additional step to remove a possible non-stationarity.

3.4 Granger Causality and Cross-Correlation

Given the three time series, the main question we are looking to answer is how much the two additional time series can help forecast future apartment prices. A structured way to verify this is by using Granger Causality [13], which is a statistical test that measures the predictive power of one time series into another and assumes they are linearly related. As the ARMA/ARIMA other approach mentioned above, this is the widely used by its simplicity, but it assumes the different datasets have a linear relationship among themselves.

4 Experiments

After generating the time series from the raw data, the ADF test is performed in R, where we see that the null hypothesis of data not being stationary cannot be rejected. To make all time series stationary, a first-order differentiation is used. After this, the ADF test is performed again, where we confirm that the data is now stationary, so the models can be applied.

4.1 Granger Causality

Once the data is stationary, we run the Granger-causality test by using the lmtest package in R. This was done with the "apartment prices"-"restaurant inspections" and "apartment prices"-"crime indices" pairs. For each pair, the Granger-causality test was performed twice: verifying the Granger-causality of the first time series into the second and vice versa. This null-hypothesis test is formulated such that a low p-value (i.e. assumed smaller than 0.05) would indicate the existence of Granger-causality (i.e. the inexistence could be rejected as null hypothesis). The results obtained are consistent with the opposite: with the exception of two zip codes (10006 and 10014) for the "apartment prices"-"crime indices" pair, all other tests returned a p-value greater than 0.05, indicating with high confidence that there is no Granger-causality.

4.2 ARIMA Model Applied to Base Time Series

The ARIMA model is commonly applied to regression problems involving time series because of its simplicity and clarity. It uses 3 parameters (P, D, Q) to determine the order of the regression: P is the number of parameters to use in the autoregressive part of the model; Q is the number of differences to use (to make sure the final time series is stationary); and Q is the number of parameters in the moving average window. Here we use a grid parameter search to show the best possible ARIMA model that could fit this problem: several values for p, d and q where tested and the MSE of the 6-month prediction was recorded. Here we present the histogram of the MSEs obtained for each of these configurations, but only showing those that obtained and MSE smaller than $2 \cdot 10^{-7}$. In this experiment, the MSE of the best ARIMA model was $1.1053 \cdot 10^{-7}$ (Fig. 1 and Table 1).

Table 1. Summary of the three best results obtained for ARIMA, with the corresponding model parameters.

P	D	Q	Mean squared error (1 × 10−7)
2	0	4	**1.1053**
0	0	2	1.1066
2	0	0	1.1113

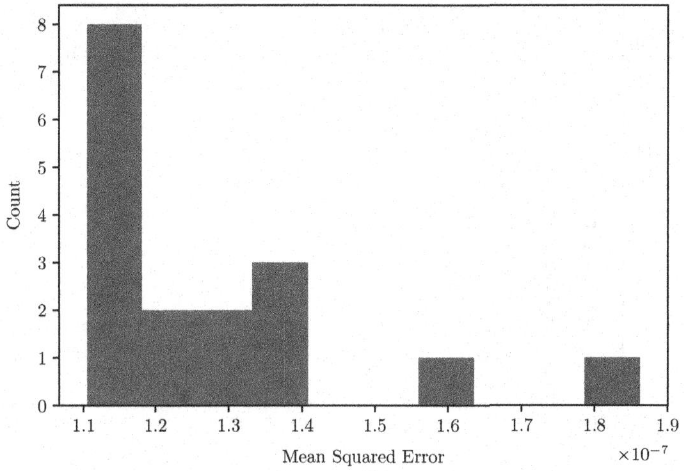

Fig. 1. Histogram showing the MSE performance of ARIMA models, which predict future apartment prices, with different hyperparameters.

4.3 LSTM Model Applied to Base Time Series

Here we use a Long Short-Term Memory neural network instead of the linear ARIMA model, following the same methodology: a large grid parameter search was used in order to find the best possible model to fit our data, based on the validation set. There were 48 different combinations of parameters and each of these was run 10 times with a different seed, totaling 480 runs. The best MSE obtained was $1.0943 \cdot 10^{-7}$. The histogram with the results is presented in the next item.

4.4 LSTM Model Applied to All Time Series

We now include the restaurant and crime time series into the LSTM model and perform the same parameter search. The best MSE obtained was $1.0906 \cdot 10^{-7}$. The histogram of the MSEs over all LSTM runs is shown in Fig. 2. Table 2 summarizes the findings.

Table 2. Summary of best results obtained for each model.

Model (best run)	Mean squared error (1 × 10−7)
ARIMA, only apt data	1.1053
LSTM, only apt data	1.0943
LSTM, all data	**1.0906**

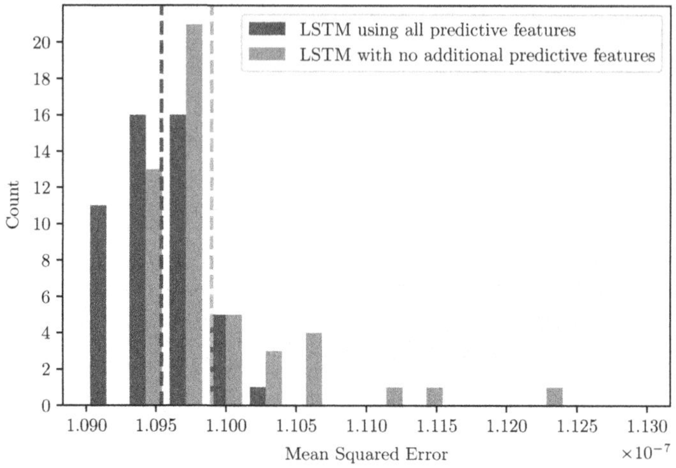

Fig. 2. Histogram showing the MSE performance of LSTM models, which predict future apartment prices, with different hyperparameters. Blue bars are the cases using all predictive features, extracted from the restaurant inspections, the crime statistics, and the apartment sales data. Bars in orange are cases when only using predictive features obtained from the apartment sales data. Dotted lines represent the average MSE of each set. Each model configuration was run 10 times and the average MSE of each is used here. A total of 98 configurations (980 runs) are shown. (Color figure online)

5 Conclusion

In this study, we presented a predictive analytics framework that provides a methodology for analyzing alternative data sources and applying recurrent neural networks to help predict real estate markets. The experiments that we conducted indicate that engineering new predictive features such from new data sources to real estate such as restaurant health inspection data and crime statistics has the potential to improve the accuracy of predictions of apartment prices using non-linear models such as recurrent neural networks. As future work, we plan to measure the predictive power of other alternative data sources related to real estate markets and to apply biologically inspired algorithms [2,6–8] for better grouping apartments. We also plan to include new features by applying Emotion Artificial Intelligence [3–5] to big data sources such as Twitter and news articles in order to build a customer index for the real estate market of NYC.

Acknowledgments. We would like thank Jing Wang, who contributed with helpful discussions to the initial analysis of this work. Also, the NYU High Performance Computing team, especially Shenglong Wang, who was always available to help with technical issues in the computer cluster.

References

1. Anderson, T.W.: The Statistical Analysis of Time Series. Wiley, Hoboken (2011)
2. Bari, A., Chaouchi, M., Jung, T.: Predictive Analytics for Dummies. Wiley, Hoboken (2016)
3. Bari, A., Liu, L.: Probing the wisdom of apple, inc., crowds using alternative data sources (2017). https://insidebigdata.com/2017/10/12/probing-wisdom-apple-inc-crowds-using-alternative-data-sources/
4. Bari, A., Peidaee, P., Khera, A., Zhu, J., Chen, H.: Predicting financial markets using the wisdom of crowds. In: 2019 IEEE 4th International Conference on Big Data Analytics (ICBDA), pp. 334–340. IEEE (2019)
5. Bari, A., Saatcioglu, G.: Emotion artificial intelligence derived from ensemble learning. In: 2018 17th IEEE International Conference on Trust, Security and Privacy in Computing and Communications/12th IEEE International Conference On Big Data Science And Engineering (TrustCom/BigDataSE), pp. 1763–1770. IEEE (2018)
6. Abdelghani Bellaachia and Anasse Bari. SFLOSCAN: a biologically-inspired data mining framework for community identification in dynamic social networks. In: 2011 IEEE Symposium on Swarm Intelligence, pp. 1–8. IEEE (2011)
7. Bellaachia, A., Bari, A.: Flock by leader: a novel machine learning biologically inspired clustering algorithm. In: Tan, Y., Shi, Y., Ji, Z. (eds.) ICSI 2012. LNCS, vol. 7332, pp. 117–126. Springer, Heidelberg (2012). https://doi.org/10.1007/978-3-642-31020-1_15
8. Bellaachia, A., Bari, A.: A flocking based data mining algorithm for detecting outliers in cancer gene expression microarray data. In: 2012 International Conference on Information Retrieval & Knowledge Management, pp. 305–311. IEEE (2012)
9. Brockwell, P.J., Davis, R.A.: Introduction to Time Series and Forecasting. Springer Texts in Statistics. Springer, Berlin (2016)
10. Campbell, C.: New York City open data: a brief history, 08 March 2017. https://datasmart.ash.harvard.edu/news/article/new-york-city-open-data-a-brief-history-991
11. The furman center for real estate & urban policy. Trends in New York City housing price appreciation (2008). https://furmancenter.org/files/Trends_in_NYC_Housing_Price_Appreciation.pdf
12. Google. Google Geocoding API. https://developers.google.com/maps/documentation/geocoding/start
13. Granger, C.W.J.: Causality, cointegration, and control. J. Econ. Dyn. Control **12**(2–3), 551–559 (1988)
14. Greene, W.H.: 1951 Econometric Analysis. Pearson Prentice Hall, Upper Saddle River (2012)
15. Hochreiter, S., Schmidhuber, J.: Long short-term memory. Neural Comput. **9**(8), 1735–1780 (1997)
16. New York City department of mental health and mental hygiene. NYC restaurant inspection results. https://data.cityofnewyork.us/Health/DOHMH-New-York-City-Restaurant-Inspection-Results/43nn-pn8j
17. New York Police department. Citywide crime statistics (2018). https://www1.nyc.gov/site/nypd/stats/crime-statistics/citywide-crime-stats.page
18. Nonko, E.: Manhattan home prices have increased dramatically in a decade, 02 February 2017. https://ny.curbed.com/2017/2/2/14483418/manhattan-home-sales-market-reports

19. NYC Department of Finance. Rolling sales data (2018). https://www1.nyc.gov/site/finance/taxes/property-rolling-sales-data.page
20. Bureau of Labor Statistics. U.S. Department of Labor. Consumer Expenditures - 2017 (2018). https://www.bls.gov/news.release/cesan.nr0.htm
21. Shumway, R.H., Stoffer, D.S.: Time Series Analysis and Its Applications: with R Examples. Springer, Berlin (2017)

Load Forecasting in District Heating Networks: Model Comparison on a Real-World Case Study

Federico Bianchi(✉), Alberto Castellini, Pietro Tarocco, and Alessandro Farinelli

Department of Computer Science, Verona University, Strada Le Grazie 15, 37134 Verona, Italy
{federico.bianchi,alberto.castellini, pietro.tarocco,alessandro.farinelli}@univr.it

Abstract. District Heating networks (DHNs) are promising technologies for heat distribution in residential and commercial buildings since they enable high efficiency and low emissions. Within the recently proposed paradigm of smart grids, DHNs have acquired intelligent tools able to enhance their efficiency. Among these tools, there are demand forecasting technologies that enable improved planning of heat production and power station maintenance. In this work we propose a comparative study for heat load forecasting methods on a real case study based on a dataset provided by an Italian utility company. We trained and tested three kinds of models, namely non-autoregressive, autoregressive and hybrid models, on the available dataset of heat load and meteorological variables. The optimal model, in terms of root mean squared error of prediction, was selected. It considers the day of the week, the hour of the day, some meteorological variables, past heat loads and social components, such as holidays. Results show that the selected model is able to achieve accurate 48-hours predictions of the heat load in several conditions (e.g., different days of the week, different times, holidays and workdays). Moreover, an analysis of the parameters of the selected models enabled to identify a few informative variables.

Keywords: Heat load forecasting · District heating network · Linear regression models · Model interpretability · Time series analysis

1 Introduction

The heating and cooling demand in the residential sector in Europe is responsible for a share of about 40% of the overall final energy usage [24]. District Heating networks (DHNs) are acknowledged as a promising technology for heat distribution, covering the demand of buildings, providing higher efficiency and better pollution control. A DHN is a network in which the heat is generated in one or more centralized locations and distributed through a pipe system via

G. Nicosia et al. (Eds.): LOD 2019, LNCS 11943, pp. 553–565, 2019.
https://doi.org/10.1007/978-3-030-37599-7_46

exchangers in buildings (see Fig. 1). The management of the system is based on seasonal heating temperature. The combination of water flow and temperature (that influence the power provided to the network) must be high enough to cover the demand of the entire network and also the heat losses [12]. Key analytical tools for optimizing the heat production for DHNs are load forecasting models. They consist of a set of statistical and artificial intelligence methods for predicting the future heat load, given the past evolution of the load and some environmental factors that influence it. Accurate models for heat load forecasting have significant advantages such as improved energy efficiency of heat production and reduced environmental impact. This topic has recently gained strong interest from the artificial intelligence community in the context of smart grids using intelligent agents to optimize their functioning.

Several predictive models [17] have been recently designed for short-term load forecasting (STLF), and the literature contains a variety of applications and models [2,4,5,7]. The application domain which mostly resembles our domain of interest is electricity demand forecasting, for which an extended literature is available [15,16]. Models can be classified into two main categories, namely, *statistical* approaches for time series forecasting, mainly based on autoregressive integrated moving average (ARIMA) models, and *machine learning/artificial intelligence* techniques, such as Gaussian processes, artificial neural networks, fuzzy logic and support vector machines. ARIMA models [15] are usually applied in univariate scenarios for short horizons. A number of studies apply seasonal ARIMA (SARIMA) techniques to STLF [8,10,18,19,25] to model seasonal patterns, and ARIMAX in multivariate scenario to simulate the effects of weather and the day of the week on the future demand. Multiple linear regression is proposed in [1,11,12,23] to consider multiple relationships between weather conditions, social components (such as holidays and weekdays) and future demand. A hybrid approach is proposed in [14], combining similar days and deep neural network to construct a model for short-term heat load forecasting. Comprehensive reviews of the literature are provided in [13,15,20,22]. A recent survey is also provided in [21], where the state-of-the-art of the last five years on load forecasting models is analyzed focusing on conventional and big data.

Load forecasting models normally consider weather and social factors. The weather is the most significant factor affecting the heat consumption, however, according to [27] most of the past publications use temperature as the main component and only a few of them consider other factor such as relative humidity, wind and rainfall. Social factors describes the behaviour of customers and can be interpreted in terms of annual, weekly and daily patterns. Other relevant social factors are calendar events [9], such as holidays and weekends in which the heat load demand is different than working days. There exist two main approaches of linear regression models, namely, *single-equation strategy* for all the hours, and *multiple-equation strategy* with specific equations for each hour of the day and each day of the week [23,25].

In this work we present a comparative study of heat load forecasting methods based on a real case of DHN managed by AGSM[1], an utility company located in Verona (Italy). The goal of our study is to improve the load forecasting performance for this network using data-driven modeling tools. The comparison concerns three modeling frameworks, namely, non-autoregressive models, autoregressive models and a hybrid method called Prophet [26], which are available in the literature. Different configurations of parameters are tested, obtaining 8 models in total. Results show that autoregressive models reach the best performance in our specific case study. An analysis is conducted to interpret the models and discover common behaviors in different weekdays and hours of the day. The main contributions of this work to the state of the art are summarized in the following points:

- we compare the performance of 8 models on a real case study of heat load forecasting and we identify an optimal model.
- we analyze the performance of the optimal model in different time instants, such as, working days, weekends, holidays.
- we analyze the parameters of the optimal model, providing a first interpretation of the most significant variables for 48-hours forecasting.

The rest of the manuscript is organized as follows. Section 2 presents the framework for model comparison, the dataset, the used methodologies and the performance measures. Result and performance are analyzed in Sect. 3. Conclusions and future directions are described in Sect. 4.

2 Materials and Methods

In this section we formalize the problem, describe our dataset of heating loads and meteorological parameters, introduce the modeling methodologies and define the performance measures used to compare different models.

2.1 System Overview and Problem Definition

The district heating, illustrated on top of Fig. 1 is a centralized network where the heat is produced and distributed through a pipe system to commercial and residential buildings. The utility company controls the variation of heating load with water flow in pipes, according to seasonal operation temperature. The supplied temperature and water flow must be high enough to cover the customers demand and provide better control of pollution, using renewable energy. The aim of the present work is to identify models which allow us to optimize the heat load forecasting in the next 48-hours and find out which variables are significant on forecasting the heat load.

The bottom of Fig. 1 is an overview of the process we used to select the optimal model for heat load forecasting. The dataset provided by utility company is used to train different statistical models. Here we consider three kinds of

[1] https://www.agsm.it/.

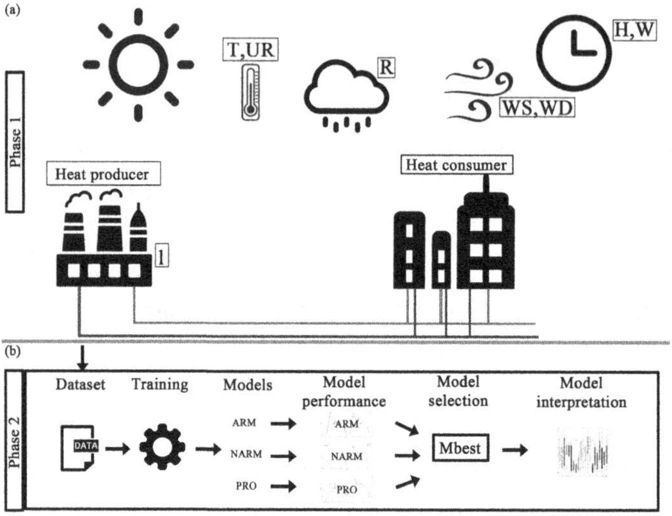

Fig. 1. Framework for model comparison: phase 1 → (a) district heating network (data acquisition). Phase 2 → (b) heat forecasting: model selection

methods, namely, non-autoregressive, autoregressive and hybrid models. After the generation of models we evaluate the performance and select the optimal model that better fits the data. The models are evaluated with standard performance measures. Finally we use the optimal model to forecast the heat load and analyze the parameters in order to identify the most significant variables.

2.2 Dataset

In this work we use a real world dataset provided by AGSM[2] an Italian energy utility company. The sampling interval of dataset is of one hour, the number of observations is 12792, including three years, namely 2016, 2017 and 2018. The public administration controls the switching on/off of DH during summer, therefore we use data concerns fall-spring period of each year with the following interval: from 2016–2017 (08.01.2016 - 11.05.2016, 11.10.2016 - 14.05.2017, and 16.10.2017 - 31.12.2017, 10296 observations in total) to train the models. Data of 2018 (08.01.2018 - 21.04.2018, 2496 observations) are used for testing. Raw dataset includes weather variables and heating load consumption of the DH. We extract new variables through a features engineering process, using domain knowledge. The complete list of variables is reported in Table 1.

[2] In accordance with the company policy, dataset and source code may not be provided at this stage of the work.

Table 1. List of variables

Symbol	Description	Symbol	Description
l	Heating load [MW]	T	Temperature [°C]
l1	Heating load of the previous day	UR	Relative humidity [%]
l2	Heating load of 2 days before	WS	Speed of wind (m/s)
l3	Heating load of 3 days before	WD	Direction of wind. Cardinal directions encoded from 0 to 9 (9 = no wind)
l4	Heating load of 4 days before	R	Rainfall. 1 = rain, 0 = otherwise
l5	Heating load of 5 days before	TV	Moving average of temperature on seven previous days
l6	Heating load of 6 days before	TM	Maximum temperature of the day
l7	Heating load of 7 days before	TS	Square temperature
lp	Peak load of previous day at 6:00 AM	T2	Square of maximum temperature
H	Holiday (0 = false, 1 = true)	TMY	Maximum temperature of the previous day
W	Weekend (0 = false, 1 = true)	TMY2	Square of maximum temperature of the previous day

2.3 Models

We propose three kinds of model[3], namely: Non-Autoregressive Model (NARM), Autoregressive Model (ARM) and Prophet (PRO) described below and summarized in Table 2.

Non-Autoregressive Model (NARM). It is built through a process known as binary recursive partitioning, which is an iterative process that splits the data into partitions (or branches) having similar properties (e.g., weekday, hour), and then continues splitting each partition into smaller groups as the method moves down each branch. The various branches refer to hour of the day, day of the week, heating on/off. The aim of this method is to analyze the fundamental components of the data in order to investigate if there are any (linear) relationships between the heat load and some independent variables in specific days of the week and hours of the day. The models are generated after splitting the data into partitions and this kind of approach is useful because, for example, if we have to make

[3] In the present work the models are written using Python in Spyder IDE.

predictions on Monday at 8:00 AM, we just need to access the model that was trained with data in the same day and time in the past. We generated two types of linear regression models, called NARM 1 and NARM 2 in the following, that use a tree structure and no autoregressive variables, such as delayed load. NARM 1 is a set of day/hour univariate linear regression models (168 models in total) where every model is related to a specific pair (weekday, hour) and has the form $l = C_1 T + C_0$, where C_0, C_1 are coefficients and T is the temperature. NARM 2 differs from NARM 1 in the independent variable, since it uses, in addition to T, variables UR, WS and WD. The learning methodology used is ordinary least squares that fits a linear model with a vector of coefficients that minimize the residual sum of squares between the observed responses in the data and the responses predicted by the linear approximation.

Autoregressive Model (ARM). It is a statistical method to characterize the impact of one or more independent variables on a dependent variable, in order to evaluate the linear dependency between them. Here we propose four types of linear regression models, in which the heat consumption l is explained by a linear model based on weather variables: T, UR, WS, WD and R. Besides the weather factors, we include social factors like annual, weekly, daily patterns of consumption and calendar events, distinguish workdays from weekend. The proposed models are called ARM 1, ARM 2, ARM 3 and ARM 4 in the following. All of them are autoregressive because they consider delayed load. In ARM 1 and ARM 2 we adopt a single-equation model strategy. The latter uses 7 delayed loads, from one day before the forecasting instant to 7 days in advance. In ARM 3 we use a multiple equations strategy with one model for each hour of the day (i.e., 24 models in total), while in ARM 4 we use a multiple equations strategy with one model for each pair (weekday, hour) (i.e., 168 models in total). In ARM 4 we forecast the load for each day and for each hour of the day with different equations, therefore we have 168 forecasting equations having the following form: $M_{ij} = f(WV, dl, social)$, where $i \in \{1, \ldots, 7\}$ refers to day of week, $j \in \{0, \ldots, 23\}$ refers to hour of the day, WV refers to weather and derived variables. The autoregressive part of the method is represented by dl and it refers to past loads. These variables measure the load a day before the prediction (at the same hour), 2 days before the prediction and so on, until 7 days before the prediction. We also consider the peak of energy load at the previous day, which is at 6:00 AM in our case. *Social* refers calendar events, such as working days or holiday. These dummy variables are important because they model the different load during different days of the week, in fact during the weekend and holiday we can observe that the heat load is lower. In ARM 3 and ARM 4 we follow the [23] idea, considering a multiple equations approach. Here we include more than one meteorological factors such as, T, UR and R. We do not consider lagged errors but 7 lagged loads to enforce the forecast accuracy.

Prophet (PRO). It is an open source library published by Facebook[4] [26] that provides the ability to make time series prediction, but it is also designed to have

[4] https://research.fb.com/prophet-forecasting-at-scale.

Table 2. Model description

Model id	Model type	Factors
ARM 1	Autoregr. mdl	dl: l1
ARM 2	Autoregr. mdl	dl: li (i ∈ [1, 7])
ARM 3	Autoregr. mdl	dl: li (i ∈ [1, 7]). WV: T, UR, WS, WD, TV, TM. Social: H, W, lp
ARM 4	Autoregr. mdl	dl: li (i ∈ [1, 7]). WV: T, UR, WS, WD, TV, TM, R, TS, TM2, TMY, TMY2. Social: H, W
NARM 1	Non-Autoregr. mdl	WV: T
NARM 2	Non-Autoregr. mdl	WV: T, UR, WS, WD
PRO 1	Prophet mdl	dl: 1
PRO 2	Prophet mdl	dl: li (i ∈ [1, 7]). WV: T, UR

intuitive parameters that can be adjusted without knowing the details of the underlying model. At its core, the Prophet procedure is similar to a generalized additive regression model (GAM) [17] with four main components: a piecewise linear or logistic growth curve trend *g(t)*, a yearly seasonal component modeled using Fourier series *s(t)*, a weekly seasonal component using dummy variables *h(t)* and the error term ε_t: $y(t) = g(t) + s(t) + h(t) + \varepsilon_t$.

The GAM formulation has the advantage to be flexible, accommodating seasonality with multiple periods, it fits very fast and parameters are easily interpretable. Here we use two types of Prophet model: PRO 1 and PRO 2. PRO 1 considers only the historical load component to predict the next 48-hours. PRO 2 also uses regressive components of load, starting from prediction instant and considering loads from a day to seven days in the past.

2.4 Performance Measures

Models are evaluated by two standard performance measures for time-series forecasting, namely, the root mean square error (RMSE) and the mean absolute percentage error (MAPE). Given an observed time-series with n observations $y_1, \ldots, y_n$ and a prediction $\hat{y}_1, \ldots, \hat{y}_n$, the RMSE is computed as formula:

$$RMSE = \sqrt{\frac{1}{N}\sum_{t=1}^{N}(\hat{y}_t - y_t)^2} \tag{1}$$

and it represents the average error that the prediction makes each time instant. The MAPE is computed as formula:

$$MAPE = (\frac{1}{N}\sum_{t=1}^{N}(\frac{|\hat{y}_t - y_t|}{y_t})) \cdot 100\% \quad (2)$$

and it represents the absolute average percentage error that the prediction makes. We are interest to compute the performance of the models on 48-hours forecasting, therefore we iterate the computation of RMSE and MAPE using a sliding window of 48-hours to evaluate the models on the entire test dataset. Starting from the point p_1 of dataset, we forecast the next 48-hours and compute RMSE and MAPE values of the interval $[p_1,p_{48}]$, then we move to the next point p_2 and repeat the previous computations. At the final step the evolution of the error on test dataset is represented by a RMSE and MAPE value for each point. Computing the average of the RMSE and MAPE points we obtain the average of root mean square error, $\overline{RMSE}$ and the average of mean absolute percentage error, $\overline{MAPE}$, that we use to compare the global performance of the models.

3 Results

Here we present a comparative evaluation of the models described in the previous section. In the first part of the analysis (see Fig. 2) we perform a comparison of the global performance (i.e., $\overline{RMSE}$ and $\overline{MAPE}$) to select the optimal model. In the second part (see Fig. 3) we observe some 48-hour predictions of the optimal model in different time instants of 2018 (e.g., working days, weekends and holidays) and compare them to real heat load in the same moments to understand how the model behaves in different situations. Finally, in the third part of the section (see Fig. 4) we analyze the parameters of the optimal model and use data visualization and model interpretation tools [6] to identify variables having stronger impact on the quality of the prediction.

Selection of Optimal Model. We compare the eight models summarized in Fig. 2a. The table shows the number of variables used in each model, the average root mean square error and the average mean absolute percentage error. The models are ordered by $\overline{RMSE}$. Model ARM 4 (highlighted in green) has the best performance. The multiple-equation model, using one equation for each pair of weekday and hour and 20 independent variables, proves to be accurate in forecasting. ARM 3 assuming a multiple-equation model with one equation for each hour of the day and 16 variables, performed equal to ARM 4 ($\overline{RMSE}$), but it achieved worse $\overline{MAPE}$. PRO 2 follows closely whereas PRO 1, highlighted in red, has the worse performance. The scatter plot in Fig. 2b, which represents the relationship between the number of variables used in a model and the RMSE, shows that models with high numbers of variables have lower error. Figure 2c shows the time evolution of the RMSE computed for 48-hour predictions by the best 3 models. Each point (x, y) represents the RMSE (coordinate y) of a 48-hours forecast starting at instant x. ARM 4 has a smaller error than ARM 3 and

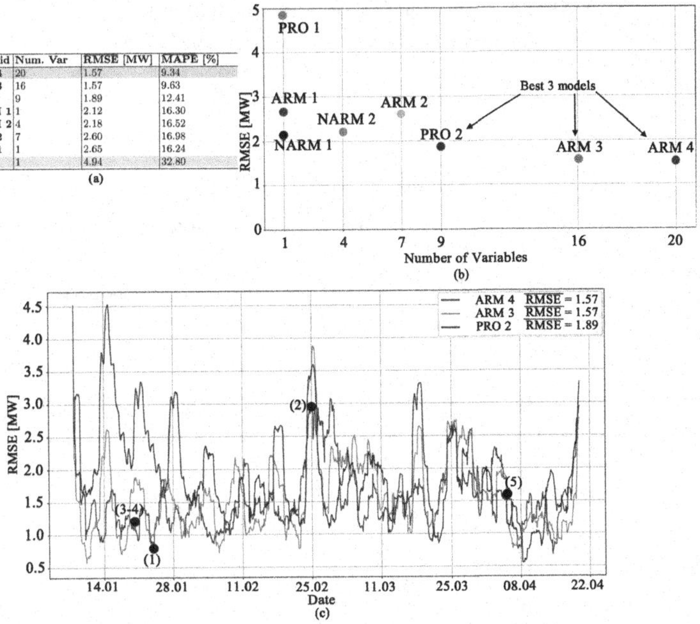

Model id	Num. Var	$\overline{RMSE}$ [MW]	$\overline{MAPE}$ [%]
ARM 4	20	1.57	9.34
ARM 3	16	1.57	9.63
PRO 2	9	1.89	12.41
NARM 1	1	2.12	16.30
NARM 2	4	2.18	16.52
ARM 2	7	2.60	16.98
ARM 1	1	2.65	16.24
PRO 1	1	4.94	32.80

(a)

Fig. 2. Comparison of model performance: (a) Table of number of variables, $\overline{RMSE}$ and $\overline{MAPE}$ for each model. (b) Relationship between number of variables and $\overline{RMSE}$ for the analyzed models. (c) RMSE of the best 3 models for each 48-hours forecast performed in 2018. Each point (x, y) represents the RMSE (y) of a 48-hours forecast starting at instant x (Color figure online)

PRO 2. In Fig. 2c we also highlight 5 significant points that allow us to focus on the model which obtained the best performance: ARM 4.

Comparison Between Real and Predicted Load. In Fig. 3 we show the comparison between real and predicted load in the 5 significant points described in the following. Point (1) represents 48-hours forecasting with minimum RMSE error which fits almost perfectly. Point (2) represents 48-hours forecasting with maximum RMSE error. Notice that even though the error in that point is the highest, ARM 4 performs better than the other methods. Points (3–4) represent 48-hours forecasting in a weekend (i.e., Friday-Saturday and Saturday-Sunday). The model is able to adapt to social behaviors, forecasting quite well working day as well as weekend, where the heat load changes. It also fits perfectly the load peak. Finally, (5) shows 48-hours forecasting on a holiday. We show 48-hours forecasting starting from a day before Easter 2018 and from Easter to Easter Monday. For each prediction, the method is able to fit load peaks, adapting to load variations very quickly. The regression line is less closest to the real load in (2) and in (5), even though the RMSE value is still low.

Informative Variables. In Fig. 4 we show the parameters of the optimal model. The bar plot (Fig. 4a) identifies the significant variables, in particular it shows the

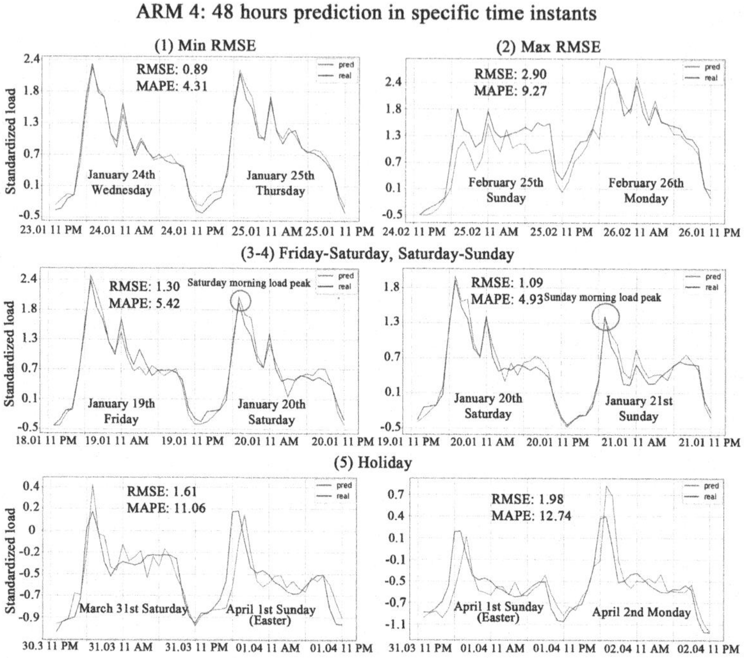

Fig. 3. ARM 4 - optimal model: (1) 48-hours forecasting with minimum RMSE error. (2) 48-hours forecasting with maximum RMSE error. (3–4) 48-hours forecasting of weekend, Friday-Saturday and Saturday-Sunday. (5) 48-hours forecasting of holiday: Easter

percentage of models in which variables has p-values less than or equal to 0.05. The graph in Fig. 4b shows which variables are statistically significant for three specific models, namely, Monday, Saturday and Sunday at 11. Figure 4a show that 3 variables are very important: T, W and l1, in fact they result significant for more than 65% of the models. Temperature T is particularly meaningful during evening and night when the temperature falls and the heat load increases in domestic buildings. The autoregressive variable l1 that refers to load of previous day is the most significant variable for the method, having a p-values ≤ 0.05 for almost 80% of the models. The use of autoregressive variables as the load of the past days, in particular the previous 24-hours prior the forecasting instant, it seems to be crucial to predict the next 48-hours. An other autoregressive variable such as the load peak of the previous day lp, it results meaningful for the model of 6 AM, when the first and the higher load peak of the days occurs. This knowledge refines the forecasting capability of the method. The other weather variables seem to have not considerable impact for a good number of models.

In Fig. 4b we analyze the significance of variables for three relevant models such as Monday, Saturday and Sunday at 11 AM, when the second load peak of the day occurs. From scatter plot we can see that lagged load l1 to forecast the load at 11 AM is very significant for each selected models. The model of Sunday

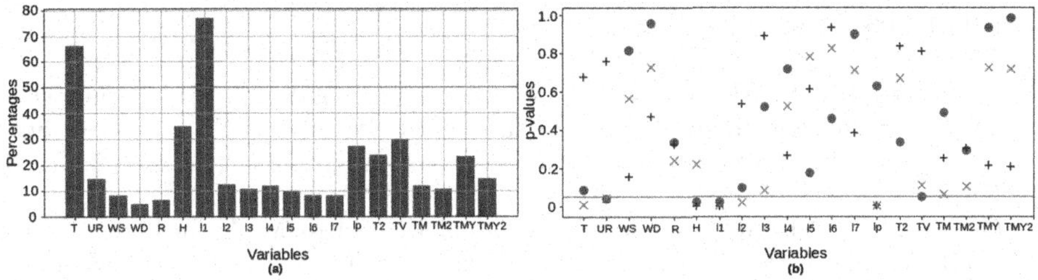

Fig. 4. Parameters of ARM 4 model: (a) the significant variables. (b) The significant variables for Monday (blue), Saturday (green) and Sunday (red) models at 11 AM (Color figure online)

considers the weather factors as temperature and relative humidity and also the autoregressive load variables. The variables l1 and lp contain information concern Saturday morning at 11:00 AM and 6:00 AM, two hours of the day when load is higher. These variables allow Sunday to have a precious knowledge about the load. The model of Saturday has almost opposite behaviour than the other two models. It considers weather variables we computed: TM, TMY, T2, TMY2 and finally TV. The model of Monday considers significant only a few variables such as UR, H, l1 and TV. In particular we can notice that the variable lp is not important for the model, because refers to Sunday at 11 AM and it is a behaviour we expected.

4 Conclusion and Ongoing Work

We proposed a comparative study of models for load forecasting in DHNs. Models were generated and analyzed on a real-world dataset. We evaluated non-autoregressive, autoregressive and hybrid models performances in terms of RMSE and MAPE to find the optimal model. Model ARM 4 performed better than the others. It is an autoregressive model with multiple-equation approach with one equation for each pair (weekday, hour), with 168 equations in total. By analyzing the parameters of model ARM 4 we discovered that weather factors, delayed load and social components are crucial for accurate forecasting. This research can be widely extended in several directions. First, we will compare the presented models with models generated by other machine learning techniques, such as, Gaussian processes and neural networks. Then, we will investigate the effects of solar radiation and other variables on heat load for improving forecast accuracy. Finally, we will consider the investigate different ways to integrate these forecasting models into planning frameworks [3] to improve operational decision making.

Acknowledgments. The research has been partially supported by the projects "Dipartimenti di Eccellenza 2018–2022, funded by the Italian Ministry of Education, Universities and Research (MIUR), and "GHOTEM/CORE-WOOD, POR-FESR 2014-2020", funded by Regione del Veneto.

References

1. Baltputnis, K., Petrichenko, R., Sobolevsky, D.: Heating demand forecasting with multiple regression: model setup and case study. In: 2018 IEEE 6th Workshop on Advances in Information, Electronic Electrical Engineering (AIEEE), pp. 1–5 (2018)
2. Castellini, A., Beltrame, G., Bicego, M., Blum, J., Denitto, M., Farinelli, A.: Unsupervised activity recognition for autonomous water drones. In: Proceedings of the Symposium on Applied Computing, SAC 2018, pp. 840–842. ACM (2018)
3. Castellini, A., Chalkiadakis, G., Farinelli, A.: Influence of state-variable constraints on partially observable monte carlo planning. In: Proceedings of 28th International Joint Conference on Artificial Intelligence (IJCAI 2019), pp. 5540–5546 (2019)
4. Castellini, A., Franco, G.: Bayesian clustering of multivariate immunological data. In: Nicosia, G., Pardalos, P., Giuffrida, G., Umeton, R., Sciacca, V. (eds.) LOD 2018. LNCS, vol. 11331, pp. 506–519. Springer, Cham (2018). https://doi.org/10.1007/978-3-030-13709-0_43
5. Castellini, A., et al.: Subspace clustering for situation assessment in aquatic drones. In: Proceedings of Symposium on Applied Computing, SAC 2019, pp. 930–937. ACM (2019)
6. Castellini, A., Masillo, F., Sartea, R., Farinelli, A.: eXplainable modeling (XM): data analysis for intelligent agents. In: Proceedings of the 18th International Conference on Autonomous Agents and Multiagent Systems (AAMAS 2019), pp. 2342–2344. IFAAMAS (2019)
7. Castellini, A., Paltrinieri, D., Manca, V.: MP-GeneticSynth: inferring biological network regulations from time series. Bioinformatics **31**, 785–87 (2015)
8. Castellini, A., Zucchelli, M., Busato, M., Manca, V.: From time series to biological network regulations: an evolutionary approach. Mol. BioSystems **9**, 225–233 (2013)
9. Dahl, M., Brun, A., Kirsebom, O.S., Andresen, G.B.: Improving short-term heat load forecasts with calendar and holiday data. Energies **11**, 1678 (2018)
10. Elamin, N., Fukushige, M.: Modeling and forecasting hourly electricity demand by sarimax with interactions. Discussion Papers in Economics and Business, pp. 17–28, Osaka University (2017)
11. Fang, T.: Modelling district heating and combined heat and power (2016)
12. Fang, T., Lahdelma, R.: Evaluation of a multiple linear regression model and SARIMA model in forecasting heat demand for district heating system. Appl. Energy **179**, 544–552 (2016)
13. Feinberg, F.A., Genethliou, D.: Load Forecasting. In: Chow, J.H., Wu, F.F., Momoh, J. (eds.) Applied Mathematics for Restructured Electric Power Systems. Power Electronics and Power Systems, pp. 269–285. Springer, Boston (2005). https://doi.org/10.1007/0-387-23471-3_12
14. Gong, M., Zhou, H., Wang, Q., Wang, S., Yang, P.: District heating systems load forecasting: a deep neural networks model based on similar day approach. Adv. Build. Energy Res., 1–17 (2019)
15. Gross, G., Galiana, F.D.: Short-term load forecasting. Proc. IEEE **75**(12), 1558–1573 (1987)
16. Hagan, M.T., Behr, S.M.: The time series approach to short term load forecasting. IEEE Trans. Power Syst. **2**, 785–791 (1987)
17. Hastie, T., Tibshirani, R.: Generalized additive models: some applications. J. Am. Stat. Assoc. **82**, 371–386 (1987)

18. Hyndman, R.J., Athanasopoulos, G.: Forecasting: Principles and Practice. Text, Melbourne (2014)
19. Kim, M.S.: Modeling special-day effects for forecasting intraday electricity demand. Eur. J. Oper. Res. **230**, 170–180 (2013)
20. Mirowski, P., Chen, S., Ho, T.K., Yu, C.N.: Demand forecasting in smart grids. Bell Labs Tech. J. **18**, 135–158 (2014)
21. Mujeeb, S., Javaid, N., Javaid, S., Rafique, A., Manzoor, I.: Big data analytics for load forecasting in smart grids: a survey (2019)
22. Muñoz, A., Sánchez-Úbeda, E.F., Cruz, A., Marín, J.: Short-term forecasting in power systems: a guided tour. In: Rebennack, S., Pardalos, P., Pereira, M., Iliadis, N. (eds.) Handbook of Power Systems II. ENERGY, pp. 129–160. Springer, Berlin (2010). https://doi.org/10.1007/978-3-642-12686-4_5
23. Ramanathan, R., Engle, R., Granger, C.W.J., Vahid-Araghi, F., Brace, C.: Short-run forecast of electricity loads and peaks. Int. J. Forecast. **13**, 161–174 (1997)
24. Buffa, S., Cozzini, M., D'Antoni, M., Baratieri, M., Fedrizzi, R.: 5th generation district heating and cooling systems: a review of existing cases in Europe. Renew. Sustain. Energy Rev. **104**, 504–522 (2019)
25. Soares, L.J., Medeiros, M.C.: Modeling and forecasting short-term electricity load: a comparison of methods with an application to Brazilian data. Int. J. Forecast. **24**, 630–644 (2008)
26. Taylor, S.J., Letham, B.: Forecasting at scale. Am. Stat. **72**, 37–45 (2017)
27. Weron, R.: Modeling and Forecasting Electricity Loads and Prices: A Statistical Approach. Wroclaw University of Technology, Hugo Steinhaus Center (2006)

A Chained Neural Network Model for Photovoltaic Power Forecast

Carola Gajek(✉), Alexander Schiendorfer, and Wolfgang Reif

Institute for Software & Systems Engineering,
University of Augsburg, Augsburg, Germany
{gajek,schiendorfer,reif}@isse.de

Abstract. Photovoltaic (PV) power forecasting is an important task preceding the scheduling of dispatchable power plants for the day-ahead market. Commercially available methods rely on conventional meteorological data and parameters to produce reliable predictions. These costs increase linearly with a rising number of plants. Recently, publicly available sources of free meteorological data have become available which allows for forecasting models based on machine learning, albeit offering heterogeneous data quality. We investigate a chained neural network model for PV power forecasting that takes into account varying data quality and follows the business requirement of frequently introducing new plants. This two-step model allows for easier integration of new plants in terms of manual efforts and achieves high-quality forecasts comparable to those of raw forecasting models from meteorological data.

Keywords: Machine learning · Neural networks · Photovoltaic power forecast

1 Motivation

In the wake of the energy revolution, more and more volatile power plants based on renewable energy sources such as wind turbines or photovoltaic (PV) plants enter the market. In Germany, for instance, solar energy accounted for 8.4% of the total electricity generated in 2018 – five years earlier it was only 5.7% [3]. This increasing ratio affects the stability of the power grid due to the intermittent generation of PV plants. Cloud movements influence the solar irradiation and consequently cause fluctuations in the generated power.

In order to guarantee balance in the energy grid, supply and demand must be approximately equal at all times. While the output generated by dispatchable plants such as gas turbines can be increased in the event of an energy decay, this is not possible for PV plants. So, if volatile plants produce too little electricity, dispatchable power plants have to compensate. In order to be able to estimate

The research presented in this paper was carried out in cooperation with and partially funded by Stadtwerke München. We thank Barnabas Kittlaus and Florian Nafz for their contributions and providing PV data.

G. Nicosia et al. (Eds.): LOD 2019, LNCS 11943, pp. 566–578, 2019.
https://doi.org/10.1007/978-3-030-37599-7_47

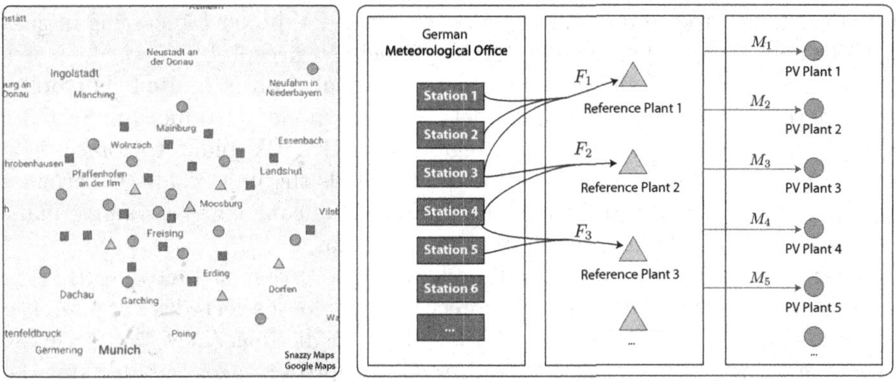

Fig. 1. Two-stage process by means of chained machine learning models: power outputs for reference PV plants (shown as blue triangles) are forecasted from raw meteorological data (at several locations, depicted as red rectangles) provided by the German Meteorological Office. The outputs of non-reference PV plants are then predicted from the output predictions of reference plants. Locations are picked for illustration purposes only and do not coincide with real plant or station locations. (Color figure online)

in advance when and how much energy will probably have to be compensated, forecasts are needed especially for volatile plants. The more accurate the power forecasts are, the less energy will have to be compensated in the short term, leading to more efficient schedules for the power plants involved [11]. Large energy operators responsible for a large number of PV systems (such as Stadtwerke München) have to create such forecasts for each individual plant, typically based on meteorological data. They can either be obtained commercially or must be created by the operators themselves based on physical equations or machine learning techniques. In order to be able to produce acceptable predictions in the latter case, usually a great amount of historical data is needed for each plant, which leads to high administrative expenses. If the underlying data consist of weather information from a local weather service, the costs of high-resolution weather information for the location of the power plant are usually quite high. Thus, costs and effort of data acquisition and management can quickly exceed the profit generated by accurate forecasting.

Consequently, in this paper, we take on the task of producing accurate PV forecasts under several business constraints for a specific application scenario:

Costs of Forecasts The driving motivation behind our study is to examine if accurate PV power predictions can be obtained from freely available meteorological data, as opposed to existing baseline predictions.

Heterogeneous Data Quality Apart from being free, the meteorological data available at relevant locations vary heavily in terms of quality (see Sect. 2).

Data Preparation and Maintenance Efforts New plants should be easily integrated into the system – optimally at little manual data cleansing costs.

Instead of trying to improve the state of the art in PV power forecasting in general, our goal is to apply machine learning to achieve acceptable power forecasts in this setting, even if the available meteorological data is limited. For doing so, we developed three prediction models: the first model introduced in Sect. 3.1 takes meteorological data from the region around a PV plant to forecast its power. Although this direct prediction model yields the best results in terms of error reduction, the data preparation effort is significant for every single plant due to the heterogeneous availability of weather data.

Therefore, we propose a chained model (Sect. 3.3) that includes precise predictions from weather data using the forecast model for a selected set of reference plants and learns mappings from the resulting predictions (Sect. 3.2) to every other remaining plant, as Fig. 1 visualizes. Our approach takes into account the organizational change processes involved with preparing and training new prediction models and offers a transformation strategy by starting with predictors for few reference plants to serve as a baseline for learned mappings and gradually converting them to have their own prediction models.

2 Heterogeneity of Data Quality

The communal energy service provider Stadtwerke München operates a *virtual power plant* which is a network of several small decentralized energy producers and consumers [12] to regulate short-term fluctuations in the energy grid. This coalition can jointly achieve production capacities matching those of ordinary power plants. In order to maintain the balance of the energy grid under the influence of many players, every provider has to forecast how much energy will be generated or consumed. For the so-called *day-ahead trading* at the energy market in Germany, these predictions are required daily in 15-minute resolution for the following day. Since unforeseen changes due to outages of some plants or changing weather conditions can quickly occur, *intraday trading* allows to compensate the deviating energy on the same day [5]. Due to their commercial relevance, we focus on day-ahead forecasts in this work.

We considered a subset of 120 PV plants of the virtual power plant throughout Germany, each of them with a maximum power between 100 and 20k kW. Historical actual power output data was available in a 15-minute resolution from December 2017 through June 2018. As the virtual power plant continues to grow, several new PV plants can accrue per month, which also need to be predicted. Due to the direct dependence of photovoltaic power on weather condition, we used numerical weather records and forecasts from the German Meteorological Office DWD[1] as a basis for the power forecasts. The hourly forecasts originate from the weather model `cosmo-d2`, which is generated for an equidistant grid of 2.2 km over Germany. By contrast, the historical data is recorded at various weather stations throughout Germany in different time resolutions. Most importantly, all of this data is distributed openly and for free which can in turn help to reduce prediction costs.

[1] Deutscher Wetterdienst https://www.dwd.de

The most natural way to obtain historical weather data for a PV plant would be to consider the records from the nearest weather station. Given the historical data from DWD, we face two problems: first, each station is only recording a small subset of all possible weather attributes, and second, multiple time gaps and series of error values occur in the recorded data – for some stations more than for others. This variation in data availability of the individual stations together with their uneven distribution over Germany lead to very heterogeneous data quality for all PV plants. For example, for some plants there could be a high-quality and close weather station, whereas others may be surrounded by many distant weather stations that simultaneously suffer from poor data quality or have recently been installed and therefore lack historical data.

Related Work

Due to their relevance for trading in the energy market, forecasting models for PV plants have been investigated thoroughly [1,7,11]. These models target several forecast horizons, ranging from a few seconds to days or weeks ahead. Similarly, they address a variety of spatial horizons, from a single site to regional forecasts. The authors of [1] provide a thorough survey of the most common approaches to PV forecasting, including analytical equations to model the irradiance of a PV system as the most important predictor for the power output (see, e.g., [4]) and statistical or machine learning models such as neural networks [8] or random forests [9]. In [8], a multilayer perceptron model is trained to forecast solar irradiance which is then converted into power outputs using a physical model. By contrast, our forecasting model (see Sect. 3.1) learns the mapping from irradiance values of multiple weather stations to a power output. This is due to the fact that we can obtain irradiance forecasts from DWD.

However, given our available (free) data, none of these forecasting models is directly applicable which is why we opted for a custom machine learning model that is trained in our specific setting and does not require, e.g., precise tilt angles and orientations of PV plants as inputs.

3 Forecasting Models

In our proposed approach (see Fig. 1), we build prediction models $P_1, \ldots, P_m$ for all m PV plants based on meteorological data from DWD. Since the generated forecasts are intended for day-ahead trading, they must be created for the following day in 15-minute resolution using current meteorological *predictions*. Consequently, estimating PV power is a regression problem where the input features $\vec{x}$ correspond to a vector of numerical meteorological parameters for a certain point in time (given as forecasts themselves) and the target quantity y is the continuous output of a PV plant at this point in time. Although a first impression might suggest that a joint model could be trained and applied to forecast the outputs of all PV plants simultaneously, there are several reasons why we decided to build a prediction model for every single plant instead:

(a) In a joint model, poor data quality of some PV plants could harm prediction accuracy for all other plants.

(b) The joint model has to be completely re-built and re-trained when new plants are added to the virtual power plant. For separate models, by contrast, only the added plants require building new forecasting models and the already existing ones can simply remain. This aspect is very important in our scenario as new plants are added frequently to the virtual power plant.
(c) Handling single models is scalable because they can be trained and evaluated in parallel.

3.1 A Power Forecasting Model from Weather Data

As mentioned before, we use publicly accessible weather data from DWD as basis for the power forecasting models F_i of a single PV plant i. Historical records of pairs of meteorological data and PV power outputs are used to train the models such that weather *forecasts* can afterwards be mapped to power *forecasts*. In order to use a weather attribute as a feature for the data set, it needs to occur both in historical and forecast data. Unfortunately, this is only the case for four attributes where we discarded the attribute `cloud coverage` due to its unavailability in more than 90% of the stations, leaving us with

- `solar diffuse irradiation`,
- `solar global irradiation` and
- `air temperature two meters above ground`

as our remaining available features. Fortunately, irradiation and temperature tend to be the most important features for PV power prediction [10,14].

The historical weather values used to train our models are measured at fixed weather stations distributed over Germany in varying time resolutions (e.g. ranging from every minute to yearly), whereas the weather forecasts are collected from the weather model `cosmo-d2` that offers points arranged in a dense, equidistant grid over Germany (called `cosmo-d2` grid in the following). However, these forecast values are only available in an hourly resolution. Since the input data need to have the same resolution, the historical values are aggregated to hourly resolution to match the available forecast data. As mentioned before, we still have to provide power forecasts in *15-minute* resolution even though our weather features only have *hourly* resolution. Therefore, we map the hourly weather input onto four consecutive quarter-hourly power forecasts for the same hour.

Mapping coarse hourly weather condition onto finer quarter-hourly power forecasts presents us with a problem: the model cannot decide how the four output values should develop. For similar weather conditions, we would expect a positive development of them in the morning while the sun is still rising, but a negative one in the afternoon due to the sunset. Therefore, we add the hour value of the data point in a one-hot-encoded way to the features.

Because only few weather stations measure the three relevant weather attributes and the records contain many gaps or error values, we include the attributes of the stations in an area around a plant into its feature set. Based on the measured average speed of clouds from [6], we choose a radius of 90 km

Table 1. Statistical values describing the minimal, maximal, mean and median number of weather attributes from DWD data that were used as features for the forecasting models of the individual PV plants.

	Min	Max	Median	Mean
`solar diffuse irradiation`	0	7	2	2.23
`solar global irradiation`	0	7	2	2.77
`air temperature two meters above ground`	8	28	20	19.03
All three features	11	41	21	24.03

around the plant that corresponds to twice the average distance traveled by clouds per hour. Since an instance can only be used productively if there exist non-error values for all features, we consider a weather attribute from a station only if it provides enough data points so that the overall amount of instances is not reduced too much. With respect to the DWD data, we first sort the stations around a plant by their distance in ascending order. Afterwards, while iterating over the ordered stations, we exclude attributes from those with a ratio of missing or error values exceeding 1% of the current amount of instances.

According to this strategy, the weather-based data sets for all 120 PV systems have been generated. Table 1 shows that due to the heterogeneity of weather data, the data sets contain much more temperature attributes than for irradiation. Moreover, some PV plants even suffer from not having any weather station in their environment measuring irradiation values. Since these features are much more important for accurate power forecasts than temperature values, the corresponding forecasting models are expected to perform poorly.

For PV power forecasts, we have to resort to `cosmo-d2` grid points as opposed to specific weather stations provided in the historical data that were used for training. Therefore, we replaced the input features of the weather stations with those of the closest `cosmo-d2` grid points during forecasting. Since the grid resolution is about 2.2 km, the distance between the forecast points and weather stations is reasonable.

In an ideal world with lots of resources available for data processing and storage, the forecasting model is a good choice to get power forecasts for all PV plants, as our evaluation in Sect. 4 shows. With limited resources, however, a few problems arise: the historical weather records for each PV system must be both continuously updated and always processed into a consistent amount of data. In addition, the available weather stations of the DWD can change, for example, by adding new stations or removing previously used stations. In the latter case in particular, the model of the affected PV plant must then be completely re-built and re-trained due to the omission of the associated feature for this station. Thus, the entire knowledge of the old model is lost.

Besides our strategy of selecting feasible weather stations for a PV plant, there is a number of other strategies to create a joint data set from the given historical records of the weather stations. Due to the heterogeneity of the available data, it may be possible that for different PV systems different merging strategies lead to the best data sets. Comparing multiple strategies for all plants to find the best one would in turn require a lot of computational or manual data preprocessing effort. The heterogeneity of the data poses another problem: As can be seen in Table 1, some considered PV plants do not have any nearby weather stations that measure irradiation attributes in sufficient quality. Unfortunately, these attributes are crucial for PV prediction. An intuitive solution would be to gradually increase the radius around the stations until an acceptable number of weather stations measuring irradiation values were found which in turn would cause increasing computational effort.

3.2 A Mapping Model Learning from Reference Power Forecasts

Instead of computing power forecasts using the weather-based forecasting model F_i for all m PV plants (including the aforementioned data preparation and cleaning efforts), we will build models $F_{r_1}, \ldots, F_{r_k}$ only for a small selection of plants, the so-called *reference plants* $\{r_1, \ldots, r_k\} \subset \{1, \ldots, m\}$, and map the resulting forecasts to the remaining $m - k$ plants via *mapping models* $M_1, \ldots, M_{m-k}$. Precisely, a PV prediction model P_i of plant i is obtained as follows:

$$P_i = \begin{cases} F_i & \text{if } i \text{ is reference plant, i.e., } i \in \{r_1, \ldots, r_k\} \\ M_i(F_{r_1}, \ldots, F_{r_k}) & \text{if } i \text{ is non-reference plant} \end{cases}$$

This approach is based on the principle of regionality, as nearby plants are expected to be usually exposed to similar weather conditions and consequently show similar power curves. The power forecasts of the reference plants are mapped to those for the remaining non-reference plants. Various factors, such as the orientation of the system with respect to both the cardinal direction and the angle of the solar panels, their location along the latitude or the topology of the environment, for example tall shading buildings, affect the output power of a PV plant. Since this information is individual, in some cases not available, and in general difficult to express analytically, we also use machine learning techniques for the mapping model. To avoid having to manually pick a subset of reference plants for each PV plant, the forecasts of all of them are used as input for the models. A resulting advantage is that the location of the target plant does not have to be known in order to obtain forecasts.

The mapping model takes the forecast values of all reference plants for a certain quarter-hourly point in time to produce the power forecast of a regarded plant for the same point. For training, historical actual power outputs of the plant as well as the corresponding forecasts of the reference plants are required in 15-minute resolution. The reference plants should be selected from all over Germany and all together provide as gapless forecast values as possible for the considered period. This approach offers the advantage that all models have the same input, only the output has to be exchanged for each plant individually. Even if a new plant is added to the virtual power plant, there is no effort for generating new input for it.

3.3 A Chained Forecasting Model

To obtain reliable power predictions for a large and increasing number of PV plants based on the available data, we combine the two methods presented so far. This means that for every non-reference PV plant, the two models are composed: for the reference plants, the forecasting models $F_{r_1}, \ldots, F_{r_k}$ predict power values based on weather data; the mapping model M_i then takes all these power forecasts and estimates the output of a non-reference plant i, as shown in Fig. 1.

The *curse of dimensionality* is the most important reason why we decided to compose the models sequentially instead of combining them into one: In the latter case, the feature set would consist of all weather attributes of all weather stations associated with reference plants. Then, too few training instances could be available to train this model due to the time gaps and error values in the historical records. By splitting the model into two successive models, in contrast, significantly fewer features are used in both steps, which we expect to improve training for both and, consequently, to lead to better predictions.

Using this stepwise approach, a small loss of accuracy is tolerable in favor of a significantly lower manual data preparation effort. The models depend on much less historical weather data from DWD, since we only need them for the small portion of reference plants. Our strategy is to pick those few reference plants carefully by evaluating available raw weather data in terms of proximity and data quality. This expenditure is well spent since once we have high-quality forecasting models for the reference plants, we can easily train additional mapping models for the remaining plants using the same inputs (forecasts of reference PV plants) but only different power targets.

4 Evaluation

For our evaluation, we investigate whether our proposed models based on freely available weather data reach the quality of the commercial baseline predictions and whether the chained model achieves acceptable results compared to the pure forecasting models – at lower development costs.

We noted that the provided actual power values of some PV plants contained days with no power output at all. Since PV plants generate a small amount of energy even on winter days, this is obviously incorrect data and was consequently excluded from the data set. These zero power days could be caused by maintenance, defects or snow on the solar panels and would significantly skew the results. After this preprocessing step, the remaining data for each plant was split into a training set with about 70% of the instances, a validation and a test set with 15%, respectively. For a better visualization on the validation and test sets, all instances of a whole day were assigned to precisely one of the three sets.

Instead of optimizing hyperparameters individually for each of the 120 PV plants considered in this paper, we opted for a uniform hyperparameter setting across all plants. By doing so, we assume that the underlying functions to be learned are similar for all PV systems and therefore it is enough to tune the capacity of a model only once for all of them. Therefore, a small subset of eleven plants was randomly selected such that they are distributed all over Germany with different maximal output, in the following referred to as $PV1$ to $PV11$. We used the root mean squared error (RMSE) as error metric, adhering to conventions about PV forecasting evaluations in the literature [1].

As a preliminary experiment in terms of model adequacy, we evaluated the three machine learning techniques *random forest*, *Gaussian process* and *feed-forward neural network* for the mapping model using given baseline forecasts for 13 reference plants. Random forests have already been successfully applied to PV forecasting in literature [14] and can represent a large number of functions without having to severely limit the inductive bias as they make no prior assumptions regarding smoothness or other properties of the function. This is an important aspect for our scenario, as the form and the complexity of the mapping between the reference predictions and the performance values of the remaining plants is not known in advance.

Gaussian processes were chosen because of their probabilistic nature, which allows us to obtain predictions and simultaneously quantify their uncertainty. Moreover, once the covariance (or kernel) function has been selected, many hyperparameters of the model are set automatically during training so that they do not need to be chosen manually [13]. Like random forests, they can describe many different functions, which can nevertheless be restricted if necessary by the kernel function. Finally, neural networks have proven to be a universal tool in practice since they are very adaptive and can fit a wide variety of functions. They have also been successfully used for PV forecasting multiple times [1].

In our implementation, we used the python library *GPy*[2] for Gaussian processes, *scikit-learn*[3] for random forests and *Keras*[4] with the *TensorFlow*[5] backend for neural networks. With random forests, the tuned hyperparameters included the `number of estimators`, the `minimum samples per leaf` and the `maximal number of features` per split. After model selection, i.e. grid searching for appropriate values of these hyperparameters, the number of estimators was set to the default value 10, the maximal number of features to 4 and the minimum samples per leaf was set to 1.

For Gaussian processes, we applied the procedure for automatic kernel selection described in [2]: At first, the best standard kernel of the framework is chosen. Afterwards, it is combined stepwise through addition or multiplication with other standard kernels until no further improvement of the resulting Gaussian process can be achieved. The best results were achieved using an additive combination of a linear and an exponential kernel for all input dimensions together after 200 optimization steps. For the neural networks, different architectures with one or two hidden layers were compared. The activation functions for these layers were taken from the standard functions of the framework, whereas the identity

[2] GPy https://sheffieldml.github.io/GPy/
[3] scikit-learn https://scikit-learn.org/
[4] Keras https://keras.io/
[5] TensorFlow https://www.tensorflow.org/

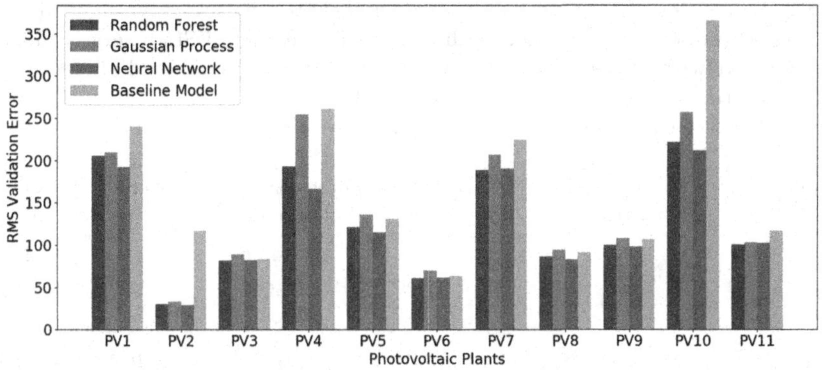

Fig. 2. Comparison of validation errors (RMSE) of all tested methods for the mapping model and the basline model.

function was selected for the output layer since we consider a multivariate regression problem. Models with one hidden layer containing ten neurons and the activation function `softplus` achieved the best results.

Figure 2 shows the error values achieved by the different models. All three techniques perform at approximately the same level for each plant, with the neural network performing best for most PV plants.

Compared to the mapping model, Gaussian processes are not equally suitable for our forecasting model due to the higher dimensionality of the weather data. The computational complexity of Gaussian processes depends cubically on the dimension [13] which makes this technique hard to apply in high-dimensional settings. Moreover, we also refrained from using random forests for our forecasting model since we expected the neural network to capture the feature interactions between the continuous weather attributes more easily. Hence, the experiments for the forecasting model were only performed with neural networks.

We used the same reference plants as in the preliminary experiments for the mapping model. Thus, no optimized selection of the reference plants has yet taken place. For each model, the number of features is different due to the heterogeneity of weather stations discussed in Sect. 3.1. For architecture tuning, we chose the number of hidden neurons as a multiple of the number of inputs. This makes the different models much more similar to each other than using a fixed number of neurons because the ratio of neurons in input and hidden layer stays the same. When tuning the forecasting models of the eleven selected power plants, three layers with the same number of neurons in the input as in the hidden layer and activation function `softplus` in the hidden layer turned out to be the most successful architecture.

Finally, for the chained model, we reused the techniques of the two basic models we identified as performing best. Whereas the weather-based forecasting models of the first part are completely adopted, the subsequent mapping models are re-tuned with the resulting reference forecasts since in the preliminary

Table 2. Evaluation of the direct forecasting model and the two-stage chained model on a shared test set in comparison with the baseline model of Stadtwerke München. Plants marked with an asterisk (*) are used as reference plants and therefore only predicted by the forecasting model. Absolute values are given as RMSE in kilowatts, relative values are error improvements compared to the basline model.

PV plants	Chained model	Improvement	Forecasting model	Improvement	Baseline model
$PV1^*$			144.6	16.2%	172.6
$PV2$	37.0	41.2%	22.5	64.3%	62.9
$PV3^*$			82.0	9.2%	90.3
$PV4$	181.8	8.0%	102.6	48.1%	197.6
$PV5$	108.7	5.2%	98.2	14.5%	114.8
$PV6$	54.5	10.6%	54.9	9.9%	61.0
$PV7$	189.3	−3.6%	146.4	19.9%	182.8
$PV8$	83.8	8.2%	71.1	22.2%	91.3
$PV9$	96.8	−0.2%	83.6	13.6%	96.7
$PV10$	199.9	44.9%	222.8	38.6%	362.8
$PV11$	110.3	14.5%	106.7	17.3%	129.0
Average	117.2	10.8%	104.9	20.2%	131.4
Sum	1288.7	10.8%	1153.4	20.2%	1445.7
Reference plants*					
Average			144.0	20.2%	180.5
Sum			1872.6	20.2%	2346.0
All 120 plants					
Average	112.0	20.0%	108.3	22.6%	139.9
Sum	13098.4	20.0%	12674.5	22.6%	16373.5

experiment they were trained on external forecast data. The resulting mapping model tuned on the selected eleven plants contains seven neurons in the hidden layer, once again applying the activation function `softplus`.

As Table 2 visualizes, both the direct forecasting and the chained model were able to achieve better results than the baseline predictions obtained commercially by Stadtwerke München. Using the forecasting model for predicting outputs of reference plants, the RMSE could be improved by 20.2%. Mapping these forecasts to the outputs of the remaining PV plants, this improvement could almost be retained. Regarding all 120 PV plants, the direct weather-based forecasting model performs better than the chained model – however, this comes at the price of a much higher manual workload, as we discussed in Sect. 3.

5 Conclusion and Future Work

We propose a chained model for PV power forecasting that takes into account the organizational change process of preparing and training new prediction models and follows requirements of business processes like easy adaption for accruing new plants. Since our models are based on meteorological data that is suffering from spatially varying data quality, we show how to deal with this heterogeneity. The evaluation shows that the chained model achieves better results than the baseline and even performs comparable to the direct forecasting model which requires significantly higher data preparation effort for every single plant.

In future work, one could use so-called *clear sky models* to interpolate hourly weather forecasts to quarter-hourly ones (like in [14]) for the forecasting model instead of mapping hourly weather forecasts to four consecutive quarter-hourly power forecasts. Moreover, transfer learning could help leverage knowledge extracted from plants with high data quality to improve models based on lower data quality. Finally, weather forecasts with short temporal horizon may serve as acceptable historical records since weather forecasts tend to be at good quality for this horizon.

References

1. Antonanzas, J., Osorio, N., Escobar, R., Urraca, R., Ascacibar, F.J., Antonanzas, F.: Review of photovoltaic power forecasting. Sol. Energy **136**, 78–111 (2016)
2. Duvenaud, D.: Automatic model construction with Gaussian processes. Ph.D. thesis, University of Cambridge (2014)
3. Frauenhofer ISE: Jährlicher Anteil der Solarenergie an der Stromerzeugung in Deutschland. https://www.energy-charts.de/ren_share_de.htm?year=all&source=solar-share&period=annual. Accessed 25 Mar 2019. (in German)
4. Inman, R.H., Pedro, H.T., Coimbra, C.F.: Solar forecasting methods for renewable energy integration. Prog. Energy Combust. Sci. **39**(6), 535–576 (2013)
5. Kraus, K.M.: Vortrag: "der regelenergiemarkt. In: Wirkungsweise und Wechselbeziehungen mit Spot-und Intra-Day-Märkten", Regelenergie-Symposium, Leipzig September (2004)
6. Lappalainen, K., Valkealahti, S.: Analysis of shading periods caused by moving clouds. Sol. Energy **135**, 188–196 (2016)
7. Liu, J., Fang, W., Zhang, X., Yang, C.: An improved photovoltaic power forecasting model with the assistance of aerosol index data. IEEE Trans. Sustain. Energy **6**(2), 434–442 (2015)
8. Mellit, A., Pavan, A.M.: A 24-h forecast of solar irradiance using artificial neural network: application for performance prediction of a grid-connected pv plant at trieste, italy. Sol. Energy **84**(5), 807–821 (2010)
9. Mohammed, A.A., Yaqub, W., Aung, Z.: Probabilistic forecasting of solar power: an ensemble learning approach. In: Neves-Silva, R., Jain, L.C., Howlett, R.J. (eds.) Intelligent Decision Technologies. SIST, vol. 39, pp. 449–458. Springer, Cham (2015). https://doi.org/10.1007/978-3-319-19857-6_38
10. Monteiro, C., Fernandez-Jimenez, L., Ramirez-Rosado, I.J., Muñoz Jiménez, A., Lara-Santillan, P.: Short-term forecasting models for photovoltaic plants: analytical versus soft-computing techniques. Math. Problems Eng. **2013**, 1–9 (2013)

11. Pierro, M., et al.: Multi-model ensemble for day ahead prediction of photovoltaic power generation. Sol. Energy **134**, 132–146 (2016)
12. Ruiz, N., Cobelo, I., Oyarzabal, J.: A direct load control model for virtual power plant management. IEEE Trans. Power Syst. **24**(2), 959–966 (2009)
13. Williams, C.K., Rasmussen, C.E.: Gaussian Processes for Machine Learning, vol. 2. MIT Press, Cambridge (2006)
14. Zamo, M., Mestre, O., Arbogast, P., Pannekoucke, O.: A benchmark of statistical regression methods for short-term forecasting of photovoltaic electricity production, part I: deterministic forecast of hourly production. Sol. Energy **105**, 792–803 (2014)

Trading-off Data Fit and Complexity in Training Gaussian Processes with Multiple Kernels

Tinkle Chugh[1,2(✉)], Alma Rahat[3], and Pramudita Satria Palar[4]

[1] University of Exeter, Exeter, UK
t.chugh@exeter.ac.uk
[2] Palacky University in Olomouc, Olomouc, Czech Republic
[3] Swansea University, Swansea, UK
A.A.M.Rahat@swanea.ac.uk
[4] Institut Teknologi Bandung, Bandung, Indonesia
pramsp@ftmd.itb.ac.id

Abstract. Gaussian processes (GPs) belong to a class of probabilistic techniques that have been successfully used in different domains of machine learning and optimization. They are popular because they provide uncertainties in predictions, which sets them apart from other modelling methods providing only point predictions. The uncertainty is particularly useful for decision making as we can gauge how reliable a prediction is. One of the fundamental challenges in using GPs is that the efficacy of a model is conferred by selecting an appropriate kernel and the associated hyperparameter values for a given problem. Furthermore, the training of GPs, that is optimizing the hyperparameters using a data set is traditionally performed using a cost function that is a weighted sum of data fit and model complexity, and the underlying trade-off is completely ignored. Addressing these challenges and shortcomings, in this article, we propose the following automated training scheme. Firstly, we use a weighted product of multiple kernels with a view to relieve the users from choosing an appropriate kernel for the problem at hand without any domain specific knowledge. Secondly, for the first time, we modify GP training by using a multi-objective optimizer to tune the hyperparameters and weights of multiple kernels and extract an approximation of the complete trade-off front between data-fit and model complexity. We then propose to use a novel solution selection strategy based on mean standardized log loss (MSLL) to select a solution from the estimated trade-off front and finalise training of a GP model. The results on three data sets and comparison with the standard approach clearly show the potential benefit of the proposed approach of using multi-objective optimization with multiple kernels.

Keywords: Machine learning · Kriging · Bayesian optimization · Multi-objective optimization · Model selection

G. Nicosia et al. (Eds.): LOD 2019, LNCS 11943, pp. 579–591, 2019.
https://doi.org/10.1007/978-3-030-37599-7_48

1 Introduction

Gaussian processes (GPs) have been widely used in machine learning and optimization communities. Some of the problems where GPs have gained their popularity are non-linear regression (also known as Kriging in geostatistics), classification [23] and Bayesian optimization [24]. The main advantage of using GPs is that they provide a predictive distribution instead of point predictions as in other models like neural networks and support vector regression. This uncertainty can be used in making the decisions [20,22] and in selecting samples by optimizing an acquisition function in Bayesian optimization [13,15].

Despite their wide applicability, little attention has been paid to the problem of selecting kernels and the hyperparameters. As mentioned in [23], multiple choices exist and it is not straightforward to select a kernel and its hyperparameters. It often requires prior knowledge about the underlying function that we are trying to model. To select the hyperparameters, the traditional approach is to maximize the marginal likelihood for a given kernel. A characteristic of this likelihood function is that it tries to balance between data-fit and complexity. For instance, the data-fit decreases monotonically with the length scale resulting in increasing the complexity of the model. A simple illustration of the model fit and complexity by varying the length scale when using a Gaussian kernel is shown in Fig. 1.

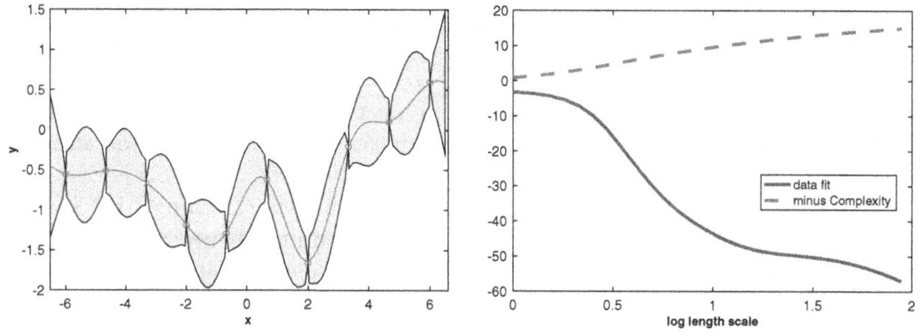

Fig. 1. Left plot: data generated with a GP realization of length scale = 1, signal variance = 1 and noise variance = 0.1, right plot: data fit and minus complexity with length scale of the data set in the left plot.

As can be seen, the data fit decreases with the increase in complexity of models with different length scales. In selecting a kernel, several options like Gaussian (or RBF), exponential, linear, matern 5/2, matern 3/2 and periodic exist. In the literature, some studies are devoted to the concern of selecting a kernel. For example, in [17], a genetic programming approach was applied to find a composite kernel and in [8], different combinations like sum and product of kernels were used. In [21], a weighted sum of kernels was used in training of GPs. Recently, in [4], different kernels were studied in the context of Bayesian optimization and different correlations were observed between a kernel and other elements in Bayesian optimization.

In applying multi-objective optimization in machine learning, some studies exist in building models like neural networks [9,12] and decision trees [10]. Approaches in these studies considered different objectives like bias and variance, model fit and complexity to find the number of hidden layers and number of nodes in neural networks and number and depth of trees in decision trees. However, to the best of our knowledge, no study exist in using multi-objective optimization in GPs, despite the fact that the inherent property of the likelihood function when building the model is to balance between model fit and complexity. Therefore, in this work, by optimizing two objectives, maximizing model fit and minimizing complexity, we find the optimal hyperparameter values and weights for different kernels.

We use a non-dominated sorting genetic algorithm, NSGA-II [6] to find approximated Pareto optimal solutions, where each solution on the Pareto front represents a model with different accuracy, complexity, hyperparameter values and weights to different kernels. It should be noted that using a multi-objective optimization algorithm does not increase the computational complexity in optimizing the hyperparameter values as both the standard and the proposed approach aim to solve an optimization problem. Also, one is free to choose another multi-objective optimizer.

We then use MSLL [23] to select a model from the approximated Pareto front. The results of the proposed approach using multi-kernel and multi-objective (MKL-MO) is compared then with standard (with single kernel and maximizing the marginal likelihood) and single-kernel and multi-objective (SKL-MO) approaches. This comparison shows the effect of using multi-objective optimization and multiple kernels. We can summarize the main contributions of this paper as follows:

- We use weighted product of multiple kernels to relieve users from selecting one or more kernels for the problem at hand.
- We use multi-objective optimization to estimate the trade-off between data fit and complexity.
- We utilize the MSLL performance metric to select a model from the approximated Pareto front, and derive predictions from a GP model.

The rest of the article is structured as follows. In Sect. 2, we provide a brief description of GPs and different kernels used in this work. In Sect. 3, we explain the proposed approach of using multi-objective optimization with single and multiple kernels. We conduct experiments and discuss the results in Sect. 4. Finally, we conclude and mention the future research directions in Sect. 5.

2 Gaussian Processes for Regression

A typical regression task is to model the relationship between some independent variables (or features) and a dependent variable. Consider a data set of M observations $\mathcal{D} = \{(\mathbf{x}_m, f(\mathbf{x}_m)) \mid m = 1, \ldots, M\}$, where $\mathbf{x} \in \mathbb{R}^n$ is a n-dimensional feature vector, and a function $f : \mathbf{x} \rightarrow \mathbb{R}$ produces a response (i.e. the dependent variable) based on $\mathbf{x}$. In the regression task, we are therefore interested in making predictions about $f_i = f(\mathbf{x}_i)$ for any arbitrary feature vector $\mathbf{x}_i$ given the data set $\mathcal{D}$.

As mentioned, GPs have grown in popularity for non-linear regression tasks in recent years. This is primarily due to its efficacy in providing a posterior probability density indicating how confident the prediction is. Essentially, a GP is a collection of random variables such that any finite number of these have a joint Gaussian distribution [23]. This means that the posterior predictive density of the function $f(\mathbf{x})$ given some data set $\mathcal{D}$ and a feature vector $\mathbf{x}$ is normally distributed:

$$p(f \mid \mathbf{x}, \mathcal{D}, \boldsymbol{\theta}) = \mathcal{N}(f \mid \mu(\mathbf{x}), \sigma^2(\mathbf{x})), \tag{1}$$

where the mean and the variance of the prediction are given by:

$$\mu(\mathbf{x}) = \boldsymbol{\kappa}(\mathbf{x}, X, \boldsymbol{\theta})(K + \sigma_e^2 I)^{-1}\mathbf{f} \tag{2}$$

$$\sigma^2(\mathbf{x}) = \kappa(\mathbf{x}, \mathbf{x}, \boldsymbol{\theta}) - \boldsymbol{\kappa}(\mathbf{x}, X, \boldsymbol{\theta})^\top (K + \sigma_e^2 I)^{-1} \kappa(X, \mathbf{x}, \boldsymbol{\theta}) \tag{3}$$

Here $X \in \mathbb{R}^{M \times n}$ is the matrix of observed feature vectors and $\mathbf{f} \in \mathbb{R}^M$ is the corresponding vector of the responses $\mathbf{f} = (f_1, \ldots, f_M)^\top$; thus $D = \{(X, \mathbf{f})\}$. The covariance matrix $K \in \mathbb{R}^{M \times M}$ represents the covariance function $\kappa(\mathbf{x}', \mathbf{x}'', \boldsymbol{\theta})$ evaluated for each pair of observations $\mathbf{x}', \mathbf{x}'' \in X$ and $\boldsymbol{\kappa}(\mathbf{x}, X, \boldsymbol{\theta}) \in \mathbb{R}^M$ is the vector of covariances between an arbitrary $\mathbf{x}$ and each of the observations. The kernel hyperparameter vector $\boldsymbol{\theta} \in \mathbb{R}^k$ is a vector of parameters that controls the shape of the kernel. σ_e^2 is a homoscadastic Gaussian noise variance that encapsulates the potential error which may occur while measuring the responses $\mathbf{f}$. The overall hyperparameter vector is therefore $\mathbf{t} = (\boldsymbol{\theta}, \sigma_e^2)^\top$.

2.1 Kernels

A kernel (or covariance function) is usually defined as $\kappa(\mathbf{x}_i, \mathbf{x}_j, \boldsymbol{\theta})$, where $\mathbf{x}_i$ and $\mathbf{x}_j$ are two feature vectors, and $\boldsymbol{\theta} \in \mathbb{R}^k$ is a vector of k hyperparameters. In essence, the kernel captures the intuition that two feature vectors that are spatially closer should have similar response, and this relationship is defined by the hyperparameters $\boldsymbol{\theta}$. A kernel with its hyperparameters thus imposes a reproducing kernel Hilbert space for all possible functions that may be represented given data set $\mathcal{D}$.

To describe the relationship between responses f_i and f_j for a pair of feature vectors x_i and x_j, typically we consider a distance measure in the feature space $r^2 = \sum_{v=1}^{n} \frac{(x_i[v] - x_j[v])^2}{l[v]^2}$ with a hyperparameter $l[v]$ that determines the length scale in the vth dimension. Here, $l[v]$ scales the vth dimension and thus controls the importance of the respective dimension in determining the response. In addition, an amplitude hyperparameter σ_f that controls how much the function response may vary with distance in the feature space. Hence, the hyperparameter vector may be constructed as $\boldsymbol{\theta} = (\sigma_f, l[1], \ldots, l[n])^\top$. With this, we can define the following five popular kernels used in this paper [23].

Radial Basis Function or Gaussian. This is the most popular kernel with infinitely many derivatives, and therefore can produce very smooth function realisations. It may be expressed as:

$$\kappa(\mathbf{x}_i, \mathbf{x}_j, \boldsymbol{\theta}_r) = \sigma_f^2 \exp\left(-\frac{r^2}{2}\right). \tag{4}$$

Exponential. Closely related to the Gaussian kernel, but can produce rougher function realisations. It can be defined as:

$$\kappa(\mathbf{x}_i, \mathbf{x}_j, \boldsymbol{\theta}_e) = \sigma_f^2 \exp\left(-r\right). \tag{5}$$

Matern. A class of functions defined by:

$$\kappa(\mathbf{x}_i, \mathbf{x}_j, \boldsymbol{\theta}_\nu) = \sigma_f^2 \frac{2^{1-\nu}}{\Gamma(\nu)} (\sqrt{2\nu} r)^{\nu+1} \beta_\nu, \tag{6}$$

where, ν is a smoothness parameter that is set to either $\frac{3}{2}$ for once differentiable functions or $\frac{5}{2}$ for twice differentiable functions in this paper and β_ν is the modified Bessel function. In the above kernels, the hyperparameter vector have the same attributes, i.e. $\boldsymbol{\theta}_r = \boldsymbol{\theta}_e = \boldsymbol{\theta}_m = (\sigma_f, l[1], \ldots, l[n])^\top$, and the number of hyperparameters is $k = n + 1$.

Periodic. To capture periodicity that may occur in a response, we may also consider the following periodic kernel [19]:

$$\kappa(\mathbf{x}_i, \mathbf{x}_j, \boldsymbol{\theta}_p) = \sigma_f^2 \exp\left[-\frac{1}{2}\sum_{v=1}^{n}\left(\frac{\sin\left(\frac{\pi}{t[v]}(x_i[v] - x_j[v])\right)}{l[v]}\right)^2\right], \tag{7}$$

where the additional hyperparameter $t[v]$ represents the distance between repetitions in the vth dimension. Thus, in this case, the hyperparameter vector is $\boldsymbol{\theta}_p = (\sigma_f, l[1], \ldots, l[n], t[1], \ldots, t[n])^\top$, and hence the number of hyperparameters is $k = 2n + 1$.

Clearly, the kernels above (and many others in the literature) can represent functions with varying degree of smoothness and periodicity. Nonetheless, a specific kernel on its own may not be appropriate for modelling all responses; no free lunch theorem applies here [27]. That is why choosing an appropriate kernel is an important step in training a GP model, and often requires domain specific knowledge.

3 Multi-objective Training of Gaussian Processes

As mentioned in the introduction, improving the data fit increases the model complexity, i.e. most complex model can fit the given data best. Therefore, data fit and complexity are conflicting objectives. This is because a very complex model may not generalise the training data well, and consequently perform poorly on unseen data set. We, therefore, want to control complexity such that it avoids over fitting without compromising performance on both training and validation data set in GP training.

Typically, training a GP model constitutes estimating the overall hyperparameter vector $\mathbf{t} = (\boldsymbol{\theta}, \sigma_e^2)^\top$ that brings together the kernel hyperparameters $\boldsymbol{\theta}$ and the Gaussian error noise variance σ_e^2 by maximising the marginal likelihood of the data:

$$\log p(\mathcal{D} \mid \mathbf{t}) = -\frac{1}{2}\mathbf{f}^\top (K + \sigma_e^2 I)^{-1}\mathbf{f} - \frac{1}{2}\log|K + \sigma_e^2 I| - \frac{M}{2}\log(2\pi) \tag{8}$$

$$= g_d(\mathcal{D}, \mathbf{t}) - g_c(\mathcal{D}, \mathbf{t}) + C, \tag{9}$$

where, $I \in \mathbb{R}^{M \times M}$ is an identity matrix. The first term is representing the *data fit* $g_d(\mathcal{D}, \mathbf{t})$ and the second term is representing *model complexity* $g_c(\mathcal{D}, \mathbf{t})$ [7,23]. The last term is a *normalisation constant* C.

Clearly, the desire to strike a balance between data fit and complexity is evident from (8): when data fit $g_d(\mathcal{D}, \mathbf{t})$ is maximised and the model complexity $g_c(\mathcal{D}, \mathbf{t})$ is minimised simultaneously, it results in maximising the marginal likelihood. Intuitively, this means we improve data fit as much as possible while penalising the complexity at the same time. Interestingly, despite the recognition of the obvious conflict between the objectives (e.g. [23]), a multi-objective optimization approach has never been adopted. Instead, the training of a GP model is posed as a single objective optimization problem for locating suitable hyperparameters and error variance:

$$\mathbf{t}^* = \underset{\mathbf{t}}{\operatorname{argmax}} \quad \log p(\mathcal{D} \mid \mathbf{t}). \tag{10}$$

The estimated optimal solution for $\mathbf{t}^* = (\boldsymbol{\theta}^*, \sigma_e^{2*})^\top$ is then used in (2) and (3) to produce the posterior predictive distribution. In this work, we propose to deal with the conflicting objectives as a multi-objective optimization problem (MOP) of maximising both data fit and complexity penalty simultaneously:

$$\max_{\mathbf{t}} \;\; g_d(\mathcal{D}, \mathbf{t}) = -\frac{1}{2}\mathbf{f}^\top (K + \sigma_e^2 I)^{-1}\mathbf{f}, \tag{11}$$

$$\min_{\mathbf{t}} \;\; g_c(\mathcal{D}, \mathbf{t}) = \frac{1}{2}\log|K + \sigma_e^2 I|. \tag{12}$$

Generally, there is not a unique solution to this multi-objective problem, but a range of solutions $\mathbf{t}$ that trade-off between the data fit and complexity. The trade-off relationship is characterised by the notion of dominance [5]. A solution $\mathbf{t}$ is said to (weakly) dominate another shape $\mathbf{t}'$, denoted as $\mathbf{t} \prec \mathbf{t}'$, iff,

$$\begin{aligned} & g_d(\mathcal{D}, \mathbf{t}) > g_d(\mathbf{t}') \text{ and } g_c(\mathbf{t}) \leq g_c(\mathbf{t}') \\ \text{or } & g_d(\mathcal{D}, \mathbf{t}) \geq g_d(\mathbf{t}') \text{ and } g_c(\mathbf{t}) < g_c(\mathbf{t}'). \end{aligned} \tag{13}$$

The set of solutions that provide an optimal trade-off between the objectives is referred to as the Pareto set:

$$\mathcal{P} = \{\mathbf{t} \mid \mathbf{t}' \not\prec \mathbf{t}\ \forall \mathbf{t}, \mathbf{t}' \in \tau \wedge \mathbf{t} \neq \mathbf{t}'\}, \tag{14}$$

where τ is the space that consists of all permissible hyperparameter vectors $\mathbf{t}$. The image of the Pareto set $\mathcal{P}$ in the objective space is known as the Pareto front $\mathcal{F}$. It may not be possible to locate the exact Pareto set within a practical time limit, even if the objective functions were computationally cheap. Therefore, the goal of an effective optimization approach is to generate a good approximation of the Pareto set, denoted as $\mathcal{P}^* \subseteq \tau$, and the associated Pareto front, denoted as $\mathcal{F}^*$. In this paper, we used the popular NSGA-II optimizer to approximate the optimal trade-off front (and one is free to chose another multi-objective optimizer).

Clearly the maximum likelihood solution $\mathbf{t}^*$ in (10) is achieved by optimizing a weighted sum of the MOP in Eqs. (11) and (12). In this case, both objectives are equally weighted, i.e. they both have the same importance. It is well-known that the optimal solution of a weighted sum must reside in the Pareto set [5], and therefore $\mathbf{t}^* \in \mathcal{P}$. However, intuitively there is no reason to believe that for all problems data fit and model complexity are equally important, and this specific set of weights will outperform others. This is precisely why, in this paper, we attempt to estimate the optimal trade-off front, and decide on which solution to select based on the estimated performance.

3.1 Training with Multiple Kernels

Thus far, we introduced a single kernel, associated hyperparameters, and how to optimize these in a multi-objective manner to train a GP model. In this section, we present how we can combine multiple kernels so that a user does not have to select a kernel for a given problem.

There are various avenues to combine multiple kernels, for instance weighted sum or weighted product of kernels [8]. In this paper, we use a composite kernel consisting of L kernels as a weighted product [25]:

$$\kappa_c(\mathbf{x}^i, \mathbf{x}^j, \boldsymbol{\Theta}) = \prod_{l=1}^{L} \omega_l \kappa_l(\mathbf{x}^i, \mathbf{x}^j, \boldsymbol{\theta}_l), \tag{15}$$

where $\boldsymbol{\omega} = (\omega_1, \ldots, \omega_L)^\top$ is a weight vector with $\sum_l \omega_l = 1$, and composite hyperparameter vector $\boldsymbol{\Theta} = (\boldsymbol{\theta}_1, \ldots, \boldsymbol{\theta}_L)^\top$. In this case, the overall hyperparameter vector becomes $\mathbf{t} = (\boldsymbol{\Theta}, \sigma_e^2, \boldsymbol{\omega})^\top$. With this, we now search over all possible $\mathbf{t}$ in Eqs. (11) and (12). Note that it is straightforward to compute the covariance matrix K using the kernel defined in Eq. (15).

In this paper, we used five kernels as described in Sect. 2.1. Thus we have $L = 5$, $|\boldsymbol{\omega}| = 5$ and $\boldsymbol{\Theta} = (\boldsymbol{\theta}_r, \boldsymbol{\theta}_e, \boldsymbol{\theta}_{m32}, \boldsymbol{\theta}_{m52}, \boldsymbol{\theta}_p)^\top$ with $|\boldsymbol{\Theta}| = 6n + 5$. Therefore the number of overall hyperparameters that we optimize is: $|\mathbf{t}| = 6n + 11$ (including parameters for noise variance).

3.2 Constructing a Model from the Estimated Pareto Front

As discussed, solving the MOP will result in a range of solutions for the overall hyperparameter vector $\mathbf{t} = (\boldsymbol{\theta}, \sigma_e^2)^\top$ for SKL-MO and $\mathbf{t} = (\boldsymbol{\Theta}, \sigma_e^2, \boldsymbol{\omega})^\top$ for MKL-MO, each of which is a potential GP model with a distinct posterior predictive distribution for $f(\cdot)$. It is, therefore, required to select one solution or combine multiple solutions to produce a single GP model for predictions. Different approaches may be adopted for this purpose: using ensemble of models [11, 18], selecting a model representing a knee point (or maximum trade-off) on the Pareto front [2] and using Bayesian information criterion [3].

In this paper, our goal is to shed light on the efficacy of SKL-MO and MKL-MO in comparison to the standard approach. To do so, intuitively, we want to estimate how good a model may be given a solution from the $\mathcal{P}^*$ and the data set. Hence, we use an performance metric called mean standardized log loss (MSLL) [23] which is defined as:

$$-\log p(\mu(x) \mid \mathcal{D}, x, \mathbf{t}) = \frac{1}{2}\log(2\pi\sigma^2(x)) + \frac{(\mu - f(x))^2}{2\sigma^2(x)} \tag{16}$$

The main benefit of using MSLL is that it is not sensitive to overall scale of the response variable values and considers both predicted values and their standard deviations.

In our approach, we split the data set into ten-folds leaving randomly chosen 90% for training and 10% for validation in each fold. For each fold, we perform multi-objective optimization to approximate the Pareto front, and select a solution with minimum MSLL value on the test set. This, of course, do not give us an idea on how to construct a model when we want to train on 100% of the data, but clearly shows which approach may yield better generalisation results. We expect to investigate this further in future.

4 Numerical Experiments

This section provides the results and discussion of numerical experiments conducted on three popular data sets. First data set used was Mauna Loa monthly mean of CO_2 concentrations (in parts per million by volume (ppmv)) from 2010–2018 [14][1] and is shown in Fig. 2. The second data set used was the concrete data set [28] in which strength of the concrete depends on the concentration of cement, furnace slag, fly ash, water, superplasticizer, coarse aggregate, fine aggregate used and the age of the concrete[2]. The third data set used was sarcos data set [26], in which 21 dimensions representing positions, velocities and accelerations map to the torque of the robot arm[3]. In this work, we used 100

[1] available from: https://www.esrl.noaa.gov/gmd/ccgg/trends/data.html.

[2] available from http://archive.ics.uci.edu/ml/datasets/concrete+compressive+strength.

[3] available from http://www.gaussianprocess.org/gpml/data/.

uniformly distributed set of points in concrete and sarcos data sets. A summary of different data bases used with number of variables and size is provided in Table 1.

Table 1. Number of variables and size of different data used

	Number of variables	Size of data set
Mauna CO_2	1	108
Concrete	8	100
Sarcos	21	100

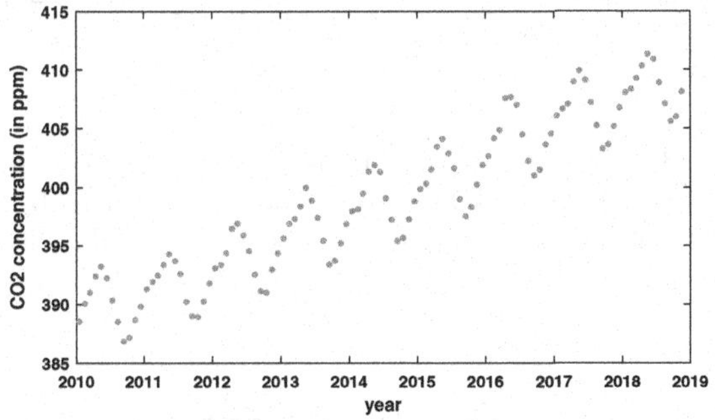

Fig. 2. The 108 observations of CO_2 concentrations in years 2010–2018

To show the potential of using multi-objective optimization and multiple kernels, we compared the proposed multi-kernel and multi-objective (MKL-MO) approach with standard and with single-kernel and multi-objective approach (SKL-MO) approaches. In both standard and SKL-MO approaches, we used the Gaussian kernel. Further, we used 10-fold cross validation and calculated the root mean square (rmse) values. In doing multi-objective optimization in SKL-MO and MKL-MO, we used NSGA-II algorithm. In using NSGA-II, we kept an archive to store all solutions and nondominated solutions from the archive were used as the final solutions. The parameter values of different elements in NSGA-II were: population size: 50, number of generations: 50, crossover: simulated binary with 0.8 probability, mutation: polynomial with 1/number of variables.

The approximated Pareto optimal solutions (of a random fold among 10 folds) representing negative of data fit and complexity of different approaches on three data sets are shown in Fig. 3. Each solution on the Pareto front has its own set of parameters i.e. kernel parameters, noise variance and weights. In solving standard approach, only one solution could be obtained which is represented with a circle in the figures. As both SKL-MO and MKL-MO solves a MOP, it is not surprising to get many solutions. However, one key observation from the results is that a much better distribution (or diversity) of solutions was obtained in MKL-MO when compared to SKL-MO approach. This is because the multi-objective optimization algorithm was able to explore in diverse regions with the help of multiple kernels. Finding a good distribution of solutions is one of the main features when solving a MOP and the proposed MKL-MO approach was able to achieve it.

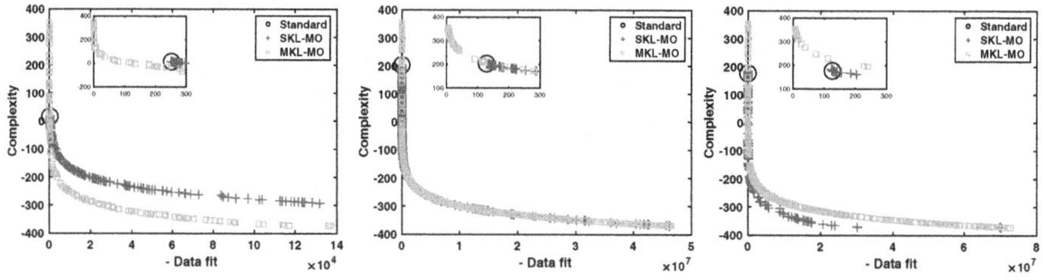

(a) (b) (c)

Fig. 3. Approximated Pareto optimal solutions representing -data fit and complexity of standard, SKL-MO and MKL-MO approaches on (a) Mauna, (b) concrete, and (c) sarcos data sets. A small part of the plot with low -data fit values is zoomed

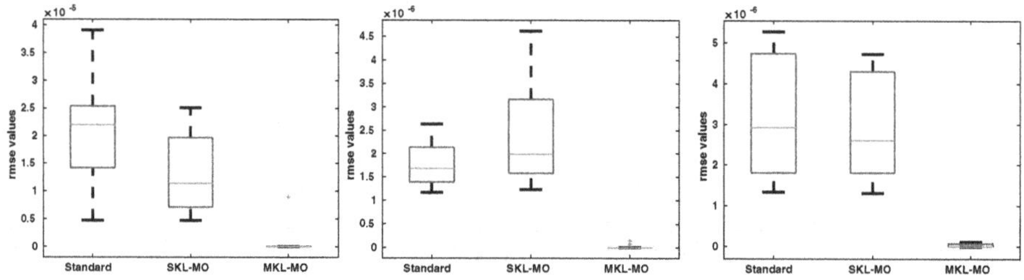

(a) (b) (c)

Fig. 4. Root mean square error (rmse) values of standard, SKL-MO and MKL-MO approaches on (a) Mauna, (b) concrete, and (c) sarcos data sets

Next, we selected a model with the least MSLL values from the $\mathcal{P}^*$ for SKL-MO and MKL-MO approaches, and calculated the rmse values. The box plots of the rmse values of all three different approaches on different data sets are shown in Fig. 4, and the corresponding MSLL values are shown in Fig. 5. To test whether one of the methods statistically significantly wins in all folds and problems, we performed Mann-Whitney-U test [16] as the folds were independently chosen. We also adjusted for multiple comparisons using Bonferroni correction [1]. The significance level was set to $\rho = 0.05$. The tests revealed that MKL-MO performed better than its competitors in concrete (rmse), mauna (rmse,

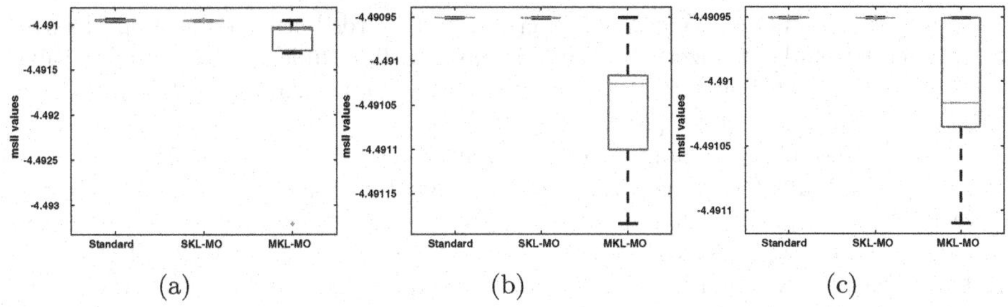

Fig. 5. Mean standardized log loss (MSLL) values of standard, SKL-MO and MKL-MO approaches on (a) Mauna, (b) concrete, and (c) sarcos data sets

MSLL), sarcos (rmse). Otherwise, we found no statistically significant results at the desired level. Visually, it is clear that MKL-MO outperforms other methods.

5 Conclusions

In this article, we focused on multi-objective optimization of two conflicting objectives, maximizing data fit and minimizing complexity when training a GP model. In addition, we combined the multi-objective approach with multiple kernels to handle the challenges of selecting a particular kernel. For this, we used the weighted product of kernels where weights and the kernel parameters were calculated during the multi-objective optimization. The mean standardized log loss values were used in selecting a model from the approximated Pareto front after solving multi-objective optimization problem. The results on three different data sets and comparison with standard and single kernel-multi-objective approach clearly showed the potential of the proposed multi-kernel multi-objective approach. In future, we will investigate more methods of combining kernels and selecting a solution from the estimated Pareto front for a diverse set of data sets from practical applications with varying sizes.

Acknowledgments. This research was partially supported by the Natural Environment Research Council, UK [grant number NE/P017436/1] and Youth and Sports of the Czech Republic under the project CZ.02.1.01/0.0/0.0/17_049/0008408: Hydrodynamic design of pumps.

References

1. Bender, R., Lange, S.: Adjusting for multiple testing: when and how? J. Clin. Epidemiol. **54**(4), 343–349 (2001)
2. Branke, J., Deb, K., Dierolf, H., Osswald, M.: Finding knees in multi-objective optimization. In: Yao, X., et al. (eds.) PPSN 2004. LNCS, vol. 3242, pp. 722–731. Springer, Heidelberg (2004). https://doi.org/10.1007/978-3-540-30217-9_73
3. Burnham, K.P., Anderson, D.R.: Multimodel inference: understanding AIC and BIC in model selection. Soc. Methods Res. **33**(2), 261–304 (2004)

4. Chugh, T., Rahat, A., Volz, V., Zaefferer, M.: Towards better integration of surrogate models and optimizers. In: Bartz-Beielstein, T., Filipič, B., Korošec, P., Talbi, E.-G. (eds.) High-Performance Simulation-Based Optimization. SCI, vol. 833, pp. 137–163. Springer, Cham (2020). https://doi.org/10.1007/978-3-030-18764-4_7
5. Coello Coello, C.A., Lamont, G.B., Veldhuizen, D.A.V.: Evolutionary Algorithms for Solving Multi-Objective Problems. Springer, Berlin (2007)
6. Deb, K., Prarap, A., Agarwal, S., Meyarivan, T.: A fast and elitist multiobjective genetic algorithm: NSGA-II. IEEE Trans. Evol. Comput. **6**, 182–197 (2002)
7. Duvenaud, D.: Automatic model construction with Gaussian processes. Ph.D. thesis, University of Cambridge (2014)
8. Duvenaud, D., Lloyd, J., Grosse, R., Tenenbaum, J., Zoubin, G.: Structure discovery in nonparametric regression through compositional kernel search. In: Proceedings of the 30th International Conference on Machine Learning, vol. 28, pp. 1166–1174. PMLR, Atlanta (2013)
9. Fieldsend, J.E., Singh, S.: Pareto evolutionary neural networks. IEEE Trans. Neural Netw. **16**(2), 338–354 (2005)
10. Fieldsend, J.E.: Optimizing decision trees using multi-objective particle swarm optimization. In: Coello, C.A.C., Dehuri, S., Ghosh, S. (eds.) Swarm Intelligence for Multi-objective Problems in Data Mining. SCI, vol. 242, pp. 93–114. Springer, Berlin (2009). https://doi.org/10.1007/978-3-642-03625-5_5
11. Friese, M., Bartz-Beielstein, T., Bäck, T., Naujoks, B., Emmerich, M.: Weighted ensembles in model-based global optimization. In: AIP Conference, vol. 2070, p. 020003 (2019)
12. Jin, Y., Sendhoff, B.: Pareto-based multiobjective machine learning: an overview and case studies. IEEE Trans. Syst. Man Cybern. Part C (Appl. Rev.) **38**, 397–415 (2008)
13. Jones, D., Schonlau, M., Welch, W.: Efficient global optimization of expensive black-box functions. J. Global Optim. **13**, 455–492 (1998)
14. Keeling, C.D., Whorf, T.P.: Atmospheric CO2 records from sites in the sio air sampling network. in trends: a compendium of data on global change. Carbon dioxide information analysis center. Oak Ridge National Laboratory, USA (2004)
15. Knowles, J.: ParEGO: a hybrid algorithm with on-line landscape approximation for expensive multiobjective optimization problems. IEEE Trans. Evol. Comput. **10**, 50–66 (2006)
16. Knowles, J.D., Theile, L., Zitzler, E.: A tutorial on the performance assesment of stochastic multiobjective optimizers. Technical report TIK214, Computer Engineering and Networks Laboratory, ETH Zurich, Zurich, Switzerland, February 2006
17. Kronberger, G., Kommenda, M.: Evolution of covariance functions for gaussian process regression using genetic programming. In: Moreno-Díaz, R., Pichler, F., Quesada-Arencibia, A. (eds.) EUROCAST 2013. LNCS, vol. 8111, pp. 308–315. Springer, Heidelberg (2013). https://doi.org/10.1007/978-3-642-53856-8_39
18. Lei, Y., Yang, H.: A Gaussian process ensemble modeling method based on boosting algorithm. In: Proceedings of the 32nd Chinese Control Conference, pp. 1704–1707 (2013)
19. MacKay, D.J.: Introduction to Gaussian processes. NATO ASI Series F Comput. Syst. Sci. **168**, 133–166 (1998)
20. Mazumdar, A., Chugh, T., Miettinen, K., López-Ibáñez, M.: On dealing with uncertainties from kriging models in offline data-driven evolutionary multiobjective optimization. In: Deb, K., et al. (eds.) EMO 2019. LNCS, vol. 11411, pp. 463–474. Springer, Cham (2019). https://doi.org/10.1007/978-3-030-12598-1_37

21. Palar, P.S., Shimoyama, K.: Kriging with composite kernel learning for surrogate modeling in computer experiments. In: AIAA Scitech 2019 Forum, pp. 2019–2209 (2019)
22. Rahat, A.A., Wang, C., Everson, R.M., Fieldsend, J.E.: Data-driven multi-objective optimisation of coal-fired boiler combustion systems. Appl. Energy **229**, 446–458 (2018)
23. Rasmussen, C.E., Williams, C.K.I.: Gaussian Processes for Machine Learning. The MIT Press, Cambridge (2006)
24. Shahriari, B., Swersky, K., Wang, Z., Adams, R.P., de Freitas, N.: Taking the human out of the loop: a review of bayesian optimization. Proc. IEEE **104**, 148–175 (2016)
25. Sonnenburg, S., Rätsch, G., Schäfer, C., Schölkopf, B.: Large scale multiple kernel learning. J. Mach. Learn. Res. **7**, 1531–1565 (2006)
26. Vijayakumar, S., Schaal, S.: Locally weighted projection regression: an O(n) algorithm for incremental real time learning in high dimensional space. In: Proceedings of the Seventeenth International Conference on Machine Learning (ICML), pp. 1079–1086 (2000)
27. Wolpert, D.H.: The lack of a priori distinctions between learning algorithms. Neural Comput. **8**(7), 1341–1390 (1996)
28. Yeh, I.C.: Modeling of strength of high-performance concrete using artificial neural networks. Cem. Concr. Res. **28**, 1797–1808 (1998)

Designing Combinational Circuits Using a Multi-objective Cartesian Genetic Programming with Adaptive Population Size

Leandro S. Lima[1], Heder S. Bernardino[1(✉)], and Helio J. C. Barbosa[1,2]

[1] Univerisdade Federal de Juiz de Fora (UFJF), Juiz de Fora, MG, Brazil
leandro.lima@engenharia.ufjf.br, heder@ice.ufjf.br, hcbm@lncc.br
[2] Laboratório Nacional de Computação Científica (LNCC), Petrópolis, RJ, Brazil

Abstract. This paper proposes a multiobjective Cartesian Genetic Programming with an adaptive population size to design approximate digital circuits via evolutionary algorithms, analyzing the trade-off between the most often used objectives: error, area, power dissipation, and delay. Combinational digital circuits such as adders, multipliers, and arithmetic logic units (ALUs) with up to 16 inputs and 370 logic gates are considered in the computational experiments. The proposed method was able to produce approximate circuits with good operational characteristics when compared with other methods from the literature.

Keywords: Combinational circuits · Multi-objective optimization · Cartesian Genetic Programming

1 Introduction

The design of electronic circuits is usually a complex task which requires knowledge of specific methodologies. The use of evolutionary algorithms (EAs) to design digital systems gave rise to the Evolvable Hardware (EH) field [9]. The reasons which brought EH to the spotlight of hardware development are its ability to [9]: (i) reach novel architectures which the conventional methods would hardly provide due to their non-flexible nature, (ii) find fair solutions for problems where the specifications are incomplete, (iii) deliver fair solutions in scenarios where there is not a perfect solution, (iv) achieve fault tolerance at the hardware level, and (v) to make the design less dependent on the expert.

EH has provided good results as in [4,7,9], but EH faces issues such as the representation of solutions, as long chromosomes are usually required to encode complex circuits which lead to large search spaces and, consequently, increased

Electronic supplementary material The online version of this chapter (https://doi.org/10.1007/978-3-030-37599-7_49) contains supplementary material, which is available to authorized users.

G. Nicosia et al. (Eds.): LOD 2019, LNCS 11943, pp. 592–604, 2019.
https://doi.org/10.1007/978-3-030-37599-7_49

difficulty in solving the problem. In addition, in the design of combinational digital circuits (CDCs), the processing time grows exponentially with the number of circuit inputs, and EH also uses techniques which are very time consuming [7].

Addressing the design of CDCs, energy efficiency, complexity and delay are features of digital circuits to be analyzed during the manufacturing process once people desire faster, simpler and energetically efficient devices. Approximate Computing (AC), a new paradigm in electronic projects, explores systems which could tolerate loss (i.e. precision) in order to reduce complexity, costs, delay and to increase the energy efficiency of the systems [4,5]. AC can be found in inherently error resilient situations [10], such as multimedia [5], machine learning [11], approximate arithmetic circuits [4], and FIR and IIR approximate filters [9].

The usage of AC and EH in the context of evolutionary design has gained attention as these approaches may lead to the conception of circuit architectures which are different and might be superior to the designs created by specialists [9]. In this scenario, digital circuits obtained via AC are classified as *approximate digital circuits* (ADCs) [10]. Their requirements are relaxed aiming at achieving savings in energy consumption, delay, and complexity. As it may be necessary to increase the complexity of a target circuit to reduce errors, energy efficiency and delay would probably be degraded as these quantities are conflicting. Thus, a multiobjective optimization problem arises [4]. A variety of trade-offs between error, delay, and energy efficiency can be found by a multiobjective approach, enabling the design of a vast number of ADCs. Applying the evolutionary approach in the scenario of intrinsic evolutionary design (e.i. in which the evolutionary process is conducted in the target device) could bring new possibilities as the final solution is already implemented in hardware.

Cartesian Genetic Programming (CGP) [7] is a genetic programming method in which programs are expressed as directed acyclic graphs (DAGs) with their nodes organized in a matrix [7]. CGP allows for a convenient representation when several inputs and outputs are required and, consequently, has become the most popular method in the evolutionary design of CDCs [7].

Here, we propose a technique to design approximate combinational digital circuits (ACDC) based on CGP where the size of the population varies according to the number of non-dominated solutions. One candidate solution starts the search process, the population is allowed to grow up to a maximum number of non-dominated solutions, and, when the population size exceeds a predefined threshold, the candidate solutions with the lowest crowding distances [3] are eliminated. We study here the trade-off between the delay, output error, and power dissipation when designing approximate digital arithmetic circuits. In particular, 8-bit adder (A8) and 8-bit multiplier (M8) are commonly used in the context of AC and EH [4]. Thus, we included them in computational experiments. In addition, the Arithmetic Logic Unit (ALU) is considered here. The results found are compared to those obtained by other methods from the literature.

2 Related Work

Vasicek et al. [10] developed ADCs in which the requirements on functional equivalences between the specifications and implementations were relaxed leading to gains in the speed of computation, area occupied on a chip, and energy consumption. That approach can also be used in multimedia and image compression applications. For instance, most users would not notice variations in the brightness degree in some pixels of an image [5].

Hrbacek et al. [4] proposed a multiobjective approach to design ADCs using CGP to represent candidate circuits, and the Non-Dominated Sorting Genetic Algorithm II (NSGA-II) [3] to explore the search space by analyzing the trade-off between error, delay and power dissipation. The initial population was composed of fully functional circuits, instead of random circuits, and approximate versions of M8 and S8 were designed with significant power consumption savings.

Kaufmann et al. [6] proposed a local search algorithm called hybrid evolutionary strategy (hES) based on the evolutionary strategy and the concepts of Pareto dominance. The hES method uses the Fast non-dominated sort and Crowding-distance of NSGA-II [3], and alternates its evolutionary process using global and local search. That method and its periodization with NSGA-II was compared with Strength Pareto Evolutionary Algorithm 2 (SPEA2) and NSGA-II when designing 2-bit multiplier and adder. CGP's representation was adopted, the functional quality of the solutions was treated as a constraint, and circuit area and delay were the objectives. The authors concluded that for the evolution of digital circuits which uses CGP as a representation model, hES and its periodization are significantly better than NSGA-II and SPEA2.

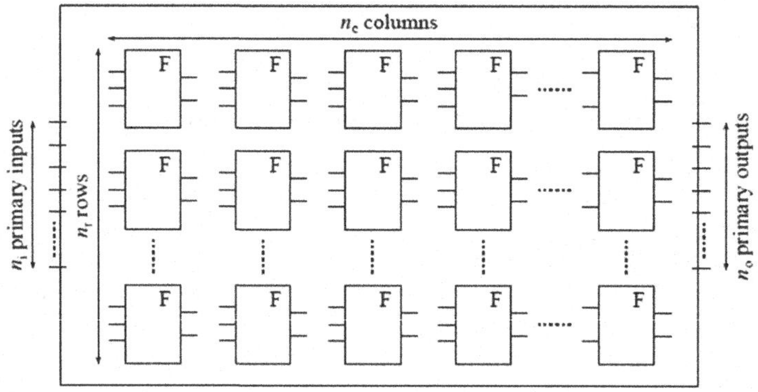

Fig. 1. Illustration of the representation used in CGP (extracted from [4]).

3 Cartesian Genetic Programming

Cartesian Genetic Programming (CGP) [7], is an evolutionary technique where a circuit is represented by a grid of $n_r \times n_c$ interconnected nodes, as in Fig. 1. The processing elements, represented by F, can be either elementary gates (e.g. AND, OR, XOR) or functional level components (e.g. adders, comparators, shifters, and multipliers). Nodes inputs can be connected either to one of the n_i primary inputs or to a node located in the previous l columns; l is defined as "levels-back", which controls the connectivity of the graphs. Each node has a fixed number of inputs, N_{ni}, and outputs, N_{no}. The nodes are able to perform one of the functions from the set Γ ($F \in \Gamma$). Each one of the n_o primary outputs can be connected either to a primary input or to an output of a node. A candidate circuit represented by CGP is described by a sequence of integers which represents the functions and the connections between the nodes. The encoding of a chromosome consists of $n_r \times n_c$ triplets (i_1, i_2, F). Thus, for the case considered here where the gates have two inputs, it is possible to represent a node with its input indices i_1 and i_2 (other nodes), and a function ψ. The last part of GP's encoding contains n_o integers which specify the primary output nodes of the circuit.

In the standard CGP, the population consists of a fixed number of individuals, commonly, randomly initialized. Also, some previous knowledge of the problem can be used. The optimization process when designing circuits is normally an evolutionary strategy $(1+\lambda)$-ES, where λ offspring are created by the application of a point mutation (i.e. $\mu_r\%$ of genes are modified) to the current solution. These $1+\lambda$ individuals are evaluated and the best one survives to the next generation. The process is repeated until a stop criterion is reached.

3.1 Objective Functions

Four objectives (to be minimized) are often considered in the design: *area*, *error*, *power dissipation*, and *propagation delay*. They are described in the sequence.

Area. Analyzing the area of a digital circuit is relevant as it affects the manufacturing cost. The larger the circuit c, the larger the area of the printed circuit board necessary for its implementation. The area occupied by c can be estimated by counting its processing elements (e.g., transistors, elementary ports, functional level components). Here, the area of c is defined as its number of gates.

Error Functions. The Hamming distance between two binary words is often adopted as the error measure in the evolutionary design of digital circuits [4]. Let $O^{(i)}_{orig}$ and $O^{(i)}_{app}$ denote, respectively, the output binary word of a fully functional digital circuit and that of an ADC generated for the input vector i. The Hamming distance is defined as $d_h = \sum_{\forall i}(O^{(i)}_{orig} \oplus O^{(i)}_{app})$, where $\oplus$ is *XOR*, i.e., the number of positions in which the bits of the two binary words differ from each other.

Power Dissipation. The main sources of power dissipation of a digital circuit are [2]: switching component of power ($P_{switching}$), short-circuit component of power ($P_{short-circuit}$), and the leakage current component of power ($P_{leakage}$). The power dissipation is then computed by $P = a_{0\to 1}.C_L.f_{clk}.V_{dd}^2 + I_{sc}.V_{dd} + I_{leakage}.V_{dd}$, where C_L is the load capacitance, f_{clk} is the clock frequency, $a_{0\to 1}$ is the node transition activity factor, I_{sc} is the short circuit current, $I_{leakage}$ is the leakage current, and V_{dd} is the supply voltage. The node transition factor quantifies the average number of times a logic gate makes a state transition that dissipates power within a period of *clock*. It can be defined as $a_{0\rightarrow 1} = p_0.p_1 = p_0.(1 - p_0)$.

Propagation Delay. The delay of a digital circuit c, the time spent for the changes in the input cause any effects in the output, is defined here as D_c and is calculated as the delay of the longest path according to $D_c = \max_{\forall p \in path} \sum_{c_i \in p} t_d(c_i)$, where t_d is the delay of a cell c_i and t_d is normally provided by the manufacturers.

4 Non-dominated Sorting Genetic Algorithm II

Most of the multiobjective EAs (MOEA) are based on the *Pareto dominance* concept which states that a solution $\mathbf{x_1}$ dominates a solution $\mathbf{x_2}$ if: (i) $\mathbf{x_1}$ is no worse than $\mathbf{x_2}$ in all objectives; and (ii) $\mathbf{x_1}$ is strictly better than $\mathbf{x_2}$ in at least one objective. NSGA-II [3] has two main features: non-dominated ranking and crowding distance. In the non-dominated ranking, the individuals in the set R_t –parent population (P_t) plus offspring population (Q_t)– are sorted according to their dominance. All Pareto optimal solutions, which are the feasible non-dominated solutions, form the first front F_1. The non-dominated solutions from $R_t \setminus F_1$[1] compose the second front F_2, and so on. A rank corresponding to the front index is assigned to each individual, i.e., i is given to the solutions in F_i. The individuals in the first fronts form P_{t+1}.

However, as the parent population size is $|R_t|/2$, where $|\cdot|$ denotes the cardinality of a set, one can determine i such that $|F_1 \cup F_2 \cup \cdots \cup F_{i-1}| \leq |R_t|/2$ and $|F_1 \cup F_2 \cup \cdots \cup F_i| > |R_t|/2$. Thus, some individuals in F_i can not be included in the next population, as the population size is fixed. The crowding-distance is calculated as the semi-perimeter of the hyperrectangle formed by the values of adjacent neighbor of each candidate solution in the objectives space and it is adopted by NSGA-II to select the candidate solutions in less populated regions.

5 The Proposed Method

The idea here is to design ACDC, considering multiple objectives, using CGP. Most of the MOEAs, such as NSGA-II, maintain a population of fixed size during the evolutionary process; the larger the population, the larger the number of trade-offs between the different objectives analyzed, as a larger population allows for better coverage of the search space. As a consequence, the chances of premature convergence is reduced. However, the computational cost for each iteration of the method associated with this population may be significant. In contrast, a small population leads to a coarser coverage, which can result in the non-exploration of promising areas of the search space increasing the probability

[1] The symbol "\" represents the operator of set difference.

that the algorithm gets stuck at a local optimum [8]. An attractive alternative is thus to permit that the size of the population varies during the search process. For this, we propose here the variation of the number of individuals in the population from one to a (user defined) maximum value. The proposed approach uses CGP to represent the candidate circuits, builds a set of Pareto solutions observing the trade-off between key circuit parameters and, when the population size becomes larger than Tam_Max, the crowding distance from NSGA-II [3] is applied to select individuals. Algorithm 1 presents a pseudocode of the proposal.

Algorithm 1. Pseudocode of the proposed technique.

```
t = 0;
P_t ← Initialize-Population(P_t);
Q_t ← ∅;
while Circuit-Evaluation ≤ N do
    R_t ← P_t ∪ Q_t;
    F_1 ← Non-dominated-Individuals(R_t);
    P_t ← F_1;
    if |P_t| > Tam_Max then
        Crowding-distance(P_t);
        Crowded-comparison-operator-sort(P_t, ≺_n);
        P_t ← P_t[1 : Tam_Max];
    end
    for i ← 1 to |P_t| do
        Q_i ← Mutate(P_i);
    end
    t = t + 1;
end
```

Initially, the population of parents (P_t) is initialized with a single (randomly generated) individual and the population of offspring (Q_t) is empty. The proposal evolves the candidate circuits while the stop criterion is not met. During the search, the parent and offspring populations are combined into a single population $R_t = P_t \cup Q_t$. Then, the non-dominated individuals from R_t, F_1, are selected to compose the population P_t. Thus, the size of the population P_t is not fixed, as the number of non dominated solutions may vary. Crowding distance [3] is applied when the population size is larger than a user-defined threshold (Tam_Max). This is used in order to avoid the population size to become very large. Crowding distance can be used to preserve the diversity of the solutions and is calculated as half of the perimeter of the cuboid formed by the nearest neighbor with respect to the objective function values. The individuals are sorted in descending order (Crowded-comparison-operator-sort) and the first ones are selected to compose P_t. Finally, new candidate circuits are generated using point mutation (*Mutate*(P_i)), the most commonly used mutation operator in CGP [7]. In this context, one allele of a gene is randomly selected and its value is replaced by another valid value chosen randomly. The point mutation rate ($\mu_r\%$) is a parameter that affects the number of genes mutated.

To deal with constrained problems, the proposed method uses the Constrained NSGA-II approach [3]. In this context, it is said that a solution $\mathbf{x_1}$ dominates a solution $\mathbf{x_2}$ if any of the following conditions is true: (i) $\mathbf{x_1}$ is feasible and $\mathbf{x_2}$ is not; (ii) $\mathbf{x_1}$ and $\mathbf{x_2}$ are both infeasible, but $\mathbf{x_1}$ has a smaller overall constraint violation; and (iii) $\mathbf{x_1}$ and $\mathbf{x_2}$ are both feasible and $\mathbf{x_1}$ dominates $\mathbf{x_2}$.

6 Computational Experiments

The experiments were conducted in the extrinsic evolutionary scenario (search occurs in software) using the truth tables of w-bit ($w = 2, 8$) adder and multiplier, and an ALU circuits. The problems are labeled here as Aw and Mw, respectively, for adder and multiplier circuits, and they can be used to construct filtering structures, such as Finite Impulse Response Filter (FIR filter) and Infinite Impulse Response Filter (IIR Filter). These filters are quite important in digital signal processing such as image processing, video processing, audio processing, and wireless communication systems [5]. Also, adders and multipliers are used in circuits dedicated to calculate hyperbolic and trigonometric functions, such as the Coordinate Rotation Digital Computer (CORDIC) [1]. The Ripple Carry Adder and Ripple Carry Array Multiplier were adopted as reference architectures for the 2- and 8-bit adder and multiplier, respectively. Finally, the architecture of ALU is important for digital signal processing, and it is present in all computing devices such as microprocessors, computers and embedded systems. The SN54/74LS181 architecture[2], a 4-bit ALU which can perform 16 logic operations and also a diversity of arithmetic operations, was adopted here. The output is encoded by a 8-bit word, 4 bits are reserved for each variable, and 6 bits are used as controllers (logical/arithmetical operator). The data provided by a current manufacturer of digital devices is used to calculate power dissipation and delay (V_{dd}, C_L and t_d in Sect. 3.1). Nexperia[3] was selected as its gates have low t_d values. The gates are: 74LVC1G08 (AND), 74LVC1G08 (NAND), 74LVC1G32 (OR), 74LVC1G86 (XOR), 74LVC1G02 (NOR), HEF4077B (XNOR). These circuits are important but simples, as we conducted the experiments as a preliminary evaluation of the proposed approach.

The proposed CGPMO+APS[4] was compared to hES and its variant with periodization (hn[10]) [6], and the CGP combined with NSGA-II [4]. The experiments were divided into three scenarios: (i) constrained multiobjective optimization (no domain information added), (ii) the design of ACDCs, (no domain information added); and (iii) the design of ACDCs with information from the domain specialist.

[2] http://pdf1.alldatasheet.com/datasheet-pdf/view/5671/MOTOROLA/SN54LS181.html.

[3] http://www.nexperia.com/products/logic/gates.

[4] The source-code of CGPMO+APS is available at https://github.com/ciml.

6.1 Scenario 1 – Constrained Multiobjective Optimization

Here, S2 and M2 are the problems, the Hamming distance is the measure of functional quality (a constraint in this scenario), and the area and delay are the objective functions to be minimized. The computational experiments were conducted as in [6], and the results obtained by the proposed CGPMO+APS are compared to those found by hES and hn^{10}. The initial population of the CGPMO+APS is composed by 1 individual randomly generated, $\Gamma = \{0, 1, a, b, \overline{a}, \overline{b}, a.b, a.\overline{b}, \overline{a}.b, \overline{a}.\overline{b}, a \oplus b, a \oplus \overline{b}, a + b, a + \overline{b}, \overline{a} + b, \overline{a} + \overline{b}, a.\overline{c} + b.c, a.\overline{c} + \overline{b}.c, \overline{a}.\overline{c} + b.c, \overline{a}.\overline{c} + \overline{b}.c\}$ is the set of Boolean functions that can be executed by processing nodes, $\mu_r = 5\%$, $n_r = 1$, $n_c = 200$, $l = 200$, 400.000 fitness evaluations were allowed for S2, and 1.600.000 fitness evaluations for M2. Finally, 20 independent runs were performed.

In [6], the Additive Epsilon Indicator (AEI) and the Kruskal-Wallis non-parametric test were used. Here we do not present a statistical comparison using the results of CGPMO+APS as the results of each independent run were not provided in [6]. However, the capacity of generating feasible circuits with various combinations of area and delay was analyzed in [6]. Thus, we provide here a comparison in terms of feasible circuits generated by the techniques considered and these values are presented in Table 1. In this scenario we analyze the number of independent runs which resulted in functionally correct solutions. The results indicate that the proposed CGPMO+APS obtained better results in terms of fully functional circuits when compared to the other methods, as it is the only technique which found functionally correct circuits in all 20 independent runs.

Table 1. Number of independent runs with functionally correct solutions.

Method	S2	M2	#correct circuits S2	#correct circuits M2
hES	12	15	–	–
hn^{10}	11	14	–	–
CGPMO+APS	**20**	**20**	**26**	**24**

6.2 Scenario 2 – Approximate Design of CLCs from Scratch

The present set of experiments is composed of three objective functions commonly employed when designing ADCs in the context of EH: (i) Hamming distance, (ii) delay, and (iii) power dissipation. The circuits S8, M8, and ALU were used as problems to be solved. The approach proposed in [4], labeled here as CGP+NSGA-II, was implemented and its results are used in the comparative analysis. Both CGP+NSGA-II and the proposed CGPMO+APS used the same parameter settings: $\Gamma = \{\text{AND, NAND, OR, XOR, NOR, XNOR}\}$, $\mu_r = 5\%$, $n_r = 1$. The number of columns were $l = n_c = 300, 200, 1000$ for the ALU, S8, and M8, respectively, and these values are those used in [4]. Also, the population size of CGP+NSGA-II is equal to 50.

Table 2. Parameters of circuits with the lowest error values when the population is randomly initialized and 30×10^6 circuit evaluations are allowed. The reference models (with error = 0) are the ALU SN54/74LS181 with delay = 22.22 ns and power = 1.14 mW, the Ripple Carry Adder with delay = 25.70 ns and power = 0.50 mW, and the Ripple Carry Array Multiplier with delay = 74.30 ns and power = 3.21 mW. Here, error is the percentage of incorrect outputs values, and delay and power are the ratio of the values observed in the circuits generated and those of the reference architecture.

	CGPMO+APS			CGP+NSGA-II		
	Error (%)	Delay (% ns)	Power (% mW)	Error (%)	Delay (% ns)	Power (% mW)
ALU	29.72	560.81	51.24	36.64	701.80	51.65
	29.81	517.57	41.41	37.35	685.14	48.87
	29.86	866.22	40.74	37.57	685.14	37.30
	29.91	371.62	27.81	37.67	685.14	36.70
	30.21	371.17	25.00	37.92	684.23	33.31
	30.38	327.93	25.76	38.14	383.33	50.62
	30.48	326.13	24.93	38.30	669.37	27.60
	30.52	326.13	21.49	38.32	373.87	32.45
	30.60	309.91	25.21	38.57	358.11	24.22
	30.79	317.12	21.42	38.96	358.11	23.19
S8	15.49	570.82	79.63	13.89	592.22	95.11
	15.54	570.82	73.98	14.18	578.21	90.20
	16.58	570.82	73.00	14.70	605.45	74.73
	16.63	564.20	64.86	14.87	578.21	87.25
	17.08	309.73	69.81	15.10	571.60	91.39
	17.32	564.20	64.27	15.44	591.05	70.79
	17.35	571.21	59.36	15.92	570.82	81.49
	17.60	296.50	68.99	16.49	310.12	97.90
	17.70	564.20	60.14	16.66	591.44	67.47
	17.95	296.50	63.68	16.88	302.72	77.83
M8	38.87	504.04	52.12	42.14	414.27	64.27
	38.92	504.04	32.81	42.36	317.50	56.22
	38.97	311.57	40.60	42.39	316.42	62.08
	38.98	308.88	38.06	42.79	411.44	47.87
	38.99	311.57	33.66	42.87	309.56	81.27
	39.02	308.88	33.67	42.91	409.15	49.12
	39.23	304.71	37.07	42.96	306.46	68.29
	39.25	219.11	47.72	42.99	411.31	39.85
	39.27	306.59	36.50	43.09	409.15	46.55
	39.28	219.11	46.86	43.17	299.87	69.49

Here we analyze the results obtained by CGPMO+APS and CGP+NSGA-II when the initial population was randomly generated and 30×10^6 objective function evaluations were allowed. Table 2 presents the operational characteristics of 10 candidate solutions for ALU, S8, and M8 with the smallest errors considering all independent runs. The techniques analyzed here were not able to design fully functional ALUs, S8, and M8 from scratch. It can also be noticed that no circuits reached delay values lower than or equal to the delay of the reference circuits. The ALU models obtained by CGPMO+APS show a smaller level of error than those found by CGP+NSGA-II. The (10) ALUs designed by CGPMO+APS and presented in Table 2 dissipate less power and have a smaller delay than those (10) obtained by CGP+NSGA-II. Regarding S8, the circuits generated dissipate less power than the reference architecture. Among the S8 circuits, that with the largest power dissipation (79.63% of the power dissipated by the reference architecture) was found by CGPMO+APS and it is superior, in terms of energy efficiency, to all 10 circuits that present the lowest error values found by CGP+NSGA-II. The error values of the S8 circuits obtained by CGPMO+APS and CGP+NSGA-II are similar. It is noted that the 10 M8 models with the smallest error values obtained by both approaches dissipate less power than the Ripple Carry Array Multiplier. Also, the M8 circuits found by CGPMO+APS dissipate less power and have smaller delay than those obtained by CGP+NSGA-II.

Table 3 presents the hypervolume found by CGPMO+APS and CGP+NSGA-II, and the p-values of the Kruskal-Wallis tests. CGPMO+APS presented the highest mean hypervolume values for all circuits analyzed. According to the Kruskal-Wallis test, the results found by the proposed CGPMO+APS are statistically different from those obtained by CGP+NSGA-II for S8 and ALU.

6.3 Scenario 3 – Approximate CLCs Using Conventional Models

Besides the design of ADCs from scratch, CGPMO+APS and CGP+NSGA-II were also applied to optimize conventional architectures by observing the trade-off between error, power, and delay. As a result, the techniques provide a set of ADCs to the specialist, who can choose the best for his/her application. Also, 30×10^6 circuit evaluations are allowed here.

The operational features of 10 circuits with the smallest errors for ALUs, S8, and M8 are shown in Table 4. CGPMO+APS and CGP+NSGA-II were able to obtain fully functional circuits in terms of error, and these circuits dissipate less power or present lower delay values than those of the base architecture for all the cases tested (ALU, S8, and M8). CGPMO+APS generated an ALU with lower values of delay and power dissipation than those of the conventional architecture. Also, with respect to S8 and M8, the fully functional circuits generated by the proposed method are better in delay or power dissipation. The results obtained by CGPMO+APS are, in general, better than those found by CGP+NSGA-II. This advantage is specifically large with respect to the delay values of ALU.

Table 3. Hypervolume values. The reference points are the highest values obtained for each objective function: (0.5221, 192.3000, 0.5861), (0.4722, 291.7000, 0.4865), and (0.5257, 374.5000, 2.6115), respectively, for ALU, S8, M8. Kruskal-Wallis test is applied and the p-values are also presented.

Circuit	Method	Best	Median	Mean	STD	Worst	p-value
ALU	CGPMO+APS	0.3532	0.3049	**0.2965**	0.0438	0.2276	3.76×10^{-7}
	CGP+NSGA-II	0.3366	0.2720	0.2700	0.0449	0.1464	
S8	CGPMO+APS	0.4059	0.3439	**0.3475**	0.0314	0.2935	9.77×10^{-4}
	CGP+NSGA-II	0.3770	0.3041	0.3049	0.0456	0.2388	
M8	CGPMO+APS	0.3192	0.1524	**0.1615**	0.0873	0.0218	6.90×10^{-2}
	CGP+NSGA-II	0.1873	0.1475	0.1406	0.0384	0.0418	

Table 5 presents the hypervolume found by CGPMO+APS and CGP+NSGA-II. CGPMO+TP presented the highest mean values of hypervolume for all circuits analyzed. According to the Kruskal-Wallis test, the results of CGPMO+APS are statistically better than those of CGP+NSGA-II for all the tested circuits.

Table 4. Parameters of circuits with the lowest error values when the population is initialized with a conventional architecture and 30×10^6 circuit evaluations are allowed. The reference models (with error = 0) are those in the caption of Table 2.

	CGPMO+APS			CGP+NSGA-II		
	Error (%)	Delay (% ns)	Power (% mW)	Error (%)	Delay (% ns)	Power (% mW)
ALU	0.00	99.55	64.80	0.00	395.50	64.34
	0.00	404.05	64.74	0.00	723.87	64.30
	0.00	411.71	64.71	0.00	746.85	64.11
	0.00	544.14	64.11	0.52	395.50	62.55
	0.33	99.55	63.46	1.04	746.85	61.24
	0.68	86.04	65.11	1.70	396.40	60.59
	0.93	404.05	62.39	2.12	395.50	62.15
	1.04	404.05	61.85	2.61	403.15	60.20
	1.09	86.04	65.04	2.73	396.40	60.42
	1.27	86.04	64.66	2.86	723.87	59.93
S8	0.00	84.82	128.18	0.00	104.28	98.54
	0.00	107.39	99.28	0.00	104.67	71.58
	0.00	114.01	85.39	1.39	95.72	99.61
	1.37	86.77	99.05	1.40	101.56	98.43
	2.02	80.93	98.75	2.02	75.88	97.45
	2.60	60.31	100.74	2.28	68.48	97.71

(*continued*)

Table 4. (*continued*)

	CGPMO+APS			CGP+NSGA-II		
	Error (%)	Delay (% ns)	Power (% mW)	Error (%)	Delay (% ns)	Power (% mW)
	2.71	81.71	97.76	2.78	95.72	91.94
	3.13	80.93	98.30	3.58	67.70	97.71
	3.32	61.09	97.51	3.82	55.25	100.08
	3.47	86.77	97.33	4.17	97.28	91.00
M8	0.00	100.40	99.95	0.00	100.13	99.98
	0.38	100.40	99.93	0.82	98.11	99.77
	0.41	98.11	99.88	2.37	98.11	99.49
	0.61	98.11	99.77	5.32	110.76	98.79
	1.09	98.11	99.65	5.39	98.11	99.37
	1.15	100.13	99.52	5.40	98.11	99.22
	1.30	95.55	99.64	5.57	100.40	98.53
	1.57	95.28	99.21	6.14	97.57	98.37
	2.03	93.53	98.39	6.41	93.53	98.28
	3.02	99.86	97.74	6.77	93.53	98.24

Table 5. Hypervolumes values and *p*-values of the Kruskal-Wallis test. The reference points are: $(0.4284, 165.8000, 1.1348)$, $(0.4722, 74.4000, 0.6369)$, and $(0.5353, 82.3000, 3.2135)$, respectively, for ALU, S8, and M8.

Circuit	Method	Best	Median	Mean	STD	Worst	p-value
ALU	CGPMO+APS	0.5354	0.4904	**0.4909**	0.0347	0.4340	2.8701×10^{-9}
	CGP+NSGA-II	0.4088	0.3397	0.3357	0.0450	0.2687	
S8	CGPMO+APS	0.5025	0.4849	**0.4687**	0.0345	0.4192	6.7183×10^{-8}
	CGP+NSGA-II	0.4422	0.4016	0.3944	0.0392	0.3326	
M8	CGPMO+APS	0.3730	0.3101	**0.2996**	0.0503	0.2028	3.1732×10^{-5}
	CGP+NSGA-II	0.2601	0.2204	0.2135	0.0368	0.1478	

7 Concluding Remarks and Future Work

A multiobjective Cartesian Genetic Programming technique with adaptive population size (CGPMO+APS) was proposed here to design approximate combinational digital circuits (ACDCs). Three situations were considered in the experiments: (i) constrained multiobjective optimization (no domain information added), (ii) the design of ACDCs (no domain information added); and (iii) the design of ACDCs with information from the domain specialist.

In (i), the results indicated that CGPMO+APS was able to design more feasible circuits than the other approaches analyzed. In (ii) and (iii), the results show that both CGPMO+APS and CGP+NSGA-II are not suitable for the construction of complex architectures such as ALUs, S8, and M8 without the introduction of the knowledge of the domain expert. On the other hand, when a conventional architecture is adopted as the initial solution, the two approaches synthesized ALUs, S8, and M8 that do not present errors with respect to the truth table and with improvement in delay or power dissipation when compared to the reference architecture. Particularly, the CGPMO+APS obtained a fully functional ALU with lower delay and power dissipation than those of the reference architecture. Also, CGPMO+APS obtained better mean values of hypervolume than those of CGP+NSGA-II.

The use of Binary Decision Diagrams to reduce the processing time and solving more complex problems are relevant research avenues.

Acknowledgments. We thanks the support provided by CNPq (312337/2017-5 and 312682/2018-2), FAPEMIG (APQ-00337-18), PPGCC, and PPGMC.

References

1. Aggarwal, S., Meher, P.K., Khare, K.: Concept, design, and implementation of reconfigurable cordic. IEEE Trans. Large Scale Integr. (VLSI) Syst. **24**(4), 1588–1592 (2016)
2. Chandrakasan, A.P., Brodersen, R.W.: Minimizing power consumption in digital CMOS circuits. Proc. IEEE **83**(4), 498–523 (1995)
3. Deb, K., Pratap, A., Agarwal, S., Meyarivan, T.: A fast and elitist multiobjective genetic algorithm: NSGA-II. IEEE Trans. Evol. Comput. **6**(2), 182–197 (2002)
4. Hrbacek, R., Mrazek, V., Vasicek, Z.: Automatic design of approximate circuits by means of multi-objective evolutionary algorithms. In: International Conference on Design and Technology of Integrated Systems in Nanoscale Era (DTIS), pp. 1–6 (2016)
5. Julio, R.O., Soares, L.B., Costa, E.A.C., Bampi, S.: Energy-efficient gaussian filter for image processing using approximate adder circuits. In: 2015 IEEE International Conference on Electronics, Circuits, and Systems (ICECS), pp. 450–453 (2015)
6. Kaufmann, P., Knieper, T., Platzner, M.: A novel hybrid evolutionary strategy and its periodization with multi-objective genetic optimizers. In: IEEE Congress on Evolutionary Computation, pp. 1–8 (2010)
7. Miller, J.F.: Cartesian Genetic Programming. Springer, Berlin (2011). https://doi.org/10.1007/978-3-642-17310-3
8. Roeva, O., Fidanova, S., Paprzycki, M.: Influence of the population size on the genetic algorithm performance in case of cultivation process modelling. In: Federated Conference on Computer Science and Information Systems, pp. 371–376 (2013)
9. Stepney, S., Adamatzky, A.: Inspired by Nature. Springer, Cham (2018). https://doi.org/10.1007/978-3-319-67997-6
10. Vasicek, Z., Sekanina, L.: Evolutionary approach to approximate digital circuits design. IEEE Trans. Evol. Comput. **19**(3), 432–444 (2015)
11. Venkataramani, S., Sabne, A., Kozhikkottu, V., Roy, K., Raghunathan, A.: SALSA: systematic logic synthesis of approximate circuits. In: DAC Design Automation Conference, vol. 2012, p. 796–801 (2012)

Multi-task Learning by Pareto Optimality

Deyan Dyankov[1], Salvatore Danilo Riccio[2,3], Giuseppe Di Fatta[1], and Giuseppe Nicosia[2(✉)]

[1] University of Reading, Reading, England
[2] University of Cambridge, Cambridge, England
gn263@cam.ac.uk
[3] Queen Mary University of London, London, England

Abstract. Deep Neural Networks (DNNs) are often criticized because they lack the ability to learn more than one task at a time: Multitask Learning is an emerging research area whose aim is to overcome this issue. In this work, we introduce the Pareto Multitask Learning framework as a tool that can show how effectively a DNN is learning a shared representation common to a set of tasks. We also experimentally show that it is possible to extend the optimization process so that a single DNN simultaneously learns how to master two or more Atari games: using a single weight parameter vector, our network is able to obtain sub-optimal results for up to four games.

Keywords: Multitask learning · Neural and evolutionary computing · Deep neuroevolution · Hypervolume · Kullback-Leibler Divergence · Evolution Strategy · Deep artificial neural networks · Atari 2600 Games

1 Introduction

In recent years, Deep Neural Networks (DNNs) [7] obtained astonishing results in Reinforcement Learning (RL) problems. DNNs outperformed humans in some tasks like in the game of chess, shogi, and Go as shown in [18].

The main limitation is their poor flexibility: although each DNN may potentially master a given individual task, there are no known benefits to re-train a DNN previously trained on a task to master a different one. This means that DNNs still lack the human-like ability to transfer knowledge between different tasks.

An emerging branch of research studies how to use the knowledge acquired for a single task to address a different one. Usually, it is required that the new task shares some common features with the one already mastered by the network [13]. This desired behaviour has been studied with different approaches: Tucker decomposition [14], Meta-learning or learning-to-learn [17], sparse coding for multitask and transfer learning [10]. A general and complete overview of Multitask learning can be found in [15,19,21].

Multitask learning (MTL) improves generalization by extracting information and shared structures between the distinct learning problems. Sometimes

G. Nicosia et al. (Eds.): LOD 2019, LNCS 11943, pp. 605–618, 2019.
https://doi.org/10.1007/978-3-030-37599-7_50

these tasks can be competing learning problems. MTL relies on training signals from multiple tasks to obtain structure that is shared across multiple learning problems. Since the shared structure must support solving multiple tasks, it is inherently more general, which leads to better generalization of the tasks [4].

Our main contribution is to study MTL problems within the framework of multi-objective optimization. We also exploit some indexes, namely the convergence direction of the Pareto front, the hypervolume index, and the Kullback-Leibler Divergence, to empirically show the robustness of the overall convergence process. Here we apply Pareto Optimality theories to analyze the performance of deep artificial neural networks on Multi-Objective Optimization Problems. The proposed approach to evolve the weights of the DNN is a gradient-free, population-based multitask evolutionary algorithm and it performs well on hard deep reinforcement learning problems such as Atari 2600 games. Instead of addressing one task at a time [9,13], we train the DNN simultaneously on more than one task. We generalize Evolution Strategies (ES) to tackle MTL problems and show the main experimental results. All final networks are deep artificial neural networks sharing the feature to play two or four Atari games. Competitive two (or four) Atari-playing agents evolve in as little as tens of generations. Our goal is to take the first steps towards a new approach to address transfer learning.

The rest of the paper is organized as follows: in Sect. 2 we introduce our algorithm and some indexes like the hypervolume and the Kullback-Leibler Divergence. In Sect. 3 we briefly discuss the experimental setup used to validate the method. In Sect. 4 we show the obtained results; finally, conclusions are drawn in Sect. 5 with some further developments proposed for future works.

2 Pareto Optimality for MultiTask Learning

Given a single DNN whose topology is a-priori defined, let θ be the weight vector parameter associated to it. Let $\mathcal{T}$ be a finite set of tasks with cardinality $t \in \mathbb{N}$. Each task $\tau_i \in \mathcal{T}$ is associated with an utility (or *score*) function $u_i \in \mathbb{R}$ that is a-priori unknown and it is stochastically dependent on the vector θ, so $u_i = f_i(\theta)$. Then the MTL problem consists in finding θ such that $\text{argmax}_\theta\{f_1(\theta), f_2(\theta), \ldots, f_t(\theta)\}$.

We propose the MultiTask Evolutionary Strategy (MTES), a generalization of Evolution Strategies, and investigate how this can make a single DNN to master two or more tasks simultaneously training the network only once.

The main idea behind the proposed approach is to model MTL in terms of Multi-Objective Optimization Problem. We then exploit indexes from Pareto optimality theory to show that the DNN has effectively learned each task.

As it will be explained in further details in Sect. 3.2, at each iteration we generate n perturbations of the weight vector parameter θ. Each perturbation θ_j $(0 < j \leq n)$ is evaluated on every task once and it is then associated with the corresponding utility $u_{ij} = u_i(\theta_j)$ for each task i and perturbation j. These values can be represented as n points in a t-dimensional space. Since we want to maximize every u_i, we can use the Pareto theory for Multi-Objective Optimization Problems.

Let Θ_k be the finite set containing all vectors of weight θ evaluated at generation k. For each $\theta \in \Theta_k$ we collect the score obtained in each game u_i. A vector of weight $\bar{\theta}$ is *Pareto optimal* with respect to iteration k if $\nexists \theta \in \Theta_k$, $\theta \neq \bar{\theta}$ such that $u_i(\theta) \geq u_i(\bar{\theta})$ for all $i = 1, \ldots, t$ and $u_j(\theta) > u_j(\bar{\theta})$ for at least one index j. Then, for each iteration k a weight vector θ belongs to the Pareto Front if it is Pareto optimal.

The Pareto front provides useful information about the evolving process. In particular, it is desirable that the Pareto front moves toward a direction suitable to every task. A measure of this property can be computed by means of the hypervolume indicator H [6]. We choose to compute the hypervolume indicator as the part of the t-dimensional space between the Pareto front and the positive semi-axle. This index penalizes solutions that are optimal for just one task and that have poor performance for the others. There are many benefits to use the hypervolume indicator as a quality measure for the objective vectors [1] and efficient heuristic algorithms can be adopted to provide an estimation of the exact value of the indicator H. We want this index H to be maximized: a greater value indicates that the overall performance of the DNN is better with respect to another strategy associated with a smaller hypervolume value.

We also exploit the Kullback-Leibler Divergence D_{KL} [8] as an index to evaluate the information gain obtained between generations [20]. This index is computed as $D_{KL}(P, Q) = \sum_P p(x) ln \frac{p(x)}{q(x)}$ where P and Q are two distributions obtained by partitioning in subintervals of same length the interval from the minimum and the maximum score achieved in a task at a given iteration. Each subinterval is associated with the fraction of perturbations that belong to that subinterval. To overcome numerical issues coming from some subintervals of Q with zero frequency, we add a small ϵ to each subinterval of both distributions. P and Q are then normalized to 1 and D_{KL} is computed. This index gives an estimate of the similarity between the two distributions: it is 0 if and only if they are identical, otherwise it grows bigger as the distributions diverge.

3 Experimental Protocol

We used some Atari 2600 games as a benchmark to train the DNN and to test the performance of the proposed approach. We started from the algorithms previously developed by Salimans et al. in [16] and by Conti et al. in [5].

Instead of evaluating each perturbation on one task only, we generate a population of distributed workers for each task we want to address. Then, each perturbation is evaluated on one among the tasks $\tau \in \mathcal{T}$. It shall be noticed that the most time consuming part of the algorithm is due to perturbations evaluation. Evaluating each perturbation only on one among the tasks results in maintaining almost the same computational time because the total amount of evaluations performed is the same. Furthermore, collecting all the scores in a single vector regardless of the task they were evaluated on introduces some cross-correlation that can help θ to evolve towards a strategy that maximizes all tasks scores.

Regarding the model of the network, we use a DNN with the same topology used in [11]. It is associated with a weight parameter vector θ with dimension 1008450×1 that is randomly initialized. Then θ is evolved using MultiTask Evolution Strategy (MTES) as explained in Sect. 3.2 and the network is trained to master the selected tasks. The algorithm runs until a chosen criterion is met. The choice of the criterion can be based on some factors such as the elapsed time from start or the total number of iterations. A common choice in the literature is to stop the simulation after 200 iterations. We try to run the simulation for 1,000 iterations to guarantee that full training is achieved. The simulations are run on a local computer, exploiting the GPU-enabled codebase.

3.1 Atari Games

Taking games from the Atari 2600 as a benchmark is a common choice in the literature to test reinforcement learning (RL) algorithms [2,3,5,12,16]. Atari games come in different variety, style and difficulty. Furthermore, each game is non-deterministic, so playing exactly the same moves twice is not a sufficient condition to achieve the same score. The low quality graphics (compared to modern videogames) limits the input parameter vector. Although there are some common features between some games, each one requires a different strategy from the others.

The full action space A is composed of 18 different actions which are given by the Cartesian product of $A_1 := \{fire, noFire\}$, $A_2 := \{up, down, noVertMov\}$, and $A_3 := \{left, right, noHorizMov\}$, i.e. $A := A_1 \times A_2 \times A_3$. Each node in the output layer of the DNN represents a single action $a \in A$. Some games accept the full action space, whereas other do not. For the sake of simplicity, in this paper we limit the experimental analysis to tasks associated with the full action space.

It is usually required to choose tasks with some common features for transfer learning to be effective. To meet this condition, we perform MTL on "River Raid" and "Zaxxon" because both of them are scrolling shooters with almost the same game dynamics. Also the maximum achievable score is almost the same. To address the 4 tasks scenario we add also "Centipede" and "Seaquest" because both of them belong to the genre *shoot 'em up* and have not so hard game dynamics. Furthermore, it is know that "Seaquest" gets stuck on local maxima where ES may converge to, so it is interesting to find out whether or not MTES is able to avoid this problem.

3.2 Evolutionary Strategy

Dealing with RL problems, Evolutionary Strategies (ES) have been shown to reach state-of-the-art performance as discussed in [16]. Among its properties, it shall be noticed that perturbations do not change during evaluation. This implies that the overall performance is evaluated, so actions that may lead to an instant small reward are not chosen if there exist other long-run strategies able to bring higher total scores. Another important difference with respect to other strategies is that perturbation is done in the parameter space instead of the action space.

The experimental analysis of this work is based on the research of Conti et al. [5], but it has been extended to address the MTL problem. First, vector $\theta_{P(0)}$ (subscript $P(0)$ stands for "Parent at generation 0") is initialized with random values. Then, at each iteration we generate a population of n perturbations by adding different random noise vectors to $\theta_{P(k)}$. Each perturbation is evaluated on one among the tasks in $\mathcal{T}$ to obtain the associated score. Note that, in previous works, $|\mathcal{T}| = 1$. We extended the algorithm so that $|\mathcal{T}| > 1$. These scores are collected together and they are used to evolve $\theta_{P(k)}$ as explained in [5]. To briefly summarize the implemented ES algorithm, the parent at iteration $k+1$ is computed as $\theta_{P(k+1)} = h\left(\theta_{P(k)}, \Delta\theta(\theta_1, \ldots, \theta_n)\right)$, where $\Delta\theta(\theta_1, \ldots, \theta_n)$ is the incremental value computed as a normalized weighted average given the scores obtained by each offspring, and $h(\cdot)$ is a chosen optimizer (e.g.: SGD or Adam). It is worth to note that we have only one parent for each iteration. Furthermore, the parent is not likely to be one of the offspring. Each vector $\theta_P(k+1)$ is evaluated on 200 episodes to validate its performance. This concludes an "iteration". The cycle is repeated, generating a new population of perturbations P_d and so on, until the desired number of generations is reached.

4 Results

We first tested our implementation of MTES on learning how to play two games: River Raid and Zaxxon. We addressed this problem first to prove experimentally whether or not our implementation of MTES would be able to learn how to play two games simultaneously. Although in the literature the common choice is to run the simulation over 200 iterations, we decided to run it through 1,000 iterations to test the convergence of Pareto front over a longer period of time.

As it is shown in Fig. 1, scores in both games grow up until almost 200 iteration. After that, Zaxxon score remains almost constant, whereas River Raid score slightly increases over time. Figures 2, 3 show minimum, mean, and maximum score for each iteration in both games respectively. Convergence for River Raid scores is better than the one for Zaxxon. This is justified by the algorithm used to evolve θ and by the fact that playing a River Raid game takes less time than playing a Zaxxon game. During the perturbation evaluation step, workers will take less time on River Raid, so finally a greater amount of perturbations will be evaluated on River Raid rather than Zaxxon, ultimately giving a better convergence of the scores.

The analysis of the Pareto front is carried out at the end of each iteration. As a post-processing step, for every iteration we compute all scores u_{ij} for each task i and perturbation j. Figure 4 only shows the feasible perturbations at the end of iteration 1, namely those that obtained a score strictly greater than zero in both tasks, among the 5,000 generated. It can be seen that only few points

Table 1. Hypervolume values computed for iterations from 1 to 1,000 in the 2 tasks scenario.

Generation	Hypervolume H $[10^6]$
1	1.83
125	42.23
250	46.08
375	49.83
500	46.39
625	51.59
750	46.14
875	55.95
1000	58.29

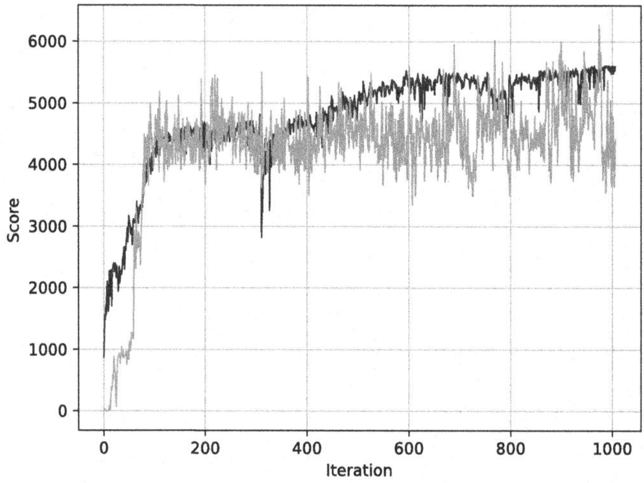

Fig. 1. Game mean score obtained in 200 episodes per iteration. Blue line: River Raid mean score; red line: Zaxxon mean score. (Color figure online)

are feasible and that the Pareto front is close to the origin, which means that choosing θ randomly will bring to poor performance on both tasks.

Figure 5 shows a fully converged network after 1,000 iterations, in which θ has been evolved according to the MTES policy. The Pareto front is far from the origin and almost every perturbation is feasible. We can also see that the network is exploring solutions on both tasks, which is what we expected. The convergence of the Pareto front is showed in Fig. 6, where it is possible to see that the front is moving in a direction common to both tasks. For a more quantitative analysis, the hypervolumes have been computed and are shown in Table 1. Although at first the index grows bigger, it may seem surprising that it does not show a monotonic behaviour. This however is not an issue because the score

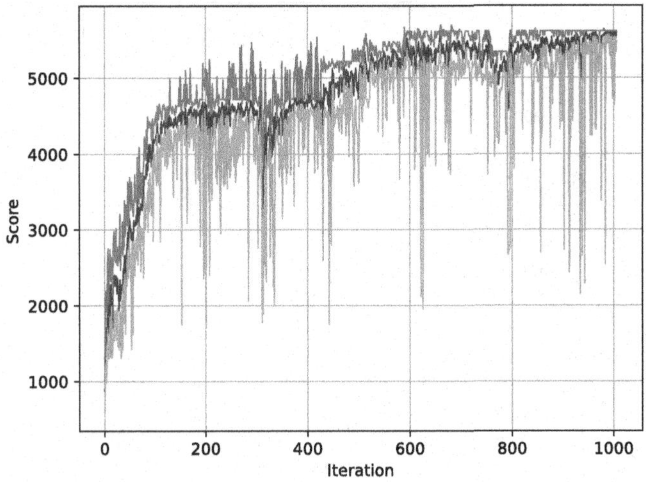

Fig. 2. River Raid scores obtained in 200 episodes per iteration. Red line: minimum score; blue line: mean score; green line: maximum score. (Color figure online)

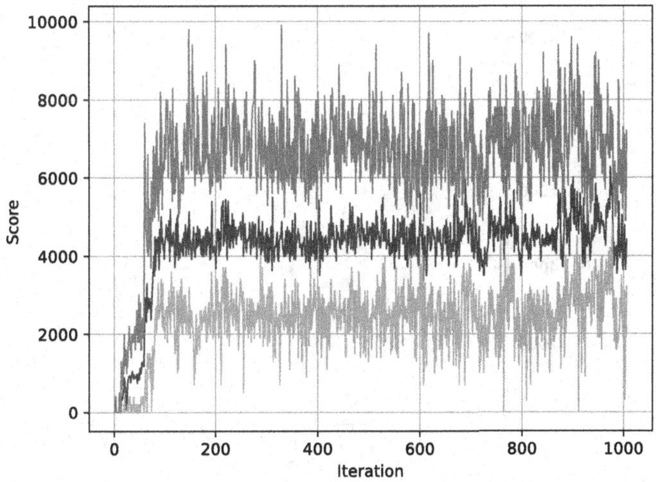

Fig. 3. Zaxxon scores obtained in 200 episodes per iteration. Red line: minimum score; blue line: mean score; green line: maximum score. (Color figure online)

obtained by each perturbation is stochastic. Furthermore, we are evaluating random perturbations of $\theta_{P(k)}$, so we are exploring parameter space without a prior knowledge of how it will affect the scores.

The resulting values for the Kullback-Leibler Divergence D_{KL} are shown in Table 2. It should not be surprising that this index is greatest between generations 1 and 125 because, as already shown, the DNN evolved fast during the first generations. After that, divergence values are almost zero for τ_2 (Zaxxon),

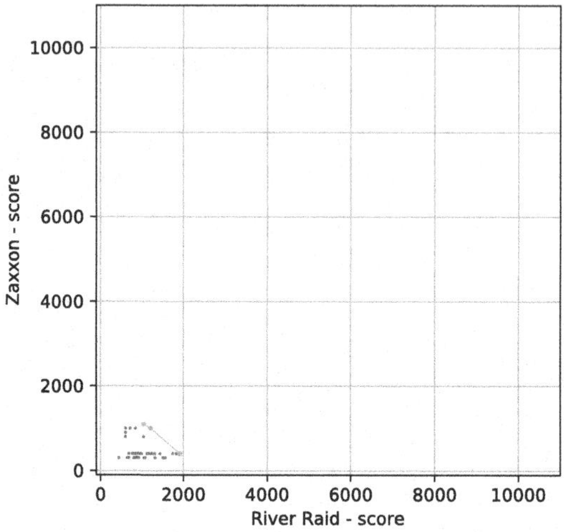

Fig. 4. Evaluation of 5000 perturbations at generation 1. Only feasible points have been plotted. Red line: Pareto front (Color figure online)

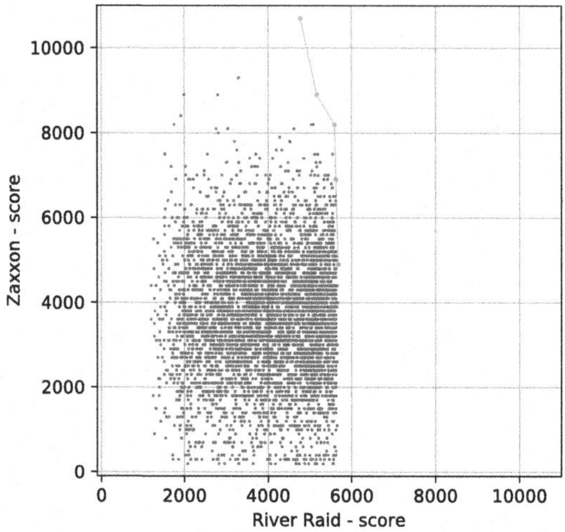

Fig. 5. Evaluation of 5000 perturbations at generation 1000. Only feasible points have been plotted. Red line: Pareto front (Color figure online)

in accordance with the fact that the DNN does not improve its performance in playing Zaxxon after iteration 200. Instead, it becomes better in River Raid, leading to D_{KL} values greater than zero for τ_1.

In the reminder the MTES results on these 4 tasks are presented: River Raid, Zaxxon, Centipede, and Seaquest. Figure 7 shows how the DNN has evolved and learned to play these games.

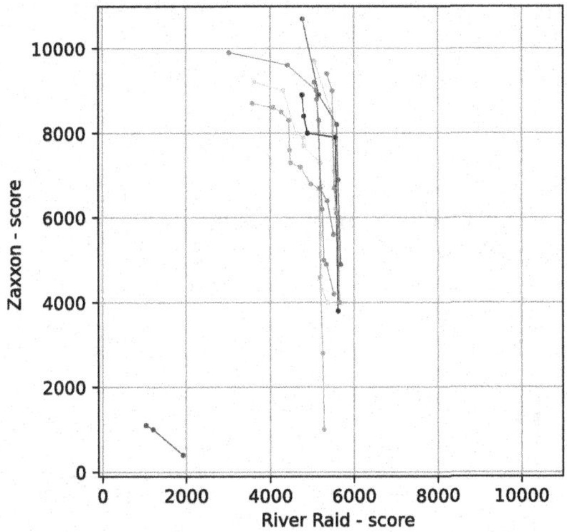

Fig. 6. Evolution of the Pareto front at different iterations. Red line: it. 1; orange line: it. 125, yellow line: it. 250, green line: it. 375, lime line: it. 500, cyan line: it. 625, blue line: it. 750, pink line: it. 875, purple line: it. 1000 (Color figure online)

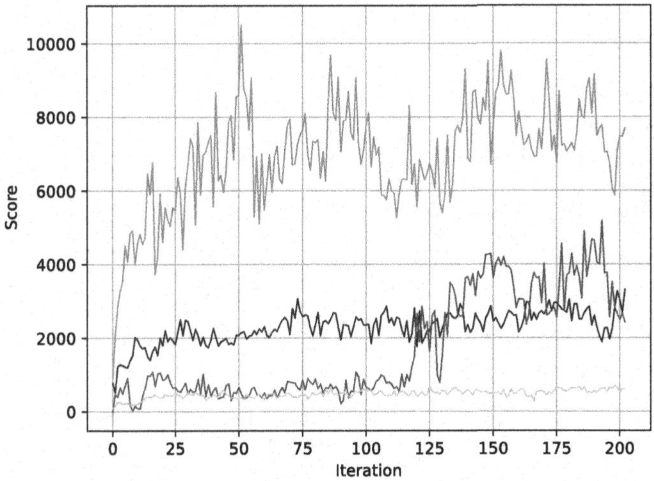

Fig. 7. Game mean score obtained in 200 episodes per iteration. Blue line: River Raid mean score; red line: Zaxxon mean score; green line: Centipede mean score; orange line: Seaquest mean score. (Color figure online)

Our implementation of MTES suffers the same problem of ES: Seaquest gets stuck on a local maxima and the DNN is unable to explore other strategies even though the other 3 games push θ to evolve into some strategies which might bring novelty to Seaquest.

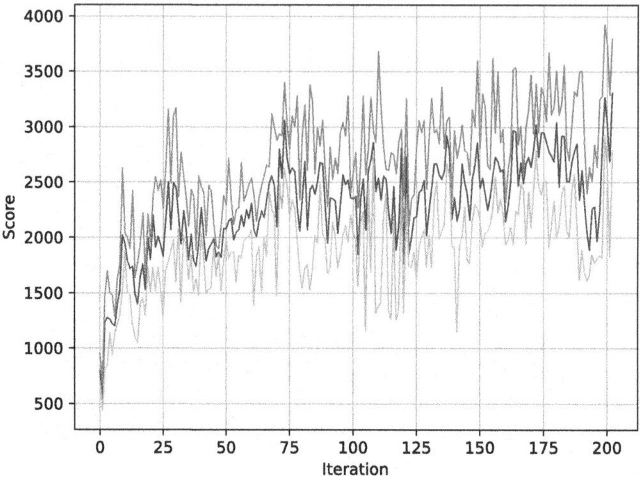

Fig. 8. River raid scores obtained in 200 episodes per iteration. Red line: minimum score; blue line: mean score; green line: maximum score. (Color figure online)

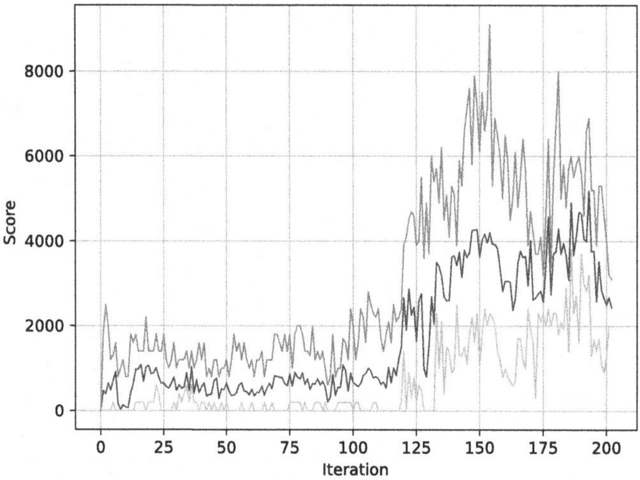

Fig. 9. Zaxxon scores in 200 episodes per iteration. Red line: minimum score; blue line: mean score; green line: maximum score. (Color figure online)

We do not show the Pareto fronts for this scenario because the points belong to a 4-dimensional space. However, hypervolumes are computed and their results are shown in Table 3. The index is almost the same for iterations 10 and 100, but then the mean score in Zaxxon has a sudden positive step around iteration 125, which leads to a higher hypervolume at iteration 200.

A further confirm of these results comes from D_{KL} indexes shown in Table 4. In the first iterations the DNN obtains the best results for Centipede: its D_{KL}

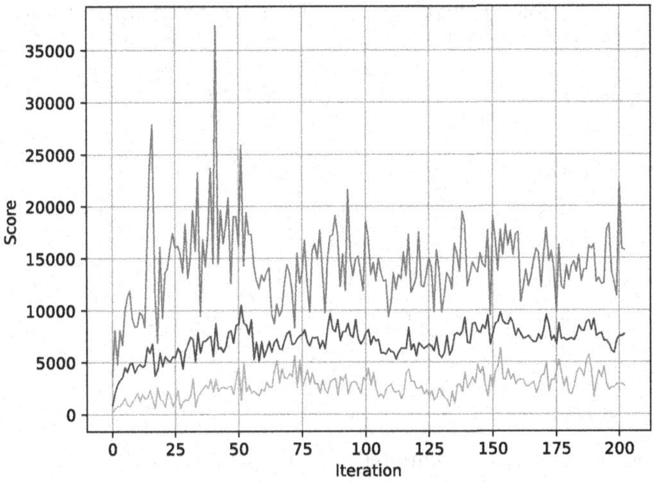

Fig. 10. Centipede scores obtained in 200 episodes per iteration. Red line: minimum score; blue line: mean score; green line: maximum score. (Color figure online)

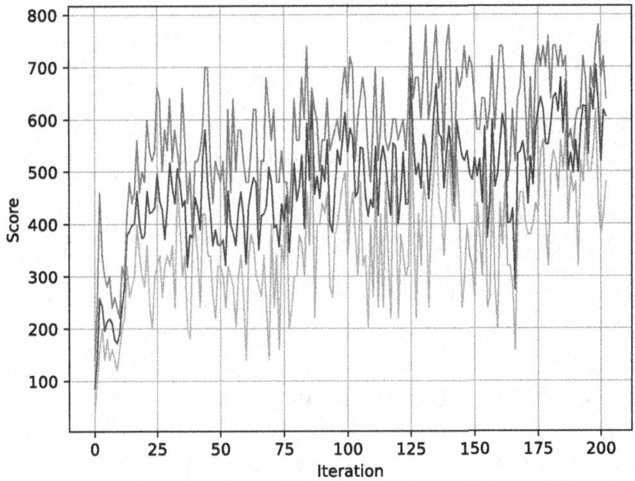

Fig. 11. Seaquest scores obtained in 200 episodes per iteration. Red line: minimum score; blue line: mean score; green line: maximum score. (Color figure online)

from iteration 1 to 100 is high. After that, Centipede score do not increase any more, accordingly to almost zero values of D_{KL} for task τ_3. The highest value for the divergence is the one relative to Zaxxon score evaluated from iteration 100 to 200: this is consistent with the positive step affecting Zaxxon score around iteration 125.

Table 2. Kullback-Leibler Divergence computed for consecutive iterations in the 2 tasks scenario. The chosen ϵ is 0.01.

From iter.	To iter.	D_{KL} for τ_1	D_{KL} for τ_2
1	125	6.84	2.43
125	250	0.13	0.04
250	375	0.41	0.02
375	500	1.04	0.02
500	625	0.17	0.03
625	750	0.05	0.03
750	875	0.09	0.04
875	1000	0.13	0.05

Table 3. Hypervolume values computed for iterations from 10 to 200 in 4 tasks scenario.

Iteration	Hypervolume H $[10^{14}]$
10	2.15
100	1.66
200	5.77

Table 4. Kullback-Leibler Divergence computed for consecutive iterations in 4 tasks scenario (River Raid, Zaxxon, Centipede, Seaquest). The chosen ϵ is 0.01.

From iter.	To iter.	D_{KL} for τ_1	D_{KL} for τ_2	D_{KL} for τ_3	D_{KL} for τ_4
1	100	0.19	0.11	1.14	0.07
100	200	0.49	1.46	0.18	0.04

5 Discussions and Conclusions

This paper lays the foundations for a new approach to transfer learning based on neuroevolution. We have investigated how to adopt MTES to extend ES for MultiTask Learning with Deep Neural Networks. Experiments on Atari games show that a single DNN trained with MTES is able to learn more than one game simultaneously in the same time required to master only one game. The Pareto front convergence towards a common direction to every addressed task indicates that the DNN is learning some features intrinsically shared between the chosen tasks. Among its other properties, it does not need temporal discounting and, as for ES, it is highly parallelizable. Hypervolumes growing trend shows that our model is effectively learning the tasks, whereas the D_{KL} index shows the amount of information extracted between two different generations.

In future works, it should be theoretically proven under which conditions the used approach leads to a convergence of the Pareto front in a direction that is common to each game. We also plan to develop and compare new evolution strategies exploiting Pareto front theory. In particular, instead of using ES, we would try evolving vector θ by means of selecting the perturbation that satisfies some properties, for example it should be selected the perturbation having the maximum 2-norm distance from the origin.

References

1. Auger, A., Bader, J., Brockhoff, D., Zitzler, E.: Hypervolume-based multiobjective optimization: theoretical foundations and practical implications. Theoret. Comput. Sci. **425**, 75–103 (2012)
2. Bellemare, M.G., Naddaf, Y., Veness, J., Bowling, M.: The arcade learning environment: an evaluation platform for general agents. J. Artif. Intell. Res. **47**, 253–259 (2013)
3. Brockman, G., et al.: Openai gym (2016)
4. Caruana R.: Multitask learning. In: Thrun S., Pratt L. (eds) Learning to Learn, pp. 95–133. Springer, Boston (1998). https://doi.org/10.1007/978-1-4615-5529-2_5
5. Conti, E., Madhavan, V., Petroski Such, F., Lehman, J., Stanley, K.O., Clune, J.: Improving exploration in evolution strategies for deep reinforcement learning via a population of novelty-seeking agents. In: NeurIPS 2018, Montreal, Canada (2018)
6. Fonseca, C.M., Paquete, L., López-Ibáñez, M.: An improved dimension-sweep algorithm for the hypervolume indicator. In: 2006 IEEE International Conference on Evolutionary Computation, pp. 1157–1163 (2006)
7. Goodfellow, I., Bengio, Y., Courville, A.: Deep Learning. MIT Press (2016) http://www.deeplearningbook.org
8. Kullback, S., Leibler, R.A.: On information and sufficiency. Ann. Math. Statist. **22**(1), 79–86 (1951). https://doi.org/10.1214/aoms/1177729694
9. Kumar, M.P., Packer, B., Koller, D.: Self-paced learning for latent variable models. In: Lafferty, J.D., Williams, C.K.I., Shawe-Taylor, J., Zemel, R.S., Culotta, A. (eds.) Advances in Neural Information Processing Systems, vol. 23, pp. 1189–1197. Curran Associates, Inc. (2010)
10. Maurer, A., Pontil, M., Romera-Paredes, B.: Sparse coding for multitask and transfer learning. In: Dasgupta, S., McAllester, D. (eds.) Proceedings of the 30th International Conference on Machine Learning. Proceedings of Machine Learning Research, vol. 28, pp. 343–351. PMLR, Atlanta, Georgia, USA, 17–19 June 2013. http://proceedings.mlr.press/v28/maurer13.html
11. Mnih, V., et al.: Asynchronous methods for deep reinforcement learning. In: International Conference on Machine Learning (2016)
12. Mnih, V., et al.: Human-level control through deep reinforcement learning. Nature **518**(7540), 529–533 (2015). https://doi.org/10.1038/nature14236
13. Murugesan, K., Carbonell, J.: Self-paced multitask learning with shared knowledge. IJCAI-17 (2017)
14. Romera-Paredes, B., Aung, H., Bianchi-Berthouze, N., Pontil, M.: Multilinear multitask learning. In: Dasgupta, S., McAllester, D. (eds.) Proceedings of the 30th International Conference on Machine Learning. Proceedings of Machine Learning Research, vol. 28, pp. 1444–1452. PMLR, Atlanta, Georgia, USA, 17–19 June 2013. http://proceedings.mlr.press/v28/romera-paredes13.html
15. Ruder, S.: An overview of multi-task learning in deep neural networks. CoRR (2017)

16. Salimans, T., Ho, J., Chen, X., Sidor, S., Sutskever, I.: Evolution strategies as a scalable alternative to reinforcement learning. arXiv e-prints arXiv:1703.03864, March 2017
17. Schmidhuber, J.: Ultimate cognition à la gödel. Cognitive Comput. **1**(2), 177–193 (2009). https://doi.org/10.1007/s12559-009-9014-y
18. Silver, D., et al.: A general reinforcement learning algorithm that masters chess, shogi, and go through self-play. Science **362**, 1140–1144 (2018)
19. Stanley, K., Clune, J., Lehman, J., Miikkulainen, R.: Designing neural networks through neuroevolution. Nat. Mach. Intell. (2019). https://doi.org/10.1038/s42256-018-0006-z
20. Stracquadanio, G., Nicosia, G.: Computational energy-based redesign of robust proteins. Comput. Chem. Eng. (2010). https://doi.org/10.1016/j.compchemeng.2010.04.005
21. Zhang, Y., Yang, Q.: An overview of multi-task learning. Nat. Sci. Rev. **5**(1), 30–43 (2018). https://doi.org/10.1093/nsr/nwx105

Vital Prognosis of Patients in Intensive Care Units Using an Ensemble of Bayesian Classifiers

Rosario Delgado[1], J. David Núñez-González[2(✉)], J. Carlos Yébenes[3], and Ángel Lavado[4]

[1] Department of Mathematics, Universitat Autònoma de Barcelona, Campus de la UAB, 08193 Cerdanyola del Vallès, Spain
delgado@mat.uab.cat

[2] Department of Applied Mathematics, Engineering School of Gipuzkoa, University of the Basque Country (UPV/EHU), Leioa, Spain
josedavid.nunez@ehu.eus

[3] Sepsis, Inflammation and Critical Patient Safety Research Group, Critical Care Department, Hospital de Mataró, Mataró, Spain
jyebenes@csdm.cat

[4] Information Management Unit, Maresme Health Consortium, Hospital de Mataró, Mataró, Spain
alavado@csdm.cat

Abstract. An Ensemble of Bayesian Classifiers (EBC) is constructed to perform vital prognosis of patients in the Intensive Care Units (ICU). The data are scarce and unbalanced, so that the size of the minority class (critically ill patients who die) is very small, and this fact prevents the use of *accuracy* as a measure of performance in classification; instead we use the *Area Under the Precision-Recall curve* (AUPR). To address the classification in this setting, we propose the use of an ensemble constructed from five base Bayesian classifiers with the weighted majority vote rule, where the weights are defined from AUPR.

We compare this EBC model with the base Bayesian classifiers used to build it, as well as with the ensemble obtained using the mere majority vote criterion, and with some state-of-the-art machine learning supervised classifiers. Our results show that the EBC model outperforms most of the competing classifiers, being only slightly surpassed by Random Forest.

Keywords: Bayesian Classifier · Ensemble · Area Under the Precision-Recall curve · Majority vote · Vital prognosis · ICU

1 Introduction

Medical attention to critically ill patients admitted in Intensive Care Units (ICU) is a costly and complex process [13,16]. Not only patient age and comorbidities

G. Nicosia et al. (Eds.): LOD 2019, LNCS 11943, pp. 619–630, 2019.
https://doi.org/10.1007/978-3-030-37599-7_51

or number of organ failures at admission are related to mortality. Delay on attention or inadequate management, are directly associated with mortality and with an increased length of stay, costs and a decrease of quality of life at hospital discharge in survivors [5,8,15]. Protocolization of care processes is essential for safety and best quality attention. But implementation of protocols can be affected by structural or situational scenarios as work overload, health care workers ratio and expertise or organizational models [17,22]. Tools to evaluate and compare deviation of results can alert about the need to monitor specific clinical pathways. The Acute Physiology and Chronic Health Evaluation (APACHE) is the gold standard scoring system to predict the probability of death over a 24-hour period from ICU admission. Although posterior versions have been introduced, it is the APACHE II [14] which is the one used mostly today. A logistic regression model validated on previous groups of patients is currently used to predict the probability of death from APACHE II [18]. However, neither this score nor its successive versions are completely satisfactory, and the search for new methods to evaluate therapeutic procedures used in critical patients that could provide more accurate estimations continues.

Besides, Bayesian networks, particular case of which are Naive Bayes, are one of the most appropriate methodologies to model situations under uncertainty, as is the case of health. The number of papers related to the use of quantitative methods to help with medical diagnosis and decision-making in health centers and hospitals has been increasing in recent years (see for instance [1] and the reviews [7] and [3]). Some of them apply methodologies of *machine learning* to deal with different diseases, and in particular a few include Bayesian networks, which are gaining popularity due to their versatility and power. Just to mention some examples, Bayesian networks have been used in public health evaluation [21], for risk assessment with emerging diseases [24], or for medical diagnosis [4]. In [9] the authors use Naive Bayes, Support Vector Machines, Gaussian Mixture Models and Hidden Markov Models, as predictive tools for the inference of lactate level and mortality risk regarding sepsis. A Dynamic Bayesian Network is applied to predict sequences of organ failures in patients admitted to ICU in [19]. A Random Forest model was used to predict mortality in emergency department of an hospital of patients with sepsis in [23], and a Support Vector Machine has been recently used in [12] to build a model for predict infection at the emergency department from free text data in addition to vital sign and demographic characteristics.

This paper aims to explore the ability of an Ensemble of Bayesian Classifiers (EBC) to estimate the risk of death of patients admitted to ICU. Indeed, for vital prognosis of critical patients, in a first step we construct a Bayesian classifier for the variable *Result* (live/die), which is an ensemble of five base Bayesian classifiers, by using the weighted majority vote rule with appropriate weights. Further, in a second step, we predict destination at hospital discharge in case the vital prognosis is *live*, while we predict the cause of death in case the vital prognosis is *die*. Up to our knowledge, this is a novel approach for estimating the risk of death of critically ill patients, as well as for predicting the destination for survivors or the cause of death, as appropriate.

As the performance of prediction models seems likely to be, in general, dependent on the context, it appears as necessary conducting specific research to compare different models for vital prognosis of patients admitted to ICU. For that, we compare our EBC model with other state-of-the-art machine learning methodologies, such as Neural Networks, Support Vector Machines and Random Forests, without intending to be exhaustive, but rather to highlight its strengths and weaknesses. The results obtained are promising and the EBC model has been successfully used to obtain some relevant conclusions from the medical point of view and the management of the ICU, and a computer application has been developed to facilitate communication between the expert system whose inference engine is the EBC model, and the human experts.

2 Materials and Methods

The Database Set

Information relating to 1,354 critical patients at the ICU unit of the Mataró Hospital (Mataró, Spain) from years 2016 and 2017, has been used in this study to associate patients' syndromic evaluation result and APACHE II, among other features of predictive relevance chosen according to medical experts, with mortality and destination at hospital discharge or cause of death, as appropriate (see Table 1). Features are grouped into "Epidemiological characteristics", "Syndromic evaluation" and "Others".

All variables were of type factor, or have been transformed into factor by a discretization procedure. In particular, the variable "Charlson", that can show integer values from 0 to 29, has been grouped into 5 categories: 0, 1, 2, 3 and > 3. Age has been categorized by using the usual intervals associated with APACHE II measurement, while APACHE II itself takes integer values between 0 and 67 and it has also been discretized. Missing values are relatively rare, and only appear in 9 of the 24 variables, none of them variables of "Syndromic evaluation". Note that coherently, patients with missing values in variable *Result* also have missing values in *Destination* and *Cause of death.*

Model Building

Bayesian networks (BN) are "white-box" graphical models representing the probabilistic relationships among variables affecting a phenomenon, which are used for probabilistic inference. For a set of random variables $V = \{X_1, \ldots, X_n\}$, a BN is a model that represents their joint probability distribution P. The graphical part of the model consists of a *directed acyclic graph (DAG)*, whose n nodes represent the random variables. The directed arcs among the nodes represent conditional dependencies governed by the **Markov condition** that establishes that each node in the DAG is independent of those who are not its descendents

Table 1. Variables of the database set, their values and percentages. Charlson: Comorbidity score [2]; SCA: Acute coronary syndrome; I_Resp: Respiratory failure; Inest_HD: Shock; Dism_Consciencia: Coma; I_Renal: Renal failure; I_HPT: Liver failure; ACR: Cardiorespiratory arrest; Postop_Electiu: Elective surgery; Alt_Ritme_Card: Cardiac arrhythmia; TCE: Cranial trauma; Policot_No_TCE: Body trauma without cranial trauma; Intox: Intoxication; SEM 1: extra-hospital attention; SEM 2: derived from other hospital. Destination = "morgue" if Result = "die". Cause of death = "not dead" if Result = "live"; moreover, we merge classes "septic complications" and "non-septic complications" (2% each) into "complications".

Variables	Values and percentages (among non-missing data)
Epidemiological characteristics	
F_1 : Sex	man (63%), woman (37%)
F_2 : Age	<45(9%), [45,55)(10%), [55,65)(18%), [65,75)(25%), [75,85)(26%), ≥ 85(12%)
F_3 : Charlson	0 (30%), 1 (25%), 2 (16%), 3 (11%), >3 (18%)
Syndromic evaluation (% of "yes")	
F_4 : SCA (18%), F_5 : I_Resp (33%), F_6 : Inest_HD (25%), F_7 : Dism_Consciencia (7%),	
F_8 : I_Renal (4%), F_9 : I_HPT (0.1%), F_{10} : ACR (5%), F_{11} : Postop_Electiu (7%),	
F_{12} : Alt_Ritme_Card (4%), F_{13} : TCE (0.1%), F_{14} : Policot_No_TCE (1%),	
F_{15} : Intox (1%), F_{16} : Other causes (7%)	
Others	
F_{17} : Origin	plant (21%), operation room (15%), SEM1 (1%), SEM2 (21%), emergencies (42%)
F_{18} : Category	elective surgery (8%), urgent surgery (10%), coronary (16%), medical (64%), trauma (2%)
F_{19} : Infection	yes (36%), no (64%)
F_{20} : Therapy requirement	coma (21%), stable (22%), unstable (25%), post-surgery observation (6%), medical surveillance (26%)
F_{21} : APACHE II	< 5 (8%), [5,10) (27%), [10,15) (24%), [15,20) (20%), [20,25) (11%), [25,30) (6%), [30,35) (3%), ≥ 35 (1%)
Consequences	
Result	live (85%), die (15%)
Destination	origin (80%), referential hospital (5%), morgue (15%)
Cause of death	reason for admission (11%), complications (4%), not dead (85%)

given its parents are known. When a Bayesian network is used to classify cases into a set of categories or classes, we term it *Bayesian classifier*.

(Bayesian) inference is the term used to refer to the update of probabilities of the network from a given evidence: we compute *a posteriori* probabilities from evidences and *a priori* probabilities. Prediction of a query variable X given the evidence E is the instantiation of X with the largest posterior probability, and this probability is said to be the *confidence level* of the prediction.

For vital prognosis of critical patients, we learn five different base Bayesian classifiers, say $\text{BC}_1, \ldots, \text{BC}_5$, from data, by considering the features and the class variable *Result*. This allows further enquiry into the causal links between the features and the vital prognosis, being this an advantage over typically "black-box" machine learning methods, such as Neural Networks, which are unable to provide explanations for their predictions. The structure of each of the five Bayesian networks is learned by maximizing a specific score function with some restrictions on the allowed directed arcs between nodes, being among them the Naive Bayes and the Tree Augmented Naive (TAN). More specifically, in Table 2 we report the traits of construction of these models, including the score function and the restrictions on the allowed directed arcs, in the form of *whitelist/blacklist* of forced/forbidden directed arcs. Maximum Likelihood Estimation is used to estimate their parameters. Note that both, BC_2 and BC_4 are Bayesian network-augmented naive Bayes [6] since the class variable is assumed to be a root node parent of every feature, and the subgraph of the features is an unrestricted Bayesian network.

Table 2. Traits of the five base Bayesian networks used to construct the Ensemble Bayesian classifier.

Model	Score	Restriction on the directed arcs
BC_1 (Naive)		Whitelist: from class variable to features. Blacklist: the rest
BC_2	BIC	Whitelist: from class variable to each feature
BC_3	AIC	Blacklist: from each feature to class variable
BC_4	AIC	Whitelist: from class variable to each feature
BC_5 (TAN)		Whitelist: from class variable to each feature
		Each feature has an extra incoming arc from other feature

Naive Bayes assumes that features are independent of each other given the class, which can be unrealistic in many applications. The other four models in Table 2 are different attempts to improve classification by relaxing this assumption and trying, at the same time, to maintain simplicity and efficiency as much as possible. In particular, TAN relaxes the feature independence assumption of the Naive Bayes through a tree structure, in which each feature only depends on the class and one other feature.

We construct the EBC model for the class variable *Result* (denoted by **M1**) from the five base classifiers $BC_1, \ldots, BC_5$, with the *weighted majority vote* criterion, which is a single-winner voting system but in which more power is given to more "competent" base classifiers.

Bearing in mind that the combination of unbalanced data (15% "die" in variable *Result*) and a small sample size (1,354) prevents the use of *accuracy* as an evaluation metric in classification, we follow [10] when considering a measure based on *Recall* (also called *Sensitivity*) and *Precision*, with "positive class" the minority class *die*, which provides a good representation of performance assessment in the binary classification: the *Area Under the Precision-Recall curve* (AUPR), being the *Precision-Recall* (PR) curve that obtained by plotting *Precision* over *Recall*. The PR curve provides a more informative picture of the performance of the classifier than the Receiver Operator Characteristic (ROC) curve when dealing with highly skewed datasets, as is our case.

Concretely, for the construction of the ensemble EBC, base classifier i has an assigned weight w_i which is obtained from its estimated AUPR, denoted by $A_i \in [0, 1]$, in the following way:

$$w_i = \frac{h_i}{\sum_{j=1}^{5} h_j}, \text{ where } h_i = \log\left(\frac{\frac{1}{2}(A_i + 1)}{1 - \frac{1}{2}(A_i + 1)}\right). \quad (1)$$

Note that $\frac{1}{2}(A_i + 1) \in [0.5, 1]$, and then $\frac{\frac{1}{2}(A_i+1)}{1-\frac{1}{2}(A_i+1)} \geq 1$ and consequently $h_i \geq 0$. This transformation is a dilatation since if $A_i < A_j$, therefore $h_j - h_i > A_j - A_i > 0$. With this assignment of weights to the component classifiers of the ensemble, we magnify their relevance based on AUPR values. Fixed a critical patient and a class j, let introduce the **discriminant** function $D_j = \sum_{i=1}^{5} w_i\, d_{i,j}$ where $d_{i,j} = 1$ if classifier i assigns class j to the patient, and 0 otherwise, that is, D_j is the sum of weights corresponding to classifiers that assign the patient to class j. The inferred class for the given critical patient by the EBC model is taken to be the one that maximizes the discriminant function. (Note that with $w_i = w_j$, $i, j = 1, \ldots, 5$, this rule corresponds to the mere majority vote criterion; we denote the corresponding ensemble model by MV).

The processing pipeline of the vital prognosis is summarized in Fig. 1. After mortality prediction with model M1, *Destination* or *Cause of death*, as appropriate, will be predicted in a second step by using models M2 or M3, respectively, which are EBC models similars to M1, but with different class variables.

Implementation

We have implemented the algorithm in R language. For structure learning of the base Bayesian networks $BC_1, \ldots, BC_4$ we used the hill-climbing score-based structure learning algorithm, implemented by function **hc** of the *bnlearn* package [20], and for BC5 we used the **tree.bayes** function of the same package, which implements the Tree-Augmented naive Bayes classifier, while we used *gRain* package [11] to carry on the Bayesian inferences. Neural Network (NN), Support Vector Machine (SVM) and Random Forest (RF) have been constructed, respectively, with the functions **mlNnet**, **mlSvm** and **mlRforest** of the *mlearning* package of R[1], by using the default values (maximum number of iterations = 1000 for NN, radial kernel for SVM, and 500 trees to generate for RF).

Validation and Comparison with Other Classifiers

To validate the EBC model, we use $k-$fold cross-validation with $k = 10$. AUPR is used as performance measure in order to compare the EBC model with its single component base classifiers $BC_1, \ldots, BC_5$, as well as with the other machine learning methods: NN, SVM and RF. A further comparison is with the ensemble

[1] Grosjean, Ph., Denis, K.: (2013) mlearning: Machine learning algorithms with unified interface and confusion matrices. R package version 1.0-0. https://CRAN.R-project.org/package=mlearning.

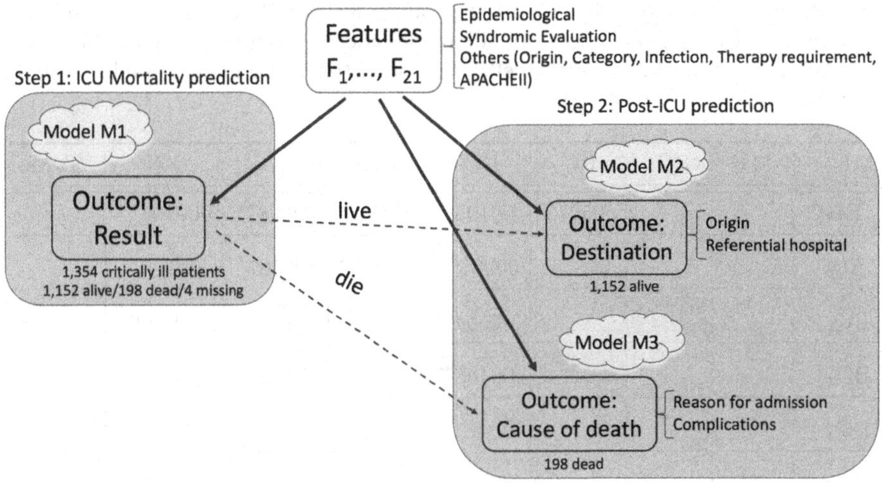

Fig. 1. Processing pipeline of vital prognosis for patients in intensive care units.

model MV, obtained from $BC_1, \ldots, BC_5$ with the mere *majority vote* criterion, for which the inferred class is the one that is the most popular among the five classifiers. Note that a tie could be possible when applying this criterion. Indeed, if the evidence consisting of the patient's features has an estimated probability equal to zero in any of the five models, that/those model/s will not be able to perform Bayesian inference to classify the patient, which could lead to an even number of class assignments and, therefore, a possible tie. The tiebreaker rule consists of simply assigning one of the categories at random, with equal probabilities.

We must highlight the fact that in the second step of the procedure, extreme imbalance of the class distribution, both in the case of *Destination* (with minority class "referential hospital") and of *Cause of death* (minority class "complications"), represents a handicap that could distort the results of the validation process. Especially dramatic is the scarcity of data in the case of *Cause of death*, since only 198 of the patients of the dataset died.

3 Results

In Table 3 below we report the statistically significant (<0.1) p-values of the Wilcoxon signed-rank non-parametric statistical hypothesis test to compare matched pairs of samples corresponding to the AUPR values obtained through the k-fold cross validation, for the EBC model M1 and the rest of classifiers. The alternative hypothesis of the test is *"the classifier of the row has a higher median than that of the column"*. This statistical test is used as an alternative to the paired Student's t-test when the population cannot be assumed to be normally distributed. As usual, * means statistical significance at 0.05 level, ** at 0.01 and *** at 0.001. Analogous results for model M2 are in Table 5.

Table 3. Statistically significant (<0.1) p-values for the Wilcoxon signed-rank test to compare AUPR medians between pairs of classifiers, for model M1 (variable *Result* with "die" as positive class). Empty cells correspond to non-significant p-values ($\geq$0.1). Completely empty rows/columns have been removed.

M1	MV	BC_1	BC_2	BC_3	BC_4	BC_5	NN	SVM
EBC			.05273	.00195**	.01367*	.00098***	.000195***	
MV			.01855*	.00098***	.04199*	.01367*	.00195**	
BC_1				.00684**		.01855*	.04199*	
BC_2				.00488**		.04199*	.00195**	
BC_3							.01367*	
BC_4				.00488**			.00195**	
BC_5				.02441*			.00293**	
SVM				.00684**			.00098***	
RF	.06543	.01855*	.00977**	.00195**	.00977**	.00684**	.00098***	.00293**

Randomization is used to alleviate the possible bias due to the (random) choice of the folds in the validation procedure. Indeed, we repeat the procedure a total of 10 times, using a different random seed in each case to carry out the partition of the database in the $k = 10$ folds. The results have been summarized in Table 4 (note that p-values in Table 3, respectively Table 5, correspond to the first repetition of model M1, respectively M2, in Table 4). We only show comparisons for EBC and MV (between them and with the other considered classifiers) that are statistically significant (p-value < 0.1), following the Wilcoxon signed-rank test for the AUPR values obtained with the k-fold cross validation. Here, ">" means that the AUPR median of the classifier "is bigger than". We can see that EBC is the one with the best overall performance except for Random Forest, which, in turn, is an ensemble learning method for classification consisting in constructing multiple decision trees and using the majority vote criterion to output the predicted class.

Finally, we can compare both ensemble classifiers, EBC and MV, between them by performing Wilcoxon signed-ranked tests for the difference (EBC − MV) in the number of significant comparisons in Table 4. p-values for these tests can be found in Table 6. We see from them that EBC outperforms MV classifier, both for prediction of variable *Result* focusing on the minority class "die", and for prediction of variable *Destination* when *Result* is "live", focusing on the minority class "reference hospital".

Table 4. Statistically significant comparisons (p-value < 0.1 for the Wilcoxon signed-rank test to compare AUPR medians) for models M1 (variable *Result*) and M2 (variable *Destination*).

Reps		M1 (*Result*)	M2 (*Destination*)
1	MV >	BC_2, BC_3, BC_4, BC_5, NN	
	EBC >	BC_2, BC_3, BC_4, BC_5, NN	BC_5
		RF > MV	
2	MV >	BC_2, BC_3, BC_4, BC_5, NN	BC_1, BC_5, NN
	EBC >	BC_2, BC_3, BC_4, BC_5, NN, SVM	BC_1, BC_2, BC_5, NN
		RF > MV, **EBC**	
3	MV >	BC_1, BC_2, BC_3, BC_5, NN	BC_5, NN
	EBC >	BC_1, BC_2, BC_3, BC_5, NN	BC_5, NN
		RF > MV	
4	MV >	BC_1, BC_2, BC_3, BC_4, BC_5, NN, SVM	BC_1, BC_5
	EBC >	BC_2, BC_3, BC_4, BC_5, NN, SVM	BC_1, BC_5
5	MV >	BC_2, BC_3, BC_4, BC_5, NN, SVM	BC_1, BC_5, NN
	EBC >	BC_2, BC_3, BC_4, BC_5, NN, SVM	BC_1, BC_4, BC_5, NN
6	MV >	BC_1, BC_2, BC_3, BC_4, BC_5, NN, SVM	BC_1, BC_4, BC_5, NN
	EBC >	BC_1, BC_2, BC_3, BC_4, BC_5, NN, SVM	BC_1, BC_2, BC_4, BC_5, NN, SVM
7	MV >	BC_3, BC_4, BC_5, NN	BC_1, BC_2, BC_4, BC_5, NN
	EBC >	BC_1, BC_2, BC_3, BC_4, BC_5, NN, SVM	BC_1, BC_4, NN
		EBC > MV	
8	MV >	BC_2, BC_3, BC_4, BC_5, NN	BC_1, NN
	EBC >	BC_2, BC_3, BC_4, BC_5, NN	BC_1, BC_5, NN
		RF > MV, **EBC**	**EBC** > MV
9	MV >	BC_2, BC_3, BC_4, BC_5, NN, SVM	BC_3, BC_5
	EBC >	BC_1, BC_2, BC_3, BC_4, BC_5, NN, SVM	BC_1, BC_3, BC_4, BC_5, NN
		EBC > MV	
10	MV >	BC_2, BC_3, BC_4, BC_5, NN	BC_5
	EBC >	BC_1, BC_2, BC_3, BC_4, BC_5, NN, SVM	BC_1, BC_2, BC_4, BC_5, NN
		EBC > MV	

Table 5. Statistically significant (<0.1) p-values for the Wilcoxon signed-rank test to compare medians for AUPR between pairs of classifiers, for model M2 (variable *Destination* with "reference hospital" as positive class). Empty cells correspond to non-significant p-values (≥0.1). Completely empty rows/columns have been removed.

M2	BC_1	BC_2	BC_4	BC_5	NN
EBC				.0801	
BC_2				.0801	
BC_4				.0290*	
SVM				.0322*	
RF	.0244*	.0654	.0801	.0244*	.0322*

Table 6. Comparison between EBC and MV models, both for M1 and for M2.

	Repetition										p-value
	1	2	3	4	5	6	7	8	9	10	
Result (M1)	+1	+1	+1	−1	0	0	+5	0	+3	+4	0.02915* ✓
Destination (M2)	+1	+1	0	0	+1	+2	−2	+3	+3	+4	0.03333* ✓

4 Conclusion

Our primary goal with this work has been to demonstrate the feasibility and benefits of routinely collecting information from critically ill patients admitted to ICU. The development of automatic tools to assist in clinical decision-making remains a challenge, although steps have already been taken in this direction. Our research is directed towards the construction of a machine learning model based on Bayesian classifiers for the vital prognosis of critical patients.

In a first step, an ensemble of Bayesian classifiers (EBC) is developed to predict the risk of death. Then, another EBC model is used to predict the destination of the patient at hospital discharge if the predicted result is *live*, or the cause of death otherwise. EBC models are constructed as ensembles of five different base Bayesian classifiers, with the weighted majority vote rule with appropriate weights, which are obtained from the estimations of the AUPR values of the base classifiers. When dealing with highly unbalanced and sparse datasets, AUPR is preferred as representation of performance assessment in the binary classification to other commonly used measures, such as *accuracy* or the *Area Under the ROC* curve (AUC).

We compare the performance of each EBC with that of the base Bayesian classifiers from which it has been constructed, and with some state-of-the-art machine learning methodologies (Neural Networks, Support Vector Machine and Random Forest), finding that EBC is the one that has best overall performance except for Random Forest. So it seems that the ensemble-based technique results in good classifiers. We also have found that EBC, which uses weights obtained from the estimated AUPR values to give more power to more "competent" base classifiers in the majority vote criterion, results in a classifier that outperforms that obtained with mere majority, MV.

In future studies we hope to enlarge the whole database set to avoid the problem of scarcity, especially for the variable *Cause of death*, resulting in a classifying model that can predict much more efficiently the cause of death for patients whose vital prognosis in the ICU is *die*. Other variables could also be included in the future. Of special interest is the variable *Duration of stay in the ICU*, which has not been addressed in this preliminary study but which will undoubtedly be included in the subsequent work.

To the extent that it can help physicians in undertaking patient-tailored therapeutic decisions, and to the health authorities to manage more optimally the available resources, the local data-driven machine learning methodology introduced in this work for estimating the risk of death and predicting the destination at hospital discharge or the cause of death, as appropriate, using an ensemble of Bayesian classifiers with weighted majority vote criterion, seems to be a useful and promising tool.

Funding Information. R. Delgado and J.D. Núñez-González are supported by Ministerio de Economía y Competitividad, Gobierno de España, project ref. MTM2015 67802-P.

References

1. Barado, J., Guergué, J.M., Esparza, L., Azcárate, C., Mallor, F., Ochoa, S.: A mathematical model for simulating daily bed occupancy in an intensive care unit. Crit. Care Med. **40**(4), 1098–1104 (2012)
2. Charlson, M., Szatrowski, T.P., Peterson, J., Gold, J.: Validation of a combined comorbidity index. J. Clin. Epidemiol. **47**(11), 1245–51 (1994)
3. Chaudhry, B., et al.: Systematic review: impact of health information technology on quality, efficiency, and costs of medical care. Ann. Intern. Med. **144**, 742–752 (2006)
4. Cruz-Ramírez, N., Acosta-Mesa, H.G., Carrillo-Calvet, H., Alonso Nava-Fernández, L., Barrientos-Martínez, R.E.: Diagnosis of breast cancer using BN: a case study. Comput. Biol. Med. **37**, 1553–1564 (2007)
5. Detsky, M.E., et al.: Six-month morbidity and mortality among intensive care unit patients receiving life-sustaining therapy. A prospective cohort study. Ann. Am. Thorac Soc. **14**(10), 1562–1570 (2017)
6. Friedman, N., Geiger, D., Goldszmidt, M.: Bayesian network classifiers. Mach. Learn. **29**(2–3), 131–163 (1997)
7. Garg, A.X.: Effects of computerized clinical decision support systems on practitioner performance and patient outcomes: a systematic review. JAMA: J. Am. Med. Assoc. **293**, 1223–1238 (2005)
8. Granholm, A., Miller, M.H., Krag, M., Perner, A., Hjortrup, P.B.: Predictive performance of the simplified acute physiology score (SAPS) II and the initial sequential organ failure assessment (SOFA) score in acutely Ill intensive care patients: post-hoc analyses of the SUP-ICU inception cohort study. PLoS ONE **11**(12), e0168948 (2016). https://doi.org/10.1371/journal.pone.0168948
9. Gultepe, E., Green, J.P., Nguyen, H., Adams, J., Albertson, T., Tagkopoulos, I.: From vital signs to clinial outomes for patients with sepsis: a machine learning basis for a clinical decision support system. J. Am. Med. Inform. Assoc. **21**, 315–325 (2014)
10. He, H., Garcia, E.A.: Learning from imbalanced data. IEEE Trans. Knowl. Data Eng. **21**(9), 1263–1284 (2009)
11. Hojsgaard, S.: Graphical independence networks with the gRain package for R. J. Stat. Softw. **46**(10), 1–26 (2012)
12. Horng, S., Sontang, D.A., Halpern, Y., Jernite, Y., Shapiro, N.I., Nathason, L.A.: Creating an automated trigger for sepsis clinical decision support at emergency department triage using machine learning. PLoS ONE **12**(4), e0174708 (2017). https://doi.org/10.1371/journal.pone.0174708
13. Kerlin, M.P., Cooke, C.R.: Understanding costs when seeking value in critical care. Ann. Am. Thorac Soc. **12**(12), 1743–1744 (2015)

14. Knaus, W.A., Draper, E.A., Wagner, D.P., Zimmerman, J.E.: APACHE II: a severity of disease classification system. Crit. Care Med. **13**(10), 818–29 (1985)
15. Li, Z., et al.: A multifactor model for predicting mortality in critically ill patients: a multicenter prospective cohort study. J. Crit. Care **42**, 18–24 (2017)
16. Lone, N.I., et al.: Five-year mortality and hospital costs associated with surviving intensive care. Am. J. Respir. Crit. Care Med. **194**(2), 198–208 (2016)
17. McGlynn, E.A., et al.: The quality of health care delivered to adults in the United States. N. Engl. J. Med. **348**(26), 2635–2645 (2003)
18. Niewiński, G., Starczewska, M., Kański, A.: Prognostic scoring systems for mortality in intensive care units. The APACHE model. Anaesthesiol. Intensive Ther. **46**(1), 46–49 (2014)
19. Sandri, M., Berchialla, P., Baldi, I., Gregori, D., De Blasi, R.A.: Dynamic Bayesian networks to predict sequences of organ failures in patients admitted to ICU. J. Biomed. Inf. **48**, 106–113 (2014)
20. Scutari, M.: Learning Bayesian Networks with the bnlearn R package. J. Stat. Softw. **35**(3), 1–22 (2010)
21. Spiegelhalter, D.J.: Incorporating Bayesian ideas into healthcare evaluation. Stat. Sci. **19**, 156–174 (2004)
22. Steinberg, E.P.: Improving the quality of care. Can we practice what we preach? N. Engl. J. Med. **348**(26), 2681–2683 (2003)
23. Taylor, R.A., et al.: Prediction of in-hospital mortality in emergency department patients with sepsis: a local big data-driven, machine learning approach. Acad. Emerg. Med. **23**, 269–278 (2016)
24. Walshe, T., Burgman, M.: A framework for assessing and managing risks posed by emerging diseases. Risk Anal. **30**(2), 236–249 (2010)

On the Role of Hub and Orphan Genes in the Diagnosis of Breast Invasive Carcinoma

Marta B. Lopes[1,2(✉)], André Veríssimo[2,3], Eunice Carrasquinha[2], and Susana Vinga[2,3]

[1] Instituto de Telecomunicações, Av. Rovisco Pais 1, 1049-001 Lisbon, Portugal
marta.lopes@tecnico.ulisboa.pt
[2] INESC-ID, Instituto de Engenharia de Sistemas e Computadores - Investigação e Desenvolvimento, Rua Alves Redol, 9, 1000-029 Lisbon, Portugal
[3] IDMEC, Instituto Superior Técnico, Universidade de Lisboa, Av. Rovisco Pais 1, 1049-001 Lisbon, Portugal

Abstract. Network information is gaining importance in the generation of predictive models in cancer genomics, with the premise that prior biological knowledge offers the models interpretability and reproducibility, an invaluable contribution in precision medicine. This work evaluates the usefulness of accounting for gene network information provided by the data correlation structure and external STRING information in the classification of Breast Invasive Carcinoma (BRCA) RNA-Seq data from The Cancer Genome Atlas (TCGA) into *tumor* and *normal* tissue, by sparse logistic regression (SLR). Within the correlation-based approaches, two directions were investigated: first, imposing smaller penalties on *hub* genes, i.e., highly connected genes in the network (hubSLR); second, favouring the selection of orphan or weakly correlated genes (orphanSLR). Without loss of predictive ability, a considerable overlap between the genes selected by the methods was achieved, with fewer genes exclusively selected by each method. Besides a consensus list of genes, the complementarity offered by sets of genes exclusively selected by each model based on different network information shall be regarded as a means to enhance biological interpretability, drawing attention to genes with a known role in the network, either hubs, orphans or highly connected genes in protein-protein interaction networks. This represents a major advantage over non network-based methods, enabling the disclosure of the relevance of known gene subnetworks in the disease under study, while boosting biomarker discovery and precision medicine.

Keywords: Gene networks · Sparse logistic regression · Regularized optimization · RNA-Seq

Partially funded by H2020 (No. 633974) and the Portuguese Foundation for Science & Technology (UID/EEA/50008/2019, UID/CEC/50021/2019, UID/EMS/50022/2019, PTDC/CCI-CIF/29877/2017, PTDC/EMS-SIS/0642/2014, and SFRH/BD/97415/2013).

G. Nicosia et al. (Eds.): LOD 2019, LNCS 11943, pp. 631–642, 2019.
https://doi.org/10.1007/978-3-030-37599-7_52

1 Introduction

Precision medicine relies on the advanced discovery of putative diagnostic and prognostic cancer biomarkers based on the analysis of patients' genetic information, therefore holding the promise of tailoring clinical decisions [12]. Although individual cancer gene mutations represent potential targets for cancer therapy, they tend to participate in common biological activities such as genome maintenance, cell differentiation or growth signaling [1]. Particularly, in complex diseases, individual differences arise not from a single gene alteration but from more subtle interactions among such alterations [12, 23]. It has also been recently hypothesized through the 'omnigenic' model that cell regulatory networks are highly interconnected to the extent that any gene expressed in disease-relevant cells is likely to affect the regulation or function of core disease-related genes, and that most heritability can be explained by effects on genes outside core pathways [2]. This fosters new research directions targeting not only differentially expressed genes, but also those that, although non-differentially expressed, play nonetheless a central role in the molecular mechanism of complex biological phenomena [26].

Data dimensionality reduction is a critical step in the analysis of high-dimensional genomic data for the selection of disease-related genes. Regularization methods such as Lasso [16], elastic net [27] and fused Lasso [17], are gold standard methods to perform feature selection. However, although promoting some grouping of correlated or neighboring genes, these methods often lack interpretability, as they do not account for the gene functional interrelationships [25]. Another limitation of variable selection in genomic data is the fact that the sets of genes selected across different methods are often not same, showing little overlap of genes [4, 7, 9]. This can be due to both technical and biological laboratory issues, or statistical reasons arising from the high-dimensionality of the data in the variables space compared to the sample size [7]. Failure to identify a set of stable and robust disease biomarkers will compromise the development of new therapies and, consequently, effective disease management.

Models incorporating knowledge on the molecular pathways involving individual alterations, as well as among multiple alterations, are expected to play a role in precision therapy. For example, the incorporation of protein-protein interaction (PPI) network information into sparse logistic regression yielded competitive modeling performance against, e.g., Lasso and elastic net, in the classification of breast cancer gene expression data [4, 22, 25]. Network-based regularization methods based on Cox regression have also been proposed in the context of survival analysis, namely Net-Cox [24] and DegreeCox [19], able to account for network information as a model constraint favouring particularly relevant genes. Either gene co-expression networks or correlation-based networks extracted from the data itself are considered in Net-Cox and DegreeCox, respectively. The two methods have shown their potential to improve accuracy of survival prediction in ovarian cancer patients based on transcriptomic data, outperforming Lasso regularization in the some of the case studies evaluated, while improving interpretability of the obtained models.

This work evaluates the use of gene networks, either extracted from protein-protein interactions available networks or extracted directly from the gene correlation data, in the identification of meaningful subnetworks regulating Breast Invasive Carcinoma (BRCA). More than increased model accuracy, we are seeking improved model interpretability and disease understanding. Within data gene correlation information, particular attention will be given to the role of highly connected genes or *hubs*, and isolated or *orphan* genes, in the classification of BRCA data, following recent work on the incorporation of network-based regularizers into sparse models when the features (e.g., genes) can be represented by a graph structure [3,18,20]. This will be accomplished by using sparse logistic regression with an additional network-based constraint, conveyed in the penalty term.

2 Methods

2.1 Classification

Sparse Logistic Regression. Binary logistic regression is a classification method that describes the relationship between one or more independent variables and a binary outcome vector $\mathbf{Y} = \{Y_i\}_{i=1,\ldots,n}$, which is given by the logistic function

$$p_i = P(Y_i = 1|\mathbf{X}_i) = \frac{\exp(\mathbf{X}_i^T \boldsymbol{\beta})}{1 + \exp(\mathbf{X}_i^T \boldsymbol{\beta})}, \tag{1}$$

where $\mathbf{X}_i$, $i = 1, ..., n$, is the vector of p covariates for observation i, p_i is the probability of success (i.e., $y_i = 1$) for observation i, and $\boldsymbol{\beta} = (\beta_1, \beta_2, \ldots \beta_p)$ are the regression coefficients associated to the p independent variables. The parameters of the *sparse logistic regression* model, herein called SLR, are estimated by maximizing the log likelihood function of the logistic model given by

$$l(\boldsymbol{\beta}) = \sum_{i=1}^{n} \{y_i \log p_i + (1 - y_i) \log[1 - p_i]\} + F(\boldsymbol{\beta}), \tag{2}$$

$F(\boldsymbol{\beta})$ denoting the regularization term, which for the elastic net penalty [27] takes the form

$$F(\boldsymbol{\beta}) = \lambda\{\alpha\|\boldsymbol{\beta}\|_1 + (1 - \alpha)\|\boldsymbol{\beta}\|_2^2\}, \tag{3}$$

with $\alpha = 1$ corresponding to Lasso and $\alpha = 0$ to ridge, and the tuning parameter λ controlling the severity of the penalty, i.e., the level of shrinkage.

Prior Network Information. It is widely recognized that the predictive ability and interpretability of regression methods can benefit from prior information on the variables under consideration, which are the genes in this particular case study. Such prior knowledge can also be introduced as constraints in the cost function. Following the recent work proposed by Veríssimo et al. on network-based regularization applied to generalized linear models [18–20], we introduce a network-based constraint into the regularization term of the SLR model, herein called hubSLR. Each vertex of the network represents a gene and the corresponding vertex centrality information is obtained from the Pearson's correlation of the gene expression dataset. Therefore, the penalty term in Eq. 3 takes the form

$$F(\boldsymbol{\beta}) = \lambda \left\{ \alpha \|\mathbf{w} \circ \boldsymbol{\beta}\|_1 + (1-\alpha)\|\mathbf{w} \circ \boldsymbol{\beta}\|_2^2 \right\}. \tag{4}$$

with vector $\mathbf{w}$ representing the factors that control how much of the penalty λ affects each coefficient, and $\circ$ standing for the elementwise product. Considering the $p \times p$ data correlation matrix $\Sigma = \{\rho_{ij}\}$, with $i, j = 1, ..., p$, the weights $\mathbf{w} = (w_1, ..., w_i, ..., w_p)$ are defined as

$$w_i = \sum_{j=1}^{p} |\rho_{ij}|, \tag{5}$$

representing the vertex weighted degree, i.e., the degree of a node will be defined as the sum of all the absolute values of the correlations with that node. By taking the inverse of w_i, hub genes, here corresponding to highly correlated genes given by the Pearson's correlation, will be less penalized. The resulting vector is scaled between 0 and 1, by dividing by the maximum value. The influence this penalization has in gene selection and in the clinical outcome is then assessed by SLR.

While favouring hub genes in the classification of gene expression data seem of particular relevance, as highly connected genes are expected to play an important role in gene regulatory pathways, the role of poorly connected or isolated genes in a disease regulatory network, herein called orphan genes, has been seldom investigated. In order to evaluate the role of orphan genes in the patient outcome (*tumor* vs. *normal* tissue), orphanSLR is also evaluated, considering the original penalty vector $\mathbf{w}$ (Eq. 5). This way, less penalized genes will be the orphan genes, i.e., nodes that show fewer and/or weaker connections with the remaining genes.

Gene network information is currently available from many databases. For instance, STRING [15] for the human genome records functional protein-protein interactions that have been observed experimentally and estimated computationally. STRING congregates information from many sources (e.g., gene co-expression, co-occurrence and experimental information) into a combined score $S = 1 - \prod_k (1 - S_k)$, computed by combining the probabilities from the different evidence channels S_k, while correcting for the probability of randomly observing an interaction [10]. The $p \times p$ score matrix $\boldsymbol{S} = \{S_{ij}\}_{i,j=1,...p}$ is transformed into an $p \times p$ adjacency matrix $A = [a_{ij}]$ ($a_{ij} = 0$ if $i = j$), which encodes the weights of the connections between pairs of proteins (nodes). The final weighted degree vector $\mathbf{w}$ is obtained by $w_i = \sum_{j=1}^{p} A_{ij}$. Having computed $\mathbf{w}$, the translation of protein-protein interaction information into networks of protein-coding genes associated with the proteins considered was performed using the `biomaRt` R package [5,14].

The usefulness of accounting for external knowledge in the classification model, herein called extSLR, encompassing known relationships between pairs of genes extracted from STRING, will also be assessed in the classification of BRCA gene expression data. Therefore, genes exhibiting a larger number of known connections will be less penalized in the network-based sparse regularization term (Eq. 4).

2.2 Data

The transcriptomic data of Breast invasive carcinoma (BRCA) patients used in this work was obtained from The Cancer Genome Atlas (TCGA) Data Portal (https://cancergenome.nih.gov/). The BRCA RNA-Seq Fragments Per Kilo base per Million (FPKM) dataset was imported using the 'brca.data' R package (https://github.com/averissimo/brca.data/releases/download/1.0/). The BRCA gene expression data is composed of 57251 variables for a total of 1222 samples from 1097 individuals. From those samples, 1102 correspond to primary solid tumor, 7 to metastases and 113 to normal breast tissue. Only samples from primary solid tumor were considered for analysis. The BRCA response variable Y is binary, coded with '1' for *tumor* samples and '0' for *normal* tissue samples. FPKM normalized BRCA gene expression data were log-transformed and Z-score normalized prior to data analysis, a standard transformation when dealing with RNA-Seq data to overcome typical skewness, mean-variance-dependency or extreme values of RNA-Seq variables [28].

Classification of BRCA Data. Sparse logistic regression based on the elastic net penalty (SLR) was used to classify RNA-Seq data from patients into *normal* and *tumor* tissue samples and to identify a set of relevant genes playing a role in the disease. Three quarters of randomly selected samples were assigned to training samples for model construction, whereas the remaining samples were assigned to test samples. As improving model accuracy was not the main focus, the choice of $\alpha = 0.3$ reflects the intention to select a reasonable number of variables yielding visual and interpretable results for the example of application chosen. The optimization of the parameters λ based on the mean squared error (MSE) of classification for the models described above was performed by 10-fold cross-validation. The same procedure was applied for evaluating the hubSLR, orphanSLR and extSLR modeling performances. A log-transformed penalization vector $\mathbf{w}$ was taken for model building, then scaled between 0 and 1. A comparison between the models was performed regarding the predictive performance given by the MSE, area under the Precision-recall curve (AUC) and the number of misclassifications. As the results obtained for a given training and test set might be data-dependent, SLR, hubSLR, orphanSLR and extSLR models were built based on 100 randomly chosen training and test sets. The median MSE, AUC, misclassfications and the identity of the variables selected in more than 50% of the runs were taken for comparison of the modeling strategies employed.

The `glmnet` R package [6] was used in our study for building the above sparse logistic regression models with elastic net regularization. The $\mathbf{w}$ vector was introduced as penalty factor in the `glmnet` function. All implementations in the R statistical software [13] can be found in a R Markdown document available at http://web.tecnico.ulisboa.pt/susanavinga/LOD2019/, which allows full reproducibility of the results presented.

3 Results and Discussion

Sparse and network-based sparse logistic regression were applied to BRCA data in the classification of breast samples into *tumor* and *normal* tissue. Besides logistic regression with elastic net penalization, two network-based approaches were considered for the construction of the penalty vector **w**: first, based on the variables' correlation matrix, either imposing a lower penalty on hub genes (hubSLR) or orphan genes (orphanSLR); second, favouring the selection of known highly connected genes in the STRING network. Overall, the three network-based models built performed similarly in terms of MSE and AUC (Table 1), with SLR and orphanSLR showing better predictive performance regarding the number of misclassifications in the train set.

Table 1. Summary of SLR, hubSLR and orphanSLR models, considering the median values obtained from 100 models built on randomly generated training and test sets (Vars, variables selected; MSE, mean squared error; AUC, area under the Precision-recall curve; Miscl, misclassifications).

	# Vars	# Miscl		MSE		AUC	
		train	test	train	test	train	test
SLR	124	1	1	0.0027	0.0048	0.99	0.99
hubSLR	117	2	1	0.0028	0.0049	0.99	0.99
orphanSLR	131	1	1	0.0026	0.0049	0.99	0.99
extSLR	116	2	1	0.0028	0.0050	0.99	0.99

From the variables selected, 72 were selected in common by the four models, 2 exclusively selected by hubSLR, 5 by orphanSLR, 24 by extSLR, and none exclusively selected y SLR (Fig. 1). The normalized variable weights of the variables selected by the three methods can be found in Fig. 2, with the higher the weight, the strongest the penalty. The vertical orange lines represent the variables selected in common by all methods, whereas coloured non-orange lines stand for the variables exclusively selected by the model under consideration. For hubSLR (Fig. 2a), all gene variables selected in common by the four methods have lower weight, indicating their role in the network as highly connected genes; the two red lines correspond to the variables exclusively selected by hubSLR, only selected by imposing a lower penalty on its associated coefficients. When moving to orphanSLR (Fig. 2b), it is expected that, as for hubSLR, less penalized genes are selected, however most genes selected in common show a higher penalty and only 2 exclusively selected by orphanSLR (yellow vertical lines) have lower weights. This is a strong indication that even imposing a stronger penalty in hub genes, these might be of special relevance in the discrimination between breast *tumor* and *normal* tissue that they are still selected. Finally, variable weights used in extSLR (Fig. 2c), namely the genes exclusively selected by the method (blue vertical lines), show the lowest weights compared to, e.g., the genes selected in common by all methods. This indicates that highly connected genes in the STRING network, i.e., functional key players in the cell, and favoured in the selection procedure, indeed play a discrimination role in the classification of *tumor* and *normal* tissue. Exploring alterations to the connection structure of such genes across classes (*tumor* vs. *normal*) might disclose new directions for therapeutical research.

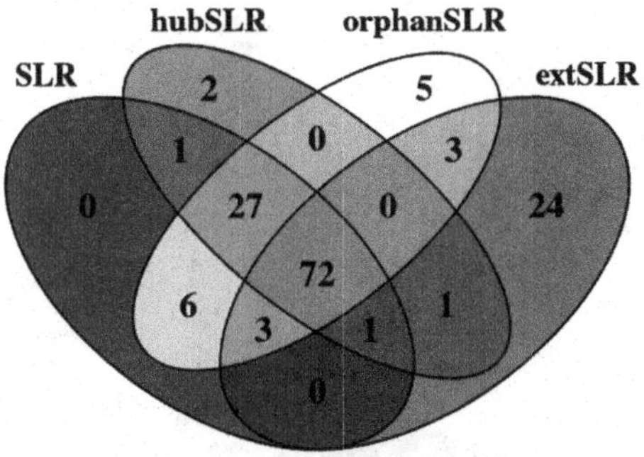

Fig. 1. Venn diagram illustrating the intersection between genes selected by SLR, hubSLR, orphanSLR and extSLR.

Table 2 represents the identity of the variables exclusively selected by each method. For a full list of the genes selected in common by the methods evaluated, please check http://web.tecnico.ulisboa.pt/susanavinga/LOD2019/. To investigate whether genes selected by hubSLR and orphanSLR have biological meaning in BRCA, and as a matter of clinical validation of our modeling approach, links to their previously described role in BRCA will be discussed next.

TMOD1 and *OR2L3* have been exclusively selected by hubSLR. While no previous report can be found relating *OR2L3* to BRCA, *TMOD1* (down regulated in BRCA) has been associated to NF-kB activation [8], which has an

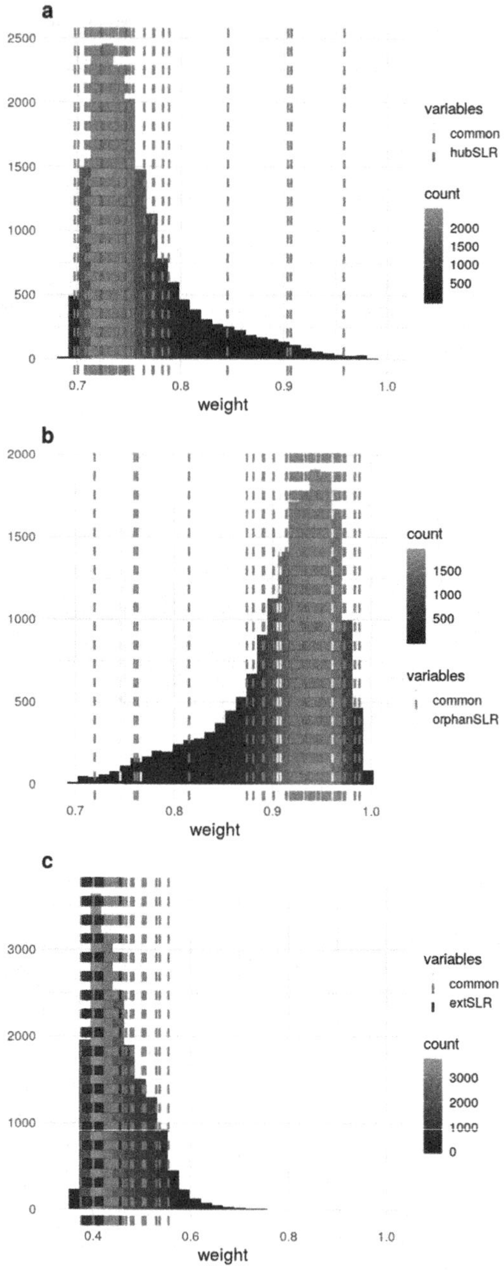

Fig. 2. Variable weights provided by correlation-based degree information where less penalized genes are (a) hubs, (b) orphans, or (c) highly connected genes provided by STRING. The weights by SLR are all equal to 1 (not represented). Coloured vertical dashed lines stand for the genes exclusively selected by the current method (hubSLR, red; orphanSLR, yellow; extSLR, blue) and in common by all methods (orange). (Color figure online)

established role in breast malignancy. Tropomodulin 1 (TMOD1) expression has been found to be regulated directly by NF-kB and significantly higher in triple-negative breast cancer (TNBC) than other breast cancer subtypes. *TMOD1* has been indicated as lead candidate molecular targets for TNBC therapy. *TMOD1* elevation was also associated with enhanced tumor growth in a mouse tumor xenograft model and in a 3D type I collagen culture [8].

Amongst the genes exclusively selected by orphanSLR, a suggestive association between SNPs in *CRHBP*, an hypothalamicpituitary-adrenal (HPA) axis gene, was found in the WISE study of Caucasians and African Americans [11]. *P4HA3* is a collagen-modifying enzyme causing 4-hydroxylation of prolines, over expressed in breast cancer tissue and associated with negative prognosis [21]. Moreover, *P4HA3* seems to be the only *P4HA* gene selectively expressed in the stroma of tumors, therefore being pointed out as a marker for malignant stroma and for stroma contributing to poor prognosis [21].

Besides the identity of the genes selected and their clinical association to BRCA, evaluating their correlation networks in the *tumor* and *normal* tissue samples is expected to provide valuable insight to disease understanding (Fig. 3). Overall, a distinct correlation pattern can be observed within tumor (Fig. 3a) and *normal* (Fig. 3b) tissue samples for the genes exclusively selected by hubSLR, orphanSR and extSLR, therefore indicating a potential role of the genes selected in the discrimination between *tumor* and *normal* tissue.

Finally, it should be noted that while the selection of previously reported genes serves the clinical validation purpose, non-reported genes, along with their gene network associations, shall be further investigated as putative target genes in BRCA, given their selection as relevant in the classification/diagnosis of breast cancer. This is particularly important in the case of orphan genes (e.g., *DEFB114* and *SCGB1D1*, showing the lowest weights; Fig. 2), as clinical relevance is often less expected from peripheral genes in the network.

Table 2. Genes exclusively selected by SLR, hubSLR, orphanSLR and extSLR.

hubSLR	*TMOD1*	*OR2L3*				
orphanSLR	*CRHBP*	*DEFB114*	*P4HA3*	*SCGB1D1*	*TCEAL6*	
extSLR	*ABCB5*	*ADH1A*	*ALDH1A2*	*ALDH1A3*	*APOB*	*CABP1*
	DMD	*F2R*	*F2RL2*	*GFAP*	*HCAR2*	*KCNE1*
	KCNIP2	*KLHL4*	*LHCGR*	*MAP1LC3C*	*MME*	*NPR1*
	PARK2	*PPARG*	*RASL11B*	*RERGL*	*SNCA*	*SPTBN1*

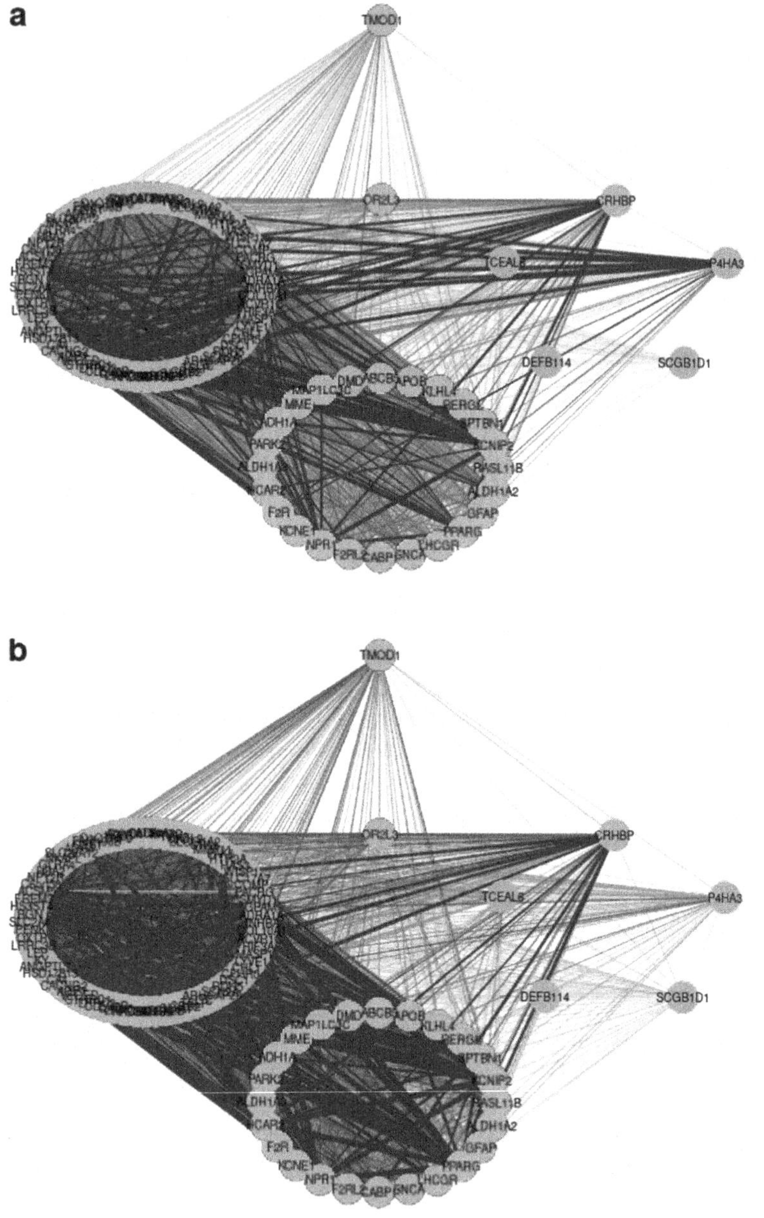

Fig. 3. Representation of the correlation-based networks obtained for the variables selected by hubSLR, orphanSLR and extSLR, for (a) tumor and (b) normal tissue samples. Coloured circles stand for the genes exclusively selected by hubSLR (red), orphanSLR (yellow), extSLR (blue) and the genes selected in common by the three models (orange). (Color figure online)

4 Conclusion

This work explores the use of prior network knowledge in the classification of breast samples into *tumor* and *normal* tissue by sparse logistic regression based on gene expression data. A penalty vector is accounted for in the model regularization term to favour the selection of either hub genes, orphan genes or known gene networks reported in the literature. Without loss of predictive ability compared to logistic regression with elastic net regularization, the results presented draw attention to the usefulness of unveiling complementary information by individual models promoting the selection of different sets of genes, regarding their role in the disease network. Such complementary information shall be regarded as a valuable contribution to disease understanding and therapy research.

References

1. Ali, M., Sjöblom, T.: Molecular pathways in tumor progression: from discovery to functional understanding. Mol. BioSyst. **5**(9), 902–8 (2009)
2. Boyle, E., Li, Y., Pritchard, J.: An expanded view of complex traits: from polygenic to omnigenic. Cell **169**, 1177–1186 (2017)
3. Carrasquinha, E., Veríssimo, A., Lopes, M., Vinga, S.: Variable selection and outlier detection in regularized survival models: application to melanoma gene expression data. In: Nicosia, G., Pardalos, P., Giuffrida, G., Umeton, R., Sciacca, V. (eds.) Machine Learning, Optimization, and Data Science, vol. 11331, pp. 431–440. Springer, Cham (2018). https://doi.org/10.1007/978-3-030-13709-0_36
4. Chuang, H., Lee, E., Liu, Y., Lee, D., Ideker, T.: Network-based classification of breast cancer metastasis. Mol. Syst. Biol. **3**, 140 (2017)
5. Durinck, S., et al.: Biomart and bioconductor: a powerful link between biological databases and microarray data analysis. Bioinformatics **21**, 3439–3440 (2005)
6. Friedman, J., Hastie, T., Tibshirani, R.: Regularization paths for generalized linear models via coordinate descent. J. Stat. Softw. **33**(1), 1–22 (2010)
7. Fröhlich, H.: Network based consensus gene signatures for biomarker discovery in breast cancer. PLoS One **6**(10), e25364 (2011)
8. Ito-Kureha, T., et al.: Tropomodulin 1 expression driven by NF-kB enhances breast cancer growth. Cancer Res. **75**(1), 62–72 (2015)
9. Kela, I., Ein-Dor, L., Getz, G., Givol, D., Domany, E.: Outcome signature genes in breast cancer: is there a unique set? Breast Cancer Res. **7**(Suppl2), 4.38 (2005)
10. Mering, C., et al.: String: known and predicted protein-protein associations, integrated and transferred across organisms. Nucleic Acid Res. **33**(Suppl 1), D433–437 (2005)
11. Nan, H., Dorgan, J., Rebbeck, T.R.: Genetic variants in hypothalamic-pituitary-adrenal axis genes and breast cancer risk in caucasians and African Americans. Int. J. Mol. Epidemiol. Genet. **675**(1), 33–40 (2015)
12. Ozturk, K., Dow, M., Carlin, D., Bejar, R., Carter, H.: The emerging potential for network analysis to inform precision cancer medicine. J. Mol. Biol. (2018). https://doi.org/10.1016/j.jmb.2018.06.016
13. R Core Team: R: A Language and Environment for Statistical Computing. R Foundation for Statistical Computing, Vienna, Austria (2017). https://www.R-project.org/

14. Durinck, S., Spellman, P.T., Birney, E., Huber, W.: Mapping identifiers for the integration of genomic datasets with the r/bioconductor package biomart. Nat. Protoc. **4**, 1184–1191 (2009)
15. Szklarczyk, D., et al.: String v10: protein-protein interaction networks, integrated over the tree of life. Nucleic Acid Res. **43**, D447–D452 (2015)
16. Tibshirani, R.: Regression shrinkage and selection via the lasso. J. Roy. Stat. Soc. B **58**(1), 267–288 (1996)
17. Tibshirani, R., Saunders, M., Rosset, S., Zhu, J., Knight, K.: Sparsity and smoothness via the fused lasso. J. Roy. Stat. Soc. B **67**(1), 91–108 (2005)
18. Veríssimo, A., Carrasquinha, E., Lopes, M., Oliveira, A., Sagot, M.F., Vinga, S.: Sparse network-based regularization for the analysis of patientomics high-dimensional survival data. bioRxiv (2018). https://doi.org/10.1101/403402
19. Veríssimo, A., Oliveira, A., Sagot, M., Vinga, S.: Degreecox - a network-based regularization method for survival analysis. BMC Bioinformatics **17**(Suppl16), 449 (2016)
20. Veríssimo, A., Vinga, S., Carrasquinha, E., Lopes, M.: `glmSparseNet` - network centrality metrics for elastic-net regularized models. Bioconductor R package version 0.99.31 (2018). https://bioconductor.org/packages/3.8/bioc/html/glmSparseNet.html
21. Winslow, S., Lindquist, K., Edsjö, A., Larsson, C.: The expression pattern of matrix-producing tumor stroma is of prognostic importance in breast cancer. BMC Cancer **16**, 841 (2016)
22. Wu, M., Zhang, X., Dai, D., Ou-Yang, L., Zhu, Y., Yan, H.: Regularized logistic regression with network-based pairwise interaction for biomarker identification in breast cancer. BMC Bioinformatics **17**, 108 (2016)
23. Zhang, Q.: A powerful nonparametric method for detecting differentially co-expressed genes: distance correlation screening and edge-count test. BMC Syst. Biol. **12**, 58 (2018)
24. Zhang, W., Ota, T., Shridhar, V., Chien, J., Wu, B., Kuang, R.: Network-based survival analysis reveals subnetwork signatures for predicting outcomes of ovarian cancer treatment. PLoS Comput. Biol. **9**(3), e1002975 (2013)
25. Zhang, W., Wan, Y., Allen, G., Pang, K., Anderson, M.: Molecular pathway identification using biological network-regularized logistic models. BMC Genomics **14**(Suppl 8), 57 (2013)
26. Zhang, W., Zeng, T., Chen, L.: Edgemarker: identifying differentially correlated molecule pairs as edge-biomarkers. J. Theorethical Biol. **362**, 35–43 (2014)
27. Zou, H., Hastie, T.: Regularization and variable selection via the elastic net. J. Roy. Stat. Soc. B **67**(2), 301–320 (2005)
28. Zwiener, I., Frisch, B., Binder, H.: Transforming RNA-Seq data to improve the performance of prognostic gene signatures. PLoS One **9**(1), e85150 (2014)

Approximating Probabilistic Constraints for Surgery Scheduling Using Neural Networks

Thomas Philip Runarsson(✉)

School of Engineering and Natural Sciences, University of Iceland, Reykjavik, Iceland
tpr@hi.is

Abstract. The problem of generating surgery schedules is formulated as a mathematical model with probabilistic constraints. The approach presented uses modern machine learning to approximate the model's probabilistic constraints. The technique is inspired by models that use slacks in capacity planning. Essentially a neural-network is used to learn linear constraints that will replace the probabilistic constraint. The data used to learn these constraints is verified and labeled using Monte Carlo simulations. The solutions iteratively discovered, during the optimization procedure, produce also new training data. The neural-network continues its training on this data until the solution discovered is verified to be feasible. The stochastic surgery model studied is inspired by real challenges faced by many hospitals today and tested on real-life data.

Keywords: Neural networks · Surgery scheduling · Stochastic mixed integer programming · Monte Carlo simulations

1 Introduction

Scheduling surgeries is undeniable a challenging task where several stakeholders [1,6,12,19], with competing objectives, are involved. The focal point of a scheduler is to keep the operating room well utilized. In general the goal is to maximize the overall throughput and so minimize the overall waiting time for patients. Patient flow from the operating rooms (ORs) impacts the workload on downstream resources such as the ward, post anesthesia care unit and intensive care unit. Scheduling too many surgeries may cause an unbalanced flow which results in resource blocking. It is of paramount important to balance the patient flow to the downstream resources to hedge against last minute surgery cancellations.

In practice, due to its complexity, nurses often create surgery schedules by hand. When scheduling they typically use the expected surgery times and the expected length of stay. This information is available from historical data and the nurse's experience. However, as pointed out by [10] using the expected surgery times will leave the ORs underutilized. Studies have revealed the importance

G. Nicosia et al. (Eds.): LOD 2019, LNCS 11943, pp. 643–654, 2019.
https://doi.org/10.1007/978-3-030-37599-7_53

of incorporating stochastic elements to the scheduling process and especially to surgery times [3,11,13–15]. The models and techniques that have been proposed vary significantly as each hospital is unique and typically requires specialized models. The flexibility of the mixed integer programming (MIP) model makes it possible to tailor the solution to the different requirements made by the hospitals. Nevertheless, many parts of the problem, remain identical [18].

Assuming that the surgery time distributions are available, for any given surgery, a number of different optimization approaches can be applied. Essentially filling the operating room can be considered a stochastic knapsack problem. In this case the capacity of the knapsack is analogous to the time available in the OR. In [14] a stochastic MIP model for the assignment of surgeries to operating days and rooms is proposed. This model is complex and considers not only the uncertainty in surgery duration but also emergency arrivals and operator's capacity. A two-stage stochastic programming model and a robust model are presented in [3]. In stochastic programming it is desirable to provide the probability distribution functions of the underlying stochastic parameters. Alternatively, realizations from these distributions, or historical data, can be used to create scenarios. In robust optimization the uncertainty is described by an upper and lower limit on stochastic parameters. The technique uses then a min-max approach guaranteeing the feasibility and optimality of the solution against all instances of the parameters within an uncertainty set. Both techniques will require various parameter settings and will not necessarily guarantee that a given OR will go over the set time limit. One way to tackle this OR time capacity constraint is by introducing a planned slack to each surgery day and room. The slack is used to hedge against uncertainty in the total surgery duration [8]. By planning a slack to each surgery day and room, the risk of overtime is minimized. The methods described above are approximate methods and will require verification using Monte Carlo simulation to validate their solution's feasibility.

Similarly to the methods described above we will be using a MIP formulation, however, all generated operating room day schedules (ORDS) will be validated using Monte-Carlo simulations. The approach improves the state-of-the art by producing guaranteed feasible solutions. Different neural-networks will be constructed in order to identify which ORDS are feasible and non-feasible. Then the neural-networks will be introduced to the MIP model in the form of linear constraints. In this manner we are able to approximate the probabilistic constraints. Our analysis is based on real life data. The data is obtained from the National University Hospital of Iceland, which contains information on all surgical activities spanning the last ten years. We will use this data to generate different instances, which will be used for our experiments. This work is the result of a close collaboration with the hospital and is inspired by the practical challenges faced by the hospital.

The paper is organized as follows. The following section provides a discussion on the sources of uncertainty faced in surgery scheduling. A complete description of the MIP model for the surgery scheduling problem is presented in Sect. 3. In

Sect. 4 the approximate ORDS classifiers using neural-networks and its implementation as MIP constraints is introduced. This is followed by the procedure used for training these models while iteratively solving the MIP model in Sect. 5. The potential and properties of this technique are investigated in Sect. 6. The paper concludes with a discussion on the main findings.

2 Stochastic Surgery Scheduling with Ward Restrictions

In practice, where schedules are frequently created manually, the expected time for the surgery is commonly used as a surrogate for the actual surgery time. This makes the process of building schedules by hand possible in practice. However, the mean is usually larger than the median, as the surgery time typically follows the log-normal distribution [20]. As a result the operating room will be underutilized. This way of scheduling effectively introduces a *planned slack* to the schedule. Nevertheless, the variance on the time of completion of the last surgery of the day is quite high. Therefore, the likelihood of going over the set time limit may also be high. Accurate Monte Carlo simulations are possible using historical data for frequently performed surgeries. The estimated mean time for any surgery is taken with respect to the operator, but there exist other causes for variations in surgery time that depend on the patient's characteristics [2]. As for the time between surgery, it may not necessarily be due to cleaning alone but the different setup required by the surgery to follow. Separating the data by these different source of variation may reduce variability.

Some surgeries may be kept open longer to accommodate for acute surgeries and so going into overtime may be acceptable for these ORs. For example, the Erasmus MC tolerates a 30% probability of going into overtime [8]. However, when many operating rooms are observed to be going over then surgeries will be canceled. Keeping the probability of going over the set time limit is, therefore, of paramount importance. The probability α of going into overtime, exceeding the time-limit T, may be bounded by the probabilistic constraint

$$\Pr\big[f(\text{ORDS}) \geq T\big] \leq \alpha \tag{1}$$

where $f(\text{ORDS})$ denotes the distribution function for the stochastic sum of all surgical procedures s, including the time between surgeries, for operator o in room r on day d.

Assuming the time distributions are available, for any given surgery, a number of different optimization approaches can be applied. However, the only exact way of solving this problem is by trying all operating room day schedules (ORDS). A technique described in [16] attempts to do precisely this. In this case a column generation approach is needed since generating all possible ORDS is intractable in general. When generating the columns a sub-problem is solved and the stochastic capacity constraint (1) is approximated using a planned slack. As a result this technique must validate the solution found using a Monte-Carlo simulation. The planned slack technique essentially assumes that the variance of the total surgery time and mean are known. The amount of slack depends on the accepted probability α that overtime occurs. When using a planned slack $\delta_{o,r,d}$ the capacity constraint is reduced to

$$\sum_{s\in\mathcal{S}^o} z_{o,r,d,s}\mu_s + \delta_{o,r,d} \leq T \quad \forall\ (o,r,d) \in \mathcal{O}\times\mathcal{R}\times\mathcal{D} \tag{2}$$

where μ_s is the mean time needed for surgery s and $z_{o,r,d,s} \in \{0,1\}$ indicates when surgery s is performed by operator o in an operating room r on day d. Determining the planned slack can be done in a straight forward manner when the distributions are known. For example in [8], the surgery times are assumed to be normally distributed. Then

$$\delta_{o,r,d} = \beta\sqrt{\sum_{s\in\mathcal{S}^o} z_{o,r,d,s}\sigma_s^2} \tag{3}$$

where σ_s^2 is the variance of surgery s. The parameter β is adjusted to achieve a suitable probability α of going into overtime. However, in general the central limit theorem will have limited application here, since the number of surgeries performed will commonly be around three or four. Furthermore, the individual surgery times are far from normal. Any reliable estimate on the planned slack, in practice, will require Monte Carlo simulations based on historical data.

In practice the ORDS selected by the scheduler is based on human experience. Although the ORDS selected are feasible they may not necessarily be optimal and this is especially true when care must be taken not to fill the wards downstream. When the wards become overcrowded surgeries will be canceled. For this an additional condition must be added, the limited ward capacity. We define a continuous variable w_d denoting the expected number of patients in the ward at any given day d. For any surgery it is assumed that it is known whether the patient will enter the ward. Historical data is then used to estimate the probability of being in the ward on day j for surgery s, denoted by $W_{j,s}$. Then w_d may be computed as follows,

$$w_d = \sum_{\substack{r\in\mathcal{R},s\in\mathcal{S}^o,j\in\{0,\ldots,m\}:\\ d-j\in\mathcal{D}}} W_{j,s}z_{o,r,d-j,s}, \quad \forall d\in\mathcal{D} \tag{4}$$

where m is some maximal number ward days, for example one week. Typically the ward is a limited resource and so an upper bound may be forced on $w_d \leq \overline{w}$ and typically also a lower bound $w_d \geq \underline{w}$, as one does not want the limited resource to be underutilized. A similar condition may be used for other costly resources in use by the hospital.

3 MIP Formulation

The flexibility of using the MIP formulation for the stochastic surgery problem will now be illustrated. In this model we will consider the elective surgery for general surgery at the National University Hospital of Iceland. Each patient is assigned to an operator (surgeon) and so there are as many patient lists as there are operators. Each operator must then be assigned to a room and day. Furthermore, patients are selected from the patient list and assigned to one of the days assigned to their operator. In general the operator room day schedule belongs to a single operator, however, it is possible to create shared ORDS in the case where two surgeons share a room during the day.

Any given surgery can only be performed once, so

$$\sum_{d\in\mathcal{D},r\in\mathcal{R},o\in\mathcal{O}:s\in\mathcal{S}^o} z_{o,r,d,s} \leq 1, \quad \forall, s \in \mathcal{S} \tag{5}$$

where $\mathcal{S}^o \subseteq \mathcal{S}$ are the surgeries on operator o's patient list. A surgeon can also not be operating in more than one room on any given day. This can be forced by the condition

$$\sum_{r\in\mathcal{R}} y_{o,r,d} \leq 1, \quad \forall d \in \mathcal{D}, o \in \mathcal{O} \tag{6}$$

where the variable $y_{o,r,d} \in \{0,1\}$ indicates that the operator is using room r on day d and is determined by the constraint

$$z_{o,r,d,s} \leq y_{o,r,d} \quad \forall (o,r,d,s) \in \mathcal{O}\times\mathcal{R}\times\mathcal{D}\times\mathcal{S}^o \tag{7}$$

Furthermore, only one operator can be in any room at any given day,

$$\sum_{o\in\mathcal{O}} y_{o,r,d} \leq 1, \quad \forall d \in \mathcal{D}, r \in \mathcal{R} \tag{8}$$

Additional conditions could be added here to force operators or patients to specific days or rooms. The flexibility of the MIP formulation is such that it will easily accommodate any additional hospital specific conditions. One such condition, we will force, is that an operator should perform at least one operation per week, or for a given week v with working days $\mathcal{D}^v$,

$$\sum_{s\in\mathcal{S}^o,d\in\mathcal{D}^v,r\in\mathcal{R}} z_{o,r,d,s} \geq 1, \quad \forall o \in \mathcal{O} \tag{9}$$

This condition can also be easily extended to allow for the operator's planned roster and days off by using operator specific days $\mathcal{D}_o^v$.

Unlike the master surgery scheduling (MSS) problem presented in [16], where the goal is to solve for the allocated time per specialty within the hospital, the formulation here focuses on master operator schedule. The problem is in essence the same in nature, however, here we will be forcing the ward capacities and limiting the number of operating rooms. Both are limited hospital resources. These conditions will have a significant effect on the schedules created since the operators have different case mixes.

The object is to maximize the throughput, or

$$\max_{z} \sum_{(o,r,d,s)\in\mathcal{O}\times\mathcal{R}\times\mathcal{D}\times\mathcal{S}^o} z_{o,r,d,s} \tag{10}$$

which is the maximum number of surgeries performed. Other equally important objectives could be considered, such as patient priority.

4 ORDS Classification

We now turn to the problem of determining which ORDS are feasible. Determining the planned slack in (2) can be almost impossible to achieve in practice and so a different technique must be devised. The closest neural-network alternative to this constraint would be the perceptron, which can be introduced as the following constraint:

$$\sum_{s\in\mathcal{S}^o} \omega_s z_{o,r,d,s} + b \geq 0 \quad \forall\ (o,r,d) \in \mathcal{O}\times\mathcal{R}\times\mathcal{D} \tag{11}$$

A natural extension to this model would be to introduce a single hidden layer, using linear activation functions, as follows:

$$\sum_{s\in\mathcal{S}^o} \omega_{s,n} z_{o,r,d,s} + b_n = h_{n,o,r,d} \quad \forall\ n \in \mathcal{N},\ (o,r,d) \in \mathcal{O}\times\mathcal{R}\times\mathcal{D} \tag{12}$$

and

$$\sum_{n\in\mathcal{N}} h_{n,o,r,d}\omega'_n + b' \geq 0 \quad \forall\ (o,r,d) \in \mathcal{O}\times\mathcal{R}\times\mathcal{D} \tag{13}$$

In the case when a non-linear activation function is used, such as the $\mathrm{ReLU}(h_{n,o,r,d})$, then constraint (12) must be written as:

$$\sum_{s\in\mathcal{S}^o} \omega_{s,n} z_{o,r,d,s} + b_n = h^+_{n,o,r,d} - h^-_{n,o,r,d} \quad \forall\ n \in \mathcal{N},\ (o,r,d) \in \mathcal{O}\times\mathcal{R}\times\mathcal{D} \tag{14}$$

where $h^+_{n,o,r,d}, h^-_{n,o,r,d} \geq 0$ and constraint (13) reduced to

$$\sum_{n\in\mathcal{N}} h^+_{n,o,r,d}\omega'_n + b' \geq 0 \quad \forall\ (o,r,d) \in \mathcal{O}\times\mathcal{R}\times\mathcal{D} \tag{15}$$

Unfortunately the solution $h_{n,o,r,d} = h^+_{n,o,r,d} - h^-_{n,o,r,d}$ is not unique, and we would like either of them to take the value zero. To achieve this one must introduce an indicator variable as proposed by [4]. That is, $q_{n,o,r,d} \in \{0,1\}$ with $h^+_{n,o,r,d} \leq Mq_{n,o,r,d}$ and $h^-_{n,o,r,d} \leq M(1 - q_{n,o,r,d})$, where M is a big positive value. Introducing these auxiliary binary variables will make the problem harder for the MIP solver, but will allow us to create more powerful classifiers.

5 Optimization and Learning ORDS

The number of feasible and infeasible ORDS can be generated using random sampling and labeling based on Monte-Carlo simulations. The training data generated will be unbalanced, with the number of infeasible solutions typically outnumbering the feasible ones. Finding feasible solutions is, however, in general easy, as assigning fewer surgeries to the ORDS will typically result in a feasible ORDS. Once the initial data set has been constructed the initial classifier may be constructed. The neural-networks learned will use Torch [17] and trained using the ADAM optimizer [9]. For all neural-network models a single output node is applied using a Tanh unit. The MIP optimization and learning procedure is described as follows:

1. Generate the initial training data $(\boldsymbol{X}, \boldsymbol{u})$ of size ℓ using randomly generated ORDS and labelling using Monte-Carlo simulations, based on the historical data available. Set counter $k \leftarrow 0$.
2. Train the initial neural-network to classify feasible ORDS and add to the MIP model as constraints, depending on the type, as described in the previous section.
3. Solve the MIP problem using a set time limit for the solver.
4. Extract the ORDS from the MIP solution and use a Monte Carlo simulation to verify that they will not result in overtime with probability α. Denote this new data by $(\boldsymbol{X}, \boldsymbol{u})^k$ (the size of this training set will be the number of rooms times planned days).
5. If the new ORDS are not all feasible continue training the classifier model using these few examples until they are satisfied. One would like them to be classified correctly, so that the infeasible ORDS do not re-appear in the next iteration. Let $k \leftarrow k + 1$, update the classifier constraints and goto step 3.

When hidden units are used dropout [21] is applied for the initial training. This was found to be essential for finding classifiers that generalize well. However, the MIP discovered ORDS must be classified correctly and so dropout is turned off when training on the MIP generated samples. Furthermore, care must be taken to not diverge too far from the initial ORDS classification model. For this reason, each time new MIP generated ORDS are discovered, the model is retrained using the initial model as the starting point (model from step 2.).

6 Experimental Study

For the experimental study schedules are created for *general surgery*. The experiments will be performed on a dual i7-5960 3.00 GHz with 4 cores and CUDA 10.0 on a GeForce RTX 2080 card. The MIP model is programmed with the Python 3.7 programming language using Gurobi version 8.1.0 [7] and PyTorch version 1.0.1.post2 [17] was used to train the neural networks.

6.1 Instance Generation

In this study a large data set was obtained from the National University Hospital of Iceland. The data consists of information on all surgical activities spanning 10 years. Only the most recent data is used, as surgery times and length of stay have been shown to either decrease or increase with time and so too do the surgeries performed.

To create a good master operator schedule, one must identify the surgeries that are most likely to occur within each planning cycle. As each patient is assigned to an operator o, each will have their own waiting list $\mathcal{S}^o$ of patients. Typically, no two surgeons will perform exactly the same set of procedures. Hence, differentiating between the surgeons is required in practice when finding the frequencies of surgeries. However, for simplification, it is assumed that each waiting list is equal in size for all surgeons. Typically the length of the operator's list is in the 100s and so one might consider the shortened lists used in this study as patients with higher and equal priority. The length of the list will be limited to twice the maximum number required to fit the available OR surgery time. In Table 1 a summary of the main characteristics of the most frequently performed surgeries, by the operators, is presented.

Table 1. Summary of main characteristics of most frequently performed surgeries by the nine operators

Operator	Number of surgery types	Mean surgery time	Mean ward probability	Mean ward length of stay
A	9	184 (min.)	0.4	2.4 (days)
B	6	137	0.5	2.9
C	9	188	0.5	2.6
D	10	167	0.7	3.2
E	11	162	0.6	3.4
F	5	214	0.4	3.7
G	7	90	0.1	2.0
H	12	143	0.3	2.5
I	13	84	0.1	2.3

For the computational experiments we will examine how different parameter settings will influence the results of the models with respect to the classifier used. Based on the scheduling practises of Landspitalinn, it is assumed that each day/room has a capacity of $T = 450$ min and that ORs are closed during the weekend. The number of operating rooms available to general surgery is two. The maximum number of ORDS created is, therefore, 10 for a 5-day working week. As discussed previously, it is not necessarily undesirable to run an OR into overtime. In the real life settings, two out of the seven ORs are kept open longer for acute surgeries that continue, when possible, after the elective program. It can be assumed here that $\alpha \approx 0.25$. The upper limit on the ward capacity will be forced at $\overline{w} = 8$, as is the case at the hospital in question. A lower bound $\underline{w} = 2$ will ensure that the ward is utilized.

Table 2. Confusion matrix using all ORDS for training perceptron.

$\ell = 544.131$	Predicted: feasible	Predicted: infeasible	
Actual: feasible	30.800	1.043	31.843
Actual: infeasible	459	511.829	512.288
	31.259	512.872	

6.2 Results

The first result presented is the classification accuracy of the three different ORDS classifiers on the complete ORDS data set. These models will then be used to model ORDS feasibility and applied to the MIP model in the final experiment that investigates their optimization performance. In these experiments models are also constructed that use only half of the complete ORDS data set for training.

Classification Accuracy. When the number of patients per operator is 20 it is possible to generate all ORDS for our test function. Each ORDS is validated using a Monte Carlo simulation using 1000 realizations. In this case there are 544.131 possible ORDS and of these the number feasible are 31.843. The training set is highly unbalanced, but this was not found to degrade the performance of the different classifiers.

The result for the first classifier, the perceptron, gives the confusion matrix presented in Table 2 for the training data. Out of the 31.843 feasible ORDS, 1.043 examples are predicted to be infeasible. This can lead to problems when these ORDS are required to construct the optimal solution. The optimal solution may no longer be reachable. The case is less serious when an infeasible ORDS is classified as feasible, these ORDS will be eliminated when the model is re-trained. To get a better understanding of the performance of this approach, to understand how critical these misclassified ORDS are, one needs to solve some problems. Clearly one would like these models to be as accurate as possible. For this multi-layered neural networks are required. In Table 3 better accuracy is achieved using a hidden layer with a linear activation function. Even greater accuracy is achieved using the non-linear ReLU activation function, rectifier

Table 3. Confusion matrix using all ORDS for training a single hidden linear layer with 100 units and 10 units.

$\ell = 544.131$	Predicted: feasible	Predicted: infeasible	Predicted: feasible	Predicted: infeasible	
	100 units		10 units		
Actual: feasible	31.370	473	31.338	505	31.843
Actual: infeasible	150	512.138	120	512.168	512.288
	31.520	512.611	31.458	512.673	

Table 4. Confusion matrix using all ORDS for training a single hidden ReLU layer with 100 units and 10 units.

$\ell = 544.131$	Predicted: feasible	Predicted: infeasible	Predicted: feasible	Predicted: infeasible	
	100 units		10 units		
Actual: feasible	31.818	25	31.594	249	31.843
Actual: infeasible	21	512.267	115	512.173	512.288
	31.839	512.292	31.709	512.422	

units, for the hidden layer. The ReLU is a popular activation function used for deep neural-networks [5]. It is essentially equivalent to the linear activation but all negative values are set to zero. Its result is given in Table 4 and achieves by far the greatest accuracy on the training data.

MIP Optimization. The models trained are now used with the MIP model and solved. In addition the models were trained with a reduced initial training set, half the size of the complete set. This will give some indication on the robustness of the classifiers. The number of patients is 20 per operator, in total 180 patients. The optimal solution to this problem may be found by searching all possible ORDS and has the value of 42 patients. The five step procedure described Sect. 5 was executed until a feasible ORDS solution is found. The procedure is repeated from step 3. as long as an optimal solution was not found within 100 s or the ORDS in the solution found were verified as being infeasible using the Monte Carlo simulation. If the ORDS are feasible no re-training is performed and the optimization continues until the optimal solution is found (MIP gap 0%). Otherwise, early stopping may speed up the generation of new infeasible ORDS data.

Table 5. The performance for the different classification models. The optimal value for this test function is 42.

Model	Hidden nodes	Training data	Best objective	MIP gap	Re-train number	Computation time (sec)
Perceptron	0	Complete	40	0%	0	21
Linear-layer	10	Complete	41	0%	0	38
Linear-layer	100	Complete	41	0%	0	41
ReLU-layer	10	Complete	40	0%	3	545
Perceptron	0	Half	40	0%	10	248
Linear-layer	10	Half	41	0%	2	78
Linear-layer	100	Half	41	0%	1	75
ReLU-layer	10	Half	40	252%	22	$\geq$10000

The result of the runs are given in Table 5. The models that require re-training are the ReLU networks and those that were trained using half the training data. With every re-training the number of misclassified ORDS is reduced. The networks using hidden linear units requires fewer re-training steps. These models also give the best objective values of 41. However, the optimal solution of 42 would appear to be unreachable. It was hoped that the more accurate the model the less likely it would be that optimal ORDS would be eliminated. That is, the more accurate ReLU network is not only more computationally expensive but does not necessarily guarantee that better solutions will be found. The computation time for the perceptron is the smallest in the case where the complete training set is used. However, the quality of the solutions found is slightly worse than that of the models using hidden linear layers. Furthermore, the perceptron requires a greater number of re-training steps when using half the training data for its initial training.

7 Discussion and Conclusion

The experimental results presented are based on one particular realization from our instance generator, however, similar results were obtained using different instances from our generator. A typical instance was chosen and the result illustrated. The known optimal solution to this problem is 42 and so the neural networks have clearly eliminated the optimal solution. Nevertheless, the quality of the solutions found are high. Typically the hospital's general surgery will schedule around 38–39 surgeries per week, which is regarded as a high number.

It was assumed that the stochastic capacity constraint for the surgeries could be replaced with a neural network classifier. Ideally one would need to know the ORDS to estimate the mean time for the stochastic sum of their surgery times and variance. From this the confidence bounds could be accurately determined or estimated using Monte Carlo simulation. Nevertheless, knowing all ORDS is intractable in general and so the problem of determining the slack for the capacity constraint remains unknown. It has been shown that the neural networks are able to classify feasible ORDS with high accuracy. The technique provides a practical approach to approximating the planned slack required in surgery scheduling. The technique was able to identify most of the quality ORDS and arrange them in such a way as to satisfy the ward constraints. Thus creating high quality surgery schedules.

Acknowledgement. The author would like to acknowledge the National University Hospital of Iceland for providing data, insights and support for this project.

References

1. Cappanera, P., Visintin, F., Banditori, C.: Addressing conflicting stakeholders' priorities in surgical scheduling by goal programming. Flex. Serv. Manuf. J. **30**(1), 252–271 (2018)

2. Cardoen, B., Demeulemeester, E., Beliën, J.: Operating room planning and scheduling: a literature review. Eur. J. Oper. Res. **201**(3), 921–932 (2010)
3. Denton, B., Miller, A.J., Balasubramanian, H.J., Huschka, T.R.: Optimal allocation of surgery blocks to operating rooms under uncertainty. Oper. Res. **58**(4-part-1), 802–816 (2010)
4. Fischetti, M., Jo, J.: Deep neural networks and mixed integer linear optimization. Constraints **23**(3), 296–309 (2018)
5. Glorot, X., Bordes, A., Bengio, Y.: Deep sparse rectifier neural networks. In: Proceedings of the Fourteenth International Conference on Artificial Intelligence and Statistics, June 2011, pp. 315–323 (2011)
6. Guido, R., Conforti, D.: A hybrid genetic approach for solving an integrated multi-objective operating room planning and scheduling problem. Comput. Oper. Res. **87**, 270–282 (2017)
7. Gurobi Optimization, LLC.: Gurobi Optimizer Reference Manual (2018). http://www.gurobi.com
8. Hans, E., Wullink, G., van Houdenhoven, M., Kazemier, G.: Robust surgery loading. Eur. J. Oper. Res. **185**(3), 1038–1050 (2008)
9. Kingma, D.P., Ba, J.: Adam: a method for stochastic optimization. arXiv:1412.6980 [cs] (2014)
10. Kroer, L.R.R., Foverskov, K., Vilhelmsen, C., Hansen, A.S., Larsen, J.: Planning and scheduling operating rooms for elective and emergency surgeries with uncertain duration. Oper. Res. Health Care **19**, 107–119 (2018)
11. Lamiri, M., Xie, X., Dolgui, A., Grimaud, F.: A stochastic model for operating room planning with elective and emergency demand for surgery. Eur. J. Oper. Res. **185**(3), 1026–1037 (2008)
12. Marques, I., Captivo, M.E.: Different stakeholders' perspectives for a surgical case assignment problem: deterministic and robust approaches. Eur. J. Oper. Res. **261**(1), 260–278 (2017)
13. Min, D., Yih, Y.: Scheduling elective surgery under uncertainty and downstream capacity constraints. Eur. J. Oper. Res. **206**(3), 642–652 (2010)
14. Molina-Pariente, J.M., Hans, E.W., Framinan, J.M.: A stochastic approach for solving the operating room scheduling problem. Flex. Serv. Manuf. J. **30**(1), 224–251 (2018)
15. Neyshabouri, S., Berg, B.P.: Two-stage robust optimization approach to elective surgery and downstream capacity planning. Eur. J. Oper. Res. **260**(1), 21–40 (2017)
16. van Oostrum, J.M., Van Houdenhoven, M., Hurink, J.L., Hans, E.W., Wullink, G., Kazemier, G.: A master surgical scheduling approach for cyclic scheduling in operating room departments. OR Spectr. **30**(2), 355–374 (2008)
17. Paszke, A., et al.: Automatic differentiation in PyTorch, October 2017
18. Riise, A., Mannino, C., Burke, E.K.: Modelling and solving generalised operational surgery scheduling problems. Comput. Oper. Res. **66**, 1–11 (2016)
19. Samudra, M., Van Riet, C., Demeulemeester, E., Cardoen, B., Vansteenkiste, N., Rademakers, F.E.: Scheduling operating rooms: achievements, challenges and pitfalls. J. Sched. **19**(5), 493–525 (2016)
20. Spangler, W.E., Strum, D.P., Vargas, L.G., May, J.H.: Estimating procedure times for surgeries by determining location parameters for the Lognormal model. Health Care Manage. Sci. **7**(2), 97–104 (2004)
21. Srivastava, N., Hinton, G., Krizhevsky, A., Sutskever, I., Salakhutdinov, R.: Dropout: a simple way to prevent neural networks from overfitting. J. Mach. Learn. Res. **15**, 1929–1958 (2014)

Determining Principal Component Cardinality Through the Principle of Minimum Description Length

Ami Tavory(✉)

Facebook Research, Core Data Science, New York, USA
atavory@fb.com

Abstract. PCA (Principal Component Analysis) and its variants are ubiquitous techniques for matrix dimension reduction and reduced-dimension latent-factor extraction. One significant challenge in using PCA, is the choice of the number of principal components. The information-theoretic MDL (Minimum Description Length) principle gives objective compression-based criteria for model selection, but it is difficult to analytically apply its modern definition - NML (Normalized Maximum Likelihood) - to the problem of PCA. This work shows a general reduction of NML problems to lower-dimension problems. Applying this reduction, it bounds the NML of PCA, by terms of the NML of linear regression, which are known.

Keywords: Minimum description length · Normalized maximum likelihood · Principal component analysis · Unsupervised learning · Model selection

1 Introduction

1.1 The Problem of Principle Component Dimension Selection

Let X be an an $n \times m$ matrix. In machine learning, it is very common to approximate it by a "simpler" product of matrices W and Z^T of lower dimensions $n \times k$ and $k \times m$, respectively (for $k \lesssim m$). Among others, these include Probabilistic Principal Component Analysis, Independent-Factor Analysis, and Non-Negative Matrix Factorization (see [3,11,25]). We will focus specifically on the simple PCA (Principal Component Analysis),

$$\arg \min_{W,Z:\ \mathrm{rank}(W)=\mathrm{rank}(Z)=k} \left\| X - WZ^T \right\|_F^2 . \tag{1}$$

The lower-dimension product is not guaranteed to losslessly approximate the original matrix. In fact, the famous Eckart-Young-Mirsky Theorem - whose properties we will use throughout - essentially guarantees some loss:

G. Nicosia et al. (Eds.): LOD 2019, LNCS 11943, pp. 655–666, 2019.
https://doi.org/10.1007/978-3-030-37599-7_54

Theorem 1. ***(Eckart-Young-Mirsky)*** *Let $X = U\Lambda V^T$ be the SVD (singular value decomposition) of X, with $\Lambda = \operatorname{diag}(\lambda_1, \ldots, \lambda_m)$, and U and V unitary. Let U_k and V_k be the matrices of the first k columns of U and V, respectively. Then*

$$\left\|X - WZ^T\right\|_F^2 \geq \|X - U_k \operatorname{diag}(\lambda_1, \ldots, \lambda_k) V_k\|_F^2 = \sum_{i=k+1}^{m} \left[\lambda_i^2\right], \tag{2}$$

and so $W = U_k \operatorname{diag}(\lambda_1, \ldots \lambda_k)$, $Z = V_k$, is optimal.

The motivation for the reduced dimension, is uncovering a structure that is, in some sense, "truer", or "more useful". To quote [14]:

"The central idea of principal component analysis is to reduce the dimensionality of a data set in which there are a large number of interrelated variables, while retaining as much as possible of the variation present in the data set. This reduction is achieved by transforming to a new set of variables, the principal components, which are uncorrelated, and which are ordered so that the first few retain most of the variation present in all of the original variables."

As the theorem shows, though, loss minimization, in itself, will not lead us to the reduced dimension - it will always favor the maximum number of components.

1.2 The Principles of MDL and NML

The MDL (Minimum Description Length) principle (see [8,9,17,20,22]) is an information-theoretic method for model selection. Probability-theory approaches to model selection - both frequentist and Bayesian - assume that there exists a true probability distribution from which the observed data were sampled. The goal is to optimize a model subject to this (indirectly-observed) distribution. MDL is similar in philosophy to Occam's Razor (see [2]). The goal is to find a model optimizing the total description length of the model and the observed data. There is no assumption that a true probability was approximated, or that it even exists. We will see that avoiding this assumption leads to a form of online optimality.

How can we objectively quantify a description length? Given a probability distribution, information theory gives an objective code length through entropy [5], but assumptions on the probability distribution are precisely what we wish to avoid. In [23], Rissanen formulated the question as a minimax problem, namely the smallest regret relative to all possible codes under mild conditions. He showed that the NML (Normalized Maximum Likelihood) (see [1,23]) is the solution to this problem.

Definition 1. ***Normalized Maximum Likelihood*** *Let X be distributed by a model specified by some parameter(s) Φ. The NML is defined as*

$$f^{\mathrm{NML}}(X) = \frac{\hat{f}\left(X ; \hat{\Phi}(X)\right)}{\int \hat{f}\left(Y ; \hat{\Phi}(Y)\right) dY}, \tag{3}$$

where

- $\hat{\Phi}(X)$ *is the maximum likelihood (ML) estimator of* Φ *given* X.
- $\hat{f}\left(Y\ ;\ \hat{\Phi}(Y)\right)$ *is the ML of* Y *assuming that the true parameters are* $\hat{\Phi}(Y)$.

The logarithm of the right-hand side of Eq. (3) is the *stochastic complexity*, and the logarithm of its denominator is the *parametric complexity*. It can be shown that choosing between different Φ based on maximizing (3), is optimal in a prequential sense (see [19]).

1.3 Main Contribution: Applying NML to PCA

Conceptually, it is possible to calculate the NML of PCA, by inserting Eq. (2) into Eq. (3). Unfortunately, evaluating the denominator requires integrating over the eigenvalues of arbitrary matrices, which is difficult. Instead, in the rest of this paper, we avoid this by bounding the NML of PCA by reducing it to the NML of linear regression (see [21]), resulting in the following theorem:

Theorem 2. *Let* $s\left(X\ ;\ k\right)$ *be the stochastic complexity of a* k*-dimensional PCA reduction of* X*. Then*

$$\begin{aligned} s(X\ ;\ k) & \\ \simeq (nm-kn)\ln\left(\sum_{i=k+1}\left[\lambda_i^2\right]\right) &+ nk\ln\left(\left\|X^TX\right\|_F^2\right) \\ +(mn-kn-1)\ln\left(\frac{mn}{mn-kn}\right) &- (nk+1)\ln(nk) + \Delta s, \end{aligned} \tag{4}$$

where

$$0 \le \Delta s \le mk\ln\left(\frac{2}{m\epsilon}\right). \tag{5}$$

This means that the number of dimensions can be chosen, by optimizing the above for k.

1.4 Outline

We continue this section with definitions and notations, and related work. Section 2 shows the main idea of NML reduction via elimination of some of the optimization parameters. We use this to reduce the problem of PCA NML to linear-regression NML. Section 3 details the specific reductions. Section 4 concludes and discusses further work. Numeric simulations appear in the full version of the paper [24].

1.5 Definitions and Notations

We will use lowercase letters (s) for scalars, underlined lowercase letters ($\underline{x}$) for column vectors, uppercase letters (X) for matrices, and calligraphic ($\mathcal{B}$) for sets. A single subscript for a matrix denotes a matrix row (X_i). $f(x)$, $f(x\ ;\ y)$, $f(x \mid y)$ denote the density of some x, the density of some x assuming some other parameter is y, and the density of some x conditional on some other random variable being y, respectively. $\|X\|_F = \left(\sum_{i,j} X_{i,j}^2\right)^{\frac{1}{2}}$ is the Frobenius norm, and $D(x \mid y)$ is the Kullback-Leibler distance.

1.6 Related Work

[3,25] contain excellent overviews of matrix factorization; in particular, PCA appears in the classic [7]. [1,8–10,17,23] describe MDL and NML, in particular, for model selection. [21,23] show closed forms of linear-regression NML. [15] uses cross validation approximations for PCA dimension estimation, [4] does so using an analysis of the conditional distribution of the singular values of a Wishart matrix, [12] uses a Bayesian approach, [26] uses patterns in the scree plots, and [13] compares statistical and heuristic approaches to this problem. To the best of my knowledge, previous works did not apply the modern form of the MDL principle to the problem of PCA dimension selection.

2 NML Reduction via Elimination of Optimization Parameters

Consider the generative form of (1), shown in the factor diagram (see [6]) in Fig. 1. In this model, $k \sim \mathcal{U}(1, m)$ determines the dimension of W_k and V_k. $X = W_k V_k^T + \Upsilon$, where $\Upsilon \sim \mathcal{N}(0, \tau I_k)$. Note that they do not appear in the original problem (at least in this form), but the problems are effectively equivalent. The distribution of k hardly affects the stochastic complexity (see [20], Chap. 5), and any distribution assigning a positive probability to any value of $1, \ldots, m$ could be used. Regarding the Gaussian additive noise Υ,

$$\arg\max_{W_k, V_k} f(X\ ;\ k) = \arg\max_{W_k, V_k} \frac{1}{(2\pi\tau)^{\frac{nm}{2}}} e^{-\frac{\left\|X - W_k V_k^T\right\|_F^2}{2\tau^2}} \overset{(a)}{=} \sum_{i=k+1}^{m} \left[\lambda_i^2\right],$$

where (a) follows from Theorem 1.

Now consider the generative model in Fig. 4 (discussed in greater detail in Sect. 3), where both the number of parameters and the loadings matrix are known. This easier problem is more similar to linear regression, whose NML is known (see [21]). Of course, in the original problem, the loadings matrix is not known, but rather optimized as well. The following Lemma, however, relates the NML of a problem depending on a number of parameters, to the the same problem where one of them is fixed.

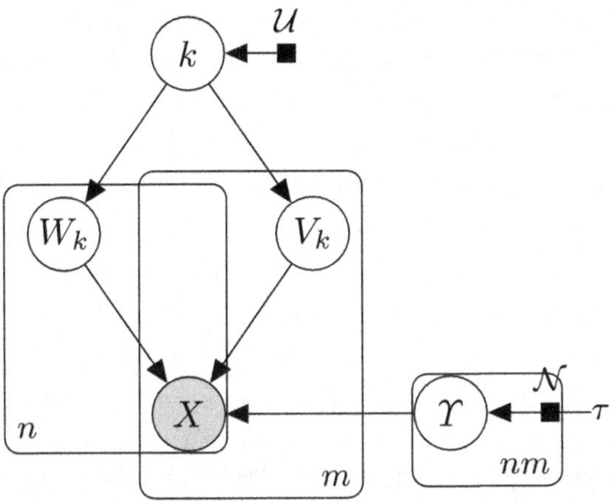

Fig. 1. Equivalent factor graph of PCA. The dimension k is a-priori uniform, and the observed matrix X is the product of the score and loadings matrices, with additive noise Υ distributed i.i.d. $\mathcal{N}(0, \tau I_k)$.

Lemma 1. *Let $\mathcal{B} = \{b_1, \ldots, b_\ell\}$ be a finite set (for some ℓ). Then*

$$\int \hat{f}\left(X \mid \hat{A}(X), \hat{b}(X)\right) dX \leq \sum_{b \in \mathcal{B}} \int \hat{f}\left(X \mid \hat{A}(X), b\right) dX. \tag{6}$$

Furthermore, if

$$\hat{b}(x) = \arg\min_{b} \hat{f}\left(X \mid \hat{A}(X), b\right), \tag{7}$$

then

$$\int \hat{f}\left(X \mid \hat{A}(X), \hat{b}(X)\right) dX \geq \max_{b \in \mathcal{B}} \int \hat{f}\left(X \mid \hat{A}(X), b\right) dX. \tag{8}$$

Proof. For inequality (6),

$$\begin{aligned}
&\int_X \hat{f}\left(X \mid \hat{A}(X), \hat{b}(X)\right) \mathrm{d}X \\
&= \sum_b \int_{X:\, \hat{b}(X)=b} \hat{f}\left(X \mid \hat{A}(X), b\right) \mathrm{d}X \\
&\overset{(a)}{\leq} \sum_b \int_X \hat{f}\left(X \mid \hat{A}(X), b\right) \mathrm{d}X,
\end{aligned}$$

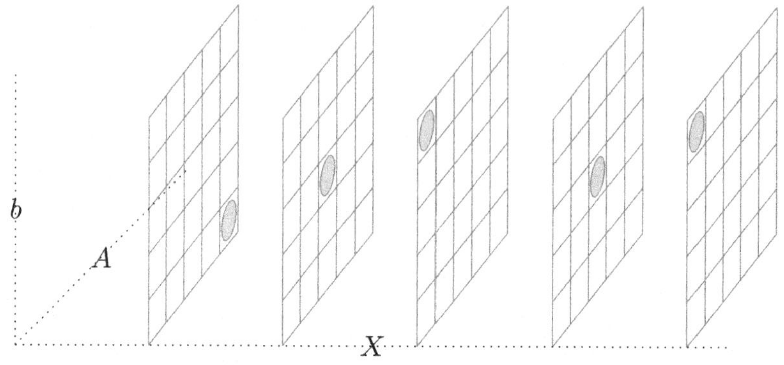

Fig. 2. Parametric complexity using only a subset of the features. For each X, there are an optimal $\hat{A}(X)$ and $\hat{b}(x)$, but we wish to bound this by expressions in which for each X, b is constant.

where (a) follows from the non-negativity of densities. In Fig. 2, this corresponds to bounding by considering the sum of all planes, then slicing them by vertical levels.

For inequality (8), consider an arbitrary $b' \in \mathcal{B}$. Then

$$
\begin{aligned}
&\int_X \hat{f}\left(X \mid \hat{A}(X), \hat{b}(X)\right) \mathrm{d}X \\
&= \sum_b \int_{X:\, \hat{b}(X)=b} \hat{f}\left(X \mid \hat{A}(X), b\right) \mathrm{d}X \\
&= \int_{X:\, \hat{b}(X)=b'} \hat{f}\left(X \mid \hat{A}(X), b'\right) \mathrm{d}X + \sum_{b \neq b'} \int_{X:\, \hat{b}(X)=b} \hat{f}\left(X \mid \hat{A}(X), b\right) \mathrm{d}X \\
&\overset{(a)}{\geq} \int_{X:\, \hat{b}(X)=b'} \hat{f}\left(X \mid \hat{A}(X), b'\right) \mathrm{d}X + \sum_{b \neq b'} \int_{\hat{b}(X)=b} \hat{f}\left(X \mid \hat{A}(X), b'\right) \mathrm{d}X \\
&= \int_X \hat{f}\left(X \mid \hat{A}(X), b'\right) \mathrm{d}X,
\end{aligned}
$$

where (a) follows from condition (7). Since this is true for an arbitrary b', it is true for the maximum. In Fig. 2, this corresponds to moving the disks until they are at the same horizontal level.

The next section formalizes the application of the lemma to PCA NML.

3 Reducing PCA NML to Linear Regression NML

Let $v_{i,j}$ be the elements of the unitary matrix V from Theorem 1. By the Cauchy-Schwartz Inequality, $|v_{i,j}| \leq 1$. Let $\epsilon \lnapprox \frac{1}{m}$ be a number such that $\frac{1}{\epsilon}$ is an integer. We can quantize $v_{i,j}$ into one of $\frac{2}{\epsilon}+1$ values, each distanced ϵ from each other, resulting in the matrix V^ϵ. By considering its Neumann series, it is clear that it is invertible, so there exists some W' such that $W'V^\epsilon = WV$.

Using Lemma 1, therefore, we can reduce the original problem to that in Fig. 3, where V_k^ϵ is a known matrix which is quantized version of a unitary matrix V_k (specifically, $V_k^\epsilon = V_k + \epsilon E_k$, where E_k has values each with absolute value at most $\frac{1}{2}$). Let $\mathcal{V}_k^\epsilon$ be the set of the quantized matrices, and let $s_i^\epsilon(X,k)$ be the stochastic complexity of Fig. 3, where the loadings matrix is known to be the ith element of $\mathcal{V}_k^\epsilon$ (according to some enumeration). Then by Lemma 1,

$$\max_{i\in\{1,\ldots,|\mathcal{V}_k^\epsilon|\}} s_i^\epsilon(X\,;\,k) \leq s(X\,;\,k) \leq \sum_{i=1}^{|\mathcal{V}_k^\epsilon|} \left[s_i^\epsilon(X\,;\,k)\right]. \tag{9}$$

Furthermore, we will see in Appendix A.1 the following lemma:

Lemma 2.

$$\ln\left(\left|\mathcal{V}_{\|}^\epsilon\right|\right) \lesssim mk\ln\left(\left(\frac{2}{\epsilon}+1\right)e^{-\left(\frac{1-\frac{1+\epsilon+\frac{\epsilon^2}{4}}{\sqrt{m}}}{2}\right)}\right) + (k-1)\ln\left(\frac{\epsilon+\frac{m\epsilon^2}{4}}{\pi}\right). \tag{10}$$

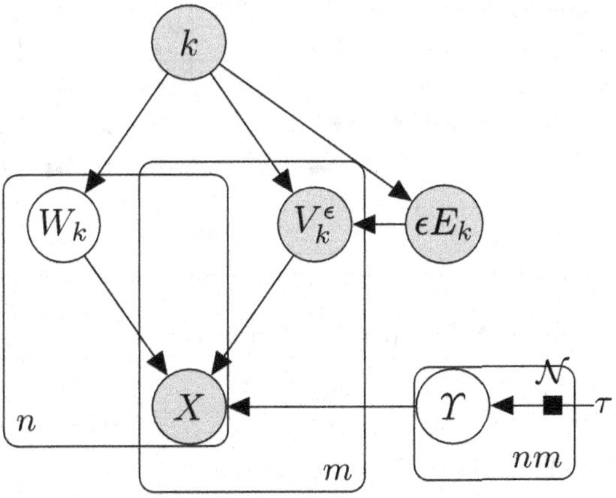

Fig. 3. Factor graph of known quantized loadings "PCA".

Let V_k^ϵ, a known quantized loadings matrix, be the ith item in $\mathcal{V}_{\|}^\epsilon$. To calculate its NML, note that Fig. 3 is very similar to linear regression (whose NML is known), except that W_k and X are matrices instead of vectors. This can be easily reduced to linear regression, though, by considering the problem

$$\begin{aligned} \underline{x} &= \tilde{V_k}^{\epsilon} \underline{w} + \underline{\upsilon} \\ &= \begin{bmatrix} X_1^T \\ \vdots \\ X_n^T \end{bmatrix} = \begin{bmatrix} V_k^{\epsilon} & \dots & 0 \\ \vdots & \ddots & \vdots \\ 0 & \dots & V_k^{\epsilon} \end{bmatrix} \begin{bmatrix} W_1^T \\ \vdots \\ W_n^T \end{bmatrix} + \begin{bmatrix} \Upsilon_1^T \\ \vdots \\ \Upsilon_n^T \end{bmatrix}, \end{aligned}$$

where $\underline{x}$ and $\underline{\upsilon}$ each have length nm, $\tilde{V_k}^{\epsilon}$ is $mn \times kn$, and $\underline{w}$ has length km. This is the dashed part of Fig. 4, and has known NML (see Equation (19) in [21])

$$\begin{aligned} s_i^{\epsilon}(X,k) &= (nm - kn)\ln(\hat{\tau}) + nk\ln\left(\left\|\tilde{V}_k^{\epsilon}\hat{w}\right\|_F^2\right) \\ &+ (mn - kn - 1)\ln\left(\frac{mn}{mn - kn}\right) - (nk+1)\ln(nk). \end{aligned} \tag{11}$$

However, we need the NML to be expressed in terms from the original problem.

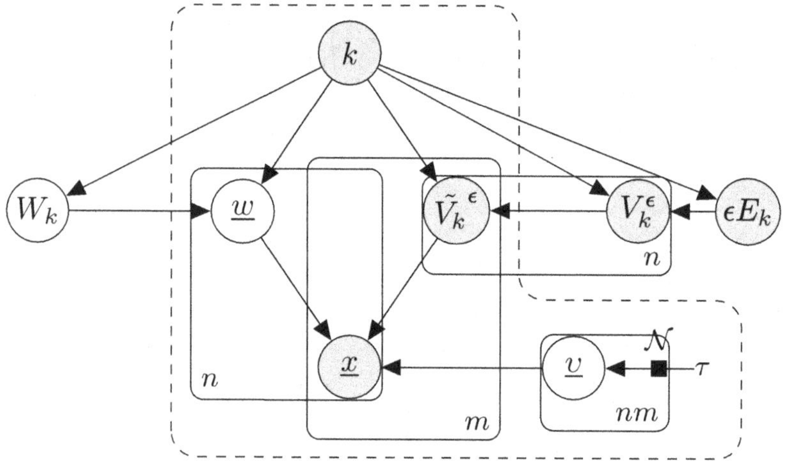

Fig. 4. Linear-regression factor graph.

It is well known (see [11]) that

$$\underline{\hat{w}} = \begin{bmatrix} \left(V_k^{\epsilon T} V_k^{\epsilon}\right)^{-1} V_k^{\epsilon T} X_1^T \\ \vdots \\ \left(V_k^{\epsilon T} V_k^{\epsilon}\right)^{-1} V_k^{\epsilon T} X_n^T \end{bmatrix}.$$

Furthermore, for the jth range,

$$\begin{aligned}\hat{W}_j^T &= \left(V_k^{\epsilon T} V_k^{\epsilon}\right)^{-1} V_k^{\epsilon T} X_j^T \\ &\simeq \left(I_k + \epsilon\left(V_k^T E + E^T V_k\right)\right)^{-1} V_k^T X_j^T \\ &\overset{(a)}{\simeq} \left(I_k - \epsilon\left(V_k^T E + E^T V_k\right)\right) V_k^T X_j^T,\end{aligned}$$

where (a) follows from [18] Equation (191). Therefore,

$$\left(V_k + \epsilon E\right) \hat{W}_j^T \simeq \left(I_k - \epsilon\left(V_k^T E + E^T V_k + E V_k^T\right)\right) X_j^T,$$

and, finally,

$$\left|\ln\left(\left\|V_k^{\epsilon} \hat{W}_j^T\right\|_F^2\right) - \ln\left(X_j^T X_j\right)\right| \lesssim 2\epsilon. \tag{12}$$

We now prove Theorem 2:

Proof. In Eq. (11), we replace $\hat{\tau}$ using Theorem 2, and $\tilde{V}_k^{\epsilon}\hat{w}$ using Eq. (12). We use the resulting expression - which is independent from i (the element of $\mathcal{V}_k^{\epsilon}$) - in Lemma 1.

4 Conclusions and Future Work

In this work we saw an NML-calculation technique based on reducing a problem through eliminating the optimization of some of its original dimensions. We saw how to use this to bound the NML of PCA. The technique is simple and general, and can be used to reduce problems in other domains, where simpler versions of the problem have a closed-form NML. Unfortunately, there are also several types of simple problems with no closed-form NML. For these cases, an MCMC evaluation of the parametric complexity (the denominator of Eq. (3)), could be a good numeric approximation. Developing an efficient algorithm for this, is a topic for further research.

A Appendix

A.1 Number of Quantized Unitary Matrices

We prove here Lemma 2. Let $\underline{v_i}, \underline{v_j}$ be two columns of a unitary matrix (perhaps the one), and $\underline{v_i^{\epsilon}}, \underline{v_j^{\epsilon}}$ be their quantized counterparts. Simple arithmetic shows that

$$\left|\underline{v_i^{\epsilon}} \cdot \underline{v_j^{\epsilon}} - \underline{v_i} \cdot \underline{v_j}\right| \le \epsilon + \frac{m\epsilon^2}{4}. \tag{13}$$

We will see that

$$\begin{aligned} P\left(\left|\underline{v_i^\epsilon}\cdot\underline{v_i^\epsilon}\right|\leq 1+\epsilon+\frac{\epsilon^2}{4}\right)&\leq e^{-mk\frac{1-\frac{1+\epsilon+\frac{\epsilon^2}{4}}{\sqrt{m}}}{2}},\\ P\left(\left|\underline{v_{i+1}^\epsilon}\cdot\underline{v_i^\epsilon}\right|\leq \epsilon+\frac{\epsilon^2}{4}\,|\,\forall_j v_j\underline{v_j^\epsilon}\leq 1+\epsilon+\frac{\epsilon^2}{4}\right)&\leq\left(\frac{\epsilon+\frac{m\epsilon^2}{4}}{\pi}\right)^{k-1}. \end{aligned} \tag{14}$$

For the first part of inequality (14),

$$\begin{aligned} &P\left(\sum_{k=1}^{m}\left[v_{i,k}^2\right]\leq 1+\epsilon+\frac{m\epsilon^2}{4}\right)\\ &\leq P\left(\left|\left\{k\mid v_{i,k}^2\geq x\right\}\right|\leq\frac{1+\epsilon+\frac{m\epsilon^2}{4}}{x}\right)\\ &\overset{(a)}{\leq} e^{-mD\left(\frac{1}{mx}\,|\,\sqrt{1-x}\right)}, \end{aligned} \tag{15}$$

where (a) follow from the Chernoff bound (see [16], Chap. 5). Using the well-known bound (see [5], [16], Chap. 5),

$$D\left(x\mid y\right)\geq\frac{(x-y)^2}{2y},\ (x\leq y),$$

and so

$$D\left(\frac{1+\epsilon+\frac{m\epsilon^2}{4}}{mx}\mid\sqrt{1-x}\right)\geq\frac{\left(\frac{1+\epsilon+\frac{m\epsilon^2}{4}}{mx}-\sqrt{1-x}\right)^2}{2\sqrt{1-x}}. \tag{16}$$

Setting $x=\frac{1}{\sqrt{m}}$, we have

$$\begin{aligned} &D\left(\frac{1+\epsilon+\frac{m\epsilon^2}{4}}{\sqrt{m}}\mid\sqrt{1-\frac{1}{\sqrt{m}}}\right)\\ &\geq\frac{\left(\frac{1+\epsilon+\frac{m\epsilon^2}{4}}{\sqrt{m}}-\sqrt{1-\frac{1}{\sqrt{m}}}\right)^2}{2\sqrt{1-\frac{1}{\sqrt{m}}}}\\ &\overset{(a)}{\gtrsim}\frac{\left(1-\frac{3+\epsilon+\frac{\epsilon^2}{4}}{2\sqrt{m}}\right)^2}{2\left(1-\frac{2}{\sqrt{m}}\right)}\\ &\overset{(b)}{\gtrsim}\frac{1-\frac{1+\epsilon+\frac{\epsilon^2}{4}}{\sqrt{m}}}{2}, \end{aligned} \tag{17}$$

where (a) and (b) follow from the Taylor expansion of $(1+x)^\alpha$.

For the second part of Inequality (14), applying Eq. (13) twice on the left side, and once on the right side, we have

$$\underline{v_i^\epsilon} \cdot \underline{v_j^\epsilon} = \left\|\underline{v_i^\epsilon}\right\| \left\|\underline{v_j^\epsilon}\right\| \cos\left(\alpha_{i,k}^\epsilon\right), \tag{18}$$

and so

$$\frac{\pi}{2} - \alpha_{i,k}^\epsilon \overset{(a)}{\simeq} \sin\left(\frac{\pi}{2} - \alpha_{i,k}^\epsilon\right) = \cos\left(\alpha_{i,k}^\epsilon\right) \leq \left|\frac{\epsilon + \frac{m\epsilon^2}{4}}{1 - \epsilon - \frac{m\epsilon^2}{4}}\right|, \tag{19}$$

with $\alpha_{i,k}^\epsilon$ the angle between the vectors, and where (a) follows from the Taylor series of $\sin(x)$. Approximating $\alpha_{i,k}^\epsilon \sim \mathcal{U}(0, 2\pi)$, we get that the probability is approximately that in the second part of Inequality (14).

References

1. Barron, A.R., Rissanen, J., Yu, B.: The minimum description length principle in coding and modeling. IEEE Trans. Inf. Theor **44**(6), 2743–2760 (1998)
2. Blumer, A., Ehrenfeucht, A., Haussler, D., Warmuth, M.K.: Occam's razor. Inf. Process. Lett. **24**(6), 377–380 (1987)
3. Bokde, D., Girase, S., Mukhopadhyay, D.: Matrix factorization model in collaborative filtering algorithms: a survey. Procedia Comput. Sci. 49, 136–146 (2015). Proceedings of 4th International Conference on Advances in Computing, Communication and Control (ICAC3 2015)
4. Choi, Y., Taylor, J., Tibshirani, R.: Selecting the number of principal components: estimation of the true rank of a noisy matrix. Ann. Statist. **45**(6), 2590–2617 (2017)
5. Cover, T.M., Thomas, J.A.: Elements of Information Theory. Wiley, New York (2006)
6. Laura, D.: Directed factor graph notation for generative models. Tech. rep. Saarbrücken, Germany (2010)
7. Eckart, C., Young, G.: The approximation of one matrix by another of lower rank. Psychometrika **1**(3), 211–218 (1936)
8. Grünwald, P.: A tutorial introduction to the minimum description length principle. In: Advances in Minimum Description Length: Theory and Applications. MIT Press, Cambridge (2005)
9. Hansen, M.H., Yu, B.: Model selection and the principle of minimum description length. J. Am. Stat. Assoc. **96**(454), 746–774 (2001)
10. Hansen, M.H., Yu, B.: Minimum description length model selection criteria for generalized linear models, vol. 40. Lecture Notes-Monograph Series, pp. 145–163. Institute of Mathematical Statistics, Beachwood (2003)
11. Hastie, T., Tibshirani, R., Friedman, J.: The Elements of Statistical Learning. SSS. Springer, New York (2009). https://doi.org/10.1007/978-0-387-84858-7
12. Hoyle, D.C.: Automatic PCA dimension selection for high dimensional data and small sample sizes. JMLR **9**, 733–2759 (2008)
13. Donald, A.J.: Stopping rules in principal components analysis: a comparison of heuristical and statistical approaches. Ecology **6**(74), 2204–2214 (1993)
14. Jolliffe, I.T.: Principal Component Analysis. Springer, New York (1986). https://doi.org/10.1007/b98835

15. Josse, J., Husson, F.: Selecting the number of components in principal component analysis using cross-validation approximations. Comput. Stat. Data Anal. **56**(6), 1869–1879 (2012)
16. Mitzenmacher, M., Upfal, E.: Probability and Computing: Randomized Algorithms and Probabilistic Analysis. Cambridge University Press, Cambridge (2005)
17. Myung, J.I., Navarro, D.J., Pitt, M.A.: Model selection by normalized maximum likelihood. J. Math. Psychol. **50**(2), 175–191 (2006)
18. Petersen, K.B., Pedersen, M.S.: The Matrix Cookbook, November 2012. Version 20121115 (2012)
19. Dawid, A.P., Vovk, V.G.: Prequential probability: principles and properties. Bernoulli **5**(1), 125–162 (1999)
20. Rissanen, J.: Stochastic complexity in statistical inquiry. Adv. Ser. Appl. Phys. **178**, 409–412 (1989)
21. Rissanen, J.: MDL denoising. IEEE Trans. Inf. Theo. **46**, 2537–2543 (1999)
22. Rissanen, J.: Stochastic complexity. J. Royal Stat. Soc. **49**(3), 223–265 (1999)
23. Rissanen, J.: Strong optimality of the normalized ml models as universal codes. IEEE Trans. Inf. Theory **47**, 1712–1717 (2000)
24. Tavory, A.: Determining principal component cardinality through the principle of minimum description length. arXiv (2019)
25. Udell, M., Horn, C., Zadeh, R., Boyd, S.: Generalized low rank models. Foundations and Trends® in Machine Learning, **9**(1), 1–118 (2016)
26. Zhu, M., Ghodsi, A.: Automatic dimensionality selection from the scree plot via the use of profile likelihood. Comput. Stat. Data Anal. **51**, 918–930 (2006)

Modelling Chaotic Time Series Using Recursive Deep Self-organising Neural Networks

Erik Berglund(✉)

Jeppesen, Gothenburg, Sweden
erik.berglund@gmail.com

Abstract. This paper proposes a Machine Learning (ML) algorithm that exhibits markedly different behaviour when trained with chaotic input data, compared to non-chaotic data. It will be demonstrated that the output of the ML system is itself chaotic when it is trained on a chaotic input. The proposed algorithm is a deep network with both direct and recurrent connections between layers, as well as recurrent connections within each layer.

Keywords: Deep network · Self-organisation · Chaos · Recursion

1 Introduction

Previous work has shown that single-layer recurrent self-organising maps can model some aspects of the behaviour of chaotic systems. The current work examines what happens when we alter the algorithm to support multi-layer networks with recurrent connections between the layers. Self-Organising Map (SOM) [10] type networks are suitable for creating deep networks, because there is no error to back-propagate, thus no problems of vanishing or exploding error signals. We show that adding more layers can greatly improve the modelling power, while reducing the total number of processing nodes and interconnections.

Chaotic systems are highly interesting due to how often they appear in nature; notably weather, markets, and animal populations. Since chaotic systems, by definition, are impossible to predict for more than short time periods, it is conceivable that instead of trying to predict them, it might be possible to model them instead, and thus learn something about their behaviour. It has also been suggested that chaos plays a role in how biological entities process information [6,7]

2 Previous Work

Chaotic time-series and chaos in neural networks has been studied by several researchers. One prominent example is Chaotic simulated annealing (CSA) [4],

G. Nicosia et al. (Eds.): LOD 2019, LNCS 11943, pp. 667–675, 2019.
https://doi.org/10.1007/978-3-030-37599-7_55

which has been shown to be able to solve combinatorial tasks like the Travelling Salesman Problem (TSP) efficiently. Unlike the present work the chaos is inherent in the network model, not learned. CSA has inspired other approaches, for example [17], which also contains a good review of similar methods. Time-series processing with SOM-type algorithms have been studied before, with the aim of predicting (not modelling) time series [3,11].

Other authors have proposed a hybrid approach using back-propagation neural networks and particle swarm optimisation [13], but this work also focused on prediction rather than modelling. Another avenue of research is the role of chaotic neural networks in memory formation and retrieval [5]. Feeding SOM output back into the input was first investigated in [16], who used the term "recursive" instead of "recurrent", since the Recurrent SOM name was already used for a SOM with leaky integrators on the inputs. The current work continues this convention.

In the standard SOM the magnitude of the weight update is dependent on the iteration number or time. For the Parameter-Less SOM 2 (PLSOM2) [1] algorithm the weight update is instead a function of what can be inferred about the state of the network.

Other recent advances in SOM algorithms include [8], which focuses on static behaviour like quantisation error, not dynamic behaviour as the current work. Deep networks constructed from layers of SOMs has been proposed earlier, for example [12], which uses layers of several parallel SOM networks interleaved with supervised learning kernels, and applied the resulting network to the task of character recognition with good results. However, this network only sends signals forwards, there are not any recurrent connections either within the layers or between the layers. One network that was not based on the SOM, but still utilised feedback weights between layers (but not within layers), was proposed by [9]. Both these networks were focused on image recognition, not time series modelling.

3 Learning Algorithm

The network consists of three layers of nodes, see Fig. 1.

A sample from the input sequence is presented to the first (leftmost) layer along with the output from the layer itself and the output of the next layer. An excitation vector for this layer is computed and used as part of the input for the next layer. The process is then repeated for the two last (rightmost) layers. The process is then repeated with the next sample from the input sequence. Recursion in SOM type neural networks works the same way as recurrence in other Neural Network algorithms: The output at time $t-1$ forms part of the input at time t. The particular method used here is adopted from [2] and [16]. In particular, each layer uses as its input not only the input from the previous layer, but also its own output and the output from the *next* layer.

The first step of the calculation carried out in each node i is given in (1),

$$\delta_{i,l,k}(t) = ||y_k(t) - w_{i,l,k}|| \tag{1}$$

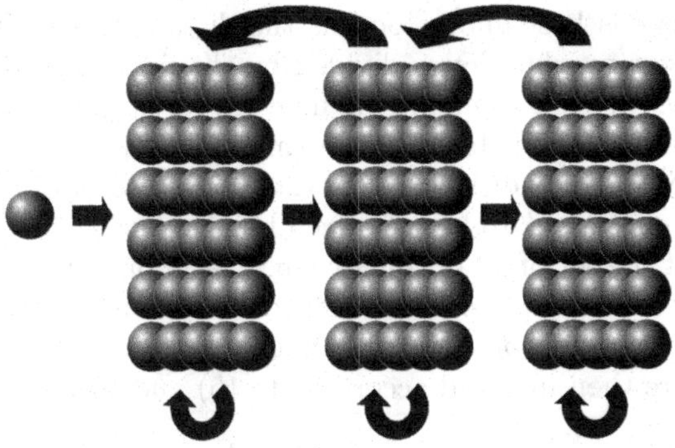

Fig. 1. Proposed network. All nodes are included. The leftmost node is the input. Connections are shown symbolically as arrows. Direct connections are shown as straight arrows between layers, recurrent intra-layer connections are shown as curved arrows below layers, and recurrent inter-layer connections are shown as curved arrows above layers.

where $y_k(t)$ is the output vector of layer k at time t or, in case of the first layer, the input vector, see (3). $w_{i,l,k}$ is the weight vector connecting node i in layer l to layer k. In short, $\delta_{i,l,k}$ decreases the closer the weight vector of node i in layer l matches the output of layer k.

The intermediate output is defined as a vector u of length n, where n is the number of nodes, where $u_{i,l}(t)$ corresponds to the excitation of the ith node of the lth layer at time-step t.

$$u_{i,l}(t) = e^{-[\alpha_l \delta_{i,l,l-1}(t) + (1-\alpha_l)(\beta_l \delta_{i,l,l}(t-1) + (1-\beta_l)\delta_{i,l,l+1}(t-1))]} \tag{2}$$

In (2), α_l is a tuning parameter that determines how much influence the new input has relative to the recurrent connections for layer l, β_l is a tuning parameter to account for the fact that two layers may have different sizes, u is a vector of length n, e is Euler's constant. In the case of the last layer, β_l is set to one, ignoring the part of the calculation based on the next layer. Following the calculation of u, the updated value of y is found by scaling and translating u so that each element lies in the range $[0, 1]$ according to (3).

$$y_{i,l}(t) = \frac{u_{i,l}(t) - \min(u_l(t))}{\max(u_l(t)) - \min(u_l(t))} \tag{3}$$

In other words, y_l is a vector where each entry $y_{i,l}$ corresponds to the normalised excitation of a node i. The entry that corresponds to the winning node, $y_{c,l}$, is equal to 1, all other entries are in the interval $[0, 1\rangle$. The vector y is the normalised excitation vector.

The weight update scaling factor ϵ_l for layer l is given by (4).

$$\epsilon_l = \frac{\alpha_l \delta_{c,l,l-1}(t) + (1-\alpha_l)(\beta_l \delta_{c,l,l}(t-1) + (1-\beta_l)\delta_{c,l,l+1}(t-1))}{\alpha_l S_{l-1} + (1-\alpha_l)(\beta_l S_l + (1-\beta_l)S_{l+1})} \tag{4}$$

The scaling factor ϵ is best understood as the distance from the current input vector to the closest weight vector, divided by the maximum distance between any two input vectors. If ϵ is small, the network already fits the current input well, and only need to adapt its weights a small amount.

The variable $\delta_{c,l,k}$ is the distance from the weight vector of the winning node c and the input from layer k to layer l, given by (1). The scalar S_l is an estimate of the diameter of the set of all outputs from layer l, or in the case of $l = 0$, the diameter of the set of all inputs. For an efficient algorithm for calculating S, see [1]. The parameters α_l and β_l are in the range $< 0, 1 >$.

The weights are then updated according to (5) and (6).

$$\eta_{i,l,k} = \frac{-D(c,i)^2}{(\gamma_l ln(1 + \epsilon_l(e-1)))^2} \tag{5}$$

$$\Delta w_{i,l,k} = \epsilon_l e^{\eta_{i,l,k}} \tag{6}$$

Here, $\Delta w_{i,l,k}$ is the amount of change for the weights of node i between layer l and k, ϵ_l is given by (4), e is Euler's constant, $D(c,i)$ is the distance between node i and the winning node c (note that this is the distance along the node lattice, not the difference between the weight vectors). γ_l is the generalisation parameter for layer l.

3.1 Idle Mode

As is clear from the above algorithm the direct input is only partially responsible for selecting the winning node, a large part is also due to the excitation vector given by (3). The map can thus be iterated without taking the input into account by calculating the winning node and the next value of the excitation vector based on the current excitation vector. This will henceforth be referred to as "idle mode".

4 Experimental Setup

As training input for time series are used:

1. A chaotic Mackey-Glass [14] series.
2. A simulated Mackey-Glass sequence, consisting of 511 sine waves with different wavelength, phase, and amplitude added together to resemble the frequency response of the real Mackey-Glass sequence.
3. The sum of two sine waves, with wavelengths 200 and 61.5.
4. A sine wave with wavelength 200.

All time series were scaled and translated to span the interval [0, 1]. The Fourier coefficients of the simulated Mackey-Glass series is indistinguishable from the Fourier coefficients of the real Mackey-Glass series, as the simulated series was created through inverse Fourier transform of the real series.

The Mackey-Glass time series is given by (7).

$$\frac{dx}{dt} = a\frac{x_\tau}{1 + x_\tau^n} - bx \tag{7}$$

here x_τ represents the value of the variable at time index $(t - \tau)$. In the present work the following values are used: $\tau = 17$, $a = 0.2$, $b = 0.1$, and $n = 10$.

Before each test the RPLSOM2 weights were initialised to a random state and trained with 50000 samples from one of the time series. The map has 30 nodes arranged in a 6 × 5 grid in three layer for a total of 90 nodes. The generalisation factor γ for the three layers were set to 24, 16, and 18, respectively. The parameter α was set to 0.8, 0.7, and 0.9 for the three layers respectively, and β was set to 0.5, 0.5, and 1, respectively. Each experiment was repeated 1000 times for each time series.

The for each time series, three key indicators of chaos were recorded:

- Repetition. One characteristic of chaotic systems is that they have very long repetition periods. The repetition period is defined as the number of samples one must draw from the series before the last 100 samples are repeated anywhere in any of the previous samples. After training the map was put into idle mode for 200000 iterations. The weight vector of the winning node will describe a one-dimensional time series, which was checked for repetition.
- Fractal dimension. To calculate the fractal dimension of the series attractor, it was embedded in 3 dimensions, using (8),

  $$X_t^T = \{x(t), x(t-1), x(t-2)\} \tag{8}$$

  where $x(t)$ is the sample drawn from the time series at time t. This results in a trajectory of vectors in 3-space. The information dimension of this trajectory was measured.
- Lyapunov exponent. The Lyapunov exponent was calculated using the excitation vector y (see Eq. 3) from the last layer of the RPLSOM2 and the numerical algorithm described in [15]. The perturbation value used was $d_0 = 10^{-12}$, and was added to the excitation vector of the last layer. The algorithm was in idle mode for 150 iterations before the divergence was measured and used to compute the Maximum Lyapunov Exponent.

5 Results

The proposed algorithm performs much better than the work it was based on [2] in terms of clearer differences in the Lyapunov exponent (see Fig. 2 and Table 1), the repetition time (see Fig. 3 and Table 2), and the fractal dimension (see Fig. 4 and Table 3).

In the present work, differences between the chaotic and non-chaotic maps are much more pronounced. The Lyapunov exponent is now positive on average, when the previous work only had a positive Lyapunov exponent in around 1 in 5 experimental runs. The repetition time is lower than for [2], but this is possibly due to the lower number of nodes in the first layer, meaning fewer potential states in the input trace, but the difference between the chaotically trained and the non-chaotically trained networks is still more pronounced in relative terms. The repetition time for the chaotic map is more than 400 times that of the best non-chaotic map, where the difference previously was less than 3 times. The fractal dimensions of the present and previous work are not directly comparable, since the present work treats stationary points as having dimension 0, which is more correct.

The proposed network is also more computationally efficient than [2], as there are 90 versus 100 nodes, and 5430 versus 10100 connections (including inter- and intra-layer recurrent connections). This is interesting because it shows that fewer nodes in more layers achieves much better results than more nodes in one layer.

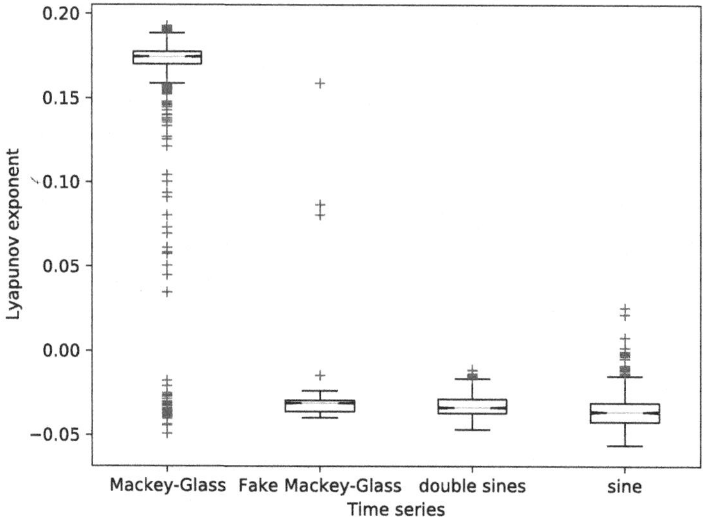

Fig. 2. Lyapunov exponent for 1000 maps trained with different inputs. Whiskers show lowest and highest outlier within 1.5 * IQR of the median.

Table 1. Mean Lyapunov exponent, with standard deviation and percentage of positive Lyapunov exponents.

Sequence	Lyapunov exponent	SD	% positive
Mackey-Glass	0.158	0.056	92.6
Fake Mackey-Glass	−0.032	0.009	0.3
Double sine	−0.033	0.006	0.0
Sine	−0.036	0.01	0.4

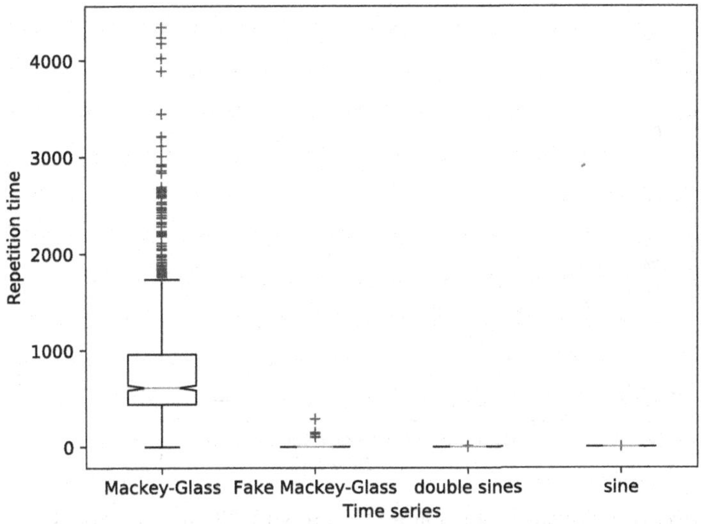

Fig. 3. Mean number of iterations of the map before repeating for 1000 maps trained with different inputs. Whiskers show lowest and highest outlier within 1.5 * IQR of the median. Notches indicate 95% confidence interval.

Table 2. Mean repetition time, with standard deviation.

Sequence	Repetition time	SD
Mackey-Glass	784.395	618.252
Fake Mackey-Glass	1.766	11.837
Double sine	0.86	0.601
Sine	0.752	0.432

Table 3. Mean information dimension, with standard deviation and percentage of non-zero values.

Sequence	Information dimension	SD	% non-zero
Mackey-Glass	0.034	0.037	91.9
Fake Mackey-Glass	0.001	0.011	0.6
Double sine	0.0	0.0	0.0
Sine	0.0	0.027	0.6

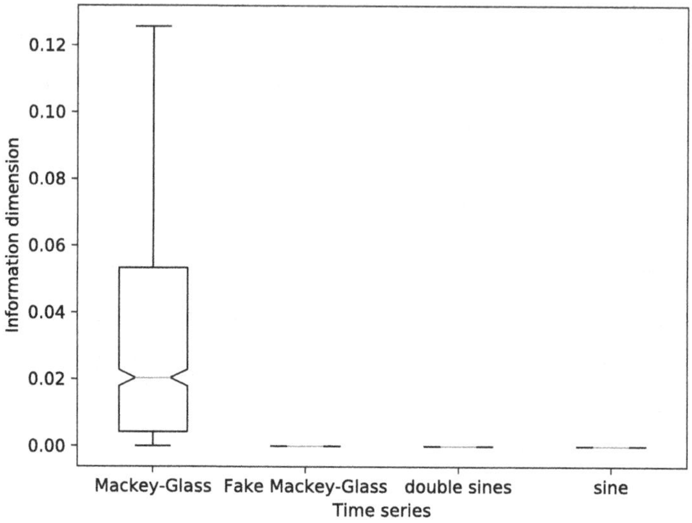

Fig. 4. Information dimension for maps trained with different inputs. Whiskers show lowest and highest outlier within 1.5 * IQR of the median. Other outliers have been omitted for clarity. Notches indicate 95% confidence interval.

6 Conclusion

It was observed that multi-layer RPLSOM2 networks trained with a chaotic time series to a significant degree exhibit the following characteristics:

- Longer repetition periods.
- Higher Lyapunov exponent.
- Higher probability of having a positive Lyapunov exponent.
- Higher fractal dimension.
- Higher probability of having non-zero fractal dimension.

when compared to networks that have been trained on non-chaotic but otherwise similar time sequences. This is consistent with chaotic behaviour. It can also be observed that by using a deep, multi-layer network, the aforementioned characteristic are more pronounced, despite the network having fewer nodes and fewer interconnecting weights. Possible future avenues of research include investigating more complex datasets such as real-world chaotic systems, for example weather and animal populations.

References

1. Berglund, E.: Improved PLSOM algorithm. Appl. Intell. **32**(1), 122–130 (2010)
2. Berglund, E.: Towards a universal modeller of chaotic systems. In: Battiti, R., Kvasov, D.E., Sergeyev, Y.D. (eds.) LION 2017. LNCS, vol. 10556, pp. 307–313. Springer, Cham (2017). https://doi.org/10.1007/978-3-319-69404-7_23

3. Chappell, G.J., Taylor, J.G.: The temporal Kohonen map. Neural Netw. **6**(3), 441–445 (1993)
4. Chen, L., Aihara, K.: Chaotic simulated annealing by a neural-network model with transient chaos. Neural Netw. **8**(6), 915–930 (1995)
5. Crook, N., Scheper, T.O.: A novel chaotic neural network architecture. In: ESANN 2001 Proceedings, April 2001, pp. 295–300 (2001)
6. Freeman, W.J.: Chaos in the brain: possible roles in biological intelligence. Int. J. Intell. Syst. **10**(1), 71–88 (1995)
7. Freeman, W.J., Barrie, J.M.: Chaotic oscillation and the genesis of meaning in cerebral cortex. In: Buzsáki, G., Llinás, R., Singer, W., Berthoz, A., Christen, Y. (eds.) Temporal Coding in the Brain. Research and Perspectives in Neurosciences, pp. 13–37. Springer, Heidelberg (1994). https://doi.org/10.1007/978-3-642-85148-3_2
8. Hameed, A.A., Karlik, B., Salman, M.S., Eleyan, G.: Robust adaptive learning approach to self-organizing maps. Knowl.-Based Syst. **171**, 25–36 (2019)
9. Hinton, G., Dayan, P., Frey, B., Neal, R.: The "wake-sleep" algorithm for unsupervised neural networks. Science **268**(5214), 1158–1161 (1995)
10. Kohonen, T.: Self-Organizing Maps. Springer Series in Information Sciences, vol. 30. Springer, Heidelberg (2001). https://doi.org/10.1007/978-3-642-56927-2
11. Koskela, T., Varsta, M., Heikkonen, J., Kaski, K.: Recurrent SOM with local linear models in time series prediction. In: 6th European Symposium on Artificial Neural Networks, pp. 167–172. D-facto Publications (1998)
12. Liu, N., Wang, J., Gong, Y.: Deep self-organizing map for visual classification. In: 2015 International Joint Conference on Neural Networks (IJCNN), pp. 1–6, July 2015
13. López-Caraballo, C.H., Salfate, I., Lazzús, J.A., Rojas, P., Rivera, M., Palma-Chilla, L.: Mackey-glass noisy chaotic time series prediction by a swarm-optimized neural network. J. Phys: Conf. Ser. **720**, 012002 (2016)
14. Mackey, M.C., Glass, L.: Oscillation and chaos in physiological control systems. Science **197**, 287 (1977)
15. Sprott, J.C.: Chaos and Time-Series Analysis. Oxford University Press, New York (2003)
16. Voegtlin, T.: Recursive self-organizing maps. Neural Netw. **15**(8–9), 979–991 (2002)
17. Wang, L., Li, S., Tian, F., Fu, X.: A noisy chaotic neural network for solving combinatorial optimization problems: stochastic chaotic simulated annealing. IEEE Trans. Syst. Man Cybern. Part B Cybern. **34**(5), 2119–2125 (2004)

On Tree-Based Methods for Similarity Learning

Stephan Clémençon[1] and Robin Vogel[1,2(✉)]

[1] Telecom Paris, LTCI, Institut Polytechnique de Paris, Paris, France
{stephan.clemencon,robin.vogel}@telecom-paris.fr
[2] IDEMIA, Courbevoie, France
robin.vogel@idemia.com

Abstract. In many situations, the choice of an adequate similarity measure or metric on the feature space dramatically determines the performance of machine learning methods. Building automatically such measures is the specific purpose of metric/similarity learning. In [21], similarity learning is formulated as a pairwise bipartite ranking problem: ideally, the larger the probability that two observations in the feature space belong to the same class (or share the same label), the higher the similarity measure between them. From this perspective, the ROC curve is an appropriate performance criterion and it is the goal of this article to extend recursive tree-based ROC optimization techniques in order to propose efficient similarity learning algorithms. The validity of such iterative partitioning procedures in the pairwise setting is established by means of results pertaining to the theory of U-processes and from a practical angle, it is discussed at length how to implement them by means of splitting rules specifically tailored to the similarity learning task. Beyond these theoretical/methodological contributions, numerical experiments are displayed and provide strong empirical evidence of the performance of the algorithmic approaches we propose.

Keywords: Metric-learning · Rate bound analysis · Similarity learning · Tree-based algorithms · U-processes

1 Introduction

Similarity functions are ubiquitous in machine learning, they are the essential ingredient of nearest neighbor rules in classification/regression or K-means/medoids clustering methods for instance and crucially determine their performance when applied to major problems such as biometric identification or recommending system design. The goal of learning automatically from data a similarity function or a metric has been formulated in various ways, depending on the type of similarity feedback available (*e.g.* labels, preferences), see [1,5,14,15,20]. A dedicated literature has recently emerged, devoted to this class of problems that is referred to as similarity-learning or metric-learning and is

G. Nicosia et al. (Eds.): LOD 2019, LNCS 11943, pp. 676–688, 2019.
https://doi.org/10.1007/978-3-030-37599-7_56

now receiving much attention, see *e.g.* [2] or [16] and the references therein. A popular framework, akin to that of multi-class classification, stipulates that pairwise similarity judgments can be directly deduced from observed class labels: a positive label is assigned to pairs formed of observations in the same class, while a negative label is assigned to those lying in different classes. In this context, similarity learning has been recently expressed as a *pairwise bipartite ranking problem* in [21], the task consisting in learning a similarity function that ranks the elements of a database by decreasing order of the posterior probability that they share the same label with some arbitrary query data point, as it is the case in important applications. In biometric identification (see *e.g.* [12]), the identity claimed by an individual is checked by matching her biometric information, a photo or fingerprints taken at an airport for instance, with those of authorized people gathered in a data repository of reference (*e.g.* passport photos or fingerprints). Based on a given similarity function and a fixed threshold value, the elements of the database are sorted by decreasing order of similarity score with the query and those whose score exceeds the threshold specified form the collection of matching elements. The ROC curve of a similarity function, *i.e.* the plot of the false positive rate *vs* the true positive rate as the threshold varies, appears in this situation as a natural (functional) performance measure. Whereas several approaches have been proposed to optimize a statistical counterpart of its scalar summary, the AUC criterion (AUC standing for Area Under the ROC Curve), see [11,18], it is pointed out in [21] that more local criteria must be considered in practice: ideally, the true positive rate should be maximized under the constraint that the false positive rate remains below a fixed level, usually specified in advance on the basis of operational constraints (see [12,13] in the case of biometric applications). If the generalization ability of solutions of empirical versions of such pointwise ROC optimization problems (and the situations where fast learning rates are achievable as well) has been investigated at length in [21], it is very difficult to solve in practice these constrained, generally nonconvex, optimization problems. It is precisely the goal of the present paper to address this algorithmic issue. Our approach builds on an iterative ROC optimization method, referred to as TreeRank, that has been proposed in [8] (see also [7] as well as [6] for an ensemble learning technique based on this method) and investigated at length in the standard (non pairwise) bipartite ranking setting. In this article, we establish statistical guarantees for the validity of the TreeRank methodology, when extended to the similarity learning framework (*i.e.* pairwise bipartite ranking), in the form of generalization rate bounds related to the sup norm in the ROC space and discuss issues related to its practical implementation. In particular, the *splitting rules* recursively implemented in the variant we propose are specifically tailored to the similarity learning task and produce symmetric tree-based scoring rules that may thus serve as similarity functions. Numerical experiments based on synthetic and real data are also presented here, providing strong empirical evidence of the relevance of this approach for similarity learning.

The paper is organized as follows. The rigorous formulation of similarity learning as pairwise bipartite ranking is briefly recalled in Sect. 2, together with the main principles underlying the TreeRank algorithm for ROC optimization. In Sect. 3, theoretical results proving the validity of the TreeRank method in the pairwise setup are stated and practical implementation issues are also discussed. Section 4 displays illustrative experimental results.

2 Background and Preliminaries

We start with recalling key concepts of similarity learning and its natural connection with ROC analysis and next briefly describe the algorithmic principles underlying the TreeRank methodology. Throughout the article, the Dirac mass at any point x is denoted by δ_x, the indicator function of any event $\mathcal{E}$ by $\mathbb{I}\{\mathcal{E}\}$, and the pseudo-inverse of any cdf $F(u)$ on $\mathbb{R}$ by $F^{-1}(t) = \inf\{v \in \mathbb{R} : \ F(v) \geq t\}$.

2.1 Similarity Learning as Pairwise Bipartite Ranking

We place ourselves in the probabilistic setup of multi-class classification here: Y is a random label, taking its values in $\{1, \dots, K\}$ with $K \geq 1$ say, and X is a random vector defined on the same probability space, valued in a feature space $\mathcal{X} \subset \mathbb{R}^d$ with $d \geq 1$ and modelling some information hopefully useful to predict Y. The marginal distribution of X is denoted by $\mu(dx)$, while the prior/posterior probabilities are $p_k = \mathbb{P}\{Y = k\}$ and $\eta_k(X) = \mathbb{P}\{Y = k \mid X\}$, $k = 1, \dots, K$. The conditional distribution of the r.v. X given $Y = k$ is denoted by μ_k. The distribution P of the generic pair (X, Y) is entirely characterized by $(\mu, \ (\eta_1, \dots, \eta_K))$. Equipped with these notations, we have $\mu = \sum_k p_k \mu_k$ and $p_k = \int_{\mathcal{X}} \eta_k(x)\mu(dx)$ for $k \in \{1, \ \dots, \ K\}$. In a nutshell, the goal pursued in this similarity learning framework is to learn from a training dataset $\mathcal{D}_n = \{(X_1, Y_1), \dots, (X_n, Y_n)\}$ composed of independent observations with distribution P a similarity (scoring) function, that is a measurable symmetric function $s : \mathcal{X}^2 \to \mathbb{R}_+$ (*i.e.* $\forall (x, x') \in \mathcal{X}^2$, $s(x, x') = s(x', x)$) such that, given an independent copy (X', Y') of (X, Y), the larger the similarity score between the input observations X and X', the higher the probability that they share the same label (*i.e.* that $Y = Y'$) should be. We denote by $\mathcal{S}$ the ensemble of all similarity functions.

Optimal Rules. Given this informal objective, the set of optimal similarity functions is obviously formed of strictly increasing transforms of the (symmetric) posterior probability $\eta(x, x') = \mathbb{P}\{Y = Y' \mid (X, X') = (x, x')\}$, namely

$$\mathcal{S}^* = \{T \circ \eta : \ \ T : Im(\eta) \to \mathbb{R}_+ \text{ borelian, strictly increasing}\},$$

denoting by $Im(\eta)$ the support of the r.v. $\eta(X, X') = \sum_k \eta_k(X)\eta_k(X')$. A similarity function $s^* \in \mathcal{S}^*$ defines the optimal preorder[1] $\preceq^*$ on the product space $\mathcal{X} \times \mathcal{X}$: for all $(x_1, x_2, x_3, x_4) \in \mathcal{X}^4$, x_1 and x_2 are more similar to each other than x_3 and x_4 iff $\eta(x_1, x_2) \geq \eta(x_3, x_4)$, and one then writes $(x_3, x_4) \preceq^* (x_1, x_2)$.

[1] A preorder on a set $\mathcal{X}$ is any reflexive and transitive binary relationship on $\mathcal{X}$. A preorder is an order if, in addition, it is antisymmetrical.

For any query $x \in \mathcal{X}$, s^* also defines a preorder $\preceq_x^*$ on the input space $\mathcal{X}$, that enables us to rank optimally all possible observations by increasing degree of similarity to x: for any $(x_1, x_2) \in \mathcal{X}^2$, x_1 is more similar to x than x_2 (one writes $x_2 \preceq_x^* x_1$) iff $(x, x_2) \preceq^* (x, x_1)$, that is $\eta(x, x_2) \leq \eta(x, x_1)$.

Pointwise ROC **Curve Optimization.** As highlighted in [21], similarity learning can be formulated as a *bipartite ranking* problem on the product space $\mathcal{X} \times \mathcal{X}$ where, given two independent realizations (X, Y) and (X', Y') of P, the input r.v. is the pair (X, X') and the binary label is $Z = 2\mathbb{I}\{Y = Y'\} - 1$, see *e.g.* [9]. In bipartite ranking, the gold standard by which the performance of a scoring function s is measured is the ROC curve (see *e.g.* [10] for an account of ROC analysis and its applications.): one evaluates how close the preorder induced by s to $\preceq^*$ is by plotting the parametric curve $t \in \mathbb{R}_+ \mapsto (F_{s,-}(t), F_{s,+}(t))$, where

$$F_{s,-}(t) = \mathbb{P}\{s(X, X') > t \mid Z = -1\},\ F_{s,+}(t) = \mathbb{P}\{s(X, X') > t \mid Z = +1\},$$

where possible jumps are connected by line segments. This P-P plot is referred to as the ROC curve of $s(x, x')$ and can be viewed as the graph of a continuous function $\alpha \in (0,1) \mapsto \mathrm{ROC}_s(\alpha)$, where $\mathrm{ROC}_s(\alpha) = F_{s,+} \circ F_{s,-}^{-1}(\alpha)$ at any point $\alpha \in (0,1)$ such that $F_{s,-} \circ F_{s,-}^{-1}(\alpha) = \alpha$. The curve ROC_s informs us about the capacity of s to discriminate between pairs with same labels and pairs with different labels: the stochastically larger than $F_{s,-}$ the distribution $F_{s,+}$, the higher ROC_s. It corresponds to the type I error *vs* power plot (false positive rate *vs* true positive rate) of the statistical test $\mathbb{I}\{s(X, X') > t\}$ when the null hypothesis stipulates that the labels of X and X' are different (*i.e.* $Y \neq Y'$) and defines a partial preorder on the set $\mathcal{S}$: one says that a similarity function s_1 is more accurate than another one s_2 when, for all $\alpha \in (0,1)$, $\mathrm{ROC}_{s_2}(\alpha) \leq \mathrm{ROC}_{s_1}(\alpha)$. The optimality of the elements of $\mathcal{S}^*$ w.r.t. this partial preorder immediately results from a classic Neyman-Pearson argument: $\forall (s, s^*) \in \mathcal{S} \times \mathcal{S}^*$, $\mathrm{ROC}_s(\alpha) \leq \mathrm{ROC}_{s^*}(\alpha) = \mathrm{ROC}_\eta(\alpha) := \mathrm{ROC}^*(\alpha)$ for all $\alpha \in (0,1)$. For simplicity, we assume here that the conditional cdf of $\eta(X, X')$ given $Z = -1$ is invertible. The accuracy of any $s \in \mathcal{S}$ can be measured by:

$$D_p(s, s^*) = ||\mathrm{ROC}_s - \mathrm{ROC}^*||_p, \tag{1}$$

where $s^* \in \mathcal{S}^*$ and $p \in [1, +\infty]$. When $p = 1$, one may write $D_1(s, s^*) = \mathrm{AUC}^* - \mathrm{AUC}(s)$, where $\mathrm{AUC}(s) = \int_{\alpha=0}^{1} \mathrm{ROC}_s(\alpha) d\alpha$ is the *Area Under the* ROC *Curve* (AUC in short) and $\mathrm{AUC}^* = \mathrm{AUC}(\eta)$ is the maximum AUC. Minimizing $D_1(s, s^*)$ boils down thus to maximizing the ROC summary $\mathrm{AUC}(s)$, whose popularity arises from its interpretation as the *rate of concording pairs*:

$$\begin{aligned}\mathrm{AUC}(s) = \mathbb{P}\left\{s(X_1, X_1') < s(X_2, X_2') \mid (Z_1, Z_2) = (-1, +1)\right\} \\ + \frac{1}{2}\mathbb{P}\left\{s(X_1, X_1') = s(X_2, X_2') \mid (Z_1, Z_2) = (-1, +1)\right\},\end{aligned}$$

where $((X_1, X_1'), Z_1)$ and $((X_2, X_2'), Z_2)$ denote independent copies of $((X, X'), Z)$. A simple empirical counterpart of $\mathrm{AUC}(s)$ can be derived from this formula, paving the way for the implementation of "empirical risk minimization" strategies, see [9] (the algorithms proposed to optimize the AUC criterion or surrogate performance measures are too numerous to be listed exhaustively here). However, as mentioned precedingly, in many applications, one is interested in finding a similarity function that optimizes the ROC curve at specific points $\alpha \in (0, 1)$. The superlevel sets of similarity functions in $\mathcal{S}^*$ define the solutions of pointwise ROC optimization problems in this context. In the above framework, it indeed follows from Neyman Pearson's lemma that the test statistic of type I error less than α with maximum power is the indicator function of the set $\mathcal{R}^*_\alpha = \{(x, x') \in \mathcal{X}^2 : \eta(x, x') \geq Q^*_\alpha\}$, where Q^*_α is the conditional quantile of the r.v. $\eta(X, X')$ given $Z = -1$ at level $1 - \alpha$. Considering similarity functions that are bounded by 1 only, it corresponds to the unique solution of the problem:

$$\max_{\substack{s\,:\,\mathcal{X}^2 \to [0,1],\\ \text{borelian}}} \mathbb{E}[s(X, X') \mid Z = +1] \quad \text{subject to} \quad \mathbb{E}[s(X, X') \mid Z = -1] \leq \alpha.$$

Though its formulation is natural, this constrained optimization problem is very difficult to solve in practice, as discussed at length in [21]. This suggests the extension to the similarity ranking framework of the TreeRank approach for ROC optimization (see [8] and [7]), recalled below. Indeed, in the standard (non pairwise) statistical learning setup for bipartite ranking, whose probabilistic framework is the same as that of binary classification and stipulates that training data are i.i.d. labeled observations, this recursive technique builds (piecewise constant) scoring functions s, whose accuracy can be guaranteed in terms of sup norm (*i.e.* for which $D_\infty(s, s^*)$ can be controlled) and it is the essential purpose of the subsequent analysis to prove that this remains true when the training observations are of the form $\{((X_i, X_j),\ Z_{i,j}) :\ 1 \leq i < j \leq n\}$, where $Z_{i,j} = 2\mathbb{I}\{Y_i = Y_j\} - 1$ for $1 \leq i < j \leq n$, and are thus far from being independent. Regarding its implementation, attention should be paid to the fact that the splitting rules for recursive partitioning of the space $\mathcal{X} \times \mathcal{X}$ must ensure that the decision functions produced by the algorithm fulfill the symmetric property.

2.2 Recursive ROC Curve Optimization - The TreeRank Algorithm

Because they offer a visual model summary in the form of an easily interpretable binary tree graph, decision trees remain very popular among practicioners, see *e.g.* [4] or [19]. In general, predictions are computed through a hierarchical combination of elementary rules comparing the value taken by a (quantitative) component of the input information (the *split variable*) to a certain threshold (the *split value*). In contrast to (supervised) learning problems such as classification/regression, which are of local nature, predictive rules for a global problem such as *similarity learning* cannot be described by a simple (tree-structured) partition of $\mathcal{X} \times \mathcal{X}$: the (symmetric) cells corresponding to the terminal leaves of the binary decision tree must be sorted in order to define a similarity function.

Similarity Trees. We define a *similarity tree* as a binary tree whose leaves all correspond to symmetric subsets $\mathcal{C}$ of the product space $\mathcal{X} \times \mathcal{X}$ (*i.e.* $\forall (x, x') \in \mathcal{X}^2$;

$(x, x') \in \mathcal{C} \Leftrightarrow (x', x) \in \mathcal{C}$) and is equipped with a 'left-to-right' orientation, that defines a tree-structured collection of similarity functions. Incidentally, the symmetry property makes it a specific *ranking tree*, using the terminology introduced in [8]. The root node of a tree $\mathcal{T}_J$ of depth $J \geq 0$ corresponds to the whole space $\mathcal{X} \times \mathcal{X}$: $\mathcal{C}_{0,0} = \mathcal{X}^2$, while each internal node (j, k) with $j < J$ and $k \in \{0, \ldots, 2^j - 1\}$ represents a subset $\mathcal{C}_{j,k} \subset \mathcal{X}^2$, whose left and right siblings respectively correspond to (symmetric) disjoint subsets $\mathcal{C}_{j+1,2k}$ and $\mathcal{C}_{j+1,2k+1}$ such that $\mathcal{C}_{j,k} = \mathcal{C}_{j+1,2k} \cup \mathcal{C}_{j+1,2k+1}$. Equipped with the left-to-right orientation, any subtree $\mathcal{T} \subset \mathcal{T}_J$ defines a preorder on $\mathcal{X}^2$: the degree of similarity being the same for all pairs (x, x') lying in the same terminal cell of $\mathcal{T}$. The similarity function related to the oriented tree $\mathcal{T}$ can be written as:

$$\forall (x, x') \in \mathcal{X}^2, \;\; s_{\mathcal{T}}(x, x') = \sum_{\mathcal{C}_{j,k}:\ \text{terminal leaf of } \mathcal{T}} 2^J \left(1 - \frac{k}{2^j}\right) \cdot \mathbb{I}\{(x, x') \in \mathcal{C}_{j,k}\}.$$

Observe that its symmetry results from that of the $\mathcal{C}_{j,k}$'s. The ROC curve of the similarity function $s_{\mathcal{T}}(x, x')$ is the piecewise linear curve connecting the knots:

$$(0, 0) \text{ and } \left(\sum_{l=0}^{k} F_-(\mathcal{C}_{j,l}), \sum_{l=0}^{k} F_+(\mathcal{C}_{j,l})\right) \text{ for all terminal leaf } \mathcal{C}_{j,k} \text{ of } \mathcal{T},$$

denoting by F_σ the conditional distribution of (X, X') given $Z = \sigma 1$, $\sigma \in \{-. +\}$. Setting $p_+ = \mathbb{P}\{Z = +1\} = \sum_k p_k^2$, we have $F_+ = (1/p_+) \sum_k p_k^2 \cdot \mu_k \otimes \mu_k$ and $F_- = (1/(1 - p_+)) \sum_{k \neq l} p_k p_l \cdot \mu_k \otimes \mu_l$. A statistical version can be computed by replacing the $F_\sigma(\mathcal{C}_{j,l})$'s by their empirical counterpart.

Growing the Similarity Tree. The TreeRank algorithm, a bipartite ranking technique optimizing the ROC curve in a recursive fashion, has been introduced in [8] and its properties have been investigated in [7] at length. Its output consists of a tree-structured scoring rule (Sect. 2.2) with a ROC curve that can be viewed as a piecewise linear approximation of ROC* obtained by a Finite Element Method with implicit scheme and is proved to be nearly optimal in the D_1 sense under mild assumptions. The growing stage is performed as follows. At the root, one starts with a constant similarity function $s_1(x, x') = \mathbb{I}\{(x, x') \in \mathcal{C}_{0,0}\} \equiv 1$ and after $m = 2^j + k$ iterations, $0 \leq k < 2^j$, the current similarity function is

$$s_m(x, x') = \sum_{l=0}^{2k-1} (m - l) \cdot \mathbb{I}\{(x, x') \in \mathcal{C}_{j+1,l}\} + \sum_{l=k}^{2^j-1} (m - k - l) \cdot \mathbb{I}\{(x, x') \in \mathcal{C}_{j,l}\}$$

and the cell $\mathcal{C}_{j,k}$ is split so as to form a refined version of the similarity function,

$$s_{m+1}(x, x') = \sum_{l=0}^{2k} (m - l) \cdot \mathbb{I}\{(x, x') \in \mathcal{C}_{j+1,l}\} + \sum_{l=k+1}^{2^j-1} (m - k - l) \cdot \mathbb{I}\{(x, x') \in \mathcal{C}_{j,l}\}$$

namely, with maximum (empirical) AUC. Therefore, it happens that this problem boils down to solve a cost-sensitive binary classification problem on the set

$\mathcal{C}_{j,k}$, see Subsect. 3.3 in [7]. Indeed, one may write the AUC increment as

$$\text{AUC}(s_{m+1}) - \text{AUC}(s_m) = \frac{1}{2} F_-(\mathcal{C}_{j,k}) F_+(\mathcal{C}_{j,k}) \times (1 - \Lambda(\mathcal{C}_{j+1,2k} \mid \mathcal{C}_{j,k})),$$

where $\Lambda(\mathcal{C}_{j+1,2k} \mid \mathcal{C}_{j,k}) \overset{def}{=} F_+(\mathcal{C}_{j,k} \setminus \mathcal{C}_{j+1,2k})/F_+(\mathcal{C}_{j,k}) + F_-(\mathcal{C}_{j+1,2k})/F_-(\mathcal{C}_{j,k})$. Setting $p = F_+(\mathcal{C}_{j,k})/(F_-(\mathcal{C}_{j,k}) + F_+(\mathcal{C}_{j,k}))$, the crucial point of the TreeRank approach is that the quantity $2p(1-p)\Lambda(\mathcal{C}_{j+1,2k} \mid \mathcal{C}_{j,k})$ can be interpreted as the cost-sensitive error of a classifier on $\mathcal{C}_{j,k}$ predicting positive label for any pair lying in $\mathcal{C}_{j+1,2k}$ and negative label fo all pairs in $\mathcal{C}_{j,k} \setminus \mathcal{C}_{j+1,2k}$ with cost p (respectively, $1-p$) assigned to the error consisting in predicting label $+1$ given $Z = -1$ (resp., label -1 given $Z = +1$), balancing thus the two types of error. Hence, at each iteration of the similarity tree growing stage, the TreeRank algorithm calls a *cost-sensitive* binary classification algorithm, termed LeafRank, in order to solve a statistical version of the problem above (replacing the theoretical probabilities involved by their empirical counterparts) and split $\mathcal{C}_{j,k}$ into $\mathcal{C}_{j+1,2k}$ and $\mathcal{C}_{j+1,2k+1}$. As described at length in [7], one may use cost-sensitive versions of celebrated binary classification algorithms such as CART or SVM for instance as LeafRank procedure, the performance depending on their ability to capture the geometry of the level sets $\mathcal{R}^*_\alpha$ of the posterior probability $\eta(x, x')$. As highlighted above, in order to apply the TreeRank approach to similarity learning, a crucial feature the LeafRank procedure implemented must have is the capacity to split a region in subsets that are both stable under the reflection $(x, x') \in \mathcal{X}^2 \mapsto (x', x)$. This point is discussed in the next section. Rate bounds for the TreeRank method in the sup norm sense are also established therein in the statistical framework of similarity learning, when the set of training examples $\{((X_i, X_j),\ Z_{i,j}\}_{i<j}$ is composed of non independent observations with binary labels, formed from the original multi-class classification dataset $\mathcal{D}_n$.

3 A Tree-Based Approach to Similarity Learning

We now investigate how the TreeRank method for ROC optimization recalled in the preceding section can be extended to the framework of similarity-learning and next establish learning rates in sup norm in this context.

3.1 A Similarity-Learning Version of TreeRank

From a statistical perspective, a learning algorithm can be derived from the recursive approximation procedure recalled in the previous section, simply by replacing the quantities $F_\sigma(\mathcal{C})$, $\sigma \in \{-,\ +\}$ and $\mathcal{C} \subset \mathcal{X} \times \mathcal{X}$ borelian, by their empirical counterparts based on the dataset $\mathcal{D}_n$:

$$\widehat{F}_{\sigma,n}(\mathcal{C}) = \frac{1}{n_\sigma} \sum_{i<j} \mathbb{I}\{(X_i, X_j) \in \mathcal{C},\ Z_{i,j} = \sigma 1\}, \tag{2}$$

with $n_\sigma = (2/(n(n-1)))\sum_{i<j}\mathbb{I}\{Z_{i,j} = \sigma 1\}$. Observe incidentally that the quantities (2) are by no means i.i.d. averages, but take the form of ratios of U-statistics of degree two (*i.e.* averages over pairs of observations, *cf* [17]), see Sect. 3 in [21]. For this reason, a specific rate bound analysis (ignoring bias issues) guaranteeing the accuracy of the TreeRank approach in the similarity learning framework is carried out in the next subsection.

The Similarity TreeRank Algorithm

Input. Maximal depth $D \geq 1$ of the similarity tree, class $\mathcal{A}$ of measurable and symmetric subsets of $\mathcal{X} \times \mathcal{X}$, training dataset $\mathcal{D}_n = \{(X_1, Y_1), \ldots, (X_n, Y_n)\}$.

1. (Initialization.) Set $\mathcal{C}_{0,0} = \mathcal{X} \times \mathcal{X}$, $\alpha_{d,0} = \beta_{d,0} = 0$ and $\alpha_{d,2^d} = \beta_{d,2^d} = 1$ for all $d \geq 0$.
2. (Iterations.) For $d = 0, \ldots, D-1$ and $k = 0, \ldots, 2^d - 1$:
   (a) (Optimization step.) Set the entropic measure:

$$\Lambda_{d,k+1}(\mathcal{C}) = (\alpha_{d,k+1} - \alpha_{d,k})\widehat{F}_{+,n}(\mathcal{C}) - (\beta_{d,k+1} - \beta_{d,k})\widehat{F}_{-,n}(\mathcal{C}) \; .$$

   Find the best subset $\mathcal{C}_{d+1,2k}$ of the cell $\mathcal{C}_{d,k}$ in the AUC sense:

$$\mathcal{C}_{d+1,2k} = \underset{\mathcal{C} \in \mathcal{A},\ \mathcal{C} \subset \mathcal{C}_{d,k}}{\arg\max}\ \widehat{\Lambda}_{d,k+1}(\mathcal{C}) \; . \tag{3}$$

   Then, set $\mathcal{C}_{d+1,2k+1} = \mathcal{C}_{d,k} \setminus \mathcal{C}_{d+1,2k}$.
   (b) (Update.) Set

$$\alpha_{d+1,2k+1} = \alpha_{d,k} + \widehat{F}_{-,n}(\mathcal{C}_{d+1,2k}),\ \beta_{d+1,2k+1} = \beta_{d,k} + \widehat{F}_{+,n}(\mathcal{C}_{d+1,2k})$$

$$\text{and } \alpha_{d+1,2k+2} = \alpha_{d,k+1},\ \beta_{d+1,2k+2} = \beta_{d,k+1} \; .$$

3. (Output.) After D iterations, get the piecewise constant similarity function:

$$s_D(x, x') = \sum_{k=0}^{2^D - 1} (2^D - k)\ \mathbb{I}\{(x, x') \in \mathcal{C}_{D,k}\}, \tag{4}$$

   together with an estimate of the curve $\mathrm{ROC}(s_D, .)$, namely the broken line $\widehat{\mathrm{ROC}}(s_D, .)$ that connects the knots $\{(\alpha_{D,k}, \beta_{D,k}) :\ k = 0, \ldots, 2^D\}$, and the following estimate of $\mathrm{AUC}(s_D)$:

$$\widehat{\mathrm{AUC}}(s_D) = \int_{\alpha=0}^{1} \widehat{\mathrm{ROC}}(s_D, \alpha)d\alpha = \frac{1}{2} + \frac{1}{2}\sum_{k=0}^{2^{D-1}-1} \widehat{\Lambda}_{D-1,k+1}(\mathcal{C}_{D,2k}).$$

The symmetry property of the function (4) output by the learning algorithm is directly inherited from that of the candidate subsets $\mathcal{C} \in \mathcal{A}$ of the product space $\mathcal{X} \times \mathcal{X}$ among which the $\mathcal{C}_{d,k}$'s are selected. We new explain at length how to perform the optimization step (3) in practice in the similarity learning context.

Splitting for Similarity Learning. As recalled in Subsect. 2.2, solving (3) boils down to finding the best classifier on $\mathcal{C}_{d,k} \subset \mathcal{X}^2$ of the form

$$g_{\mathcal{C}|\mathcal{C}_{d,k}}(x,x') = \mathbb{I}\{(x,x') \in \mathcal{C}\} - \mathbb{I}\{(x,x') \in \mathcal{C} \setminus \mathcal{C}_{d,k}\} \text{ with } \mathcal{C} \subset \mathcal{C}_{d,k},\ \ \mathcal{C} \in \mathcal{A},$$

in the empirical AUC sense, that is to say that minimizing a statistical version of the cost-sensitive classification error based on $\{((X_i, X_j),\ Z_{i,j}) :\ \ 1 \leq i < j \leq n,\ (X_i, X_j) \in \mathcal{C}_{d,k}\}$

$$\Lambda(\mathcal{C} \mid \mathcal{C}_{d,k}) = \frac{\mathbb{P}\{g_{\mathcal{C}|\mathcal{C}_{d,k}}(X,X') = 1 \mid Z = -1\}}{\mathbb{P}\{(X,X') \in \mathcal{C}_{d,k} \mid Z = -1\}} + \frac{\mathbb{P}\{g_{\mathcal{C}|\mathcal{C}_{d,k}}(X,X') = -1 \mid Z = 1\}}{\mathbb{P}\{(X,X') \in \mathcal{C}_{d,k} \mid Z = 1\}}.$$

Notice that, equipped with the notations previously introduced, the statistical version of $\Lambda(\mathcal{C} \mid \mathcal{C}_{d,k})$ is $\Lambda_{d,k+1}(\mathcal{C})/\left((\alpha_{d,k+1} - \alpha_{d,k})(\beta_{d,k+1} - \beta_{d,k})\right)$. In [7], it is highlighted that, in the standard ranking bipartite setup, any (cost-sensitive) classification algorithm (*e.g.* Neural Networks, CART, Random Forest, SVM, nearest neighbours) can be possibly used for splitting, whereas, in the present framework, classifiers are defined on product spaces and the symmetry issue must be addressed. For simplicity, assume that $\mathcal{X}$ is a subset of the space $\mathbb{R}^q$, $q \geq 1$, whose canonical basis is denoted by $(e_1, \ldots, e_q)$. Denote by $P_V(x, x')$ the orthogonal projection of any point (x, x') in $\mathbb{R}^q \times \mathbb{R}^q$ equipped with its usual Euclidean structure onto the subspace $V = Span((e_1,\ e_1), \ldots, (e_q,\ e_q))$. Let W be V's orthogonal complement in $\mathbb{R}^q \times \mathbb{R}^q$. For any $(x, x') \in \mathcal{X}^2$, denote by $f(x, x') = (f_1(x, x'), \ldots, f_{2q}(x, x'))$ the $2q$-dimensional vector, whose first q components are the coordinates of the projection $P_V(x, x')$ of (x, x') onto the subspace V in an orthonormal basis of V (say $\{(1/\sqrt{2})(e_1,\ e_1), \ldots, (1/\sqrt{2})(e_q,\ e_q)\}$ for instance) and whose last components are formed by the *absolute values* of the coordinates of the projection $P_W(x, x')$ of (x, x') onto W expressed in a given orthonormal basis (say $\{(1/\sqrt{2})(e_1,\ -e_1), \ldots, (1/\sqrt{2})(e_q,\ -e_q)\}$ for instance). Observing that, by construction, $f(x, x') = f(x', x)$ for all $(x, x') \in \mathcal{X}^2$, our proposal relies on the following result (whose proof is straightforward and left to the reader).

Lemma 1. *Let* $S : \mathcal{X}^2 \to \mathbb{R}$. *Then,* S *is symmetric iff there exists* $s : \mathbb{R}^q \times \mathbb{R}^q_+ \to \mathbb{R}$ *such that:* $\forall (x, x') \in \mathcal{X}^2,\ \ S(x, x') = (s \circ f)(x, x')$.

In order to get splits that are symmetric w.r.t. the reflection $(x, x') \mapsto (x', x)$, we propose to build directly classifiers of the form $(G \circ f)(x, x')$. In practice, this splitting procedure referred to as Symmetric LeafRank and summarized below simply consists in using as input space $\mathbb{R}^q \times \mathbb{R}^q_+$ rather than $\mathbb{R}^{2q}$ and considering as training labeled observations the dataset $\{(f(X_i, X_j),\ Z_{i,j}) :\ \ 1 \leq i < j \leq n,\ \ (X_i, X_j) \in \mathcal{C}_{d,k}\}$ when running a cost-sensitive classification algorithm. Just like in the original version of the TreeRank method, the growing stage can be followed by a pruning procedure, where children of a same parent node are recursively merged in order to produce a similarity subtree that maximizes an estimate of the AUC criterion, based on cross-validation usually, one may refer to Sect. 4 in [7] for further details. In addition, as in the standard bipartite ranking context, the Ranking Forest approach (see [6]), an *ensemble learning* technique based on TreeRank that combines aggregation and randomization, can be implemented to dramatically improve stability and accuracy of similarity tree models both at the same time, while preserving their advantages (*e.g.* scalability, interpretability).

Symmetric LeafRank

- **Input.** Pairs $\{((X_i, X_j),\ Z_{i,j}) : 1 \le i < j \le n,\ (X_i, X_j) \in \mathcal{C}_{d,k}\}$ lying in the (symmetric) region to be split. Classification algorithm $\mathcal{A}$.
- **Cost.** Compute the number of positive pairs lying in the region $\mathcal{C}_{d,k}$

$$p = \frac{\sum_{1\le i<j\le n} \mathbb{I}\{(X_i, X_j) \in \mathcal{C}_{d,k},\ Z_{i,j} = +1\}}{\sum_{1\le i<j\le n} \mathbb{I}\{(X_i, X_j) \in \mathcal{C}_{d,k}\}}$$

- **Cost-sensitive classification.** Based on the labeled observations

$$\{(f(X_i, X_j),\ Z_{i,j}) : 1 \le i < j \le n,\ (X_i, X_j) \in \mathcal{C}_{d,k}\},$$

run algorithm $\mathcal{A}$ with cost p for the false positive error and cost $1-p$ for the false negative error to produce a (symmetric) classifier $g(x, x')$ on $\mathcal{C}_{d,k}$.
- **Output** Define the subregions:

$$\mathcal{C}_{d+1,2k} = \{(x, x') \in \mathcal{C}_{d,k} : g(x, x') = +1\} \text{ and } \mathcal{C}_{d+1,2k+1} = \mathcal{C}_{d,k} \setminus \mathcal{C}_{d+1,2k}.$$

3.2 Generalization Ability - Rate Bound Analysis

We now prove that the theoretical guarantees formulated in the ROC space equipped with the sup norm that have been established for the TreeRank algorithm in the standard bipartite ranking setup in [8] remain valid in the similarity learning framework. The rate bound result stated below is the analogue of Corollary 1 in [8]. The following technical assumptions are involved:

- the feature space $\mathcal{X}$ is bounded;
- $\alpha \mapsto \mathrm{ROC}^*(\alpha)$ is twice differentiable with a bounded first order derivative;
- the class $\mathcal{A}$ is intersection stable, *i.e.* $\forall(\mathcal{C},\ \mathcal{C}') \in \mathcal{A}^2$, $\mathcal{C} \cap \mathcal{C}' \in \mathcal{A}$;
- the class $\mathcal{A}$ has finite VC dimension $V < +\infty$;
- we have $\{(x, x') \in \mathcal{X}^2 : \eta(x, x') \ge q\} \in \mathcal{A}$ for any $q \in [0, 1]$;

Theorem 1. *Assume that the conditions above are fulfilled. Choose $D = D_n$ so that $D_n \sim \sqrt{\log n}$, as $n \to \infty$, and let s_{D_n} denote the output of the* Similarity TreeRank *algorithm. Then, for all $\delta > 0$, there exists a constant λ s.t., with probability at least $1 - \delta$, we have for all $n \ge 2$: $D_\infty(s_{D_n}, s^*) \le \exp(-\lambda\sqrt{\log n})$.*

Proof. The proof is based on the following lemma, proved in [21] (in a more general version, the present one being a restriction to classes of indicator functions), which provides upper confidence bounds for the suprema of collections of ratios of U-statistics.

Lemma 2. *(Lemma 1, [21]) Suppose that Theorem 1's assumptions are fulfilled. Let $\sigma \in \{-, +\}$. For any $\delta \in (0,1)$, we have with probability at least $1-\delta$,*

$$\sup_{\mathcal{C}} \left| \widehat{F}_{\sigma,n}(\mathcal{C}) - F_\sigma(\mathcal{C}) \right| \leq 2C\sqrt{\frac{V}{n}} + 2\sqrt{\frac{\log(1/\delta)}{n-1}},$$

where C is a universal constant, explicited in [3] (see page 198 therein).

This crucial result permits to control the deviation of the progressive outputs of the SIMILARITY TREERANK algorithm and those of the nonlinear approximation scheme (based on the true quantities) investigated in [8]. The proof can be thus derived by following line by line the argument of Corollary 1 in [8]. □

This *universal* logarithmic rate bound may appear slow at first glance but attention should be paid to the fact that it directly results from the hierarchical structure of the partition induced by the tree construction and the *global* nature of the similarity learning problem. As pointed out in [8] (see Remark 14 therein), the same rate bound holds true for the deviation in sup norm between the empirical ROC curve $\widehat{\mathrm{ROC}}(s_{D_n}, .)$ output by the TREERANK algorithm and the optimal curve ROC*.

4 Illustrative Numerical Experiments

To begin with, we study the ability of similarity ranking trees to retrieve the optimal ROC curve for synthetic data, issued from a random tree of depth D_{gt} with a noise parameter δ. Our experiments illustrate three aspects of learning a similarity s_D with TreeRank of depth D: the impact of the class asymmetry $p_+ \ll 1 - p_+$ as seen in the bounds of [21], the trade-off between number of training instances and model complexity, see Theorem 1, and finally the impact of model biais. Results are summarized in the table below, with 95%-confidence intervals between parenthesis based off the normal approximation obtained on 400 runs. Details about the synthetic data experiments and real data experiments can be found in the appendix.

Class asymmetry			Model complexity			Model bias		
p_+	$D_1(s_D, s^*)$	$D_\infty(s_D, s^*)$	D_{gt}	$D_1(s_D, s^*)$	$D_\infty(s_D, s^*)$	D	$D_1(s_D, s^*)$	$D_\infty(s_D, s^*)$
0.5	0.07(±0.07)	0.30(±0.07)	1	0.00(±0.01)	0.06(±0.01)	1	0.21(±0.13)	0.65(±0.13)
10^{-1}	0.08(±0.08)	0.31(±0.08)	2	0.03(±0.04)	0.20(±0.04)	2	0.11(±0.10)	0.43(±0.10)
10^{-3}	0.42(±0.17)	0.75(±0.17)	3	0.07(±0.07)	0.30(±0.07)	3	0.07(±0.07)	0.30(±0.07)
$2 \cdot 10^{-4}$	0.45(±0.08)	0.81(±0.08)	4	0.12(±0.09)	0.43(±0.09)	8	0.06(±0.06)	0.28(±0.06)
Parameters: $D = D_{gt} = 3$.			$D_{gt} = D$, $p = 0.5$.			$D_{gt} = 3$, $p = 0.5$.		
Shared parameters: $\mathcal{X} = \mathbb{R}^3$, $\delta = 0.01$, $n_{\text{test}} = 100,000$, $n_{\text{train}} = 150 \cdot (5/4)^{D_{gt}^2}$.								

5 Conclusion

In situations where multi-class data are available, the objective of *similarity learning* can be naturally formulated as a ROC curve optimization problem, whose solutions are given by similarity functions yielding a maximal true positive rate with a false positive rate below a fixed value of reference, when thresholded at an appropriate level. Given the importance of this learning task, that finds its motivation in many practical problems, related to biometrics applications in particular, the present paper proposes an extension of the recursive approach TreeRank for ROC optimization to the similarity framework. Precisely, from an algorithmic viewpoint, it is shown how to adapt it in order to build *symmetric* scoring functions and, from a theoretical angle, the accuracy properties are proved to be preserved in spite of the complexity of the data functional that is optimized by the algorithm in a recursive manner. Experimental results supporting the approach promoted are also presented.

References

1. Bellet, A., Habrard, A.: Robustness and generalization for metric learning. Neurocomputing **151**(1), 259–267 (2015)
2. Bellet, A., Habrard, A., Sebban, M.: Metric Learning. Morgan & Claypool Publishers, San Rafael (2015)
3. Bousquet, O., Boucheron, S., Lugosi, G.: Introduction to statistical learning theory. In: Advanced Lectures on Machine Learning, pp. 169–207 (2004)
4. Breiman, L., Friedman, J., Olshen, R., Stone, C.: Classification and Regression Trees. Wadsworth and Brooks, Monterey (1984)
5. Cao, Q., Guo, Z.C., Ying, Y.: Generalization bounds for metric and similarity learning. Mach. Learn. **102**(1), 115–132 (2016)
6. Clémençon, G., Depecker, M., Vayatis, N.: Ranking forests. J. Mach. Learn. Res. **14**, 39–73 (2013)
7. Clémençon, S., Depecker, M., Vayatis, N.: Adaptive partitioning schemes for bipartite ranking. Mach. Learn. **83**(1), 31–69 (2011)
8. Clémençon, S., Vayatis, N.: Tree-based ranking methods. IEEE Trans. Inf. Theory **55**(9), 4316–4336 (2009)
9. Clémençon, S., Lugosi, G., Vayatis, N.: Ranking and empirical minimization of U-statistics. Ann. Stat. **36**(2), 844–874 (2008)
10. Fawcett, T.: An introduction to ROC analysis. Lett. Pattern Recogn. **27**(8), 861–874 (2006)
11. Huo, J., Gao, Y., Shi, Y., Yin, H.: Cross-modal metric learning for AUC optimization. IEEE Trans. Neural Netw. Learn. Syst. **29**(10), 4844–4856 (2018)
12. Jain, A., Hong, L., Pankanti, S.: Biometric identification. Commun. ACM **43**(2), 90–98 (2000)
13. Jain, A.K., Ross, A., Prabhakar, S.: An introduction to biometric recognition. IEEE Trans. Circuits Syst. Video Technol. **14**(1), 4–20 (2004)
14. Jain, L., Mason, B., Nowak, R.: Learning low-dimensional metrics. In: NIPS (2017)
15. Jin, R., Wang, S., Zhou, Y.: Regularized distance metric learning: theory and algorithm. In: NIPS (2009)

16. Kulis, B.: Metric learning: a survey. Found. Trends Mach. Learn. **5**(4), 287–364 (2012)
17. Lee, A.J.: U-statistics: Theory and practice. Marcel Dekker Inc., New York (1990)
18. McFee, B., Lanckriet, G.R.G.: Metric learning to rank. In: ICML (2010)
19. Quinlan, J.: Induction of decision trees. Mach. Learn. **1**(1), 1–81 (1986)
20. Verma, N., Branson, K.: Sample complexity of learning mahalanobis distance metrics. In: NIPS (2015)
21. Vogel, R., Clémençon, S., Bellet, A.: A probabilistic theory of supervised similarity learning: pairwise bipartite ranking and pointwise ROC curve optimization. In: ICML (2018)

Active Learning Approach for Safe Process Parameter Tuning

Stefano De Blasi[1,2](✉)

[1] Bosch Rexroth AG, 97816 Lohr am Main, Germany
Stefano.DeBlasi@boschrexroth.de
[2] University of Applied Sciences Fulda, 36037 Fulda, Germany

Abstract. The amount of sensor data in industry increases dramatically due to the digitization of plants and the Industrial Internet of Things (IIoT) evolution. This rapid technological revolution creates new opportunities in intelligent automation, but also challenges. As the complexity of the manufacturing processes increases accordingly, it becomes increasingly difficult to understand how a process is affected by physical conditions, not to mention optimizing process parameters with traditional methods.

In order to deal with the latter problem, an active learning approach is presented that can be used during production processes and that guarantees a user-defined minimum performance during learning. Based on Gaussian processes it is also possible to contribute already existing domain expertise to the learning process. We demonstrate that an additional solid expertise prevents an increase in the risk of failure throughout the automated optimization process.

With the proposed method manufacturers are able to evaluate their expertise based policy or even gain further knowledge about correlations by made performance improvements. The approach is generic and suitable for applications in mass production or frequently recurring tasks. Since the test results are promising, the next step is to create initial industrial prototypes.

Keywords: Safe optimization · Active learning · Parameter tuning

1 Introduction

Process parameters are often adjusted manually in order to achieve the best product quality, process performance or operating time of plants. Due to the changing situation, the so-called environment, experts have to make continuous adjustments. An industrial environment can be defined by measured temperature, humidity, weight of products, but also by calculated metrics based on several different sensor data. As the complexity increases accordingly, it becomes increasingly difficult to optimize process parameters with traditional methods based on physical conditions.

G. Nicosia et al. (Eds.): LOD 2019, LNCS 11943, pp. 689–699, 2019.
https://doi.org/10.1007/978-3-030-37599-7_57

To automate the adjustment decisions or actions, optimization algorithms explore a mapping of how the environment affects performance. Taking into account the parameters used, a policy can be created and used to determine the predicted optimal parameters for the future. This exploring task is called policy search. The learning agent interprets the environment, suggests an action and learns how its action affected the environment. Therefore, the quality of its decisions has to be evaluated which requires expertise of the environment. In industrial use-cases, domain experts and data scientists work together to enable the agent to evaluate the process results based on sensor data. Here, reinforcement learning approaches are worth mentioning, which mostly use simulations to allow mistakes during training or at least during pretraining. However, the focused scenario does not require a complex model, e.g. digital twin, which takes into account all physical effects of the process.

Since agent-based systems pave the way for active optimization, they gain growing attention also in manufacturing industry [14]. Such active learning algorithms choose the data from which they learn [13]. Learning an optimal policy of a system through a minimal number of experiments is still a problem [5], that is particularly relevant for real-world interacting applications of machine learning. An efficient global optimization for expensive experiments approximates the black-box behavior function of the real system based on Gaussian process theory and a set of observation points [8]. To achieve this, the so-called Kriging surrogate model [12] estimates the uncertainty of each predicted point and updates its model by adding gained knowledge about new observations for the most promising candidates. Such surrogate-based methods are derivative-free and widely used [16], and extensions are also available that take constraints into account [2].

For industrial applications, however, it is not the learning rate that is most decisive, but the minimization of risks during learning. Therefore, approaches with safety guarantees are particularly interesting in this application field, which lead to increased research relevance [1,6,15]. To address this problem, this paper presents an iteration- and surrogate-based approach that can be used during production processes and that guarantees a user-defined minimum performance during learning. Based on the assumptions and methods described in Sect. 2, an active learning approach ensuring safety is proposed and evaluated through the use of experiments with different complexities in Sect. 3. Finally, the results are critically interpreted and an outlook is given on further research activities.

2 Assumptions and Methods

The problem is defined mathematically in the Sect. 2.1 based on assumptions made in Sect. 2.2. Theoretical background about used methods of the presented approach to optimize the parameters of our experiments is content of Sect. 2.3.

2.1 Problem Definition

The general optimization problem of a multidimensional nonlinear function $f_{(x)}$ is formulated by:

$$\max_{x \in \mathcal{X}^d} f_{(x)}. \tag{1}$$

In this work the approach focuses on industrial processes. Their performance (e.g. product quality or cycle time) is described by $f_{(x)}$ and depends on multidimensional sensor data z (e.g. product material or weight) and process parameters a (e.g. control parameter or conveyor speed). While the sensor data is given by the industrial environment, the process parameters are adjustable by the agent. For the unknown performance function $f_{(x)}$, the variable x is defined as (a, z) and the general problem (1) can be reformulated:

$$\max_{a \in \mathcal{A}^d} f_{(a,z)}. \tag{2}$$

In detail, a is the action of adjusting the parameters and z is the state variable of the industrial environment defined by sensor data. For each value of z there exist an optimized adjustment of a to obtain the maximum achievable performance of the experiment. The scenario discussed is the automated acquisition of knowledge about this optimal mapping of states and actions without negatively affecting the process performance during and after the learning.

2.2 Assumptions

The proposed approach is based on assumptions that generally do not apply to all production lines. It is therefore recommended to check the assumptions mentioned below before applying the method in the field to avoid undesirable behavior.

Continuity: Even thought the presented approach is derivative-free, the dynamics of the system $f_{(a,z)}$ must be continuous in the domain space $(\mathcal{A}, \mathcal{Z})$ to ensure safe active learning. Any discontinuity, even with cautious exploration, could lead to an unexpected observation of unsafe behavior.

Domain Expertise: For reliable active learning during processes, domain experts already know how to adjust the parameters a depending on the environmental conditions of the plant to prevent a downtime. There is no need for optimal performances with this given policy, which is to be improved as learning progresses.

Performance Assignment: The performance for each executed experiment can be assigned to a known parameter setting and environmental condition to include the observation knowledge for the model update. Especially in mass production, this assumption cannot always be fulfilled.

2.3 Background on Used Methods

In this section the necessary knowledge is provided to understand the workflow of active process learning in the field with minimal risk of performance failures.

Gaussian Processes: For the simple regression of an unknown function $f_{(x)}$, Gaussian processes are often used in machine learning [11]. Based on the previously made observations, mean values for each point x_P can be calculated to predict $f_{(x_P)}$. In addition, the variances for each x_P is also available, which indicate the uncertainty and are often used to explore the function in as few steps as possible. Depending on the problem, different kernel functions $k(x, x')$ can be used to handle high frequencies of $f_{(x)}$.

Bayesian Optimization: Bayesian optimization [10] is a method that explores the unknown function $f_{(x)}$ in pursuit of the global maximum, which is the best process performance in our case, by selecting new observation points of x. By modeling $f_{(x)}$ with random functions the most informative observation points are determined. Here, $f_{(x)}$ is modeled by a Gaussian process, which is the most common probabilistic model used in Bayesian optimization [11]. This method can be used to solve the mathematical problem (1). The main advantage of Bayesian optimization as optimization method is the low number of necessary experiments, while the higher computational cost per iteration is the drawback. Process experiments in manufacturing industry take often seconds (e.g. welding) till hours (e.g. 3D printing). Bayesian optimization is therefore suitable for process optimization in industrial environments, especially for learning on the edge. In addition, Bayesian optimization does not need to take into account derivatives.

Contexts: External variables, which are considered as fixed and cannot be influenced directly for the next experiment, are defined as contexts z. Since flexible manufacturing processes play an important role in modern factory concepts, incoming features for the next process step can be interpreted as possible contexts. Gaussian processes can also be used to model the contextual influence by the usage of the kernel $k(z, z')$.

Contextual Bayesian Optimization: Contextual Bayesian optimization [9] is an conceptual extension of Bayesian optimization to optimize the performance of complex systems with environmental contexts [3]. By multiplying the kernel functions of the context and the action space (3), we assume that a performance is dependent on these variables and there are no other variable influences that should change the agents behavior. In this way each action results in a similar way if contexts were similar. Here, squared exponential kernels are used because of their diversity.

$$k(a, z, a', z') = k(a, a') \cdot k(z, z') \tag{3}$$

Since the context z is given for each iteration, it can not be optimized but affects the aiming to find the optimal performance for this z. The modification is generally able to solve the problem (2). However, it does not meet the defined limitations, as the method would select *greedy* observation points and accept temporary failures during exploration.

Safe Bayesian Optimization: SAFEOPT [15] is a safe Bayesian optimization algorithm enabling efficient, automatic and global optimization of parameters a without the risk of failures [4], which is important, as the plant and its components can be expensive. Furthermore, industrial customers are usually not willing to take any risks of a lower production quality. For each iteration n the optimization is restricted to a safe set S_n, which is extended during the exploration. Depending on the safe set and the performance uncertainty the next observation point is selected [3].

3 Safe Active Learning

Based on the Gaussian process framework GPY [7] and the safe Bayesian optimization framework SAFEOPT [4], the approach and all experiments are implemented in Python as described in Sect. 3.1 to meet industrial requirements such as manufacturing quality assurance. Two experiments with different complexity are described in Sect. 3.2 and finally used as proof of concept to evaluate the method in Sect. 3.3.

3.1 Approach

An initial safe parameter a_0 with known context z_0 defines the safe set as $S_0 \ni (a_0, z_0)$. This set is extended step-wise based on observations and the confidence intervals. Whenever SAFEOPT fails to suggest safe parameters a_n for the next iteration n based on the obtained context z_n, the algorithm redefines a_n thought a user-defined function $s_{(z)}$, which enables a continuous learning during industrial daily routine.

During the exploration with SAFEOPT, performances of the system can vary. Since manufacturers have to guarantee a defined quality all the time, a safe performance threshold J_{min} is chosen. Assuming that performance can be measured or calculated by sensor data, the threshold should be based on the minimum performance achieved before the learning approach. If it is not measured directly, it should be calculated using the same formula as later.

3.2 Experiments

For reasons of illustration and comprehensibility, one-dimensional policy functions are optimized in the following experiments. Since the proposed method focuses on low-dimensional problems and becomes exponentially harder when

Algorithm 1. Active Learning Approach

Data:

- Number of iterations N,
- Gaussian process regression $\mathcal{GP}$ with kernel $k(a, z, a', z')$,
- Standard parameter function $s_{(z)}$,
- Domain $\mathcal{A}$ for parameters a,
- Domain $\mathcal{Z}$ for parameters z,
- Initial safe set $S_0 \subseteq (\mathcal{A}, \mathcal{Z})$,
- Minimum performance threshold J_{min},
- Properties of SafeOpt.

Result: trained $\mathcal{GP}$.

```
begin
    initialize SafeOpt(GP, A, J_min);
    for n < N do
        wait until new process starts;
        get z from environmental sensor data;
        get a_n(z) from SafeOpt with context z;
        if (a_n(z), z) ∉ S_n then
            a_n(z) ← s_(z);
        end
        set process parameters;
        while process do
            collect necessary sensor data;
        end
        calculate f_(a_n(z), z) from collected data;
        update GP and S_n+1 ⊆ (A, Z);
    end
    return GP;
end
```

dimensions increase, it is recommended to use a subspace of $\mathcal{A}$ or $\mathcal{Z}$ for higher dimensional problems [17].

The first simulated process behavior is defined by a simple function with a very dangerous minimum, see Fig. 1 top left contour plot:

$$f_{1(a,z)} = 0.5\sin(2a) + 0.25\cos(3z) + 0.25\cos(3az) + 1. \tag{4}$$

For the second experimental setup, a more complex problem with several local maxima and minima was randomly modeled by a Gaussian process, see Fig. 1 bottom left contour plot.

In each setup domains $\mathcal{A}$ and $\mathcal{Z}$ are chosen such that $a \in [-1, 1]$ and $z \in [0, 2]$ are true. Here, the process parameter a can be interpreted as a percentage change of the standard value. Hence, without expertise, the process will be executed with $a = 0$ for each z like in the first experiment (see green dashed line of Fig. 1 top left contour plot). By domain expertise, the expert-defined parameter function $s_{(z)}$ can also be set up more complex than $s_{1(z)}$ for the first example,

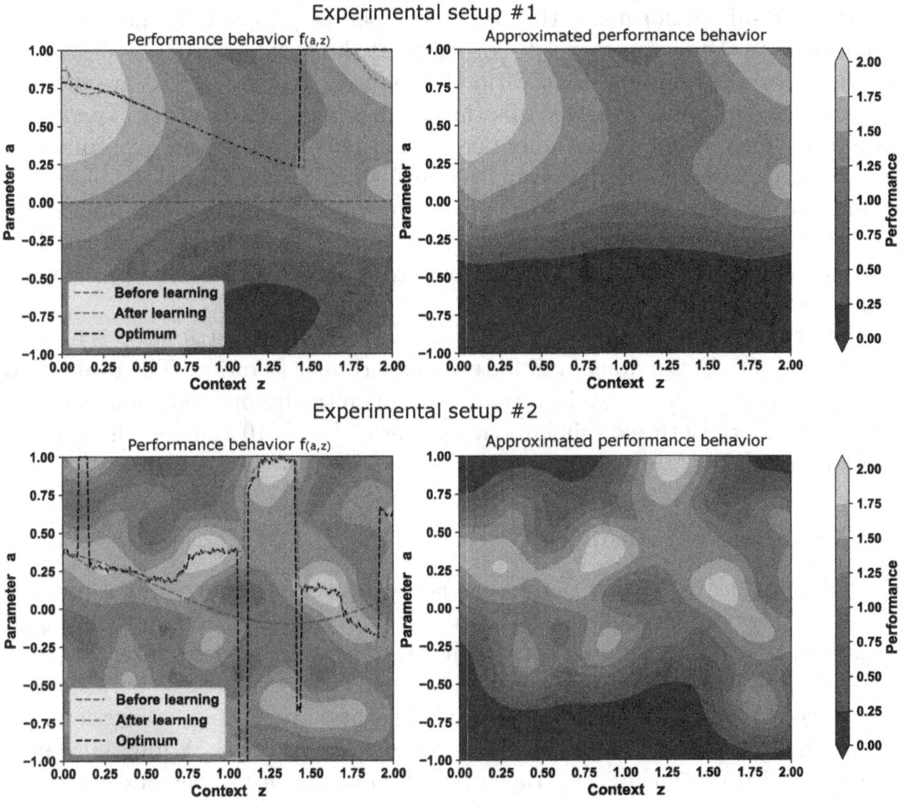

Fig. 1. Experimental setups. Left-hand contour plots: The true performance behavior of two experimental setups depending on the one-dimensional setting and the given context are illustrated. Dashed lines indicate the standard setting before optimization (green), which is declared as safe, the theoretical optimum in our selected space (black) and the optimized setting after 200 iterations (red). Right-hand contour plots: The safe explored and approximated performance behavior of one out of ten exemplary runs after 200 iterations. (Color figure online)

$$s_{1(z)} = 0. \tag{5}$$

For the second example, a sinusoidal function

$$s_{2(z)} = 0.25 \sin(2z + 2) + 0.15 \tag{6}$$

represents the prior expertise, see green dashed line of Fig. 1 bottom left contour plot. In addition, S_0 can be extended with previous expert experimental results, but no S_0 was used for these experiments.

Each parameter selection for a context leads to a normalized performance between 0 and 2. For both experiments, a minimum performance of 1.0 is gained by the usage of the standard policy $s_{(z)}$. To ensure safety during the active learning, the safe performance threshold is chosen as $J_{min} = 0.95$. Lower values would increase the learning speed, but decrease the provided safety.

The theoretical optimal policies in the selected domain ($\mathcal{A}$, $\mathcal{Z}$) are marked as black dashed lines in Fig. 1. While the optimal policy for the first experiment has almost linear sections and only one discontinuity, the second experiment is quite complex with several discontinuities and difficult to define mathematically. In order to model such an optimal policy by a human expert, comprehensive process knowledge would be required. The goal of the optimization problem is to create a good policy by carrying out experiments while not reducing the obtained experiment performances.

The random number generator of NumPy with uniform distribution is used to select contexts z in the domain $\mathcal{Z}$ for each experiment iteration. The evaluation is carried out after $N = 200$ iterations, i.e. theoretically one iteration per 0.01 of the context z. Both experiments are performed $n = 10$ times with different random seeds.

3.3 Experimental Results

The determined observation points did not approach the dangerous minimum area of the first experimental setup because of the large defined threshold J_{min}, which leaded to an unexplored spot of the systems behavior even after 200 iterations, see top right contour plot in Fig. 1. In this way, the process performance during the exploration was not significantly negatively affected, which can also be identified by the upper performance distribution in Fig. 2. A minimum performance of 1.0 was ensured for the entire 200 reaction iterations of each of the 10 experiments performed per setup. In the industrial context, the plant could therefore produce high quality, even if it learns more about its behavior.

In fact, the qualities have already improved during exploration, with less frequent occurrences of nearly 1.0 and more frequent occurrences of up to 2.0 cases. However, the upper right performance behavior maximum of the first experiment was still not detected, which leads to the differences between optimal (black) and learned policy (red), see top left contour plot in Fig. 1. This maximum is taken into account after 50 additional iterations for the illustrated exemplary run in Fig. 1. Depending on the used random seed for the experiment, it is also possible to archive this knowledge within the first 200 iterations.

In the upper left plot of Fig. 2, the performance depending on the context before and after learning is compared with the optimal achievable one. Again, the only big difference of our learned model, depending on the random seed used, is caused by a small number of experiments with a context close to 2.

For the second experimental setup, several local maxima were detected, see bottom right contour plot in Fig. 1. A complex policy with several discontinuities was learned by the agent, see bottom left contour plot in Fig. 1. However, similar to the first experiment also here are still some differences to the optimal policy. This can also be seen in the bottom left plot of Fig. 2. Also for the second experimental setup, no performance below 1.0 was achieved during the learning process. The high demands on the performance guarantee also have a downside

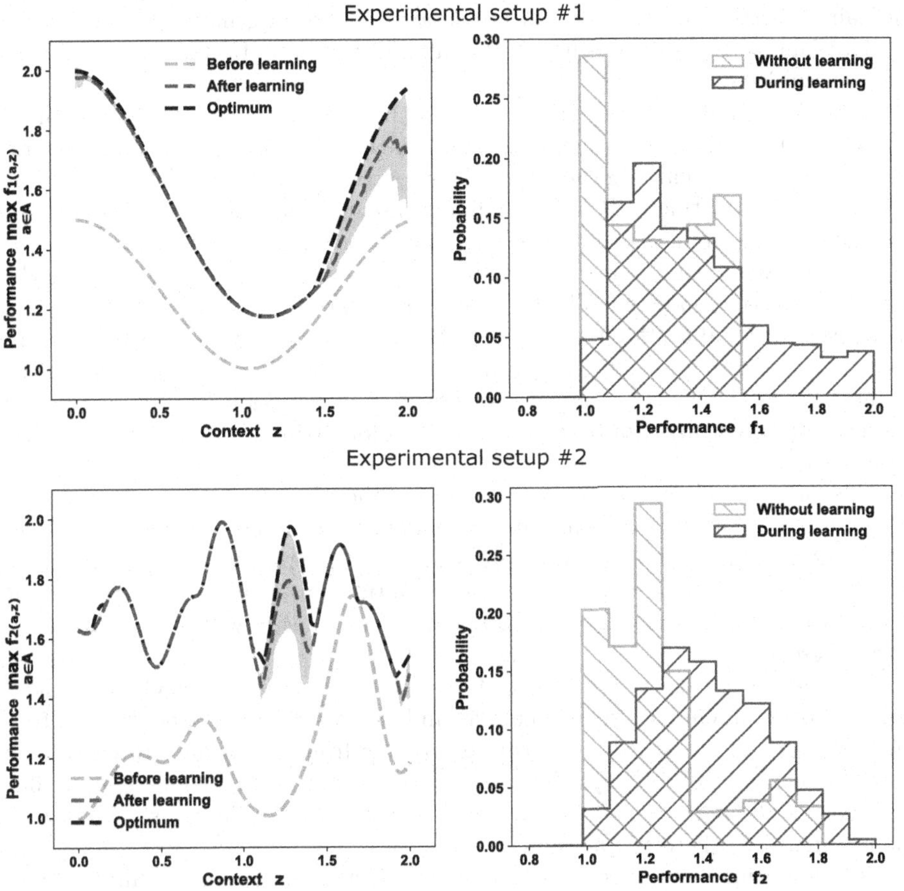

Fig. 2. Left-hand plots: The performance behaviors of the experiments against the given context is illustrated before (green) and after learning (red) with standard deviation (light red colored area). For almost all context values the performances are improved. However, after 200 iterations there are still differences to the optimum (black). Right-hand histograms: During 200 iterations for each experiment, the obtained performance probability (red) was improved in contrast the performance with standard policy (green), without causing failures (no cases with a lower performance than 1.0). (Color figure online)

as the agent will never explore the local maximum at $f_{2(1.0,0.15)}$ without leaving the domain space due to the fact that it is surrounded by prohibited areas in the domain ($\mathcal{A}$, $\mathcal{Z}$).

4 Conclusion and Outlook

The presented approach enables iteration-based learning in industrial daily routine whilst not increasing the risk of failure. It was possible to include prior domain expertise to improve and accelerate the learning method. Even if the evaluation results are promising, they should be interpreted with caution because of three major drawbacks.

First, the method is not able to obtain the optimal policy when prohibited minima surround a local maximum, as shown with the second experimental setup. Second, the learning rate is small compared with other methods due to the conservative adjustment. It is reasonable to assume that the method is suitable for applications in mass production or frequently recurring tasks with assignable performances. Third, the method requires increasing computing time during the extending of the modeling. The usage of batch learning methods could improve the computational costs but complicate the safety concerns of the approach. In addition, it is essential to forget redundant information in order to minimize the computing costs caused by increased observation information. Therefore, we plan to study the correlation of batch methods and forgetting redundant information approaches.

An continuous optimization should include the adaptation to changing conditions such as aging sensors. This need is reinforced by the changing behavior over time due to wear or long-term changes in environmental conditions that have not been taken into account. The discrimination problem between redundant information to forget and important findings to memorize will be the subject of future research.

As the concept has turned out to be feasible, research will also be continued with hardware including experiments for industrially relevant applications. This includes various tasks such as feature extraction from sensor data to reflect the environmental condition with the lowest possible dimensional context. A similar challenge is the performance measurement of the influenced process, which must be assignable to the made action.

The long-time developed expertise of manufactures can lead to complex functions depending on sensor data to obtain good performance without machine learning approaches. This domain expertise can be used for the definition of the standard parameter function $s_{(z)}$. With the proposed method manufacturers are able to evaluate their expertise based policy or even gain further knowledge about correlations by made performance improvements without increased risk.

References

1. Akametalu, A.K., Fisac, J.F., Gillula, J.H., Kaynama, S., Zeilinger, M.N., Tomlin, C.J.: Reachability-based safe learning with Gaussian processes. In: 53rd IEEE Conference on Decision and Control, pp. 1424–1431. IEEE (2014)
2. Audet, C., Denni, J., Moore, D., Booker, A., Frank, P.: A surrogate-model-based method for constrained optimization. In: 8th Symposium on Multidisciplinary Analysis and Optimization, p. 4891 (2000)
3. Berkenkamp, F., Krause, A., Schoellig, A.P.: Bayesian optimization with safety constraints: safe and automatic parameter tuning in robotics. arXiv preprint arXiv:1602.04450 (2016)

4. Berkenkamp, F., Schoellig, A.P., Krause, A.: Safe controller optimization for quadrotors with Gaussian processes. In: 2016 IEEE International Conference on Robotics and Automation (ICRA), pp. 491–496. IEEE (2016)
5. Chatzilygeroudis, K., Vassiliades, V., Stulp, F., Calinon, S., Mouret, J.B.: A survey on policy search algorithms for learning robot controllers in a handful of trials. arXiv preprint arXiv:1807.02303 (2018)
6. Fisac, J.F., Akametalu, A.K., Zeilinger, M.N., Kaynama, S., Gillula, J., Tomlin, C.J.: A general safety framework for learning-based control in uncertain robotic systems. IEEE Trans. Autom. Control **64**(7), 2737–2752 (2019). https://doi.org/10.1109/TAC.2018.2876389
7. GPy: a Gaussian process framework in Python (2012). http://github.com/SheffieldML/GPy
8. Jones, D.R., Schonlau, M., Welch, W.J.: Efficient global optimization of expensive black-box functions. J. Glob. Optim. **13**(4), 455–492 (1998)
9. Krause, A., Ong, C.S.: Contextual Gaussian process bandit optimization. In: Advances in Neural Information Processing Systems (NIPS), pp. 2447–2455 (2011)
10. Mockus, J.: Bayesian Approach to Global Optimization: Theory and Applications, vol. 37. Springer, Heidelberg (2012)
11. Rasmussen, C.E., Williams, C.: Gaussian Processes for Machine Learning, vol. 1. MIT press, Cambridge (2006). **39**, 40–43
12. Sacks, J., Welch, W.J., Mitchell, T.J., Wynn, H.P.: Design and analysis of computer experiments. Stat. Sci. **4**, 409–423 (1989)
13. Settles, B.: Active learning literature survey. Computer Sciences Technical report 1648, University of Wisconsin-Madison Department of Computer Sciences (2009)
14. Shen, W., Hao, Q., Yoon, H.J., Norrie, D.H.: Applications of agent-based systems in intelligent manufacturing: an updated review. Adv. Eng. Inf. **20**(4), 415–431 (2006)
15. Sui, Y., Gotovos, A., Burdick, J., Krause, A.: Safe exploration for optimization with Gaussian processes. In: International Conference on Machine Learning, pp. 997–1005 (2015)
16. Vu, K.K., D'Ambrosio, C., Hamadi, Y., Liberti, L.: Surrogate-based methods for black-box optimization. Int. Trans. Oper. Res. **24**(3), 393–424 (2017)
17. Wang, Z., Zoghi, M., Hutter, F., Matheson, D., De Freitas, N.: Bayesian optimization in high dimensions via random embeddings. In: Twenty-Third International Joint Conference on Artificial Intelligence, pp. 1778–1784 (2013)

Federated Learning of Deep Neural Decision Forests

Anders Sjöberg[1,2(✉)], Emil Gustavsson[1,2], Ashok Chaitanya Koppisetty[3], and Mats Jirstrand[1,2]

[1] Fraunhofer-Chalmers Centre, 412 88 Gothenburg, Sweden
{anders.sjoberg,emil.gustavsson,mats.jirstrand}@fcc.chalmers.se
[2] Fraunhofer Center for Machine Learning, Gothenburg, Sweden
[3] Volvo Car Corporation, 405 31 Gothenburg, Sweden
ashok.chaitanya.koppisetty@volvocars.com
http://www.fcc.chalmers.se/
https://www.volvocars.com/

Abstract. Modern technical products have access to a huge amount of data and by utilizing machine learning algorithms this data can be used to improve usability and performance of the products. However, the data is likely to be large in quantity and privacy sensitive, which excludes the possibility of sending and storing all the data centrally. This in turn makes it difficult to train global machine learning models on the combined data of different devices. A decentralized approach known as federated learning solves this problem by letting devices, or clients, update a global model using their own data and only sending changes of the global model, which means that they do not need to communicate privacy sensitive data.

Deep neural decision forests (DNDF), inspired by the versatile algorithm random forests, combine the divide-and-conquer principle together with the property representation learning. In this paper we further develop the concept of DNDF to be more suited for the framework of federated learning. By parameterizing the probability distributions in the prediction nodes of the forest, and include all trees of the forest in the loss function, a gradient of the whole forest can be computed which some/several federated learning algorithms utilize. We demonstrate the inclusion of DNDF in federated learning by an empirical experiment with both homogeneous and heterogeneous data and baseline it against a convolutional neural network with the same architecture as the DNDF. Experimental results show that the modified DNDF, consisting of three to five decision trees, outperform the baseline convolutional neural network.

Keywords: Federated learning · Deep neural decision forests · Parameterization

G. Nicosia et al. (Eds.): LOD 2019, LNCS 11943, pp. 700–710, 2019.
https://doi.org/10.1007/978-3-030-37599-7_58

1 Introduction

An almost endless amount of data is stored in modern technical products, in everything from vehicles to mobile phones and tablets [8]. These products are equipped with sensors, e.g. cameras, microphones and GPS, and this together with the fact that the products are frequently used and widely spread [15] creates this plenitude of data. Modern machine learning models can utilize this massive amount of data to improve the usability and user experience of the product. Moreover, the automotive industry benefits greatly of data and machine learning to improve active safety and self driving cars [1]. However, the nature of the data collected from technical products can be highly private, which means that machine learning models cannot be trained in a centralized location due to the increased risk of exposing personal information.

A learning technique that allows the users to utilize the benefits of machine learning models trained from rich data, without storing the data centrally, is called *Federated Learning* [13]. The reason behind the name is that the learning task is solved by a loose federation of *clients* (e.g., vehicles and mobile phones) that are coordinated by a central *server*. The idea is that each client has its own local data set which is never uploaded to the server. Instead, the server constructs a global model and distribute it to all the clients and then the clients update the model with their own data and only the update of the model is communicated. There is no need for direct access of the data for the server, which is the advantage of this approach. The updates of the model generally contain much less information than the raw data itself [13]. In other words, federated learning considerably decreases security and privacy risks since the server never stores any raw data.

Random forests [2] are versatile machine learning models [7] that are widely used through history, especially in computer vision [4]. While working with high dimensional data it has been empirically shown that random forests are competitive to many of the efficient learners [3]. Furthermore, random forests can be distributed on parallel hardware architectures without difficulty and they are also considered to be close to an ideal learner [17]. These reasons together with many computationally attractive properties make random forests convenient to use for both commercial products [16] and for research questions [5].

It would be highly desirable to adapt the versatile and efficient algorithm random forests to fit the framework of federated learning. Unfortunately, this is not easily implemented since federated learning emphasizes that the training process of the model is gradual, through communication, and that is not the case for random forests. Once the forests are established there is no clear way to further train them. However, an intuitive and naive approach of adapting random forests to fit the framework of federated learning is for the forests to share their decision trees from different clients, or by modifying the nodes in the trees after each communication, or by some sort of pruning. A more advanced approach is instead to introduce *Deep Neural Decision Forests* [11]. Deep neural decision forests combine the divide-and-conquer principle of decision trees together with the attractive property representation learning as known from deep architectures.

In this paper we introduce some modification of deep neural decision forests to make them well-suited for federated learning.

1.1 Contributions

Our main contribution in this paper is that we incorporate the concept of deep neural decision forests in the federated learning framework. We modify the deep neural decision forest so the whole forest can be trained with gradient descent methods. This is convenient because several federated learning algorithms are derived from or based on gradient updates, such as *Federated Averaging* [13] and *Federated Stochastic Variance Reduced Gradient* [9]. This modification is done by parameterizing the distributions in the prediction nodes of the deep neural decision forest. This enables the deep neural decision forests to be completely described by parameters. Lastly, we include all the decision trees in the loss function so one can compute a gradient for the whole forest.

2 Deep Neural Decision Forests

This section is based on the theory in [11]. Let us consider a classification problem where the input and the output space are given by $\mathcal{X}$ and $\mathcal{Y}$, respectively. An ensemble of decision trees is a decision forest $\mathcal{F} = \{T_i : i \in \mathcal{I}\}$, where T_i denotes decision tree $i \in \mathcal{I}$. The internal decision nodes in decision tree i are indexed by $\mathcal{N}_i$, while the prediction nodes, the terminal nodes of the tree, are indexed by $\mathcal{L}_i$. In each prediction node $l \in \mathcal{L}_i$ there is a distribution $\boldsymbol{\pi}_l(\boldsymbol{\Theta}_\pi)$, parameterized with $\boldsymbol{\Theta}_\pi$, over $\mathcal{Y}$. The probability of a sample point that has reached leaf l will take on class y is denoted $\pi_{ly}(\boldsymbol{\Theta}_\pi) = \boldsymbol{\pi}_l(y; \boldsymbol{\Theta}_\pi)$, which is parameterized with a softmax function, i.e.,

$$\pi_{ly}(\boldsymbol{\Theta}_\pi) = \frac{e^{\Theta_{\pi_{ly}}}}{\sum_{i \in \mathcal{Y}} e^{\Theta_{\pi_{li}}}}. \tag{1}$$

Each internal decision node $n \in \mathcal{N}_i$ is given a *decision function* $d_n(\cdot; \boldsymbol{\Theta}_d) : \mathcal{X} \to [0, 1]$, parameterized by $\boldsymbol{\Theta}_d$, which describes how the sample points route through the decision tree. When a sample point $\boldsymbol{x} \in \mathcal{X}$ arrives at a decision node n it will either move into the left or right subtree based on the output of the decision function $d_n(\boldsymbol{x}; \boldsymbol{\Theta}_d)$. In a regular decision tree the function d_n is binary and completely deterministic. However, the routing through a decision tree in the deep neural decision forests is probabilistic. The direction of the routing is decided by the output of a Bernoulli stochastic variable with mean $d_n(\boldsymbol{x}; \boldsymbol{\Theta}_d)$. Whenever a sample point in a standard decision tree ends up in a terminal node l the prediction is given by the leaf distribution $\boldsymbol{\pi}_l$. However, in a probabilistic routing the leaf predictions are averaged by the probability of reaching that leaf. This means that the final prediction of sample point $\boldsymbol{x}$ from decision tree T_i with decision and prediction nodes parameterized by $\boldsymbol{\Theta} = (\boldsymbol{\Theta}_\pi, \boldsymbol{\Theta}_d)$ is given by

$$P_{T_i}(y|\boldsymbol{x}, \boldsymbol{\Theta}) = \sum_{l \in \mathcal{L}_i} \pi_{ly}(\boldsymbol{\Theta}_\pi) \mu_l(\boldsymbol{x}; \boldsymbol{\Theta}_d), \ i \in \mathcal{I}, \tag{2}$$

where $\mu_l(\boldsymbol{x}; \boldsymbol{\Theta}_d)$ is the *routing function*, which gives the probability that sample point $\boldsymbol{x}$ reaches leaf l.

The explicit form of the routing function is given by

$$\mu_l(\boldsymbol{x}; \boldsymbol{\Theta}_d) = \prod_{n \in \mathcal{N}_i} d_n(\boldsymbol{x}; \boldsymbol{\Theta}_d)^{\mathbb{1}_{l \swarrow n}} \bar{d}_n(\boldsymbol{x}; \boldsymbol{\Theta}_d)^{\mathbb{1}_{n \searrow l}},$$

where $\bar{d}_n(\boldsymbol{x}; \boldsymbol{\Theta}_d) = 1 - d_n(\boldsymbol{x}; \boldsymbol{\Theta}_d)$, and the indicator functions are defined as

$$\mathbb{1}_{l \swarrow n} = \begin{cases} 1, & \text{if } l \text{ belongs to the left subtree of node } n, \\ 0, & \text{otherwise,} \end{cases}$$

and

$$\mathbb{1}_{n \searrow l} = \begin{cases} 1, & \text{if } l \text{ belongs to the right subtree of node } n, \\ 0, & \text{otherwise,} \end{cases}$$

see Fig. 1 for an illustration.

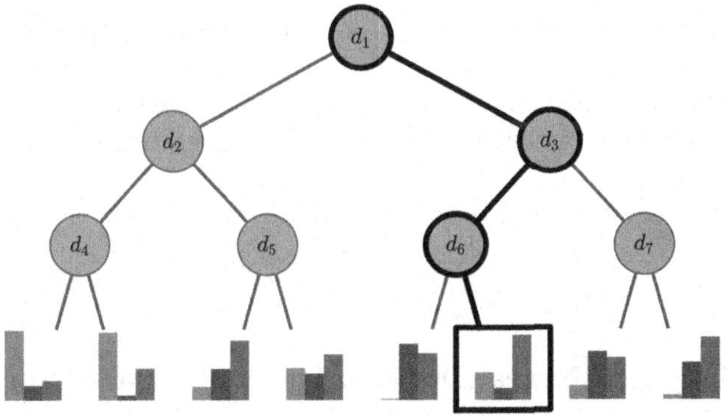

Fig. 1. Each node $n \in \mathcal{N}_i$ of tree i decides the routing decision with the decision function $d_n(\cdot; \boldsymbol{\Theta}_d)$. The black path shows an example of route that a sample point $\boldsymbol{x}$ can take in a tree, which has the probability $\bar{d}_1(\boldsymbol{x}; \boldsymbol{\Theta}_d) d_3(\boldsymbol{x}; \boldsymbol{\Theta}_d) \bar{d}_6(\boldsymbol{x}; \boldsymbol{\Theta}_d)$.

In this paper the decision functions, that are providing a stochastic routing, are defined as:

$$d_n(\boldsymbol{x}; \boldsymbol{\Theta}_d) = \sigma(f_n(\boldsymbol{x}; \boldsymbol{\Theta}_d)), \tag{3}$$

where $\sigma(x) = (1 + e^{-x})^{-1}$ is the sigmoid function, and $f_n(\cdot; \boldsymbol{\Theta}_d) : \mathcal{X} \to \mathbb{R}$ is a real-valued function depending on the sample and the parameterization $\boldsymbol{\Theta}_d$. Furthermore, we consider the functions $f_n(\cdot, \boldsymbol{\Theta}_d)$ to be embedded within a deep convolutional neural network with parameters $\boldsymbol{\Theta}_d$. More precisely, we regard each function f_n as an output of a deep convolutional neural network that is turned into a stochastic routing decision by the action of d_n, which due the sigmoid function is in the range $[0, 1]$. Note that we can construct a deep neural

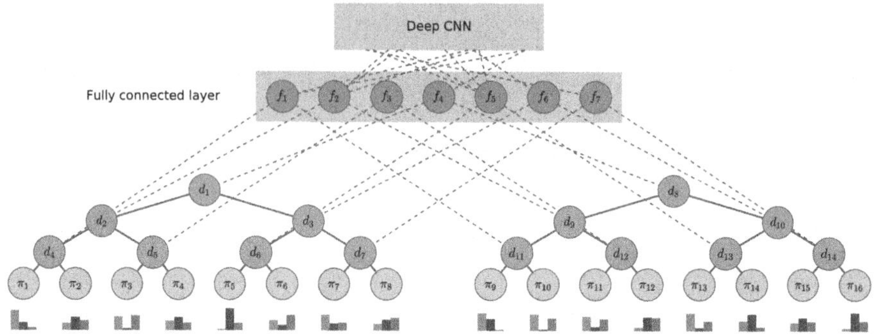

Fig. 2. An illustration of a deep neural decision forest. On the top there is a deep convolutional neural network, CNN, with an arbitrary number of layers, subsumed via $\boldsymbol{\Theta}_d$. Below that is a fully connected layer that is used to yield the functions $f_n(\cdot;\boldsymbol{\Theta}_d)$, see Eq. (3). Each output of f_n corresponds to a decision node for each decision tree, i.e., $d_n(\boldsymbol{x};\boldsymbol{\Theta}_d) = \sigma(f_n(\boldsymbol{x};\boldsymbol{\Theta}_d))$. The orange circles at the bottom represent prediction nodes, each with a probability distribution $\boldsymbol{\pi}_l(\boldsymbol{\Theta}_\pi)$ (Color figure online)

decision forest so that for every $(i,j) \in \mathcal{I}\times\mathcal{I}$ there exist $n \in \mathcal{N}_i$ and $m \in \mathcal{N}_j$ such that $f_n \equiv f_m$, see Fig. 2. In other words, an output from a deep convolutional neural network can be connected to many different decision trees.

The final prediction of the decision forest $\mathcal{F}$ is given by averaging the output of all the trees, i.e.,

$$P_{\mathcal{F}}(y|\boldsymbol{x},\boldsymbol{\Theta}) = \frac{1}{|\mathcal{I}|}\sum_{i\in\mathcal{I}} P_{T_i}(y|\boldsymbol{x},\boldsymbol{\Theta}).$$

2.1 Learning Deep Neural Decision Forests by Back-Propagation

Training deep neural decision forests requires estimating the parameter $\boldsymbol{\Theta} = (\boldsymbol{\Theta}_\pi, \boldsymbol{\Theta}_d)$. By using the minimum empirical risk principle we can estimate the parameter $\boldsymbol{\Theta}$ with respect to a given data set $\mathcal{T} \subset \mathcal{X}\times\mathcal{Y}$ under log-loss, i.e., we want to minimize the following risk sum:

$$R(\boldsymbol{\Theta};\mathcal{T}) = \frac{1}{|\mathcal{T}||\mathcal{I}|}\sum_{(\boldsymbol{x},y)\in\mathcal{T}}\sum_{i\in\mathcal{I}} L_i(\boldsymbol{\Theta};\boldsymbol{x},y),$$

where $L_i(\boldsymbol{\Theta};\boldsymbol{x},y)$ is the log-loss term for the sample $(\boldsymbol{x},y)\in\mathcal{T}$, which is defined as

$$L_i(\boldsymbol{\Theta};\boldsymbol{x},y) = -\log(P_{T_i}(y|\boldsymbol{x},\boldsymbol{\Theta})), \quad i \in \mathcal{I},$$

where P_{T_i} is defined in Eq. (2).

Minimizing the risk $R(\boldsymbol{\Theta};\mathcal{T})$ can definitely be a difficult and large-scale optimization problem, therefore we choose to use gradient based methods, e.g., *Adam* [6] and Stochastic Gradient Descent. The gradient of $R(\boldsymbol{\Theta};\mathcal{T})$ can be decomposed into

$$\frac{\partial R}{\partial \boldsymbol{\Theta}}(\boldsymbol{\Theta}; \mathcal{T}) = \frac{1}{|\mathcal{T}||\mathcal{I}|} \sum_{(\boldsymbol{x},y)\in\mathcal{T}} \sum_{i\in\mathcal{I}} \begin{pmatrix} \frac{\partial L_i}{\partial \boldsymbol{\Theta}_d}(\boldsymbol{\Theta}; \boldsymbol{x}, y) \\ \frac{\partial L_i}{\partial \boldsymbol{\Theta}_\pi}(\boldsymbol{\Theta}; \boldsymbol{x}, y) \end{pmatrix}.$$

The gradient of the loss L_i with respect to $\boldsymbol{\Theta}_d$ can be decomposed by the chain rule to

$$\frac{\partial L_i}{\partial \boldsymbol{\Theta}_d}(\boldsymbol{\Theta}; \boldsymbol{x}, y) = \sum_{n\in\mathcal{N}_i} \frac{\partial L_i(\boldsymbol{\Theta}; \boldsymbol{x}, y)}{\partial f_n(\boldsymbol{x}; \boldsymbol{\Theta}_d)} \frac{\partial f_n(\boldsymbol{x}; \boldsymbol{\Theta}_d)}{\partial \boldsymbol{\Theta}_d}.$$

Furthermore, the gradient term that depends on the decision tree is given by

$$\frac{\partial L_i(\boldsymbol{\Theta}; \boldsymbol{x}, y)}{\partial f_n(\boldsymbol{x}; \boldsymbol{\Theta}_d)} = d_n(\boldsymbol{x}; \boldsymbol{\Theta}_d) A_{n_r} - \bar{d}_n(\boldsymbol{x}; \boldsymbol{\Theta}_d) A_{n_l}, \tag{4}$$

where n_l and n_r indicate the left and right child of node n, respectively, and we define A_q for an arbitrary node $q \in \mathcal{N}_i$ as

$$A_q = \frac{\sum_{l\in\mathcal{L}_q} \pi_{ly}(\boldsymbol{\Theta}_\pi)\mu_l(\boldsymbol{x}; \boldsymbol{\Theta}_d)}{P_{T_i}(y|\boldsymbol{x}, \boldsymbol{\Theta})}.$$

The set $\mathcal{L}_q \subseteq \mathcal{L}_i$ denotes the set of leaves held by the subtree rooted in node q. A detailed derivation of Eq. (4) can be found in Supplementary Material in [11]. Finally, the gradient of the lost L_i with respect to $\boldsymbol{\Theta}_\pi$ is

$$\frac{\partial L_i}{\partial \boldsymbol{\Theta}_\pi}(\boldsymbol{\Theta}; \boldsymbol{x}, y) = -\frac{1}{P_{T_i}(y|\boldsymbol{x}, \boldsymbol{\Theta})} \sum_{l\in\mathcal{L}_i} \mu_l(\boldsymbol{x}; \boldsymbol{\Theta}_d) \frac{\partial \pi_{ly}(\boldsymbol{\Theta}_\pi)}{\partial \boldsymbol{\Theta}_\pi},$$

where the last factor is the gradient of the softmax function defined in Eq. (1).

3 Federated Learning

In this paper we assume a synchronous update scheme that communicates in rounds. Furthermore, we assume that there is a fixed set of K clients, each with a fixed local data set. In the beginning of each round of communication, a random fraction C of clients is selected, and the server sends the current global model (e.g., its parameters) to each of these clients. The reason for selecting a fraction of clients is computational efficiency, since adding more clients beyond a certain point only gives diminishing returns [13]. Each selected client updates its model locally by using its own client data set, thereafter the client sends the update to the server. The server then aggregates the models into an updated global model, and the process repeats until termination criteria is reached, see Fig. 3 for an illustrative description.

The aggregation method on the server that we choose to use in this paper is called *Federated Averaging* [13], which has shown good empirical results [14]. The idea is to average the incoming updated parameters of the model, weighted

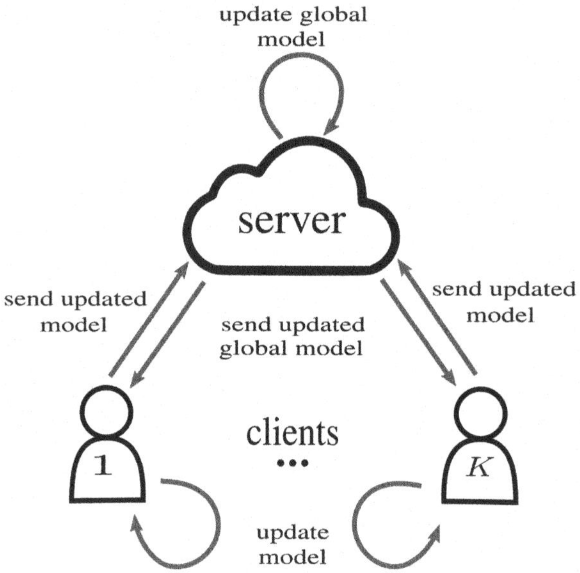

Fig. 3. The server distributes the global model to a fraction C of the K clients, where in turn they update the model with their local data and send the updated model back to the server. The server aggregate all models into one new updated global model and the process repeats, until termination criteria is reached.

according to the clients' relative data size. That means, if the total data is partitioned over K clients, with $\mathcal{P}_k$ as the set of indexes of sample points on client k, with $n_k = |\mathcal{P}_k|$, and $n = \sum_{k=1}^{K} |\mathcal{P}_k|$ as the total number of sample points, then the parameters are updated in the following way:

$$\boldsymbol{\Theta}^{(t)} \leftarrow \sum_{k \in S^{(t)}} \frac{n_k}{n} \boldsymbol{\Theta}_k^{(t)},$$

where t is the current iteration, and $S^{(t)}$ is a random set of $\max(C \cdot K, 1)$ clients at iteration t. Furthermore, let E be the number of epochs each client makes over its local data set for each communication round. Lastly, let B denote the local minibatch size used for the client updates. Federated averaging is summarized in Algorithm 1.

4 Experimental Results

The purpose of this empirical experiment is to show that the modified DNDF works in the frame work of federated learning. We consider an image classification problem, more precisely, MNIST the digit recognition task [12]. Furthermore, we partition the MNIST data set over the clients in two ways as in [13]. The first way is **IID**, independent and identically distributed, where the sample points are

Algorithm 1. *Federated Averaging.* The K clients are indexed by k; E is the number of local epochs, and B is the local minibatch size.

```
Server executes:
  Initialize the parameters Θ^(1)
  for each round t = 1, 2, ... do
    S^(t) ← (random set of max(C · K, 1) clients)
    for each client k ∈ S^(t) in parallel do
      Θ_k^(t+1) ← ClientUpdate(k, Θ^(t))
    Θ^(t+1) ← Σ_{k∈S^(t)} (n_k/n) Θ_k^(t+1)

ClientUpdate(k, Θ):
  𝓑 ← (split 𝓟_k into batches of size B)
  for each local epoch i from 1 to E do
    for batch b ∈ 𝓑 do
      Θ ← Adam(b, Θ)[6]
  return Θ to server
```

shuffled and then partitioned into 20 clients each receiving 3000 sample points. The second way is **Non-IID**, where we first sort the data by digit label, and divide into 40 shards of size 1500, and assign 2 shards to each of the 20 clients. Note that Non-IID sampling is extreme, a client can at most posses examples of two digits. The reason behind Non-IID sampling is that it reflects the real world where different users will supply with different type of data.

We construct a deep neural decision forest consisting of a convolutional neural network with three 3×3 convolution layer (the first with 32 channels, the second with 64, the third with 128, each followed with 2×2 max pooling and ReLu activations) followed with two fully connected layers (625 and 15 neurons), and this is in turn connected with two to five decision trees that have a depth of four. Each of the outputs from the convolutional neural network are connected to each of the decision trees, as in Fig. 2. Furthermore, we want to compare the result with a convolutional neural network, CNN, without any decision forest connected to it, as a baseline. This convolutional neural network is constructed in the same fashion: three 3×3 convolution layer (the first with 32 channels, the second with 64, the third with 128, each followed with 2×2 max pooling and ReLu activations) followed with two fully connected layers (625 and 10 neurons) completed with a softmax function.

Let the parameters in federated averaging, Algorithm 1, be the following: batch size $B = 10$, number of training rounds locally $E = 5$, and fraction of clients $C = 1.0$.

Table 1 shows the average number of communication rounds that is necessary to achieve a test set accuracy of 99% for the different number of decision trees in the forest and for the CNN for Non-IID and IID data over 20 and 60 runs, respectively. Figure 4 shows the average test accuracy over 20 runs corresponding to the number of communication rounds for Non-IID data.

Table 1. The data set MNIST is distributed on 20 clients. The parameters in Algorithm 1 are the following: $E = 5$, $C = 1.0$, and $B = 10$. A convolutional neural network, CNN, and a deep neural decision forest, DNDF, with four different architecture setting, are tested. The entries in the columns Non-IID and IID give the number of communication rounds needed to achieve an accuracy of 99% on a test set, averaged over 20 and 60 runs, respectively.

Algorithms	Architecture	Non-IID	IID
CNN	Fully connected layer	200.1	5.4
DNDF	2 Decision trees	120.0	5.9
	3 Decision trees	80.5	5.3
	4 Decision trees	65.6	5.2
	5 Decision trees	57.9	5.1

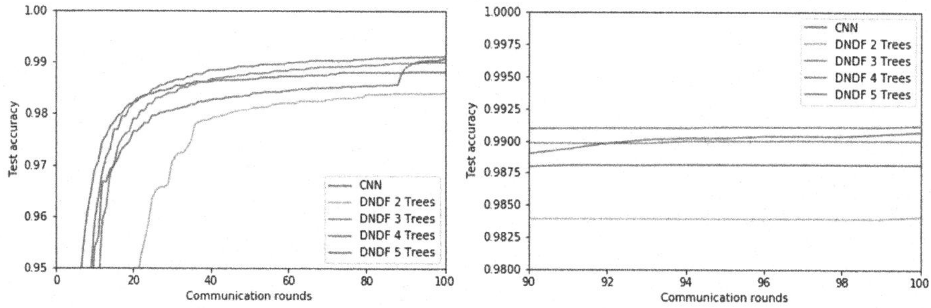

Fig. 4. The left figure visualizes the average test accuracy of 20 runs for the deep neural decision forest with four different settings, two to five decision trees, and for the convolutional neural network with fully connected layer. The algorithms are trained and evaluated on the classification data set MNIST for 100 communication rounds with 20 clients, where the data was Non-IID. The parameters in Algorithm 1 are $E = 5$, $C = 1.0$, and $B = 10$. The right figure shows the last five communication rounds in higher resolution.

5 Conclusion and Future Work

In Table 1 we can see that including decision trees at the end of the convolutional neural network rather than only having fully connected layers decreases the number of communication rounds needed for both Non-IID and IID data, at least for three to five decision trees. It is also clear that increasing the number of decision trees improves the results. Figure 4 further strengthen the impression that the deep neural decision forest outperforms the convolutional neural network. We can clearly see that three to five decision trees are more efficient than the convolutional neural network with fully connected layers at communication round 100, while the CNN with fully connected layers performs better than the deep neural decision forest with only two decision trees. However, the number of communication rounds needed to achieve accuracy of 99% is considerably higher for the CNN, see Table 1.

In this paper we describe how to modify deep neural decision forests to be well-suited for federated learning. This modification is done by parameterizing the terminal nodes and include all decision trees in the loss function. The experimental results do not only show that deep neural decision forest works in the setting of federated learning, it also indicates good performance compared to the convolutional neural network. However, the purpose of our empirical experiment is not to show that a deep neural decision forest is superior a convolutional neural network.

Future work includes analyzing and experimenting with more realistic scenarios, such as having a much larger pool of clients, where high communication efficiency is needed [10], updates that can occur asynchronously, and the amount of data is unevenly distributed between the clients. CO-OP [14] is one example of an asynchronous federated learning algorithm that could be used. Another interesting idea is to incorporate different types of monitoring algorithms on the server which can, for example, detect different groups of clients depending on the updates of the global model the server receive, or identify anomalous behavior of certain clients. We also aim at experimenting with the architecture in deep neural decision forests to see how that influence the result, e.g., if increasing the number of decision trees consistently improves the result and to what extent. Lastly, our goal is to implement the proposed idea in the OODIDA framework [18], which is an Erlang-based framework developed for distributed machine learning on edge devices.

Acknowledgements. This research was supported by the project Onboard/Offboard Distributed Data Analytics (OODIDA) in the funding program FFI: Strategic Vehicle Research and Innovation (DNR 2016-04260), which is administered by VINNOVA, the Swedish Government Agency for Innovation Systems.

This work was developed in Fraunhofer Cluster of Excellence Cognitive Internet Technologies.

References

1. Bojarski, M., et al.: End to end learning for self-driving cars. arXiv preprint arXiv:1604.07316 (2016)
2. Breiman, L.: Random forests. Mach. Learn. **45**(1), 5–32 (2001)
3. Caruana, R., Karampatziakis, N., Yessenalina, A.: An empirical evaluation of supervised learning in high dimensions. In: Proceedings of the 25th International Conference on Machine Learning, pp. 96–103. ACM (2008)
4. Criminisi, A., Shotton, J.: Decision Forests for Computer Vision and Medical Image Analysis. Springer, London (2013). https://doi.org/10.1007/978-1-4471-4929-3
5. Díaz-Uriarte, R., De Andres, S.A.: Gene selection and classification of microarray data using random forest. BMC Bioinform. **7**(1), 3 (2006)
6. Diederik, P., Kingma, J.B.: Adam: a method for stochastic optimization. arXiv preprint arXiv:1412.6980 (2014)

7. Fernández-Delgado, M., Cernadas, E., Barro, S., Amorim, D.: Do we need hundreds of classifiers to solve real world classification problems? J. Mach. Learn. Res. **15**(1), 3133–3181 (2014)
8. Harari, G.M., Lane, N.D., Wang, R., Crosier, B.S., Campbell, A.T., Gosling, S.D.: Using smartphones to collect behavioral data in psychological science: opportunities, practical considerations, and challenges. Perspect. Psychol. Sci. **11**(6), 838–854 (2016)
9. Konečný, J., McMahan, H.B., Ramage, D., Richtárik, P.: Federated optimization: distributed machine learning for on-device intelligence. arXiv preprint arXiv:1610.02527 (2016)
10. Konečný, J., McMahan, H.B., Yu, F.X., Richtárik, P., Suresh, A.T., Bacon, D.: Federated learning: strategies for improving communication efficiency. arXiv preprint arXiv:1610.05492 (2016)
11. Kontschieder, P., Fiterau, M., Criminisi, A., Rota Bulo, S.: Deep neural decision forests. In: Proceedings of the IEEE International Conference on Computer Vision, pp. 1467–1475 (2015)
12. LeCun, Y., Bottou, L., Bengio, Y., Haffner, P., et al.: Gradient-based learning applied to document recognition. Proc. IEEE **86**(11), 2278–2324 (1998)
13. McMahan, B., Moore, E., Ramage, D., Hampson, S., Arcas, B.A.: Communication-efficient learning of deep networks from decentralized data. In: Artificial Intelligence and Statistics, pp. 1273–1282 (2017)
14. Nilsson, A., Smith, S., Ulm, G., Gustavsson, E., Jirstrand, M.: A performance evaluation of federated learning algorithms. In: Proceedings of the Second Workshop on Distributed Infrastructures for Deep Learning (DIDL 2018), New York, NY, USA, vol. 18, pp. 1–8 (2018)
15. Poushter, J., et al.: Smartphone ownership and internet usage continues to climb in emerging economies. Pew Res. Cent. **22**, 1–44 (2016)
16. Shotton, J., et al.: Real-time human pose recognition in parts from single depth images. In: CVPR, vol. 2, p. 3 (2011)
17. Hastie, T., Tibshirani, R., Friedman, J.: The Elements of Statistical Learning: Data Mining, Inference, and Prediction, vol. 2, p. 745. Springer, New York (2009). https://doi.org/10.1007/978-0-387-84858-7
18. Ulm, G., Gustavsson, E., Jirstrand, M.: Functional federated learning in erlang (`ffl-erl`). In: Silva, J. (ed.) WFLP 2018. LNCS, vol. 11285, pp. 162–178. Springer, Cham (2019). https://doi.org/10.1007/978-3-030-16202-3_10

Conditional Anomaly Detection for Quality and Productivity Improvement of Electronics Manufacturing Systems

Eva Jabbar[1,2(✉)], Philippe Besse[1], Jean-Michel Loubes[1], and Christophe Merle[2]

[1] IMT - Institut de Mathématiques, Université de Toulouse, CNRS UMR5219, Toulouse, France
eva.jabbar1@gmail.com
[2] Continental Powertrain France SAS, Toulouse, France

Abstract. Today the integration of Artificial Intelligence (AI) solutions is part of the strategy in the industrial environment. We focus on anomaly detection in the framework of manufacturing electronic cards manufacturing under mass production conditions (24/7). Early anomaly detection is critical to avoid defects. Researches and applications of anomaly detection techniques in the industry have been published but when they face production constraints success is not guaranteed. Today's manufacturing systems are complex and involve different behaviors. We propose and evaluate a new realistic methodology for detecting conditional anomalies that could be successfully implemented in production. The proposed solution is based on Variational Autoencoders (VAEs) which provide interesting scores under the near real-time constraints of the production environment. The results have been thoroughly evaluated and validated with the support of expert process engineers.

Keywords: Smart factory · Electronic circuit manufacturing · Artificial Intelligence · Deep conditional anomaly detection

1 Introduction

Within the electronic board assembly industry additional challenges are also arising such as the increase in product complexity and global competitiveness. The goal is to produce high quality products in the shortest time at the lowest cost. To tackle these challenges the integration of Artificial Intelligence (AI) solutions is becoming a substantial part of the innovative industries strategy.

This paper focuses on anomaly detection as a part of a zero-defect strategy. Concretely, the detection and prevention of anomalies at the earliest stage in surface mount technology (SMT) production lines. SMT production lines include three main steps: solder paste printing (SPP), component pick and place (P&P),

Supported by Continental Powertrain SAS France.

G. Nicosia et al. (Eds.): LOD 2019, LNCS 11943, pp. 711–724, 2019.
https://doi.org/10.1007/978-3-030-37599-7_59

and reflow soldering. The production flow starts when printed circuit boards (PCBs) are introduced to the SPP process in which solder paste is pressed through an stencil mask to create a film over the card that serves as attachment medium between the components and connection pads. PCBs then continue through the P&P process in which components are placed on the pads covered with solder paste. With the main components placed, the PCBs pass through the reflow oven in which the solder paste film is melt to permanently bind the components to the card. For optimization purposes these processes are performed on a panel of several PCBs which is separated into sub-panels at the end of the production process.

Optical inspection systems are used for in-process inspection and quality control, which includes Solder Paste Inspection (SPI) and Automated Optical Inspection (AOI). Inspection is automated via hard coded optical processing fine tuned through statistical analysis. In case of detected failure, a manual diagnosis, performed by an operator, remains mandatory to make the final judgement.

Most of the research and applications on SMT line are focused on product defect detection at AOI based on the AI image recognition [1–4] and with a lesser extent they focus on the SPI stage [5,6]. The AOI-based innovative solutions are significant but are coming at the end of production line, while there is a need to identify and anticipate defects at the earliest stages. Additionally, in the case of cyber-physical approaches [7,8], most of solutions either use supervised algorithms with available labels in offline phase, or assume that the training dataset is produced by the same generative process (i.e, the random observations are *independently and identically distributed* (i.i.d)). However, the production environment displays different behaviours according to several conditions. At the best of our knowledge, almost none of these solutions provide conditional anomaly detection in unsupervised manner or/and handle the deployment aspect.

The framework proposed in this paper is focusing on unsupervised solder paste anomaly detection using physical measurements from real-world SPI data. We take into account the variety, velocity and volume properties of the production data. Based on this, we have established a methodology covering historical data preprocessing, model learning through the application of adequate deep learning algorithm and finally, result evaluation and display. Altogether in a close collaboration with the expert engineers as part of transparency, explainability and acceptability of AI.

2 Problem Statement

The SPI equipment is programmed to measure and inspect physical characteristics of solder paste printed on the pads of the PCB. The SPI judgment is based on pre-specified tolerance limits selected to make a compromise between the false alarms rate and the defects slipped through. In fact, most of the printing related defects are detected later at AOI (around 40% of the total defects), which is considered as a waste of time and money, especially after placing components that can be expensive. The "10X Rule" can be used for estimating rework costs at each stage of the assembly line [9]. In addition, some of these defects may be detected at the end of the process or even after delivery to the customer. In a context of increasing expectations from the market (single digit ppm quality), such inspection processes are not sufficient anymore and result in increased costs at a manufacturing level but also huge penalties in case of customer returns.

A further technical aspect is the high-volume production. In one production line, the solder paste printing process performs more than 12 million operations per day and records measurements of more than 60 variables. In addition, it is assumed that this system operates in different conditions/behaviors depending on the complexity of the product, the component category and the production phase. In this case, the definition of what is an anomaly will depend on these conditions. Therefore, we are dealing with a conditional anomaly detection in high dimension framework.

Although the frequency of abnormal events is very low compared to good ones, defining the boundary between normal and abnormal operations is an important and challenging problem. The aim is to learn 'normal' regions depending on each condition and to identify behaviours that are not consistent with these ideal expectations.

3 Overall Methodology

We assume that we have a sample $X = (x_1, \cdots, x_{d_m}, r_1, \cdots, r_{d_c}) = (x, r)$ that have two sets of features (x, r), where x represents the set of product measurement features, and r represents the condition features. In our case, the r features, selected by the process expert engineer, are the product ID and the component package. Each row in the data represents a vector of SPI measurements at the pad level of a component from a package k within a panel of product ID P_i. Note that k depends on each P_i. We denote by$C_i^k = (P_i, k_i)$ the homogeneous condition in our case (Fig. 1).

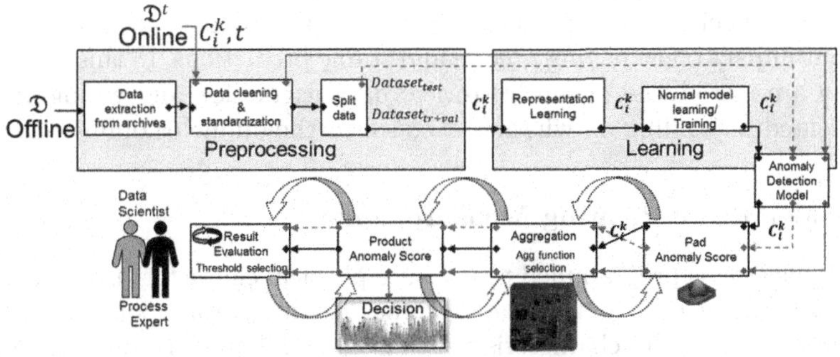

Fig. 1. Pipeline of the proposed methodology for detecting anomalies in electronic industry

The main idea is to first learn a robust and stable anomaly detection model based on high quality data. For that, big historical data are extracted, preprocessed (see Sect. 4.2) and then splitted into training, validation and test datasets. As the number of dimension has a significant effect on the machine learning operation (concept of *curse of dimensionality*), we learn first the so-called manifold or representation by performing non-linear dimensionality reduction techniques to focus on compacted sub-space conditionally to C_i^k. The Conditional normal model is then trained. The aim is to find the region where most of points are located on the low-dimensional manifold space. Using validation data, the anomaly score is calculated for each C_i^k. For a quick and global identification of the anomaly, it is necessary to aggregate all these conditional scores at panel level to enable the process expert to first identify the deviated product and then to bring the appropriate corrective action. The most suitable threshold is then selected depending on quality requirements. The optimal configuration is to detect the highest number of anomalies (critical or/and potential ones) with a lower false call rate. This step is iterative as it requires an in-depth evaluation and analysis of results. After validating a first version of the model, it is then tested on different test datasets. The performance is partially evaluated in terms of the accuracy, recall and specificity, then followed by a thorough analysis of unsanctioned anomalies. Once, the final model is validated by the expert in the offline phase, we switch then to the online phase, by first collecting, cleaning and standardizing as in the offline phase, then to directly perform the final anomaly detection algorithm.

In this work we present, test and evaluate the performance of a deep conditional anomaly detection approach based on Variational AutoEncoders (VAEs) on a real-world SMT environment. In the following sections, we will detail each step according to the studied approach.

4 Data Preprocessing

A preprocessing pipeline is needed before using the data for model learning. This step requires special handling, as learning from high-quality business data has a significant impact on achieving precise and stable predictions. In this section, the SPI data are introduced. Then, historical data extraction and wrangling steps are presented depending on whether it concerns the offline or online phase.

4.1 Solder Paste Printing Measurements

The SPI machine is programmed to measure physical characteristics of all solder depots printed on each panel. Pad volume, height, surface, offsets and PCB warping are among the characteristics that can be found. In one day a single production line could inspect more than 12 million pads. In the dataset each panel is presented in multiple rows, one per pad, with 153 features.

In our environment, the data is collected at the SPI station through a data collection agent and pushed into the cloud. These data are then archived at

x-minutes interval. Therefore, we have two kind of data: archive data including historical records and hot data generated in the last minutes.

4.2 Data Extraction and Wrangling

For the offline phase, and given the large volume of SPI data archived, we extract, clean and standardize the data using Apache Spark, an open-source unified analytics engine for big data processing. Steps within this phase are often made in an iterative way.

The extracted data include numerical and categorical features. Within the exploratory analysis step, we have identified a set of constant/null features. These features are typically non-activated SPI parameters. By variance calculation, we have filtered 69 features (from the 153 features in the original datasets), which gives a total of 84 features. These features include variables related to the pad measurements such as*volume*, *height*, *size*, *offset* and*bridge*, and also variables related to the panel such as *shrinkage*, *warping* and *Mask height*. For each feature, additional information can be given, as for instance for the feature *bridge*, there is also: its *position*, its *lengh min/max*, *area*, *offset*, and its*volume*.

Among the categorical features, we have kept only two features namely *product ID* and *component package ID* as they present the condition features, where the remaining features represent measurement attributes. The techniques we utilize in this paper tend to use numerical data as input. As the 84 features have different scales, we standardize to zero mean (μ) and unit variance (σ).

While in the online phase, we proceed only to clean and standardize the data by making use of work previously made.

5 Deep Conditional Anomaly Detection

Here we describe in detail the conditional anomaly detection approach thoroughly evaluated to handle the industrial problem previously stated: the deep anomaly detection based on VAEs.

5.1 Background of Variational AutoEncoders

Variational AutoEncoders (VAEs) are a deep Bayesian network [10]. They model the relationship between a latent variable z and the visible variable x independent and identically distributed (i.i.d) using two neural networks: an "encoder" and a "decoder" based on the reparameterization trick. The encoder network will output parameters describing a distribution for each latent dimension z. A prior is chosen for z, which is usually multivariate unit Gaussian $N(0, I)$ [10]. We consider a probabilistic model $p_{C_i^k}(x, \theta)$ where $x \in \mathbb{R}^{N_{C_i^k}}$ is the data of the condition C_i^k and $z \in \mathbb{R}^{N_z}$ is the latent variable for $N_z < N_{C_i^k}$

$$p_{C_i^k}(x, \theta) = \int p_{C_i^k}(z) p_{C_i^k}(x/z; \theta) \, \mathrm{d}z \tag{1}$$

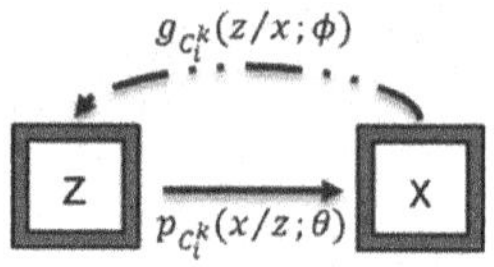

Fig. 2. Architecture of VAE

where θ is a set of network parameters. The output is then two vectors including the mean and variance of the latent variable distributions. The aim is to compute $p_{C_i^k}(z/x;\theta)$ as the mapping from z to x with the network parameter θ:

$$p_{C_i^k}(z/x;\theta) = \frac{p_{C_i^k}(x/z;\theta)\; p_{C_i^k}(z;\theta)}{p_{C_i^k}(x;\theta)} \tag{2}$$

where $p_{C_i^k}(z,\theta)$ is the prior conditionally to C_i^k. Since computing $p_{C_i^k}(x,\theta)$ is quite difficult, the encoder network is used to approximate it by the posterior distribution $g_{C_i^k}(z/x;\phi)$, with network parameter ϕ. The aim is to find the optimal parameters ϕ so that $g_{C_i^k}(z/x;\phi)$ is very similar to $p_{C_i^k}(z/x;\theta)$ while simultaneously maximizing the reconstruction likelihood. The Kullback-Leiber divergence is often used as a measure of difference between two probability distributions [10], therefore, this term has to be minimum. The formulation of the problem is then to maximize the negative evidence lower bound (ELBO) (Fig. 2):

$$\begin{aligned} -\mathcal{L}(x) &= \mathbb{E}_{g_{C_i^k}(z/x;\phi)} \log(p_{C_i^k}(x/z;\theta)) - KL(g_{C_i^k}(z/x;\phi) \| p_{C_i^k}(z;\theta)) \\ &:= R(x) - K(x) \end{aligned} \tag{3}$$

where $R(x)$ is referred to as a reconstruction error and $K(x)$ as a regularization term. $R(x)$ is calculated by Monte Carlo sampling from the variational posterior which is selected to be $g_{C_i^k}(z/x;\phi) \sim \mathcal{N}(\mu_\phi(x), \Sigma_\phi(x))$, where $\mu_\phi(x), \Sigma_\phi(x) = \sigma_\phi^2(x)\,\mathbb{I}d$ are obtained by the encoder model $e_{C_i^k}(x;\phi)$. Therefore, calculating $K(x)$ could be computed in closed form as follows:

$$\begin{aligned} K(x) &= KL(\mathcal{N}(\mu_\phi(x), \Sigma_\phi(x)) \| \mathcal{N}(0,1)) \\ &= \frac{1}{2}\Big(tr(\Sigma_\phi(x)) + \mu_\phi(x).\mu_\phi(x) - m - log(det(\Sigma_\phi(x))\Big) \end{aligned} \tag{4}$$

where $tr(.)$ is the trace function which can be computed as the sum of the diagonal of $\Sigma(X)$, $det(.)$ is the determinant function and m is the dimension of the Gaussian. As the determinant of a diagonal matrix could be computed as product of its diagonal, we have then:

$$\begin{aligned} K(x) &= \frac{1}{2}\Big(\sum_m \Sigma_\phi(x) + \sum_m \mu_\phi^2(x) - \sum_m 1 - log(\prod_m \Sigma_\phi(x)\Big) \\ &= \frac{1}{2}\Big(\sum_m \Sigma_\phi(x) + \sum_m \mu_\phi^2(x) - \sum_m 1 - \sum_m log(\Sigma_\phi(x)\Big) \\ &= \frac{1}{2}\Big(\sum_m \Sigma_\phi(x) + \mu_\phi^2(x) - 1 - log(\Sigma_\phi(x)\Big) \end{aligned} \quad (5)$$

Given the sample z, and the parameters of $p_{C_i^k}(z/x;\theta)$ obtained by the decoder model $d_{C_i^k}(z;\theta)$, the reconstruction $\hat{x}$ are sampled from $p_{C_i^k}(x; e_{C_i^k}(z;\theta))$. This phase gives to the VAEs the property of being a stochastic generative model. In fact, VAEs use a stochastic process to generate new data by modeling the parameters of a posterior distribution.

5.2 Anomaly Flagging Criteria

Within the training phase, the VAEs learn the compact information from the training dataset. As discovered in [11], VAE recognizes normal regions while calculating the distribution of training data in the latent feature space. When applying the model on new data including normal and abnormal points, the abnormal points first cannot be mapped to the compact subspace computed by the encoder network and second, the decoder network cannot restore them correctly. We define then the anomaly flagging score as the difference between the input and generated output. For that, we compute the residual error (reconstruction error) as : $L_{(x,\hat{x})} = \sum_j \|x_i - \hat{x}_i\|^2$.

In the validation phase, an extensive analysis is required to select a threshold α whereby if $L_{(x,\hat{x})} > \alpha$ then x_i is abnormal else x_i is normal. In the offline phase, α is determined according to expert process and the partially anomaly labels we have. Then, the expert can decide to adjust it according to his need: detecting critical anomalies or potential ones.

6 Experimental Setup

The purpose of this study is to develop an anomaly detection model that can detect abnormalities from SPI measurements on SMT line. The SPI machine performs around 140 operations in a second which generates a large amount of data. These data are collected and pushed in real time into the cloud. They remain as hot data for several minutes before being archived. This cloud architecture makes the model learning more convenient using the big historical data and especially facilitate the implementation of the model in production.

6.1 Datasets Description

In this section we provide a detailed description of the real world datasets used to build a fully anomaly detection model pipeline in the earliest stage of the

Table 1. Statistics of offline phase datasets: Total number of rows for each datasets, with the total number of PCBs, the studied product ID and the total number of component packages (representing C_i^k condition).

Datasets	Total rows	Total panel	Total Prod ID	Total CPack ID
Training	8 155 786	1 956	4	161
Validation	9 342 406	2 413	4	161
Test	8 717 490	2 555	4	161

SMT line. The Table 1 shows general statistics of the datasets used in the offline phase. Note that all datasets include more than 150 features. After performing data Extraction and wrangling steps described in Sect. 4.2, the resulting datasets contain 84 features. We decided to train the model on one day data including 4 product IDs. We chose 2-days additional datasets for the validation and test steps. With VAEs model there is no need to train the model only on normal data. In fact, for VAEs, it has been proven in [11] that they recognize "normal" regions in the latent feature space and then eliminate outliers in the decoder step (sampling step).

The target is to classify the validation and test datasets into normal and abnormal data. According to the anomaly score threshold, we define two anomaly categories: potential (warning) and critical anomaly. The selection of this threshold depends on the quality requirements as well as on the evaluation results.

6.2 Environmental Setup

In order to extract big historical data from archives, we developed a PySpark script on a cluster with 5 nodes using Apashe Spark. Each cluster node is a Linux PC with the Intel Xeon Platinum 8000 series (Skylake-SP) processor (4 CPUs, 16 GiB of DRAM host memory).

For VAEs model learning, we develop a python script using publicly available code based on Keras [12], Tensorflow-GPU [13]. Training is performed on GPU graphics machines with access to NVIDIA Tesla M60 GPUs delivering up to 2,048 parallel processing cores, 8 GiB of GPU memory and 16 vCPUs based on custom 2.7 GHz Intel Xeon E5 2686 v4 processors and 122 GiB of DRAM host memory.

6.3 Network Optimization and Hyperparameter Selection

We setup the hyperparameter for the prior as a unit normal Gaussian distribution. Both the encoder and decoder have two hidden layers with 14 neurons. We use Rectified Linear unit (ReLu) non-linearities activation [14] for the encoder layer and sigmoid activation for the decoder layer. We chose $z = 3$ for the dimension of the latent space. We assume that three latent variables are enough to obtain accurate compressed representation of our problem. The latent layer has three nodes each for encoding mean and variances of the Gaussian distribution. The latent representation z is then sampled from that distribution using the reparameterization trick [10]. We use ADAM [15] optimizer as the standard techniques to speed convergence. The learning rate was chosen at 0.001. Batch size was tested from 20 to 100 and set to 32 as it has been shown that small batch size provides improved generalization performance and allows a significantly smaller memory footprint in [16]. This architecture with the above-mentioned hyperparameter setting have yielded significant results in terms of anomaly detection.

6.4 Aggregation

Considering that we are seeking to identify anomalies, i.e. panels with a high reconstruction error. We have decided to use the *maximum* of this error as an aggregation function in order to specifically target extreme cases. The *mean* or *median* functions could also be considered but rather for the purpose of identifying a deviation from the overall production behaviour.

6.5 Performance Metrics

In the unsupervised anomaly detection, the evaluation of the model remains very challenging. Labels are occasionally available although far from complete and accurate. We use AOI verification sanctions and traceability web application (with the whole process sanctions). Particularly, as the AOI verification sanctions are human-made, some uncertainties or errors may be included. As nouns the difference between defect and anomaly is that defect is a fault or malfunction while anomaly is a deviation from a rule or from what is regarded as normal. Consequently, some critical anomalies may be missed. In light of these challenges, the analysis of the model results was conducted on two parts.

We have first measured the model performances on identifying known anomalous and normal panels for different anomaly thresholds. For that, the following performances metrics are considered as detailed in [17]:

Recall. True Positive (TP) rate $\frac{TP}{TP+FN}$. This measures how well we identify all anomalous panels.

Specificity. True negative (TN) rate $\frac{TN}{TN+FP}$. This measures how well we identify all normal panels.

Accuracy. Proportion of correctly identified panels $\frac{(TP+TN)}{(TP+TN+FP+FN)}$.

In our context of anomaly detection, positives refer to anomalous panels and negatives to normal panels.

Then, according to the selected threshold, we have thoroughly analyzed a number of cases considered as "false alarms" whereas some may be real anomalies. The aim is to enable the process expert engineer to adjust α according to the quality requirement. The best practice would be to define a first threshold indicating a warning situation and a second threshold for critical situations.

We also consider **the computation time** C_t metric. This measures the inference time of the one panel and needs to satisfy the condition $C_t < 3(s)$.

7 Experimental Results

In this section, we evaluate the results of CAD experiments according to the performance metrics described in the previous section.

7.1 Performance Analysis

Table 2 details the performances of the conditional VAEs on the validation datasets. Different anomaly score thresholds α are tested for each product ID. The first remark is that suitable α depends on each product. For lower α, recall reaches its maximum value except for **P4**, but gives the lowest specificity and accuracy values. This is due to the presence of false positives. By increasing α, the false positives rate decrease leading to the rise of the specificity and accuracy. However, if α is overly increased, the recall can then decrease. The best performance is then selected according to a good tradeoff between the three measures. One exception concerns **P4** which gives low recall scores regardless of the alpha value. We chose the suitable α providing the high specificity and accuracy scores. The Fig. 3 shows the aggregated reconstruction error histograms for each product on validation dataset. We represent the selected threshold α_{sel} with a vertical dashed line.

Once the optimal threshold α_{sel} is chosen for each product ID, we have performed the model on test dataset. As shown in Table 4, the performance metrics provide relevant scores. The recall achieves its **maximum values of 1** for three product ID **P2**, **P3** and **P4** which means all printing-related defect identified later on the SMT line are detected correctly. While for **P1**, only **66,7%**of known anomalies are detected. The performance in terms of accuracy and specificity are

Table 2. Performance metric on validation dataset for each product ID and anomaly score thresholds α. The best values of performance metrics are highlighted in bold

Product ID	α	Accuracy	Recall	Specificity
P1	0,075	0,936	1	0,935
	0,079	**0,974**	**1**	**0,974**
	0,081	0,986	0,5	0,989
P2	0,085	0,962	1	0,961
	0,089	**0,976**	**1**	**0,976**
	0,095	0,982	0	0,986
P3	0,068	0,984	1	0,983
	0,07	**0,988**	**1**	**0,988**
	0,074	0,993	0,5	0,995
P4	0,07	0,925	0,333	0,931
	0,074	0,961	0,333	0,967
	0,08	**0,980**	**0,333**	**0,987**

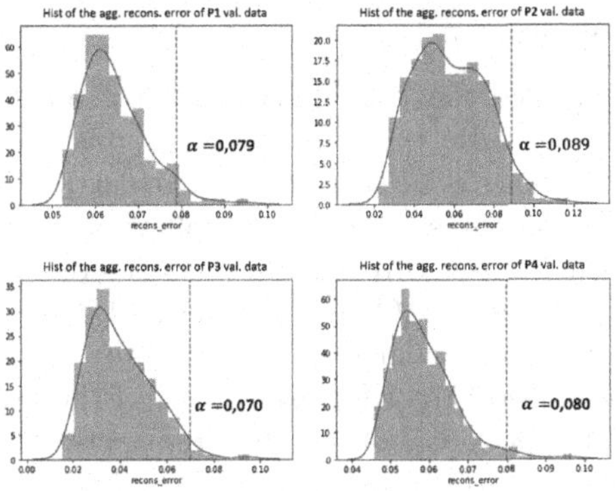

Fig. 3. Reconstruction error histograms for each product ID. The vertical dashed line represents the selected anomaly threshold.

significantly high for **P1** but slightly lower for the other product IDs. After investigating these results, we highlighted two possible reasons for this low score: a possible mislabelling of the error type or insufficient information in the learning data used in this study. A further step is to include additional data (from a large period) in the learning phase.

Furthermore, we add a final performance in terms of computational time and learning efficiency. Looking at Table 3, it can be seen that the learning time is relevant for the data size and complexity of the VAEs model used. A special emphasis is given to the C_t results which is considered to be excellent compared to the production constraint time (less than 3 min).

Table 3. The total computational time for each product ID on GPU machine

Product ID	Learning time	Inference time(s) Val	Inference time(s) Te	Inference time(s) C_t
P1	2 H	78	81	0,18
P2	21 m	78	79	0,08
P3	1 H 03 m	25	79	0,13
P4	2 H 45 m	201	153	0,23

Table 4. Results of performance metrics on test datasets for each product ID with the selected anomaly threshold α_{sel}.

Product ID	α_{sel}	Accuracy	Recall	Specificity
P1	0,079	0,991	0,667	0,994
P2	0,089	0,952	1	0,951
P3	0,07	0,895	1	0,894
P4	0,08	0,989	1	0,989

7.2 Anomaly Analysis

Besides the significant performance results, an in-depth analysis has to be carried out of cases considered as false positives under the current inspection system. We remind that the purpose is to surpass the current production performance on identifying anomalies at the earliest stage. Following this perspective, among the cases exceeding the chosen threshold, we considered the cases labelled as false defects by the operator at AOI verification and if no sanctions are given, we investigated the physical parameter values case-by-case. As an illustration, the aggregated reconstruction error for each panel of the product **P1** is plotted over time in Fig. 4. The horizontal dashed line represents the selected threshold as previously detailed.

The red dots represent the real defect panels, the orange dots represent the cases considered as false defects by the operator and the green ones are normal panels without any inspection flags (i.e. passed directly good). We observe that the two highest peaks are panels which were considered as false defects. The two real defects identified during this production period have indeed high reconstruction error values. After a thorough analysis of the physical measurement values of the two panels exceeding the threshold but considered as normal (green dots), we found that they have an outlying volume values and bridge length that can lead to short circuits. These panels are in fact atypical but not presenting a critical incident according to the expert's view. Nonetheless, from an expert point of view, having such an 'anomaly' metric would be extremely valuable.

Consequently, a key question arises about the quality requirement for these extreme cases. Our method proposes to store an anomaly score for each panel in a local database that can be consulted in case of a later incident. This procedure is very crucial to first improve the quality of products delivered to the customer and especially to continually enhance our AI solution.

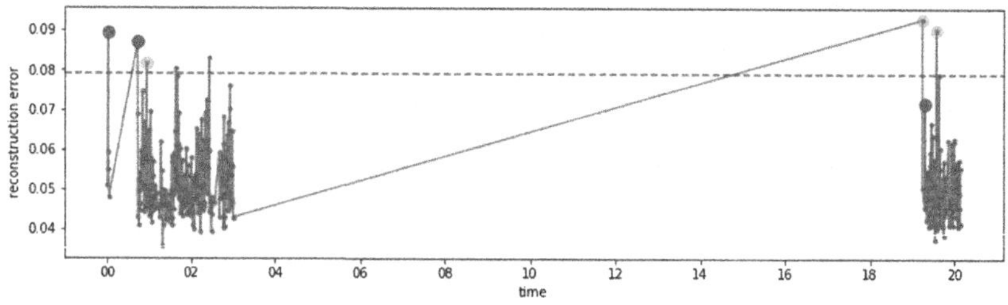

Fig. 4. The aggregated reconstruction error on P1 test dataset. The horizontal dashed line represents the selected anomaly threshold. (Color figure online)

8 Conclusion and Future Work

Among the most challenging problematic in the electronics manufacturing industry, we focus in this work on the detection and prevention of anomalies at the earliest stage in SMT lines. As a response to the actual limitation of SPI process, a rigorous and realistic methodology was proposed that take into account the production conditions. This methodology relies on cloud architecture and involves the overall AI process based on Variational Autoencoders that use reconstruction error as an anomaly score. The effectiveness of our solution was demonstrated with real-world data, in which not only known defects were identified but also labelling errors and atypical products were discovered. This leads to a very good performance in term of accuracy, recall and specificity as well as computational time. Besides, a close collaboration was made with the process expert engineers as part of transparency, explainability and acceptability of AI. The aim is to enable the process expert to adjust a tradeoff according to the quality requirement. The best practice would be to define a first threshold indicating warning situations and a second threshold for critical ones.

Accordingly, further investigations are necessary for the deployment in production, with a special attention on the acceptance aspect of our solution. If an anomaly is detected, the user must be informed of the precise root cause and the location on the PCB. Furthermore, a backgroud-learning step needs to be integrated into our solution in case of a model deviation or when non-learned conditions are introduced.

Acknowledgments. This work is supported by Continental Powertrain France. Special thanks goes to Minds team members: Nathalie Barbosa Roa, Alain Le Grand and Jean-Noël Tomasini.

References

1. Hao, W.: Solder joint defect classification based on ensemble learning. Solder. Surf. Mt. Technol. **29**, 06 (2017)
2. Acciani, G., Brunetti, G., Fornarelli, G.: A multiple neural network system to classify solder joints on integrated circuits. Int. J. Comput. Intell. Res. **2**, 337–348 (2006)
3. Hao, W., Xianmin, Z., Yongcong, K., Gaofei, O., Hongwei, X.: Solder joint inspection based on neural network combined with genetic algorithm. Optik **124**(20), 4110–4116 (2013)
4. Song, J.-D., Kim, Y.-G., Park, T.-H.: SMT defect classification by feature extraction region optimization and machine learning. Int. J. Adv. Manuf. Technol. **101**, 1303–1313 (2018)
5. Jiang, J., Cheng, J., Tao, D.: Color biological features-based solder paste defects detection and classification on printed circuit boards. IEEE Trans. Componen. Packag. Manuf. Technol. **2**(9), 1536–1544 (2012)
6. Tsai, D., Huang, C.: Defect detection in electronic surfaces using template-based Fourier image reconstruction. IEEE Trans. Componen. Packag. Manuf. Technol. **9**(1), 163–172 (2019)
7. Jabbar, E., Besse, P., Loubes, J.M., Roa, N.B., Merle, C., Dettai, R.: Supervised learning approach for surface-mount device production. In: Nicosia, G., Pardalos, P., Giuffrida, G., Umeton, R., Sciacca, V. (eds.) LOD 2018. LNCS, vol. 11331, pp. 254–263. Springer, Cham (2019). https://doi.org/10.1007/978-3-030-13709-0_21

8. Kim, D., Koo, J., Kim, H., Kang, S., Lee, S.H., Kang, J.T.: Rapid fault cause identification in surface mount technology processes based on factory-wide data analysis. Int. J. Distrib. Sens. Netw. **15**(2), 1550147719832802 (2019)
9. Riddle, M.E.: Solder paste measurement: a yield improvement strategy that helps improve profits. SMT Express (2004)
10. Kingma, D.P., Welling, M.: Auto-encoding variational bayes. arXiv preprint arXiv:1312.6114 (2013)
11. Xu, H., et al.: Unsupervised anomaly detection via variational auto-encoder for seasonal KPIs in web applications. In: Champin, P.-A., Gandon, F.L., Lalmas, M., Ipeirotis, P.G., (eds.) WWW, pp. 187–196. ACM (2018)
12. Chollet, F., et al.: Keras (2015). https://keras.io
13. Abadi, M., et al.: TensorFlow: Large-scale machine learning on heterogeneous systems. Software available from tensorflow.org (2015)
14. Krizhevsky, A., Sutskever, I., Hinton, G.E.: ImageNet classification with deep convolutional neural networks. In: Advances in Neural Information Processing Systems, pp. 1097–1105 (2012)
15. Kingma, D.P., Ba, J.: Adam: a method for stochastic optimization. arXiv preprint arXiv:1412.6980 (2014)
16. Masters, D., Luschi, C.: Revisiting small batch training for deep neural networks. CoRR, abs/1804.07612 (2018)
17. Avati, A.: Evaluation metrics, April 2019. http://cs229.stanford.edu/section/evaluation_metrics.pdf

Data Anonymization for Privacy Aware Machine Learning

David Nizar Jaidan[1(✉)], Maxime Carrere[2], Zakaria Chemli[3], and Rémi Poisvert[4]

[1] Innovation L@B Scalian France, Labège, France
david.jaidan@gmail.com
[2] Centre d'Excellence Datascale Scalian France, Le Haillan, France
[3] Innovation L@B Scalian France, Paris, France
[4] Innovation L@B Scalian France, Rennes, France

Abstract. The increase of data leaks, attacks, and other ransom-ware in the last few years have pointed out concerns about data security and privacy. All this has negatively affected the sharing and publication of data. To address these many limitations, innovative techniques are needed for protecting data. Especially, when used in machine learning based-data models. In this context, differential privacy is one of the most effective approaches to preserve privacy. However, the scope of differential privacy applications is very limited (e. g. numerical and structured data). Therefore, in this study, we aim to investigate the behavior of differential privacy applied to textual data and time series. The proposed approach was evaluated by comparing two Principal Component Analysis based differential privacy algorithms. The effectiveness was demonstrated through the application of three machine learning models to both anonymized and primary data. Their performances were thoroughly evaluated in terms of confidentiality, utility, scalability, and computational efficiency. The PPCA method provides a high anonymization quality at the expense of a high time-consuming, while the DPCA method preserves more utility and faster time computing. We show the possibility to combine a neural network text representation approach with differential privacy methods. We also highlighted that it is well within reach to anonymize real-world measurements data from satellites sensors for an anomaly detection task. We believe that our study will significantly motivate the use of differential privacy techniques, which can lead to more data sharing and privacy preserving.

Keywords: Privacy · Anonymization · Machine learning · Text encoding · Natural language processing · Time series · Anomaly detection

1 Introduction and Related Work

With the advent of large-scale data collection, the ability to analyze data while maintaining privacy became a challenging task. Moreover, when we collect personal data and/or confidential data in some specific industries such as health-care

G. Nicosia et al. (Eds.): LOD 2019, LNCS 11943, pp. 725–737, 2019.
https://doi.org/10.1007/978-3-030-37599-7_60

or defense, data-privacy becomes a crucial consideration. It is therefore difficult to access this type of data to analyze it with machine learning algorithms. Anyway, if the data is available, we are subject to a risk of disclosure and leakage of sensitive information.

The new European Union's General Data Protection Regulation (GDPR) [1] states that personal data controllers must set up appropriate organizational and technical measures to ensure the data protection principles. Therefore, business processes that handle personal data must be designed and built with consideration of these principles and must provide guarantees to protect data (e.g., using anonymization). This by using default highest-possible privacy settings so that the data is not available publicly without explicit informed consent and cannot be used to identify a person without combining external data. The key idea is that anonymized data are not concerned by data storage restrictions.

Several studies have been conducted to ensure the protection of individual sensitive information. Among the developed approaches we find the k-anonymity model [2] which led to the development of new models, such as, l-diversity [3], (α; k)-anonymity [4], t-closeness [5], (k; e)-anonymity [6] and (c; k)-safety [7]. These models can be very effective, but without formal mathematical guarantees. To deal with this lack, Dwork [8] introduced a new approach named ϵ-differential privacy. Unlike the previous methods, it provides formal privacy guarantees that do not depend on a background knowledge, computational power or post-processing on the anonymized data. It is considered as one of the most efficient approach for privacy-preserving data publishing. This approach has received a lot of attention in recent years in the fields of machine learning and data mining [9–11]. Differential privacy is able to preserve relevant information for machine learning models while preserving privacy. Several studies investigate the effectiveness of this approach to anonymize different types of data in the area of Natural language processing (NLP) [12,13] and time series analysis [14,15]. There are several approaches and implementations to construct differential privacy [16,17]. The aim of this study is to evaluate the performances of two Principal Component Analysis (PCA)-based differential privacy algorithms (PPCA [18] and DPCA [19]) in terms of utility and privacy preserving trade-off.

In this paper, we firstly describe briefly the different studied approaches, anonymization, text encoders and machine learning tools in Sect. 2. Then, we focus on the evaluation procedure and results analysis in Sect. 3. Finally, conclusions and an outlook of future work are given in Sect. 4.

2 Methodology

In this study, we investigate and evaluate the performances of two PCA-based differential confidentiality algorithms. To that end, we tested these methods on two similar textual data-sets with different sizes. Both data-sets contain texts movie reviews categorized as positive or negative. The Fig. 1 shows a schematic view of the different steps of our study. Firstly, using Skip-Thought vectors model, we encode movie reviews from the Cornell and Imdb data-sets. We also

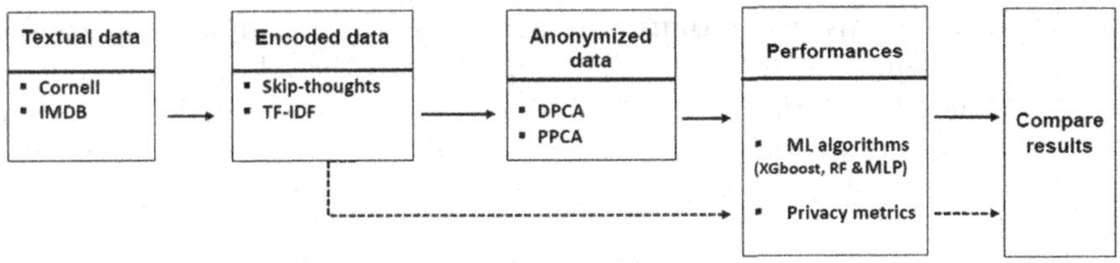

Fig. 1. Schematic view of study methodology.

tested our algorithms with a representation based on TF-IDF to be able to compare results and validate our interest for Skip-Thought. These text representations are described in Sect. 2.1. Then, to anonymize the encoded datasets, we use two differential privacy algorithms, PPCA and DPCA (detailed in Sect. 2.2). We classify negative and positive reviews using three machine learning algorithms, Multilayer Perceptron, XGBoost and Random Forest (detailed in Sect. 2.3). After this, we develop and use some informative metrics to evaluate the anonymization quality. Finally, we evaluate the ML algorithms performances (utility) and the privacy preserving quality trade-off.

2.1 Data Encoding Techniques

The process of extracting meaning and learning from text data is an active and complex topic. Among several text encoding methods, we selected Skip-Thought vector model for several reasons: (1) it is customized for texts, and not just for single words like word2vec [20], (2) Skip-Thought reduces the size of the data input about 10 times compared to the Term Frequency-Inverse Document Frequency (TF-IDF) method, and (3) it offers a semantic representation of words. We also tested our algorithms with a representation based on TF-IDF to compare results and validate our interest for Skip-Thought.

Skip-Thought Vectors: Skip-Thought is a generalized encoding technique introduced by [21]. The Skip-Thought consist to train an encoder-decoder model where the encoder maps the input sentence to a sentence vector and the decoder generates the sentences surrounding the original sentence. A large collection of novels was used to build this model, namely the Book-Corpus data-set [22]. The data-set contains 16 different kinds of books: Romance (2,865 books), Fantasy (1,479), Science Fiction (786), and more. In addition to the narratives, the books contain dialogue, emotion and a wide range of interaction between characters. Thus, with a sufficiently large collection, the construction of the model is not biased in favor of a specific field or context. Two separate models were trained on the book corpus. One is a unidirectional encoder (uni-skip) with 2400 dimensions. The other is a bidirectional model (bi-skip) with forward and backward encoders of 1200 dimensions each.

Term Frequency Inverse Document Frequency: TF-IDF is a way to score how a word is important to a document (in a corpus) based on how frequently they appear across multiple documents. Two terms are computed to determine TF-IDF: the term of frequency (TF) that represents the occurrence of the word in the document, and the inverse document frequency (IDF) that measures the importance of a given term in the entire document.

2.2 Differential Privacy Techniques

Differential privacy is a powerful approach to preserving privacy that, unlike most privacy-preserving technologies, do not rely on encryption. It is a promising way to collect, use and preserve confidentiality of data. Differential privacy techniques deprive data of its identifying characteristics in order to prevent it from being used by malicious entities that may compromise a user's privacy. We chose PPCA and DPCA algorithms because both methods focus on optimizing the application of noise and maximizing the usefulness of the output data. PPCA algorithm returns a condensed data representation, while DPCA returns transformed data in its original dimension space.

Privacy-Preserving PCA (PPCA): Unlike a standard PCA algorithm integrating differential privacy, PPCA is not performed by adding noise to the PCA subspace, but by randomly generating a subspace that provides high data utility. This is done by sampling from the Bingham distribution matrix using a Markov Chain Monte Carlo (MCMC) procedure due to [23]. Once this matrix is generated, simply applies it to raw data, as in the case of a standard PCA. The sampling process is complex and requires simulating the MCMC convergence from a random matrix over a number of iterations, making the computation very time-consuming. The author does not publish his source code, so we implemented it in Python, using an optimized Numpy version with OpenBLAS to perform calculations. We chose to use the same R package as used by the author, using a python module to execute R code.

Differential-Private Data Publishing Through Component Analysis (DPCA): DPCA is a mechanism that achieves privacy preserving data publishing (PPDP) satisfying ϵ-differential privacy with improved utility through component analysis, by combining it with Laplacian noise during the projection and recovery steps. This consists of decomposing a noisy variance matrix instead of the exact matrix. Then, adding noise to the projected matrix before recovery. As for the PPCA, the author did not publish his source code, so we did it using the python research article.

2.3 Machine Learning Techniques

In this study, we used three different classifiers to evaluate the utility of data:

eXtreme Gradient Boosting (XGBoost): XGBoost [24] is a scalable algorithm that has recently been dominating applied machine learning (known to provide better solutions than other ML algorithms) and considered as the "state-of-the-art" for structured data. It is an implementation of gradient boosted decision trees designed for speed and performance. It consists to create new models that predict the residues or errors of previous models, then combine them to make the final prediction. It uses a gradient descent algorithm to minimize the loss when adding new models.

Multilayer Perceptron (MLP): MLP [25] is a class of feed-forward artificial neural network. An MLP is structured in several layers (at least, an input layer, a hidden layer and an output layer) in which information flows only from the input layer to the output layer. Each node is a neuron that uses a nonlinear activation function (except for the input nodes). MLP uses a supervised learning technique called back-propagation for training. Each layer is composed of a flexible number of neurons; the neurons of the last layer are the outputs of the global system. It can distinguish data that is not linearly separable.

Random Forest (RF): RF [26] is an ensemble learning method for classification, regression and other tasks. RF operates by aggregating a multitude of decision trees at training phase and outputting the class that is the mode of the classes of the individual trees. In other words, the decision of each tree is taken as a vote to make a final decision based on the majority rule. The RF algorithm is often more powerful in terms of prediction accuracy than a single decision tree. The use of several decision trees reduces variance by aggregating classifiers. This makes RF more robust to imbalanced data and less subject to over-fitting.

2.4 Performance Metrics

In this section, we define some informative metrics to evaluate the anonymization quality. Given a data set (or matrix) X_{np} of [0,1] bounded entries. We identified two criteria to evaluate the anonymization method performances:
(1) The utility of the anonymous data-set: this is usually done by comparing the ML algorithms performances with a Baseline. However, ML algorithms can have different performances and some of them may be more sensitive to noise and/or dimension reduction than others. Therefore, we tested different classifiers simultaneously and we estimated the loss of utility P, that we defined as follows:

$$P = \max_{m \subset EM} Accuracy_m(X_{np}) - \max_{m \subset EM} Accuracy_m(\hat{X}_{np}) \tag{1}$$

with X_{np} the encoded data-set, $\hat{X}_{np}$ the anonymized data-set and accuracy score $(\frac{\text{Number of correct predictions}}{\text{Total number of predictions made}})$ obtained by classifiers, EM is the ensemble of models (XGBoost, RF and MLP). In other words, the loss of utility is calculated as the difference between the highest performance reached by classifiers on the raw and the anonymized data-sets.

(2) The anonymization quality: a small ϵ-value is supposed to guarantee a high-level of anonymization, but how to measure anonymization in practice? It can only be guaranteed by a mathematical formal proof, already performed using the differential privacy algorithm. We have chosen to use three metrics that are good indicators of anonymization quality because they offer, at least, necessary conditions for a better anonymization. We defined the first metric M_1 as follows:

$$M_1 = \frac{\mathrm{MSE}(X_{np}, \hat{X}_{np})}{\frac{1}{np}\sum_{i=0}^{n-1}\sum_{j=0}^{p-1}|\hat{x}_{ij}|} \tag{2}$$

where MSE is defined as:

$$\mathrm{MSE}(X_{np}, \hat{X}_{np}) = \frac{1}{np}\sum_{i=0}^{n-1}\sum_{j=0}^{p-1}(x_{ij} - \hat{x}_{ij})^2.$$

M_1 allows us to estimate the relative difference between X_{np} and $\hat{X}_{np}$. The closer M_1 is to zero the weaker is anonymization. We defined the second metric M_2 as follows:

$$M_2 = \frac{\frac{1}{n}\sum_{i=0}^{n-1}\mathrm{dist}(x_i, \hat{x}_i)}{\frac{1}{n}\sum_{i=0}^{n-1}\min_{\substack{j=0\\ j\neq i}}^{p-1}\mathrm{dist}(\hat{x}_i, \hat{x}_j)} \tag{3}$$

where $\mathrm{dist}(x_i, \hat{x}_i)$ measures the Euclidean distance between an original sample and its anonymization. $\min_{\substack{j=0\\ j\neq i}}^{p-1}\mathrm{dist}(\hat{x}_i, x_j)$ calculates the distance between an anonymized sample and its nearest anonymized neighbor sample. M_2 indicate how well anonymized samples can be correlated with their corresponding initial samples. If M_2 is larger than 1, the original and anonymized samples are very difficult to associate because the anonymized samples are very close to each other. Finally, we defined the third metric M_3 as:

$$M_3 = \frac{\frac{1}{n}\sum_{i=0}^{n-1}\mathrm{dist}(x_i, \hat{x}_i)}{\frac{1}{n}\sum_{i=0}^{n-1}\min_{\substack{j=0\\ j\neq i}}^{p-1}\mathrm{dist}(\hat{x}_i, x_j)} \tag{4}$$

where $\min_{\substack{j=0\\ j\neq i}}^{p-1}\mathrm{dist}(\hat{x}_i, x_j)$ the distance between an anonymized sample and the nearest raw sample (excluding his original sample). If M_3 is larger than 1, this means that there is a potential candidate closer to an anonymized sample than his original sample, and this makes the identification of the original sample extremely challenging.

3 Evaluation and Analysis

This study is composed of two parts. The first part consists to evaluate the performances of the DPCA and PPCA models using a text representation based on Skip-Thought. The second part is dedicated to the application of DPCA on real-world measurements data (time series) from satellites sensors for an anomaly detection task.

3.1 Anonymization of Textual Movie Reviews

In order to evaluate the impact of data-set size variations on each of the implemented anonymization methods, we evaluated PPCA and DPCA methods on two similar textual data-sets with different sizes. Both data-sets contain texts movie reviews, and categorized as positive or negative:

- Data-set Cornell (10,662 samples) were used in the experiments described in [27]. The data-set contains 5331 positive and 5331 negative review. Each line in these data correspond to a single review.

- Data-set Imdb [28] for binary sentiment classification containing substantially longer averages of reviews, and more samples (50,000) than Cornell data-set (10,000). The Imdb data-set contain 25,000 highly polar movie reviews for training, and 25,000 for testing. Movie-review data is used in sentiment-analysis experiments, labeled with respect to its overall sentiment polarity (positive or negative).

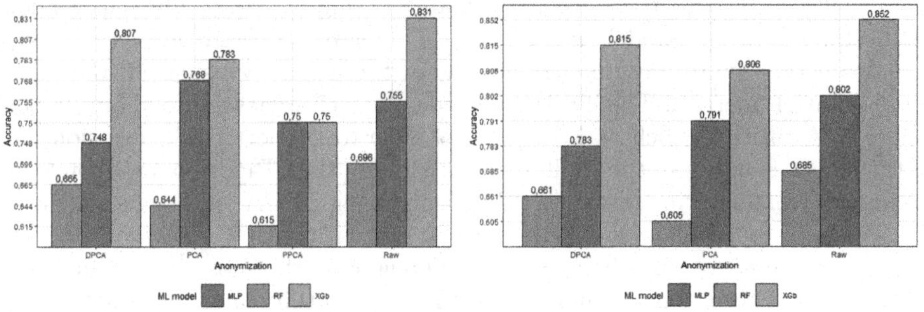

Fig. 2. Accuracy obtained from different models (XGBoost, RF and MLP) applied to the Cornell (left) and Imdb (right) data-set, and based on Skip-Thought encoding.

Performance Comparison and Results Interpretation: The first finding is that the XGBoost model is systematically better than the other two models for both raw and anonymized data (see Fig. 2). Note that PPCA was not tested on Imdb (very time-consuming and estimate at least 250 h). The use of a PCA (reducing dimensions by 50%) results in a decrease of about 5% of accuracy compared to the raw data, on both data-sets. Compared to PCA, the anonymization algorithms result in a slight additional 3% decrease of accuracy using PPCA and

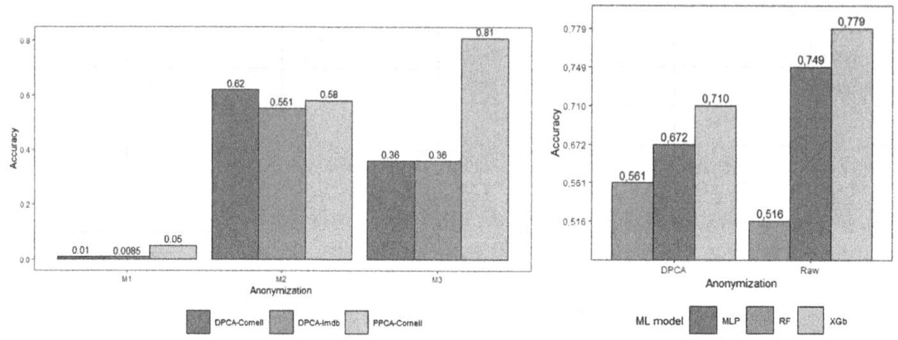

Fig. 3. (left) Anonymization quality obtained using different metrics (M_1, M_2 and M_3), and based on Skip-Thought encoding. (Right) Accuracy obtained from different models (XGBoost, RF and MLP) applied to the Cornell dataset, and based on TF-IDF encoding.

1–2% increase using DPCA (depending the data-set). We note that the utility is the same for both data-sets and is not significantly influenced by their size differences. Figure 3 (left) shows the anonymization quality as measured by the three metrics: M_1, M_2 and M_3 (defined in Sect. 2.4). We observe consistent results between the three metrics, with a better anonymization quality for PPCA than DPCA in terms of metrics M_1 and M_3. In terms of M_2, both are equivalents. Note that we obtained the same performances using bi-skip or uni-skip version of the Skip-Thought model (not shown). Therefore, we used only the bi-skip version in this study. Figure 3 (right) shows the accuracy using TF-IDF text representation method. The TF-IDF encodes Cornell reviews into about 16,000 features. Algorithms such as MLP are not able to support this dimension out of the 50,000 samples in the Imdb data-set. In addition, PPCA cannot converge (memory error), limiting tests of TF-IDF only using DPCA on the Cornell data-set. We set $\epsilon = 1$ billion to ensure data utility within DPCA method. In this context, we approve our choice of Skip-Thought as the most interesting text representation according to anonymization quality, utility, and time-consuming.

Effect of the Dimension Reduction: Figure 4 shows the k-parameter (dimension) variation as a function of the utility metric P. Note that the point $k = 25\%$ was reported for PPCA (corrupted results). In order to avoid the impact of differences in classifier performances, we consider $k = 50\%$ as the origin of each pattern. We note that standard PCA is more relevant with a small data size (<50%). Unlike PCA, DPCA is more relevant in terms of utility on higher dimensions (>50%) but the M_1 metric is twofold lower when $k = 90\%$ (not shown here). For this reason, DPCA is best used with k in the vicinity of 50%. Generally, DPCA and PCA show consistent results for both data-sets (Imdb and Cornell). PPCA shows a different behavior from other methods, we tested

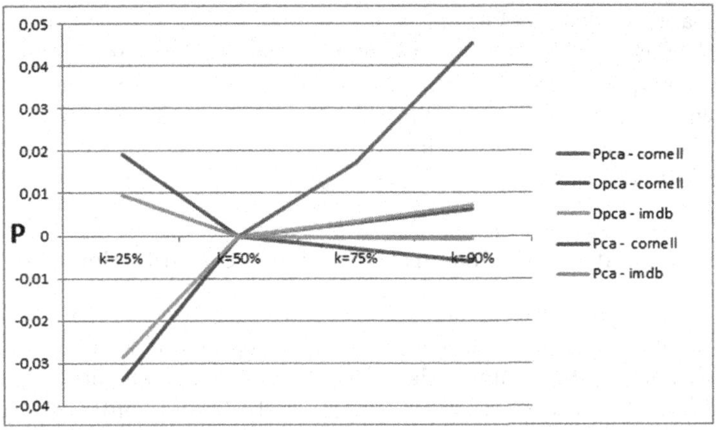

Fig. 4. k-parameter variation as a function of the utility P. The utility at $k = 50\%$ is considered as the origin.

$k = 75\%$ specifically to check this behavior. This difference is likely due to the projection subspace (not directly based on a PCA). In addition, for $k = 90\%$, the metric values M_1 and M_2 are still relevant ($M_2 = 1.1$ and $M_1 = 5$).

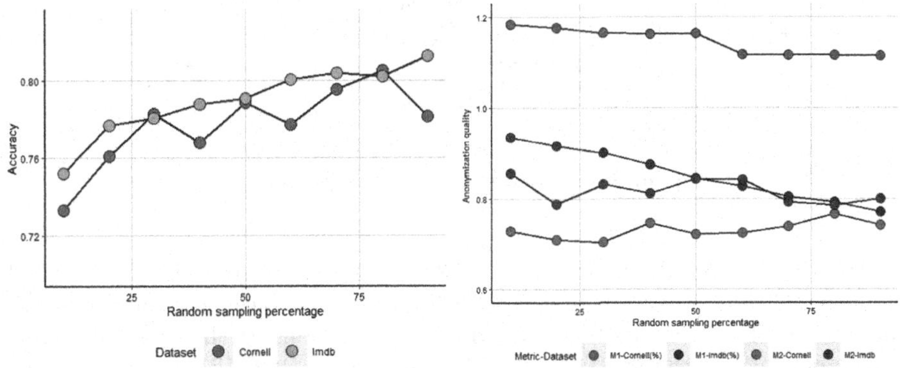

Fig. 5. Random sampling percentage as a function of (left) utility (XGBoost accuracy), and (right) anonymization quality. M_1 Metric values are displayed in percentages.

Effect of the Sub-sampling Data: In this section, we evaluate the effect of data size changes on the Cornell and Imdb data-sets. We chose random sampling because it is simple to use and more effective on large data-sets. Figure 5 (left) shows the utility as a function of the training data-set size (the validation data-set is unchanged). We note that the utility increases with the data size, and tends to stabilize. The performances are similar for the Cornell data-set, but less smooth. Figure 5 (right) shows the effect of the random sampling size on the anonymization quality. We note that M_1 decreases very slightly and M_2 remains constant whatever the sampling. Thus, the number of selected samples doesn't influence the anonymization quality. We also performed the PPCA tests with 20, 40, 60 and 80% sampling (not shown), and we observed the same trend.

3.2 Anonymization of Measurements from Satellite Sensors

We supported a leading European manufacturer to implement an anomaly detection model for satellite-sensors. The resulting model was able to detect most of the anomalies within the scope of the study. The following questions were raised: is it possible to perform an anomaly detection study on anonymized real-word data (time series)? In this section, we investigate if it is possible to perform this study. The first step was to split our data-set into days, in order to train and evaluate our model on how well it could detect anomalies from a single day's measurement. Each day was then labeled anomaly or not. Before any data transformation, we split the data-set into two parts: training and test.

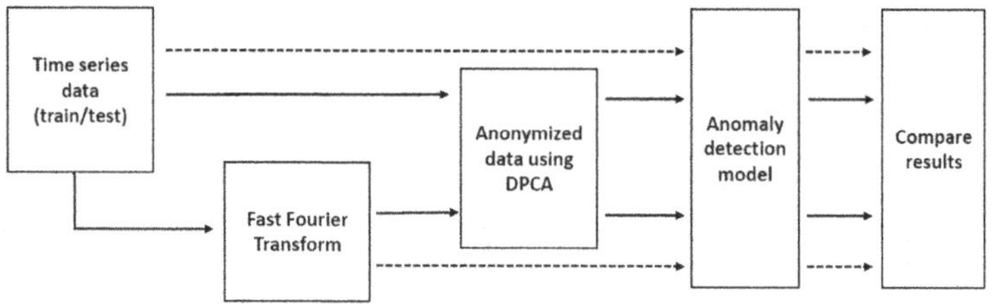

Fig. 6. Schematic view of study methodology.

The Fig. 6 shows the design methodology of our approach. First, we applied the Fast Fourier Transform (FFT), which allows us to keep the most relevant information. Then, we applied anonymization before using the anomaly detection model. We use the AUC (Area Under The Curve) ROC (Receiver Operating Characteristics) curve to check the performance of our anomaly detection problem. The reason of this choice is to provide the capability to adjust and calibrate the behavior of the model (for example, reduce false positive rate).

Table 1 shows the M_3 values according to the ϵ-parameter and the type of data-set. For raw data, we observe a high anonymization quality for $\epsilon = 1$ and $\epsilon = 1000$ but with a limited utility for $\epsilon = 1$ (not shown). For FFT data, good anonymization is only possible with $\epsilon = 1$. Therefore, we keep $\epsilon = 1000$ for raw data and $\epsilon = 1$ for FFT data. Table 2 shows the AUC values performed using our anomaly detection model. We note that by applying an FFT to our raw data, AUC value decreases by 2.5%. While anonymizing the FFT data with $\epsilon = 1$, the AUC value decrease by 2%. By the way, training and validating a model on

Table 1. Anonymization quality (using the M_3 metric) on the various datasets according to different $\epsilon - value$

	Train	Test	Train_{fft}	Test_{fft}
$\epsilon = 1$	1	0.99	0.34	0.92
$\epsilon = 1000$	0.96	0.90	0.14	0.04

Table 2. AUC scores obtained from the black-box anomaly detection model applied to the measurements sensors data-set. The application of FFT is shown in brackets.

Train	Test	ϵ (FFT)	AUC (FFT)
Raw	Raw	– (–)	0.993 (0.968)
Anon	Anon	1000 (1)	0.996 (0.948)
Anon	Raw	1000 (1)	0.991 (0.950)
Raw	Anon	1000 (1)	0.992 (0.878)

anonymized data and testing it on raw data is considered as the most expected and realistic anonymization use case. In this case, we note that if applied DPCA anonymization to FFT data (with $\epsilon = 1$), the AUC decreases only by 1.8% and if we apply only anonymization (with $\epsilon = 1000$), the AUC decreases by 0.2%. In conclusion, we showed that we can well perform an anomaly detection task on anonymized time series from satellites sensors.

4 Conclusions and Future Work

This paper was carried out to investigate and evaluate the application of differential privacy to textual data and time series from real-world use cases. We compared two differential privacy algorithms based on PCA and we extensively evaluate their performances in terms of utility and anonymization quality. Three informative metrics were defined in order to evaluate the anonymization quality. The PPCA algorithm provides a high anonymization quality at the expense of a high time-consuming. While, the DPCA is more able to support bigger data-set, but offers lower privacy guarantees. We showed that before stabilizing, the utility grows as the data-set size increases, ensuring that the anonymization quality is not affected by the data size variations. One the one hand, we demonstrate the possibility of combining a text representation approach with anonymization methods that ensure differential privacy. On the other hand, we showed that it is possible to perform an anomaly detection task on anonymized time series data from satellites sensors. As future work, we plan to apply anonymization on image and video-based data combining Convolutional Neural Network (CNN) with differential privacy methods. In order to preserve privacy and motivate data sharing, we emphasize the need to apply, explore and compare other anonymization methods on real-world data.

Acknowledgments. This work is supported by Scalian.

References

1. Albrecht, J.P.: How the GDPR will change the world. Eur. Data Prot. L. Rev. **2**, 287 (2016)
2. Samarati, P., Sweeney, L.: Protecting privacy when disclosing information: k-anonymity and its enforcement through generalization and suppression. Technical report, SRI International (1998)
3. Machanavajjhala, A., Gehrke, J., Kifer, D., Venkitasubramaniam, M.: l-diversity: privacy beyond k-anonymity. In: 22nd International Conference on Data Engineering (ICDE 2006), pp. 24–24. IEEE (2006)
4. Wong, R.C.-W., Li, J., Fu, A.W.-C., Wang, K.: (α, k)-anonymity: an enhanced k-anonymity model for privacy preserving data publishing. In: Proceedings of the 12th ACM SIGKDD International Conference on Knowledge Discovery and Data Mining, pp. 754–759. ACM (2006)
5. Li, N., Li, T., Venkatasubramanian, S.: t-closeness: privacy beyond k-anonymity and l-diversity. In: 2007 IEEE 23rd International Conference on Data Engineering, pp. 106–115. IEEE (2007)
6. Zhang, Q., Koudas, N., Srivastava, D., Yu, T.: Aggregate query answering on anonymized tables. In: 2007 IEEE 23rd International Conference on Data Engineering, pp. 116–125. IEEE (2007)
7. Martin, D.J., Kifer, D., Machanavajjhala, A., Gehrke, J., Halpern, J.Y.: Worst-case background knowledge for privacy-preserving data publishing. In: 2007 IEEE 23rd International Conference on Data Engineering, pp. 126–135. IEEE (2007)
8. Dwork, C.: Differential privacy. In: van Tilborg, H.C.A., Jajodia, S. (eds.) Encyclopedia of Cryptography and Security, pp. 338–340. Springer, Boston (2011). https://doi.org/10.1007/978-1-4419-5906-5
9. Friedman, A., Schuster, A.: Data mining with differential privacy. In: Proceedings of the 16th ACM SIGKDD International Conference on Knowledge Discovery and Data Mining, pp. 493–502. ACM (2010)
10. Mohammed, N., Chen, R., Fung, B., Yu, P.S.: Differentially private data release for data mining. In: Proceedings of the 17th ACM SIGKDD International Conference on Knowledge Discovery and Data Mining, pp. 493–501. ACM (2011)
11. Sarwate, A.D., Chaudhuri, K.: Signal processing and machine learning with differential privacy: Algorithms and challenges for continuous data. IEEE Signal Process. Mag. **30**(5), 86–94 (2013)
12. Fernandes, N., Dras, M., McIver, A.: Generalised differential privacy for text document processing. In: Nielson, F., Sands, D. (eds.) POST 2019. LNCS, vol. 11426, pp. 123–148. Springer, Cham (2019). https://doi.org/10.1007/978-3-030-17138-4_6
13. Fernandes, N., Dras, M., McIver, A.: Processing text for privacy: an information flow perspective. In: Havelund, K., Peleska, J., Roscoe, B., de Vink, E. (eds.) FM 2018. LNCS, vol. 10951, pp. 3–21. Springer, Cham (2018). https://doi.org/10.1007/978-3-319-95582-7_1
14. Zhang, X., Hamm, J., Reiter, M.K., Zhang, Y.: Statistical privacy for streaming traffic. In: Proceedings of the ISOC Network and Distributed System Security Symposium (2019)
15. Beaulieu-Jones, B.K., et al.: Privacy-preserving generative deep neural networks support clinical data sharing, p. 159756. BioRxiv (2018)

16. Chaudhuri, K., Monteleoni, C., Sarwate, A.D.: Differentially private empirical risk minimization. J. Mach. Learn. Res. **12**(Mar), 1069–1109 (2011)
17. McSherry, F., Talwar, K.: Mechanism design via differential privacy. In: Null, pp. 94–103. IEEE (2007)
18. Chaudhuri, K., Sarwate, A.D., Sinha, K.: A near-optimal algorithm for differentially-private principal components. J. Mach. Learn. Res. **14**(1), 2905–2943 (2013)
19. Jiang, X., Ji, Z., Wang, S., Mohammed, N., Cheng, S., Ohno-Machado, L.: Differential-private data publishing through component analysis. Trans. Data Priv. **6**(1), 19 (2013)
20. Goldberg, Y., Levy, O.: word2vec Explained: deriving Mikolov et al'.s negative-sampling word-embedding method. arXiv preprint arXiv:1402.3722 (2014)
21. Kiros, R., et al.: Skip-thought vectors. In: Advances in Neural Information Processing Systems, pp. 3294–3302 (2015)
22. Zhu, Y., et al.: Aligning books and movies: towards story-like visual explanations by watching movies and reading books. In: Proceedings of the IEEE International Conference on Computer Vision, pp. 19–27 (2015)
23. Hoff, P.D.: Simulation of the matrix Bingham-von Mises-Fisher distribution, with applications to multivariate and relational data. J. Comput. Graph. Stat. **18**(2), 438–456 (2009)
24. Tianqi, C., Carlos, G.: XGBoost: a scalable tree boosting system. In: Proceedings of the 22nd ACM SIGKDD International Conference on Knowledge Discovery and Data Mining, KDD 2016, New York, NY, USA, pp. 785–794. ACM (2016)
25. Gardner, M.W., Dorling, S.R.: Artificial neural networks (the multilayer perceptron)—a review of applications in the atmospheric sciences. Atmos. Environ. **32**(14–15), 2627–2636 (1998)
26. Breiman, L.: Random forests. Mach. Learn. **45**(1), 5–32 (2001)
27. Pang, B., Lee, L.: Seeing stars: exploiting class relationships for sentiment categorization with respect to rating scales. In: Proceedings of the 43rd Annual Meeting on Association for Computational Linguistics, pp. 115–124. Association for Computational Linguistics (2005)
28. Maas, A.L., Daly, R.E., Pham, P.T., Huang, D., Ng, A.Y., Potts, C.: Learning word vectors for sentiment analysis. In: Proceedings of the 49th Annual Meeting of the Association for Computational Linguistics: Human Language Technologies-Volume 1, pp. 142–150. Association for Computational Linguistics (2011)

Exploiting Similar Behavior of Users in a Cooperative Optimization Approach for Distributing Service Points in Mobility Applications

Thomas Jatschka[1(✉)], Tobias Rodemann[2], and Günther R. Raidl[1]

[1] Institute of Logic and Computation, TU Wien, Vienna, Austria
{tjatschk,raidl}@ac.tuwien.ac.at
[2] Honda Research Institute Europe, Offenbach, Germany
tobias.rodemann@honda-ri.de

Abstract. In this contribution we address scaling issues of our previously proposed cooperative optimization approach (COA) for distributing service points for mobility applications in a geographical area. COA is an iterative algorithm that solves the problem by combining an optimization component with user interaction on a large scale and a machine learning component that provides the objective function for the optimization. In each iteration candidate solutions are generated, suggested to the future potential users for evaluation, the machine learning component is trained on the basis of the collected feedback, and the optimization is used to find a new solution fitting the needs of the users as good as possible. While the former concept study showed promising results for small instances, the number of users that could be considered was quite limited and each user had to evaluate a relatively large number of candidate solutions. Here we deviate from this previous approach by using matrix factorization as central machine learning component in order to identify and exploit similar needs of many users. Furthermore, instead of the black-box optimization we are now able to apply mixed integer linear programming to obtain a best solution in each iteration. While being still a conceptual study, experimental simulation results clearly indicate that the approach works in the intended way and scales better to more users.

Keywords: Cooperative optimization · Facility location problem · Matrix factorization

1 Introduction

There exists a vast amount of literature regarding setting up service points for mobility applications such as bike sharing systems [1] or charging stations for

Thomas Jatschka acknowledges the financial support from Honda Research Institute Europe.

G. Nicosia et al. (Eds.): LOD 2019, LNCS 11943, pp. 738–750, 2019.
https://doi.org/10.1007/978-3-030-37599-7_61

electric vehicles [2]. A fundamental ingredient for optimizing the locations of service points is the distribution of existing customer demand to be potentially fulfilled in the considered geographical area. An estimation of this existing demand distribution is usually obtained upfront by performing customer surveys, considering demographic data, information on the street network and public transport, and not that seldom including human intuition and political motives. Unfortunately, this estimation is frequently imprecise and a system built on such assumptions might not perform as well as it was hoped for. Therefore, we have recently proposed the concept of a cooperative optimization algorithm (COA) [3,4], which, instead of estimating customer demand upfront, directly incorporates potential users in the optimization process by iteratively suggesting them solution scenarios and asking for feedback. Based on this user feedback a machine learning (ML) model is trained, which is used as evaluation function by an optimization component. This optimization core is responsible for generating new promising solution candidates, from which scenarios to be presented to the users are again derived. A major bottleneck in this previous approach is the large ML model consisting of many smaller components—one per considered user and potential service point location— which need to be trained in each iteration, and the used black-box optimization at the core.

In this contribution, we aim to improve the scalability of COA by replacing the ML model as well as the optimization core in a way that allows to exploit similar behavior of users. We refine the user interaction of COA by assuming that each potential user has certain use cases for the system, such as going to work, to a recreational facility, or shopping. The demand of these individual cases can be satisfied by different service points to different degrees, depending on the customer's preferences about the locations of these service points. It is unlikely that two customers have the same needs in all respect, i.e., they have the very same use cases with the same demands; however, given a sufficiently large number of users, it is safe to assume that some customers share some use cases and then have similar opinions on the suitability of service point locations w.r.t. such a use case. Our goal is to exploit these similarities using collaborative filtering techniques, in particular matrix factorization [5], to predict a customer's preferences of service point locations.

Concerning the optimization core in COA, we investigated in [4] a variable neighborhood search and a population-based iterated greedy algorithm, but both act as black-box methods, which do not exploit any structural features except of the ML model used to evaluate candidate solutions. Their scalability to larger instances therefore also is rather limited. Using now the matrix factorization based ML model allows to formulate the optimization problem as mixed integer linear program, which we are able to solve sufficiently fast to proven optimality.

This article is structured as follows. In Sect. 2 related work is discussed, while Sect. 3 formalizes the considered service point location problem. Section 4 presents our new approach. In Sect. 5 we experimentally evaluate the new COA variant based on a user simulation and discuss obtained results. Section 5.2 concludes this work with an outlook on future work.

2 Related Work

The Service Point Distribution Problem (SPDP) we consider here can generally be classified as a variant of the uncapacitated Facility Location Problem (FLP) [6]. For a survey on FLPs see [7]. Although the SPDP is quite generally phrased, we specifically have mobility applications in mind, especially the distribution of charging stations for electric vehicles. While there exists a vast amount of literature for setting up such systems, see e.g. [8–11], to the best of our knowledge all existing work essentially assumes customer demand to be estimated upfront. In our approach we substantially deviate from this traditional way of solving the SPDP by resorting to an interactive approach. Potential future customers are incorporated in the optimization process as an integral part by iteratively providing feedback on meaningfully constructed solution scenarios. In this way we learn user demands on-the-fly and may avoid errors due to unreliable a priori estimations. For a survey on interactive optimization algorithms see [12].

As we cannot expect a user to evaluate hundreds of solutions, a common way to unburden the users is to train a surrogate function [13] with the user feedback which is then used to evaluate intermediate solutions. In this contribution we use matrix factorization [5] as ML model to realize the surrogate function. Matrix factorization is a collaborative filtering technique which is frequently used in recommender systems [14]. The idea of collaborative filtering is to make recommendations for users based on the preferences of similar users, which means in our context to estimate some user demand for a use case by the feedback already provided by other users for similar use cases.

Matrix factorization is based on singular value decomposition which decomposes a matrix into two smaller matrices. Unknown values can then be estimated my multiplying the corresponding rows and columns of the decomposed matrices [14]. The two most popular techniques for decomposing a matrix with missing values are stochastic gradient descent (SGD) [15] and alternating least squares (ALS) [16]. ALS is usually only preferred over SGD for parallelization [5].

3 The Service Point Distribution Problem

The SPDP was originally defined in [3] as follows. We are given a set of locations $V = \{1, \ldots, n\}$ at which service points may be built and a set of potential users $U = \{1 \ldots, m\}$. The fixed costs for setting up a service point at location $v \in V$ are $z_v^{\text{fix}} \geq 0$, and this service point's maintenance over a defined time period is supposed to induce variable costs $z_v^{\text{var}} \geq 0$. The total construction costs must not exceed a maximum budget $B > 0$. Erected service stations may satisfy an arbitrary amount of customer demand, and for each unit of satisfied customer demand a prize $p > 0$ is earned.

A solution to the SPDP is a binary incidence vector $x = (x_v)_{v \in V}$, where $x_v = 1$ indicates that a service point is to be set up at location v. A solution x is feasible if its total fixed costs do not exceed the maximum budget B, i.e.,

$$z^{\text{fix}}(x) = \sum_{v \in V} z_v^{\text{fix}} x_v \leq B. \tag{1}$$

The objective function $f(x)$ of the problem is not explicitly given but only implicitly by allowing solutions to be evaluated by the users. In the original problem definition a user provides as feedback the estimated amount of demand (e.g., per week) that would be satisfied for him at each service point included in the solution x.

We now refine this user feedback by asking users already initially to specify use cases by a name and the demand each of them induces. Hence, we are also given for each user $u \in U$ the set of use cases E_u and the demand $D_{u,e}$ for each use case $e \in u$. Note, however, that we do not know which users share which use cases, their names have no meaning to us. The number of service points required to satisfy a use case e in general depends on the underlying application scenario. In our experiments in Sect. 5, we only consider scenarios where a use case requires one suitable service point to be satisfied, such as setting up charging stations for electric vehicles. Our approach, however, is in principle more general. For example when setting up rental stations for a bike sharing system, a use case will typically require two suitable service stations, one close to the origin and one close to the destination of a trip.

The objective is to find a feasible solution that maximizes the expected prizes earned for satisfied customer demands reduced by the variable costs for maintaining the service points, which is in our case

$$f(x) = q \cdot \sum_{u \in U} \sum_{e \in E_u} D_{u,e} \cdot \max_{v \in V} w(u, e, v)\, x_v - \sum_{v \in V} z_v^{\text{var}}\, x_v, \tag{2}$$

where function $w(u, e, v) \in [0, 1]$ denotes the suitability of a service point at location v to satisfy the needs of user u concerning his use case e. This objective function assumes that a user chooses for a use case always a location that is most suitable. The objective function $f(x)$ further interprets the determined suitability value for each use case as probability of the actual usage of the system to satisfy the demand $D_{u,e}$.

Note that $w(u, e, v)$ is not known upfront, but respective values can only be partially obtained from the users by providing them sample scenarios for evaluation. The evaluation of scenarios is discussed in more detail in Sect. 4.2.

As we are in general only able to obtain a small portion of all relevant values for $w(u, e, v)$ from the users, we exploit user behavior similarities and replace $w(u, e, v)$ by an approximation $\tilde{w}(u, e, v)$, yielding the surrogate objective function $\tilde{f}(x)$. This approximation will be realized by a ML model.

4 Cooperative Optimization Algorithm

The basic procedure of our COA remains almost the same as presented in [3], i.e., the framework consists of an evaluation component (EC) (containing the ML model), an optimization component (OC), a feedback component (FC), and

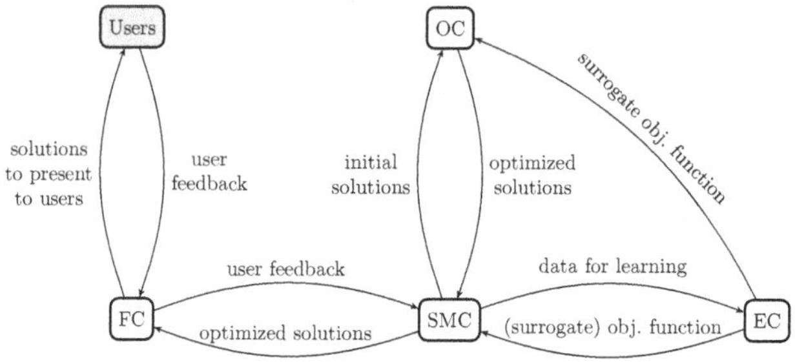

Fig. 1. Components of the COA framework and their interaction.

Algorithm 1. Basic Framework

Input : an instance of the SPDP
Output: a solution $x = (x_v)_{v \in V} \in \{0, 1\}^n$

while *no termination criterion satisfied* **do**
    Feedback Component:
        for $u \in U$ **do**
            for $e \in E_u$ **do**
                determine set of scenarios $S_{u,e}$ to be evaluated by user u;
                let user u evaluate $S_{u,e}$;
                update SMC with ratings obtained from $S_{u,e}$;
            end
        end
    Evaluation Component:
        train ML model with ratings in R, yielding surrogate obj. func. $\tilde{f}(x)$;
        re-evaluate all solutions stored in the SMC with new $\tilde{f}(x)$;
    Optimization Component:
        $x^{\mathrm{OC}} \leftarrow$ generate optimal solution w.r.t. the EC's $\tilde{f}(x)$;
        update SMC with x^{OC};
end
return *overall best found solution* $\tilde{x}^*$;

---

a solution management component (SMC). Figure 1 illustrates the communication between the components, and Algorithm 1 shows the main procedure in pseudo-code. We now use, however, different algorithms in these components as explained in the following.

4.1 Solution Management Component

The SMC stores and manages so far considered solutions and evaluations by the users. This includes in particular the set of tuples $R = \{(u, e, v) \mid w(u, e, v)$ is known from user feedback, $u \in U,\ e \in E_u,\ v \in V\}$ with the respective ratings $w(u, e, v)$. Moreover, the SMC also maintains the set X of all solutions obtained from the OC over all major iterations with their current surrogate objective values and, if available, their exact objective values. The current best solution is the solution in X with the highest surrogate objective value, denoted by $\tilde{x}^*$. With $V(u,e)$ the SMC also keeps track of the set of all locations $v \in V$ for which $(u,e,v) \in R$, with $u \in U$, $e \in E_u$. Last but not least, through the FC we are also able to obtain upper bounds on ratings $w(u,e,v)$, with $v \in V$, $u \in U$, $e \in E_u$, as explained in the next section. These upper bounds are stored in the SMC as $w^{\text{UB}}(u,e,v)$.

4.2 Feedback Component

The FC generates location scenarios for users to evaluate. Similar to solutions these scenarios are binary incidence vectors $s = (s_1, \ldots, s_n) \in \{0,1\}^n$, however they are not restricted by the budget constraint (1) and can therefore contain an arbitrary number of service points. In each COA iteration we present a set of scenarios to each user $u \in U$ for each of his use cases E_u for evaluation. If a user u selects in a scenario s for a use case e a suitable service point location v, he grades it with a rating $w(u,e,v) \in (0,1]$. If a user u decides that for a use case e there is no suitable service point location in scenario s, he indicates this by selecting no service point location, and we then know that $w(u,e,v) = 0$ for all $v \in V(s)$. Note that the user is required to select a best suited service point in the scenario if not all service points are unsuitable.

The obtained ratings are used in the EC for training the surrogate function. Moreover, the obtained ratings also serve as upper bounds for unknown ratings. As each user selects the best suited service point v in the presented scenario x w.r.t. a use case e, it must hold that $w(u,e,v) \geq w(u,e,k)\ \forall k \in x$. Hence, $w(u,e,v)$ serves as upper bound $w^{\text{UB}}(u,e,k)$ of $w(u,e,k)$. Moreover, $w^{\text{UB}}(u,e,k)$ is updated in the SMC whenever a lower upper bound is obtained.

We use two approaches to generate scenarios that are presented to a user $u \in U$ w.r.t. a use case $e \in E_u$. First, a scenario $s^{\text{V}} = \{v \in V \mid w(u,e,v) \notin R\}$ containing all locations that have not been rated yet w.r.t. u and e is presented to the user. Then, the user is also asked to evaluate the scenario $s^* = \{v \in \tilde{x}^* \mid w(u,e,v) \notin R\}$ containing all locations from the current best solutions that have not been rated yet w.r.t. u and e.

A main goal is to keep the number of presented scenarios per use case as low as possible. For this purpose, we exploit that users may show similar preferences for single use cases, hence, not every user needs to evaluate every location for a use case. Therefore, the scenario s^{V} is presented to u with a probability of 90% and s^* is shown to u with a probability of 20%.

4.3 Evaluation Component

The EC provides the means for evaluating solutions, in particular also within the OC. The real objective function $f(x)$, cf. (2), which contains many unknown user ratings, is approximated by the surrogate objective function $\tilde{f}(x)$ that is defined in accordance to $f(x)$ but makes use of *estimated ratings*

$$\tilde{w}(u,e,v) = \begin{cases} w(u,e,v) & \text{if } (u,e,v) \in R \\ \min\{w^{\mathrm{UB}}(u,e,v),\, \tilde{g}(u,e,v)\} & \text{else,} \end{cases} \tag{3}$$

where $\tilde{g}(u,e,v)$ is an approximate rating of location v for user u w.r.t. use case e.

We use matrix factorization [5] in order to predict unknown ratings. Given matrix $W = (W_{(u,e),v})_{u \in U, e \in E_u, v \in V}$ with $W_{(u,e),v} = w(u,e,v)$ for $(u,e,v) \in R$ and the other values unknown, matrix factorization identifies for each row (u,e), $u \in U$, $e \in U_e$ a vector $\xi_{u,e} \in \mathbb{R}^{\phi}$ and for each column $v \in V$ a vector $\nu_v \in \mathbb{R}^{\phi}$, respectively, with a space of features $F = \{1,\ldots,\phi\}$. The number of features ϕ is hereby a parameter that is chosen, e.g., in dependence of an estimation of the overall number of different use cases. An unknown value in W is approximated via the dot product $W_{(u,e),v} = \xi_{u,e}{}^{\mathsf{T}}\nu_v$, and $\tilde{g}(u,e,v) = W_{(u,e),v}$. The vectors $\xi_{u,e}$ and ν_v are learned by minimizing the loss function

$$\min \sum_{(u,e,v) \in R} \left(W_{(u,e),v} - (\mu + b_{u,e} + b_v + \xi_{u,e}{}^{\mathsf{T}}\nu_v)\right)^2 + \lambda(\|\xi_{u,e}\|^2 + \|\nu_v\|^2 + b_{u,e}^2 + b_v^2), \tag{4}$$

where λ is a regularization parameter which is set to 0.001 in our experiments, $b_{u,e} \in \mathbb{R}$ and $b_v \in \mathbb{R}$ are biases for users and locations, respectively, and μ is the average over all known values in R. For this minimization, stochastic gradient descent is used. In the first iteration of COA, the weights of the model are initialized randomly, while in later iterations, the model is re-trained starting with the values from the previous iteration.

4.4 Optimization Component

The OC solves the following mixed integer programming (MIP) formulation to determine an optimal solution w.r.t. the current surrogate objective function with ratings $\tilde{w}(u,e,v)$ provided by the EC. We use a binary variable x_v to indicate whether or not a location $v \in V$ is in the solution. Continuous variable $y_{u,e} \in [0,1]$ represents the expected degree to which a use case $e \in E_u$ is satisfied for user $u \in U$. Binary variable $z_{u,e,v} \in \{0,1\}$ indicates whether or not a user u would use a service point at location v to satisfy the demand of a use case $e \in E_u$.

$$\max q \cdot \sum_{u \in U} \sum_{e \in E_u} D_{u,e}\, y_{u,e} - \sum_{v \in V} z_v^{\mathrm{var}}\, x_v \tag{5}$$

$$y_{u,e} \le \sum_{v \in V} \tilde{w}(u,e,v) \cdot z_{u,e,v} \qquad \forall u \in U,\ e \in E_u \tag{6}$$

$$z_{u,e,v} \le x_v \qquad \forall u \in U,\ v \in V,\ e \in E_u \tag{7}$$

$$\sum_{v \in V} z_{u,e,v} \le 1 \qquad \forall u \in U,\ e \in E_u \tag{8}$$

$$\sum_{v \in V} c_{v,1}^{\mathrm{fix}}\, x_v \le B \tag{9}$$

$$x_v \in \{0,1\} \qquad \forall v \in V \tag{10}$$

$$y_{u,e} \in [0,1] \qquad \forall u \in U,\ e \in E_u \tag{11}$$

$$z_{u,e,v} \in \{0,1\} \qquad \forall u \in U,\ e \in E_u,\ v \in V \tag{12}$$

Inequalities (6) determine the expected degrees of satisfying the use cases in dependence of the user ratings and the location selection variables. Inequalities (7) express that a location can only satisfy demand if it contains a service point. According to (8), the demand of a service point for a user can only be satisfied at a single location. Finally, inequality (9) ensures that a solution does not exceed the budget.

5 Experimental Evaluation

As this contribution is only a conceptual study, we do not test with real users but simulate the user interaction in an idealized manner in certain benchmark scenarios. For this purpose we adopt the user simulation from [3] and extend it to our new needs.

5.1 Benchmark Scenarios

The n possible locations for service stations are randomly distributed in the Euclidean plane with coordinates coord(v), $v \in V$ chosen uniformly from the grid $\{0, \ldots, L-1\}^2$, with $L = \lceil 10\sqrt{n} \rceil$. The fixed costs c_v as well as the variable costs z_v for setting up a service station at each location $v \in V$ are uniformly chosen at random from $\{50, \ldots, 100\}$. The budget is assumed to be $B = \lceil 7.5 \cdot n \rceil$ so that about 10% of the stations with average costs can be set up.

The number of use cases for each user $u \in U$ is chosen randomly according to a shifted Poisson distribution with offset one and expected value three. Each of these use cases $e \in E_u$ is associated with an individual demand $D_{u,e}$ chosen at random from $\{5, \ldots, 50\}$ and a particular geographic location $r_{u,e} \in \{0, \ldots, L-1\}^2$. In order to model similarities in the users' use cases, these locations are generated in the following dependent way. We first select α *attraction points* A with uniform random coordinates from $\{0, \ldots, L-1\}^2$. Then, each use case location is derived by choosing one of these attraction points $(a_x, a_y) \in A$ and adding an individual deviation, i.e.,

$$r_{u,e} = (\lfloor \mathcal{N}(a_x, \sigma_{\mathrm{v}}) \rfloor, \lfloor \mathcal{N}(a_y, \sigma_{\mathrm{v}}) \rfloor), \tag{13}$$

where $\mathcal{N}(\cdot, \cdot)$ denotes a random value sampled from a normal distribution with the respectively given mean value and standard deviation. Note that coordinates beyond the grid are re-sampled.

A service point location $v \in V$ is generally considered suitable for the use case e if its Euclidean distance to the use case location does not exceed 15.

In this case v receives a positive rating that decreases exponentially with the distance but is also perturbed by a Gaussian noise:

$$w(u, e, v) = \mathcal{N}(e^{-||r_{u,e} - \mathrm{coord}(v)||/10}, \sigma_{\mathrm{r}}). \tag{14}$$

If $w(u, e, v) \notin (0, 1]$, the random sampling is repeated in order to obtain a valid rating.

In our experiments we consider benchmark scenarios with $n = 100$ locations and $m \in \{500, 1000, 1500\}$ users. For each combination we derive three groups of 30 independent instances with different parameters $\alpha \in \{10, 17, 25\}$, $\sigma_\mathrm{v} \in \{5, 7, 10\}$, and $\sigma_\mathrm{r} \in \{0.03, 0.1, 0.15\}$. All benchmark instances are available at https://www.ac.tuwien.ac.at/research/problem-instances#spdp.

5.2 Computational Results

The whole approach was implemented in Python 3.7. The matrix factorization has been realized with Keras 2.2.4 and TensorFlow 1.13.1 without GPU support. The number of features ϕ of the matrix factorization was set in accordance to the number of attraction points α of the test instances. At each iteration, the model was trained with the SGD optimizer to minimize loss function (4). Each training was done over 300 epochs with a batch size of 32, or until the loss function did not improve within 10 epochs. We use 20% of the training data as validation data with which the loss of the model is calculated.

The MIP is solved with Gurobi 8.1.0. All test runs have been executed on an Intel Xeon E5-2640 v4 2.40 GHz machine in single-threaded mode. COA was terminated after five major iterations or when a CPU-time limit of 7200 s has been reached and returned as the overall best solution $\tilde{x}^*$, i.e., the solution with the highest surrogate objective value at the end.

We compare our results to optimal solutions obtained by solving the MIP in the OC with exact values $w(u, e, v)$ provided by the user simulation, and with our previous COA variant from [3], here denoted as COA_0. In order make the comparison to COA_0 as fair as possible, the same termination criteria were applied, but otherwise all parameters of COA_0 were set as described in [3].

Table 1 shows the obtained results. Each line lists, for COA as well as COA_0, the average number of iterations $\overline{n_\mathrm{it}}$, the average optimality gap $\overline{\%\text{-gap}}$ between the objective value of $\tilde{x}^*$ and the optimal solution, the average percentage error of the surrogate function values of the final solutions $\overline{\%\text{-}\Delta\tilde{f}}$, with $\%\text{-}\Delta\tilde{f} = 100\% \cdot |\tilde{f}(\tilde{x}^*) - f(\tilde{x}^*)|/f(\tilde{x}^*)$, the average ratio of locations the users had to rate during the course of the algorithm per use case and their relevant locations per use case $\overline{\rho}$, and the median computation times in seconds $t[s]$.

The results clearly show that COA is able to converge to very reasonable solutions with small remaining optimality gaps of typically less than 2.3% within only five major iterations. For $\%\text{-}\Delta\tilde{f}$, we can observe that the percentage errors decrease as the number of users increases. This is especially evident for the hardest instance groups C, F, and I where $\%\text{-}\Delta\tilde{f}$ decreases from 8.17% to 4.42% on average. This documents that, given a sufficient amount of users, the surrogate function is able to approximate the real objective function at the end well in the relevant parts w.r.t. the returned solution. The table also shows that not all runs have been completed with five iterations, i.e., COA was aborted due to the time limit for 9 instances from the instance groups H and I. Column $\overline{\rho}$ of COA also shows that in general users do not need to rate more locations than their total number of relevant locations for each of their use cases.

COA_0 is significantly outperformed by COA in all aspects. COA is able to generate better solutions in less time for all instance groups. In many cases COA_0 exceeded the time limit of 7200 s already in the first or second iteration which explains the large difference in performance between COA and COA_0. It is not quite easy to compare $\overline{\rho}$ between COA and COA_0 since COA_0 was not able to perform as many iterations as COA. However, in general we can observe that users are required to evaluate significantly more locations with COA_0 than with COA.

Table 1. Average results of COA and COA_0.

						COA					COA_0				
Inst.	m	α	σ_v	σ_r	ϕ	$\overline{n_{it}}$	%-gap	%-$\overline{\Delta\tilde{f}}$	$\overline{\rho}$	t[s]	$\overline{n_{it}}$	%-gap	%-$\overline{\Delta\tilde{f}}$	$\overline{\rho}$	t[s]
A	500	10	5	0.03	10	5.00	**0.35**	**2.28**	0.86	**751**	1.97	16.40	28.07	0.82	7172
B	500	17	7	0.10	17	5.00	**1.18**	**5.19**	0.88	**888**	2.43	18.37	21.44	1.24	7168
C	500	25	10	0.15	25	5.00	**2.23**	**8.17**	0.84	**1033**	2.07	14.61	26.54	0.89	7190
D	1000	10	5	0.03	10	5.00	**0.39**	**1.94**	0.84	**1540**	2.90	16.93	22.63	1.53	7180
E	1000	17	7	0.10	17	5.00	**1.61**	**4.73**	0.83	**2407**	2.30	13.34	21.91	1.07	7181
F	1000	25	10	0.15	25	5.00	**1.52**	**5.72**	0.86	**3383**	2.53	16.98	20.86	1.32	7191
G	1500	10	5	0.03	10	5.00	**0.26**	**1.73**	0.85	**2579**	2.83	14.78	14.81	1.50	7189
H	1500	17	7	0.10	17	4.90	**1.18**	**3.81**	0.82	**4478**	1.77	17.78	28.88	0.65	7179
I	1500	25	10	0.15	25	4.73	**1.63**	**4.42**	0.80	**5605**	1.97	18.08	26.13	0.83	7189

In Fig. 2 we take a closer look at the distributions of the optimality gaps of the obtained solutions and how well our surrogate function is able to learn the behavior of the users. Considering a fixed number of users, the obtained optimality gaps deteriorate as the complexity of the instances (i.e., α, σ_v, σ_r) increases. Interestingly, increasing the number of users does not have a substantial impact on the optimality gaps when the complexity parameters stay the same. For %-$\overline{\Delta\tilde{f}}$, however, we can observe that the medians of the percentage errors slightly improve as the number of users increases. The large outliers of the instances groups H and I are from runs that have been aborted due to the time limit.

Generally, Fig. 2 indicates that the new approach scales now much better to larger numbers of users, and instead of the users, the number of actually different use cases is now what matters primarily. Thus, the similarity among users is indeed effectively exploited.

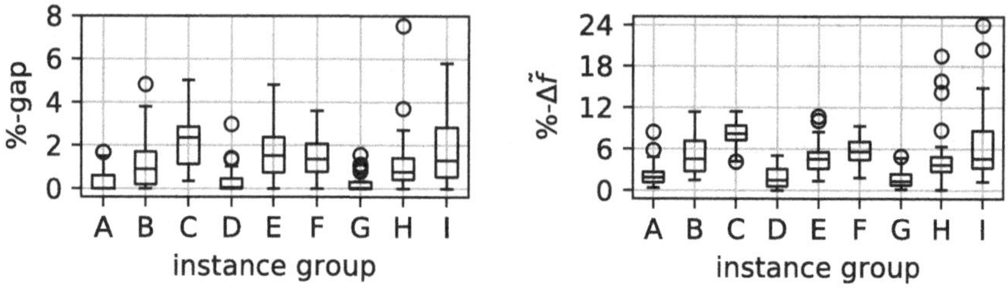

Fig. 2. Distributions of the optimality gaps and surrogate percentage errors of the best found solutions.

In Fig. 3 we analyze the computation times of the individual components of COA. Note that we omitted the computation times of the FC in Fig. 3 as they are negligible in comparison to the computation times of the EC and the OC. We see that the number of users has the strongest impact on the overall times. However, with an increasing complexity of the test instances, the OC quickly becomes the main bottleneck of our COA, as it generally requires more computation time than the other two components together.

While for COA_0 the EC was a major bottleneck, it now scales very well with an increasing number of users w.r.t. our benchmark instances. Hence, matrix factorization turns out to be an excellent choice as underlying model of our surrogate function.

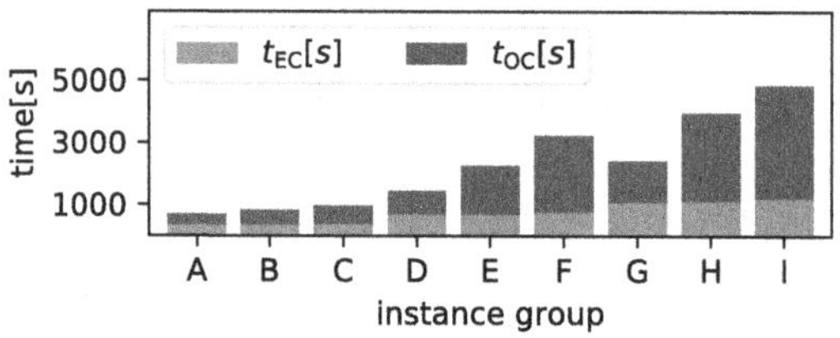

Fig. 3. Computation times of COA grouped by its framework components.

6 Conclusion and Future Work

In this contribution we have made major progress in improving the scalability of our previously presented COA [3] by using a matrix factorization model as our new surrogate function in the EC. Due to this change we were also able to abandon our previous black box optimization model of the OC and use a MIP instead. The new surrogate function as well as the new optimization core resulted in a major speedup and improvement in the scalability of our COA. Moreover, our new approach also requires a significantly lower number of user interactions.

In future work we aim at improving the approach further by refining in particular the feedback component to further reduce the number of user evaluations that are necessary to obtain reliable results. Moreover, for larger instances solving the MIP becomes the major bottleneck, as we have seen. Hence, a natural step to further improve the scalability is to replace the exact MIP with a reasonable heuristic approach. The loss of the proven optimality does not seriously matter in our application as enough other uncertainties remain. Last but not least, remember that COA was designed with more general applications in mind, and one of our next steps will be to apply it to more complex scenarios like bike sharing station planning, where we have to deal with trips instead of single locations in the use cases.

References

1. Kloimüllner, C., Raidl, G.R.: Hierarchical clustering and multilevel refinement for the bike-sharing station planning problem. In: Battiti, R., Kvasov, D.E., Sergeyev, Y.D. (eds.) LION 2017. LNCS, vol. 10556, pp. 150–165. Springer, Cham (2017). https://doi.org/10.1007/978-3-319-69404-7_11
2. Frade, I., Ribeiro, A., Gonçalves, G., Antunes, A.: Optimal location of charging stations for electric vehicles in a neighborhood in Lisbon, Portugal. Transp. Res. Rec. J. Transp. Res. Board **2252**, 91–98 (2011)
3. Jatschka, T., Rodemann, T., Raidl, G.R.: A cooperative optimization approach for distributing service points in mobility applications. In: Liefooghe, A., Paquete, L. (eds.) EvoCOP 2019. LNCS, vol. 11452, pp. 1–16. Springer, Cham (2019). https://doi.org/10.1007/978-3-030-16711-0_1
4. Jatschka, T., Rodemann, T., Raidl, G.R.: VNS and PBIG as optimization cores in a cooperative optimization approach for distributing service points. In: Computer Aided Systems Theory - EUROCAST 2019. LNCS, Springer (2019, to appear). https://www.ac.tuwien.ac.at/files/pub/jatschka_19a.pdf
5. Bell, R.M., Koren, Y., Volinsky, C.: Matrix factorization techniques for recommender systems. Computer **42**(08), 30–37 (2009)
6. Cornuéjols, G., Nemhauser, G.L., Wolsey, L.A.: The uncapacitated facility location problem. In: Mirchandani, P.B., Francis, R.L. (eds.) Discrete Location Theory, pp. 119–171. Wiley, Hoboken (1990)
7. Farahani, R.Z., Hekmatfar, M.: Facility Location: Concepts, Models Algorithms and Case Studies. Springer, London (2009). https://doi.org/10.1007/978-3-7908-21
8. Awasthi, A., Venkitusamy, K., Padmanaban, S., Selvamuthukumaran, R., Blaabjerg, F., Singh, A.K.: Optimal planning of electric vehicle charging station at the distribution system using hybrid optimization algorithm. Energy **133**, 70–78 (2017)
9. Cavadas, J., Homem, G.D.A.C., Gouveia, J.: A MIP model for locating slow-charging stations for electric vehicles in urban areas accounting for driver tours. Transp. Res. Part E Logistics Transp. Rev. **75**, 188–201 (2015)
10. Chung, S.H., Kwon, C.: Multi-period planning for electric car charging station locations: a case of korean expressways. Eur. J. Oper. Res. **242**(2), 677–687 (2015)

11. Kameda, H., Mukai, N.: Optimization of charging station placement by using taxi probe data for on-demand electrical bus system. In: König, A., Dengel, A., Hinkelmann, K., Kise, K., Howlett, R.J., Jain, L.C. (eds.) KES 2011. LNCS (LNAI), vol. 6883, pp. 606–615. Springer, Heidelberg (2011). https://doi.org/10.1007/978-3-642-23854-3_64
12. Meignan, D., Knust, S., Frayret, J.M., Pesant, G., Gaud, N.: A review and taxonomy of interactive optimization methods in operations research. ACM Trans. Inter. Intell. Syst. **5**(3), 17:1–17:43 (2015)
13. Koziel, S., Ciaurri, D.F., Leifsson, L.: Computational Optimization, Methods and Algorithms. Studies in Computational Intelligence, vol. 356, pp. 33–59. Springer, Heidelberg (2011). https://doi.org/10.1007/978-3-642-20859-1
14. Ekstrand, M.D., Riedl, J.T., Konstan, J.A.: Collaborative filtering recommender systems. Found. Trends Hum.-Comput. Inter. **4**(2), 81–173 (2011)
15. Robbins, H., Monro, S.: A stochastic approximation method. Ann. Math. Stat. **22**(3), 400–407 (1951)
16. Bell, R.M., Koren, Y.: Scalable collaborative filtering with jointly derived neighborhood interpolation weights. In: Seventh IEEE International Conference on Data Mining, pp. 43–52 (2007)

Long Short-Term Memory Networks for Earthquake Detection in Venezuelan Regions

Sergi Mus[1], Norma Gutiérrez[1], Ruben Tous[1], Beatriz Otero[1(✉)], Leonel Cruz[1], David Llácer[1], Leonardo Alvarado[2], and Otilio Rojas[3,4]

[1] Universitat Politècnica de Catalunya (UPC), Barcelona, Spain
botero@ac.upc.edu
[2] Fundaciòn Venezolana de Investigaciones Sismológicas, Caracas 1070, Venezuela
[3] Facultad de Ciencias, Universidad Central de Venezuela, Caracas, Venezuela
[4] Barcelona Supercomputing Center (BSC), Barcelona, Spain

Abstract. Reliable earthquake detection and location algorithms are necessary to properly catalog and analyze the continuously growing seismic records. This paper reports the results of applying Long Short-Term Memory (LSTM) networks to single-station three-channel waveforms for P-wave earthquake detection in western and north central regions of Venezuela. Precisely, we apply our technique to study the seismicity along the dextral strike-slip Boconó and La Victoria - San Sebastián faults, with complex tectonics driven by the interactions between the South American and Caribbean plates.

Keywords: Earthquake detection · Neural networks · Deep learning · LSTM

1 Introduction and Related Works

Most earthquake detection methods are designed for moderate and large earthquakes, and fail to detect low-magnitude events, buried in seismic noise. However, correctly detecting these earthquakes through the existing seismic records is key to understanding their causes and to mitigate the seismic risk. This paper reports the results of an approach to apply Long Short-Term Memory (LSTM) networks over seismic data collected by broadband stations at western, central and northern Venezuela, during the time period of 2015 to 2018. The seismicity in the region results from the right-lateral strike-slip faulting experienced along the interface between the Caribbean and South American plates, as the former moves to the east with respect to the latter. A review of the seismic history and tectonics of our study area and related regions can be found in [3]. Artificial Neural Networks have been actively applied to earthquake detection since mid-late 90s. In particular recurrent networks, well suited for recognition and inference of temporal patterns, have been used for small-event detection in noisy data in [7]

G. Nicosia et al. (Eds.): LOD 2019, LNCS 11943, pp. 751–754, 2019.
https://doi.org/10.1007/978-3-030-37599-7_62

and [6], and for early warning systems and earthquake forecasts in [4]. The network architectures in [6] and [4], present few preprocessing convolutional layers, preceeding the core LSTM structure, in a similar way to our current approach.

2 Methodology

2.1 Dataset and Preprocessing

The Venezuelan Foundation of Seismological Research (FUNVISIS) network counts with 40 broadband stations recording three-channel continuous data at 100 Hz. The input dataset includes waveforms from seismic events with magnitudes in the range [1.7, 5.2] Mw that occurred between 2015 and 2018. Few of these events took place on the western state of Tachira, while hypocenters of the remaining bigger set are located across the Northcentral states of Carabobo, Aragua and Miranda.

Input waveforms are first normalized and divided into single-station streams. In P signal detection, the Z component is the most relevant and that is why the rest of the components are deleted [2]. After that, we cut the signals in 50 seconds windows, which gives us windows of 5001 samples. Then, we compare this window's starting and final times to the times specialized analysts label as containing an event, to determine if a window contains an event or not. After that, we put the windows containing events in one folder and the ones not containing on another one. From this folders we load the same number of events and noise windows so we have a balanced dataset. Then, the dataset is splitted into 20% for validation and the rest for training.

2.2 Network

In our first experiment, we tried to use only LSTM layers with some fully connected layers. This first experiments did not give positive results, since the accuracies did not exceed 65%. We hypothesized that this could be due to the erratic nature of seismographic signals. In order to solve this problem, we added some convolutional layers before the LSTM layers. The reasoning behind these layers that we added is to first extract the features using these layers and afterwards to feed this features to the LSTM.

Our final network model (see Fig. 1) starts with three convolutional layers, each one of them followed by a max-pooling layer. Convolutional layers had 128 filters each, and a window size of 3 samples. As said previously, the purpose of these layers is to extract all the important features from the wave signal. After these layers, three LSTM layers were added with 128 units each one. Furthermore, we have used the CuDNN implementation [1]. This CUDA library enables to use GPU acceleration to train our neural network, which makes the training process remarkably faster. The main goal of this step is to extract time related features from the input signal and to extract the important data from it. The last phase consists of two fully connected layers with 128 units and a rectified linear activation function (ReLU). The purpose of this step is to make the final classification after having extracted all the features. And after the two dense layers a final output layer with a single perceptron, that has a sigmoid activation function which classifies the signal, as containing an event or containing just noise. The neural network was trained with the ADAM optimizer [5] with a learning rate of 0.001. The loss function used is binary cross entropy, which was chosen because the problem can be summed up to be a binary classification problem.

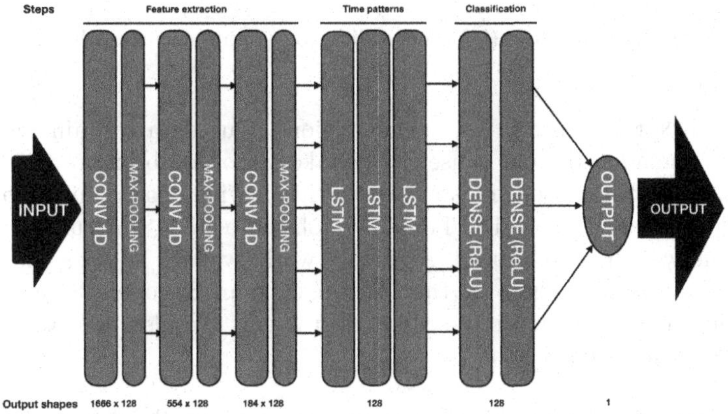

Fig. 1. Network architecture with 3 convolutional+max_pooling layers, 3 LSTM layers and 2 fully connected layers.

3 Experiments and Results

The network has been trained using a system with an Intel Core i7-4770 CPU @ 3.40GHz, 16 GB of RAM memory and an NVIDIA GeForce GTX 1060 6 GB. The version of Tensorflow used is 1.12.0 with CUDA 9.0.176. During the training process a Tensorboard callback was used, which enabled to monitor the accuracy and loss over each epoch (see Fig. 2). This way the networks could be compared to each other. After some networks were trained, we determine which architectures and parameters gave the biggest benefits and those were kept while the underperforming ones were rejected. Finally, the accuracy has been topped out at 97.61%, with a difference in loss between the training and validation datasets smaller than 1%. This indicates that no overfitting takes place.

4 Discussion and Conclusions

In this paper we have reported the results of an approach to apply LSTM networks over single-station waveforms for P-wave based earthquake detection in

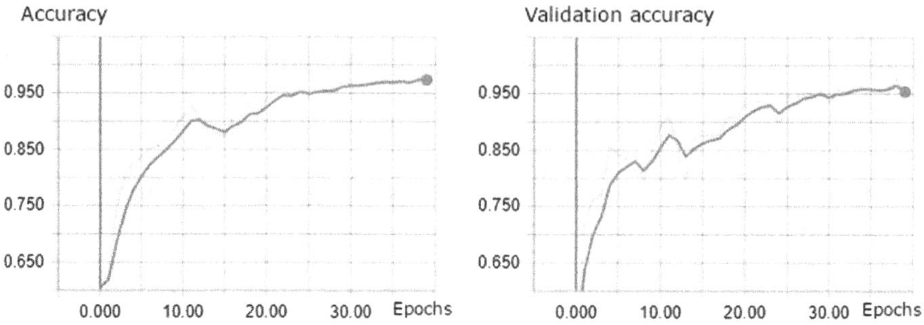

Fig. 2. Evolution of training (left) and validation (right) accuracy during training.

western and North Central Venezuelan regions. Our data contains waveforms from diverse magnitude and cause earthquakes recorded by more than 40 heterogeneous and distant seismic stations, within a wide geographic area encompassing four different geological faults. We obtain a 97.61% accuracy when we train the network for 40 epochs. As future work, we will address the problem of estimating the pick time for the P-wave with an Encoder-Decoder LSTM, approaching it as a sequence-to-sequence problem. We also plan to apply our approach to S-wave detection.

Acknowledgements. This work is partially supported by the Spanish Ministry of Economy and Competitivity under contract TIN2015-65316-P and by the SGR programmes (2014-SGR-1051 and 2017-SGR-962) of the Catalan Government. We thank FUNVISIS for providing the seismic data subject of our current studies.

References

1. NVIDIA cuDNN. https://developer.nvidia.com/cudnn. Accessed: 15 Jun 2018
2. Aki, K., Richards, P.G.: Quantitative Seismology. University Science Books, 2 edn. (2002). http://www.worldcat.org/isbn/0935702962
3. Audemard, F.A.: Revised seismic history of the El pilar fault, Northeastern Venezuela, from the cariaco 1997 earthquake and recent preliminary paleoseismic results. J. Seismol. **11**(3), 311–326 (2007). https://doi.org/10.1007/s10950-007-9054-2
4. Ibranhim, M.A., Park, J., Athens, N.: Earthquake warning system: detecting earthquake precursor signals using deep learning networks. In: Proceedings of the University of Stanford, CS230, Stanford, California, USA, pp. 1–7 (2018)
5. Kingma, D.P., Ba, J.: Adam: a method for stochastic optimization. In: 3rd International Conference on Learning Representations, 2015, San Diego, CA, USA (2015)
6. Mousavi, S.M., Zhu, W., Sheng, Y., Beroza, G.C.: CRED: a deep residual network of convolutional and recurrent units for earthquake signal detection. CoRR abs/1810.01965, pp. 1–23 (2018).
7. Wiszniowski, J., Plesiewicz, B., Trojanowski, J.: Application of real time recurrent neural network for detection of small natural earthquakes in Poland. Acta Geophys. **62**(3), 469–485 (2014)

Zero-Shot Fashion Products Clustering on Social Image Streams

Jonatan Poveda(✉) and Rubén Tous

Universitat Politècnica de Catalunya (UPC), Barcelona, Spain
jonatan.poveda@estudiant.upc.edu, rtous@ac.upc.edu

Abstract. Computer Vision methods have been proposed to solve the problem of matching photographs containing some products from users in social media to products in retail catalogues. This is challenging due to the quality of the photographies, difficulties in dealing with garments and their category taxonomy. A *N-Shot Learning* approach is required as retail catalogues may contain hundreds of different products for which, in many cases, only one image is provided. This framework can be solved by means of *Deep Metric Learning (DML)* techniques, in which a metric to discriminate similar than dissimilar samples is learnt. The performance of different authors tackling this problem varies a lot but even if they perform reasonably well, the set of elements they need to return in order to include the exact product is large. As after the query there is a person curating the results, it is important to return the smallest set of elements possible, being ideally just to return only one: the related product. This paper proposes to solve the image-to-product image matching problem through a product retrieval system using DML and Zero-short Learning, focusing on garments, and applying some of the last advances on clustering techniques.

Keywords: Computer Vision · Deep Metric Learning · Zero-shot Learning · Clustering · Supervised learning

1 Introduction

Nowadays, there is a growing interest in deriving benefit from the photos that users share on social networks such as Instagram or Twitter. Computer Vision is essential for many tasks: image classification can be applied to pick images that match a *visual brand identity* [2], object recognition to find brand logos [1,2], or to find the relationship between photos generated online by users and real world events [3]. Another desired application, specially in e-commerce, is to match street images to products. This problem, where an image of a product is

This work is partially supported by the Spanish Ministry of Economy and Competitivity under contract TIN2015-65316-P and by the SGR programme (2014-SGR-1051 and 2017-SGR-962) of the Catalan Government.

G. Nicosia et al. (Eds.): LOD 2019, LNCS 11943, pp. 755–758, 2019.
https://doi.org/10.1007/978-3-030-37599-7_63

being used to find that product, is approached using a variety of techniques in computer vision: from feature engineering to deep learning.

As dealing with street images, we have to consider some complications: only part of the image could be important, the photo may have bad light conditions, multiple items may appear on the scene, or the item we are interested on could be occluded. Moreover, in the case of garments the problems is even tougher as a result of non-rigid transformations between instances of the same piece of clothing. In addition, we may be dealing with a single image as a reference of an item or discriminating among different items with near-inappreciable differences. In terms of garment categories, no one agrees with a taxonomy, and it use to change over time within the same brand.

2 Related Work

We use *N-Shot learning* to tackle this problem. *N-Shot learning* is a framework where there are only few samples available for each class. *Zero-Shot Learning* is the special case where no samples are available for some of the classes. Therefore, the architecture has to be able to differentiate between classes in a way that, a new class is not confused with already seen classes, and to cluster unseen samples of the same class close. These methods are a particular case of supervised learning, as we have labels for all data, even though they are *loose* ('similar', 'dissimilar'), and where all instances of some classes are held out of training data. This method avoids learning garment categories and produces a taxonomy-independent solution. *Deep Metric Learning* group techniques where a discriminative embedding function, suited for a clustering or retrieval task, is learnt. In this framework, the learnt feature embedding function embeds input images into vectors, so-called *embeddings*, in such a way that the embeddings of semantically similar images are close to each other in the *embedding space*. These techniques are perfectly suit for solving *N-Shot Learning* problems.

3 Methodology

ModaNet dataset [5] is used to train and evaluate our system. It is a curated set of 166,132 fashion images from chictopia.com. Each sample contains an image, the user post and its social interactions, bounding boxes of items, their segmentation, and their category. Categories are grouped into 13 *meta-categories* by semantic similarity (Fig. 1). We instantiate the model *ProxyNCA* described in [4] to solve the problem. It is based on a triplet network built using BN-Inception. ProxyNCA tackles the problem of speed in mining techniques by using *proxies*: virtual points near actual data points in the embedding space that represent them. Proxies are used to learn a distance metric. This idea helped to cut off the convergence time to up to a third compared to their state-of-the-art. The original authors compare two proxy-assignment methods: a *static* one when class labels are available, and a *dynamic* when only *loose labels* are available. Their implementation used can be found in https://github.com/dichotomies/proxy-nca.

In our tests we only use images and meta-categories, employing the *static* proxy-assignment method. We build balanced datasets by randomly picking 13,653 samples per class for training and 385 per class for validation. This ensures performance metrics are representative of all classes. For training we use the classes *'top'*, *'outer'*, *'pants'*, *'footwear'*, *'bag'*, *'dress'*, *'belt'*, *'skirt'*, and for evaluating *'sunglasses'*, *'shorts'*, *'headwear'*, *'scarf/tie'*, *'boots'* (see Fig. 1). The selection of these classes is picked to maximize the number of training samples. The base model, BNInception, is initialized using pre-trained weights on Imagenet and fine-tuned along 40 epochs in batches of 32 samples. Data augmentation and data adaptation transformations are adopted from original authors: random crop of 227 × 227, resize to 256 × 256, random horizontal flip and an input normalization stage at the end. Images are in RGB colour space. For input normalization only samples from the train set are considered to avoid data leakage. The embedding size is set to 64. Adam optimization algorithm is used for back-propagation with a learning rate of 6.25×10^{-4} and a weight decay of 0.94 per epoch.

Fig. 1. Example images of all categories (from upper-left to lower-right): bag, belt, boots, outer, dress, sunglasses, pants, top, shorts, skirt, headwear, scarf/tie

4 Evaluation and Results

We use Recall@k to analyse the model performance on retrieving the correct class in a subset of k samples. In other words, given a query image the model returns a ranked list of similar images. When any of the first returned k-images is from the same class as the input it counts as a positive. To analyse performance on clustering we use the *Normalized Mutual Information (NMI)* [4], also called *symmetric uncertainty*, for its properties: invariant to label permutation and independent of clustering' cardinality. Its value lies between 0 (random clustering) and 1 (perfect clustering). We evaluate the architecture with two different labels. One using the bounding boxes to crop the objects from the image and the other using the segmentation to additionally mask the background. As we are normalizing samples, note that on using segmentation, we are taking into account all the zeroes of the removed background, then computing a lower mean. In order to fix it, we modified the input normalization to take into account only the item pixels. Results are shown on Fig. 2.

We should be cautions with the results as the classes split may introduce some bias. Note that performance may be different on a more realistic case when dealing with imbalanced categories.

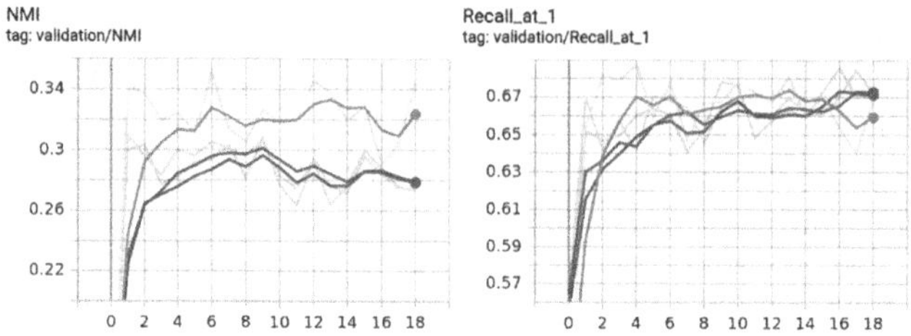

Fig. 2. Performance on clustering and retrieval using bounding boxes (orange), segmentation with default input normalization (blue), and segmentation using a mask-aware input normalization (red) using Tensorboard. An smoothing factor of 0.6 is applied to show its tendency (vivid lines). (Color figure online)

5 Conclusions

We observe that using the context of the items helps for clustering. Even if the tendency seems to indicate the same for retrieval, it is not statistically significant (Fig. 2). Therefore, extracting a segmentation from the objects not only could add complexity to the system but also seems not to help. We also note that correcting the input normalization to be effective on masked items does not help neither for clustering nor for retrieval. Even though, we should take into account that on other datasets, as it could lead the model to underperform. We conclude that there is a low clustering quality, therefore classes could not be separated successfully. We hypothesize that selecting a more diverse set would lead to a more generalized solution, therefore improving model performance. Further work should be done to improve class clustering, specially when using segmented images, and additional evaluations selecting a different class split to cross-validate the current results.

References

1. Orti, O., Tous, R., Gomez, M., Poveda, J., Cruz, L., Wust, O.: Real-time logo detection in brand-related social media images. In: 15th International Work-Conference on Artificial Neural Networks, IWANN 2019, Gran Canaria, Spain. 12–14 June 2019 (2019)
2. Tous, R., et al.: Automated curation of brand-related social media images with deep learning. Multimedia Tools Appl. **77**(20), 27123–27142 (2018). https://doi.org/10.1007/s11042-018-5910-z
3. Tous, R., Torres, J., Ayguadé, E.: Multimedia big data computing for in-depth event analysis. In: Proceedings of the 2015 IEEE International Conference on Multimedia Big Data (BigMM), 20–22 April 2015, Beijing, China, pp. 144–147. IEEE (2015)
4. Ying, X., Zha, H.: No fuss distance metric learning using proxies. In: Proceedings - International Conference on Pattern Recognition, vol. 1, pp. 535–538 (2017). https://arxiv.org/pdf/1703.07464.pdf
5. Zheng, S., Yang, F., Kiapour, M.H., Piramuthu, R.: ModaNet: a large-scale street fashion dataset with polygon annotations (2018). http://arxiv.org/abs/1807.01394. https://github.com/eBay/modanet

Treating Artificial Neural Net Training as a Nonsmooth Global Optimization Problem

Andreas Griewank[1(✉)] and Ángel Rojas[2]

[1] Research Center on Mathematical Modelling (MODEMAT), Escuela Politécnica Nacional, Quito, Ecuador
griewank@math.hu-berlin.de
[2] School of Mathematical and Computational Sciences, Yachay Tech, Urcuquí, Ecuador
angel.rojas@yachaytech.edu.ec

Abstract. We attack the classical neural network training problem by successive piecewise linearization, applying three different methods for the global optimization of the local piecewise linear models. The methods are compared to each other and steepest descent as well as stochastic gradient on the regression problem for the Griewank function.

Keywords: Abs-normal/linear form · Dynamic search trajectory · Proximal term · Coordinate descent · Mixed integer linear optimization

1 Introduction

In machine learning objective functions that are only piecewise smooth and should be globally minimized abound. The standard method of dealing with them is to apply a stochastic gradient method disregarding the nonsmoothness and hoping for the best as far as global optimality of the computed solution is concerned. Without doubting that this optimistic approach often works very well in practice, we explore in this paper a more traditional deterministic, but partially also heuristic approach. All piecewise smooth functions in so-called abs-normal form can be locally approximated by a corresponding abs-linear model, such that the discrepancy is of second order in the distance to the reference point [7]. We consider three methods for globally minimizing these local models: Coordinate Global Descent (CGD) (see Wright [12]), the Target Oriented Averaging Search Trajectory (TOAST) proposed in [9] and the solution of an equivalent Mixed Integer Bilinear Optimization Problem [11] by modern branch and cut solvers like Gurobi (see www.gurobi.com). Only the latter choice guarantees global optimality of the model minimizers, but in any case, when generated by successive abs-linearization, their cluster points need not be globally optimal on the underlying nonlinear problem. Nevertheless we find on some test problems that these cluster points have typically much lower function values than the stationary points reached by variants of steepest descent or a local successive abs-linear minimization algorithm [10].

G. Nicosia et al. (Eds.): LOD 2019, LNCS 11943, pp. 759–770, 2019.
https://doi.org/10.1007/978-3-030-37599-7_64

2 Preliminary Observations on Neural Net Learning

A learning problem for a neural net with a single intermediate layer of size n and hinge activation [6] can be mathematically described by

$$\min_w |f(w;x) - y| \text{ where } f(w;x) \equiv p^\top \max(0, Wx + b) \text{ with } w \equiv (W, b, p). \quad (1)$$

Here $x \in \mathbb{R}^d$ is a feature vector from a training set, $y \in \mathbb{R}$ the corresponding label, $W \in \mathbb{R}^{n\times d}$ the weight matrix between the sample input and the intermediate layer, $b \in \mathbb{R}^n$ the shift inhomogeneity and $p \in \mathbb{R}^n$ the output weight vector.

The empirical risk objective to be minimized is the average of the losses $|f(w;x) - y|$ for $(x, y) \in \mathbb{R}^{d+1}$ ranging over a training set. In other words we are looking for an optimal l_1 fit over a set of sample points rather than just minimizing the discrepancy at one particular point. This summation does not affect the properties of the objective with respect to the optimization variables $w = (W, b, p) \in \mathbb{R}^{n\times(d+2)}$ to be discussed here. Firstly we observe that due to the positive homogeneity of the hinge function for a scalar $0 < \rho \in \mathbb{R}$

$$\begin{aligned} f(\rho W, \rho b, p/\rho; x) &= (p/\rho)^\top \max(0, (\rho W)x + (\rho b)) \\ &= p^\top \max(0, Wx + b) = f(W, b, p; x). \end{aligned}$$

This invariance implies immediately that all local and global minimizers cannot be isolated and their Hessians must be singular if they exist at all.

Of course a similar reciprocal scaling of weights and shifts can be applied between successive layers in more general networks leaving the resulting prediction function invariant and thus the risk function minimization problem singular. To remove the singularity in the single layer case considered here, we can fix the components of the output weight vector to be -1 or $+1$ so that the prediction function is effectively split into

$$f(W_+, W_-, b_+, b_-; x) = \underbrace{e_+^\top \max(0, W_+x + b_+; x)}_{\equiv f_+(W_+, b_+; x)} - \underbrace{e_-^\top \max(0, W_-x + b_-)}_{= f_-(W_-, b_-; x)}$$

where e_+ and e_- denote vectors of ones in suitable dimensions n_- and n_+. If we choose $n_+ = n = n_-$ we have twice as many weights and shifts and all possible prediction functions of the original model can be reproduced. Even for this extremely simple model we have universality in the sense that every continuous function can be approximated up to an arbitrarily small absolute error $\varepsilon > 0$ as shown by Yarotsky in [16]. A similar result was proven via the representation of B-splines by neural nets in [3]. Of course the necessary size $2n$ of the intermediate layer will grow rapidly as ε becomes small. We also note that in this normalized single layer case the prediction function $f(w; x)$ and the averages of the losses $|f(w; x) - y|$ are piecewise linear w.r.t to the weights $w = (W, b)$. Hence the learning problem is then itself a global piecewise linear optimization problem. For more general nets we may apply successive piecewise linearization.

It is interesting to note that with respect to the sample point x both f_+ and f_- are maxima of affine functions and thus in particular convex. This common property makes the difference f a so-called DC function [2] for which various methods to compute the global minimum are available. However, for neural network problems we are normally not interested in optimizing the prediction function f itself with respect to the sample x but rather in minimizing the average of the discrepancies $|f_+(W_+, b_+; x) - f_-(W_-, b_-; x) - y|$ over the samples $(x, y) \in \mathbb{R}^{n+1}$ with respect to the weights and shifts. This problem is not DC so that the given decomposition of f into f_+ and f_- does not seem to help for our global optimization effort.

Apart from singularity the neural network training problems based on hinge activation exhibit another rather undesirable property from the optimization point of view. The average of the discrepancies $|f(w; x) - y|$ is multi-piecewise-linear in the weights w in that for any Cartesian basis vector e_j the function $f(w + te_j; x) - y$ and thus its absolute value is piecewise linear in t. Then the following result follows easily.

Proposition 1 (Minimizers are generically nonsmooth) . *Suppose a locally Lipschitz continuous function* $\varphi : \mathbb{R}^{\tilde{n}} \mapsto \mathbb{R}$ *has a nonempty bounded level set* $\{x \in \mathbb{R}^{\tilde{n}} : \varphi(w) \leqslant \varphi(w_0)\}$ *for some* $w_0 \in \mathbb{R}^{\tilde{n}}$ *and is multi-piecewise linear as defined above. Then* φ *has global minimizers where it is not differentiable, which applies also at all geometrically isolated local minimizers.*

In general a point w_* which is *coordinate minimal* in that all univariate restrictions $\varphi_j(t) = \varphi(w_* + te_j)$ are first order minimal need not be first order minimal for $\varphi(w)$ itself. This is true even for piecewise linear φ like for example the lemon squeezer described in [8]. However, there is some hope that in the special case of neural nets, where the variables represent weights coordinate minimality might have some implications for the behavior of φ in the full space.

Of course, the observation that machine learning problems are not everywhere differentiable is not new. However, it is often suggested that this non-smoothness is only transitionary and does not matter in the end, which we believe to be unduly optimistic, even if one uses a smooth loss function. It was recently documented by Kakade and Lee [13] that standard ML packages yield inconsistent derivative values at nonsmooth points. In the book [2] several classes and many individual examples of nonsmooth problems are described and locally optimized by bundle and other general purpose nonsmooth optimization methods. The globality issue is not specifically addressed except in the DC case.

The essential message of the above observations is that all global and even local minimizers of multi-piecewise functions are nonsmooth. Theoretically and with some luck this problem can be overcome by smoothing of the activation function, as for example in [3]. However, it seems rather difficult to find the right smoothing scale such that the algorithm behavior is markedly improved without the objective and its minimizers being altered significantly. We pursue the strategy of accounting for the nonsmoothness and its combinatorial consequences explicitly. Near a kink steepest descent and all other algorithms designed for smooth problems are likely to chatter back and forth across the kink and thus severely limiting their ability to move tangentially along a valley towards a minimizer that also has directions with smooth growth of second order. It was shown in [8] that under the Linear Independent Kink Qualification such a so-called $\mathcal{V}-\mathcal{U}$ decomposition always exists. The usual recipe for dealing with this chattering or zig-zagging problem is to successively reduce the step-size at a suitable rate. Loosely speaking, the stepsize, also known as learning rate, must be smaller than twice the local Lipschitz constant, which is of course difficult to determine.

There is a forth difficulty, namely for all bounded x and sufficiently large negative b the prediction function $f(x)$ and its gradients with respect to all variables will vanish identically, so that gradient based methods including stochastic variants can not move away at all. In conclusion of this preliminary section we note that from a classical optimization point of view neural net training problems represent the worst of two worlds, namely nonsmoothness and various singularities. In some other contexts these two properties can be traded off against each other, but here they occur jointly in a generic fashion. This means of course that the frequently presented convergence analyses [1], [4] for variants of steepest descent assuming smoothness and strict convexity simply do not apply to neural networks with piecewise linear activation. To deal with such problems we pursue the strategy of piecewise linearization.

3 The Generalized Abs-Normal and Abs-Linear Forms

All objective functions in machine learning and other applications are evaluated by a sequence of arithmetic operations, smooth intrinsic functions and the nonsmooth elements abs, min and max. Mathematically we can interpret such an evaluation procedure as the generalized abs-normal form

$$\text{Min} f(x, z, |z|) \quad s.t. \quad z = F(x, z, |z|) \quad \text{where} \tag{2}$$

$$f : \mathbb{R}^{n+s+s} \mapsto \mathbb{R} \quad \text{and} \quad F : \mathbb{R}^{n+s+s} \mapsto \mathbb{R}^s$$

This is a slight generalization of the usual abs-normal form [7] where z does not occur explicitly as arguments of f and F. We must require that the matrices

$$M \equiv \frac{\partial F}{\partial z} \in \mathbb{R}^{s \times s} \quad \text{and} \quad L \equiv \frac{\partial F}{\partial w} \in \mathbb{R}^{s \times s}$$

are strictly lower triangular. That means we can compute for any x the piecewise smooth functions $z(x)$ which finally yields the objective

$$\varphi(x) = f(x, z(x), |z(x)|) : \mathbb{R}^d \mapsto \mathbb{R}$$

Given code for evaluating the functions in (2) every AD tool will be able to compute for given x and z the vectors y and the partitioned Jacobian

$$\frac{\partial[F-z,f]}{\partial[x,z,w]} \quad \equiv \quad \begin{bmatrix} Z & M-I & L \\ a^\top & b^\top & c^\top \end{bmatrix} \quad \in \quad \mathbb{R}^{(s+1)\times(n+s+s)}.$$

Now the generalized abs-normal form evaluated with its derivatives at some consistent point $(\mathring{x}, \mathring{z})$ yields the piecewise linear model

$$\begin{aligned} z &= \mathring{x} + Z(x-\mathring{x}) + M(z-\mathring{z}) + L(|z|-|\mathring{z}|) \\ &= \underbrace{(\mathring{x} - Z\mathring{x} - M\mathring{z} - L|\mathring{z}|)}_{=\mathring{d}} + Zx + Mz + L|z|, \end{aligned} \tag{3}$$

$$\begin{aligned} y &= \mathring{y} + a^\top(x-\mathring{x}) + b^\top(z-\mathring{z}) + c^\top(|z|-|\mathring{z}|) \\ &= \underbrace{(\mathring{y} - a^\top\mathring{x} - b^\top\mathring{z} - c^\top|\mathring{z}|)}_{=\mathring{\mu}} + a^\top x + b^\top z + c^\top|z|. \end{aligned} \tag{4}$$

where the constant shift $\mathring{\mu}$ does not matter since we are doing optimization. Note that due to triangular structure of M and L we can unambiguously evaluate the piecewise linear functions $z(x)$ and $y(x)$ for any $x \in \mathbb{R}^d$ by forward substitution. We will denote y as a function of x and the development point as the piecewise linearization

$$\Delta\varphi(\mathring{x}; x-\mathring{x}) \equiv y(x) - \mathring{\mu}\,.$$

The key property on which the local minimization method SALMIN [10] is based is the generalized Taylor approximation [7]

$$|\varphi(x) - \varphi(\mathring{x}) - \Delta\varphi(\mathring{x}; x-\mathring{x})| \le \tfrac{\bar{q}}{2}\|x-\mathring{x}\|^2. \tag{5}$$

We will generally refer to a quadratic regularization term in the Euclidean norm as the **prox**imal term and label functions that include this term accordingly. For a theoretical properties of piecewise smooth and piecewise linear functions see the book by Scholtes [15]. There exists also some software for piecewise linear functions, but it does not utilize the abs-linear form (see e.g. [5]). Regularization by a strictly proximal term has been utilized for a long time in machine learning applications (see e.g. [14]).

4 Three Optimization Strategies for the Inner Loop

The local minimization of the model objective

$$\Delta\varphi(x_k; \Delta x) + \tfrac{q_k}{2}\|\Delta x\|^2 \tag{6}$$

can be performed by an active signature method SALMIN described in [10], which is very similar to convex quadratic optimization. To increase the chance of finding a global minimizer one can perform bi-directional global line-searches along coordinate axes, which is quite cheap on the GANF. Various kinds of coordinate descent methods have received a lot of attention in recent years [12]. We call this first option Coordinate Global Descent (CGD) because the optimization along the coordinates will be global. It is not clear yet whether in the nonconvex situations that we are interested in, CGD will be guaranteed to come close to at least a stationary point. The second option called TOAST for Target Oriented Averaging Search Trajectory will again be more ambitious than SALMIN in that we try to compute a global minimizer of the model objective. This will be done using the continuous search directory proposed in [9], which is defined by a second order ODE that can be solved in closed form for the model objective occurring on the right hand side of (6). As the third and final option we consider the reformulation as a mixed integer bilinear optimization problem, which can be reformulated as a mixed integer linear optimization problem (MILOP) and then solved by modern branch and cut solvers. In contrast to CGD and TOAST this last option is actually guaranteed to yield global minima. However, it must be noted that solving the inner model problem globally does not ensure that one approximates a global or even local minimizer of the underlying nonlinear problem.

In the remainder of the paper we consider and compare the three options CGD, TOAST and MILOP.

4.1 Coordinate Global Descent

Suppose that given a point $\mathring{x} \in \mathbb{R}^d$ and a direction $0 \neq e \in \mathbb{R}^d$ we try to solve the problem

$$\mathring{t} \;=\; \underset{0\leqslant t\in\mathbb{R}}{\text{arglobmin}}\; \varphi(\mathring{x}+te) \in [0,\infty).$$

Here it is assumed that φ is piecewise linear and unbounded below so that it always attains a global directional minimizers. Moreover, we assume that

$$y(x) = \varphi(x) = \varphi(\mathring{x}) + \Delta\varphi(\mathring{x}, x - \mathring{x})$$

and the corresponding $z = z(x)$ can be evaluated by the abs-linear form (3) and (4). Restricting the arguments x to the ray $x(t) = \mathring{x} + te$ with $t \geqslant 0$ we can consider the $z(t) = z(x(t))$ and $y(t) = y(x(t))$ also as univariate functions. Denoting their derivatives in the directions $x' = e$ by superscript $'$ we obtain

$$z'(t) = Ze + Mz'(t) + L\Sigma(t)z'(t) \tag{7}$$

$$y'(t) = a^\top e + b^\top z'(t) + c^\top \Sigma(t)z'(t). \tag{8}$$

where $\Sigma(t) = \text{diag}(\text{sgn}(z(t))$. Due to the strict lower triangularity of M and L these uniquely defined directional derivatives can be evaluated by forward substitution. In some cases it may be advantageous to not compute the abs-linear form explicitly, but to compute the pair $(z(t), y(t)), (z'(t), y'(t))$ in the forward mode of automatic differentiation applied directly to (2). In fact this is what we have done in our preliminary numerical experiments because as we will see in the last subsection for the neural network learning problem, the abs-linear form with respect to the weights w is not quite as naturally available as with respect to the sample point x. The key common property is that we still have the same hierarchy of switching variable z_ℓ irrespective of whether we consider x as variable and w as constant or the other way round.

We need two bounds t_{lo} and t_{hi} which are supposed to have the property that $\varphi(x(t))$ has no kink and thus no local minimizers below t_{lo} and above t_{hi}, In the following algorithmic description we will treat $\mathring{x}$ and e as global variables and perform a recursion on an interval $[\check{t}, \hat{t}\,) \subset [0, \infty)$, in which the global minimizer is sought. At level ℓ we will ensure that none of the switching functions $z_j(t) = z_j(\mathring{x} + te)$ for $j < \ell \leqslant s$ has a root inside $[\check{t}, \hat{t}\,)$. Then we evaluate $\check{z}_\ell = z_\ell(\check{t})$ and $z'_\ell = z'_\ell(\check{t})$. Now if $\check{z}_\ell z'_\ell < 0$ and $\mathring{t} = \check{t} - \check{z}_\ell / z'_\ell < \hat{t}$ we must split the interval $[\check{t}, \hat{t}\,)$ into $[\check{t}, \mathring{t})$ and $[\mathring{t}, \hat{t})$. On each one of these two subintervals the search can be repeated, which leads to the following pseudo-code.

Algorithm 1. Recursive minimization along ray $x(t) = \check{x} + te$

Ensure: $\check{t} = 0,\ \check{z} = z(\check{t}),\ \check{z}' = z'(\check{t}),\ \check{y} = y(\check{t}),\ \hat{t} = \infty,\ \hat{y} = \infty,\ \ell = 1,\ 0 \leq s \in \mathbb{N}$
  function GLOMIN$(\ell, \check{t}, \check{z}, \check{z}', \check{y}, \hat{t}, \hat{y})$ ▷ returns minimizer and minimal value
    if $k > s$ **then** ▷ we have reached a leaf, $s = 0$ allowed
      $(t, y)_{\min} = \min\{(\check{t}, \check{y}), (\hat{t}, \hat{y})\}$ ▷ min passes minimizer and value
    else
      $\ell = \ell + 1$ ▷ we go to the next level
      if $\check{z}_\ell \check{z}'_\ell < 0$ and $\mathring{t} = \check{t} - \check{z}_\ell / \check{z}'_\ell < \hat{t}$ **then** ▷ we have to split interval
        evaluate $\mathring{z} = z(\mathring{t}), \mathring{z}' = z'(\mathring{t}), \mathring{y} = y(\mathring{t})$
        $(t, y)_{\min} = \min\{\text{GLOMIN}(k, \check{t}, \check{z}, \check{z}', \check{y}, \mathring{t}, \mathring{y}), \text{GLOMIN}(\ell, \mathring{t}, \mathring{z}, \mathring{z}', \mathring{y}, \hat{t}, \hat{y})\}$
      else
        $(t, y)_{\min} = \text{GLOMIN}(\ell, \check{t}, \check{z}, \check{z}', \check{y}, \mathring{t}, \mathring{y})$ ▷ no split necessary
      end if
    end if
    return $(t, y)_{\min}$
  end function

The sequence of coordinate directions $e = \pm e_j$ can be selected in various ways. In our preliminary numerical tests we chose $j \in \{1 \dots n\}$ uniformly at random.

4.2 Target Oriented Averaging Trajectory Search

The TOAST option of SALGO is based on a continuous search trajectory originally proposed ODE in [9] under the assumption that the objective $\varphi \in C^2(\mathbb{R}^n)$ is twice differentiable and that its gradient $\nabla\varphi(w)$ is nonzero at all w where $f(w) = c$.

Here the method parameter c is the *target* or *c-level* that one wants to reach from above at any particular stage of the optimization procedure. Once it has been reached, the target c can be lowered to a more ambitious level, unless one is satisfied with what has been found. In neural network training the objective is usually some positive experimental risk that one wishes to get down as close to zero as possible. Therefore we have adopted a halving strategy whenever a certain level has been reached. Of course one may find during the minimization calculation that the target appears to be unattainable, and one then has to move it up towards the values that have already been attained during the current run. This adjustment of c is the main heuristic aspect of the proposed method and will be discussed later. Under conditions of machine learning the smoothness assumption is only likely to be satisfied on domain segments that decompose the Euclidean space $\mathbb{R}^d$. So the corresponding numerical methods and theoretical results from [9] need to be spliced together. This is fairly easy if we assume transversality, i.e that the search trajectory crosses the boundary between the smooth pieces always at an angle. In [9] various considerations led to the autonomous second order ODE

$$\ddot{w}(t) = -e\left[I - \frac{\dot{w}(t)\,\dot{w}(t)^\top}{\|\dot{w}(t)\|^2}\right]\frac{\nabla\varphi(w(t))}{[\varphi(w(t)) - c]} \quad \text{with} \quad \|\dot{w}(t_0)\| = 1. \tag{9}$$

In addition to the target level c we have here the *sensitivity parameter* $e > 0$, which needs to be chosen adequately. The basic idea behind the equation is that, given the current point x and the search direction $\dot{w}$, we adjust the latter towards the direction of steepest descent $-\nabla\varphi(w)$ scaled by the reciprocal $1/[\varphi(w) - c]$. As long as we are high above the target level, the reciprocal will be small and thus the search direction more or less constant, hopefully ignoring small local wiggles in the objective. Once $\varphi(w)$ gets closer to c, the adjustment of $\dot{w}$ towards steepest descent will be more drastic and in the limit as $\varphi(w) - c$ tends to zero $\dot{w}$ reduces to the steepest descent direction $-\nabla\varphi(w)/\|\nabla\varphi(w)\|$.

Throughout the tangent $\dot{w}$ is scaled to have the Euclidean norm $\|\dot{w}\| = 1$ so that the dependence on the independent variable t is really the arclength parametrization. The projector $I - \dot{w}\dot{w}^\top/\|\dot{w}\|^2 = I - \dot{w}\dot{w}^\top$ implies $2\dot{w}^\top\ddot{w} = \frac{d}{dt}\|\dot{w}\|^2 = 0$ so that $\|\dot{w}\| \equiv 1$ analytically. Of course, if one solves the ODE numerically this will not hold exactly and the denominator in the projector cannot be dropped. Before discussing that let us regard the integrated equivalent of (9), namely

$$\frac{\dot{w}(t)}{[\varphi(w(t)) - c]^e} = \frac{\dot{w}_0}{[\varphi_0 - c]^e} - e\int_0^t \frac{\nabla\varphi(w(\tau))}{[\varphi(w(\tau) - c]^{e+1}}d\tau. \tag{10}$$

Here $w(\tau)$ for $0 \leqslant \tau \leqslant t$ is the analytical solution path, which is unique as long as the $\nabla\varphi$ and thus the right hand side of (9) is Lipschitz continuous, and of course $\varphi > c$. As we can see $\dot{w}(t)$ is a weighted average of the initial $\dot{w}_0$ and the negative gradients along the way. That justifies the label *target oriented averaging search* or *generalized descent* though the latter terminology might suggest the use of generalized gradients. As we can see the sensitivity e occurs here mainly as an exponent of $(\varphi - c)$. The convergence statements proven in the original paper suggest that e should be selected as the reciprocal of the growth rate of $\varphi(w) - \varphi(\mathring{w})$ within the catchments of its local minimizers $\mathring{w}$ such that

$$\varphi(w) - \varphi(\mathring{w}) \;\sim\; \|w - \mathring{w}\|^{1/e}.$$

We can also think of $1/e$ as an average degree of positive homogeneity. This would suggest the choice $e = \frac{1}{2}$ for essentially quadratic and otherwise smooth functions and $e = 1$ functions with linear growth. If φ is only piecewise smooth so that the right hand side can have jumps, then the trajectory is still unique and the integrated equation still valid, provided the kink surfaces of φ are reached and transitioned in a transversal fashion. Thus, baring severe degeneracies the averaging descent philosophy can still be applied in the piecewise smooth case. Of course the finitely many transitions must be handled numerically.

Rather than following the ODE on the original φ we perform averaging descent on its piecewise linearization with a proximal term. As already observed in [9] the trajectory then reduces to a sequence of circle segments between the boundaries of the polyhedra own which the prox-abs-linear model is smooth. Due to the generalized abs-linear representation (3) one can explicitly compute the circles and handle the transition from one polyhedra to a neighbor, so that we effectively generate a sequence of iterates until the target level is reached. The details of this method will be described in a future.

4.3 Mixed Integer Bilinear Optimization

The minimization of a function in abs-linear from can be written as

$$\text{Min}_{x,z,\sigma}\, a^\top x + b^\top z + c^\top \Sigma z \quad s.t. \quad z = d + Zx + Mz + L\Sigma z \quad \text{and} \quad \Sigma z \geqslant 0, \quad (11)$$

where the elements $\sigma_i \in \{-1, 1\}$ of the diagonal matrix $\Sigma = \text{diag}(\sigma)$ are binary variables. Equation (11) is a Mixed Integer Bilinear Optimization Problem. It can be transformed [11] into a Mixed Integer Linear Optimization (MILOP), provided we have a uniform bound γ on the components of $|z|$. We may then even add a proximal term and get the MILOP

$$\text{Min}_{x,z,h,\sigma}\, \left(a^\top x + b^\top z + c^\top h + \tfrac{q}{2}\|x\|^2\right) \quad s.t. \quad z = Zx + Mz + Lh, \quad (12)$$

$$-h \leqslant z \leqslant h \text{ and } h + \gamma(\sigma - e) \leqslant z \leqslant -h + \gamma(\sigma + e), \quad (13)$$

which can be solved by very effective modern solvers like Gurobi.

However, there is the slight problem that for learning we actually want to minimize not with respect to x but with respect to the coefficients

$$w = (Z, M, L, J, N, Y) \in \mathbb{R}^{(s\times n, s\times s, s\times s, m\times n, m\times s, m\times s)} \simeq \mathbb{R}^{s(n+s-1)+m(n+2s)}.$$

Whereas in general we have only multi piecewise linearity w.r.t. w, we have noted that in the single layer case with fixed weighting, this problem is also piecewise linear and it can be rewritten as the mixed bilinear programming problem

$$\min_w \varphi(w) = \tfrac{1}{K}\sum_{k=1}^{K} u_k \quad s.t. \quad z_k = Wx_k + b, \;\; g_k = \tfrac{1}{2}p^\top(z_k + h_k) - y_k\,,$$

$$\mu_k g_k = u_k \geqslant 0 \in \mathbb{R}, \quad \text{and} \quad h_k = \Sigma_k z_k \geqslant 0 \in \mathbb{R}^n, \quad (14)$$

where the $\mu_k \in \{-1, 1\}$ and the diagonal elements $\sigma_k \in \{-1, 1\}^n$ of the matrices Σ_k are binary variables. The bilinear terms can be replaced in the usual fashion by the system of linear inequalities for $k = 1 \dots K$

$$-u_k \leqslant g_k \leqslant u_k \text{ and } u_k + \delta_k(\mu_k - 1) \leqslant g_k \leqslant -u_k + \delta_k(\mu_k + 1) \in \mathbb{R}, \quad (15)$$

$$-h_k \leqslant z_k \leqslant h_k \text{ and } h_k + \gamma_k(\sigma_k - e) \leqslant z_k \leqslant -h_k + \gamma_k(\sigma_k + e) \in \mathbb{R}^n. \quad (16)$$

where $\delta_k \in \mathbb{R}$ must be an upper bound on the $g_k \in \mathbb{R}$ and $\gamma_k \in \mathbb{R}^n$ on all components of $z_k \in \mathbb{R}^n$. Here e represents the corresponding vector of ones.

5 Numerical Experiments and Comparison

Given the limitations of space and time we have not been able to conduct and report numerical results that are in any way conclusive. We simply minimized the averaged losses $|f(w;x) - y(x)|$ defined in (1) for the Griewank function [9]

$$y(x) = 1 + \frac{1}{40000}\sum_{k=1}^{d} x_k^2 - \prod\nolimits_{k=1}^{d} \cos\left(\frac{x_k}{\sqrt{k}}\right)$$

over K different sample points $x \in \mathbb{R}^d$ chosen uniformly at random in the cube $[-8, 8]^d$. The results reported are for $d = 4$, the number of nodes $n = 10$ and $K = 20$ sample points. This results in a mixed integer linear optimization problem specified in (14, 15, 16) with 220 binary variables and equations, 484 real variables, and 880 equality constraints. Larger choices of d, n and K were not acceptable for our AMPL model using the solver Gurobi via the NEOS server. The good news is that the MILOP system was solved exactly with an objective value of zero so that the regression problem reduced actually to an interpolation problem.

The bad news is that Gurobi reached this result by solving 2529 branch and bound nodes using a total number of 95309 Simplex iterations. Though a direct complexity comparison appears difficult, it seems certain that each of them is much more costly than evaluating the empirical risk and its gradient by back propagation. That is the cost of each steepest descent iteration of which we allowed 10000. We allowed the same number of iterations to TOAST, whose steps are a little more costly than those of steepest descent. Each step of stochastic gradient costs about one Kth of that of the for steepest descent which is why we allowed a total of 20000 iterations for SG. At every 200th iteration of SG we evaluated and plotted the full empirical risk and computed the minimum of all these values over the trajectory. For the other methods including the CGD we plotted every tenth value but took the minimal value over all iterations. The numbers in the second and the third row of the following table give the minimal value and the iteration counter at which it was achieved, respectively.

Method	CGD	TOAST	MILOP	SD	SG
minvalue	0.3005248	0.0000006	0.0000000	0.069400	0.0017503
iteration	9510	9999	195309	7036	189400

SD and SG were run with the step sizes given respectively by 0.35 and 0.01 divided by the square root of the iterations count. The vertical axis in Fig. 1 represents the decimal logarithm of the ratio between the current and the initial empirical risk. As one can see the coordinate search gets stuck in the neighborhood of a local minimizer after some 3000 iterations. Except for Gurobi TOAST achieved the lowest value of $6\,10^{-6}$, which involved halving the target about 19 times. Every time the trajectory picks up speed and goes uphill for a while before dropping down to the desired lower value. The average rate of convergence appear to be linear, i.e. what in machine learning is sometimes called exponential. Steepest descent and stochastic gradient appear to get stuck or at least slow down near a local minimizer or saddle point. Their performance most probably can be improved by a more sophisticated step size selection heuristic.

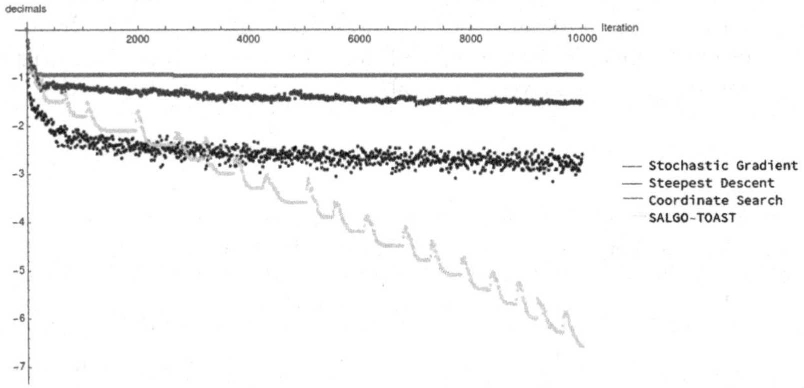

Fig. 1. Decimal digits gained by four methods on single layer regression problem.

6 Summary and Outlook

We observed that many machine learning problems and in particular the training of neural networks lead to global optimization problems for piecewise smooth objectives. Moreover, these functions have a piecewise linear local approximation, which can be obtained in abs-linear form. This motivates our strategy of successive abs-linear global optimization (SALGO). We consider three methods for solving the inner problem of globally minimizing an abs-linear function. Firstly, global minimization along randomly chosen coordinate directions (CGD). Secondly, a Target Oriented Averaging Search Trajectory (TOAST), which can be computed explicitly as a sequence of circular segments. Finally, the reformulation as a Mixed Integer Linear Optimization Problem (MILOP), which is solvable exactly by Gurobi or other branch and bound solvers. Very preliminary numerical result on a low dimensional regression problem for a single layer network show that CGD and to a lesser extend the standard methods of steepest descent and stochastic gradients may get stuck or slow down near local minimizers or saddle points. MILOP yields the exact global minimum, but may take much more computing time than the other essentially gradient based methods.

The negative result for the coordinate descent method CGD seems pretty conclusive and it is not apparent how it can be improved. We lack the sophistication and experience to imagine how the mixed integer optimization approach

MILOP could be accelerated by orders of magnitude to make it suitable for training problems of significant size. It certainly should be used as a reference for testing the other, more heuristic approaches. So the real interesting issue is whether and how TOAST can be made competitive with or superior to advanced implementations of stochastic gradient variants on problems of significant size. Apart from a refinement of the target adjustment and restarting strategy, the linear algebra needs to be improved and adapted to parallel computers and the use of graphical processing units (GPU)s. A key observation is that at every iteration of TOAST the weight matrices only undergo a rank-one change, which can be used to update all other relevant quantities at a significantly reduced cost. All methods would probably benefit from replacing the ℓ_1 norm by an ℓ_2 or other smooth loss function. That then calls for successive piecewise linearization, which is required for multi-layered neural nets anyhow.

References

1. Arora, S., Cohen, N., Golowich, N., Hu, W.: A convergence analysis of gradient descent for deep linear neural networks. CoRR, abs/1810.02281 (2018)
2. Bagirov, A., Karmitsa, N., Mäkelä, M.: Introduction to Nonsmooth Optimization: Theory, Practice and Software. Springer, Cham (2014). https://doi.org/10.1007/978-3-319-08114-4
3. Bölcskei, H., Grohs, P., Kutyniok, G., Petersen, P.: Optimal approximation with sparsely connected deep neural networks. ArXiv:abs/1705.01714 (2019)
4. Bottou, L., Curtis, F.E., Nocedal, J.: Optimization methods for large-scale machine learning. SIAM Rev. **60**, 223–311 (2018)
5. Fourer, R., Kernighan, B.W.: AMPL: a modeling language for mathematical programming (2003)
6. Glorot, X., Bordes, A., Bengio, Y.: Deep sparse rectifier neural networks. J. Mach. Learn. Res. **15**, 315–323 (2011)
7. Griewank, A.: On stable piecewise linearization and generalized algorithmic differentiation. Optim. Methods Softw. **28**(6), 1139–1178 (2013)
8. Griewank, A., Walther, A.: First and second order optimality conditions for piecewise smooth objective functions. Optim. Methods Softw. **31**(5), 904–930 (2016)
9. Griewank, A.: Generalized descent of global optimization. J. Optim. Theory Appl. **34**, 11–39 (1981)
10. Griewank, A., Walther, A.: Finite convergence of an active signature method to local minima of piecewise linear functions. Optim. Methods Softw. **34**, 1035–1055 (2019)
11. Gupte, A., Ahmed, S., Cheon, M., Dey, S.: Solving mixed integer bilinear problems using MILP formulations. SIAM J. Optim. **23**(2), 721–744 (2013)
12. Wright, S.J.: Coordinate descent algorithms. Math. Program. **151**, 3–34 (2015)
13. Kakade, S.M., Lee, J.D.: Provably correct automatic sub-differentiation for qualified programs. ArXiv:abs/1809.08530 (2018)
14. Kärkkäinen, T., Heikkola, E.: Robust formulations for training multilayer perceptrons. Neural Comput. **16**, 837–862 (2004)
15. Scholtes, S.: Introduction to Piecewise Differentiable Equations. Springer, New York (2012). https://doi.org/10.1007/978-1-4614-4340-7
16. Yarotsky, D.: Error bounds for approximations with deep ReLU networks. Neural Netw. Off. J. Int. Neural Netw. Soc. **94**, 103–114 (2017)

Author Index